BIOLOGY
of the cell, floweri

BIOLOGY

of the cell, flowering plant and mammal

J. Simpkins BSC DipEd MIBiol
Lecturer in Biology
People's College of Further Education, Nottingham

and

J. I. Williams PhD BSC DipEd MIBiol
Senior Lecturer in Biology
People's College of Further Education, Nottingham

Series Editor

M. K. Sands BSC MIBiol
Lecturer in Education
University of Nottingham

Mills and Boon Limited, London

First published in Great Britain 1980 by Mills & Boon
Limited, 15–17 Brooks Mews, London W17 1LF

ISBN 0 263 06423 9

Printed in Great Britain by
Fletcher & Son Ltd, Norwich

List of contents

Acknowledgements

We gratefully acknowledge the assistance of:

Our editor, Margaret Sands, for constant encouragement and helpful guidance.

Joan Bryant of Mills and Boon Ltd., for her endless patient and thoughtful advice in ensuring completion of the project.

Dr. R. Jones and Dr. P. M. Davies of the Department of Zoology, University of Nottingham, Dr. I. Mackenzie of the Department of Botany, University of Nottingham and Dr. A. Booth of the Department of Botany, University of Sheffield who devoted much time to reading the typescript. Their comments were immensely helpful.

Dr. D. Brindley of the Queen's Medical Centre, University of Nottingham for his useful advice on biochemical nomenclature.

The Joint Matriculation Board, the Schools Examination Department of the University of London and the Joint Committee for Ordinary National Certificates and Diplomas for their permission to use recent examination questions.

Mr. J. Kugler of the Queen's Medical Centre, University of Nottingham, without whose expertise and hard work the task of obtaining most of the photographs would have been immensely more demanding. Nearly all of the photographs not acknowledged below or in the text were provided by Mr. Kugler.

Mrs. M. Hollingsworth of the Department of Anatomy, University of Sheffield, Mrs. A. Tomlinson of the Queen's Medical Centre, University of Nottingham, and Mr. B. Case of the Department of Botany, University of Nottingham, who kindly gave their time and skills in preparing some of the photographs.

Mr. P. Bailey and Mr. A. Bezear of the Queen's Medical Centre, University of Nottingham who supplied the photographs and drawings for Figures 7.15(b), 9.7(a), 9.8, 10.17(a), 15.36, 18.8, 18.29, 18.30 and 22.16.

Philip Harris Biological Ltd., for contributing photographs for Figures 1.21(b), 2.12, 7.1, 7.11, 7.13(a), 7.14(a), 7.14(b), 7.15(a), 8.7, 12.15(a) and 19.2. Also for supplying microscopic preparations from which many other figures were produced.

Professor R. M. H. McMinn of the Department of Anatomy, the Royal College of Surgeons of England, for kindly providing Figures 9.3(c) and 15.38 as well as the photograph used on the front cover.

Dr. M. Davey of the Department of Botany, University of Nottingham for his generosity in providing Figures 1.5(a), 1.16(a), 1.19, 1.20, 1.21(a), 3.11(a), 3.14, 6.24(a), 5.5(a), 12.16 and 21.28.

Dr. M. A. Tribe of the School of Biological Sciences, University of Sussex for his help in acquiring Figure 5.18.

Mr. D. Gould, of the Department of Dental Surgery, General Hospital, Nottingham for providing Figure 14.3.

Mrs. S. York, of the EEG department, Leicester Royal Infirmary, for providing Figure 15.45(a).

The many other individuals and institutions who, provided or who gave permission to use illustrative material. They have been credited in the text alongside the appropriate figures.

Any errors in the book are entirely the responsibility of the authors.

Preface

In the past ten years or so there has been a shift in emphasis in the content of Advanced Level syllabuses in biological science. As well as traditional topics such as anatomy and physiology students are now expected to cope with the complexities of molecular genetics and biochemistry. Most Examination Boards have in recent years also changed the style of assessment. Less weight is now placed on memorising facts and more on an understanding of principles. It is with these changes in mind that this book has been written.

We have consciously omitted certain topics which are part of many Advanced Level syllabuses. To do justice to a study of the variety of life, ecology, evolution and genetics is impossible in a volume of this size. We have concentrated on a study of cells and the functional anatomy of the mammal and flowering plant which are the core of all Advanced Level syllabuses in Biology. Such topics will figure prominently in syllabuses which may in future replace GCE Advanced Level syllabuses. The contents of the book are also relevant to courses leading to the Ordinary National Diploma in Sciences. Technician students enrolled for Technician Education Council (TEC) Certificate programmes will find the objectives of Level II and III Standard Units in Cell Biology, Biochemistry and Mammalian Physiology well covered in this volume.

The text has been written for student use rather than as a reference book. The chapters can be read in any order. However, in our opinion, the content is more intelligible if the chapters are studied in the order in which they are set out. The first seven chapters provide a foundation in biological, biochemical and physical knowledge and principles which should make it easier to understand the topics in the rest of the book.

Biology is a practical subject but in the space available we could not include details of practical work. However, second-hand evidence from experimental work is quoted extensively in the text and in a number of the questions in Appendix 3. It is hoped that this will encourage students to realise that factual evidence is essential if valid conclusions are to be arrived at in biological science. Where we have written about debatable issues an attempt has been made to strike a reasonable balance in the conflicting evidence available. Where appropriate, the practical uses to which biological knowledge can be put have been mentioned.

J.S.
J.I.W.
Nottingham, 1980.

Chapter 1

The structure of cells

What is known about cell structure has largely depended on the development of microscopes and microscopical techniques. A simple microscope was invented by Galileo in 1610 but there are no reports that he used it to examine living organisms. In 1676 a Dutch draper, Antonie van Leeuwenhoek, whose hobby was the grinding of lenses, used one of his **simple microscopes** (Fig. 1.1) to examine rainwater in which grains of pepper had been soaked. He observed a variety of unicellular organisms which he called animalcules. It is now known that the organisms he saw were bacteria. A decade earlier Robert Hooke in England had made a **compound microscope** with which he examined, among other things, thin slices of cork tissue. He saw that the cork was porous, rather like a honeycomb, consisting of a great many small compartments which he called **cells**. Such are the origins of cell biology.

FIG. 1.1 *Van Leeuwenhoek's simple microscope*

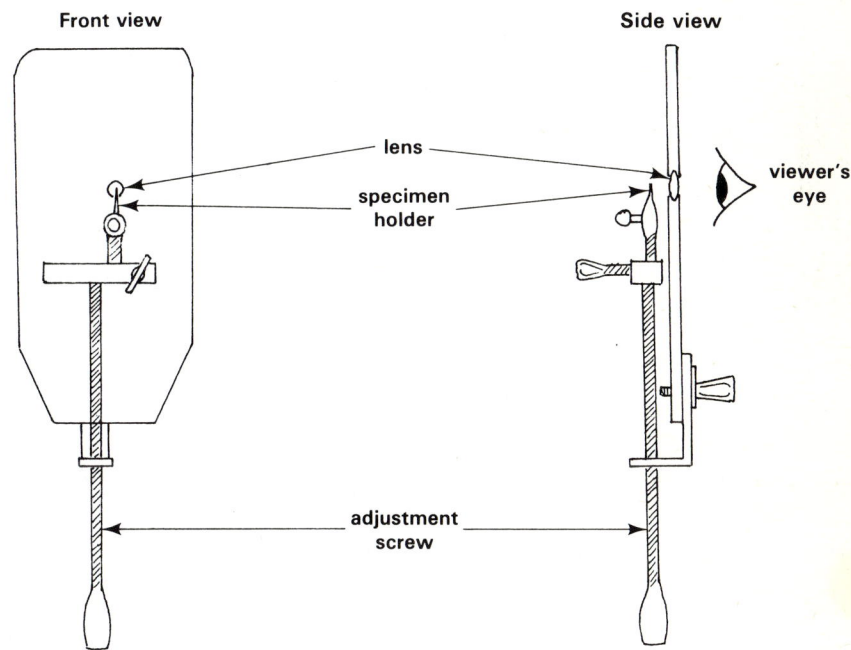

During the next 150 years, with further improvement of the compound microscope a great deal was learned of the structure of cells and tissues from many plants and animals. Gradually it was recognised that all living organisms are composed of cells, some organisms consisting of a single cell, others being multicellular. In 1839 the Belgian botanist Mathias Schleiden concluded that cells are the basic structural units of higher plants, each cell contributing to the integrated life of the organism. At the same time a German zoologist, Theodor Schwann, came to the same opinion about higher animals. These deductions were later formulated into what is called the **cell theory**. For some time after the theory was first proposed there was a tendency for biologists to emphasise the importance of individual cells, the structure and activity of which were thought to mirror that of the whole organism. Recently there has been more interest in the ways in which different types of cells interact in the functioning and development of multicellular organisms.

1.1 Cell structure as interpreted by the light microscope

The compound microscope is still used by most biologists to examine cells and tissues (Fig. 1.2(a)). With this instrument it is possible to observe living material. Good images showing much detail can be obtained, especially if the microscope is of the **phase-contrast** type (Fig. 1.2(b)). A phase-contrast microscope exaggerates small differences in refractive index of cell components to create an image in which the components are more clearly distinguished by the human eye. Nevertheless, most of what was known about the structure of cells up to the 1940s was obtained from observations on dead tissue which had been treated with preservatives. The technique, which had been developed for over a hundred years, first involves immersing the tissue in a fixative such as methanal or ethanol in order to prevent deterioration and to keep the structure as life-like as possible. Following **fixation** the tissue is **dehydrated** with ethanol and then **cleared** with an organic solvent such as dimethylbenzene which is miscible with paraffin wax. The next stage is to **embed** the tissue in molten wax which on hardening supports the tissue while thin sections, 2–10 μm thick, are cut using a microtome. After attaching the sections to microscope slides the wax is removed before the tissue is **stained**. Staining changes the refractive index of

FIG. 1.2 (a) *A modern compound microscope* (courtesy Vickers Instruments)

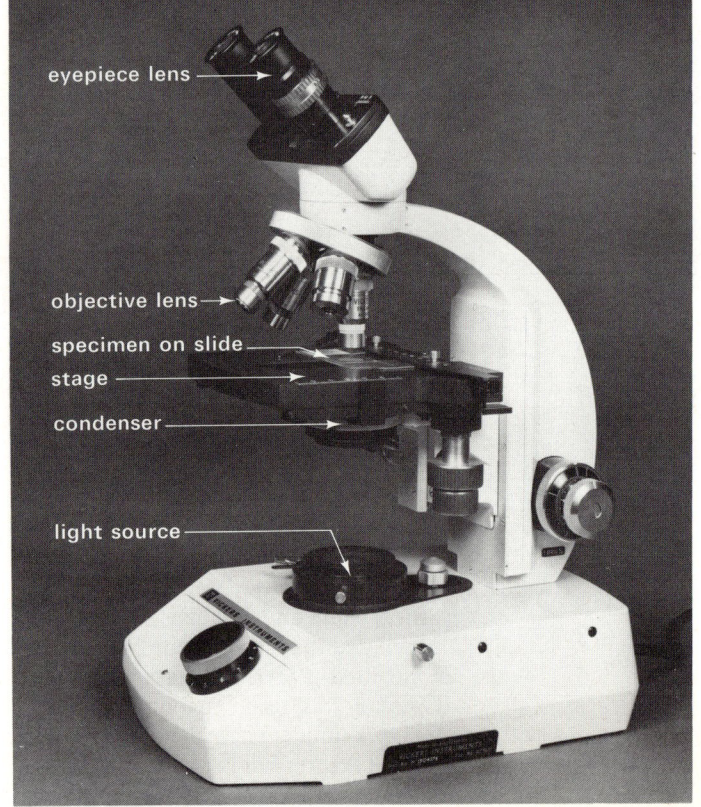

eyepiece lens

objective lens

specimen on slide

stage

condenser

light source

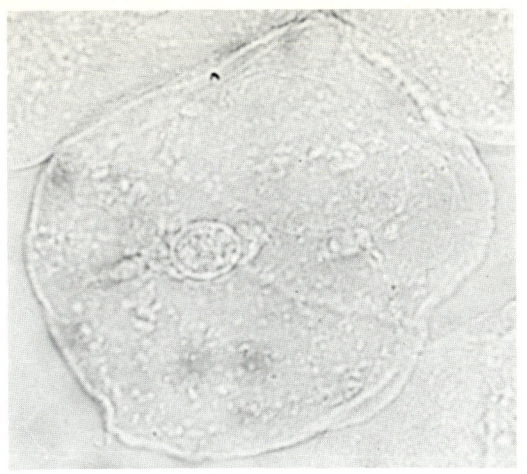

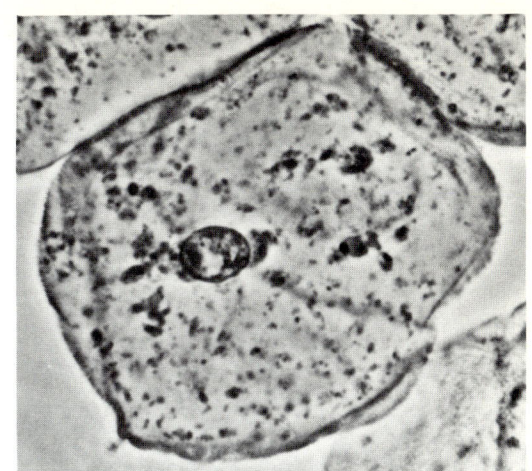

FIG. 1.2 (b) *An epithelial cell from the mouth lining*

(i) *as seen with a conventional light microscope, × 900*

(ii) *as seen with a phase-contrast microscope, × 900*

cell components so that they are more easily seen when the material is examined microscopically. Any of the above steps can lead to distortion of the specimen, so it is necessary to guard against artificial structures, called **artefacts**, which appear in the specimen during treatment but are not present in living cells.

Using such methods, coupled with observations on living cells, it is possible to build up a fairly detailed picture of cell structure (Fig. 1.3(a) and (b)).

The living material of cells is called **protoplasm** and is enclosed in a **plasma membrane**, an elastic outer layer. In plant cells the protoplast, the unit of protoplasm found in a single cell, is surrounded by an inert **cell wall**. The main ingredient of plant cell walls is a glucose polymer called cellulose. Adjacent plant cells are held together by a thin layer composed mainly of calcium pectate and known as the **middle lamella**. Another distinctive feature of many plant cells is the presence of pigment-containing bodies called plastids, the most common of which are the green **chloroplasts**. Large sap-filled **vacuoles** are also frequent.

FIG. 1.3 *Structure of cells as revealed by a compound microscope*

(a) (i) *section of a leaf mesophyll cell as seen with a light microscope × 1200*

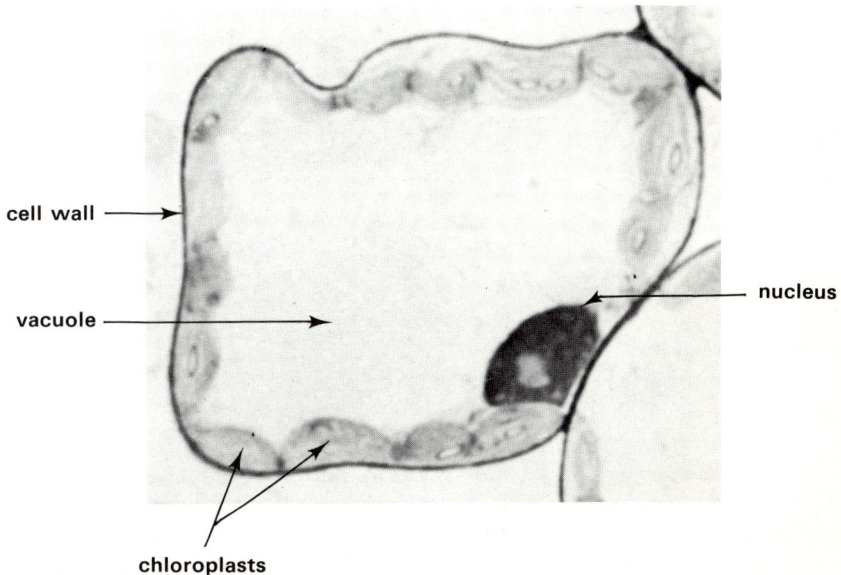

cell wall

nucleus

vacuole

chloroplasts

3

(ii) *Leaf cell of* Elodea

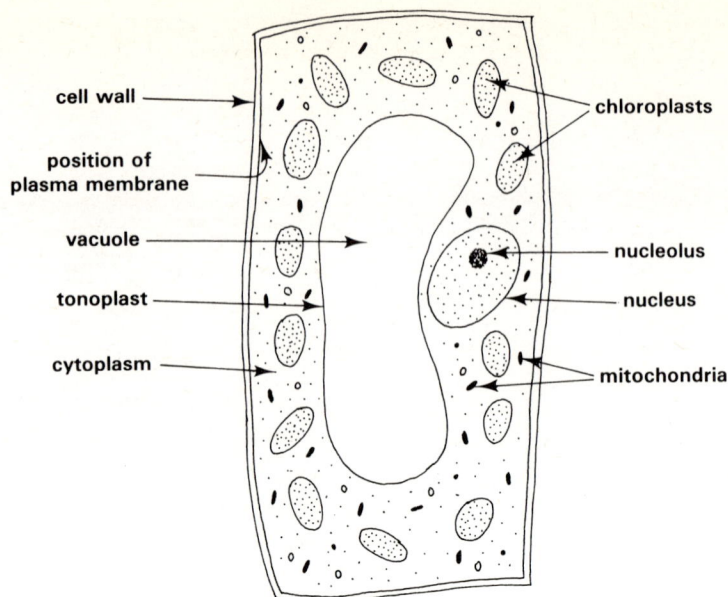

cell wall

chloroplasts

position of
plasma membrane

vacuole

nucleolus

tonoplast

nucleus

cytoplasm

mitochondria

(b) (i) *section of a mammalian
liver cell as seen with a
light microscope,* × *1200*

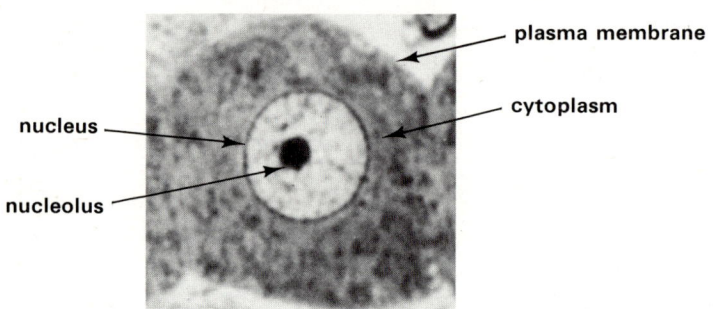

plasma membrane

cytoplasm

nucleus

nucleolus

(ii) *cell from mammalian
gastric gland (fixed and
stained)*

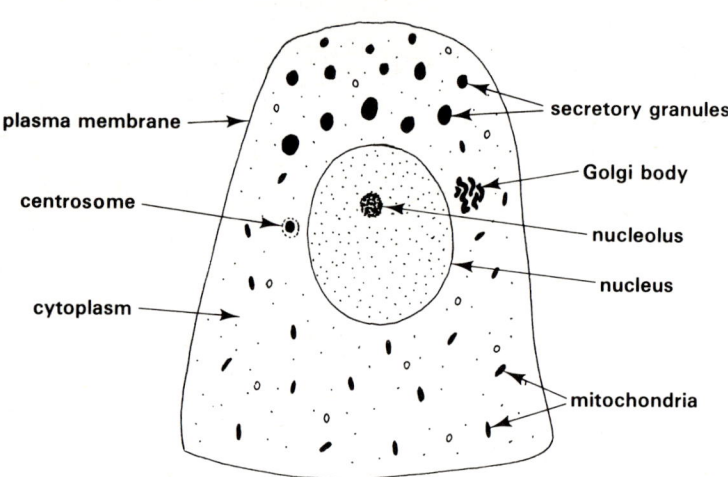

plasma membrane

secretory granules

Golgi body

centrosome

nucleolus

nucleus

cytoplasm

mitochondria

Both plant and animal cells contain a **nucleus** at some stage in their development. When not dividing the nucleus consists of a mass of granular material called **chromatin** which contains one or more dense areas known as **nucleoli**, the whole being surrounded by a **nuclear membrane**. During nuclear division the nucleoli disappear and the chromatin appears as thread-like **chromosomes** which distribute the hereditary material to the new nuclei (Chapter 7).

The protoplasm outside the nucleus is called **cytoplasm**. Using nerve cells stained in a special way, Camillo Golgi at the end of the nineteenth century observed a unique cytoplasmic structure which looked like a tiny net. This structure was subsequently called the **Golgi body**.

Other inclusions of the cytoplasm common to most forms of life and just visible as tiny granules with a compound microscope are **mitochondria**, the centres of respiration (Chapter 5). **Storage materials** are often seen in the cytoplasm of cells. In plant cells **starch** is the main storage substance while in animal cells granules of **glycogen** are commonly found.

Although all living cells have many common features there is no such thing as a typical or generalised cell. Attention has already been drawn to the main differences between plant and animal cells. It is also important to realise that multicellular plants and animals consist of a variety of cell types. The precise structure of a cell is related to the functions it performs.

1.2 Cell ultrastructure as revealed by the electron microscope

Since the 1950s biologists using the electron microscope (Fig. 1.4(a)) have made tremendous strides in our knowledge of the detailed structure of cells. The preparation of specimens for sectioning prior to examination with an **electron microscope** is in some ways similar to the method used for light microscopy but there are important differences. Firstly it is not possible to observe living material with an electron microscope because the beam of electrons directed on to the specimen consists of minute negatively charged particles which are readily scattered by atmospheric atoms and molecules. It is therefore necessary to place the specimen in a vacuum in which live

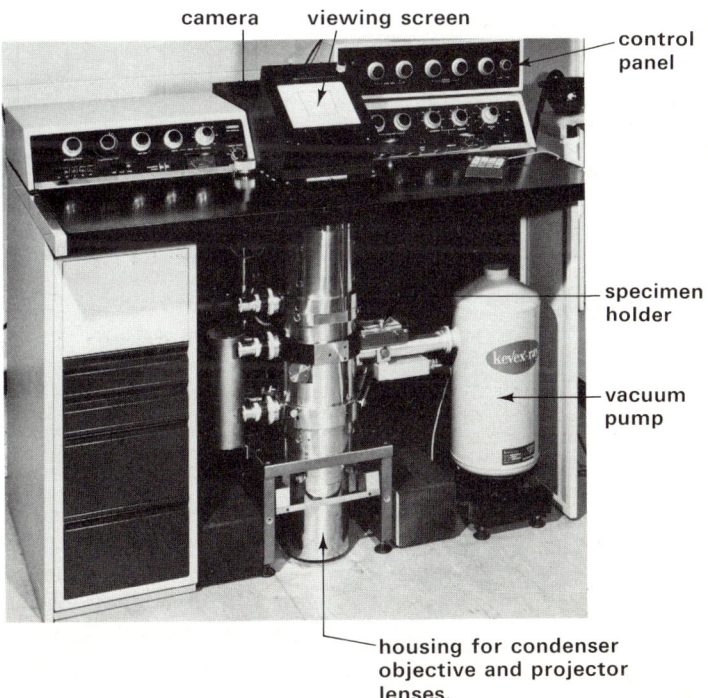

FIG. 1.4 (a) *A modern electron microscope* (courtesy Kratos Ltd.)

camera   viewing screen   control panel

specimen holder

vacuum pump

housing for condenser objective and projector lenses.

cells immediately die. Secondly, sections of cells and tissues must be extremely thin (0·01–0·5 μm thick) otherwise electrons will not pass through them. For **fixation** osmium(VIII) oxide which stains lipids and proteins is often used. Otherwise gluteraldehyde which fixes the material without rendering it electron-dense is used. The sample is next **dehydrated** with ethanol as described for light microscopy, then **cleared** ready for embedding. Paraffin wax fragments if cut into very thin sections so is unsuitable as an embedding substance for electron microscopy. Instead, clear plastic epoxy resins such as *Araldite* are used. The material is cleared in liquid resin which is hardened by gentle heat. The sample is now embedded in a tough, clear supporting substance which can be cut into very thin slices.

Sectioning is carried out with an **ultra-microtome**, the blade of which is usually made by breaking a thick piece of glass to give a hard cutting edge. The sections are then transferred to tiny circular copper grids on which they can be **stained**. Solutions of lead salts are taken up by lipid components in the specimen whilst uranyl salts react with protein and nucleic acid. The effect is to make the lipids, proteins and nucleic acids electron-dense so that they contrast with other cell components in the final image. A special holder is used to place the sections, still supported on their grids, in the electron microscope which is evacuated of air before the electron beam is switched on (Fig. 1.4(b)).

FIG. 1.4 (b) *Stages in the preparation of thin sections for electronmicroscopy*

(i) *cutting the sections with an ultramicrotome*

(ii) *mounting the sections on to copper grid*

specimen embedded in plastic resin

knife edge

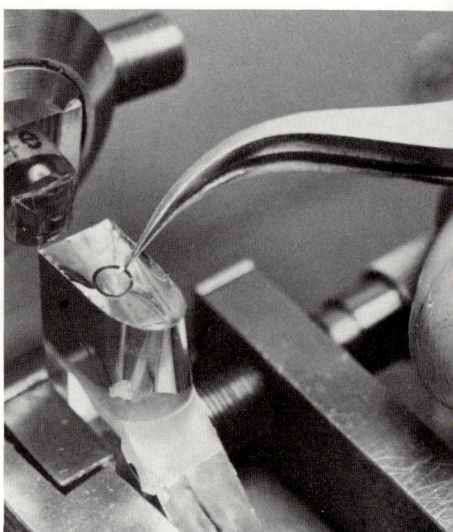

water bath for sections to float in

When electron microscope techniques were used for the first time both plant and animal cells were shown to have a detailed structure previously undreamed of. Not only were some components discovered for the very first time, others already well-known were found to be incredibly complex in structure (Fig. 1.5). Furthermore the subcellular components called **organelles** common to plant and animal cells were seen to be remarkably similar in appearance.

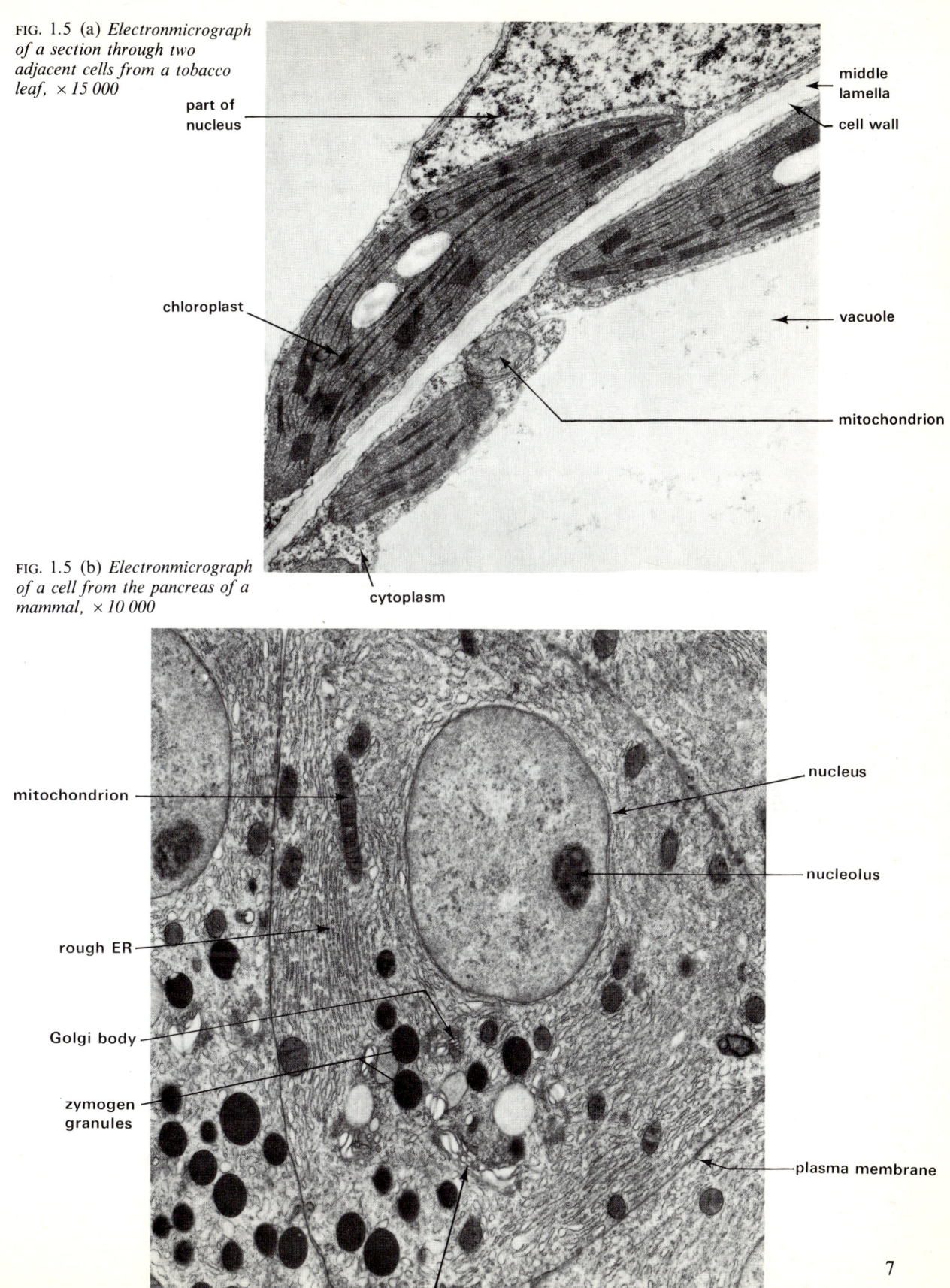

FIG. 1.5 (a) *Electronmicrograph of a section through two adjacent cells from a tobacco leaf,* × 15 000

part of nucleus

middle lamella

cell wall

chloroplast

vacuole

mitochondrion

FIG. 1.5 (b) *Electronmicrograph of a cell from the pancreas of a mammal,* × 10 000

cytoplasm

mitochondrion

nucleus

nucleolus

rough ER

Golgi body

zymogen granules

plasma membrane

smooth ER

7

1.2.1 Cell membranes

The outer boundary of the protoplast, the **plasma membrane**, is invisible with the light microscope. Even so, its presence was inferred towards the end of the last century because protoplasm was seen to leak out of animal cells when the cell surface was punctured. Overton in 1895 suggested that the membrane was made of fatty substances. Other workers later deduced that two layers of lipid were present in the plasma membrane. In 1935 Danielli and Davson suggested a model for membrane structure in which the **lipid bilayer** was coated on either side with **protein** (Fig. 1.6). Mutual attraction between the hydrocarbon chains of the lipids, and electrostatic forces between the protein and the glycerol part of the lipid molecules, were thought to maintain the stability of the membrane (Chapter 3).

FIG. 1.6 *The Danielli–Davson model of the cell membrane*

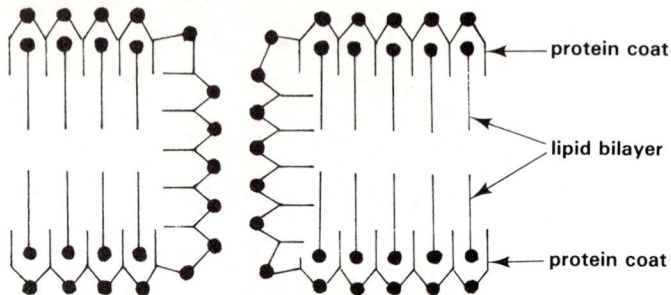

protein coat

lipid bilayer

protein coat

Using evidence from electronmicrographs Robertson, in 1960, proposed a **unit membrane hypothesis** based on the observation that all membranes of cells have a comparable appearance when viewed with the electron microscope. The two outer layers of protein are each about 2 nm thick and appear densely granular. They enclose a clear central area about 3·5 nm wide consisting of lipid (Fig. 1.7). The term **unit membrane** is now used to denote all cell membranes which are similar to this in structure.

FIG. 1.7 *Electronmicrograph of a section through microvilli from intestinal epithelium, × 180 000. The unit membrane can be clearly seen around each microvillus*

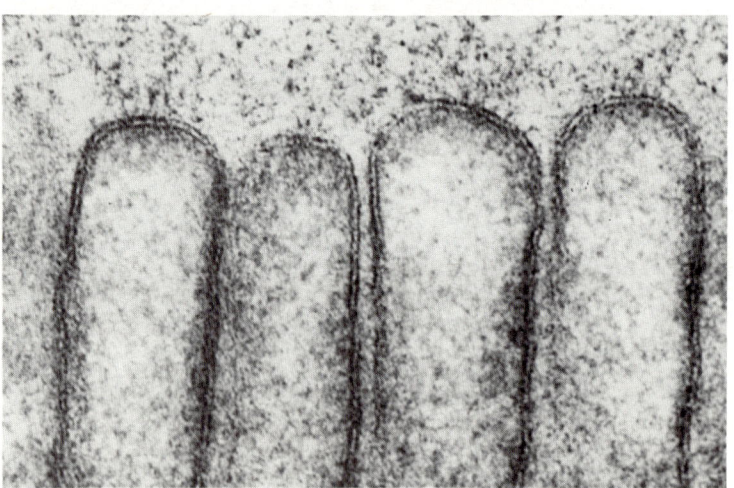

Much has since been learned about the composition and probable organisation of cell membranes. The lipids are mainly **phospholipids** which are polar molecules (Chapter 3). Considerable variation in fatty acid content occurs in membranes from the cells of different species. The amount of **protein** relative to the quantity of lipid also varies from one cell type to another. Proteins in the outer part of the membrane may even differ from those of the inner part. Many are **carrier proteins** which transport substances across the membrane. The way in which the components of cell membranes are now thought to be arranged is shown in Figure 1.8.

FIG. 1.8 *Modern interpretation of the structure of the cell membrane*

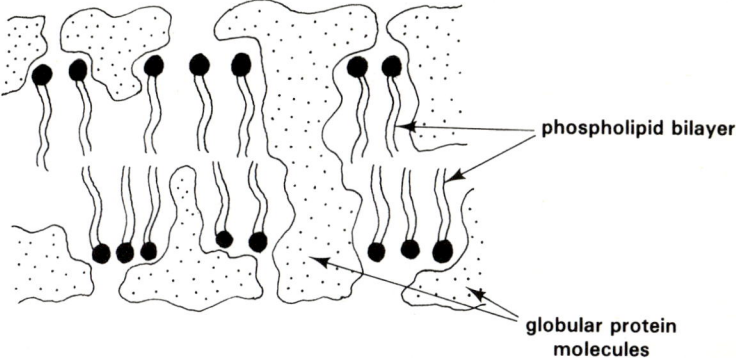

phospholipid bilayer

globular protein molecules

FIG. 1.8 *Modern interpretation of the structure of the cell membrane*

The plasma membrane is much more than just a protoplasmic boundary. It provides a means of controlling the passage of materials both into and out of the cell. Some materials are taken in by **phagocytosis** and **pinocytosis**. Phagocytosis is a mechanism whereby large suspended particles are taken wholesale into cells (Chapter 9). Pinocytosis is the intake of droplets of liquid by the formation of tiny pockets in the membrane. The plasma membrane also behaves as a **selectively permeable** membrane thereby controlling the passage of water and dissolved substances into or out of the cell. Water passes through the membrane by **osmosis** (Chapter 2). Water-soluble substances cross the membrane by **diffusion** or by **active transport**. It is now generally agreed that such substances are transported through the membrane by carrier proteins (Chapter 13). Lipid-soluble compounds can pass through membranes by dissolving in the phospholipid layer.

1.2.2 Endoplasmic reticulum

Based on observations with the light microscope cell biologists once regarded the cytoplasm as a homogeneous jelly. This view was radically changed when thin sections of cells were examined with electron microscopes. Extending throughout the cytoplasm is a three-dimensional network of sac-like and tubular cavities bound by a unit membrane. This structure is called the **endoplasmic reticulum** (**ER**) (Fig. 1.9). Here and there the membranes are covered on the cytoplasmic side with ribosomes, the **rough ER**. In other places the ribosomes are lacking, the **smooth ER**. The total area of the ER membranes in a cell of volume 5000 μm^3 can be as much as 40 000 μm^2.

If labelled amino acids are introduced into cells, radioactivity first appears in the ribosomes; then within a few minutes it is found in the membrane-enclosed sacs of the rough ER. The reason for this is that the rough ER is concerned with the production and storage of protein molecules before they are used inside the cell or are secreted to the exterior. It is thus not surprising that the rough ER is very prominent in **enzyme-secreting** cells such as those of the pancreas. The enzymes enter vesicles formed by the Golgi body before they are moved through the cell membrane by **reverse pinocytosis**.

The smooth ER is prominent in **steroid-secreting** cells such as the interstitial cells of the testes and the adrenal cortex and also in cells concerned with lipid metabolism such as the epithelial cells of the intestine. It also gives rise to the Golgi body (section 1.2.3). Both types of ER are continuous with the nuclear membrane.

The main functions of the ER are to partition the cytoplasm thus allowing metabolic reactions to be separated, to provide a relatively large surface area for the sites of metabolic reactions, and to act as a transport system for conveying materials from the inside to the outside of the cell.

FIG. 1.9 (a) *Electronmicrograph of a section through rough ER, × 100 000*

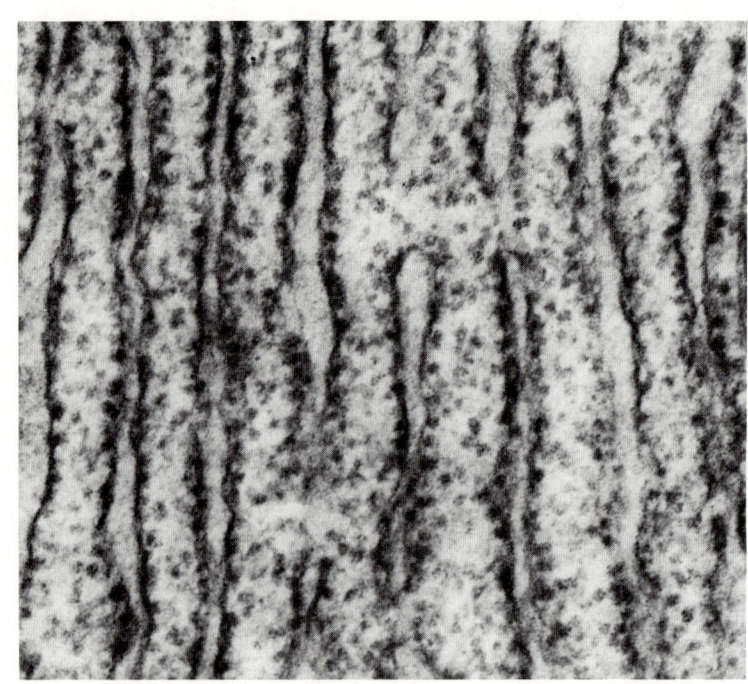

FIG. 1.9 (b) *Three-dimensional drawing of part of the rough endoplasmic reticulum*

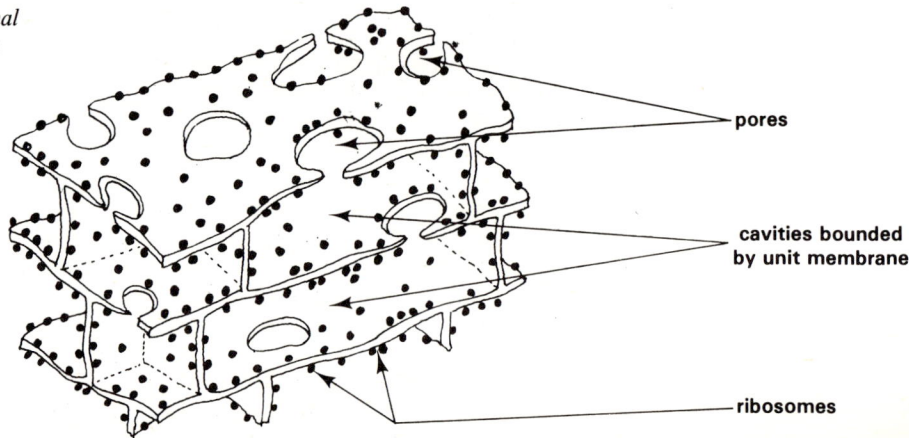

pores

cavities bounded by unit membrane

ribosomes

1.2.3 The Golgi body

As mentioned earlier in the chapter the **Golgi body** was discovered in mammalian nerve cells. Since then it has been seen in cells from almost every group of living organisms. In transverse section the Golgi body usually appears as a series of closely packed, parallel curved pockets (Fig. 1.10). The pockets are bound by unit membranes and are called **cisternae**. From the edges of the cisternae tiny vesicles arise. Some of the vesicles become lysosomes (section 1.2.5), some fuse with and enlarge the plasma membrane, others carry secretions to the plasma membrane for release to the exterior. Recently it has been shown that the cisternae are net-like in appearance (Fig. 1.11).

FIG. 1.10 *Electronmicrograph of a thin section through a Golgi body,* × *24 000*

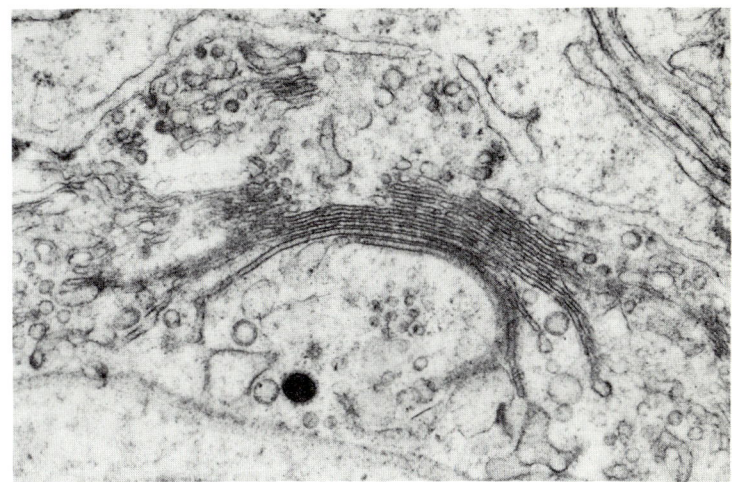

FIG. 1.11 *Three-dimensional drawing of part of the Golgi body*

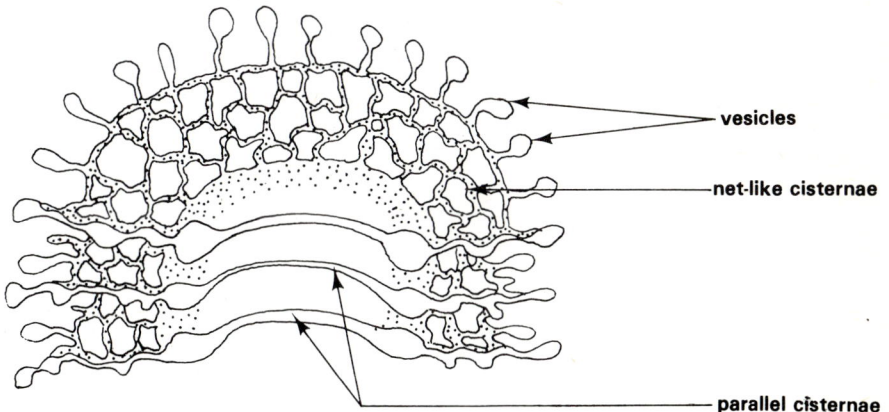

vesicles

net-like cisternae

parallel cisternae

Like the ER, Golgi bodies are well developed in **secretory cells**. Many animal cell secretions are **glycoproteins** which are polysaccharides joined to protein molecules. In the Golgi body carbohydrate is added to protein coming from the ER and the product is used internally by the cell or secreted at the cell surface. **Mucus** is a typical glycoprotein secreted by goblet cells which abound in the respiratory and gastro-intestinal tracts of mammals. Granules of **zymogen**, an inactive form of several enzymes such as trypsin and chymotrypsin, are found in the Golgi body of pancreatic cells. In plant cells the Golgi body may be concerned with the synthesis of cell wall materials such as hemi-celluloses and pectic compounds and possibly cellulose too.

11

1.2.4 Vacuoles

Large fluid-filled cytoplasmic cavities called **vacuoles** are typical of many plant cells. Each vacuole is bounded by a unit membrane known as the **tonoplast**, which is probably an extension of the ER. The permeability of the tonoplast often differs from that of the plasma membrane. In immature plant cells vacuoles are often inconspicuous but as a result of cell growth the vacuoles may eventually occupy up to 90 per cent of the cell volume. Vacuoles contain an aqueous solution of salts, sugars, pigments and other substances, often in very high concentrations. The negative water potential of the vacuolar contents is largely responsible for drawing water into plant cells thereby creating a pressure potential (Chapter 2).

1.2.5 Lysosomes

The term **lysosome** was given by Christian de Duve in 1955 to tiny organelles containing hydrolytic enzymes lying near the nucleus of liver cells. Electronmicrographs show **primary lysosomes** as small vesicles bounded by a unit membrane arising from the edges of the Golgi body. Larger **secondary lysosomes** are formed by fusion of primary lysosomes with small vacuoles. The vacuoles arise by infolding of the plasma membrane (Fig. 1.12). Primary lysosomes may also fuse with **autophagosomes** which are small membraneous pockets enclosing worn-out organelles such as mitochondria and ribosomes.

The main function of lysosomes is the digestion of materials made in the cell or taken into the cell from outside. A large variety of enzymes has been demonstrated in lysosomes from different sources. Many are lipases, carbohydrases and peptidases which hydrolyse lipids, carbohydrates and proteins respectively. The enzymes are sometimes secreted to the exterior and catalyse biochemical reactions outside the cell. In other cases they work inside the cell. The destruction of foreign particles such as bacteria engulfed by phagocytes depends on lysosomal enzymes, as does the breakdown of ageing organelles in all cells. Absorption of the tails of amphibian tadpoles,

FIG. 1.12 (a) *Electronmicrograph of a section through lysosomes, × 28 000*

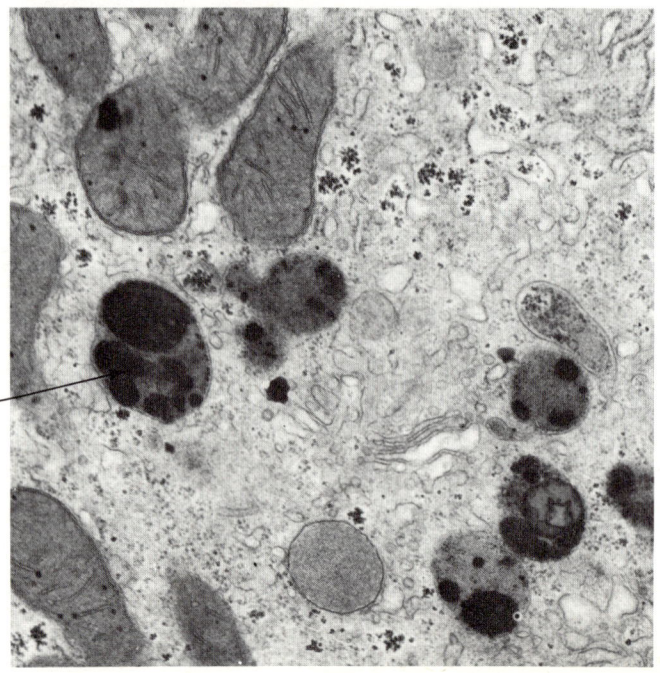

lysosome

once thought to be brought about by phagocytic blood cells, is now known to be due to the activity of lysosomes inside the tail cells. Lysosomal enzymes are also involved in the metamorphosis of insects.

Only recently have lysosomes been identified in plant cells. Enzymes secreted by lysosome-like vesicles are responsible for the breakdown of stored food in the endosperm of germinating seeds (Chapter 20).

In old and diseased cells, enzymes released internally by lysosomes bring about self-destruction, **autolysis**, of the protoplast. For this reason lysosomes are sometimes called 'suicide bags'.

FIG. 1.12 (b) *Summary of the functions of lysosomes*

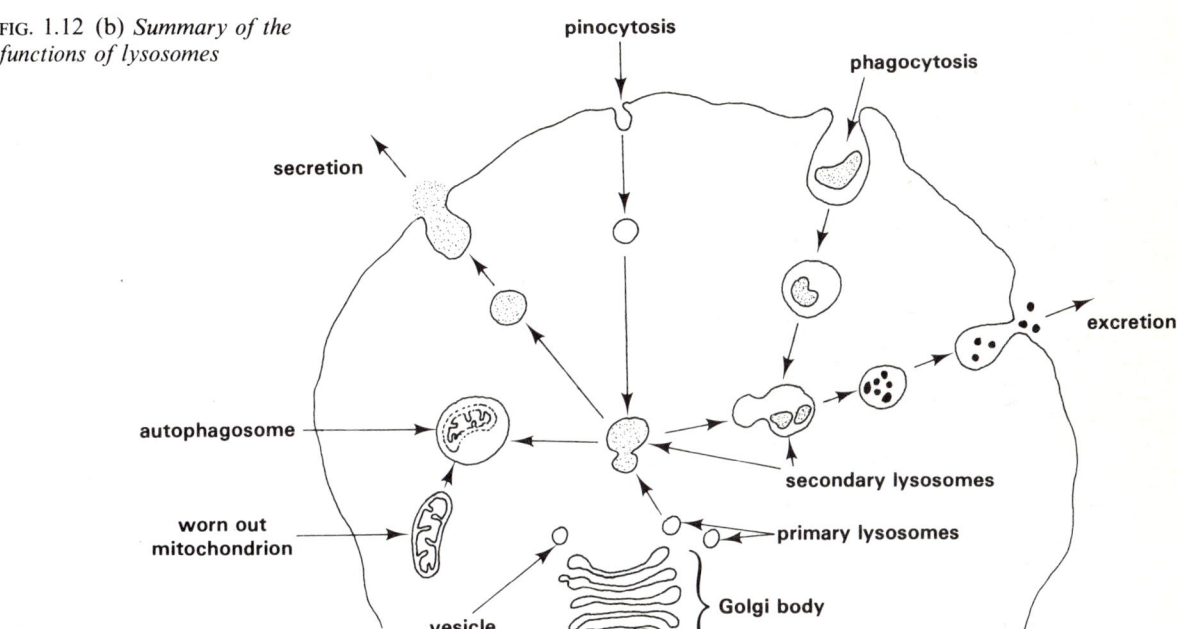

1.2.6 Ribosomes

In electronmicrographs **ribosomes** appear as small, dense, spherical bodies about 15 nm in diameter. Chemically they consist of **protein** and **ribonucleic acid (RNA)**. Frequently clusters of 2, 3, 4 or 5 ribosomes called **polysomes** are connected by a common strand of messenger RNA. The function of ribosomes is the synthesis of proteins under the control of the genetic code (Chapter 6). Proteins which are secreted at the cell surface are made by ribosomes which form part of the rough endoplasmic reticulum. Ribosomes floating freely in the cytoplasm make proteins for use inside the cell.

1.2.7 Centrosomes

Centrosomes are characteristic of animal and fungal cells but are not found in plants. A centrosome consists of a pair of **centrioles** which in non-dividing cells lie near the nucleus with their long axes at 90° to one another. A centriole is made of nine tubular filaments about 0·2 μm in length arranged in the form of a cylinder. At very high resolution each filament can be seen to consist of three fused hollow fibrils (Fig. 1.13(a) and (b)).

During nuclear division the centrosome divides and a pair of centrioles moves to opposite poles of the cell. They produce a system of microtubules called **spindle fibres** radiating towards the equator of the cell. Chromosomes become attached to the spindle equator before migrating to the poles of the cell, seemingly connected to the microtubules (Chapter 7).

In motile cells centrioles divide to produce basal bodies from which flagella and cilia develop.

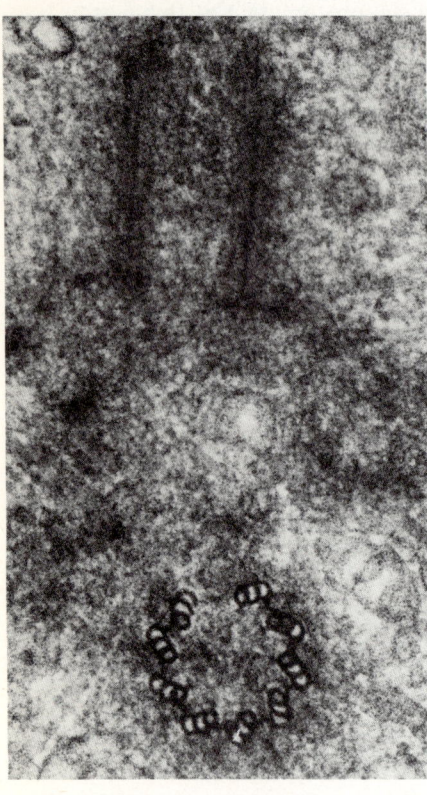

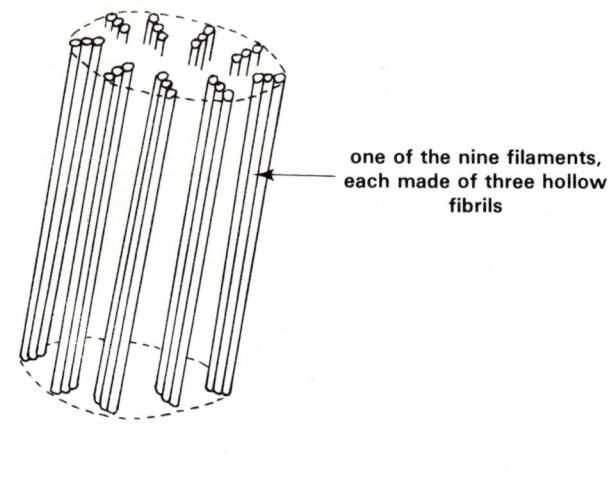

FIG. 1.13 (b) *Diagrammatic representation of the structure of a centriole*

one of the nine filaments, each made of three hollow fibrils

1.2.8 Mitochondria

The presence of small elongate bodies in the cytoplasm of plant and animal cells was first reported in the 1850s. The name **mitochondria**, meaning thread-granules, was later given to these structures which are just visible as very small grains with a light microscope. It was not until a hundred years later that thin sections of mitochondria examined under the electron microscope showed that they had a structural complexity previously unsuspected. Mitochondria from different organisms are similar in structure. They have a smooth outer unit membrane and a much folded inner unit membrane. The folds, which are called **cristae**, increase the surface area of the inner membrane thus providing a large number of sites for biochemical activity. Between the two membranes is a fluid-filled **intracristal space** while the inner membrane encloses a space containing a mixture of protein, lipid and nucleic acids called the **matrix**. In some electronmicrographs small stalked spherical bodies about 9 nm in diameter have been seen attached to the matrix side of the cristae (Fig. 1.14).

The shape of mitochondria is variable, from rod-shaped to spherical, spiral and even cup-shaped. They also vary in size depending on their source. The mitochondria from mammalian liver cells measure 1–2 μm long and 0·3–0·7 μm wide, while those from pancreatic cells, although of the same width, are up to 10 μm long. Cells which are metabolically very active contain large numbers of mitochondria. Insect flight muscles for example have numerous mitochondria packed between the muscle fibres (Fig. 1.15). The companion cells of phloem (Chapter 12) and the tubule cells of kidney nephrons (Chapter 10) are also packed with mitochondria.

FIG. 1.14 (a) *Electronmicrograph of a section through a mitochondrion*, × 81 000

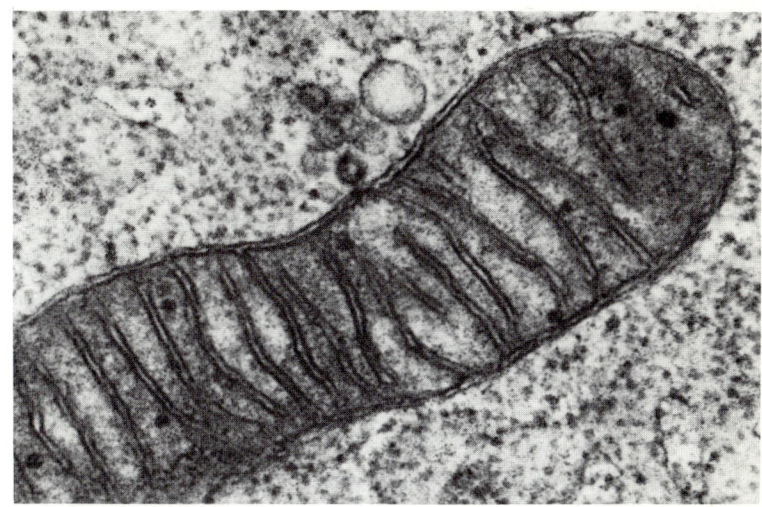

FIG. 1.14 (b) *The internal structure of a mitochondrion*

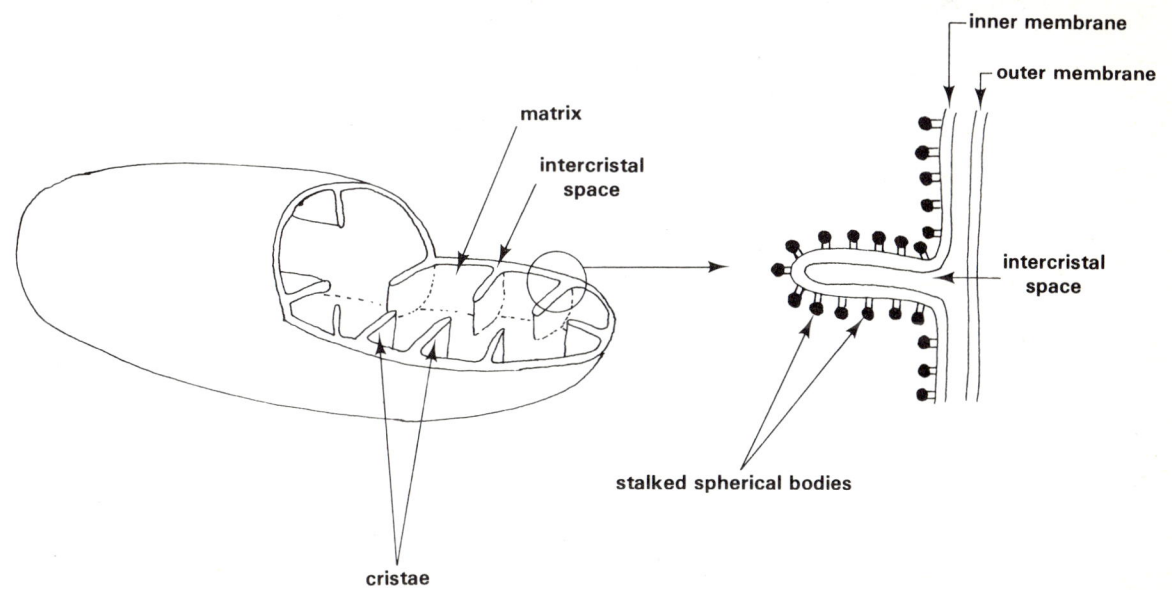

inner membrane

outer membrane

matrix

intercristal space

intercristal space

stalked spherical bodies

cristae

Structurally, mitochondria provide sites isolated from the rest of the cytoplasm on which enzyme-catalysed reactions occur. Their main function is the production of **adenosine 5′-triphosphate (ATP)** from adenosine 5′-diphosphate (ADP) and inorganic phosphate ions. The conversion of ADP to ATP is an energy-consuming reaction. Energy for the reaction is obtained from the oxidation of energy-rich substrates such as pyruvic acid which is derived from sugars metabolised in the cytoplasm. The oxidation process is called **respiration**, the biochemistry of which is described in detail in Chapter 5.

In some cells it is possible to observe mitochondria dividing. In others they appear to arise from immature mitochondria called **promitochondria**. The evolution of mitochondria is the subject of great controversy. Some cell biologists believe they were originally formed from infolding of the plasma membrane while others consider mitochondria to be degenerate symbiotic bacteria!

FIG. 1.15 *Electronmicrograph of a section through the flight muscles of a housefly,* × 58 000

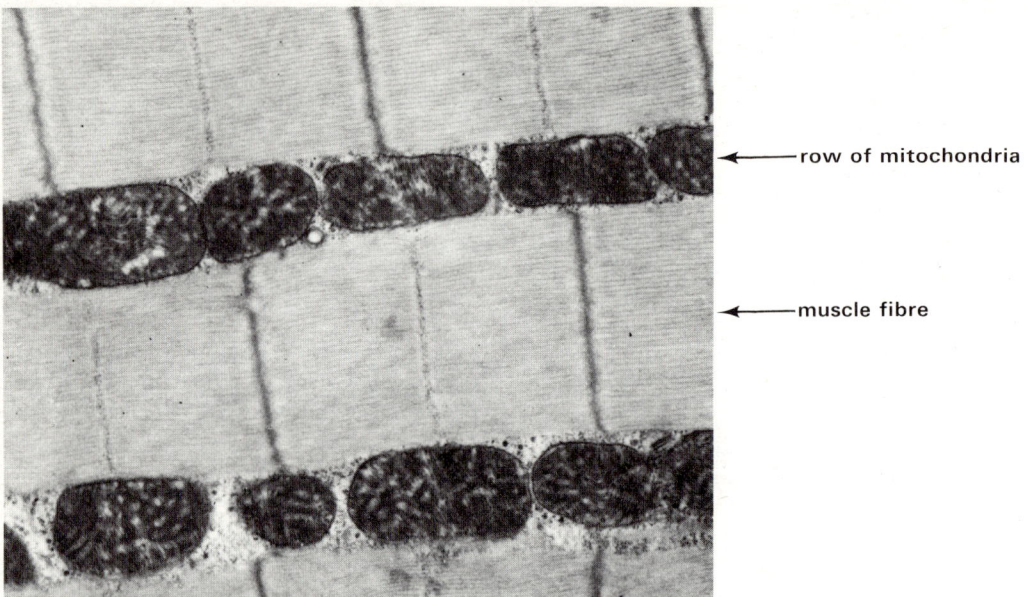

← row of mitochondria

← muscle fibre

1.2.9 Chloroplasts

In all photosynthetic plants apart from the blue-green algae and photo-autotrophic bacteria the light-absorbing pigments are housed in complex organelles called **chloroplasts**. In the cells of higher plants chloroplasts are shaped like biconvex lenses 4–10 μm in diameter and 2–3 μm thick. They are found chiefly in the mesophyll cells of leaves. Being so large chloroplasts can be seen with a light microscope and they have been studied from the mid-seventeenth century onwards.

Thin sections of chloroplasts viewed under an electron microscope show them to be bounded by a double unit membrane. The outer membrane is smooth while the inner is extended inwards as a system of layers called **lamellae** in which the photosynthetic pigments are located. In places the lamellae appear as flat discs known as **grana** piled on top of each other. These are connected by **intergrana lamellae**. The entire system of internal membranes is suspended in an aqueous matrix called the **stroma** which contains protein and nucleic acids. Following a period of illumination, photosynthetic end-products such as **starch grains** and **lipid globules** appear in the stroma (Fig. 1.16).

The arrangement of chlorophyll and other photosynthetic pigments in the chloroplast membranes is highly organised and concentrated in the grana which are just visible with a light microscope as darker green spots inside chloroplasts. Figure 1.17 shows the way in which the pigments are probably distributed in a granum. It has been suggested that the pigment layers consist of particles called **quantasomes** which are thought to be the basic photosynthetic units.

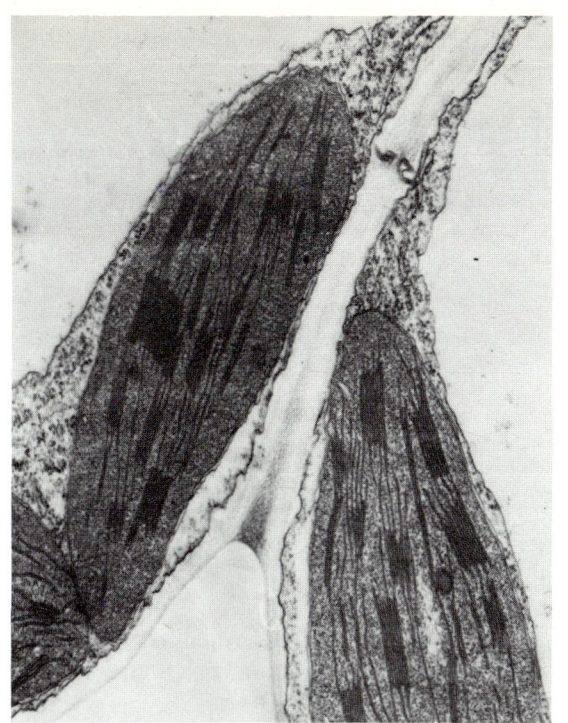

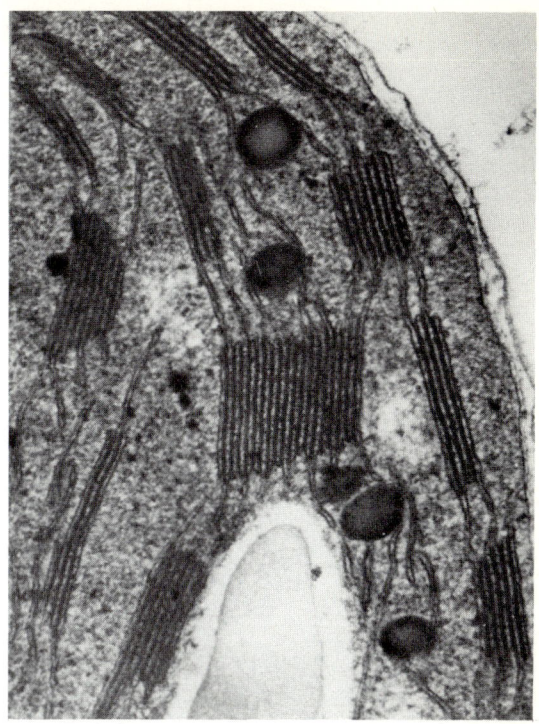

FIG. 1.16 (a) *Electronmicrographs of sections through chloroplasts*

(i) *whole chloroplast from a tobacco leaf cell, ×30 000*

(ii) *part of a chloroplast from a* Petunia *leaf cell, ×75 000*

FIG. 1.16 (b) *Drawing of a thin section of a chloroplast*

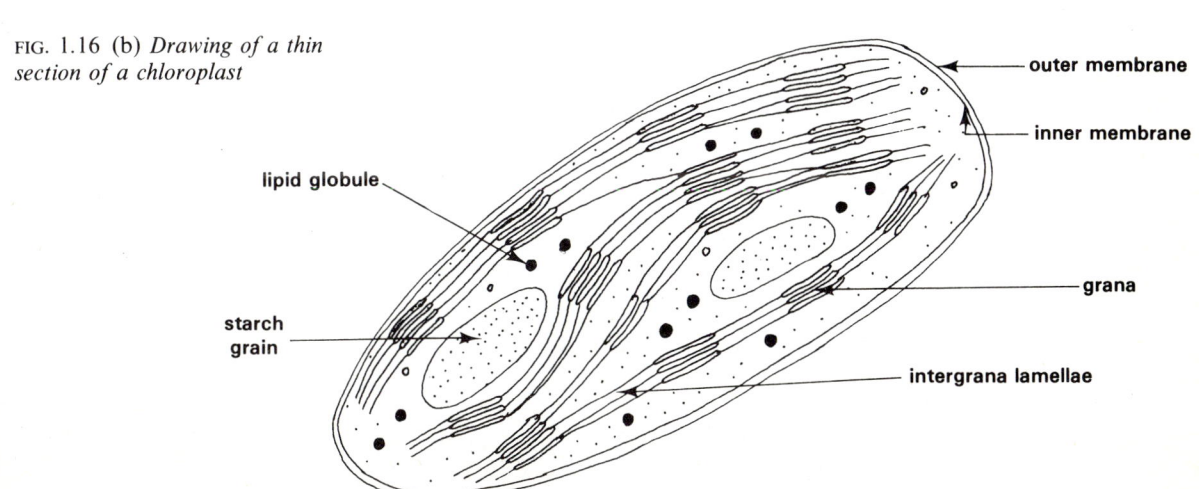

lipid globule

starch grain

outer membrane

inner membrane

grana

intergrana lamellae

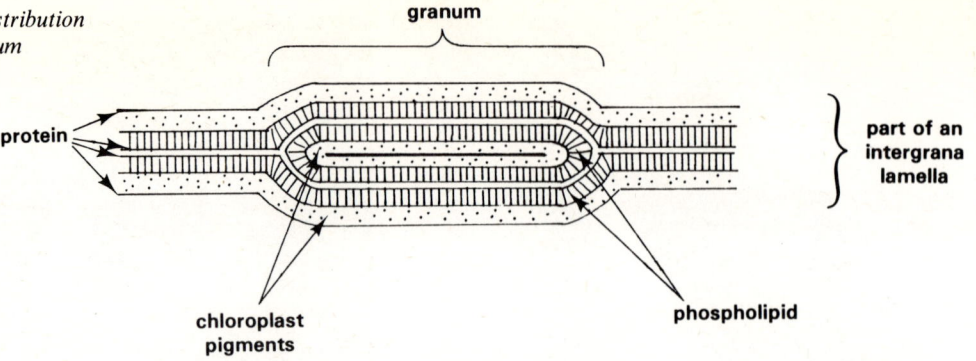

FIG. 1.17 *Probable distribution of pigments in a granum*

Chloroplasts provide sites on which the biochemical and photochemical reactions of photosynthesis can proceed without interference from those going on in the rest of the cytoplasm. Details of the reactions are given in Chapter 5. In the pigment-containing membranes, and especially in the grana, solar energy is used to produce reduced **nicotinamide-adenine dinucleotide phosphate (NADPH)** and ATP in the light-dependent reactions of photosynthesis. The stroma contains the enzymes necessary for the light-independent reactions in which NADPH, carbon dioxide and energy from ATP are used to synthesise energy-rich organic compounds such as sugars and starch. Chloroplasts are thus the organelles in which energy from sunlight is converted to chemical bond energy.

The chloroplasts of higher plant cells originate from simple organelles called **proplastids** which are abundant in actively dividing cells. The proplastids divide and during mitosis (Chapter 7) pass into newly formed cells. Later they differentiate into chloroplasts. There is just as much speculation about the evolution of chloroplasts as there is with mitochondria. A recent hypothesis favours the idea that chloroplasts are the vestiges of primitive algae which once lived symbiotically in the cells of non-green organisms.

1.2.10 The nucleus

The **nucleus** is usually the largest of the cell's organelles and can thus be seen easily with a light microscope. Van Leeuwenhoek is credited with the discovery of the nucleus in the red cells of salmon blood towards the end of the seventeenth century. A distinct nucleus is present at some stage in the cells of all forms of life apart from bacteria, blue-green algae and viruses. Although usually more or less spherical the nucleus can be more complex in shape. For example, the nucleus of a phagocytic white blood cell is lobed (Chapter 9).

In electronmicrographs the nucleus is seen to be bounded by a double-layered **nuclear membrane** (Fig. 1.18). Each layer is a unit membrane, the outer one often covered with ribosomes and continuous with the endoplasmic reticulum. Between the two layers is a **perinuclear space** about 20 nm wide. A prominent feature of the nuclear membrane not seen in the covering membranes of mitochondria and chloroplasts is the presence of numerous **pores**. These may occupy up to 15 per cent of the membrane's surface area, each pore being approximately 50 nm in diameter (Fig. 1.19). The pores provide routes for the passage of materials from the nucleus to the cytoplasm and vice versa.

Inside the nuclear membrane are found two main ingredients, **nucleic acids** and **protein**. Both **ribonucleic acid (RNA)** and **deoxyribonucleic acid (DNA)** are present, the nucleus being the main store of the cell's DNA.

Nuclear DNA is bound to a number of proteins collectively called **histone**. When the nucleus is not dividing the nucleic acid-protein complex appears as tiny granules of **chromatin**. Among the chromatin granules are one or more densely granular bodies called **nucleoli** composed mainly of DNA.

FIG. 1.18 *Electronmicrograph of a section through a nucleus, × 14 000*

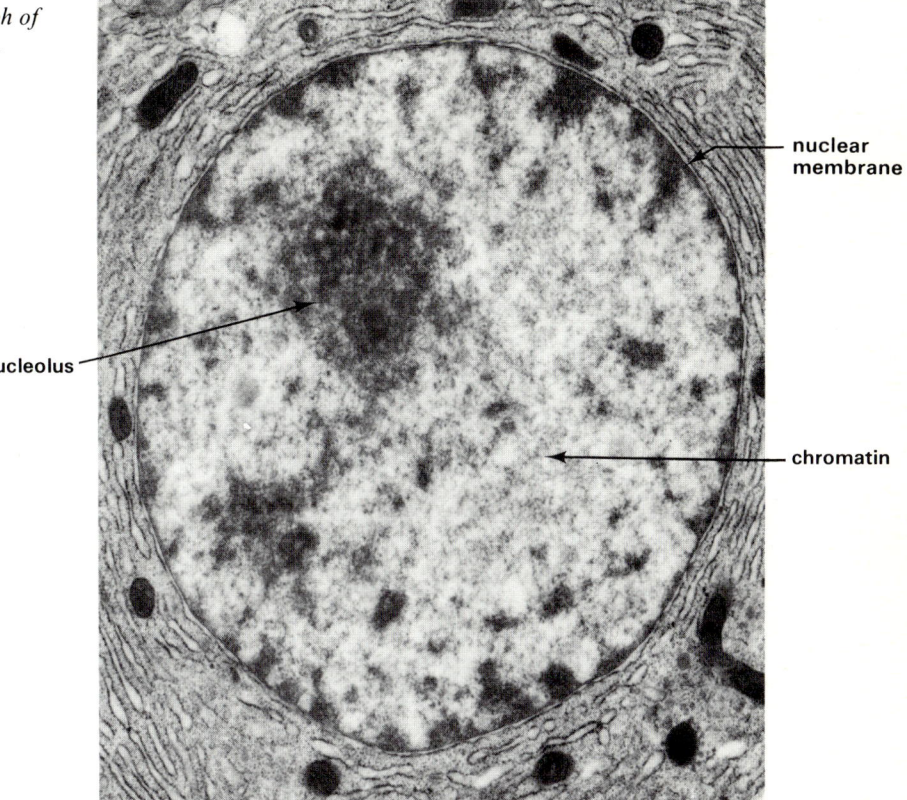

nucleolus

nuclear membrane

chromatin

FIG. 1.19 *Electronmicrograph of freeze-etched nuclear membrane of a carrot cell, × 37 000*

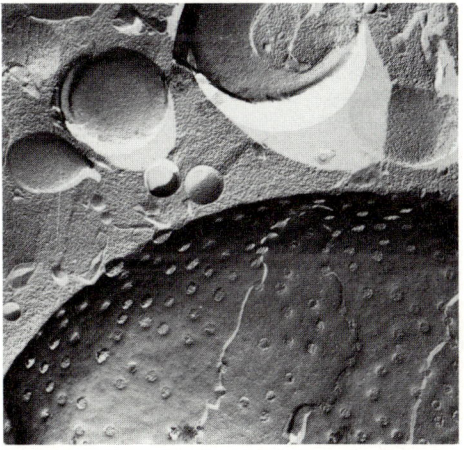

During nuclear division the nuclear membrane and nucleoli disappear and the chromatin becomes visible as thread-like bodies called **chromosomes**. At telophase (Chapter 7) the nucleoli reappear near the **nucleolar organiser**, a secondary constriction of one of the pairs of homologous chromosomes. It is thought that nucleoli code the manufacture of tiny globular bodies consisting of protein and RNA. After passing through the nuclear pores the bodies are changed into ribosomes.

The way in which the nucleus transmits hereditary materials is described more fully in Chapter 7. In Chapter 6 you can read about the way in which the nucleic acids provide the genetic code for protein synthesis at the ribosomes.

1.2.11 The plant cell wall

A characteristic of plant cells is that they have a **cell wall**. Structurally the cell wall is like fibre-glass, consisting of fibres embedded in an amorphous matrix. The fibres are of **cellulose**, an unbranched polymer of up to 10 000 β-D-glucose units (Chapter 3). Each fibre is made of several hundred microfibrils in which about 2000 cellulose molecules are held together by hydrogen bonds. The matrix consists of **pectic acid** and its salts calcium and magnesium pectate, and **hemicelluloses** which are polymers of various pentose and hexose sugars. Pectic substances also make up most of the **middle lamella** which binds adjacent plant cells to one another.

The cell wall is secreted by the protoplast it encloses. In young undifferentiated cells, such as those cut off by root and shoot apical meristems, the wall is thin and is called the **primary wall**. The cellulose fibres are orientated at random (Fig. 1.20) so that the primary wall is highly porous. This arrangement allows for considerable stretching when cells grow. Adjacent protoplasts are joined by **plasmodesmata**, threads of cytoplasm which extend through the pores. As a cell grows more cellulose fibres are laid down on the inside of the primary wall forming a thicker **secondary wall**. The cellulose fibres of the secondary wall are closely packed and laid down in an orderly way giving considerable rigidity to the cell wall. Plasmodesmata still exist but the pores through which they pass are now deeper and are called **pits** (Fig. 1.21).

FIG. 1.20 (a) *Electronmicrograph of freeze-etched primary wall of a carrot cell, × 37 000. Note the random arrangement of the cellulose fibres*

FIG. 1.20 (b) *Electronmicrograph of primary walls of adjacent cells of deadly nightshade, × 35 000*

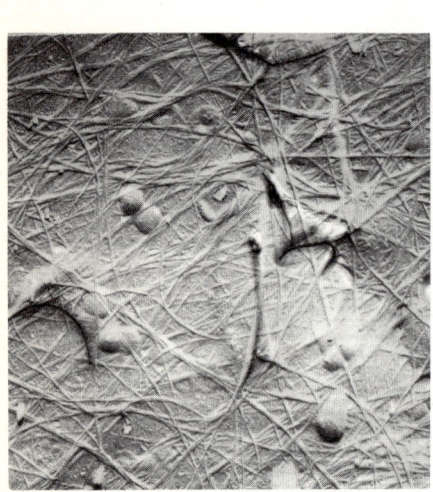

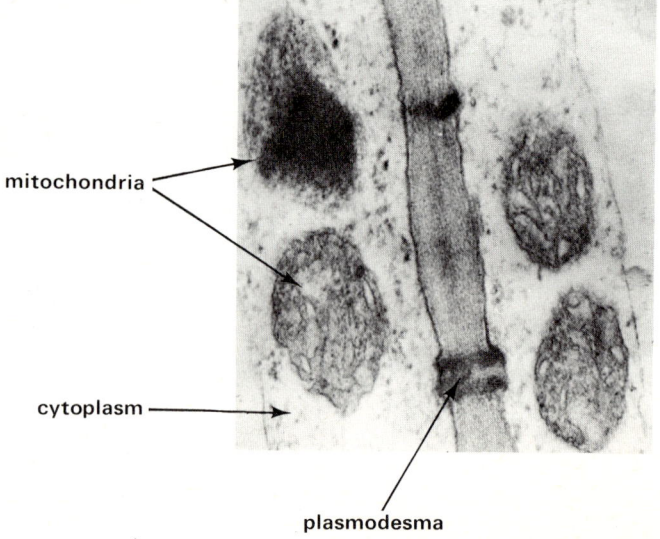

mitochondria

cytoplasm

plasmodesma

FIG. 1.21 (a) *Electronmicrograph of freeze-etched secondary wall of a carrot cell, × 37 500. Note the orderly arrangement of the cellulose fibres*

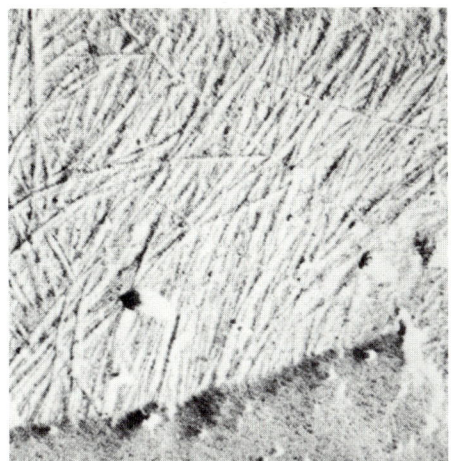

FIG. 1.21 (b) *Electronmicrograph of a freeze-etched bordered pit* (courtesy Philip Harris Biological Ltd.). *The torus is the thickened remains of the plasma membrane. It can be pushed against the border to close the pit* (see Fig. 11.20)

torus — border

During the differentiation of specialised cells such as xylem vessels, tracheids and sclerenchyma fibres the secondary wall often becomes extremely thick. Together with the primary wall it becomes impregnated with an alcohol polymer called **lignin** which gives the wall great strength. In cork tissue **suberin** impregnates the cell wall, while in the outer walls of epidermal cells of leaves and young stems **cutin** appears instead. Both suberin and cutin are waxy materials which provide an effective waterproof covering to the aerial surfaces of plants (Chapter 11).

1.3 A comparison of light and electron microscopes

Why is it that so much more detail can be made out when specimens are examined with an electron microscope as opposed to a light microscope? Before we can answer this question it is necessary to compare the principles on which light and electron microscopes work.

Figure 1.22 shows the paths of radiation in the two instruments. In both microscopes a tungsten filament lamp is used as a **source of radiation** but whereas the light microscope uses **visible light** to create the final image, the electron microscope uses a **beam of electrons**. The radiation is focused on to the specimen by a **condenser**, which in the light microscope consists of thick glass lenses mounted beneath the stage. Glass is opaque to electrons so in electron microscopes the condenser is a vertical magnetic field produced by a large cylindrical electromagnet which straightens and intensifies the electron beam. The light rays or electrons, as the case may be, now pass through the **specimen** after which the radiation is focused by an **objective lens**. An **eyepiece lens** in the light microscope further enlarges the image. The human eye cannot see electrons so in the electron microscope the final image is projected on to a viewing screen coated with a fluorescent compound such as zinc sulphide. When irradiated with electrons the fluorescent substance emits light visible to the human eye. A comparable process is used to create a picture on a television screen. With both microscopes it is possible to photograph the final image to produce **photomicrographs** and **electronmicrographs**.

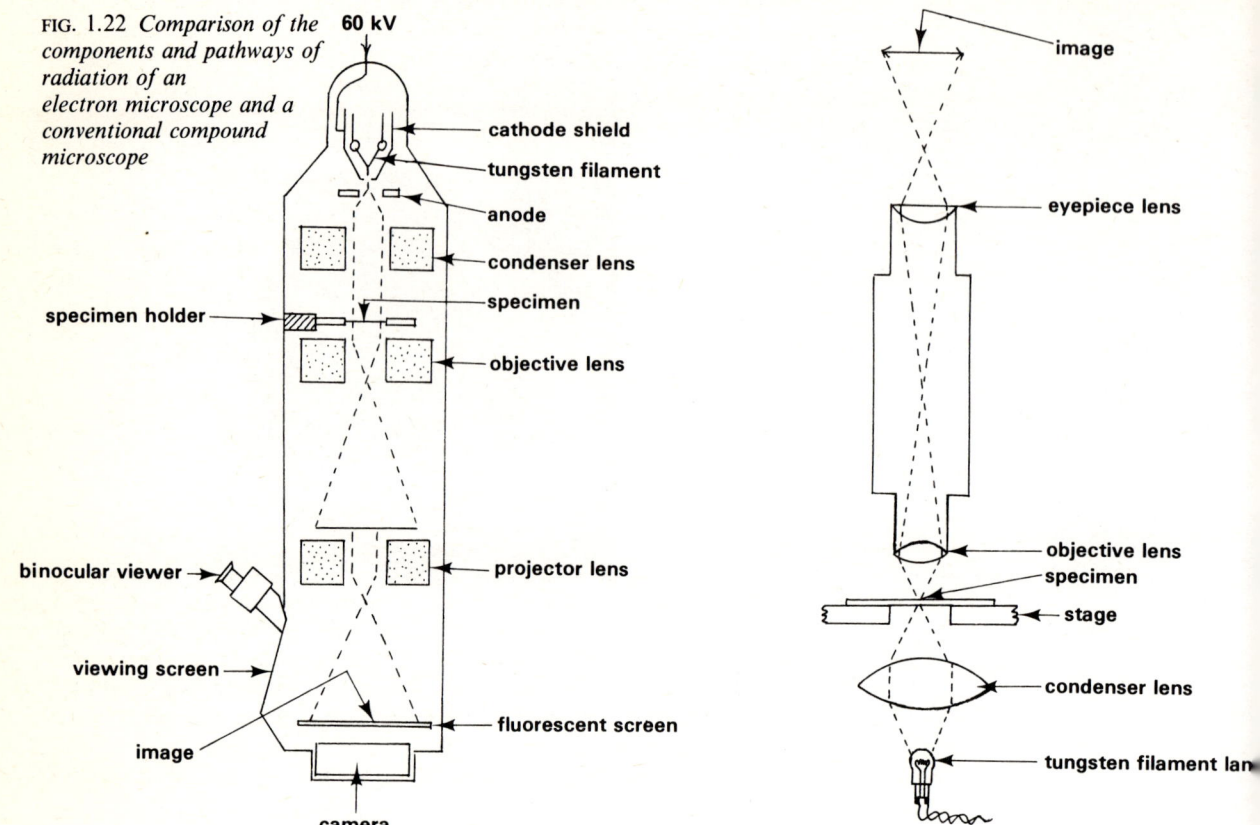

FIG. 1.22 *Comparison of the components and pathways of radiation of an electron microscope and a conventional compound microscope*

Perhaps the most obvious difference in the final images is the extent to which the specimens are magnified. With a high quality compound microscope fitted with an oil-immersion objective a magnification of about 1500 times is possible. An electron microscope can magnify up to 500 000 times. However, it is the **resolving power** of the electron microscope rather than its magnifying power which enables it to produce images containing so much more detail. Resolving power or **resolution** is the ability to make out as separate entities bodies which lie close to one another.

The power of resolution (R) of a microscope depends mainly on the wavelength of the radiation used and on the numerical aperture of the objective lens.

$$R = \frac{0.5\,\lambda}{n\,.\,sin\,\theta}$$

Where λ = wavelength of radiation

n = refractive index (**RI**) of medium between specimen and objective lens

$\theta = \frac{1}{2}$ angle of aperture (Fig. 1.23(a))

The expression n . $sin\,\theta$ is called the **numerical aperture** (**NA**) of the objective. The smaller the value of R, the better is the resolving power of the microscope. Using a compound microscope fitted with an oil-immersion objective, R = 210 nm (Fig. 1.23(b)). In other words this instrument is theoretically capable of resolving particles lying as close as 210 nm to one another. In practice it is difficult to achieve a value of R as small as this when viewing unstained preparations with a conventional compound microscope. The phase-contrast microscope provides better powers of contrast but does not improve resolution.

FIG. 1.23 *Angle of aperture and resolution. Calculate the resolution R of each of the objectives*

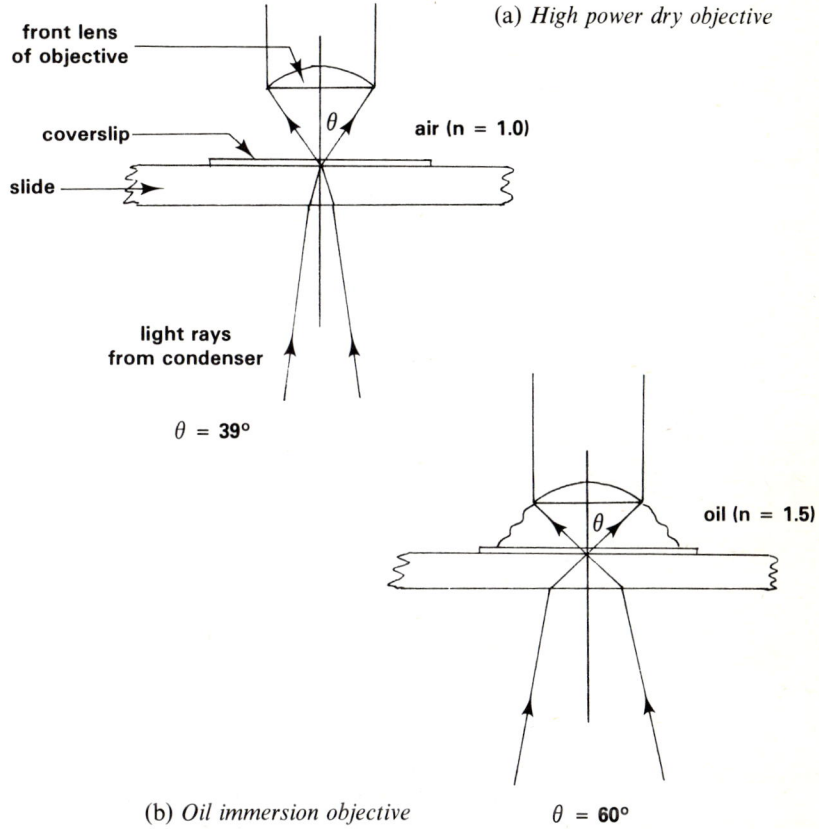

(a) *High power dry objective*

front lens of objective

coverslip

slide

air (n = 1.0)

θ

light rays from condenser

$\theta = 39°$

oil (n = 1.5)

θ

(b) *Oil immersion objective* $\theta = 60°$

With an electron microscope λ is 0·05 nm which is 10 000 times shorter than the average wavelength of white light. Other things being equal this should produce a corresponding increase in resolution but there are technical difficulties preventing this from being realised in practice. Present-day electron microscopes have resolving powers of about 0·5 nm, about 400 times greater than the light microscope. Compare these figures with the resolving power of the human eye which is about $1·0 \times 10^8$ nm.

The major disadvantage of the electron microscope is that it cannot be used to examine live specimens. Because it is necessary to process cells and tissues before they are used for electronmicroscopy, artificial bodies called **artefacts** may appear in electronmicrographs. With this instrument we see mainly the static aspects of cell biology.

How then is so much known about the functions of the various components of cells? One technique which has been of enormous help involves separating or **fractionating** the organelles. The activities of the organelles can then be studied without interference from all of the other reactions which take place in whole cells.

1.4 Cell fractionation

Live tissue is first chopped up in a cold **isotonic buffer** solution. The isotonic solution prevents distortion of the organelles. The chopped tissue is then ground up in an **homogeniser**. A domestic blender can be used for this purpose but it usually breaks most of the organelles. For sophisticated work a motor driven ground glass pestle which fits into a tube is used. This type of homogeniser develops shearing forces just sufficient to rupture the cells. Cells can also be ruptured using ultrasonic waves. The homogenate is then transferred to a **centrifuge** in which the mixture is spun at known speeds whereby the organelles are sedimented separately. The main factors governing sedimentation are the magnitude of the centrifugal force, which depends on the spinning speed, and on the size and density of the suspended particles relative to the medium in which the organelles are suspended. Exact times and speeds of centrifugation vary from one tissue to another and are determined by trial and error.

Centrifugal forces of up to 1000 times the force of gravity (1000 × g) can be attained with simple bench centrifuges used in school laboratories. Large organelles such as nuclei and chloroplasts can be sedimented by spinning at 500–600 × g for 5–10 minutes. If the supernatant liquid is spun at 10 000–20 000 × g for 15–20 minutes, mitochondria and lysosomes are sedimented. Fragmented endoplasmic reticulum with attached ribosomes, collectively called **microsomes**, can be sedimented from the supernatant liquid by spinning it for 60 minutes in an **ultracentrifuge** (Fig. 1.24) in which

FIG. 1.24 *An ultracentrifuge (courtesy Beckman–RIIC Ltd.). B shows one of the buckets in which the homogenised cell suspension is placed for centrifugation*

rotor

B

control panel

forces of 100 000 × g and above are developed. Figure 1.25 summarises the various steps in fractionating cell organelles by differential centrifugation.

Of course, the separation of organelles in this way can lead to their physical and chemical damage. They may not then behave as they normally do in the intact cell where their activities are closely co-ordinated. Nevertheless by suspending the separated components in a medium which closely resembles the intracellular environment some of the functions of organelles can be investigated.

FIG. 1.25 *Flow diagram showing the major stages in the fractionation of organelles by differential centrifugation*

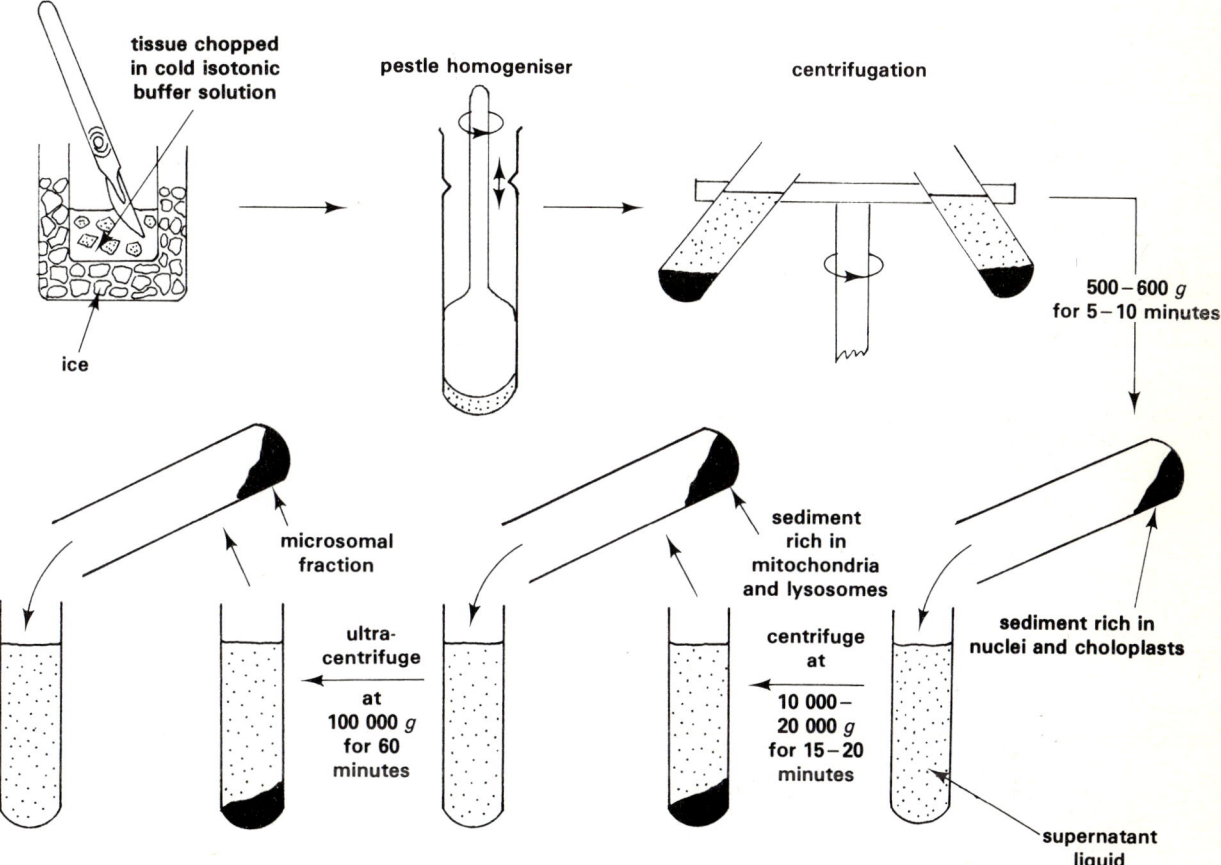

The wealth of information we have about cell structure has clearly depended on technological improvements in microscopy. The electron microscope is the most sophisticated of instruments available to date for probing the fine structure of cells. Useful as it is to know the detailed structure of cells, electronmicrographs tell us little about the functions of the various cellular components. Cell biologists therefore make use of a variety of techniques in investigating the relationship between structure and function. Some of the techniques such as cell fractionation have already been described in this chapter. Details of other techniques are given at appropriate places in the following chapters.

Chapter 2

Water and aqueous solutions

Water is the most common substance on earth. Vast quantities of frozen water make up the ice caps of the north and south poles. Even larger volumes of liquid water fill rivers, lakes and oceans. The atmosphere contains an enormous volume of water vapour. Wherever there is water there is life. Yet the importance of water to living organisms is often overlooked.

Life is thought to have originated in water (Chapter 3) and many present-day species of plants and animals live either in fresh water or in the sea. Irrespective of whether a creature lives in water or on land its life processes cannot occur without water. The living cells of all organisms contain a high proportion of water (Table 2.1) and it is in aqueous solution that metabolic reactions take place. However as we shall see in this chapter there are many other ways in which water is indispensable to all living creatures.

Table 2.1 Water content of living organisms

SPECIMEN	WATER CONTENT PER CENT FRESH MASS
lettuce leaf	93–95
carrot taproot	89–91
strawberry fruit	88–90
jellyfish	95–98
earthworm	82–84
mammal	63–68

In order to appreciate why water is so important to living organisms it is first necessary to know some of the characteristics of water molecules.

2.1 The water molecule

A water molecule consists of two atoms of hydrogen and one atom of oxygen. The atoms are joined in such a way that the single electron of each of the hydrogen atoms is shared with electrons in the outer shell of the

FIG. 2.1 *Formation of a molecule of water*

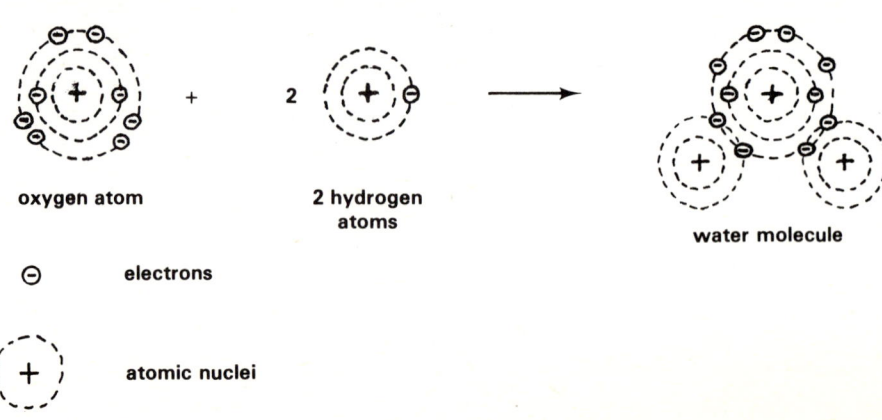

oxygen atom   2 hydrogen atoms   water molecule

⊖   electrons

+   atomic nuclei

oxygen atom (Fig. 2.1). The unshared pairs of electrons repel the shared pairs so that the hydrogen nuclei are pushed towards one another. As a consequence the molecule is bent (Fig. 2.2). The end of the molecule where the hydrogen nuclei are located is positively charged while the other end where the oxygen atom is found is negatively charged.

Such a molecule is said to be **polar** because the positively charged end differs from the negatively charged end in the way it reacts with ions and charged molecules. The polarity of water molecules accounts for many of the properties of water which are of importance to living organisms.

FIG. 2.2 *The non-linear water molecule*

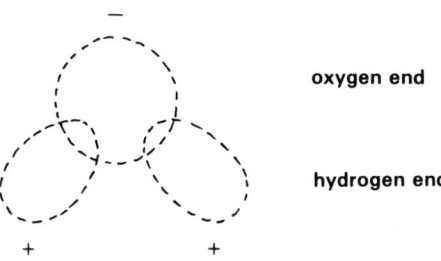

2.2 Properties of water

2.2.1 Thermal properties

Compared with other compounds of about the same relative molar mass, water has higher melting and boiling points and is liquid over a wider temperature range (Table 2.2). The reason for this can be understood by looking at the way in which water molecules react with each other. Because particles of opposite charge attract each other, the negative pole of a water molecule is drawn towards the positive pole of another. Weak **hydrogen bonds** are formed between linked water molecules. In this way a water molecule can bind with up to four others, two attracted to the oxygen atom and one to each of the hydrogen atoms (Fig. 2.3). Water therefore exists as **molecular clusters** rather than as individual molecules.

Table 2.2 Some thermal properties of water and compounds of comparable relative molar mass

SUBSTANCE	FORMULA	RELATIVE MOLAR MASS	MELTING POINT/°C	BOILING POINT/°C
water	H_2O	18	0	100
ammonia	NH_3	17	-78	-33
methane	CH_4	16	-184	161

Temperature has an effect on the amount of bonding between water molecules. At 0°C, the freezing point of water, the molecules are arranged in a regular, hexagonal, crystalline network in which each water molecule is hydrogen-bonded to four others (Fig. 2.4(a)). At temperatures just above 0°C some of the bonds are broken producing clusters of molecules which fit together more compactly than the evenly spaced water molecules in ice (Fig. 2.4(b)). This is why liquid water is denser than ice. Water is most dense at 4°C.

Changes in the density of water are important for the survival of many aquatic organisms. The temperature of a body of open water such as a pond or a lake is affected by the temperature of the air above the water.

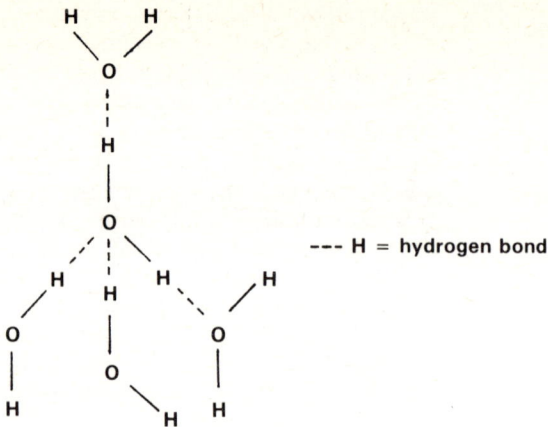

FIG. 2.3 *Hydrogen bonding between water molecules*

--- H = hydrogen bond

Very cold air causes a body of water to cool from the surface downwards. When the temperature of the upper layers falls to 4°C the dense water sinks bringing warmer water to the surface. Convection currents of this sort delay the freezing of a body of water. Ultimately a point may be reached when the water near the surface freezes. The ice so formed, being less dense than the water beneath, floats on the surface where it insulates the water from further heat loss. Consequently liquid water, in which aquatic life can survive, remains beneath the ice.

FIG. 2.4 (a) *Crystalline structure of ice*

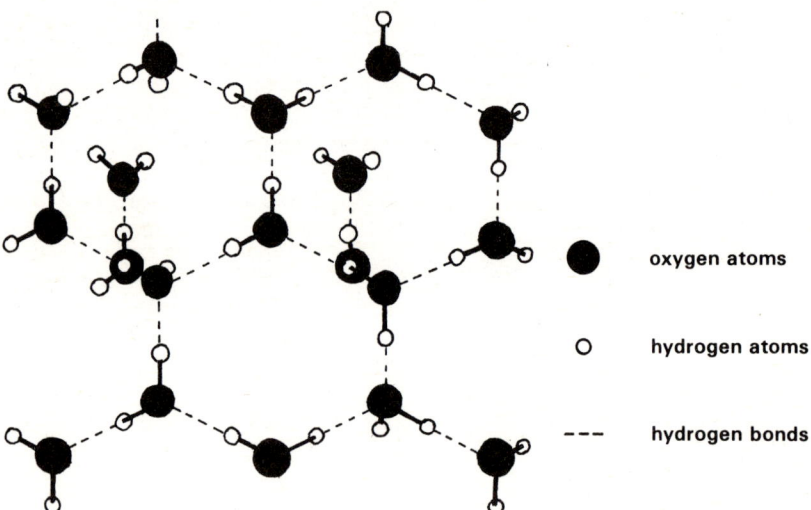

● oxygen atoms

○ hydrogen atoms

--- hydrogen bonds

FIG. 2.4 (b) *Packed molecules in liquid water*

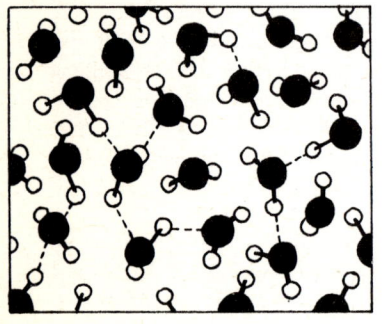

It should by now be evident that heat energy can break the bonds between water molecules. Because of the extensive hydrogen-bonding in ice, a relatively large amount of heat, called the **latent heat of melting**, is required to convert ice to liquid water. This is why water has a relatively high melting point. The formation of water vapour involves the breaking of hydrogen bonds between molecules of liquid water. Evaporation uses a lot of heat energy called the **latent heat of vaporisation**. This is why water has a relatively high boiling point. It also explains why a considerable cooling effect occurs when water evaporates from the bodies of living organisms.

28

Where do you suppose most of the heat energy for this evaporation comes from? The evaporation of water from terrestrial organisms plays an important role in preventing them from becoming overheated (Chapters 11 and 17).

The amount of heat required to raise the temperature of a given mass of a compound is called the **specific heat capacity**. Water has a much higher specific heat capacity than other compounds of similar relative molar mass. Why do you think this is so? The significance of water's high specific heat capacity to living organisms is that water does not quickly heat up or cool down as air does. It is for this reason that large bodies of water such as oceans and lakes which are inhabited by living organisms have a steady temperature. The ability of water to act as a **thermal buffer** also helps to prevent rapid changes in body temperature of terrestrial organisms. Remember that land-dwelling creatures, like all forms of life, contain a large volume of water.

2.2.2 Density and viscosity

Compared with many other substances water has a high **relative density** (specific gravity). It is for this reason that many living organisms readily float in water. Water is thus a means of external support for aquatic organisms. The buoyancy of water also helps the swimming of motile gametes and in the dispersal of fruits, seeds and spores.

Viscosity is a measure of the difficulty with which molecules slide over one another. You may be excused for thinking that the hydrogen bonds between water molecules should cause water to be highly viscous. However, this is not so. The hydrogen bonds are continually broken and re-formed so that water molecules slide over each other with relative ease. The viscosity of water is thus relatively low and enables water to flow rapidly through the bodies of terrestrial plants (Chapter 11).

Aqueous solutions on the other hand can be very viscous. The high viscosity of blood plasma and lymph is of importance in the smooth flow of body fluids in mammals (Chapter 9).

2.2.3 Surface tension, capillarity and tensile strength

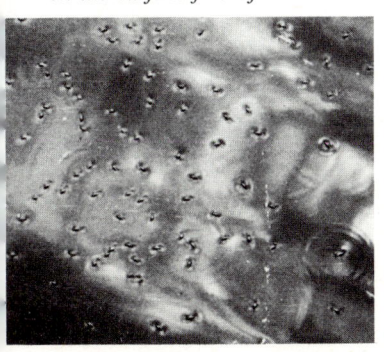

FIG. 2.5 *Pond skaters suspended on the surface film of water*

Yet another property of water attributable to the polarity of its molecules is its high **surface tension**. At the surface of an aqueous solution water molecules are pulled downwards and inwards by the hydrogen bonds which join them to water molecules just below the surface. The result is that the surface area of the water shrinks and forms an elastic skin. Some aquatic organisms such as pond skaters and duckweed are sufficiently light to be supported by the elastic skin (Fig. 2.5). Others, such as mosquito larvae, can remain suspended from the underside of the skin.

Surface tension is also mainly responsible for the capillary movement of liquids through fine tubes. **Capillarity** is one of the forces which helps bring about the transport of water through terrestrial plants (Chapter 11).

The individual hydrogen bonds between two linked water molecules are not very strong. However, the combined hydrogen bonding between a large number of water molecules can be strong enough to prevent a column of water from breaking when the column is stretched. Water thus has a high **tensile strength**. This property can be important in the upward movement of water through the vascular tissue of terrestrial plants (Chapter 11).

2.2.4 Penetration by light

Light rays can penetrate water with relative ease because water is transparent. The depth to which light can penetrate is of course reduced if the water is turbid due to suspended particles. In clear water red and yellow light can reach to a depth of 50 m while blue and violet rays can go down to 100 m.

The ability of light to penetrate water enables photosynthetic organisms to inhabit the vast surface volumes of lakes and seas. It also means that light can easily penetrate the transparent water-filled epidermis of leaves and reach the underlying cells which contain the light-absorbing pigments (Chapter 12).

2.2.5 Dissociation of water

The way in which electrons of two hydrogen atoms are shared with electrons of the outer shell of an oxygen atom when a water molecule is formed is described in section 2.1. The nucleus of a hydrogen atom consists of a single positively charged particle called a **hydrogen ion** (H^+), also called a proton. In an aqueous solution small numbers of water molecules lose their hydrogen ions. The remainder of the water molecule is called a **hydroxyl ion** (OH^-). Such molecules are said to be **dissociated**. The process is usually written as:

$$H_2O \rightleftharpoons H^+ + OH^-$$

water   hydrogen ion   hydroxyl ion

In reality the hydrogen ions become attached to the oxygen ends of other water molecules to form **oxonium ions** (H_3O^+):

$$2H_2O \rightleftharpoons H_3O^+ + OH^-$$

oxonium ion

What do you think causes the hydrogen ions to bind with the oxygen end rather than the hydrogen end of other water molecules?

In pure water the concentration of hydrogen ions $[H^+]$, or more correctly the oxonium ions, is 10^{-7} mol dm^{-3} at 298 K. The product of the concentrations of hydrogen and hydroxyl ions, which is called the **ionic-product of water K_w**, is always 10^{-14} mol dm^{-3} in any aqueous solution at 298 K:

$$K_w = [H^+] \times [OH^-]$$

ionic-product   concentration   concentration
of water   of H^+ of OH^-

Thus for pure water at 298 K:

$$K_w = 10^{-7} \times 10^{-7} = 10^{-14} \text{ mol dm}^{-3}$$

Remember to add the indices when multiplying numbers expressed in this way.

K_w is the basis of the **pH scale**, a means of indicating the concentration of hydrogen ions in an aqueous solution. The pH of a solution is defined as:

the logarithm to the base 10 of the reciprocal of the hydrogen ion concentration of the solution.

$$pH = \log_{10}\frac{1}{[H^+]}$$

$$\text{or} \quad pH = -\log_{10}[H^+]$$

A solution of pH 7·0 is a **neutral** solution. If the pH is less than 7·0 the solution is **acidic** but if the pH is greater than 7·0 the solution is **alkaline**. The pH of pure water at 298 K is:

$$-\log_{10} 10^{-7} = 7\cdot0, \text{ so pure water is neutral.}$$

Table 2.3 Concentrations (mol dm^{-3}) of H$^+$ and OH$^-$ in solutions of pH ranging from 0–14

pH	0·0	1·0	2·0	3·0	4·0	5·0	6·0	7·0	8·0	9·0	10	11	12	13	14
[H$^+$]	1·0	10^{-1}	10^{-2}	10^{-3}	10^{-4}	10^{-5}	10^{-6}	10^{-7}	10^{-8}	10^{-9}	10^{-10}	10^{-11}	10^{-12}	10^{-13}	10^{-14}
[OH$^-$]	10^{-14}	10^{-13}	10^{-12}	10^{-11}	10^{-10}	10^{-9}	10^{-8}	10^{-7}	10^{-6}	10^{-5}	10^{-4}	10^{-3}	10^{-2}	10^{-1}	1·0

In all solutions $K_w = [H^+] \times [OH^-] = 10^{-14} \text{ mol dm}^{-3}$

Table 2.3 gives the concentrations of hydrogen and hydroxyl ions in solutions ranging from pH 0–14, the full range of the pH scale. How many times more concentrated are the hydrogen ions in solutions of pH 4·0 and 5·0 respectively compared to a solution of pH 7·0?

The pH of cellular fluid is normally between 6·5–8·0. The body fluids of higher animals also have a pH within this range. Even more significant, the hydrogen ion concentration in living organisms is kept fairly constant. Examples of the way in which this is achieved are described later in the chapter. One of the advantages of maintaining a relatively fixed pH inside living organisms is that it enables enzymes to catalyse metabolic reactions efficiently. Should the pH of body fluids go outside the pH range of 6·5–8·0 most enzymes will not work (Chapter 4).

2.2.6 Water as a solvent

A wide range of inorganic and organic substances readily dissolve in water. The reason why water acts as a **solvent** for so many inorganic compounds can be understood by considering the way in which a simple salt such as sodium chloride reacts with water. In a crystal of sodium chloride the sodium and chloride ions are held together by electrovalent bonds. When placed in water the positively charged sodium ions (Na$^+$) are attracted to the negatively charged oxygen ends of water molecules. Conversely the negatively charged chloride ions (Cl$^-$) are pulled towards the positively charged hydrogen ends of water molecules. As a consequence the sodium and chloride ions become separated by clusters of water molecules and an **aqueous solution** of sodium chloride is formed (Fig. 2.6). The polarity of water molecules is clearly a key factor in the ability of water to dissolve inorganic substances.

FIG. 2.6 *Reaction between sodium chloride and water*

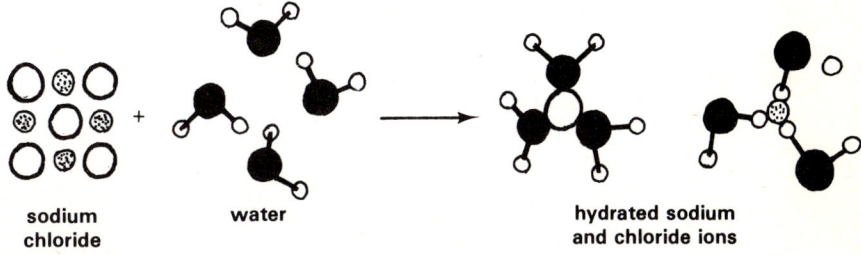

sodium chloride + water → hydrated sodium and chloride ions

Many biologically important organic substances do not ionise yet they also dissolve in water. This is because they form hydrogen bonds with water. The hydroxyl (—OH) groups of sugar molecules, imino ($>$NH) groups of amino acids and proteins, and carbonyl ($>$C $=$ O) groups of organic acids react with water in this way. Molecules of such organic compounds thus become surrounded by water molecules and go into solution.

1. TRUE SOLUTIONS

For the reasons outlined above many inorganic and organic particles smaller than 10^{-5} cm in diameter dissolve quickly in water to form what is called a **true solution**. The dissolving power of water is particularly important in the uptake by, and transportation of substances inside the bodies of plants and animals (Chapters 9, 11, 13 and 14). Metabolic reactions catalysed by enzymes also take place in aqueous solution (Chapter 4).

2. THE COLLOIDAL STATE

Particles between 10^{-3} and 10^{-5} cm in diameter such as polysaccharides and proteins of high relative molar mass do not form true solutions with water. Such particles attract water owing to hydrogen-bonding between water molecules and hydroxyl, imino and carbonyl groups. A colloidal state then exists with water acting as a **dispersion medium** in which the particles, known as the **disperse phase**, are permanently suspended. Substances which react with water in this way are called **hydrophilic colloids**. Cytoplasm is a good example of a colloid, suspended protein molecules providing cells with a vast number of fixed sites on which metabolic reactions can occur (Chapter 1). The plasma of mammalian blood is another example. Here the blood proteins which form the disperse phase play important roles in blood clotting, combating infection and in the formation of tissue fluid (Chapter 9). Hydrophilic colloids cling to water and so help to prevent the evaporation of water from organisms which live on land.

3. SUSPENSIONS

Particles larger than 10^{-3} cm in diameter do not dissolve in water but can be temporarily dispersed in water to form a **suspension**. On standing, the disperse phase of a suspension gradually separates from the dispersion medium. A suspension is formed when fats and oils are emulsified in the mammalian gut (Chapter 14). The disperse phase has a large surface area on which fat-splitting enzymes can work.

The dissolving power of water is very important for many of the activities of living organisms. Nevertheless, we should not overlook the fact that the inability of some substances to dissolve readily in water sometimes poses serious problems to plants and animals. For example, if we compare the volumes of some common gases in air and in water in direct contact with air we immediately see some startling differences (Table 2.4). Whereas carbon dioxide dissolves in water with ease, oxygen and nitrogen do not. Oxygen for respiration is thus much less available to aquatic as opposed to terrestrial organisms. It also means that the transport of oxygen in aqueous solutions inside the bodies of living organisms is inefficient unless an oxygen-carrying pigment is present (Chapter 9). Even though the relative volumes of carbon dioxide in air and in aerated water are similar, the rate at which gases diffuse through water is much slower than through air. For this reason the rate at which carbon dioxide diffuses through the protoplasm of photosynthesising cells before reaching the chloroplasts is a factor which limits the productivity of plants (Chapter 12).

Table 2.4 Amounts of some common gases in air and in water in direct contact with air

| | PERCENTAGE COMPOSITION | | |
	OXYGEN	NITROGEN	CARBON DIOXIDE
air	20·95	78·0	0·03
water (at 10°C)	0·64	1·2	0·03

2.3 Some properties of aqueous solutions

Some of the properties of aqueous solutions have already been touched on in this chapter. It is worthwhile exploring a number of these properties more fully. Two properties in particular are important to living organisms. Firstly there are those properties governed by the nature of the solute. Of special importance in this context are **acids**, **bases** and **buffers**. Then there are the **colligative properties** which are governed by the number of solute particles per unit volume of water, that is the concentration of an aqueous solution.

2.3.1 Acids, bases and buffers

An **acid** is a substance which dissociates in solution to release **hydrogen ions** (H^+). Most organic acids are called **weak acids** because few of their molecules dissociate. Conversely, many inorganic acids are **strong acids** because most of their molecules dissociate. Both weak and strong acids are found in living cells. Carboxylic acids are weak acids produced by all living organisms. Hydrochloric acid produced by cells in the stomach lining is a strong acid.

Substances which accept H^+ are called **bases**. A reaction between an acid and a base involves a **conjugate acid-base pair** which consists of a H^+ donor and a H^+ acceptor. Acid-base pairs are found in solutions containing weak acids and their salts. Let us consider a mixture of ethanoic acid and sodium ethanoate. The acid, being a weak acid, dissociates to yield a small number of ions:

$$CH_3COOH \rightleftharpoons H^+ + CH_3COO^-$$
ethanoic acid   hydrogen ion   ethanoate ion

However, the salt dissociates to produce a large number of ethanoate ions:

$$CH_3COONa \rightleftharpoons CH_3COO^- + Na^+$$
sodium ethanoate   ethanoate ion   sodium ion

Because the ethanoate ions have a high affinity for H^+, they remove a large proportion of H^+ added to the mixture so the pH of the solution remains almost constant. This effect is called the **buffering capacity** of the acid-base pair.

The main buffer which prevents large fluctuations of pH in cells is the dihydrogenphosphate and hydrogenphosphate acid-base pair. The blood and tissue fluids of higher animals are buffered by the carbonic acid–hydrogencarbonate acid-base pair (Chapters 8 and 9). The importance of buffers in living organisms cannot be overstressed. They help to stabilize the pH of cellular and body fluids, so helping to maintain suitable conditions for enzymes to efficiently catalyse metabolic reactions (Chapter 4).

2.3.2 Colligative properties

Biologically, the colligative properties of most interest are **depression of the freezing point** and **elevation of the boiling point** of aqueous solutions and the fact that aqueous solutions have **osmotic properties**.

1. DEPRESSION OF THE FREEZING POINT

When a mole of a non-electrolyte such as a sugar is dissolved in 1 dm³ of water the solution so formed of concentration 1.0 mol dm^{-3}, has a freezing point of $-1.86°C$. More concentrated solutions freeze at lower temperatures.

Because the water in living cells contains dissolved substances it follows that cellular fluid freezes only when exposed to temperatures below the freezing point of pure water. The body fluids of fish living in seas near the north and south poles are rich in solutes. Some of the solutes are similar to anti-freeze compounds used to stop the cooling system of a motor car engine freezing during winter. Even when the sea in which these fish live becomes frozen solid their body fluids remain liquid. A similar effect occurs to a lesser extent in all living organisms. The chances of damage to body tissues due to the formation of razor-sharp ice crystals inside living cells is thus minimised.

2. ELEVATION OF THE BOILING POINT

A solution of a non-electrolyte of concentration 1.0 mol dm^{-3} has a boiling point of $100.54°C$ compared with $100°C$ for pure water. The more concentrated the solution the higher is the temperature at which it boils.

The higher boiling point of a solution compared with pure water is of some importance to land-dwelling organisms. It means that slightly more heat is required if water is to evaporate from their bodies. As a result the chance of desiccation is lessened.

3. OSMOTIC PROPERTIES

When pure water is separated from an aqueous solution by a selectively permeable membrane there is a net flow of water into the solution (Fig. 2.7). The process by which the net flow of water occurs is called **osmosis**. Entry of water into the solution can be stopped if pressure is applied at C. The pressure required to prevent water passing from A to B is called the osmotic pressure of the solution. The Dutch chemist Van't Hoff discovered that the size of the osmotic pressure depends on the concentration of the solution. He found that solutions of non-electrolytes such as sugars of concentration 1.0 mol dm^{-3} have an osmotic pressure of -2270 kPa (-22.4 atm) at STP compared with pure water which has an osmotic pressure of 0 under similar conditions. Today the term **osmotic potential** is used instead of osmotic pressure.

In the arrangement shown in Figure 2.7 the water molecules on either side of the membrane have kinetic energy and move about at random. The total amount of kinetic energy of water on either side of the membrane is called the **water potential**. Pure water at STP has a water potential of 0. Water molecules moving near the membrane may hit the solid parts of the membrane and bounce off or they may slip through the pores. In this way water moves either way through the membrane from A to B and from B to A. The movement of water from A to B is impeded only by other water molecules whereas movement from B to A is impeded by dissolved solute molecules to which water molecules are attracted. The water molecules in A therefore have more kinetic energy than those in B. In other words the water potential in A is higher than in B. Because diffusible substances move from areas of high potential to areas of low potential, water passes from A to B. One way of defining osmosis is therefore:

the net diffusion of water through a selectively permeable membrane from a place of high water potential to a place of low water potential.

It should be apparent that a net osmotic flow of water from A to B will occur if a solution is placed in A instead of pure water providing the solution in A is less concentrated than in B. Attempt to explain why this happens in terms of a difference in water potential.

FIG. 2.7 *A simple osmometer*

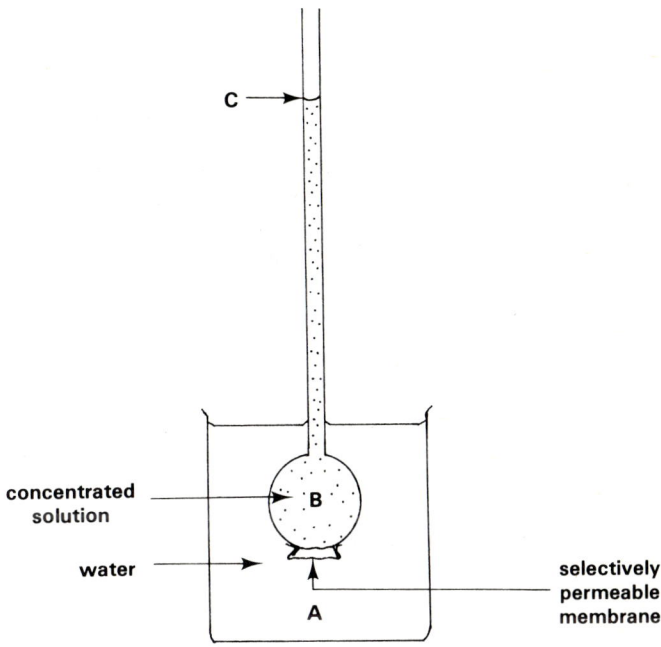

2.4 Uptake and loss of water by living cells

2.4.1 Osmotic behaviour of plant cells

One of the ways in which water enters and leaves living cells is by osmosis. It is convenient to look at the osmotic behaviour of plant and animal cells separately.

The main part of the body of most terrestrial plants consists of a tissue called parenchyma (Chapter 21). Parenchyma cells have a **wall** made of a jelly-like mixture of hemicelluloses and pectic compounds in which fibres of cellulose are enmeshed. The wall is porous and permeable to aqueous solutions. In contrast the outer boundary of the protoplast called the plasma membrane is **selectively permeable**. Whereas water can move freely through the plasma membrane, dissolved solutes cannot. The protoplast contains **sap**, an aqueous solution of inorganic and organic substances.

35

Much of the sap is localised in one or more conspicuous vacuoles each bounded by a selectively permeable membrane called the tonoplast (Fig. 2.8).

FIG. 2.8 *A parenchyma cell*

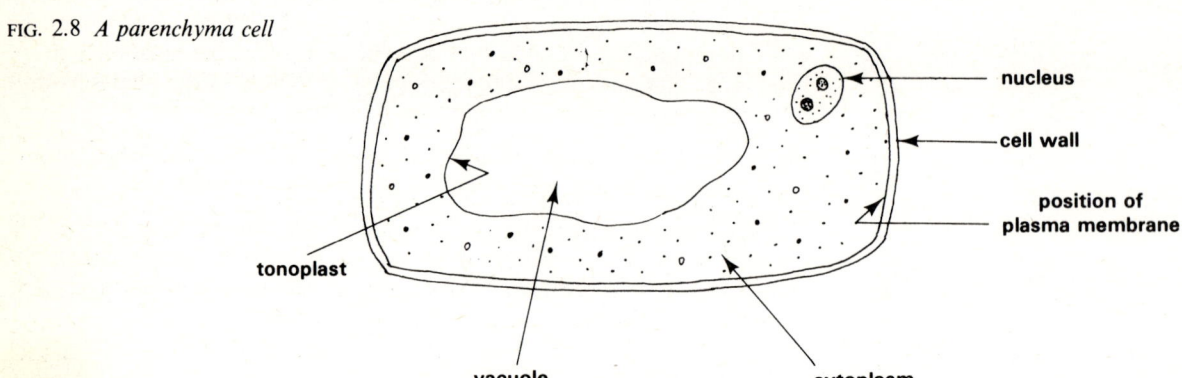

If the protoplast is sufficiently swollen to press firmly against the wall the cell is **turgid**. Should the protoplast be too small to create such a pressure the cell is **flaccid**. When a flaccid cell is immersed in water there is a net inward osmotic flow of water because the selectively permeable plasma membrane separates a place of low water potential, the cell sap, from a place of higher water potential, the water outside the cell. As water enters the cell the protoplast enlarges until it presses against the cell wall with a force called **turgor pressure**. The cell wall opposes the turgor pressure with a force of equal magnitude known as **wall pressure**. If water continues to enter the cell the wall pressure ultimately becomes so great that any further net osmotic uptake of water stops. At this stage the cell is fully turgid. The term **pressure potential** is used to indicate the tendency for water to be forced out of an expanding cell due to the difference between turgor pressure and wall pressure. Pressure potential is usually a positive force.

The ability of a cell to absorb water by osmosis is called its **water potential**. Remember that pure water has a potential of 0 at STP. Also keep in mind that cell sap has a lower, that is a negative water potential compared with water. The two factors which determine the size of the water potential of a plant cell are the osmotic potential of its sap (OPi) and the pressure potential (PP) of the protoplast. The following equation shows how the factors are related:

$$\underset{\text{water potential of cell}}{\text{WP}} \quad = \quad \underset{\substack{\text{osmotic potential} \\ \text{of cell sap}}}{\text{OPi}} \quad + \quad \underset{\substack{\text{pressure} \\ \text{potential}}}{\text{PP}}$$

For example, if
OPi $= -1013\,k\text{Pa}\,(-10\,\text{atm})$ and PP $= +506 \cdot 5\,k\text{Pa}\,(+5\,\text{atm})$:
WP $= -1013 + 506 \cdot 5 = -506 \cdot 5\,k\text{Pa}\,(-5\,\text{atm})$

The osmotic behaviour of a cell immersed in an aqueous solution depends on the difference between the water potential of the cell and the water potential and thus the osmotic potential of the solution (OPe). A non-turgid cell absorbs water so long as its water potential is more negative than that of the solution. A solution which has this effect is said to be **hypotonic**.

For example, if
OPi $= -1013$ kPa $(-10$ atm$)$, PP $= +506.5$ kPa $(+5$ atm$)$
and OPe $= -303.9$ kPa $(-3$ atm$)$:
WP $= -1013 + 506.5 - (-303.9) = -202.6$ kPa $(-2$ atm$)$,
so the cell absorbs water.

If, however, the water potential of the solution is lower than that of the cell, water is withdrawn from the cell. As water is lost the protoplast shrinks and soon ceases to press against the cell wall. This condition is called **incipient plasmolysis**. Further withdrawal of water causes the protoplast to display **total plasmolysis**. When this happens the protoplast becomes a spherical mass at the centre of the cell. A solution which has this effect on the cell is said to be **hypertonic**.

FIG. 2.9 *Epidermal cells of beetroot petiole,* $\times 250$

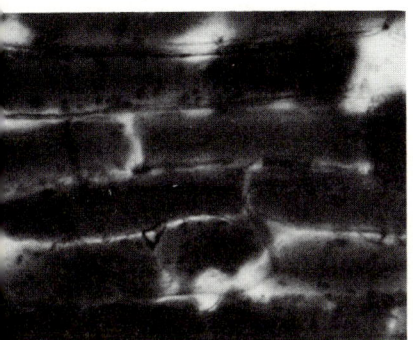

(i) *partial turgidity*

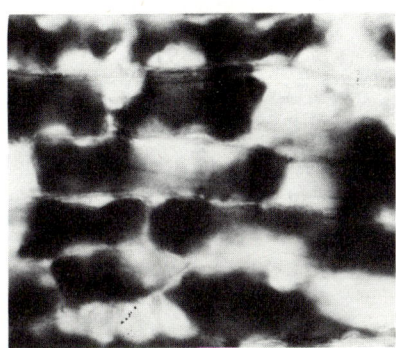

(ii) *incipient plasmolysis*

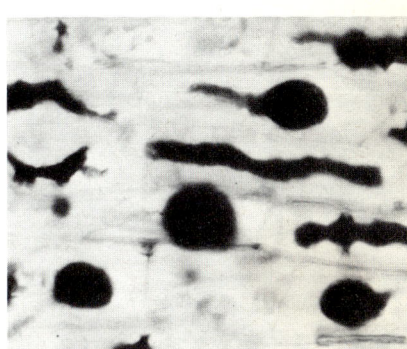

(iii) *total plasmolysis*

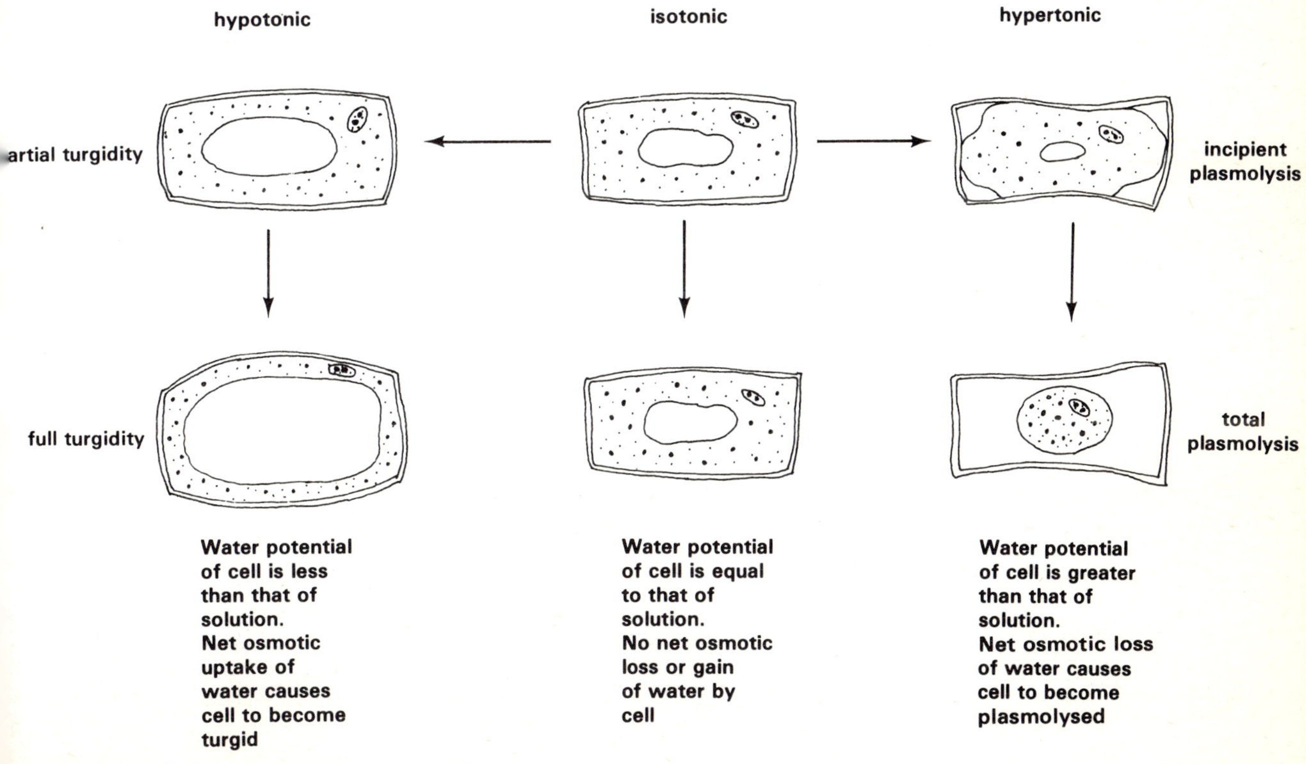

hypotonic isotonic hypertonic

partial turgidity   incipient plasmolysis

full turgidity   total plasmolysis

Water potential of cell is less than that of solution. Net osmotic uptake of water causes cell to become turgid

Water potential of cell is equal to that of solution. No net osmotic loss or gain of water by cell

Water potential of cell is greater than that of solution. Net osmotic loss of water causes cell to become plasmolysed

37

For example, if

$$OPi = -1013\ kPa\ (-10\ atm),\ PP = +506.5\ kPa\ (+5\ atm)$$
$$\text{and } OPe = -1215.6\ kPa\ (-12\ atm):$$
$$WP = -1013 + 506.5 - (-1215.6) = +709.1\ kPa\ (+7\ atm),$$
$$\text{so the cell loses water.}$$

The third possibility is that the osmotic potential of the solution is such that there is no net loss or gain of water by the cell. In other words the water potential of the cell and the external solution are equal. Such a solution is said to be **isotonic**.

For example, if

$$OPi = -1013\ kPa\ (-10\ atm),\ PP = +506.5\ kPa\ (+5\ atm)$$
$$\text{and } OPe = -506.5\ kPa\ (-5\ atm):$$
$$WP = -1013 + 506.5 - (-506.5) = 0\ kPa\ (0\ atm),$$
$$\text{so there is no net loss or gain of water by the cell.}$$

The events outlined above are illustrated in Figure 2.9. Figure 2.10 shows the relationship between the various factors which affect the osmotic behaviour of plant cells. The way in which osmosis participates in the movement of water through terrestrial plants is described in Chapter 11.

FIG. 2.10 *Graphical relationship between the factors affecting the osmotic behaviour of plant cells*

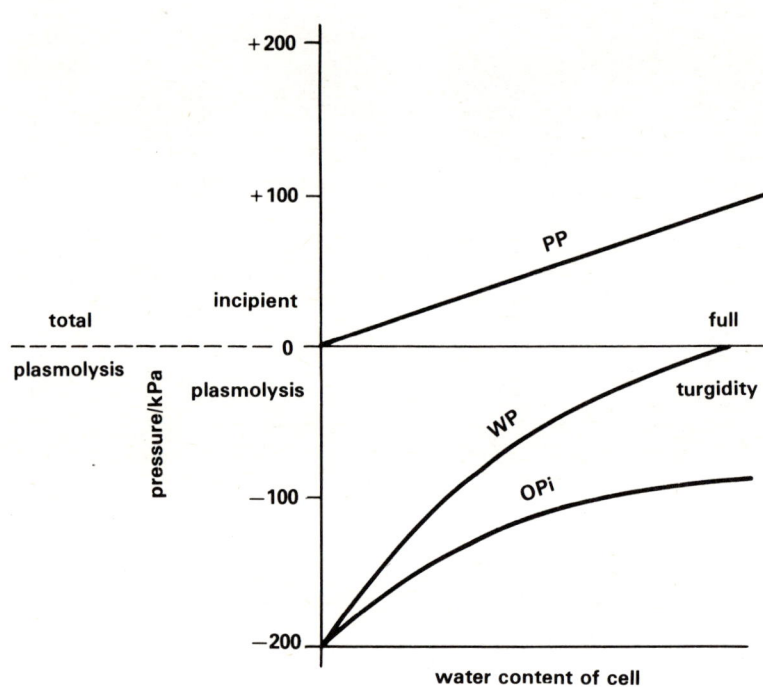

2.4.2 Osmotic behaviour of animal cells

Animal cells do not have a cell wall so wall pressure is not among the factors affecting their water potential. The osmotic potential of the cell sap (OPi) is the main factor which determines the water potential of animal cells.

When immersed in an aqueous solution animal cells behave in three different ways depending on the osmotic potential (OPe) of the external solution. If placed in a hypotonic solution there is a net osmotic uptake of water by the cell which therefore becomes turgid. If a large volume of water is absorbed the plasma membrane bursts and the contents of the cell are spilled into the solution. The bursting of red blood cells in this way is called **haemolysis**. If immersed in a hypertonic solution there is a net osmotic loss of water by the cell which therefore shrinks. Red blood cells take on a **crenated** appearance when this happens. Finally when placed in an isotonic solution there is no net loss or gain of water by the cells. In this solution the cell's water content remains steady. Figure 2.11 illustrates the osmotic behaviour of red blood cells when placed in solutions of different osmotic potential.

FIG. 2.11 *Osmotic behaviour of a red blood cell following immersion in different solutions*

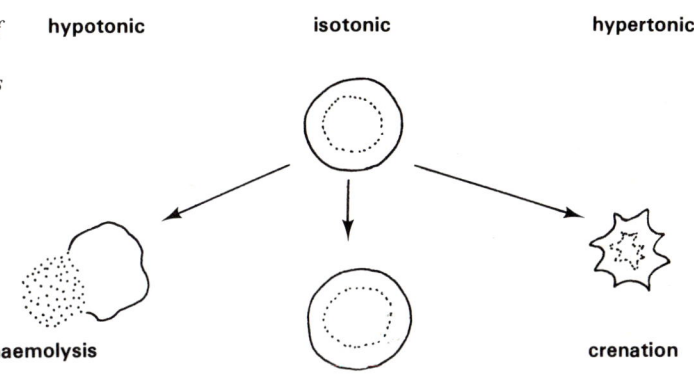

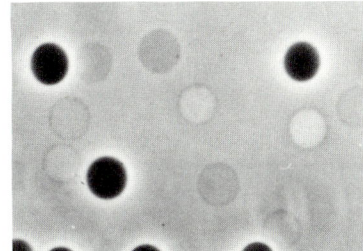

(i) *haemolysis*

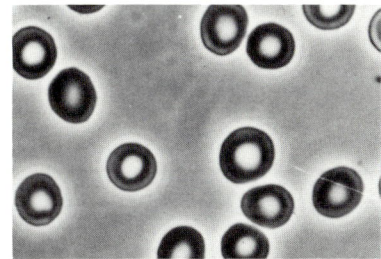

(ii) *red blood cells, ×800*

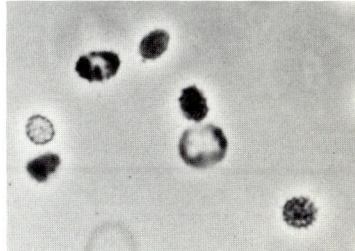

(iii) *crenation*

A variety of mechanisms has evolved among animals which protect their cells from excessive intake or loss of water by osmosis. One of the simplest devices is seen in fresh-water protozoans such as *Amoeba* and *Paramecium* where the **contractile vacuole** eliminates excess water absorbed from the environment (Fig. 2.12). In mammals the means of keeping the tissue fluid isotonic with cellular fluid is an elaborate process in which the kidneys play a key role (Chapter 10).

2.4.3 Non-osmotic movement of water

Osmosis is not the only process responsible for water movement in and out of the cells. Water is taken in by cells when pockets are formed in the plasma membrane during **pinocytosis** and **phagocytosis** (Chapter 1). Cells also contain hydrophilic substances such as polysaccharides and proteins which chemically attract water, a phenomenon called **imbibition** (section 2.2.6). The absorption of water by dry seeds during the early stages of

FIG. 2.12 *The protozoan* Paramecium *as seen with a phase-contrast microscope* (courtesy Philip Harris Biological Ltd.) × *700*

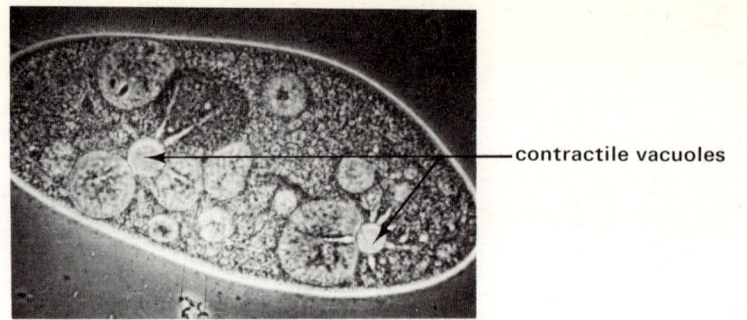

contractile vacuoles

germination is largely due to imbibition (Chapter 20). The imbibing power of the polysaccharides in cell walls also probably plays a substantial role in the transport of water through terrestrial plants (Chapter 11).

Diffusion is the name given to the process whereby particles move from a place where they are highly concentrated to places where they are less concentrated. Osmosis is a special case of diffusion because of the involvement of a selectively permeable membrane. However, there are instances when water diffuses without passing through membranes. The **evaporation** of water and its subsequent diffusion into the atmosphere is the main cause of water loss from terrestrial organisms. Such processes are also of significance in the transport of water through land-dwelling plants. These and other aspects of water movement in whole organisms are described more fully in Chapters 9 and 11.

Chapter 3

The chemical constituents of cells

Little is known of the origin of life on this planet. Fossil evidence suggests that the first forms of life were bacteria-like organisms which evolved some three to three and a half thousand million years ago. The results of experimental studies indicate that the earth's atmosphere may have provided the materials for the beginning of life. In the early 1950s Stanley Miller in America exposed gaseous mixtures of ammonia, methane, hydrogen and water to electrical discharges. Miller believed that these gases were important constituents of the earth's early atmosphere. The electrical discharges were used to simulate the effect of lightning. Among the end-products which appeared were amino acids and other organic compounds such as ethanoic acid.

There is now some doubt whether the gases used by Miller were in fact the main atmospheric constituents of the primordial earth. Recent evidence points to carbon dioxide, carbon monoxide, nitrogen and hydrogen as the atmospheric gases present in significant volumes. When mixtures of these gases were subjected to electrical discharges and other physical phenomena such as ultraviolet and visible light, a much wider range of end-products was formed. They included sugars, amino acids and carboxylic acids as well as nitrogenous bases such as those in the nucleic acids. It is possible therefore that the energy from natural physical forces was used in the formation of simple organic molecules in the earth's early atmosphere. On entering the oceans in rainfall the simple molecules could have served as the building blocks for more complex substances such as polysaccharides, proteins, lipids and nucleic acids. These organic substances, simple and complex, are the 'stuff of life' as we know it today.

3.1 Carbohydrates

Sugars, starch, glycogen and cellulose are well-known examples of a vast group of compounds called **carbohydrates**. They all consist of just three elements: carbon, hydrogen and oxygen.

3.1.1 Monosaccharides

Monosaccharides are carbohydrates of low relative molar mass. They have an empirical molecular formula $(CH_2O)_n$. All monosaccharides are white, crystalline solids which are sweet to taste and dissolve readily in water. They can be classified according to the number of carbon atoms present in each molecule (Table 3.1).

Table 3.1 Some examples of monosaccharide sugars

	MOLECULAR FORMULA	EXAMPLES
trioses	$C_3H_6O_3$	glyceraldehyde
pentoses	$C_5H_{10}O_5$	ribose, ribulose
hexoses	$C_6H_{12}O_6$	glucose, fructose

The molecules of all monosaccharides have an unbranched backbone of carbon atoms, all but one of these atoms having an attached hydroxyl ($-OH$) group. The exceptional carbon atom is part of a carbonyl ($>C=O$) group. In **aldose sugars**, for example **glucose**, the carbonyl group is situated at one end of the molecule where it forms part of an aldehyde ($-\overset{\displaystyle H}{\underset{\displaystyle}{C}}\!=\!O$) group. Where the carbonyl group occurs at any other position, as in **fructose** molecules, it is part of a ketone group and the substance is called a **ketose sugar**. The carbonyl group has a strong tendency to donate electrons and it is this property which is responsible for the **reducing powers** of monosaccharide sugars. In Fehling's test the carbonyl group reduces copper(II) ions to copper(I) ions.

Reaction on mixing Fehling's solutions 1 and 2:

$$CuSO_4 + 2NaOH \longrightarrow Cu(OH)_2 + Na_2SO_4$$

| copper(II) sulphate | sodium hydroxide | copper(II) hydroxide | sodium sulphate |

Reaction with reducing sugar:

$$2Cu(OH)_2 + R - \overset{\displaystyle H}{\underset{\displaystyle}{C}} = O \longrightarrow Cu_2O + R.COOH + 2H_2O$$

copper(I) oxide (brick red precipitate)   sugar acid

Molecules of nearly all the monosaccharide sugars have one or more **asymmetric carbon atoms**, that is carbon atoms bonded to four different functional groups. Glyceraldehyde has one such carbon atom and because of this can exist in two forms (Fig. 3.1). The D- and L-forms differ only in the spatial disposition of the functional groups attached to the asymmetric carbon atom. The two forms are called **stereoisomers** and the D-form is a non-superimposable mirror image of the L-form. Only D-isomers are commonly found in living organisms.

FIG. 3.1 *Stereoisomers of glyceraldehyde*

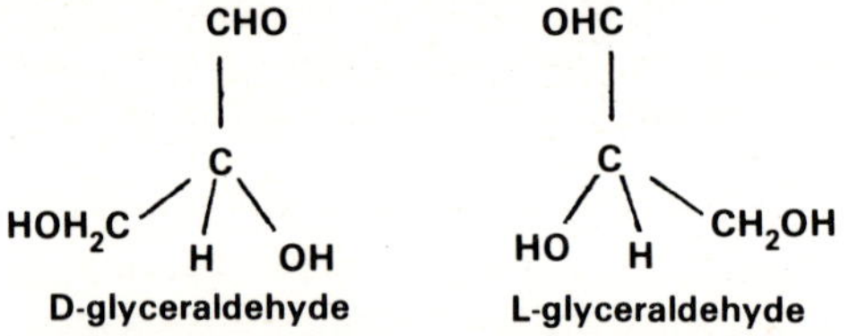

D-glyceraldehyde   L-glyceraldehyde

The number of isomers of any compound is determined by the number of asymmetric carbon atoms in one of its molecules. Some hexose sugars have 4 such atoms so 16 isomers are possible (derived from the formula 2^n, where n = no. of asymmetric carbon atoms). The molecular structures of some biologically important D-sugars are shown in Figure 3.2.

FIG. 3.2 *Molecular structure of some D-sugars*

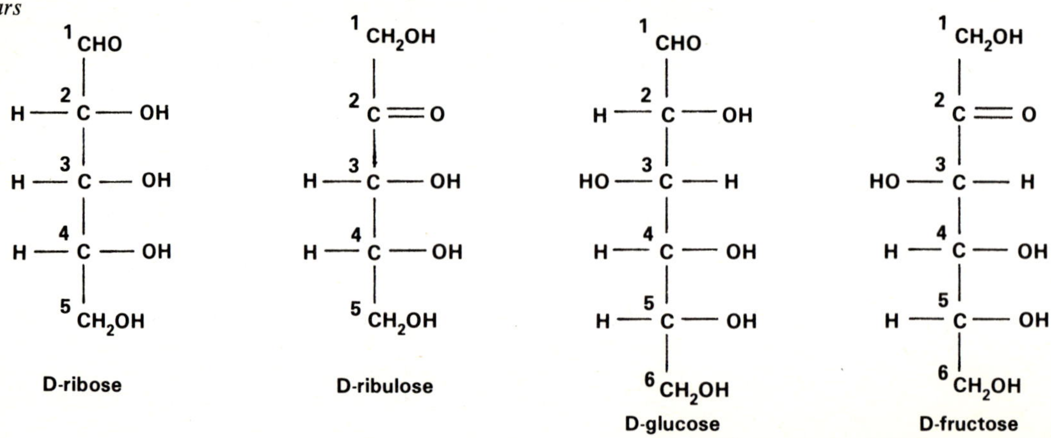

D-ribose   D-ribulose   D-glucose   D-fructose

42

When dissolved in water many monosaccharides behave as though they have one more asymmetric centre. Evidently in these conditions they no longer exist in the straight-chain form depicted in Figure 3.2. For instance, when D-glucose is dissolved in water it occurs in two different forms called α- and β-D-glucose. These are six-sided ring structures which appear because the carbonyl group at carbon atom *1* reduces the hydroxyl group attached to carbon atom *5*. The only difference between α- and β-D-glucose is in the spatial orientation of the hydrogen and hydroxyl groups at carbon atom *1* (Fig. 3.3).

FIG. 3.3 *Ring forms of* D-*glucose*

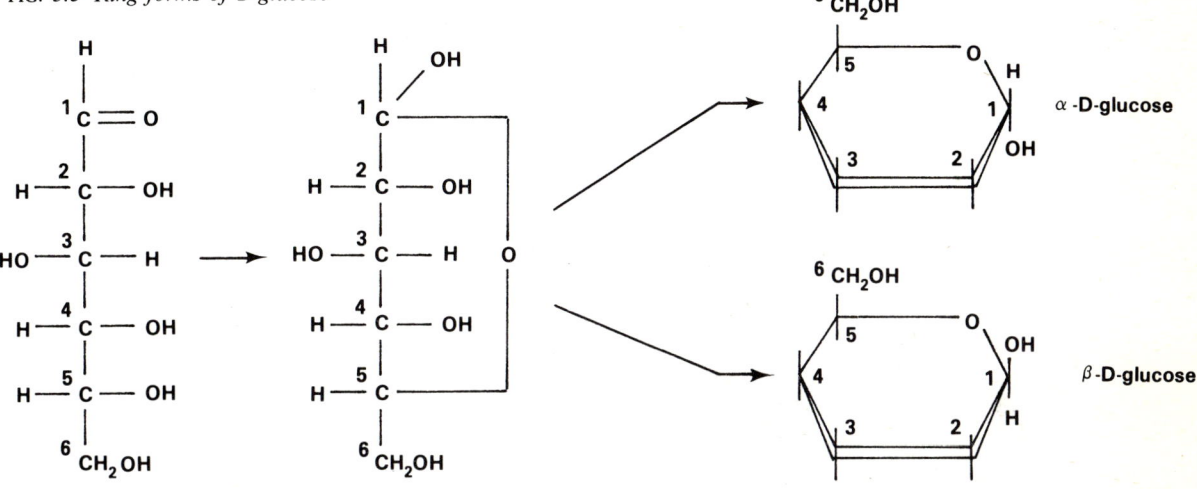

Fructose can also exist in a six-sided ring form but it more frequently exists as a more stable five-sided molecule in which the carbonyl group at carbon atom *2* reduces the hydroxyl group at carbon atom *5* (Fig. 3.4).

FIG. 3.4 *Ring forms of* D-*fructose*

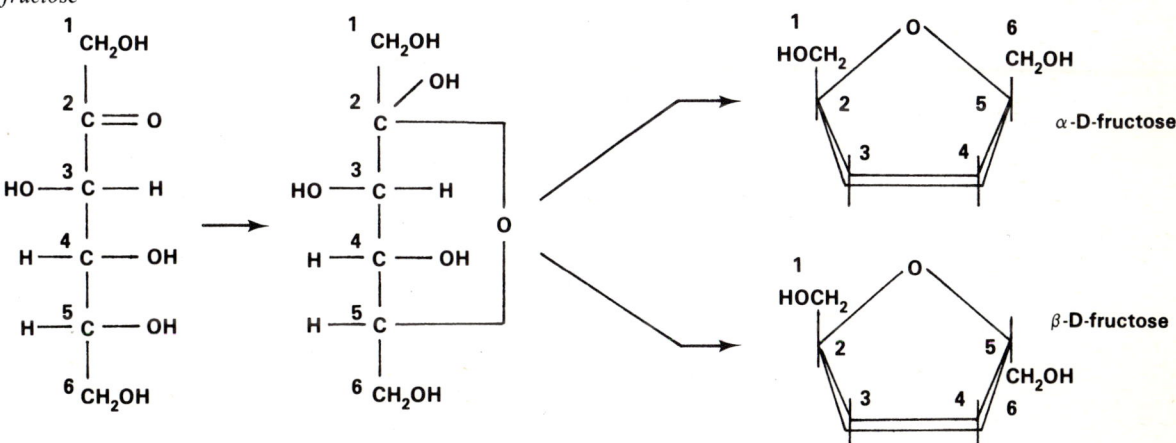

The ways in which the ring-forms of glucose and fructose have been drawn may lead you to think that the molecules are flat. This is not so. Normally the molecules have an elaborate three-dimensional structure, which in the case of glucose is often described as chair-shaped (Fig. 3.5). This has important implications in the chemistry of the more complex carbohydrates described later.

FIG. 3.5 *Chair-shaped glucose molecule*

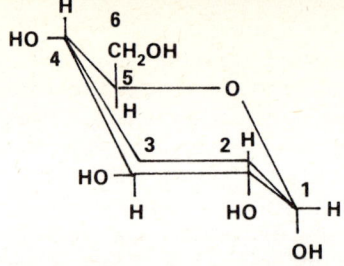

3.1.2 Disaccharides

Sucrose and maltose are the two **disaccharide sugars** most often found in living organisms. Disaccharides are formed when two hexose sugar molecules unite in a type of chemical reaction called condensation. In condensation reactions water is often one of the products:

$$2C_6H_{12}O_6 \longrightarrow C_{12}H_{22}O_{11} + H_2O$$

$$\text{hexose} \qquad \text{disaccharide} \quad \text{water}$$

Sucrose is found in most plants. It is stored in large amounts in sugar cane and sugar beet which are cultivated on a large scale to provide our supply of sucrose (Fig. 3.6).

FIG. 3.6 *A sugar-beet plant*

swollen tap root
containing sucrose

Sucrose is formed when a molecule of α-D-glucose and one of β-D-fructose join by condensation. The reaction involves the creation of a bond between carbon atom *1* of the glucose ring and carbon atom *2* of the fructose ring (Fig. 3.7). It is at these two carbon atoms that a carbonyl group occurs in the glucose and fructose molecules. This means that there are no free carbonyl groups in a molecule of sucrose which is therefore a **non-reducing sugar**. However, sucrose can be readily hydrolysed into its constituent monosaccharide reducing sugars. Hydrolysis can be achieved by boiling sucrose with dilute acid or mixing it with the enzyme sucrase (invertase).

Maltose is an intermediate product released from starch which has been hydrolysed by the complex of enzymes called amylase. Starch hydrolysis occurs in the mammalian gut (Chapter 14) and in germinating seeds (Chapter 20). A maltose molecule consists of two α-D-glucose rings joined together by an α(1→4)glycosidic linkage (Fig. 3.8). The linkage involves the aldehyde group of only one of the glucose rings. Maltose therefore has one free carbonyl group and is a **reducing sugar**.

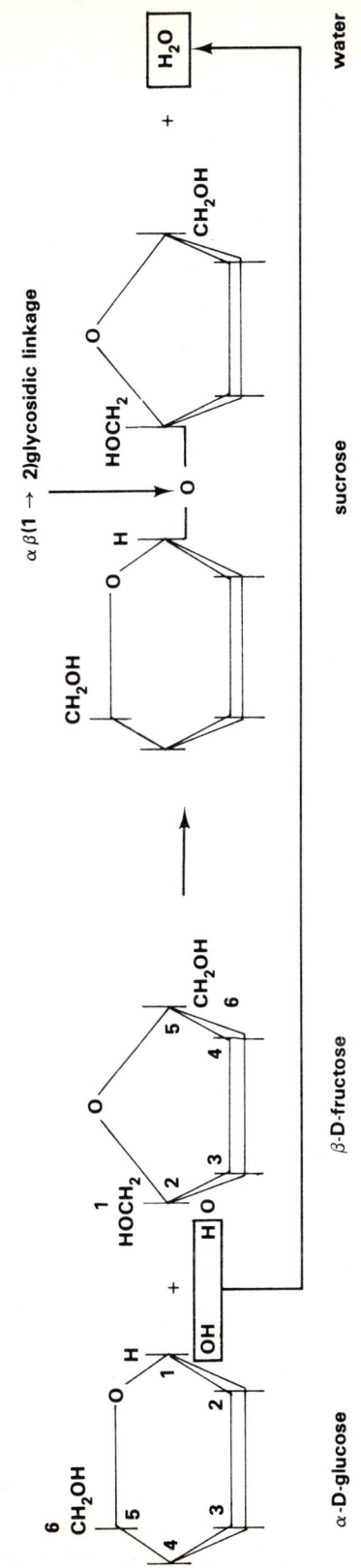

FIG. 3.7 *Formation of a molecule of sucrose*

FIG. 3.8 *Formation of a molecule of maltose*

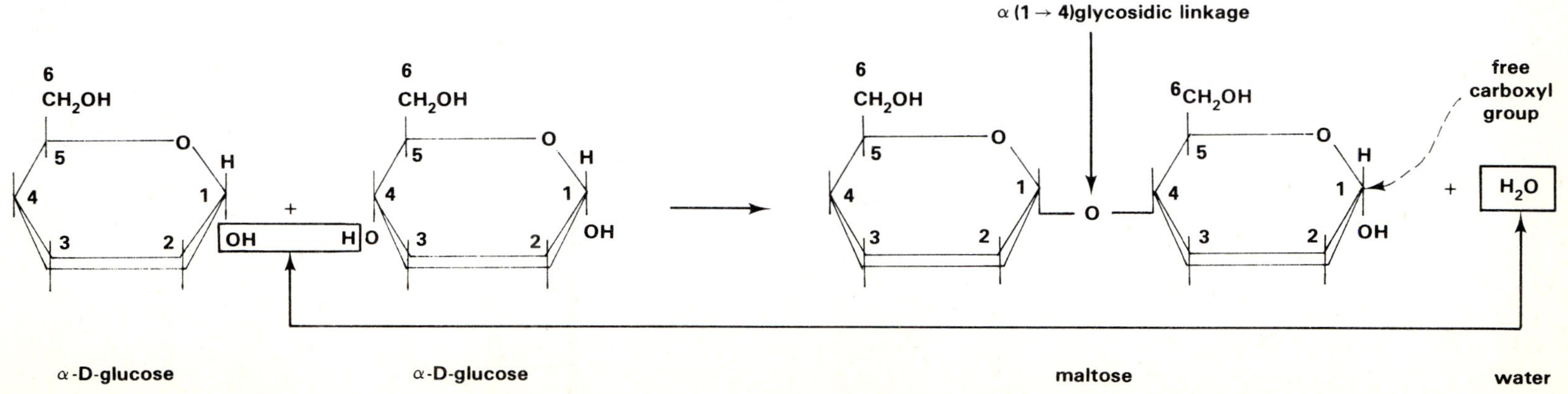

45

Simple chemical tests such as Fehling's test can be used to distinguish between pure solutions of reducing and non-reducing sugars. How is it possible to identify a number of different sugars in a mixture? Chromatography has revolutionised the art of separating and identifying many naturally occurring compounds. Descriptions of chromatographic and other techniques are given in section 3.4 of this chapter.

3.1.3 Poly-saccharides

FIG. 3.9 *Helix of amylose*

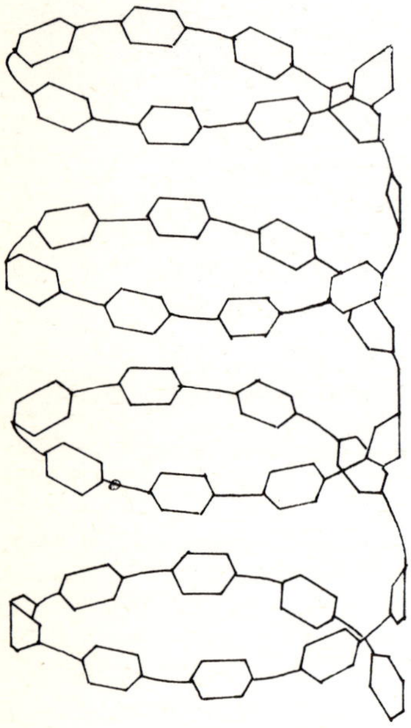

A **polymer** is a substance of large relative molar mass and is formed as a result of the joining together of a large number of basically similar smaller molecules. Polysaccharides are polymers formed by the condensation of many monosaccharide molecules. D-glucose is the main sugar involved and the products serve as storage and structural materials in plants and animals.

Starch is the main storage material of green plants. Its molecules have two components, **amylose** and **amylopectin**. The proportion of each varies from one type of starch to another. Soluble starches consist mainly of amylose. Amylose does not truly dissolve in water but forms a colloidal suspension. Amylose is a polymer of several hundred to a few thousand α-D-glucose rings joined by $\alpha(1\rightarrow4)$glycosidic linkages. The resulting molecule is unbranched and wound into a helix (Fig. 3.9). The bore of the helix is just big enough to trap iodine molecules. It is this reaction which produces the blue coloured complex when starch is mixed with iodine in potassium iodide solution. The iodine test is frequently used to investigate the presence of starch in materials of plant origin.

In contrast amylopectin is a branched molecule though again it is a polymer of α-D-glucose. The backbone of the molecule is held together by $\alpha(1\rightarrow4)$glycosidic linkages as in amylose. Branches arise about every tenth glucose ring where $\alpha(1\rightarrow6)$glycosidic bonds occur (Fig. 3.10). The branches which consist of 15–20 glucose rings joined by $\alpha(1\rightarrow4)$linkages may themselves be branched. Altogether several thousand glucose rings are present. Like amylose, amylopectin forms a colloidal suspension when mixed with water. The suspension reacts with iodine in potassium iodide solution giving a red-violet colour.

FIG. 3.10 *Part of a molecule of amylopectin*

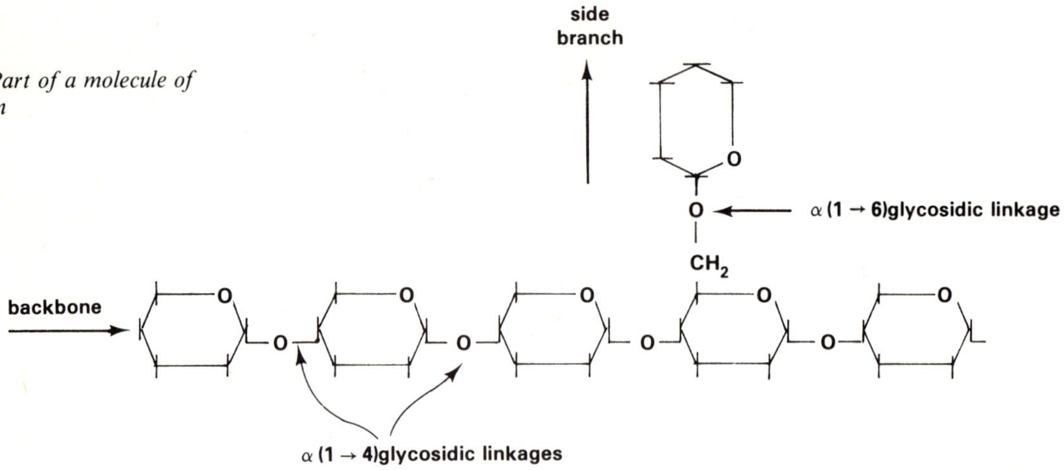

The two components of starch fit together to form a complex three-dimensional structure in which amylose helices are entangled in the branches of amylopectin molecules. Because starch is insoluble in water, it can be stored in large amounts without having any great effect on the osmotic potential of cells. It is usually stored as grains which appear in chloroplasts and in the tissues of storage organs such as potato tubers (Fig. 3.11).

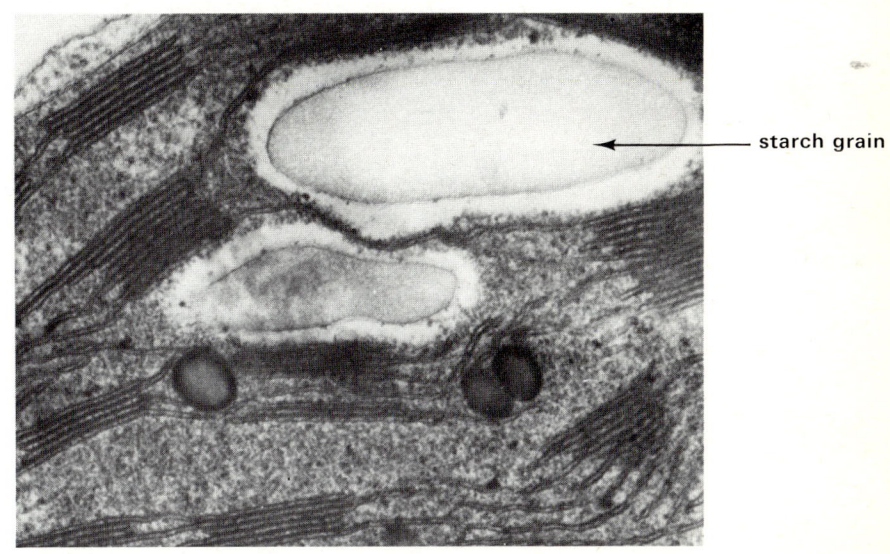

starch grain

FIG. 3.11 (b) *Starch grains in a section of potato tuber,* × 200

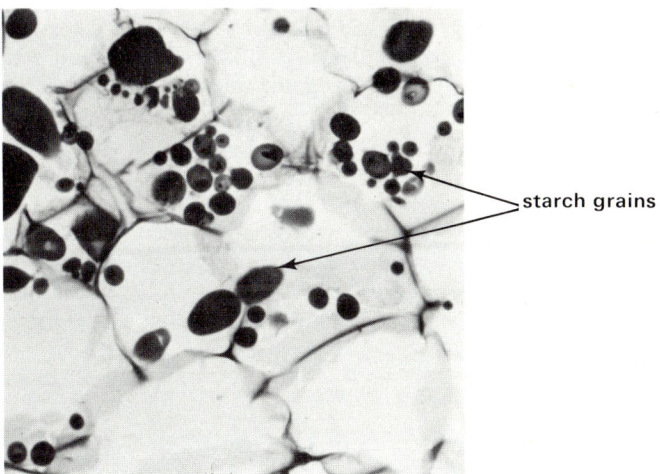

starch grains

Glycogen is a storage polysaccharide found mainly in muscle and liver tissues of animals. In molecular structure it is very similar to amylopectin except that side branches occur more frequently and are somewhat longer. Glycogen is insoluble in water and therefore is ideal as a storage material because it has little effect on the osmotic potential of cellular fluid. It is usually stored as granules in animal cells (Fig. 3.12) and its reaction with iodine is similar to that of amylopectin.

FIG. 3.12 *Electronmicrograph of a liver cell containing glycogen granules, × 28 000*

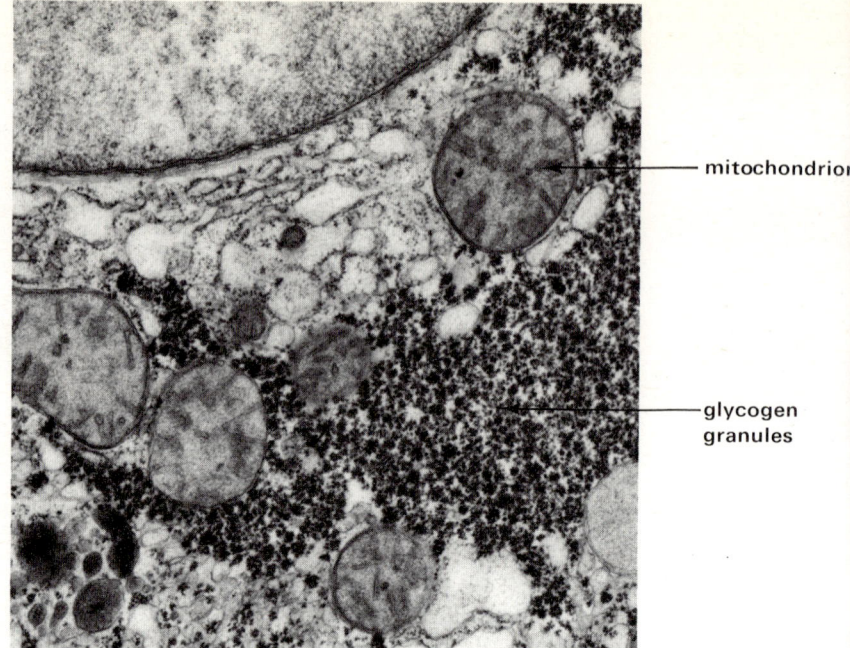

mitochondrion

glycogen granules

Cellulose is the most abundant structural polysaccharide in living organisms, although it is found only in plants. Unlike starch and glycogen it is a polymer of β-D-glucose. Several hundred to a few thousand glucose rings are joined by $\beta(1 \rightarrow 4)$glycosidic linkages in the long, unbranched molecules of cellulose (Fig. 3.13(a)). Attraction between hydroxyl ($-OH$) and hydrogen ($-H$) groups of the glucose rings of adjacent cellulose molecules results in the formation of hydrogen bonds which bind the molecules in a regular crystal-like lattice (Fig. 3.13(b)). The crystalline property of cellulose has been known for a long time but the way in which cellulose is laid down in the walls of plant cells was not known until the electron microscope came into use. Electronmicrographs show that plant cell walls consist of many fine **fibres** of cellulose which in some instances are laid down in parallel bundles but in others are distributed at random (Fig. 3.14). Each fibre contains a large number of cellulose molecules.

FIG. 3.13 *Structure of cellulose*
(a) *part of a cellulose molecule*

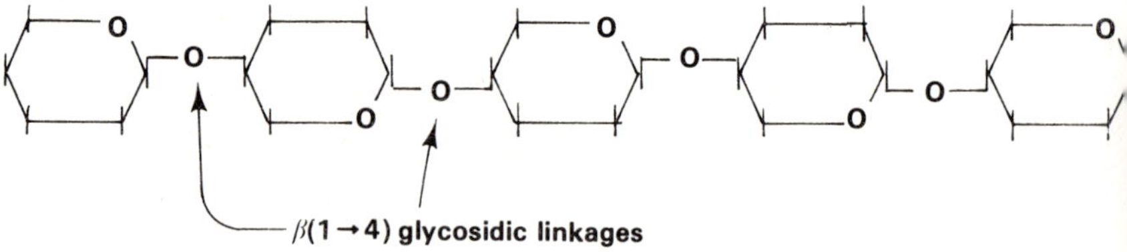

$\beta(1 \rightarrow 4)$ **glycosidic linkages**

(b) *crystalline lattice of cellulose molecules*

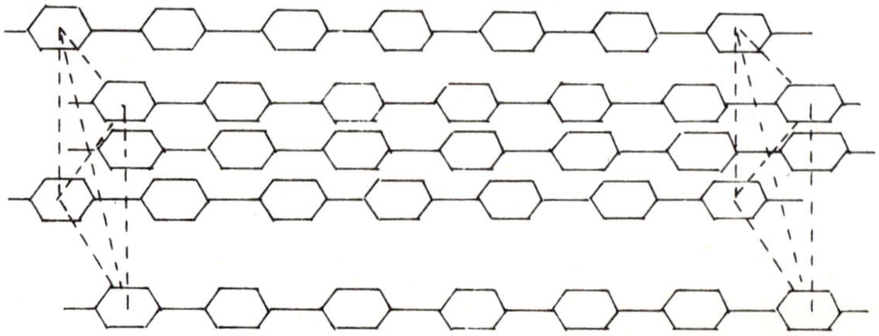

Hemicelluloses and pectic compounds hold the fibres together. Such a complex of materials is elastic enough to allow for cell growth yet has sufficient tensile strength to withstand turgor pressure which pushes the protoplast against the cell wall (Chapter 2). In sclerenchyma fibres and in xylem elements the walls are thickened with extra cellulose and stiffened with deposits of an alcohol polymer called lignin. This gives the tissues considerable rigidity enabling them to help support the shoot system in the air.

FIG. 3.14 *Electronmicrograph of freeze-etched primary wall of a carrot cell, showing cellulose fibres, × 37 500*

Cellulose reacts with iodine to form a yellow-brown complex. With chlor-zinc iodide cellulose is stained a violet colour. Lignified cell walls stain a yellow-orange colour when treated with phenylamine dyes.

3.2 Lipoids

The word **lipoid** is a collective term used to describe a range of organic compounds which can be extracted from living tissue using non-polar organic solvents such as benzene and trichloromethane. Lipoids are insoluble in polar solvents such as water and they include such diverse substances as fats, oils, phospholipids, waxes, steroids and sterols.

3.2.1 Fats and oils

Fats and **oils** are made from two ingredients, glycerol and monocarboxylic acids which are generally called **fatty acids**. Glycerol is an alcohol derivative of the triose sugar glyceraldehyde. Each molecule of glycerol has three hydroxyl ($-OH$) groups (Fig. 3.15(a)). Over seventy different fatty acids have been isolated from natural sources. Unlike glycerol, fatty acids are insoluble in water. All fatty acid molecules have a long hydrocarbon chain and a terminal carboxyl ($-C\underset{\displaystyle O}{\overset{\displaystyle OH}{}}$) group. The length of the hydrocarbon chain differs from one fatty acid to another. In some fatty acids the outer electron shell of each carbon atom in the chain is completed by sharing electrons with an adjacent carbon atom and with attached hydrogen atoms. In this way the valency of each of the carbon atoms is satisfied by the formation of four single covalent bonds. These are the **saturated fatty acids**. In contrast the hydrocarbon chain of **unsaturated fatty acids** has one or more double covalent bonds formed by the sharing of electron pairs between adjacent carbon atoms. Fatty acids therefore differ in the length of the hydrocarbon chain and in the number and positions of the double bonds, if any, in the chain. The fatty acids most frequently found in living organisms have hydrocarbon chains containing 15 to 17 carbon atoms (Fig. 3.15(b)).

FIG. 3.15 (a) *Molecular structure of glycerol*

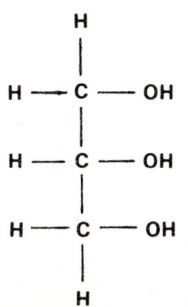

49

FIG. 3.15 (b) *Molecular structures of two fatty acids*

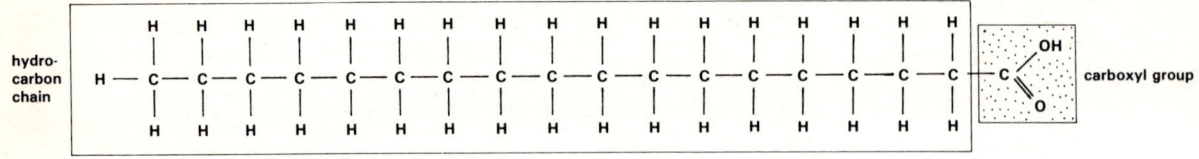

stearic acid (a saturated fatty acid)

double covalent bond

oleic acid (an unsaturated fatty acid)

Condensation reactions between the carboxyl groups of fatty acid molecules and the hydroxyl groups of glycerol produce **acylglycerols** and water (Fig. 3.16). The oxygen bridges joining the two components are called ester linkages and this type of reaction is known as **esterification**. Fats and oils are triacylglycerols in which all three of the hydroxyl groups of glycerol are used up in this way. Acylglycerols formed from unsaturated fatty acids have relatively low melting points and are liquid at 15–20°C. They are generally called oils. Most vegetable oils such as olive oil, peanut oil and corn oil are derived from unsaturated fatty acids. Many acylglycerols of animal origin are derived from saturated fatty acids and have higher melting points. They are called fats. Butter and lard are animal products which consist almost entirely of fats.

Fats and oils are non-polar compounds and are therefore insoluble in water. For this reason they can be stored without affecting the osmotic potential of plant and animal cells. In mammals, fats are stored in adipose

FIG. 3.16 *Formation of a triacylglycerol*

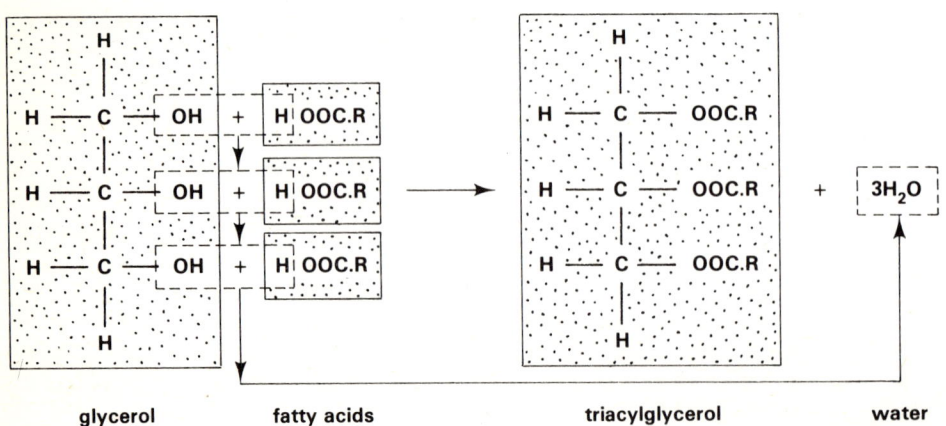

glycerol   fatty acids   triacylglycerol   water

tissue beneath the dermis of the skin and around the internal organs (Fig. 3.17). The seeds of flowering plants are often the sites of oil storage.

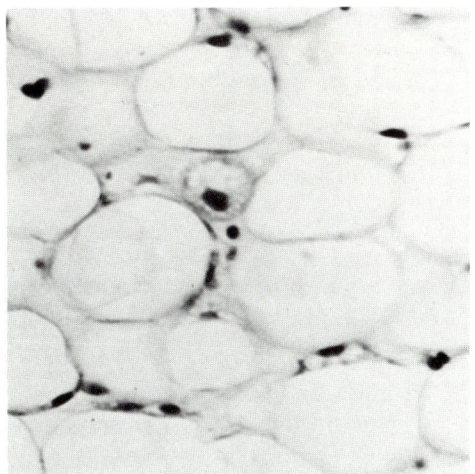

FIG. 3.17 *A section through adipose tissue, × 400. Each cell is virtually filled with fat so that the cytoplasm and nucleus are pushed close to the cell membrane*

Crude demonstrations of the presence of acylglycerols in tissue extracts or slices involve the use of fat-soluble dyes such as Sudan III. More sophisticated procedures such as chromatography are required for analysing the acylglycerol content of natural products (section 3.4).

3.2.2 Phospholipids

Phospholipids are triacylglycerols in which one of the hydroxyl groups of glycerol is esterified by **phosphoric acid** (H_3PO_4) rather than by a fatty acid. Also one of the hydroxyl groups of the phosphoric acid forms an ester linkage with an **amino alcohol**. Choline is one of the most common amino alcohols used for this purpose. The phospholipid so formed is called glycerophosphocholine (Fig. 3.18). An important property of a phospholipid molecule is its polarity. The phospho-choline 'head' is charged and

FIG. 3.18 *Formation of a phospholipid*

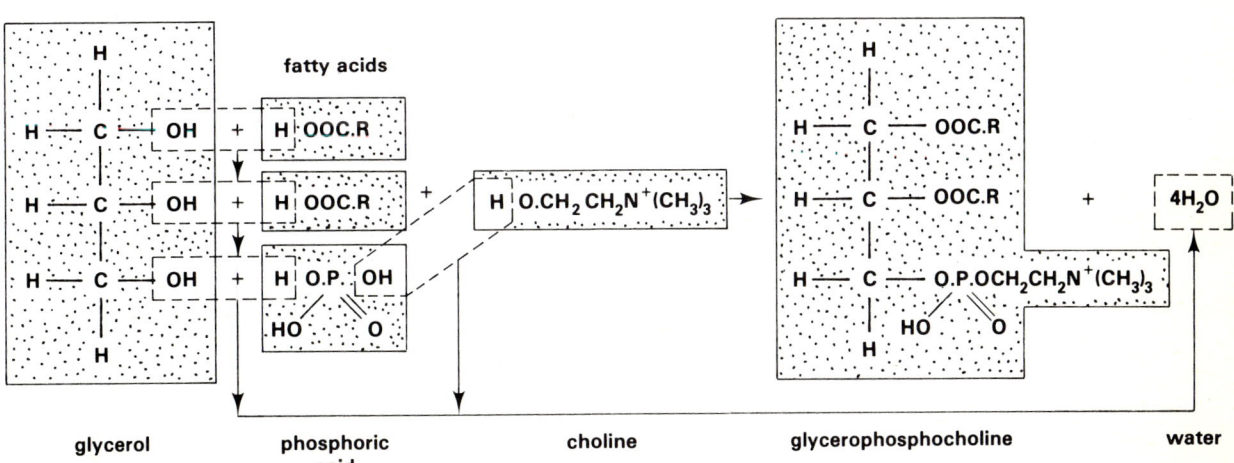

51

is therefore soluble in polar solvents such as water while the 'tail' of two hydrocarbon chains is non-polar. Such molecules do not truly dissolve in water but in certain conditions become arranged as a double layer called a **bimolecular leaflet** (Fig. 3.19). This arrangement is seen in the ultrastructure of unit membranes (Chapter 1).

FIG. 3.19 *A bimolecular leaflet*

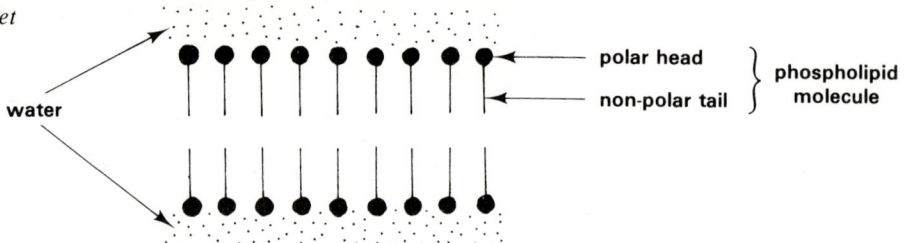

There is no simple procedure whereby the presence of phospholipids can be demonstrated in cell membranes. Osmic(VIII) acid, a solution of osmium(VIII) oxide, is used as a phospholipid fixative in preparing material for electron miscroscopy (Chapter 1). The oxide is reduced by the carbonyl groups in the hydrocarbon chains to lower oxides of osmium making the phospholipid component of membranes electron-dense.

3.2.3 Waxes

Waxes are esters of fatty acids with long hydrocarbon chains and alcohols of high relative molar mass. The alcohols have only one hydroxyl group compared with the three hydroxyl groups of glycerol. For example:

$$C_{23}H_{47}COO\boxed{H} + C_{24}H_{49}\boxed{OH} \rightarrow C_{23}H_{47}COOC_{24}H_{49} + \boxed{H_2O}$$

carnaubic acid   carnaubyl   wax   water
                     alcohol   (carnaubyl
                                  carnaubate)

Waxes are thus compounds of very high relative molar mass. They form waterproof coverings on the exterior surfaces of higher plants and animals. In terrestrial plants the **cuticle** is a waxy coating secreted by the epidermis of leaves and stems which prevents excessive evaporation from the shoot system. The sebaceous glands of mammals produce a waxy product called **sebum** which is released on to the surface of the skin (Chapter 17). Sebum helps to reduce evaporative losses from the bodies of terrestrial mammals. It also acts as a barrier which prevents waterlogging of the fur of aquatic mammals such as the otter.

3.2.4 Steroids and sterols

This group of compounds has little in common with the lipoids so far described except that like fats, oils and waxes they are soluble in non-polar solvents. **Steroids** have a complex molecular structure. They include the bile acids, hormones of the adrenal cortex and the male and female sex hormones (Chapters 14, 18 and 22). **Sterols** are alcohol derivatives of steroids. They include cholesterol, a component of the plasma membranes of cells and a precursor in the formation of steroid hormones.

3.3 Proteins

Proteins are compounds of extremely high relative molar mass. As much as 50 per cent of the dry mass of living organisms consists of protein. All proteins are polymers of **amino acid** molecules.

3.3.1 Amino acids

Nearly all amino acids consist of just four elements: carbon, hydrogen, oxygen and nitrogen. A few amino acids contain sulphur too. Every amino acid molecule has at least one amino ($-N\underset{H}{\overset{H}{\diagdown}}$) group and one carboxyl ($-C\underset{OH}{\overset{O}{\diagup}}$) group. The empirical formula for an amino acid may be written thus:

$$R.NH_2.CH.COOH$$

The structural formula is given in Figure 3.20. Because the amino group is bonded to the carbon atom next to the carboxyl group such compounds are called α-**amino acids**. Twenty different α-amino acids have been isolated from the proteins of higher plants and animals (Table 3.2). In glycine, the simplest amino acid, R is a hydrogen atom. In alanine it is a methyl ($-CH_3$) group, while in tyrosine it is a derivative of benzene. Cysteine is an example of an amino acid having a thiol ($-SH$) group (Fig. 3.21).

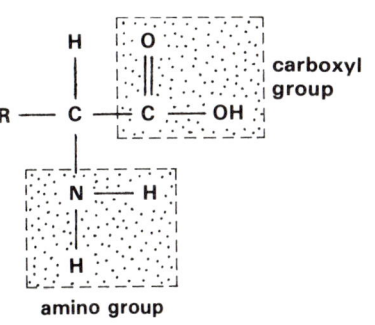

FIG. 3.20 *Structural formula of an α-amino acid*

Table 3.2 α-amino acids commonly found in plant and animal protein

alanine	leucine
arginine	lysine
asparagine	methionine
aspartic acid	phenylalanine
cysteine	proline
glutamine	serine
glutamic acid	threonine
glycine	tryptophan
histidine	tyrosine
isoleucine	valine

FIG. 3.21 *Molecular structures of some amino acids*

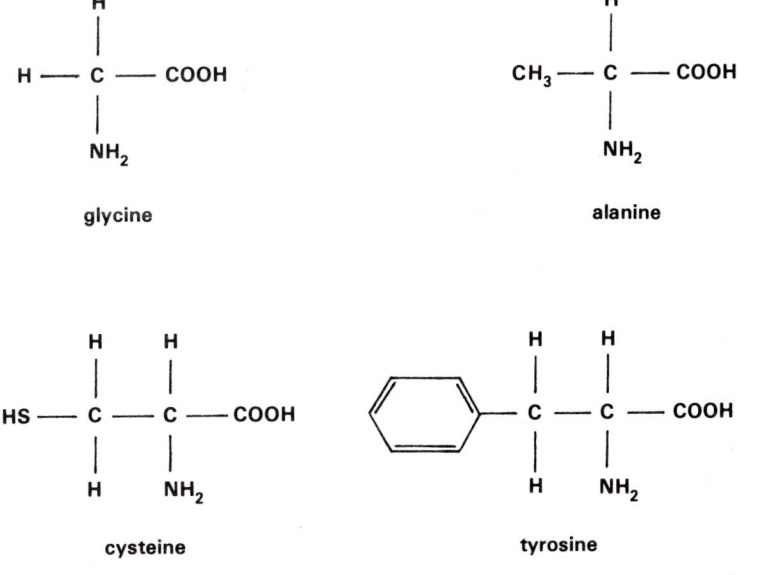

With the exception of glycine all amino acids have an asymmetric carbon atom and thus display stereoisomerism. Alanine, for example, can exist in two forms, D- and L-alanine (Fig. 3.22). As with glyceraldehyde (section 3.1.1), the two forms are non-superimposable mirror images. Interestingly, all of the amino acids which exist in proteins are L-isomers.

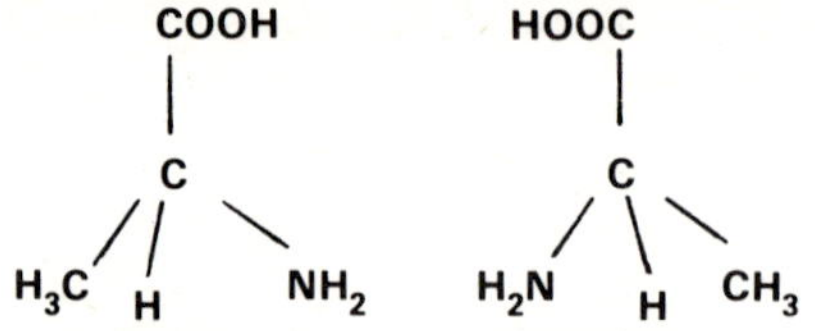

FIG. 3.22 *Comparison of the molecular structures of the stereoisomers of alanine and glyceraldehyde*

Note that the carboxyl group of alanine is in the same position as the carboxyl group of glyceraldehyde. Which groups in glyceraldehyde are in the same positions as the amino and the methyl group of alanine?

Table 3.3 Classification of α-amino acids by polarity of R group

POLARITY OF R	EXAMPLE	POLARITY DUE TO
non-polar	alanine	non-polarity due to CH_3 group
polar, but neutral	tyrosine	polarity due to OH group
	cysteine	polarity due to SH group
negative	aspartic acid	polarity due to COOH group
	glutamic acid	polarity due to COOH group
positive	lysine	polarity due to NH_2 group
	arginine	polarity due to NH_2 group

A useful way of classifying amino acids is based on the polarity of R. According to such a scheme there are four main groups of amino acids (Table 3.3). A knowledge of polarity is important in understanding some of the methods which can be used to separate a mixture of amino acids (section 3.4) and in appreciating some of the properties of proteins. In neutral aqueous solutions amino acids with non-polar R groups exist as dipolar ions called **amphions** in which the carboxyl group donates a hydrogen ion (H^+) to the amino group:

$$R - \overset{\overset{\displaystyle H}{|}}{\underset{\underset{\displaystyle NH_2}{|}}{C}} - COOH \rightleftharpoons R - \overset{\overset{\displaystyle H}{|}}{\underset{\underset{\displaystyle \underset{\oplus}{NH_3}}{|}}{C}} - COO^-$$

Amino acids can therefore act as donors and as acceptors of hydrogen ions and are said to be **amphoteric**. In acid solutions, where the hydrogen ion concentration is high, amino acid molecules are able to bind some of the hydrogen ions, while in alkaline solutions amino acids can donate hydrogen ions to the excess hydroxyl ions (OH^-):

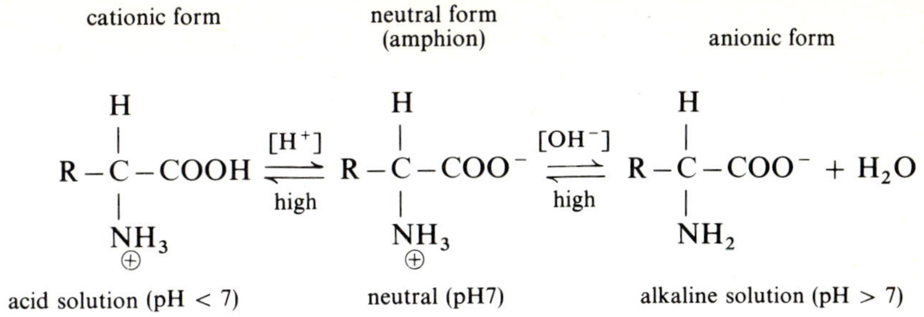

cationic form | neutral form (amphion) | anionic form

$$R-\underset{\underset{\oplus}{NH_3}}{\overset{H}{\underset{|}{C}}}-COOH \underset{\text{high}}{\overset{[H^+]}{\rightleftharpoons}} R-\underset{\underset{\oplus}{NH_3}}{\overset{H}{\underset{|}{C}}}-COO^- \underset{\text{high}}{\overset{[OH^-]}{\rightleftharpoons}} R-\underset{NH_2}{\overset{H}{\underset{|}{C}}}-COO^- + H_2O$$

acid solution (pH < 7) | neutral (pH7) | alkaline solution (pH > 7)

The ability of substances such as amino acids to control fluctuations in pH is called **buffering capacity** (Chapter 2).

The equilibrium point for the reactions described above depends on the hydrogen ion concentration, $[H^+]$. Where R is polar the situation is rather more complicated. However for every amino acid a particular $[H^+]$ causes amphions to predominate. The pH of the solution at which this occurs is called the **isoelectric point**, which for most amino acids is between pH 6 and 7. At this point the amino acids, being neutral, will not move towards the anode or to the cathode when an electric current is passed through the solution.

3.3.2 Polypeptides and proteins

In suitable conditions amino acids polymerise. They react in such a way that the α-amino group of one molecule becomes joined to the carboxyl group of another. The condensation reaction results in the formation of **peptide linkages** (Fig. 3.23). The free amino and carboxyl groups at each end of the dipeptide can form peptide linkages with other amino acid molecules so building up a chain of amino acid residues. Some short-chain peptides are of biological importance. Vasopressin (antidiuretic hormone) and ocytocin, two mammalian hormones, are each made from just nine amino acid molecules (Chapter 18). Long chains of amino acid residues are called **polypeptides**. One or more polypeptide chains are used in the formation of a protein molecule. The number of amino acid molecules ranges from about forty in the smallest proteins to several thousand in the largest. The relative molar mass of proteins ranges from 6000 to over 1 000 000.

FIG. 3.23 *Formation of a peptide linkage*

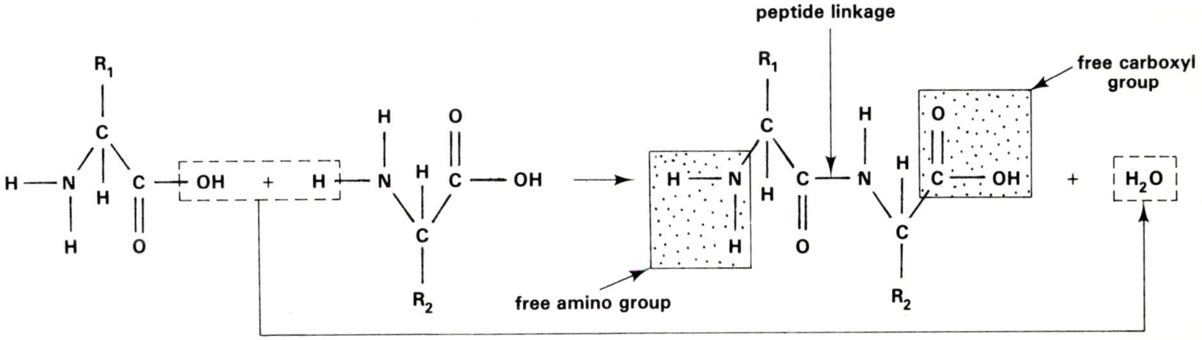

The amino acid composition of a few proteins was known in the 1940s. In the 1950s chromatographic techniques made it possible to identify the variety of amino acids in many other proteins. However, the sequence of amino acids in any protein molecule was a mystery until 1953 when Frederick Sanger published the findings of his research on the structure of the mammalian hormone insulin. Sanger showed that insulin consists of two polypeptide chains bonded to each other by **disulphide bridges** (Fig. 3.24). One chain contains 21 amino acid molecules the other 30. In the late 1950s and during the 1960s, the amino acid sequences of other small protein molecules were determined. Corticotropin (ACTH) was found to contain 39 amino acid molecules, ribonuclease has 124 while the α- and β-globins of haemoglobin have 141 and 146 respectively. More recently proteins of much higher relative molar mass have been analysed and a great deal of information has been compiled on the composition and amino acid sequences of a range of proteins from many different sources.

Such studies have shown that few proteins contain all twenty α-amino acids. Insulin, for example, has seventeen different amino acids. Some proteins consist of a small range of amino acids, in others a much wider range is found. The results of the sequence analyses revealed two general principles. Firstly, the sequence of amino acids differs from one type of protein to another. Secondly, the sequence in any one type of protein is precisely fixed. The findings have had a dramatic impact in molecular genetics (Chapter 6).

FIG. 3.24 *Structure of insulin*

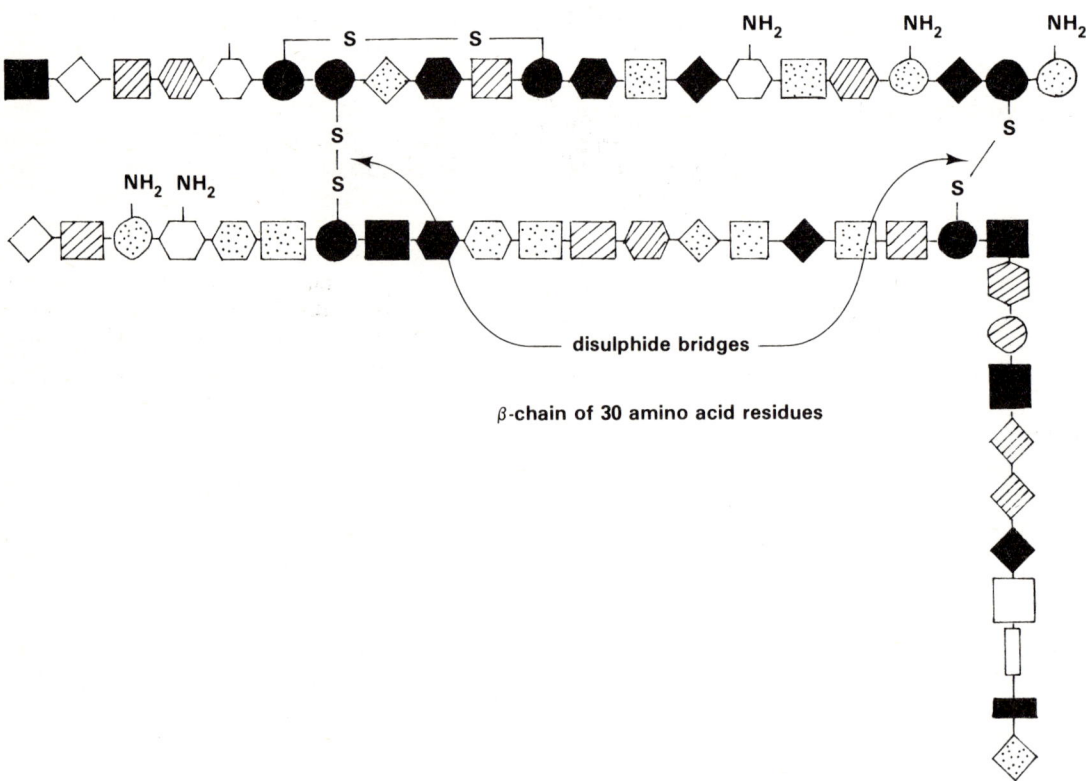

Without question, amino acid composition and sequence studies have yielded much valuable information about proteins. Yet proteins display many properties which cannot be accounted for from the results of such studies. In order to understand some of these properties it is necessary to know something about protein structure.

1. PROTEIN STRUCTURE

The name **primary structure** is given to the sequence and three-dimensional arrangement of amino acids in a polypeptide chain. Protein molecules have a very complex three-dimensional shape which cannot be explained in terms of the primary structure of their polypeptide chains. **X-ray diffraction analysis** has been a very important tool in unravelling the spatial arrangement of the atoms in protein molecules. In this technique molecules are bombarded with a beam of X-rays. Some of the rays are deflected by atoms in the molecules and are then passed through a photographic film to give a **diffraction pattern**. From the pattern, which is characteristic for a particular compound, the shape of the molecules can be deduced. It all sounds very simple but the interpretation of the pattern necessitates the use of a computer.

In 1939 Astbury obtained an X-ray diffraction pattern for keratin, a fibrous protein found in hair. The pattern indicated that the polypeptide chains in keratin were twisted or folded in a regular manner. Pauling and Corey in 1951 showed that the chains were twisted into a right handed helix. In this α-helix the peptide linkages form the backbone from which the R groups of the amino acids jut out in all directions. Stability of the helix is maintained by hydrogen bonds which occur between the keto ($>C=O$) and imino ($>N-H$) groups of every fourth peptide link (Fig. 3.25). It is now generally agreed that all fibrous proteins have a **secondary structure** of this kind.

FIG. 3.25 *The α-helix (simplified)*

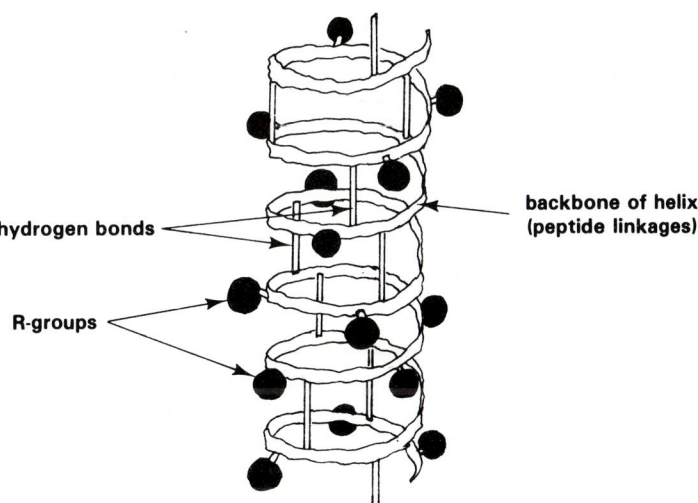

hydrogen bonds

backbone of helix (peptide linkages)

R-groups

In keratin three to seven α-helices are coiled around each other like the strands of a rope. The strands are held together by disulphide ($-S-S-$) cross linkages between the thiol ($-SH$) groups of cysteine molecules. Earlier we saw that this type of bond holds together the two polypeptide chains of insulin (Fig. 3.24). The rope-like arrangement has considerable strength and elasticity. The helices are spring-like and are capable of extending and contracting. Proteins of this kind perform useful supporting and contractile functions in the mammalian body. They include myosin of striped muscle (Chapter 15) and elastin of yellow elastic tissue. In collagen of white fibrous tissue the polypeptide chains are permanently extended. Three such chains are wound around each other and are bound to one another by hydrogen bonds. This arrangement gives less elasticity but has considerable tensile strength (Fig. 3.26).

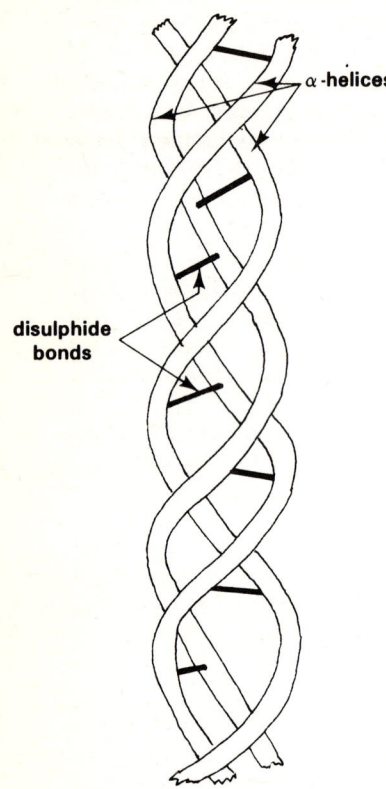

FIG. 3.26 *Tertiary structure of collagen*

α-helices

disulphide bonds

The grouping together of several α-helices gives a protein its **tertiary structure**. Globular proteins such as myoglobin, haemoglobin and enzyme proteins have a less regular tertiary structure. Again using X-ray diffraction analysis, J. C. Kendrew in the late 1950s showed that myoglobin, the oxygen-carrying pigment in muscles, consists of a single polypeptide chain folded asymmetrically. The chain has eight relatively straight segments in which the polypeptide is coiled in an α-helix. The segments are united by bent peptide bonds and the chain bonded to an iron-containing haem group (Fig. 3.27). Disulphide, hydrogen and electrovalent bonds help to stabilise the three-dimensional shape of such molecules (Fig. 3.28).

FIG. 3.27 *Tertiary structure of myoglobin*

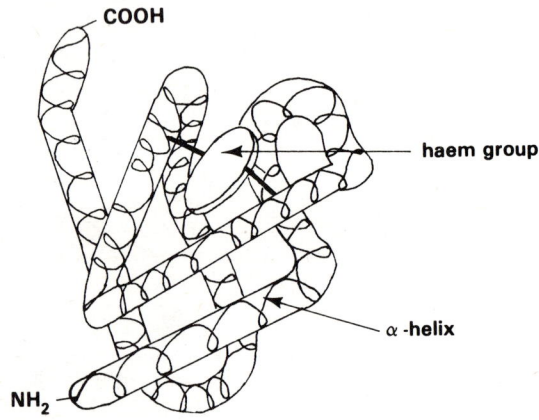

COOH

haem group

α-helix

NH_2

Most globular proteins with a relative molar mass of more than 50 000 consist of two or more polypeptide chains. The way in which they fit together is called the **quaternary structure** of the protein. Haemoglobin for example consists of four chains, two of α-globin and two of β-globin, each bonded to a haem group. The α- and β-chains are very similar in structure to myoglobin (Chapter 8).

FIG. 3.28 *Some of the bonds which maintain the three-dimensional structure of a globular protein*

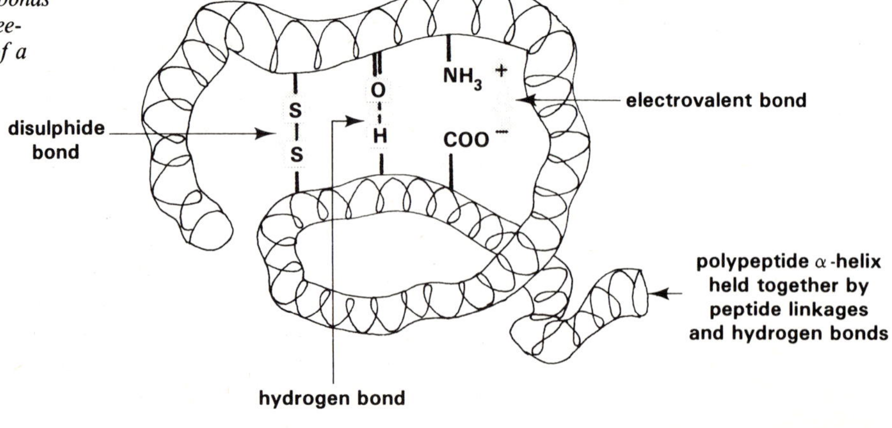

disulphide bond

NH_3^+

electrovalent bond

COO^-

polypeptide α-helix held together by peptide linkages and hydrogen bonds

hydrogen bond

2. SOME PROPERTIES OF PROTEINS IN AQUEOUS SOLUTIONS

(i) SOLUBILITY. Globular proteins are generally **soluble in water** because the polar R groups jutting outwards from the α-helix attract water molecules. Enzymes are globular proteins which function in aqueous solutions. Some globular proteins of high relative molar mass form **colloidal suspensions** in water. They include the proteins which are present in cell membranes, ribosomes and in chromosomes (Chapter 7). The **insolubility** of fibrous proteins in water is accounted for by the large number of non-polar hydrophobic R groups on the outside of the α-helix.

(ii) BUFFERING CAPACITY. The ability of a protein to accept or donate hydrogen ions (H^+) depends on the number and charge of ionisable R groups on the surface of its molecules. The most important R groups in this respect are the amino ($-NH_2$) and carboxyl ($-COOH$) groups. Amino groups bind H^+ in acid solutions (pH < 7) while carboxyl groups donate H^+ in alkaline solutions (pH > 7). In this way proteins act as **buffers** in stabilising the pH of their surroundings.

(iii) ISOELECTRIC POINT. At a precise pH in aqueous solution each protein exists as an **amphion** which carries no net electric charge. This is called the **isoelectric point** of the protein. If the protein has a large number of free amino groups the isoelectric point is at a pH above 7 while proteins with a large number of free carboxyl groups form amphions at a pH below 7. Why is this so? The electrical properties of proteins are exploited in separating mixtures of proteins by electrophoresis and ion-exchange chromatography (section 3.4).

3. PROTEIN DENATURATION

Most proteins function effectively only within a limited range of temperature and pH. At extremes of temperature and pH proteins undergo physical changes called **denaturation** which generally result in their precipitation from solution. Further details of protein denaturation are given in Chapter 4.

It is an interesting fact that denatured protein molecules often spontaneously return to their original form when placed in ideal conditions of temperature and pH, providing denaturation was not too extreme.

4. FUNCTIONS OF PROTEINS

Proteins are used for an extraordinary range of activities in living organisms. Some of their more important functions are summarised as follows:

(i)	Enzymes	e.g. pepsin, amylase, lipase
(ii)	Contraction	e.g. actin and myosin in muscle
(iii)	Protection	e.g. antibodies, fibrinogen and prothrombin in mammalian blood
(iv)	Hormones	e.g. insulin, somatotropin, corticotropin
(v)	Transport	e.g. haemoglobin and myoglobin which carry oxygen in mammalian blood and muscle
(vi)	Support	e.g. collagen, elastin
(vii)	Storage	e.g. albumin in egg white, gliadin in wheat seeds.

5. BIOCHEMICAL TESTS FOR PROTEINS AND AMINO ACIDS

One of the most common biochemical tests used to demonstrate the presence of amino acids in extracts from living organisms is the **ninhydrin**

reaction. When heated with ninhydrin the α-amino group forms a blue coloured complex. Proline gives a yellow colour in this reaction.

The **biuret test** is frequently used to investigate materials of living origin for the presence of compounds with peptide links. Proteins and polypeptides form a violet complex when their peptide linkages react with alkaline copper(II) sulphate solution.

Millon's reagent is a solution of mercury(II) nitrate in nitric acid. Mercury reacts with thiol groups of amino acids such as cysteine which is found in many proteins. At the same time electrostatic forces in the protein molecules break due to the high concentration of hydrogen ions from the acid. The protein becomes denatured and a white precipitate is formed. When heated the precipitate takes on a red coloration.

3.4 Biochemical techniques

Breakthroughs in scientific knowledge often flow from the development of new procedures and techniques. Advances in biochemical knowledge have been no exception. Some technological developments used in biochemistry such as X-ray crystallography have already been mentioned earlier in this chapter. Let us now look at other techniques which have been used even more widely during the past twenty-five years or so to provide a wealth of information about the organic compounds found in living organisms. Some of the techniques are simple and you can use them for yourself. Scientific progress does not rely exclusively on sophisticated equipment and complicated technology. The techniques described here are mainly concerned with the separation and identification of biological compounds in mixtures and the analysis of complex molecules.

3.4.1 Chromatography

Chromatography is a technique used to separate the components of mixtures. It is based on the partitioning of compounds between a **stationary phase** and a **moving phase**. In **partition chromatography** the stationary phase is a liquid while in **adsorption chromatography** the stationary phase is a solid.

Partition chromatography depends mainly on the relative solubility of substances in two or more solvents. When a solute is added to a mixture of equal volumes of two immiscible solvents it may dissolve entirely in one or other of the solvents. If the solute dissolves in both solvents it may not dissolve to the same extent in each solvent.

The ratio: $\dfrac{\text{concentration of solute in solvent 1}}{\text{concentration of solute in solvent 2}}$ at equilibrium is called the **partition coefficient**. Use is made of differences in the partition coefficient to separate components of mixtures and to identify the components of mixtures extracted from living organisms.

Adsorption chromatography relies mainly on electrostatic interaction between solutes and the stationary phase. In partition chromatography the liquid stationary phase is held in position by a solid **support medium**. During separation of the components of a mixture some of the components may be adsorbed to the support medium. Thus in practice there is often no clear distinction between partition and adsorption chromatography.

Paper, column, thin-layer, gas-liquid and ion-exchange are examples of the various forms of chromatography commonly used by biochemists.

1. PAPER CHROMATOGRAPHY
Paper chromatography is used extensively in biological research yet it is a

simple technique and is often practised in school laboratories. A **starting line** is drawn in pencil about half an inch from one edge of a sheet of absorbent paper which is the support medium. On the line a number of **origins** about half an inch apart are marked. A drop of solution of the mixture to be analysed is placed on one of the origins. On each of the remaining origins is placed a drop of a pure solution of a substance suspected to be in the mixture. The technique is extremely sensitive and can be used to detect minute quantities of unknown substances. For this reason a micropipette is used to spot the origins.

The paper is placed vertically in a tank so that the starting line touches a solvent which runs up (in **ascending chromatography**) or down (in **descending chromatography**) the support medium by capillarity. The known compounds and the components of the unknown mixture are carried different distances according to their partition coefficients between the solvent and water present in the paper. Adsorption to the support medium also affects movement of the solutes.

FIG. 3.29 *Flow diagram showing the main stages in ascending paper chromatography*

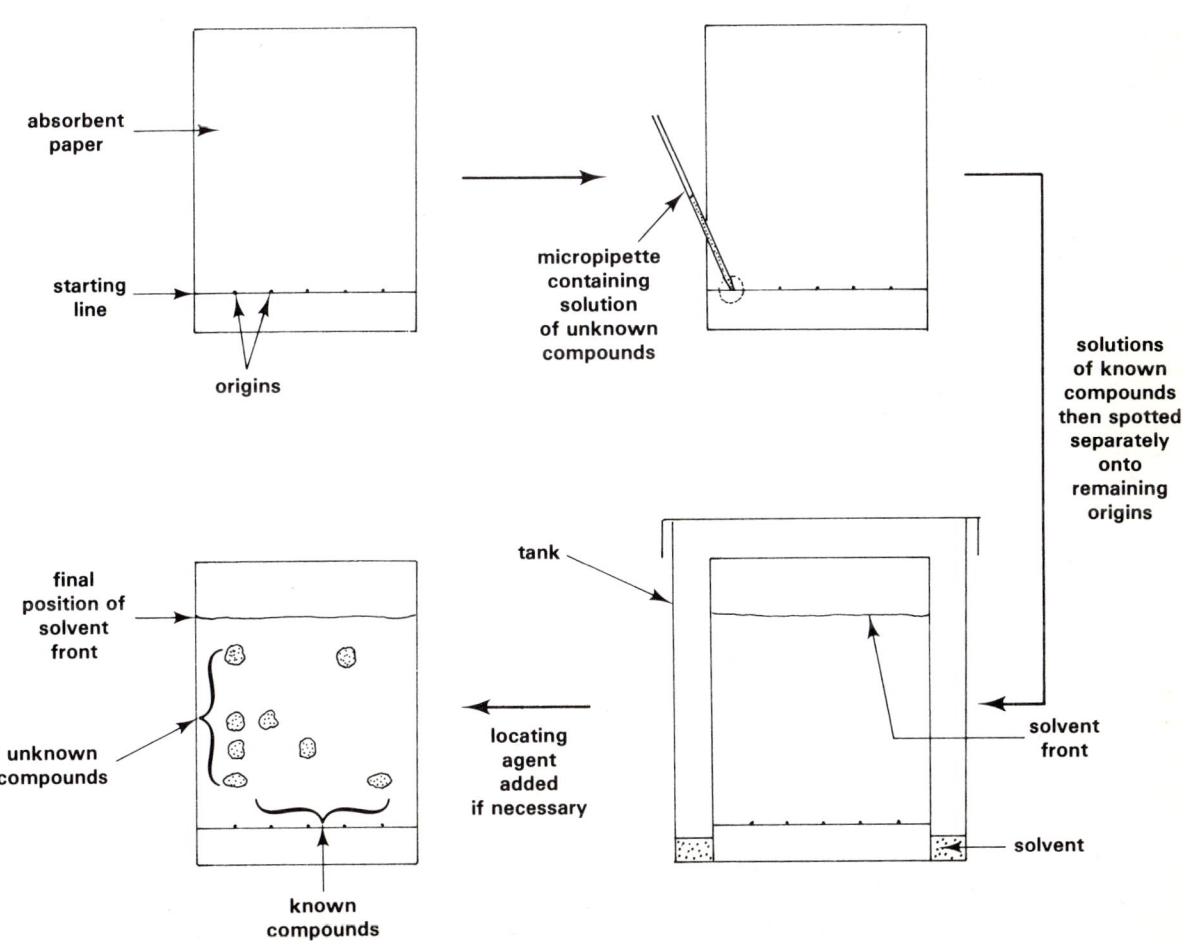

When the solvent has travelled most of the length of the paper, the distance it has moved, the **solvent front**, is marked and the paper is dried. If the solutes are coloured, as in the case of leaf pigments, the positions of the components can be seen without further steps. The positions of colourless substances have to be detected using a **locating agent** which reacts with the compounds under investigation to form coloured end products. Ninhydrin, for example, is used to locate amino acids (section 3.3.2).

The ratio: $\dfrac{\text{distance travelled by a compound}}{\text{distance travelled by solvent front}}$ is called the **retardation factor (R_f value)**. For a given solvent at a given temperature every compound has a characteristic R_f value. Thus by comparing the R_f values of the components of a mixture with those of the known compounds it usually is possible to identify some if not all of the substances in the mixture (Fig. 3.29). Should the first solvent not produce a satisfactory separation of all the compounds in the mixture the absorbent paper may then be turned through 90° and a second solvent run along it. This is **two-dimensional chromatography.**

The amounts of substances which have been separated can also be determined. The first step is to cut out the coloured spots and place them separately in a standard volume of a solvent. This procedure is called **elution**. After a standard period of time the paper discs are removed and the intensity of colour in the solvent is measured using a colorimeter.

FIG. 3.30 (a) *Stages in the separation of the components of a mixture by column chromatography*

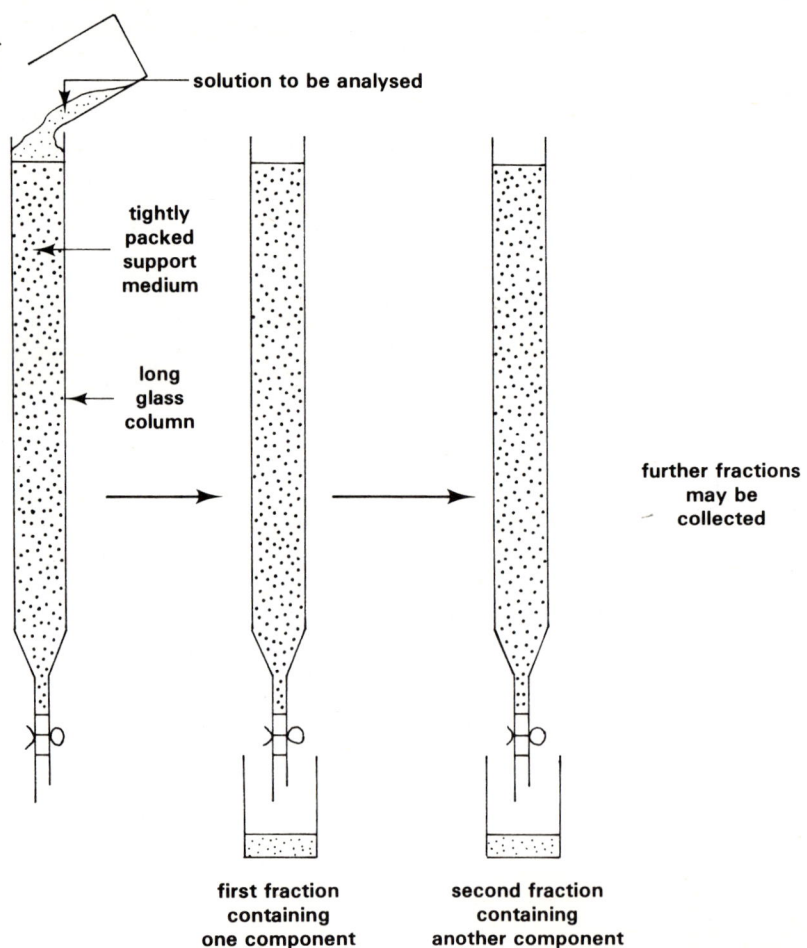

solution to be analysed

tightly packed support medium

long glass column

further fractions may be collected

first fraction containing one component

second fraction containing another component

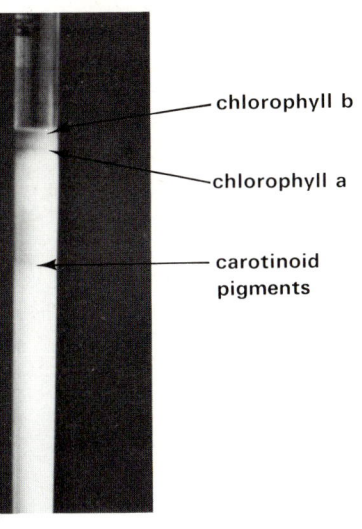

FIG. 3.30 (b) *A column chromatogram showing separated photosynthetic pigments*

chlorophyll b

chlorophyll a

carotinoid pigments

2. COLUMN CHROMATOGRAPHY

A long glass tube is packed with a support medium such as hydrated starch, silica gel or diatom shells called kieselguhr. A solution of the mixture to be analysed is poured into the top of the column. The components of the mixture move down the column at different rates. Differences in partition coefficients between the solvent and water in the support medium and adsorption to the support medium account for the rates at which the solutes travel. Again, the separation of coloured compounds such as leaf pigments can be seen without further steps. When the solvent begins to run out at the bottom of the column, fractions collected at staggered intervals of time will contain different components of the mixture. The separated components can be identified and their amounts can be measured using a colorimeter (Fig. 3.30).

3. THIN-LAYER CHROMATOGRAPHY

A glass plate is covered with an aqueous slurry of a support medium, for example, silica gel or cellulose powder. The plate is dried in a hot air oven and the mixture to be analysed, together with known compounds, spotted on to separate origins on the starting line. The plate is placed vertically in a tank so that the solvent almost touches the starting line. As in paper chromatography the solvent rises by capillarity and the components of the mixture become separated according to their partition coefficients and adsorption to the support medium. When the solvent front has moved a sufficient distance to separate the components of the mixture the plate is removed from the tank, dried, and if necessary a locating agent sprayed on to the support medium. The procedure can be made two-dimensional if the first solvent gives an incomplete separation. Once more the amounts of substances in the mixture can be determined. After scraping off the coloured spots and eluting the coloured complex, the intensity of colour can be determined with a colorimeter.

Sugars, amino acids and pigments in biological extracts can be separated and identified using any of the methods of chromatography described above. Thin-layer chromatography is often used to analyse the acylglycerols in natural fats and oils.

4. GAS–LIQUID CHROMATOGRAPHY

Analysis of the fatty acids in a mixture is normally carried out using **gas–liquid chromatography**. In this technique the vaporised methyl esters of the fatty acids are transported along a heated capillary tube using an inert carrier gas such as nitrogen. The tube is packed with a finely powdered support medium coated with paraffin grease or other similar high melting point substance. Because they have different partition coefficients between vaporised and liquid grease, the esters are separated and are carried out of the tube to a detector at different times. The time between the introduction of a sample into the tube and the appearance of a component in the detector is called the **retention time**. Each component has a characteristic retention time. Changes in potential difference between two electrodes when the component is burned in a hydrogen flame is one of several methods used to detect the amount of the component in the sample (Fig. 3.31).

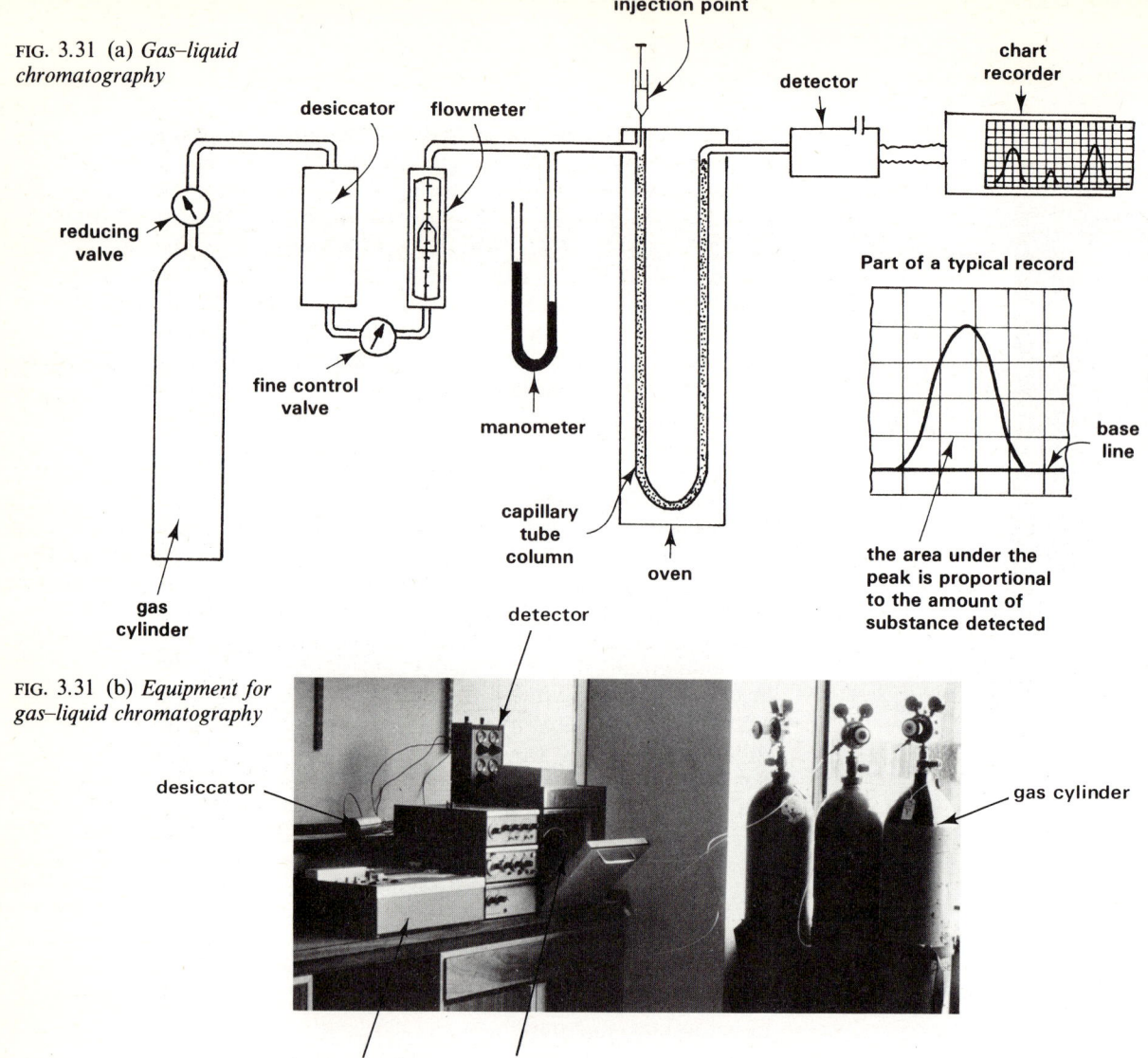

FIG. 3.31 (a) *Gas–liquid chromatography*

injection point

chart recorder

detector

desiccator   flowmeter

reducing valve

fine control valve

manometer

capillary tube column

oven

gas cylinder

Part of a typical record

base line

the area under the peak is proportional to the amount of substance detected

FIG. 3.31 (b) *Equipment for gas–liquid chromatography*

detector

desiccator

gas cylinder

chart recorder   oven

5. ION-EXCHANGE CHROMATOGRAPHY

Ion-exchange chromatography depends on the polarity of molecules and is therefore used to separate charged molecules. A long glass column is packed with a synthetic resin possessing charged functional groups. If the groups are positively charged the packing material is called a **cation exchange resin**, if negatively charged it is called an **anion exchange resin**.

Mixtures of amino acids are often separated using cation-exchange resins. Columns filled with sulphonated polystyrene beads are suitable. The polystyrene resin is first saturated with sodium hydroxide solution so that the beads are fully charged with sodium ions (Na^+). The amino acid mixture is adjusted to pH3 so that the acids are cationic (positively charged). Why does this happen? On pouring the mixture into the column the amino acids displace Na^+ ions from the resin. The amount of displacement depends on the polarity of the acids. Basic amino acids with an additional amino group such as lysine and arginine displace two Na^+ ions and are bound more firmly to the resin than acidic amino acids such as glutamic

FIG. 3.32 (a) *An amino acid analyser* (courtesy LKB Instruments Ltd.)

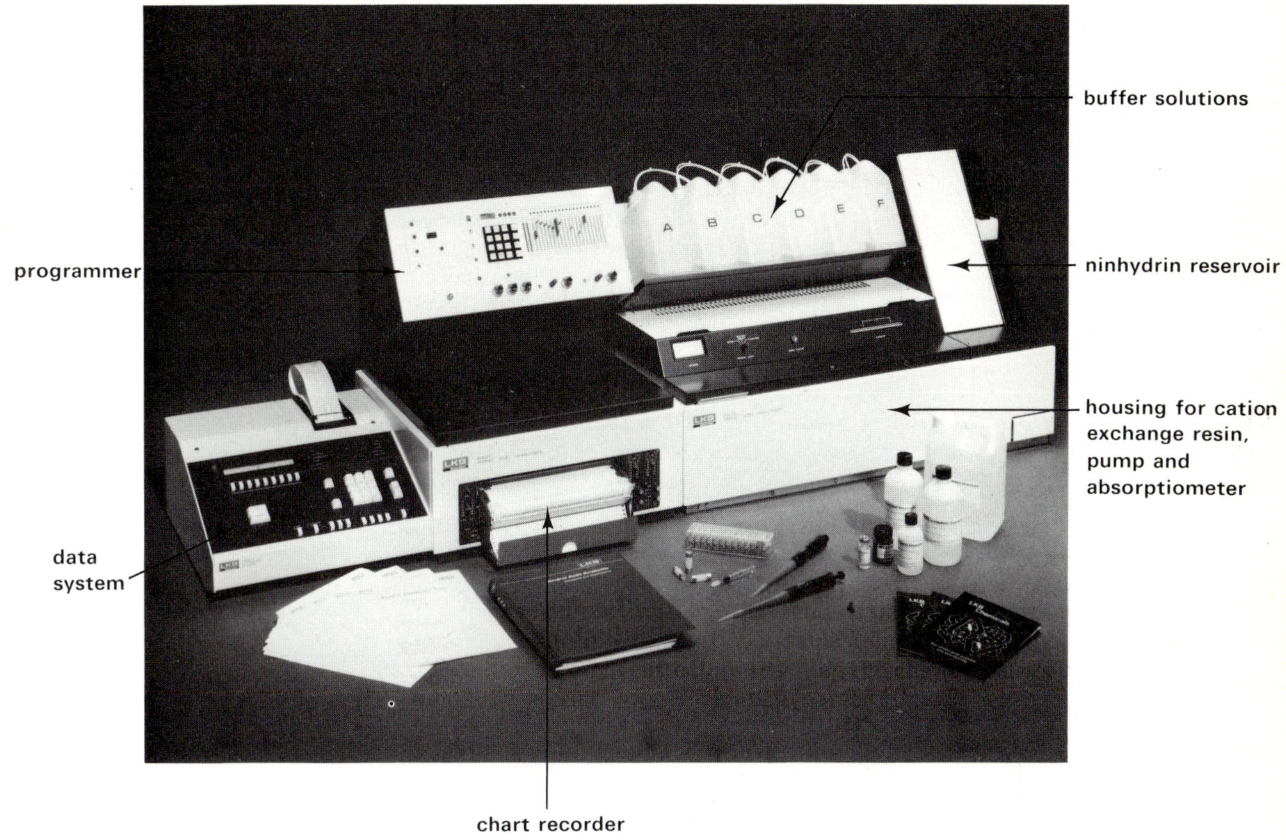

programmer

buffer solutions

ninhydrin reservoir

housing for cation exchange resin, pump and absorptiometer

data system

chart recorder

FIG. 3.32 (b) *Trace obtained from an amino acid analyser. The area under each peak is a measure of the amount of each amino acid in the mixture*

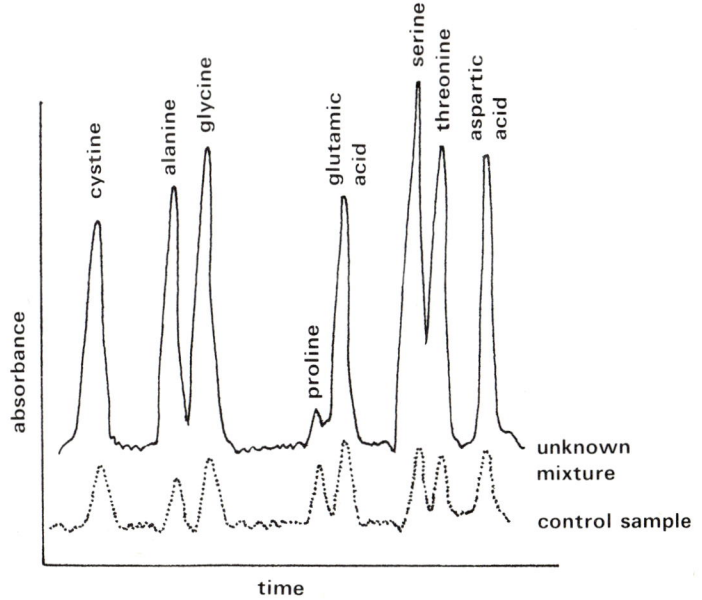

cystine

alanine

glycine

glutamic acid

proline

serine

threonine

aspartic acid

absorbance

time

unknown mixture

control sample

and aspartic acids which displace one Na^+ ion. For these reasons some amino acids move more quickly down the column than others. Samples collected from the bottom of the column at staggered time intervals contain different amino acids the amounts of which can then be measured by colorimetry. The collection of samples and their colorimetric determination can be performed automatically in **amino acid analysers** (Fig. 3.32).

3.4.2 Electrophoresis

The polarity of amino acids and proteins is exploited in separating mixtures of the compounds by **electrophoresis**. At its isoelectric point (section 3.3.2) a protein or an amino acid molecule is neutral and will not move to the positive pole (anode) or to the negative pole (cathode) if an electric current is passed through a solution in which the molecules are dissolved or suspended. At a pH above the isoelectric point amino acids and protein molecules have a net negative charge and move towards the anode. At a pH below the point they have a net positive charge and move towards the cathode.

In **paper electrophoresis** an aqueous solution or suspension of amino acids or proteins is applied to a starting line marked across the centre of a strip of filter paper or cellulose acetate paper soaked in a buffer solution.

FIG. 3.33 *Equipment for paper electrophoresis*

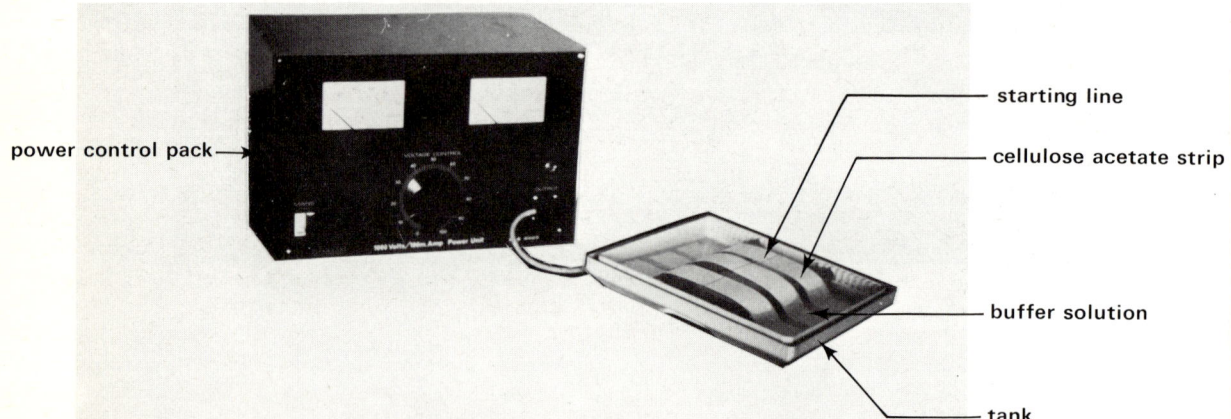

power control pack

starting line

cellulose acetate strip

buffer solution

tank

An electric current is passed through the buffer and the molecules move to either the anode or to the cathode at a rate depending mainly on the pH of the buffer, the net electrical charge of the molecules, and the voltage applied (Fig. 3.33). After a period of time the current is switched off and the positions of the separated components of the mixture identified using a locating agent. The pattern of separation is compared with that of known mixtures to identify the amino acids or proteins in the unknown mixture. Electrophoresis is used extensively in medicine to analyse blood proteins such as the immunoglobulins (Chapter 9).

It is impossible in a text of this nature to give more than a brief survey of some of the more useful biochemical techniques available today. Reference is made to other procedures at appropriate places in the book. Even now technical difficulties deny to man much biochemical knowledge and the search for new and improved techniques continues.

Chapter 4

Chapter 4

Enzymes

In living cells hundreds of different biochemical reactions take place rapidly and simultaneously. The reactions go on at relatively low temperatures (4°C to 60°C) and are normally controlled in such a way that useful products are made and wastes removed at rates which satisfy the metabolic needs of cells. How is it possible for there to be such orderliness in what must be a potentially chaotic situation? How can reactions take place so rapidly at such modest temperatures? The answers to these questions come from a study of **enzymes**.

Enzymes were discovered almost accidentally in 1897 by Hans and Eduard Buchner. They were interested in making extracts of yeast cells for medical purposes. One of the problems the Buchner brothers were faced with was how to preserve the yeast juice. They tried various methods one of which involved adding large amounts of sucrose to the juice. To their surprise they found that the sugar was fermented even though there were no live yeast cells in the mixture. Up to this time it was thought that fermentation could only take place in the presence of living organisms.

The word enzyme, meaning 'in yeast', was used to describe the substance or substances in yeast juice which brought about the fermentation of sucrose.

4.1 The structure of enzymes

To appreciate the structure of enzymes it is necessary to have a knowledge of protein structure (Chapter 3). Over 90 per cent of enzymes are simple, **globular proteins**. The remainder are conjugated proteins which have a non-protein fraction called the **prosthetic group**. Most enzymes have relative molar masses of between 50 000 and 200 000.

FIG. 4.1 *Three-dimensional structure of chymotrypsinogen* (after P. B. Sigler *et al.*, J. Mol. Biol., 35:143 (1968))

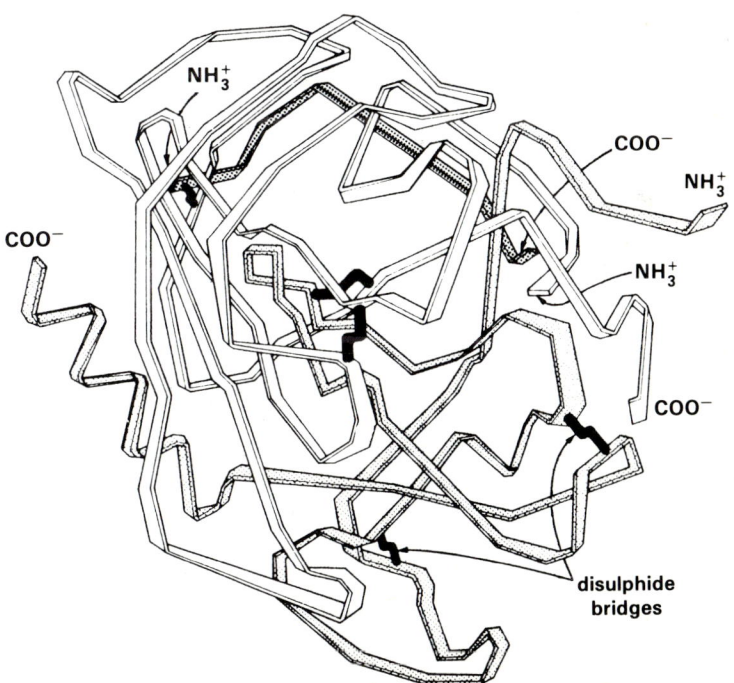

COO⁻

NH₃⁺

COO⁻

NH₃⁺

NH₃⁺

COO⁻

disulphide bridges

Amino acid analysis, X-ray diffraction studies and other techniques have in recent years provided a lot of information about the structure of enzymes (Chapter 3). The three-dimensional structure of a number of enzymes is now known in detail. Chymotrypsinogen, an inactive precursor of the pancreatic enzyme chymotrypsin, consists of 245 amino acid residues which are bonded to form three polypeptide chains (Fig. 4.1 and Chapter 14).

4.2 Enzyme action

Enzymes are soluble in water and work in aqueous solution in living cells. They are sometimes described as **organic catalysts**. A catalyst is a substance which affects the rate of a chemical reaction. The biochemical reactions catalysed by enzymes are mainly **reversible reactions**. An example of a reversible reaction is as follows:

$$\underset{\text{reactants}}{A + B} \rightleftharpoons \underset{\text{product}}{C}$$

At equilibrium the rate at which A and B are converted to C is equal to the rate at which C is converted to A and B. The position of the equilibrium depends on the energy difference between the reactants and the product. Enzymes do not alter the direction of a reaction, they speed up the rate at which equilibrium is reached. In so doing they can catalyse reversible reactions in either direction providing it is energetically feasible. Let us look at the way in which enzymes can speed up biochemical reactions.

FIG. 4.2 *Energetics of a biochemical reaction*

(a) *Uncatalysed*

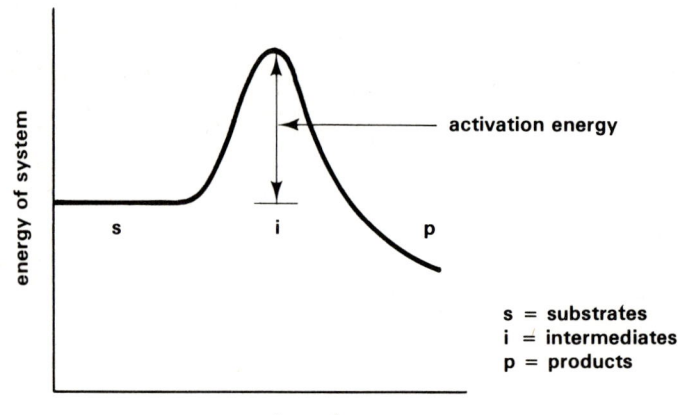

s = substrates
i = intermediates
p = products

(b) *Catalysed by an enzyme*

4.2.1 Enzyme catalysis

Biochemical reactions involve the formation or the destruction of chemical bonds. When two or more reactants are joined, chemical bonds are formed. When a complex molecule is split into simpler components chemical bonds are destroyed. In either case energy is required to bring about the changes. The energy required for a chemical reaction to proceed is called **activation energy** (Fig. 4.2(a)). Heat can be used as a source of activation energy. Indeed some reactions do not proceed quickly unless the reactants are raised to relatively high temperatures. In living cells however, reactions take place rapidly at relatively low temperatures. Enzymes lower the amount of activation energy needed, making it possible for reactions to occur at temperatures which are otherwise energetically unfavourable (Fig. 4.2(b)).

4.2.2 The lock and key mechanism

For chemical reactions to take place the reactants must be brought into contact. The quicker the reacting molecules collide the quicker is the reaction rate. One of the characteristics of enzymes is that they increase the chances of molecular collisions by drawing reactants together. They do this by what is called a **lock and key mechanism**.

Emil Fischer in 1894 proposed the lock and key hypothesis to explain how enzymes and reactants called **substrates** behave when mixed. Fischer suggested that enzyme and substrate molecules combine to form an **enzyme-substrate complex** before the products of the reaction are released (Fig. 4.3).

FIG. 4.3 *The lock and key mechanism of enzyme action*

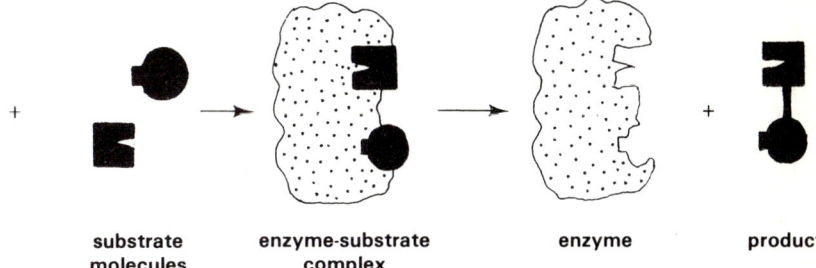

| enzyme molecule | substrate molecules | enzyme-substrate complex | enzyme | product |

The sites on enzyme molecules where substrates fit are called **active centres**. Spectroscopic evidence supports the lock and key hypothesis (Fig. 4.4) as do the results of studies on enzyme inhibition (section 4.3.2) and the effect of substrate concentration on enzyme activity (section 4.3.6). Additional support comes from studies on enzyme specificity.

FIG. 4.4 *Change in absorption spectrum of NADH on addition of alchohol dehydrogenase*

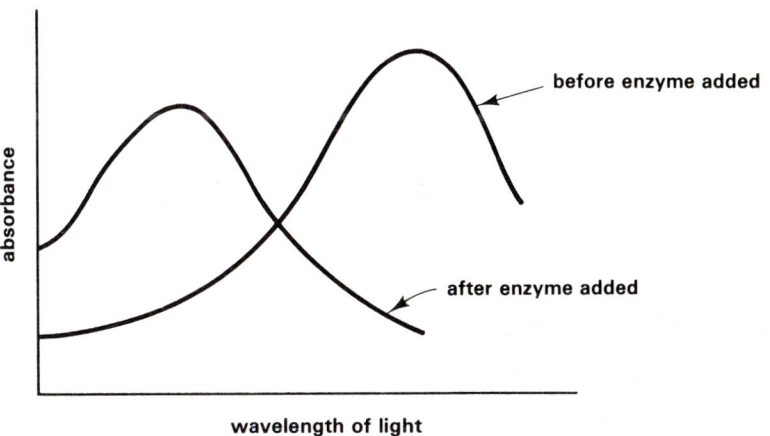

69

4.2.3 Enzyme specificity

Because enzyme and substrate molecules are three-dimensional structures the formation of an enzyme-substrate complex requires that the shapes of the reactants and the active centres of enzymes are **complementary**. Otherwise the enzyme and substrate cannot unite (Fig. 4.5). The phenomenon whereby most enzymes work with only one or with a limited range of substrates is called **enzyme specificity**.

One of the systems used to name enzymes is based on the substrates with which the enzyme combines. For example lipases react with lipids. What types of substrates do you think carbohydrases and peptidases react with?

FIG. 4.5 *Enzyme specificity*

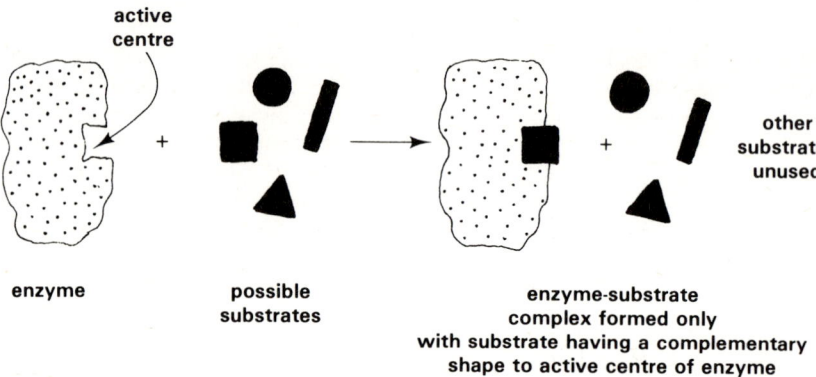

Biochemical pathways usually involve a number of linked reactions. Each reaction is catalysed by a specific enzyme:

$$A \xrightleftharpoons{enzyme\ X} B \xrightleftharpoons{enzyme\ Y} C \xrightleftharpoons{enzyme\ Z} D$$

4.2.4 Control of enzyme action

In some instances accumulation of one of the products formed near the end of the chain inhibits the action of an enzyme used in one of the earlier reactions. In the hypothetical example shown above the presence of a high concentration of product D for example may slow down the rate at which enzyme X converts A to B. This is an example of **negative feedback inhibition**, an important device in the co-ordination of the metabolism of cells.

$$A \rightleftharpoons B \rightleftharpoons C \rightleftharpoons D$$

inhibition by negative feedback

Negative feedback ensures that reactants are used efficiently and prevents the excess manufacture of end products. Like other homeostatic devices described elsewhere in this book the control of enzyme action helps to maintain a stable internal environment in living organisms.

4.3 Factors affecting enzyme action

To function at all some enzymes need the presence of **co-factors**. To function efficiently enzymes need a suitable **temperature** and **pH** and there must be enough **substrate** available. Enzyme action can be stopped partially or completely if **inhibitors** are present in the reaction mixture.

4.3.1 Co-factors

A **prosthetic group** is an essential co-factor attached to the protein part of a conjugated enzyme. If the prosthetic group is removed the enzyme fails to function. The prosthetic group sometimes contains a heavy metal such as iron (Fe), molybdenum (Mo), copper (Cu) or manganese (Mn).

Mineral ions must be mixed with the reactants before some enzymes will work. Many trace elements behave as **enzyme activators** in this way. It is partly for these reasons that minerals are essential to living organisms (Chapters 13 and 14).

In other instances enzymes do not function unless **co-enzymes** are present. Co-enzymes are organic, non-protein substances which are not bonded to enzyme molecules like prosthetic groups. Several important co-enzymes are vitamin-derivatives. **Nicotinamide-adenine dinucleotide (NAD)** and its phosphate ester **nicotinamide-adenine dinucleotide phosphate (NADP)** for example, contain a derivative of nicotinamide, a B-group vitamin (Chapter 14). Riboflavin (vitamin B_2) forms part of **flavin-adenine dinucleotide (FAD)**. NAD, NADP and FAD function as hydrogen acceptors in reactions catalysed by dehydrogenase enzymes (section 4.4).

4.3.2 Inhibitors

Many substances inhibit the activity of enzymes. **Inhibitors** fall into two categories, **reversible** and **non-reversible**.

1. REVERSIBLE INHIBITORS

Reversible inhibitors are substances which prevent enzymes from combining with substrates. The enzymes are not denatured and full activity is restored when the inhibitor is removed.

Competitive inhibitors affect enzyme action by becoming attached to active centres, so stopping the substrate from binding to the enzyme. A well-known example of this behaviour is inhibition of the enzyme succinic dehydrogenase by malonic acid. Succinic dehydrogenase catalyses the oxidation of succinic acid to fumaric acid in the Krebs' tricarboxylic acid cycle (Chapter 5). In the presence of malonic acid the reaction rate is slowed down. Malonic acid has a molecular structure which is very similar to that of succinic acid. Inhibition happens because some of the dehydrogenase enzyme's active centres become occupied by malonic acid rather than by the normal substrate succinic acid. In effect malonic acid and succinic acid are competing for the same active centres of the enzyme (Fig. 4.6). The degree of inhibition by a competitive inhibitor is less if the ratio of substrate to inhibitor molecules is high. Why is this so?

FIG. 4.6 *Competitive inhibition of an enzyme*

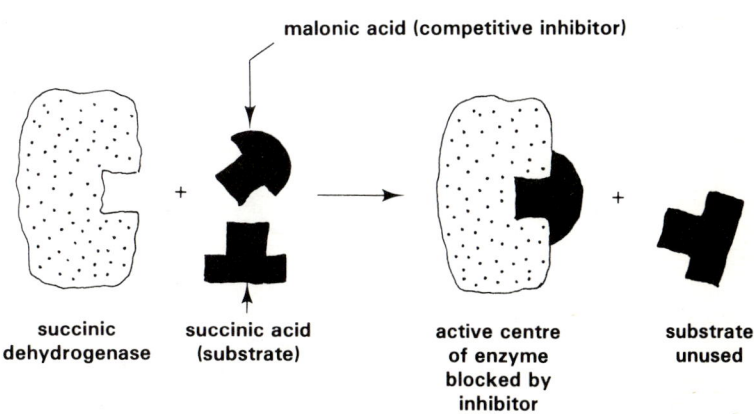

malonic acid (competitive inhibitor)

succinic dehydrogenase   succinic acid (substrate)   active centre of enzyme blocked by inhibitor   substrate unused

The degree of inhibition by **non-competitive inhibitors** cannot be reduced by increasing the number of substrate molecules. Here the inhibitor becomes attached to the enzyme at a position other than the active centre. Nevertheless the enzyme, the substrate, or possibly both become changed so that enzyme activity stops. Disulphide bridges are important in maintaining the tertiary structure of enzyme molecules (Chapter 3). If the disulphide bridges are broken the three-dimensional shape of the enzyme changes. The ions of heavy metals such as mercury (Hg), silver (Ag) and copper (Cu) affect enzymes in this way. Hg^{2+}, Ag^+ and Cu^{2+} ions combine with thiol ($-SH$) groups in enzymes, so denaturing enzyme molecules and inhibiting enzyme activity. Cyanide is another non-competitive inhibitor. It blocks the action of some enzymes by combining with iron which may be present in a prosthetic group or which may be required as an enzyme activator. It is not surprising therefore that the salts of heavy metals and cyanide are potent poisons to living organisms.

2. NON-REVERSIBLE INHIBITORS

Organophosphorus compounds which are used as insecticides are good examples of non-reversible inhibitors. They bind with enzymes in such a way that activity of the enzyme is permanently stopped. Insecticides of this type inactivate the enzyme cholinesterase which is essential for the functioning of the nervous system (Chapter 15).

4.3.3 Temperature

Heat provides activation energy and therefore chemical reactions are more likely to occur at high temperatures. Secondly, heat supplies kinetic energy to reacting molecules, causing them to move more rapidly. The chances of molecular collisions taking place are increased at high temperatures so it is more likely that enzyme-substrate complexes will be formed. However, heat energy also increases the vibration of the atoms which make up enzyme molecules. If the vibrations become too violent chemical bonds in the enzyme break and the precise three-dimensional structure, so essential for enzyme activity, is lost. At high temperatures therefore enzymes become **denatured**.

When the effect of temperature on enzyme activity is investigated ex-

FIG. 4.7 *The effect of temperature on the rate of an enzyme-catalysed reaction*

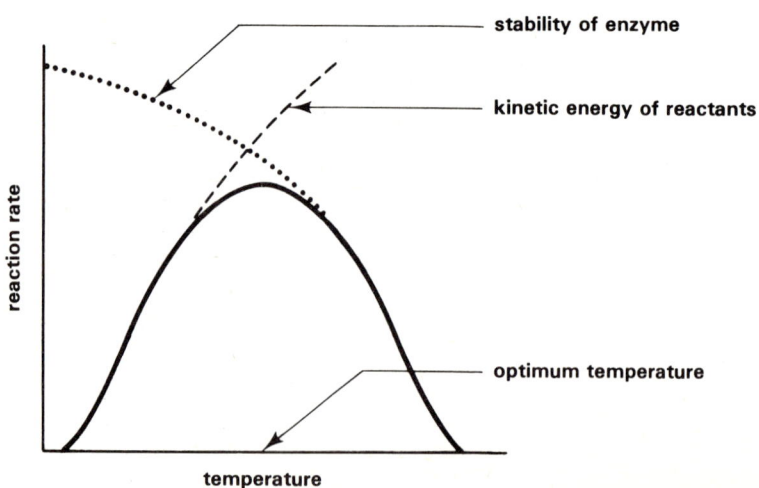

stability of enzyme

kinetic energy of reactants

optimum temperature

reaction rate

temperature

perimentally a temperature usually called the **optimum temperature** is observed at which the reaction proceeds most rapidly. This temperature is not necessarily that at which the enzyme is most stable. It is the resultant of the contrary effects of temperature on the movement of reactants and enzyme denaturation (Fig. 4.7).

The term **temperature coefficient** (Q_{10}) is used to express the effect of a 10°C rise in temperature on the rate of a chemical reaction.

$$Q_{10} = \frac{\text{rate of reaction at t} + 10°C}{\text{rate of reaction at t°C}}$$

Between 4°C and the optimum temperature the Q_{10} for enzyme-catalysed reactions is 2.

4.3.4 pH

The symbol **pH** refers to the concentration of hydrogen ions in solution (Chapter 2). The concentration of hydrogen ions [H^+] affects the stability of the electrovalent bonds which help to maintain the tertiary structure of protein molecules (Chapter 3). Extremes of pH cause the bonds to break resulting in enzyme denaturation. For every enzyme there is an **optimum pH** at which the reaction it catalyses proceeds most rapidly (Fig. 4.8). Many enzymes work within a pH range of 5–9 and catalyse reactions most efficiently at pH 7. There are exceptions. For example, pepsin and rennin secreted in the mammalian stomach work best at pH 1·5–2·5 (Chapter 14).

FIG. 4.8 *The effect of pH on the rates of enzyme-catalysed reactions*

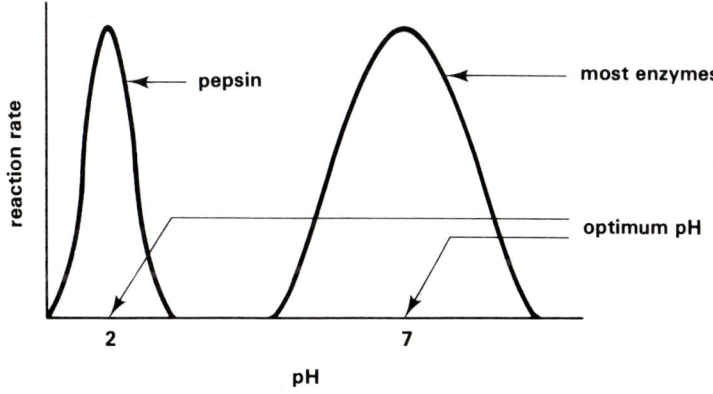

It must be remembered that even small changes in pH mean relatively large changes in [H^+]. A change of 1 on the pH scale involves a ten-fold change in [H^+] while a change in pH of 2 represents a hundred-fold increase in [H^+] (Chapter 2). Thus even small changes in pH can have a great effect on enzyme activity.

Apart from the effect in denaturing enzymes, changes in [H^+] can change the ionisation of the amino acid residues at the active centres of enzymes. Ionisation of the reactants can also be affected. The formation of enzyme-substrate complexes sometimes depends on the active centres and substrate molecules having opposite electrostatic charges. If the charges are altered by changes in pH some enzymes fail to function (Fig. 4.9).

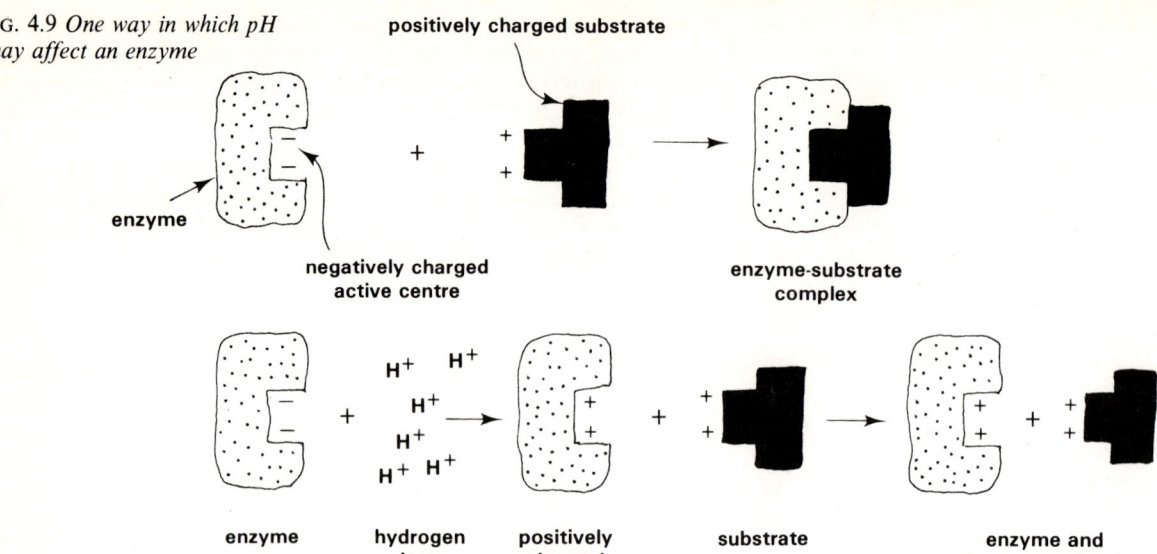

FIG. 4.9 *One way in which pH may affect an enzyme*

positively charged substrate

enzyme

negatively charged
active centre

enzyme-substrate
complex

enzyme   hydrogen   positively   substrate   enzyme and
ions   charged   substrate molecules
active centre   repel each other

4.3.5 Enzyme concentration

Enzymes catalyse reactions rapidly at very low **enzyme concentrations**. This is because enzyme molecules form complexes with substrates only very briefly. The products of the reaction are quickly released and the enzyme is then available for further activity.

The rate at which enzymes use substrates is described as the **turn-over number**. For some enzymes the turn-over number is very high. Catalase for example breaks down 40 000 molecules of hydrogen peroxide into water and oxygen every second! Even the slowest of enzymes have turn-over numbers of $100 \, \text{sec}^{-1}$. The larger the number of enzyme molecules present the greater is the amount of substrate used in a given period of time provided that there is an excess of substrate available (Fig. 4.10).

FIG. 4.10 *The effect of enzyme concentration on the rate of an enzyme-catalysed reaction*

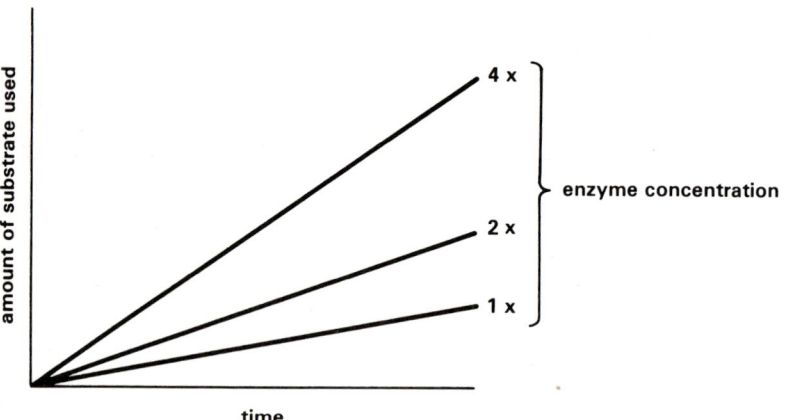

amount of substrate used

4 x

2 x

1 x

enzyme concentration

time

4.3.6 Substrate concentration

Figure 4.11 shows the effect of **substrate concentration** on the rate of a biochemical reaction when the amount of enzyme is limited. At low concentrations of substrate there is a linear relationship between the reaction rate and substrate concentration. In these conditions the ratio of enzyme to substrate molecules is high. Consequently some active centres are always free for reactants to bind with the enzyme. However, a point is reached when a further increase in substrate concentration does not cause the reaction to go any faster. The enzyme to substrate ratio is then lower and there are more substrate molecules present than there are free active centres with which to bind. Adding more substrate will not make the reaction go more quickly.

The concentration of substrate at which the reaction rate is half of its maximum is characteristic for each combination of enzyme and substrate. It is called **Michaelis' constant**.

FIG. 4.11 *The effect of substrate concentration on the rate of an enzyme-catalysed reaction*

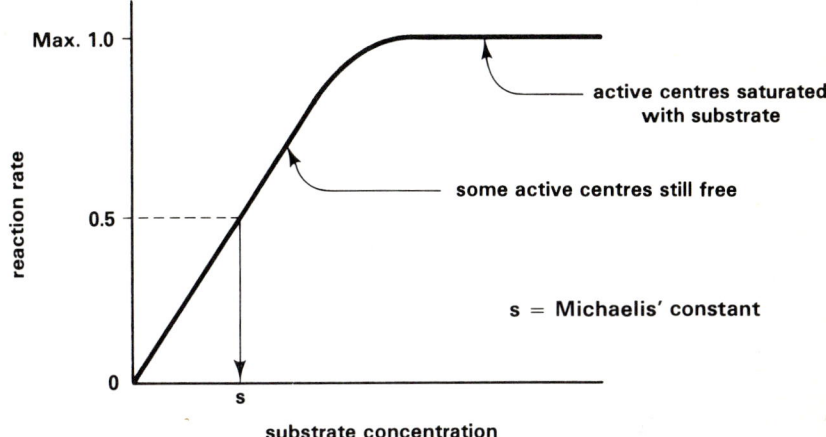

4.4 The classification of enzymes

In the past twenty years many new enzymes have been discovered. Older methods of classifying enzymes have become unpractical and for this reason, in the early 1960s the International Union of Biochemistry (IUB) recommended a new system for the classification and naming of enzymes. According to the IUB system there are six groups of enzymes.

4.4.1 Hydrolases

These enzymes catalyse reactions in which substrates are hydrolysed into simpler products. During **hydrolysis** hydrogen atoms from water enter one of the products while the hydroxyl groups end up in the other product. The reaction is shown in a simplified form as follows:

$$\text{AB} + \text{H.OH} \rightleftharpoons \text{AH} + \text{B.OH}$$

substrate   water   products

Figure 4.12 shows how some important substrates found in living organisms are hydrolysed. Hydrolases bring about the breakdown of materials in lysosomes (Chapter 1), the digestion of food in the guts of mammals (Chapter 14) and the mobilisation of food reserves in plants (Chapter 20).

FIG. 4.12 *Hydrolysis of some important biological compounds*

(a) *Glycosidic linkage in a disaccharide*

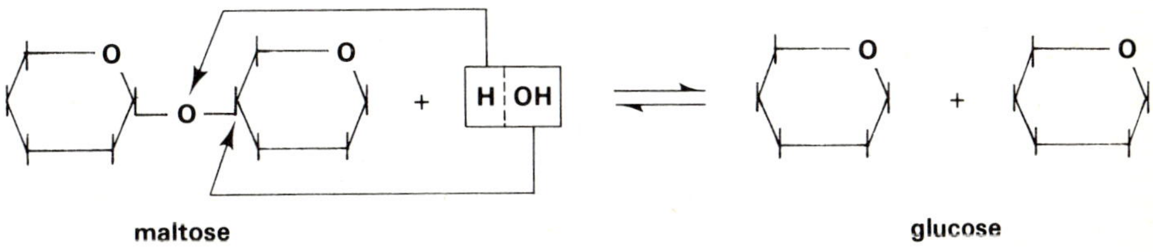

maltose   glucose

(b) *Peptide linkage in a protein*

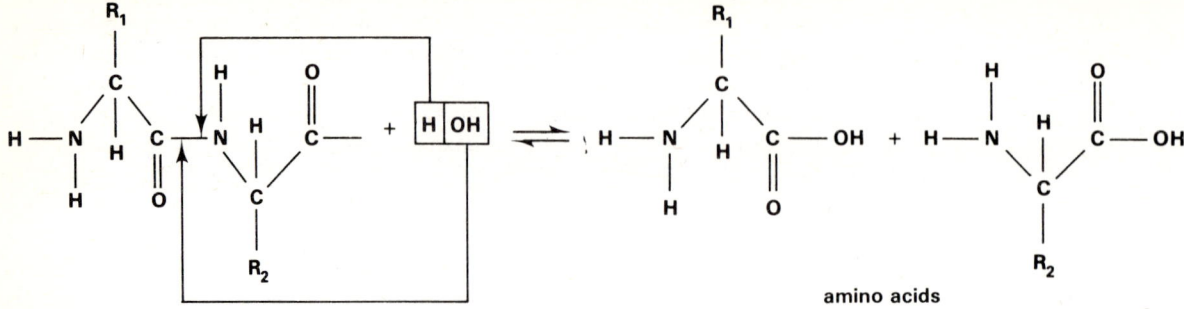

amino acids

(c) *Ester linkages in a fat or oil*

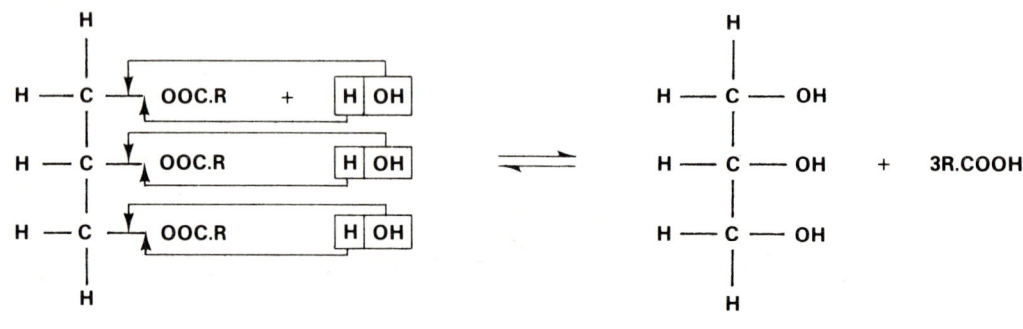

glycerol   fatty acids

4.4.2 Oxido-reductases

The **oxidation** of substrates is catalysed by this group of enzymes. There are two kinds of oxido-reductase enzymes:

1. OXIDASES catalyse the transfer of hydrogen atoms to molecules of oxygen. A simplified way of expressing the reaction is as follows:

$$AH_2 + \tfrac{1}{2}O_2 \rightleftharpoons A + H_2O$$

substrate · oxygen · oxidised substrate · water (reduced oxygen)

Hydrogen peroxide is sometimes formed instead of water.

2. DEHYDROGENASES catalyse the oxidation of substrates by transferring hydrogen atoms to co-enzymes such as NAD and NADP:

$$AH_2 + 2NAD^+ \rightleftharpoons A + 2NADH$$

substrate · co-enzyme · oxidised substrate · reduced co-enzyme

Alternatively, electrons are transferred. In reactions catalysed by oxido-reductases, substrates are oxidised while oxygen or co-enzymes are reduced. Oxidation-reduction reactions usually release energy. The importance of these reactions is described more fully in Chapter 5.

4.4.3 Transferases

Transferases catalyse the transfer of functional groups from one substrate to another:

$$AB + C \rightleftharpoons A + BC$$

Two important kinds of transferases are **transaminases** which are concerned with the transfer of amino groups (Fig. 4.13 and Chapter 14) and **kinases** which control the transfer of energy-rich phosphate groups (Chapter 5).

FIG. 4.13 *Transamination*

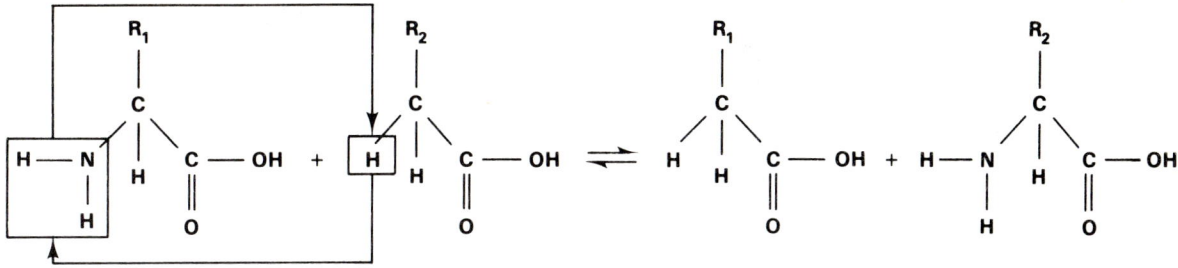

4.4.4 Isomerases

Isomerases convert one isomer of a compound to another isomer of the same compound:

$$ABC \rightleftharpoons ACB$$

Isomerases are involved in the interconversion of sugar isomers in glycolysis (Chapter 5).

4.4.5 Ligases

This group of enzymes catalyses reactions in which new chemical bonds are formed. **Ligases** control the synthesis of macromolecules such as proteins and nucleic acids (Chapter 6). ATP provides the energy to make the new chemical bonds.

4.4.6 Lyases

Lyases catalyse the breakdown of complex substrates into simpler products but water is not used as in hydrolytic reactions:

$$AB \rightleftharpoons A + B$$

Carboxylases which regulate the release from or the addition of carbon dioxide to substrates are examples of lyase enzymes. The roles of carboxylases in respiration and photosynthesis are described in Chapter 5.

Much of what is known about enzymes has come from investigations carried out *in vitro* on purified enzyme extracts. In vitro means in glass, indicating that it is easier to obtain information from test-tube studies than it is *in vivo*, in living cells. The way in which enzymes control vast numbers of different reactions taking place simultaneously in microscopic packets of protoplasm is therefore all the more remarkable. Compartmentalisation of cells (Chapter 1) undoubtedly helps. Each of the organelles of a living cell has a specific range of enzymes so that different kinds of metabolic reactions are separated. Even so, the way in which enzymes regulate the many, rapid and complex metabolic reactions in living cells surely gives cause for wonder.

Chapter 5

Energy changes in living cells

The composition of living organisms is very different from their surroundings. Whereas the environment consists of relatively simple substances such as gases, water and minerals, living organisms are made up of very complex molecules (Chapter 3). Most of the atoms in the environment are distributed in energy-poor compounds. In contrast, the atoms which make up living organisms are mainly found in energy-rich compounds.

Living organisms constantly use energy to synthesise the compounds of which they are made. They also use energy to perform osmotic, mechanical and electrical work (Chapters 2, 10, 11, 13 and 15 respectively). Where does the energy come from and how is it used?

The sun is the ultimate source of energy available to the living world. Energy in sunlight is changed to **chemical energy** in green plants in a process called **photosynthesis**. The products of photosynthesis are complex energy-rich molecules from which energy can be released to do work in living organisms. Green plants are called the **producer organisms** of ecosystems. Organisms which cannot photosynthesise are called **consumers**. Primary consumers acquire energy-rich compounds by feeding on green plants. Secondary consumers feed on primary consumers. Whatever the trophic level an organism belongs to, energy is made available when energy-rich substances are oxidised in living cells during **respiration**. Thus photosynthesis and respiration bring about the conversion of sunlight into energy which can be used to do work in living organisms.

5.1 Respiration

Respiration is the oxidation of energy-rich substrates. Carbohydrates, sugars especially, are the **respiratory substrates** most used by living organisms although lipids and proteins can also be oxidised. Polysaccharides are first hydrolysed to the hexose sugar glucose, lipids to glycerol and fatty acids, proteins to amino acids.

Most forms of life are **aerobes**. They respire using oxygen. Many aerobic organisms can respire for short periods of time in the absence of oxygen. They are called **facultative anaerobes**. A few species of bacteria are **strict anaerobes** and respire only in the absence of oxygen. Whatever the kind of respiration it is important to understand what energy is provided and how this energy is used in processes which could not occur without respiratory energy. In this context it is useful to know about a substance called adenosine 5′-triphosphate.

5.1.1. Adenosine 5′-triphosphate

Adenosine 5′-triphosphate (ATP) has the structure shown in Figure 5.1. ATP was first isolated by Lohmann from muscle extracts in 1929. Ten years later Lipmann suggested that ATP transfers energy from energy-rich substrates to energy-requiring processes in living cells.

According to the First Law of Thermodynamics every chemical reaction involves energy changes among the reactants. In some reactions a lot of heat energy is made. Heat energy is of little use to living organisms except that it can help speed up enzyme activity (Chapter 4). Heat cannot be used to synthesise proteins and other essential macromolecules, nor can it bring about the contraction of muscle or a host of other energy-requiring reactions. Energy which is used to do work in living organisms is called **free energy**.

It has been calculated that when the bond holding the terminal phosphate group of ATP is broken by hydrolysis, $30 \cdot 66$ kJ mol^{-1} of free energy is released. The same amount of free energy is released when the terminal phosphate bond of **adenosine 5′-diphosphate (ADP)** is hydrolysed. However, when the phosphate bond of **adenosine 5′-phosphate (AMP)** is hydrolysed only $14 \cdot 28$ kJ mol^{-1} of free energy is released. The amounts of free energy released in living cells may be even higher than the figures quoted here.

FIG. 5.1 *The structure of ATP*

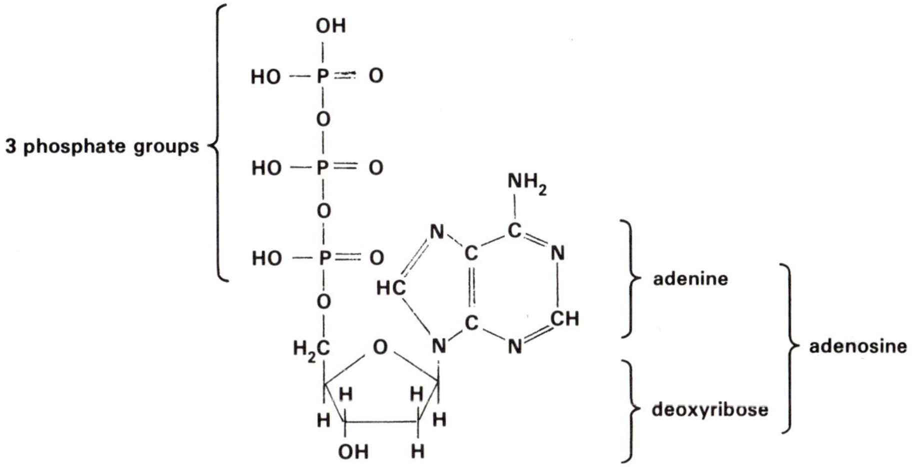

Similar amounts of energy are required to make the phosphate bonds. For example, when ADP and inorganic phosphate are converted to ATP, $30 \cdot 66$ kJ mol^{-1} of free energy is needed. The bonds between the first, second and third phosphate groups of ATP are called **high energy bonds** and adenosine 5′-triphosphate is sometimes written as:

$$A - P \sim P \sim P$$

where A = adenosine
P = phosphate group
$\sim$ = high energy bond
$-$ = low energy bond

The bonds in other substances found in living cells yield even more free energy than the high energy bonds of ATP. **Phosphocreatine** in striped muscle is an example of such a compound (Chapter 15). However, ATP is the only substance present in sufficient quantities in most living cells to be of general use as a provider of free energy. ATP is therefore indispensable in the transfer of free energy from energy-rich substrates to reactions in which free energy is used.

But how is ATP made in living cells and where does the energy for its synthesis come from?

5.1.2 Oxidation of respiratory substrates

In anaerobic conditions energy-rich substrates such as glucose are oxidised to lactic acid or to ethanol and carbon dioxide. Lactic acid is a product of anaerobic respiration in some bacteria and in animal cells. Ethanol and carbon dioxide are made by yeasts and higher plant cells when oxygen is absent. The overall processes can be summarised as follows:

lactic fermentation: $C_6H_{12}O_6 \longrightarrow 2CH_3CHOHCOOH$

glucose   lactic acid

alcoholic fermentation: $C_6H_{12}O_6 \longrightarrow 2CH_3CH_2OH + 2CO_2$

glucose   ethanol   carbon dioxide

The equations give us no idea of the amount of free energy released nor do they tell us anything about the way in which the substrate is oxidised. What is more they disguise the fact that the two processes are very similar, each sharing a number of common steps. **Glycolysis** (glyco = sugar; lysis = to split) is the name given to the common steps. Glycolysis also takes place when sugars are respired aerobically.

1. GLYCOLYSIS

Figure 5.2(a) shows the more important common steps in the oxidation of glucose, whether in anaerobic or aerobic conditions. Far from producing free energy, some of the earlier reactions need an input of free energy. The energy comes from the hydrolysis of high energy bonds of ATP. Without a supply of free energy from ATP the reactions could not proceed. The

FIG. 5.2 (a) *Glycolysis*

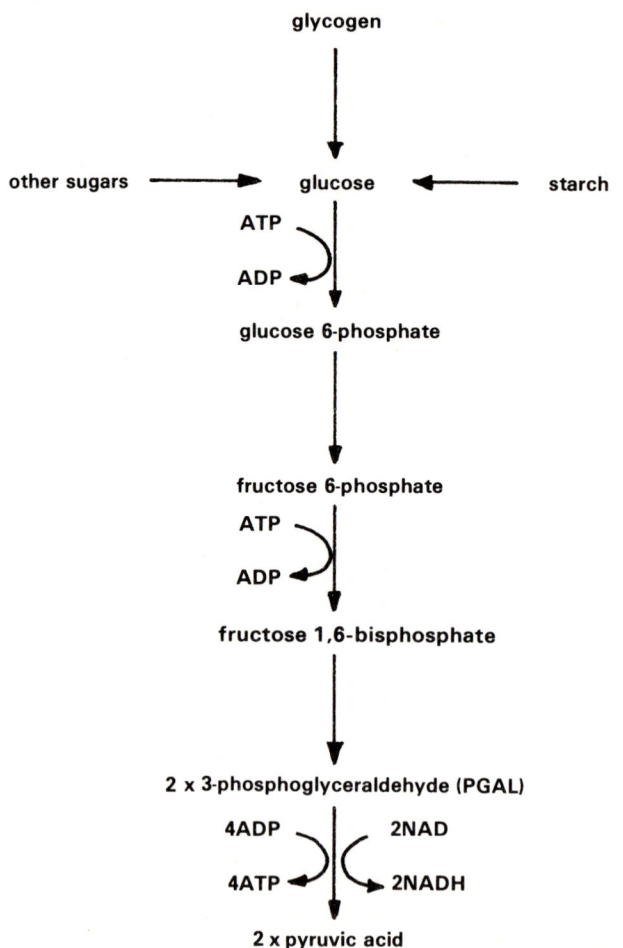

splitting of each **fructose 1, 6-bisphosphate** molecule yields two molecules of the triosephosphate called **3-phosphoglyceraldehyde (PGAL)** which is then oxidised to **pyruvic acid**. Oxidation of PGAL involves the removal of hydrogen and is catalysed by dehydrogenase enzymes (Chapter 4). At the same time the co-enzyme nicotinamide-adenine dinucleotide (NAD) becomes reduced to NADH. Oxidation of each PGAL molecule produces enough free energy for two ATP molecules to be synthesised from ADP and inorganic phosphate.

The fate of the pyruvic acid depends on the organism in which it was produced and on whether oxygen is available or not. If lactic acid is a product of anaerobic respiration the NADH formed in glycolysis transfers its hydrogen to pyruvic acid and **lactic acid** is formed. This is what happens in mammalian muscle cells when they have an oxygen debt (Chapter 8). Bacteria which sour milk also make lactic acid when oxygen is not available. If yeasts and higher plant cells are kept without oxygen, pyruvic acid is split into **carbon dioxide** and **ethanal**. The ethanal is then reduced to **ethanol** by hydrogen from NADH (Fig. 5.2(b)). Either way NAD is remade and once again acts as a hydrogen acceptor when more PGAL is oxidised.

FIG. 5.2 (b) *The fate of pyruvic acid in the absence of oxygen*

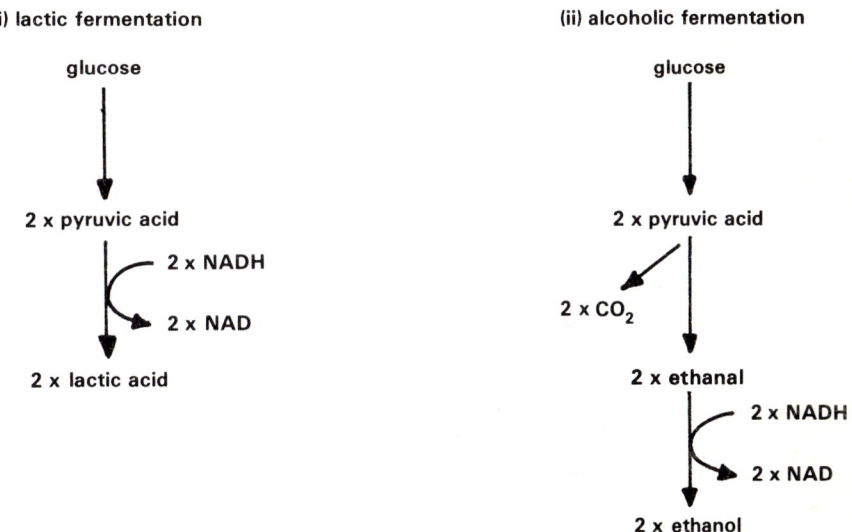

How has it been possible to unravel the steps of a complicated series of events such as glycolysis? The Buchner brothers in 1897 showed that extracts from yeast cells could ferment the disaccharide sugar sucrose (Chapter 4). In 1905 Harden and Young discovered that fructose 1,6-bisphosphate was formed in alcoholic fermentation which was stimulated by phosphate ions. During the 1930s Meyerhof observed that extracts of striped muscle broke down glucose to lactic acid. About the same time Embden accounted for the splitting of the hexose sugar into PGAL while Warburg found out how ADP and NAD are used in respiration. By discoveries such as these the intermediate compounds and end products were identified and a step by step scheme for glycolysis was pieced together. The enzymes which control glycolysis are found in the cytoplasm.

What happens to pyruvic acid if oxygen is available? In aerobic conditions the pyruvic acid is taken into the **tricarboxylic acid cycle**.

2. THE TRICARBOXYLIC ACID CYCLE

At the beginning of this century Thunberg discovered that animal tissues contain **dehydrogenase enzymes** which catalyse the transfer of hydrogen from carboxylic acids (Fig. 5.3). In 1935 Szent-Gyorgyi noticed that dehydrogenation of succinic acid in muscle is blocked by the competitive inhibitor malonic acid. Furthermore the inhibition stopped respiration of the muscle. Two years later Hans Krebs showed that respiration of pigeon breast muscle was stimulated by a specific variety of carboxylic acids. Krebs went on to carry out a series of brilliant experiments from which he

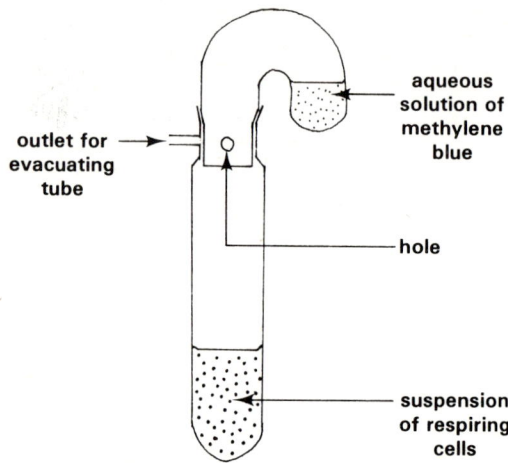

FIG. 5.3 *A Thunberg tube. The top of the tube is rotated so that the hole coincides with the outlet, and the air in the tube is then sucked out using a vacuum pump. The top is then turned to seal off the contents from the outside air. Finally the tube is inverted so that the blue dye mixes with the respiring cells and the dye is gradually reduced to a colourless compound. If air is left in the tube oxygen is reduced instead of the dye*

deduced that, in aerobic conditions, **pyruvic acid** combines with **oxaloacetic acid** to form **citric acid**. The citric acid is then dehydrogenated via a number of intermediate compounds back to oxaloacetic acid. One of the intermediates is **succinic acid**. Carbon dioxide is released in these reactions. It was some time later before it was found that pyruvic acid is converted to **acetyl coenzyme A** before it enters the cycle (Fig. 5.4). Today the citric acid cycle is called **Krebs' tricarboxylic acid (TCA) cycle**.

In the late 1940s mitochondria, separated from other cell components by differential centrifugation (Chapter 1), were found to break down pyruvic acid to carbon dioxide and water. The dehydrogenase enzymes which catalyse the oxidations in the TCA cycle are thus in the mitochondria. Biochemists are now fairly sure that the enzymes are in the matrix enclosed by the inner membrane of mitochondria.

So far so good, but what happens to the hydrogen removed from the carboxylic acids in the TCA cycle? Where does oxygen fit into the sequence of events?

In the TCA cycle oxidation of carboxylic acid molecules is catalysed by dehydrogenase enzymes. Dehydrogenation of the acids is accompanied by reduction of NAD to NADH. The fate of NADH in aerobic conditions is, however, different from NADH formed in anaerobic conditions. When oxygen is available the hydrogen from NADH is passed via a succession of hydrogen acceptors until it finally reduces oxygen to water. The succession is called an **electron transport chain** (Fig. 5.5). After NADH the hydrogen is passed to **flavin-adenine dinucleotide (FAD)**. The hydrogen atoms now split into electrons and protons. FADH transfers the electrons to the next acceptors called **cytochromes**. Reduced cytochromes then pass the elec-

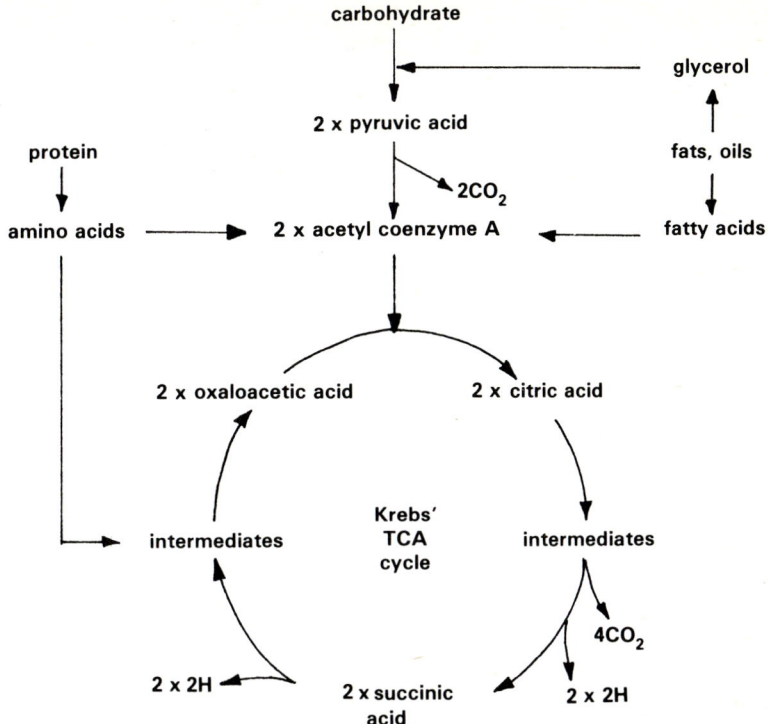

trons to **cytochrome oxidase**. Finally reduced cytochrome oxidase transfers the electrons back to the protons. The hydrogen so formed reduces oxygen to water. Each of the reactions in the chain is an oxidation-reduction reaction. As the hydrogen atoms and later the electrons move along the chain the reduced acceptors are oxidised and more hydrogen can be taken into the chain.

FIG. 5.5 *An electron transport chain*

What is particularly important is that three of the oxidation-reduction reactions in an electron transport chain yield enough free energy to make ATP from inorganic phosphate ions and ADP. This type of reaction is called **oxidative phosphorylation**. It takes place in the stalked bodies projecting from the cristae of mitochondria. How many ATP moles are formed from each mole of glucose oxidised aerobically? How does this figure compare with the number obtained when glucose is oxidised anaerobically?

5.1.3 Energy yields from respiratory substrates

The breakdown of a mole of glucose in glycolysis uses the free energy from the terminal high energy bonds of two moles of ATP. On the other hand enough free energy is released in glycolysis to produce four moles of ATP from ADP and inorganic phosphate (Pi):

$$2ATP \longrightarrow \underset{\text{glucose}}{\overset{\text{glucose}}{\bigg|}} \longrightarrow 4ADP + 4Pi$$
$$2ADP + 2Pi \longleftarrow \underset{\text{pyruvic acid}}{\bigg\downarrow} \longrightarrow 4ATP$$

The net gain in free energy from a mole of glucose broken down to pyruvic acid therefore is the free energy released when the terminal high energy bonds of two moles of ATP are later hydrolysed. Thus living cells respiring anaerobically make 2×30.66 kJ of free energy for every mole of glucose oxidised.

In aerobic conditions ATP is made in the electron transport chains into which NADH from the TCA cycle is fed. Altogether 38 ATP moles are made from each mole of glucose oxidised aerobically, 2 from glycolysis and 36 from the TCA cycle (Table 5.1). Living cells respiring aerobically thus release 38×30.66 kJ of free energy per mole of glucose. Aerobic respiration is clearly much more efficient than anaerobic respiration in releasing free energy.

Table 5.1 Origin of ATP formed in aerobic respiration of glucose

SOURCE	NO. OF ATP MOLECULES
glycolysis	2 (net gain)
NADH formed in glycolysis	6 (2 × 3)
NADH formed in Krebs' TCA cycle	30 (2 × 15)
Total	38

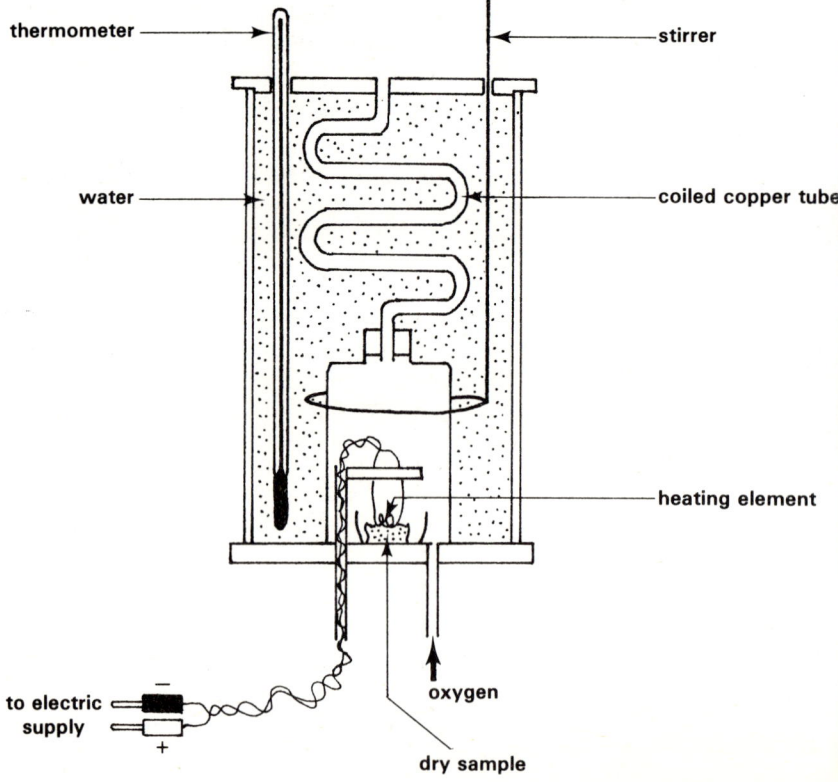

FIG. 5.6 *A simple calorimeter. When the supply of electricity is switched on the heating element becomes red hot and the sample ignites in the stream of oxygen. The heat given off by the sample causes the temperature of the water to rise. The amount of temperature rise, the volume of water and the mass of the sample are used to work out the calorific value of the sample*

thermometer — stirrer

water — coiled copper tube

heating element

to electric supply

oxygen

dry sample

What proportion of the energy in the respiratory substrate is used to do work in living cells? The amount of heat energy produced when glucose is burned to carbon dioxide and water can be measured using a **calorimeter** (Fig. 5.6). One mole of glucose releases 2880 kJ of heat energy. This is the amount of energy potentially available to do work. We can now calculate the efficiency of energy change in living cells when glucose is respired aerobically:

$$\text{Efficiency of energy change} = \frac{38 \times 30.66}{2880} \times 100 = 44 \text{ per cent}$$

This means that less than half of the potential energy in glucose is used to do work.

You may wonder what has happened to the rest of the energy in the respiratory substrate. The Second Law of Thermodynamics states that whenever one form of energy is changed into another, some of the energy is converted into heat. In aerobic respiration over half of the energy in the substrate is released as heat. Heat energy helps to maintain the constant body temperature of warm-blooded animals (Chapter 17). However, remember that heat energy cannot be used to do work in living organisms.

5.1.4 Measuring the rate of respiration

In the 1930s Otto Warburg introduced a constant volume manometer to measure the rate of oxygen uptake by slices of living tissue. The apparatus is often called a **Warburg manometer** (Fig. 5.7). **Simple manometers** are often used in school laboratories for measuring rates of oxygen uptake by living organisms (Fig. 5.8).

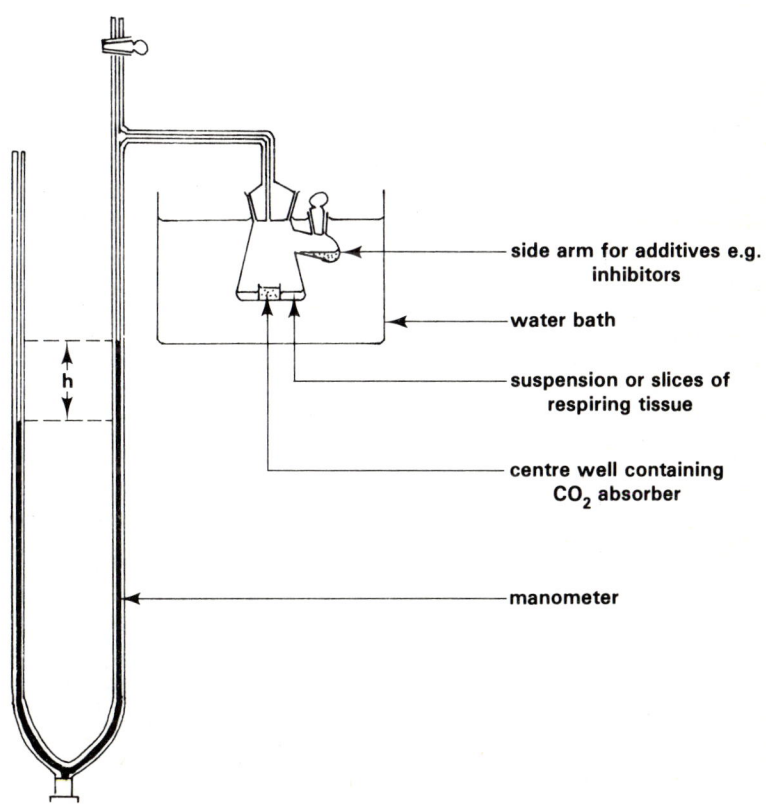

FIG. 5.7 *A Warburg respirometer. As the tissue respires the CO_2 it evolves is taken up by the CO_2 absorber. Consumption of oxygen by the tissue is then registered by a change in height* (h) *of the manometer fluid*

- side arm for additives e.g. inhibitors
- water bath
- suspension or slices of respiring tissue
- centre well containing CO_2 absorber
- manometer

h

FIG. 5.8 *A simple respirometer*

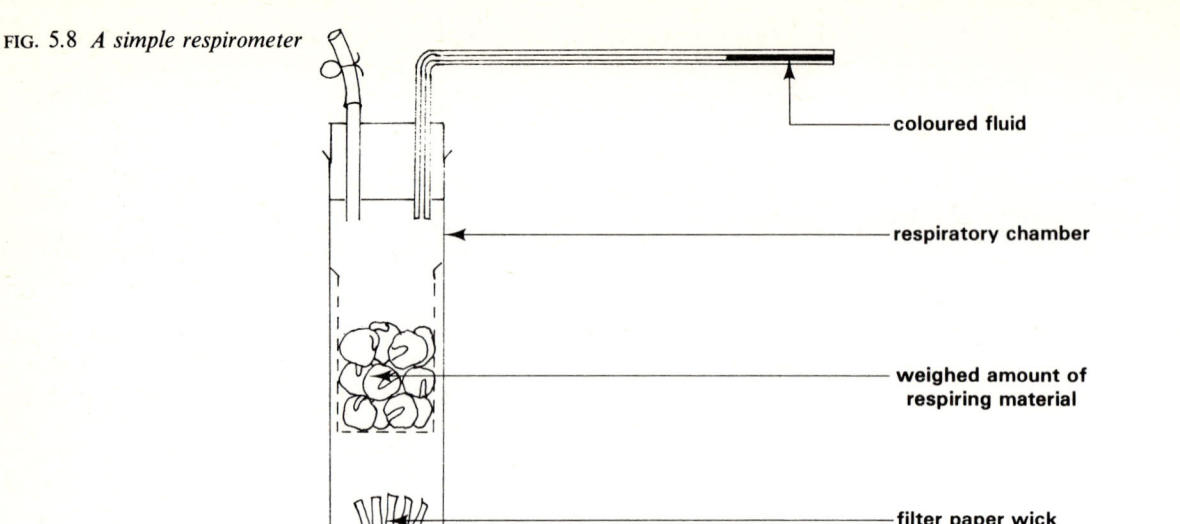

- coloured fluid
- respiratory chamber
- weighed amount of respiring material
- filter paper wick
- potassium hydroxide solution

A known mass of live material is placed in the respiratory chamber. The apparatus is kept at a fixed temperature and the valve is closed. The potassium hydroxide absorbs the carbon dioxide given off in respiration. The volume of air in the respiratory chamber decreases as oxygen is used up. Consequently the coloured fluid moves towards the respiratory chamber. Knowing the distance moved by the meniscus of the coloured fluid and the bore of the tube the rate of oxygen uptake can be calculated. The results are normally expressed as mm^3 oxygen consumed g^{-1} tissue h^{-1} at t°C, where t is the temperature at which the measurements were made.

FIG. 5.9 *Haldane's method for measuring the rate of oxygen uptake*

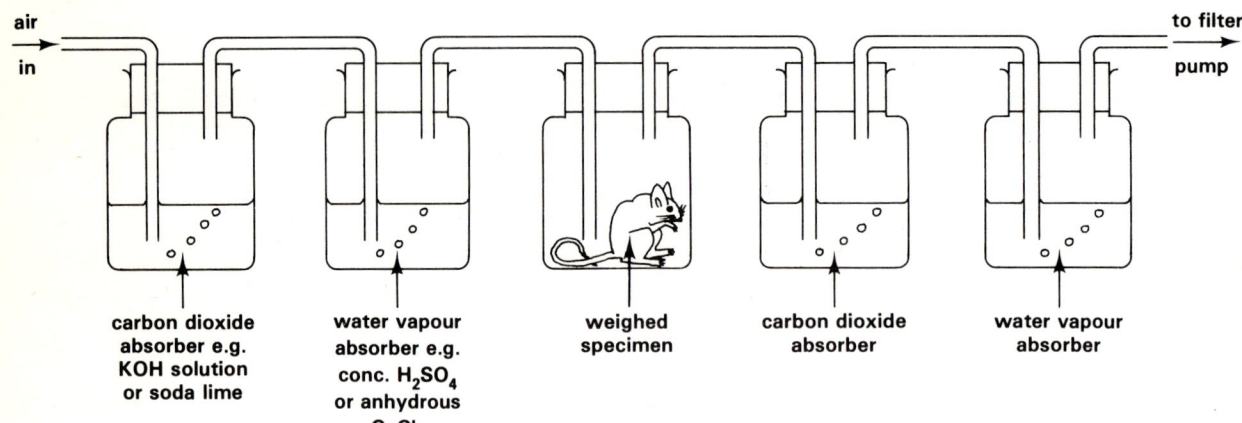

air in

to filter pump

| carbon dioxide absorber e.g. KOH solution or soda lime | water vapour absorber e.g. conc. H_2SO_4 or anhydrous $CaCl_2$ | weighed specimen | carbon dioxide absorber | water vapour absorber |

A different approach was adopted by J. B. S. Haldane (Fig. 5.9). Again a known mass of live material is placed in the respiratory chamber. Air from which carbon dioxide and water vapour have been removed is drawn through the respiratory chamber. The air then passes through the reagent bottles in which expired carbon dioxide and water vapour are separately trapped. The live material and the reagent bottles are reweighed after a fixed period of time. The rate at which oxygen was consumed is calculated on the assumption that any loss in mass of the material is due to expiration of the carbon dioxide and water vapour formed when the respiratory substrate is oxidised:

$$\text{Respiratory substrate} + O_2 \longrightarrow CO_2 + H_2O$$
$$\qquad\qquad (1) \qquad\qquad\qquad\qquad (2) \qquad (3)$$

Knowing factors (1), (2) and (3) it is possible to calculate the rate of oxygen uptake.

Which of the two methods do you think provides the more accurate measurement of oxygen uptake?

Another method used to calculate the rate of respiration is to measure the rate at which living tissue gives off carbon dioxide. The method involves passing air devoid of CO_2 through the respiratory chamber in which a known mass of live material is placed. The expired air is bubbled through a solution of calcium hydroxide of known concentration (Fig. 5.10). The volume of carbon dioxide given off is determined after titrating the hydroxide with standard hydrochloric acid.

FIG. 5.10 *A method for measuring the rate of production of carbon dioxide by a living organism*

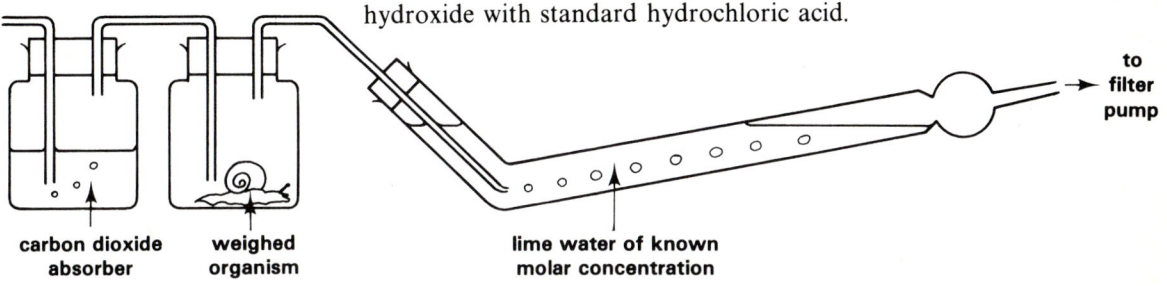

carbon dioxide absorber | weighed organism | lime water of known molar concentration | to filter pump

In each of the methods outlined above the respiratory chamber must be light-proof if the experimental subject can photosynthesise. Why do you think it is necessary to do this?

5.1.5 Respiratory quotient

The volume of carbon dioxide given off divided by the volume of oxygen taken up in a fixed period is called the **respiratory quotient (RQ)**:

$$RQ = \frac{\text{volume } CO_2 \text{ given off}}{\text{volume } O_2 \text{ taken up}}$$

RQ values indicate the type of respiration, aerobic, anaerobic or both. They also tell us which substrates are oxidised.

In aerobic conditions an RQ of 1·0 indicates that the respiratory substrate is a carbohydrate:

$$\underset{\text{carbohydrate}}{C_6H_{12}O_6} + 6O_2 \longrightarrow 6CO_2 + 6H_2O$$

For each volume of oxygen taken up a similar volume of CO_2 is given off:

$$RQ = \frac{6}{6} = 1\cdot0$$

When a lipid is oxidised aerobically the RQ is 0·7:

$$\underset{\text{tristearoylglycerol}}{C_{57}H_{110}O_6} + 81\cdot5\ O_2 \longrightarrow 57CO_2 + 55H_2O$$

$$RQ = \frac{57}{81\cdot5} = 0\cdot7$$

For the aerobic oxidation of protein an RQ of 0·99 is obtained. The aerobic breakdown of a mixture of carbohydrates, lipids and proteins gives an RQ of 0·8–0·9. Animals fed a balanced diet use carbohydrates and lipids as respiratory substrates and have an RQ of about 0·85. The RQ of starving animals is between 0·9 and 1·0. Can you explain why this is so?

A mixture of aerobic and anaerobic respiration takes place when there is a shortage of oxygen. Here the RQ is greater than 1·0. The exact value depends on the respiratory substrates used and on the relative rates of the two types of respiration. For example, when anaerobic and aerobic respiration of glucose occur at similar rates, an RQ of 1·33 is obtained.

anaerobic respiration $\qquad C_6H_{12}O_6 \qquad\qquad \longrightarrow 2CO_2 + 2CH_3CH_2OH$
aerobic respiration $\qquad\quad C_6H_{12}O_6 + 6O_2 \longrightarrow 6CO_2 + 6H_2O$

$$\text{Total} \qquad\qquad 6O_2 \longrightarrow 8CO_2$$

$$RQ = \frac{8}{6} = 1·33$$

What might the RQ be if aerobic respiration takes place more rapidly, or more slowly than anaerobic respiration of glucose?

If oxygen is totally lacking the RQ is infinity. Why?

5.2 Photosynthesis

Photosynthesis consists of two main phases, the **light-dependent** and **light-independent** (dark) reactions.

5.2.1 Light-dependent reactions

White light is a mixture of light of different wavelengths (Fig. 5.11). Einstein and Planck suggested that light energy is transmitted as units called photons. Because the energy of light is inversely proportional to its wavelength it follows that light of short wavelength has more energy than light of long wavelength. Green plants convert the energy of light into chemical bond energy.

FIG. 5.11 *The visible spectrum*

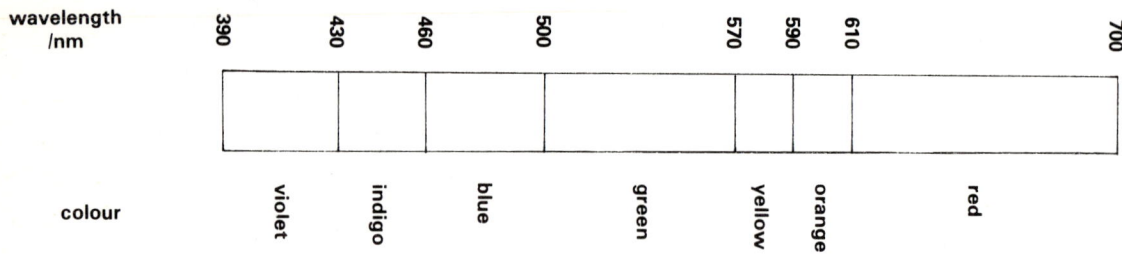

1. ABSORPTION AND ACTION SPECTRA

Green plants do not use energy from all of the components of white light for photosynthesis. This was neatly demonstrated by the German plant physiologist H. T. Engelmann in 1882. He placed filaments of the green alga *Cladophora* in a drop of water on a microscope slide. The filaments were illuminated with light of different wavelengths. He then watched the distribution of aerobic bacteria in the water. Engelmann noted that the bacteria clustered near to the filaments when blue light (450 nm) or red light (650 nm) was used (Fig. 5.12). Knowing that oxygen is given off during photosynthesis, Engelmann deduced that light of mean wavelength 450 nm and 650 nm is the most effective for photosynthesis. It has since been shown that photosynthesis in nearly all green plants occurs most rapidly when they are illuminated by blue and red light.

FIG. 5.12 *Engelmann's experiment*

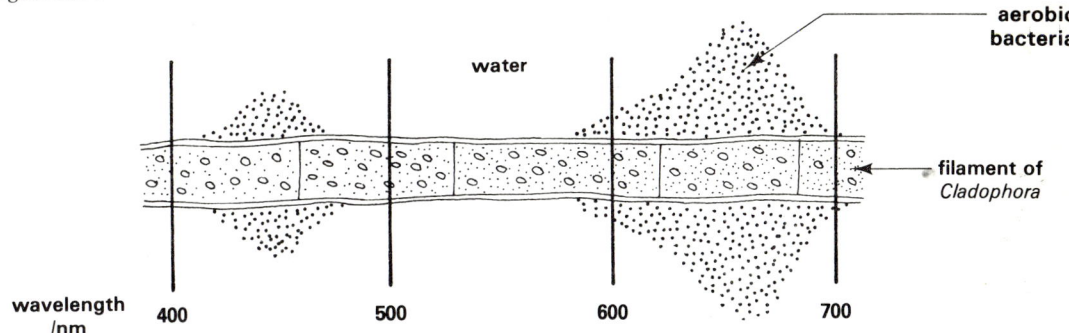

When the rate of photosynthesis is plotted against wavelength of light an **action spectrum** is produced (Fig. 5.13). The reason why light from the blue and red parts of the spectrum is effective is that it is absorbed efficiently by the pigments of green plants. The photosynthetic pigments most commonly present are chlorophylls and carotinoids (Table 5.2). Chlorophylls belong to a group of chemicals called porphyrins which includes the cytochromes and haemoglobin (Chapter 8). They all have four nitrogen-containing pyrrole rings (Fig. 5.14). In the chlorophylls a magnesium atom is held between the rings. In the cytochromes and haemoglobin there is an iron atom instead. Chlorophylls are green because they reflect green light.

FIG. 5.13 *An action spectrum for photosynthesis*

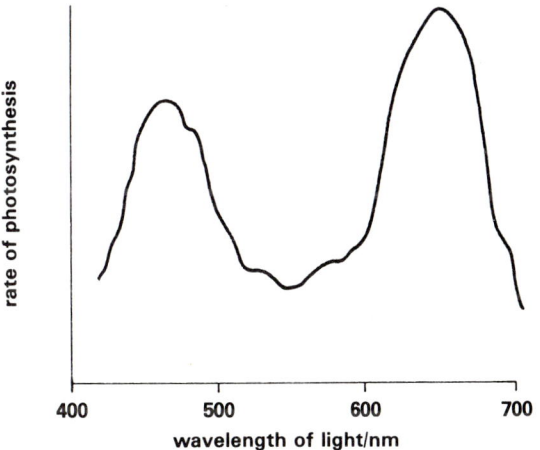

Table 5.2 Photosynthetic pigments commonly found in leaves of higher plants

GROUP	PIGMENT	FORMULA	COLOUR
chlorophylls	chlorophyll a	$C_{55}H_{72}O_5N_4Mg$	green
chlorophylls	chlorophyll b	$C_{55}H_{70}O_6N_4Mg$	green
carotinoids	xanthophyll	$C_{40}H_{56}$	yellow
carotinoids	carotene	$C_{40}H_{56}O_2$	orange

FIG. 5.14 (a) *Structure of chlorophyll*

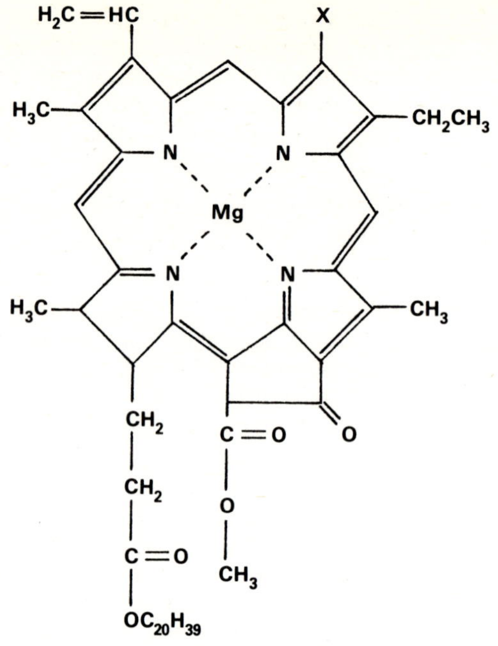

$H_2C{=}HC$
X
H_3C
CH_2CH_3
N
N
Mg
N
N
H_3C
CH_3
CH_2
CH_2
C=O
$OC_{20}H_{39}$
C=O
O
CH_3
O

In chlorophyll a X = CH_3; in chlorophyll b X = CHO

FIG. 5.14 (b) *Structure of β-carotene*

H_3C
CH_3
CH_3
CH_3
CH_3
CH_3
CH_3
H_3C
H_3C
CH_3

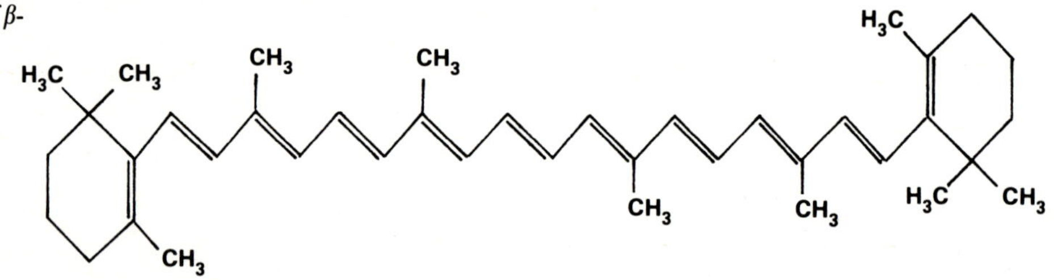

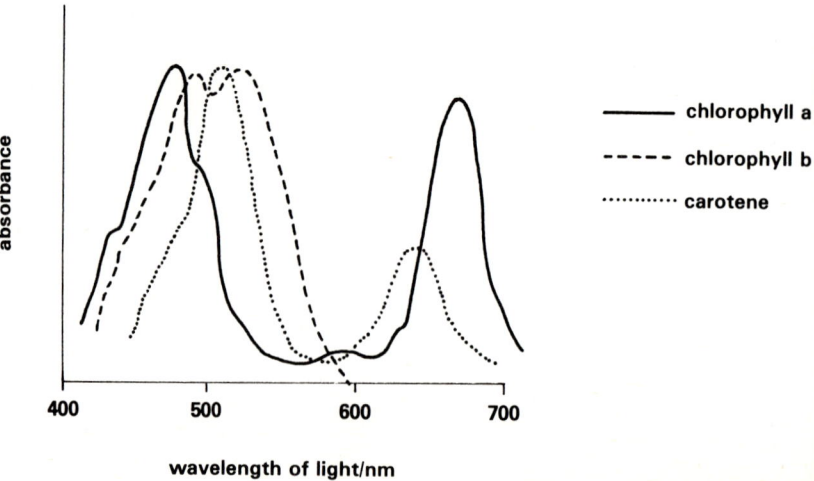

FIG. 5.15 *Absorption spectra of chloroplast pigments*

absorbance
chlorophyll a
chlorophyll b
carotene
400
500
600
700
wavelength of light/nm

The wavelengths they absorb most strongly are in the blue and red parts of the visible spectrum. Carotene and xanthophyll are long chain hydrocarbons. Their colours indicate that they reflect orange and yellow light respectively. However, they absorb blue light effectively. If the percentage light absorption is plotted against wavelength of light an **absorption spectrum** is obtained for each pigment (Fig. 5.15). Notice the close correlation between the absorption and action spectra.

2. ELECTRON TRANSFER IN CHLOROPLASTS

What happens when chloroplast pigments absorb photons of light? One clue comes from the behaviour of the pigments extracted from leaves in warm ethanol or in propanone. When the solution is placed in white light the pigments give off red light. This is called fluorescence. It occurs because light energy temporarily raises the energy level of electrons in the outer shells of magnesium atoms in the chlorophyll molecules causing the electrons to become displaced. In their high energy state the electrons are said to be **excited**. Red light appears during fluorescence because shortly after being displaced the electrons fall back into position. The extra energy they held is emitted as red light.

What has just been described is of course an experiment. In living tissues the solvent is water. Green plants do not emit red light when they photosynthesise. Green plants also give off oxygen during photosynthesis. These facts strongly suggest that in living plants the energy of excited electrons is linked to reactions involving the release of oxygen. Where does the oxygen come from and what happens to the excited electrons?

Some idea of what happens to the excited electrons was established in 1937 by the English biochemist Robin Hill. He observed that an illuminated suspension of chloroplasts in water, while evolving oxygen, reduced iron(III) ions to iron(II) ions. This behaviour, called the **Hill reaction**, suggests that chloroplasts contain electron acceptors. In the 1950s it was discovered that the normal acceptor of excited electrons from chlorophyll molecules was **nicotinamide-adenine dinucleotide phosphate (NADP)**. Transfer of the electrons was later found to be catalysed by an iron-containing protein called **ferredoxin**. As we shall see later, other electron acceptors are also involved.

3. PHOTOLYSIS

Having given electrons to NADP, chlorophyll molecules are in an electron-deficient (oxidised) state. How is the pigment reduced to its former stable condition? One way it could be reduced is to receive electrons from another source. Water is the most likely source of the electrons. Water is a weak electrolyte and a small proportion of its molecules dissociate into hydrogen ions (H^+) and hydroxyl ions (OH^-):

$$H_2O \rightleftharpoons H^+ + OH^-$$

If electrons are removed from hydroxyl ions by oxidised chlorophyll, oxygen is given off:

$$2OH^- + \text{oxidised chlorophyll} \longrightarrow \tfrac{1}{2}O_2 + H_2O + \text{reduced chlorophyll}$$

Equilibrium is re-established as more water molecules dissociate. Thus one result of light absorption by chlorophyll is a rapid splitting of water molecules into hydrogen ions and oxygen. This is called **photolysis** (photo = light; lysis = to split). Using isotopes it is possible to prove that all of the oxygen given off in photosynthesis comes from water. When heavy water, that is water containing the heavy isotope of oxygen ^{18}O, is supplied to photosynthesising plants instead of normal water containing the isotope ^{16}O, heavy oxygen is given off.

$$CO_2 + 2H_2{}^{18}O \longrightarrow {}^{18}O_2 + (CH_2O) + H_2O$$

If a control experiment is also run in which the ^{18}O is supplied in carbon dioxide, while normal water is provided, the evolved oxygen is of the ^{16}O type.

What happens to the hydrogen ions from photolysed water is explained in the next section.

4. PHOTOPHOSPHORYLATION

Excited electrons displaced from chlorophyll b pass through an electron transport chain in which **cytochromes** act as electron carriers (Fig. 5.16(a)). As excited electrons pass through the chain some of their energy is used to make ATP. When they leave the chain the electrons have a normal, **ground-state** amount of energy. The ground-state electrons reduce chlorophyll a which has also lost excited electrons. The excited electrons from chlorophyll a pass through a second electron transport chain in which **ferredoxin** is the main electron carrier. Once again ATP is made. The production of ATP in the two electron transport chains is called **non-cyclic photophosphorylation.**

FIG. 5.16 (a) *Non-cyclic photophosphorylation*

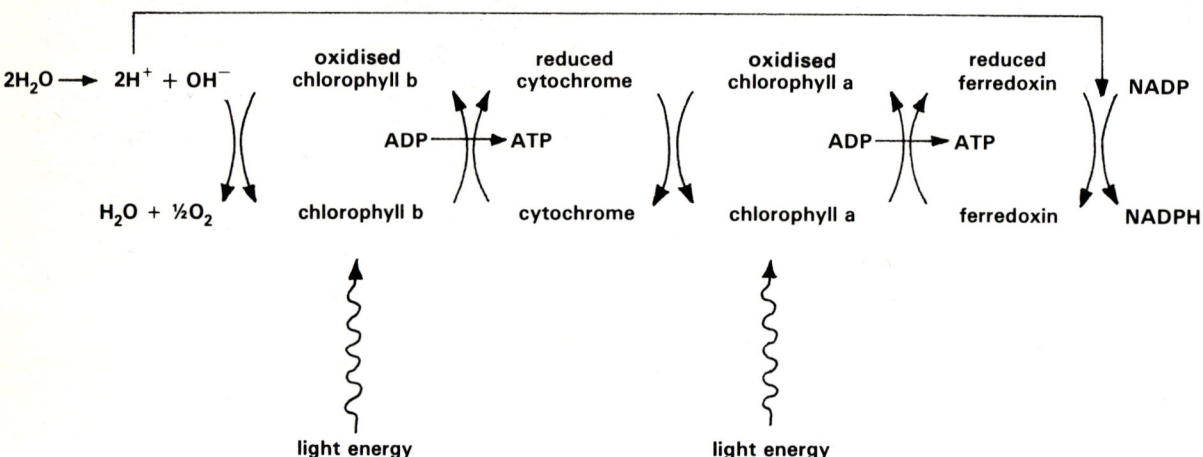

NADPH is another important product of the second electron transport chain. Hydrogen ions from photolysed water combine with ground-state electrons coming out of the chain. The hydrogen atoms so formed reduce NADP to NADPH. Oxygen is given off when hydroxyl ions from photolysed water donate electrons to oxidised chlorophyll b.

ATP is also made in a third electron transport chain called **cyclic photophosphorylation**. Excited electrons from chlorophyll a pass first to ferredoxin, then to cytochromes. The reduced cytochromes then pass the electrons back to oxidised chlorophyll a (Fig. 5.16(b)).

The important products of the light-dependent reactions of photosynthesis are ATP, NADPH and oxygen. ATP is a source of free energy required for the dark reactions. NADPH is also used in the dark reactions to reduce 3-phosphoglyceric acid (section 5.2.2). Oxygen given off in photosynthesis is used by living organisms for respiration.

No mention has yet been made of the role of the carotinoid pigments.

FIG. 5.16 (b) *Cyclic photophosphorylation*

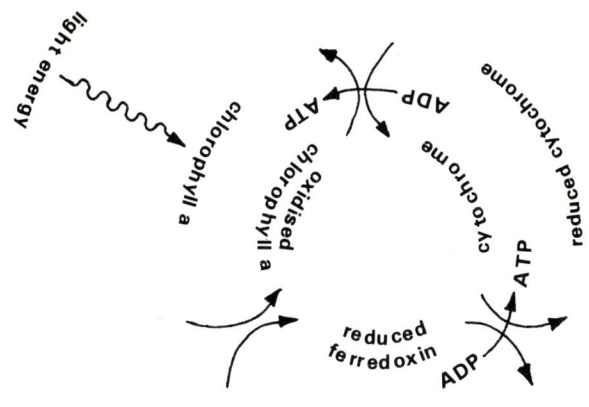

They are thought to shield the chlorophylls from excessive oxidation in intense light. The carotinoids are not involved in electron transfer. This is why photosynthesis occurs only in the green parts of variegated leaves.

5.2.2 Light-independent reactions

1. FIXATION OF CARBON DIOXIDE

For a long time plant scientists had little idea of what happens to carbon dioxide in photosynthesis. In the late 1940s Professor Melvin Calvin of the University of California decided that one way to find out was to allow plants to photosynthesise using carbon dioxide labelled with the radio-isotope ^{14}C. Calvin experimented with unicellular green algae such as *Chlorella* and *Scenedesmus* which he grew in a mineral solution held in flat glass containers he called lollipops (Fig. 5.17). The isotope was added as sodium hydrogencarbonate solution which breaks down to form $^{14}CO_2$. After a fixed period of illumination the algae were rapidly killed by running the mineral solution into a flask of hot ethanol. The cell contents were then analysed by **autoradiography** to find out which substances were labelled.

FIG. 5.17 *Calvin's experiment*

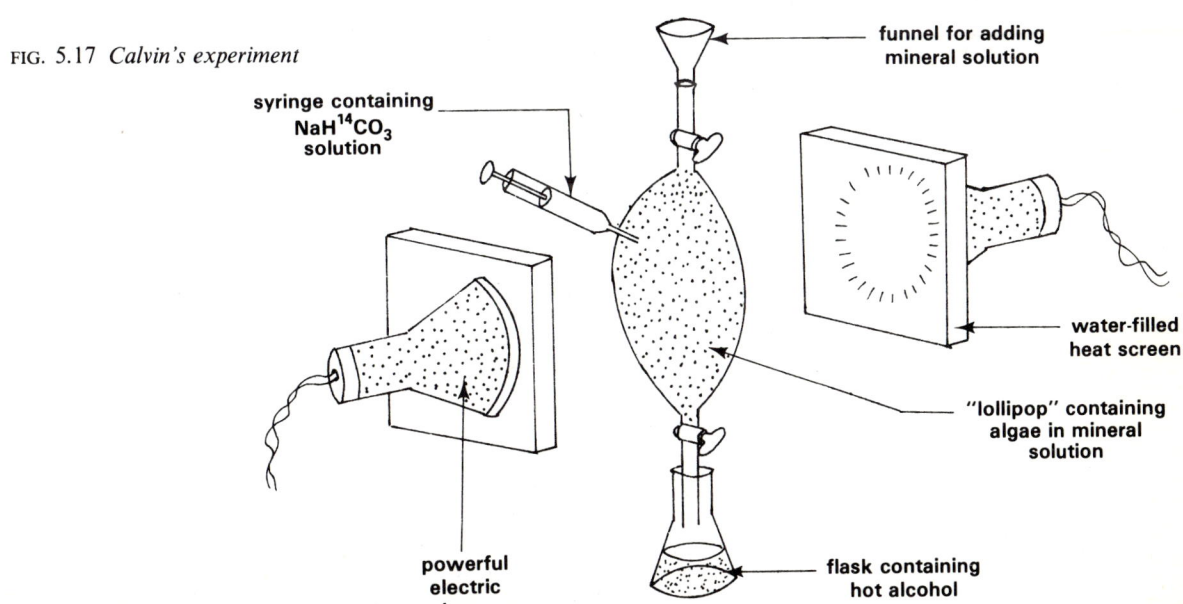

First the cell extracts were separated by **paper chromatography**. A variety of known organic compounds was run on the chromatogram at the same time in order to identify the components of the cell extracts. The chromatogram was then placed on a sheet of X-ray film. Radiations emitted by radioisotopes cause black spots called **fogging** to appear on the film (Fig. 5.18). When the fog marks on the autoradiogram are checked against the positions of the known compounds on the chromatogram it is possible to identify which of the components of the cell extracts are labelled with the isotope.

FIG. 5.18 *Autoradiographs of photosynthetic products containing ^{14}C in* Scenedesmus (courtesy Dr. M. Tribe)

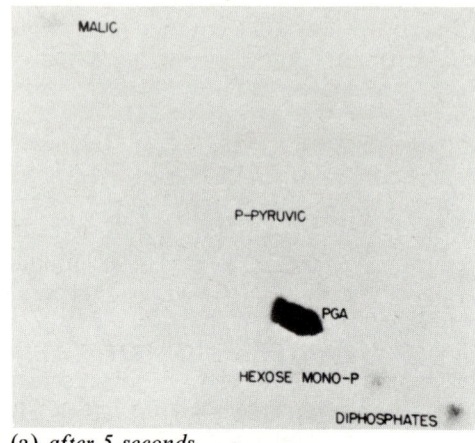

(a) *after 5 seconds*

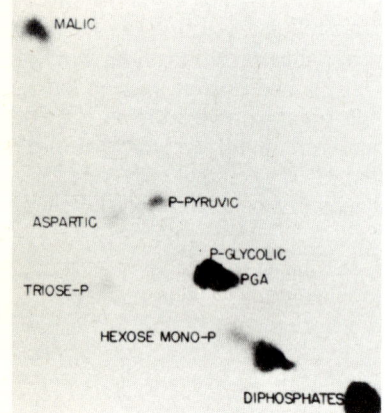

(b) *after 15 seconds*

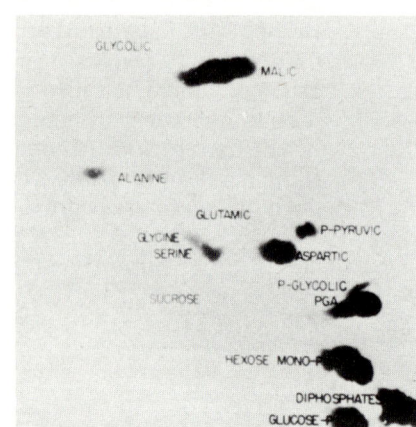

(c) *after 60 seconds*

Calvin deduced that after only a short period of illumination the radioisotope should appear in the first products of carbon dioxide fixation. With longer exposures to light, the intermediate and end products of photosynthesis should be labelled. Table 5.3 shows a summary of his findings. These reactions take place in the stroma of chloroplasts (Chapter 1).

3-phosphoglyceric acid (**PGA**) is the first product of carbon dioxide fixation. But what does CO_2 combine with to form PGA? Because each PGA molecule contains three carbon atoms Calvin thought it was logical to search for a CO_2-acceptor molecule containing two carbon atoms. Despite much effort, no such compound was found in green plants. Later a pentose sugar, **ribulose 1,5-bisphosphate** (**RBP**) was shown to be the main CO_2-acceptor in many plant species. RBP combines with carbon dioxide

Table 5.3 Components of algal cell extract labelled with radioactive carbon after different periods of photosynthesis in the presence of $^{14}CO_2$

TIME OF EXPOSURE TO LIGHT AFTER ISOTOPE ADDED/SECS	MAIN SUBSTANCES CONTAINING ^{14}C
5	3-phosphoglyceric acid (PGA)
15	PGA, hexose phosphates
60	PGA, hexose phosphates, sucrose, amino acids
300	PGA, hexose phosphates, sucrose, starch, amino acids, proteins, lipids

to form an unstable intermediate which quickly breaks down into two molecules of PGA. But how is PGA built up into higher carbohydrates, proteins and lipids? Furthermore, how is a constant supply of RBP maintained so that CO_2 fixation can continue indefinitely?

2. LINK BETWEEN LIGHT AND DARK STAGES

So far there seems to be no connection between the light-dependent reactions and carbon dioxide fixation which is light-independent. It is now possible to explain the ways in which NADPH and ATP, the products of the light stage, are used. PGA is reduced by NADPH to form the triose sugar **3-phosphoglyceraldehyde (PGAL)**. The reaction uses free energy provided by ATP. One sixth of the PGAL is built into **hexose sugars**, the remainder is used to resynthesise the carbon dioxide acceptor, RBP. These events are sometimes called the **Calvin cycle** (Fig. 5.19).

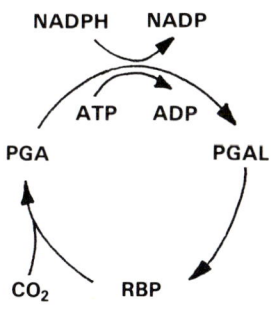

FIG. 5.19 *The Calvin cycle*

FIG. 5.20 *Summary of the fate of PGAL*

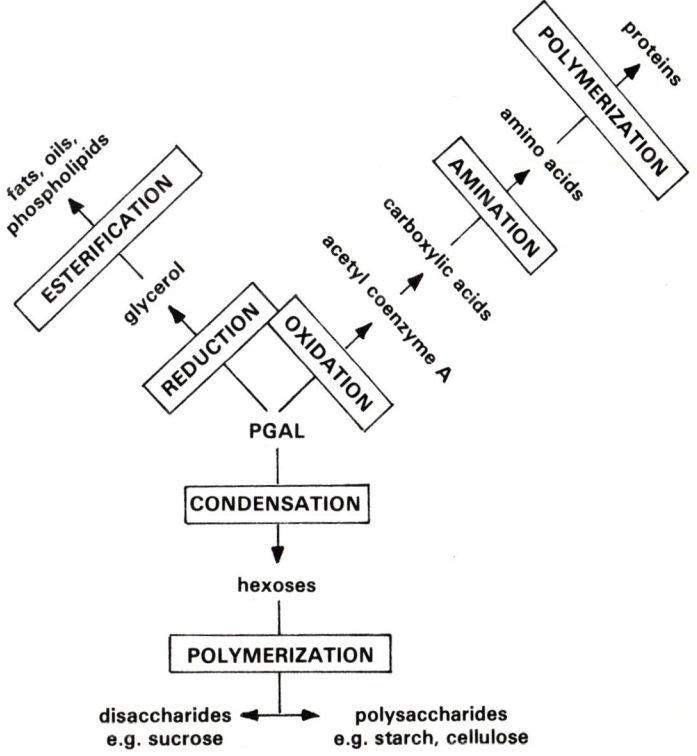

Hexoses are condensed into **disaccharides** and **polysaccharides**. Further reduction of PGAL yields glycerol which can be esterified to form **lipids**. Some PGA is fed into the Krebs' tricarboxylic acid cycle. Amination of some of the acids formed in the cycle results in the production of **amino acids** which are condensed into **protein** molecules. The polymerisation of higher carbohydrates and proteins and esterification of glycerol use the free energy provided when high energy bonds of ATP are hydrolysed. Thus a wide variety of complex organic substances, rich in chemical bond energy, results from photosynthesis (Fig. 5.20).

The synthesis of one mole of a hexose sugar uses 2880 kJ of energy

$$6CO_2 + 12H_2O + 2880 \text{ kJ} = C_6H_{12}O_6 + 6O_2 + 6H_2O$$

How much free energy is made available for work when a mole of glucose is oxidised in respiration?

3. OTHER PATHWAYS FOR CARBON DIOXIDE FIXATION

In the past few years it has been found that the fixation of carbon dioxide in some green plants occurs differently from that described above. The plants include some important tropical crops such as sugar-cane, maize and sorghum. When given $^{14}CO_2$ the carbon radioisotope first appears in **oxaloacetic acid**. Oxaloacetic acid contains four carbon atoms compared with PGA which contains three. This pathway, sometimes called the **Hatch-Slack** or **C4 pathway**, is outlined in Figure 5.21.

FIG. 5.21 *The Hatch-Slack (C_4) pathway*

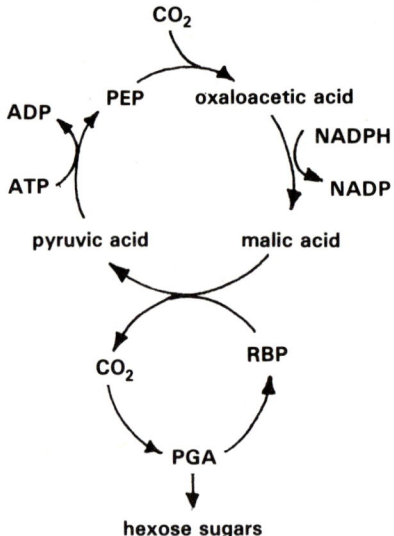

The source of energy for photosynthesis in C4 plants is sunlight, and ATP and NADPH are again the products of the light-dependent reactions. Some ATP is used to phosphorylate pyruvic acid with the formation of **phosphoenolpyruvic acid (PEP)**. On losing its phosphate group PEP releases free energy which is used to fix CO_2. Oxaloacetic acid is the first product of carbon dioxide fixation in C4 plants. NADPH reduces the oxaloacetic acid to **malic acid** which reacts with RBP to form **pyruvic acid** and **PGA**. The fate of PGA is the same as that described for C3 plants.

A similar process occurs in succulent plants including desert cacti. Here the events are called **Crassulacean acid metabolism (CAM)**. The main difference is that CAM plants take in and fix carbon dioxide at night rather than during the daytime. This is of great importance to the water economy of such plants (Chapter 11). Malic acid is again the main product of CO_2 fixation. During the daytime it reacts with RBP and PGA is formed as described above.

Among the products of CO_2 fixation in all green plants is **hydroxyethanoic acid**. The acid is immediately oxidised in C3 plants in a process called **photorespiration** and carbon dioxide is given off. Up to 30 per cent of carbon dioxide fixed in photosynthesis can be recycled in this way. Consequently a considerable amount of the solar energy used to fix carbon dioxide is wasted. C4 plants do not photorespire so their photosynthesis is more efficient in making raw materials for growth. Furthermore, with a rise in temperature the rate of photorespiration increases more rapidly than carbon dioxide fixation. C4 plants are therefore much more productive than C3 plants, especially in tropical climates.

In this chapter we have concentrated on the energetics and biochemistry of respiration and photosynthesis. There are many questions about the two processes which we have not yet answered. How, for example, do cells respiring aerobically obtain a supply of oxygen? What happens to the carbon dioxide made in respiration? What is the source of carbon dioxide for photosynthesising cells and how does it reach the chloroplasts? How does light energy reach green cells? These and other aspects of respiration and photosynthesis are described in Chapters 8 and 12.

Chapter 6

Nucleic acids and protein synthesis

Progress in scientific knowledge normally comes a little at a time. There are occasions, however, when advances occur in leaps and bounds. This is aptly illustrated by a study of the structure and functions of the nucleic acids.

In 1869 a German physician called Meischer reported on a chemical analysis he had made of the nuclei of human pus cells collected from bandages. He found the nuclei to be rich in nitrogen, sulphur and phosphorus. Meischer set a pattern for investigating the composition of the nuclei of cells using the methods of the analytical chemist.

During the next eighty years analytical chemists provided small but significant facts about the composition of nuclei. Crude extracts of nuclei were found to be acidic in reaction, hence the term **nucleic acids** to describe the most important of substances present in nuclei. Gradually the building blocks which make up the nucleic acids were discovered. But the way in which the building blocks were put together and the roles of the nucleic acids in living organisms were still poorly understood at the middle of this century. It was then that giant strides in our knowledge and understanding of the nucleic acids were made.

The turning point came in the early 1950s when James Watson and Francis Crick proposed a structure for **deoxyribonucleic acid** (**DNA**). The discovery of how exact copies of DNA are made in living cells soon followed and its biological significance was quickly realised. The momentum was maintained by the unravelling of the **genetic code**, the 'secret of life' as it has been called. We shall see in this chapter that great strides in unlocking the secrets of nucleic acids are still being made, one of which has recently confronted biologists with a moral dilemma.

6.1 The nucleic acids

Deoxyribonucleic acid (DNA) and **ribonucleic acid (RNA)** are the two main kinds of nucleic acids. DNA is found mainly in the nuclei of cells, small amounts occurring in mitochondria and chloroplasts. RNA is located mainly in the cytoplasm, particularly at the ribosomes.

Both DNA and RNA are polymers. The building blocks of which they are constructed are called nucleotides. For this reason the nucleic acids are described as **polynucleotides**.

6.1.1 Deoxyribonucleic acid

A nucleotide is made up of three parts, a **pentose sugar**, a **nitrogenous base** and **phosphate**. In DNA the sugar is **deoxyribose** (Fig. 6.1). Four different nitrogenous bases occur in DNA (Fig. 6.2). **Adenine (Ade)** and **guanine (Gua)** belong to a group of compounds called **purines**. **Cytosine (Cyt)** and **thymine (Thy)** are **pyrimidines**. The sugar, nitrogenous base and phosphate are bonded as shown in Figure 6.3 to form a **nucleotide**. In a DNA molecule the phosphate groups act as bridges joining together a chain of nucleotides by their sugar molecules (Fig. 6.4). During the early 1950s biochemists established that the number of adenine molecules in DNA from different organisms is the same as the number of thymine molecules. The ratio of cytosine and guanine molecules is also 1 : 1. However, the ratio of (Ade + Thy) : (Cyt + Gua) varies according to the source of the DNA. At this

FIG. 6.1 *Molecular structure of deoxyribose*

FIG. 6.2 *The nitrogenous bases of DNA*

5
HOCH₂

O

OH

4

1

H H

H H

3

2

OH

H

deoxyribose

adenine

guanine

cytosine

thymine

FIG. 6.3 *A nucleotide*

FIG. 6.4 *Part of a polynucleotide*

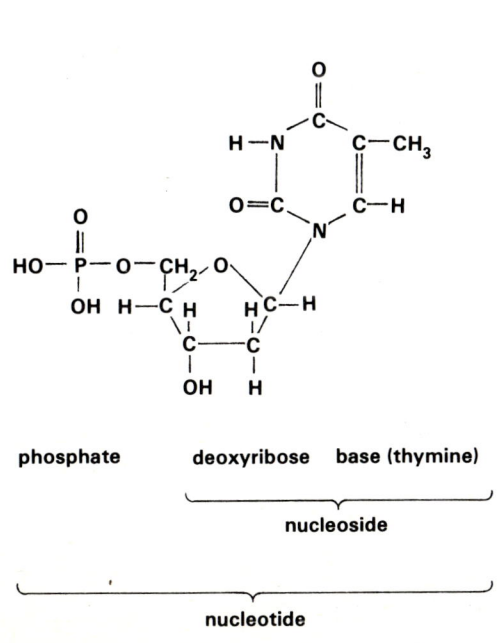

phosphate   deoxyribose   base (thymine)

nucleoside

nucleotide

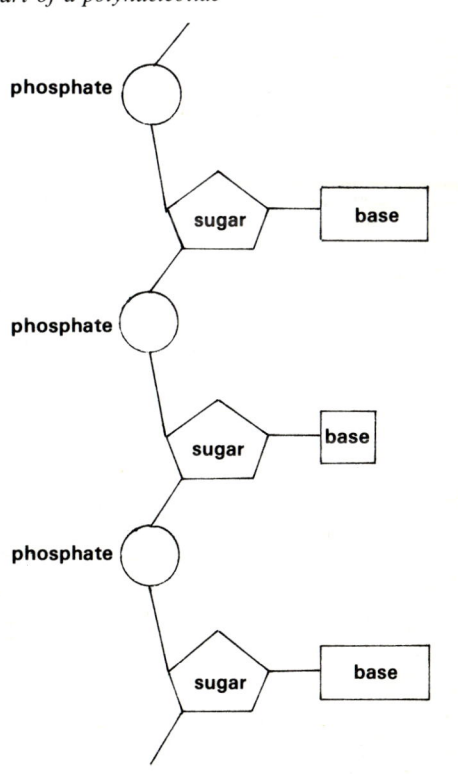

phosphate

sugar   base

phosphate

sugar   base

phosphate

sugar   base

99

FIG. 6.5 (a) *Atomic model of DNA*

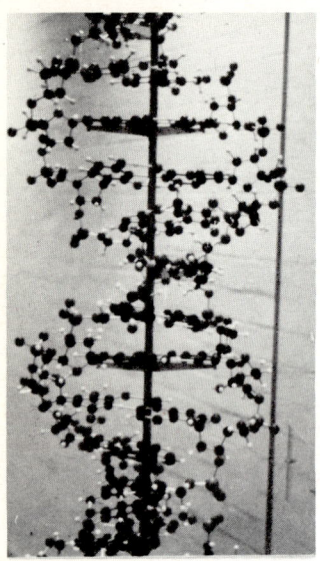

FIG. 6.5 (a) *Atomic model of DNA*

time Rosalind Franklin was working on the structure of DNA using X-ray crystallographic techniques (Chapter 3). Her results were part of the evidence which, in 1953, led James Watson and Francis Crick to propose that a DNA molecule consists of **two polynucleotide strands** each coiled in a **right-handed helix**. The two strands of the double helix are held together by hydrogen bonding between the nitrogenous bases of adjacent nucleotides. For each complete twist of the double helix there are 10 pairs of nucleotides, each twist measuring 3·4 nm in length (Fig. 6.5).

It is important to appreciate that the two polynucleotide strands of a DNA molecule are not identical. They are **complementary**. Where Ade occurs in one strand Thy is found in the other. Where Cyt appears in a strand Gua is present in the complementary strand. Watson and Crick were aware that such a structure could explain how exact copies of a DNA molecule can be made (section 6.1.4).

Altogether there can be several thousand pairs of nucleotides in a single molecule of DNA. The exact number depends on the origin of the DNA. The sequence in which the pairs of bases occurs also differs in DNA from different species. In theory the number of permutations of the bases is infinite.

During nuclear division the DNA of the nucleus is organised into visible threads called chromosomes (Chapter 7). The DNA double helix then has a coil of histone proteins wound around it. Under the electron microscope the DNA-histone complex making up the chromosomes has a woolly appearance (Fig. 6.6).

FIG. 6.5 (b) *The Watson–Crick model of DNA*

The double helix

The pairing of the bases in the polynucleotides

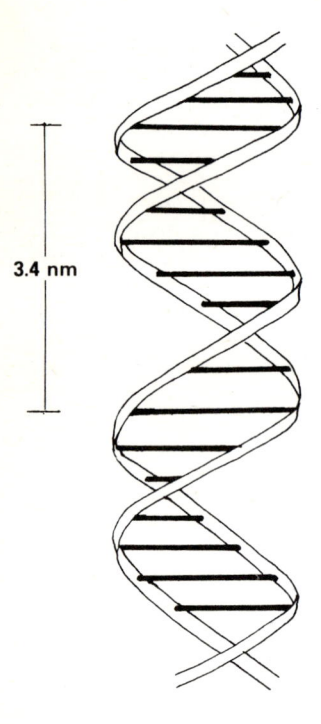

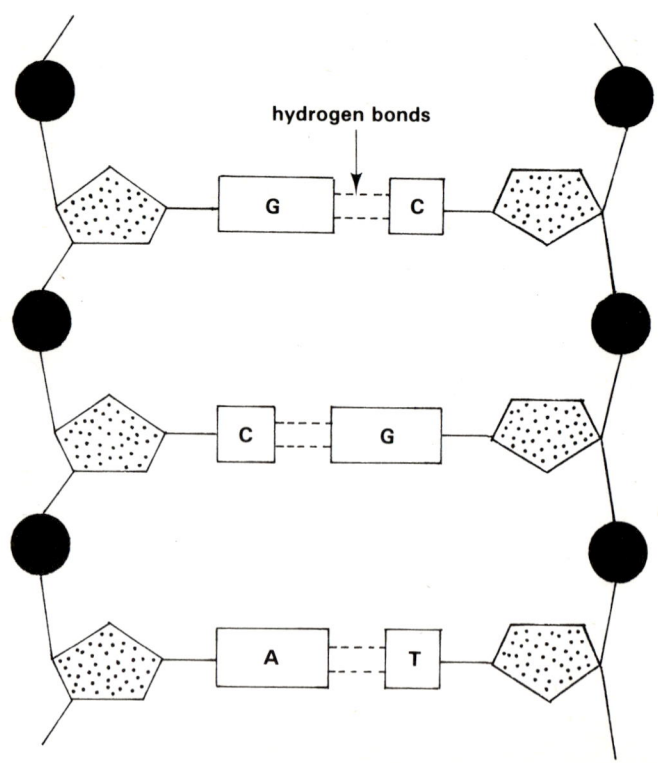

A = adenine   T = thymine
G = guanine   C = cytosine

6.1.2 Ribonucleic acid

FIG. 6.6 *Ultrastructure of a human chromosome (based on an electronmicrograph)*

The two chromatids consist of long, entangled chromatin fibres. The fibres contain about 75 per cent protein, 15 per cent DNA and 10 per cent RNA

There are three different kinds of ribonucleic acid, **messenger RNA (mRNA), transfer RNA (tRNA)** and **ribosomal RNA (rRNA)**. The nucleotides of which they are made contain the sugar **ribose** not deoxyribose as in DNA. Another difference is that thymine is not found in RNA. The nitrogenous base **uracil (Ura)** is present instead (Fig. 6.7(a)).

1. MESSENGER RNA

Between 3 and 5 per cent of the RNA in a cell is mRNA. The molecules of mRNA are single, helical strands made up of several thousand nucleotides. Messenger RNA is made in the nucleus by transcription (section 6.2.1). Its sequence of bases is complementary to that of one of the helices of DNA. The mRNA passes to the ribosomes where triplets of bases in the mRNA act as **codons** in the synthesis of proteins. The way in which the codons function is explained in section 6.2.1.

2. TRANSFER RNA

Transfer RNA makes up between 10 and 15 per cent of a cell's RNA content. The single strand of 75–90 nucleotides which make up a tRNA molecule is wound into a double helix which usually has three prominent bulges. In outline the molecule resembles a clover-leaf (Fig. 6.7(b)). One of the free ends of every tRNA molecule ends with nucleotides containing the following order of bases A ← C ← C ← with A at the very end. There are at least twenty different kinds of tRNA. They differ in the sequence of base triplets making up the **anticodon** by which tRNA binds to the codons of mRNA during the synthesis of proteins (Fig. 6.7(c)).

FIG. 6.7 (a) *Some components of RNA which are not found in DNA*

FIG. 6.7 (b) *Transfer RNA*

FIG. 6.7 (c) *An anticodon. The important difference between the 20 or so forms of tRNA is the sequence of bases in the anticodon triplet*

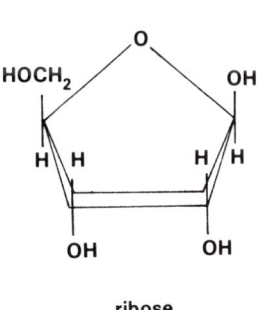

ribose

uracil

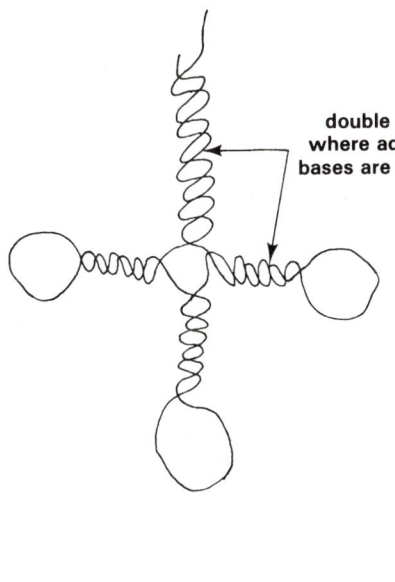

double helix where adjacent bases are bonded

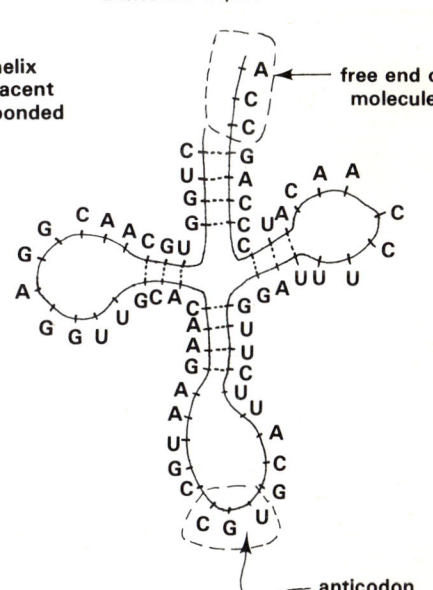

free end of molecule

anticodon

3. RIBOSOMAL RNA

The many thousands of nucleotides which make up a molecule of rRNA are wound into a complex structure consisting partly of single and partly of double helices (Fig. 6.8). Ribosomal RNA is made in the nucleus under the control of the nucleoli. It enters the cytoplasm and binds with protein molecules to become ribosomes. Over half the mass of a ribosome consists of rRNA and more than 80 per cent of the total RNA in a cell is rRNA. Even so, the precise function of rRNA is still not known.

FIG. 6.8 *Part of a molecule of rRNA*

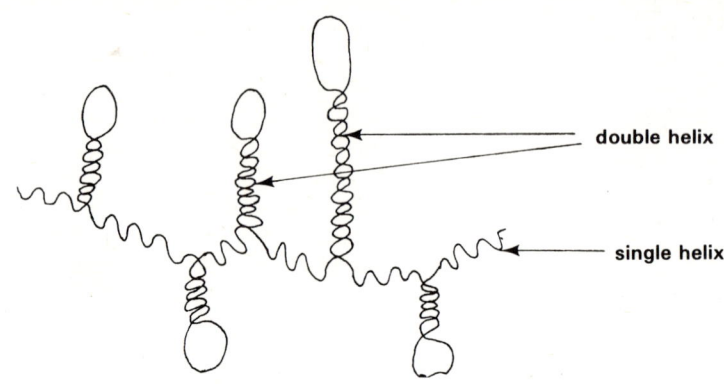

6.1.3 DNA as the hereditary substance

In 1928 Fred Griffith, an English bacteriologist, noted two distinct strains of the bacterium *Pneumococcus*. One called the **S-strain** has capsules, and when inoculated on to an agar-based medium it grows into smooth, glistening colonies. The other, called the **R-strain**, lacks capsules and its colonies have a rough, dull appearance. Griffith also observed that mice injected with the S-strain soon developed pneumonia and died. The R-strain is non-pathogenic and mice injected with it show no symptoms of illness. However, if heat-killed S-pneumococcus is mixed with the R-strain

FIG. 6.9 *Griffith's experiment*

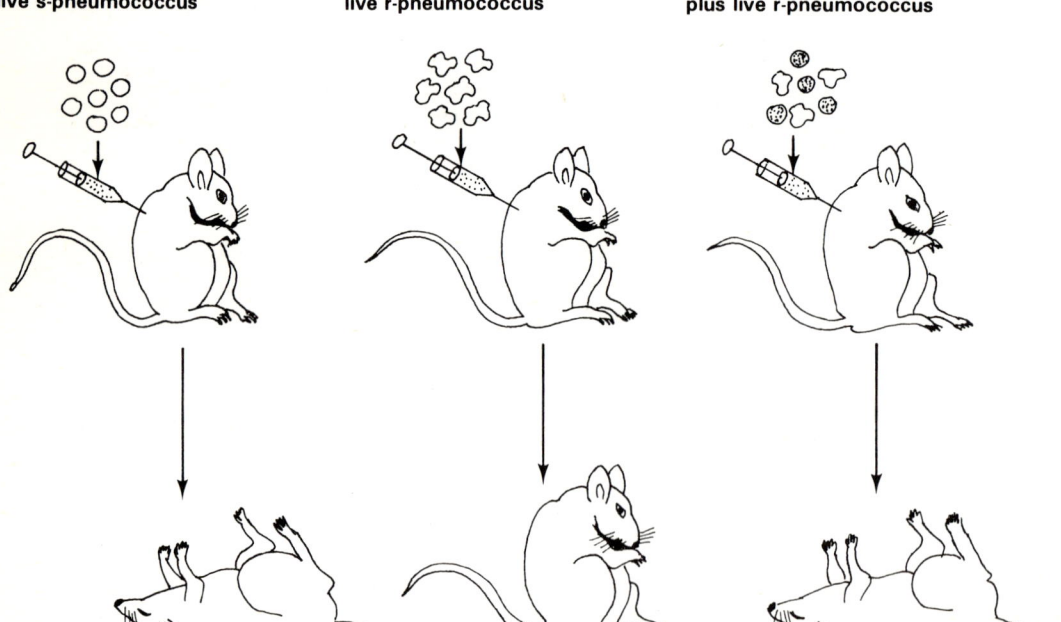

live s-pneumococcus

live r-pneumococcus

heat-killed s-pneumococcus plus live r-pneumococcus

and then injected into the mice, the animals die of pneumonia. Moreover Griffith was able to isolate live S-pneumococcus from the mice subjected to this treatment (Fig. 6.9). He was, however, unable to explain his findings.

The reason for the peculiar behaviour of Griffith's *Pneumococcus* was given by Avery in 1943. Avery reported that the non-pathogenic R-strain could be converted to the S-strain by simply adding DNA from the S-strain to the medium in which the R-strain was growing. Furthermore, the newly-acquired characteristic was permanent and was passed to subsequent generations when the bacterium reproduced. Avery gave the name **transformation** to the process whereby the R-strain was converted to the S-strain. Here was a convincing piece of experimental evidence to suggest that DNA is the substance responsible for the characteristics of living organisms.

Another very important early piece of evidence confirming the role of DNA in inheritance came from experiments using bacteriophages. Bacteriophages are viruses which parasitise bacteria. Like all viruses they consist mainly of just two substances, protein and nucleic acid. A 'phage first becomes attached to the outer surface of a bacterium then injects its nucleic acid into its host leaving its protein coat outside. After some time complete new 'phages appear inside the bacterium which then bursts open. The entire sequence of events is called the **lytic cycle** (Fig. 6.10). In 1952 Hershey and Chase wrote about an experiment they had carried out using the T_2 'phage which attacks the bacterium *Escherichia coli*. The bacterium was first grown in a medium containing radioactive sulphur as $^{35}SO_4^{2-}$, or phosphorus as $^{32}PO_4^{3-}$. The T_2 'phage was then added and allowed to multiply as described in the lytic cycle. ^{35}S was used by *E. coli* to make sulphur-containing amino acids which were built into the protein coats of one of the sets of 'phages. The DNA of the other set of 'phages became labelled with ^{32}P. The labelled 'phages were then added separately to fresh cultures of *E. coli* growing in a non-radioactive medium. Shortly after the 'phages had become attached to the bacteria a homogeniser was used to chop off the protein coats of the 'phages adhering to the host cells. The cultures were then spun in a centrifuge. The bacteria formed a pellet at the bottom of the centrifuge tube, the 'phage coats remained in the supernatant liquid.

FIG. 6.10 *Lytic cycle of a bacteriophage*

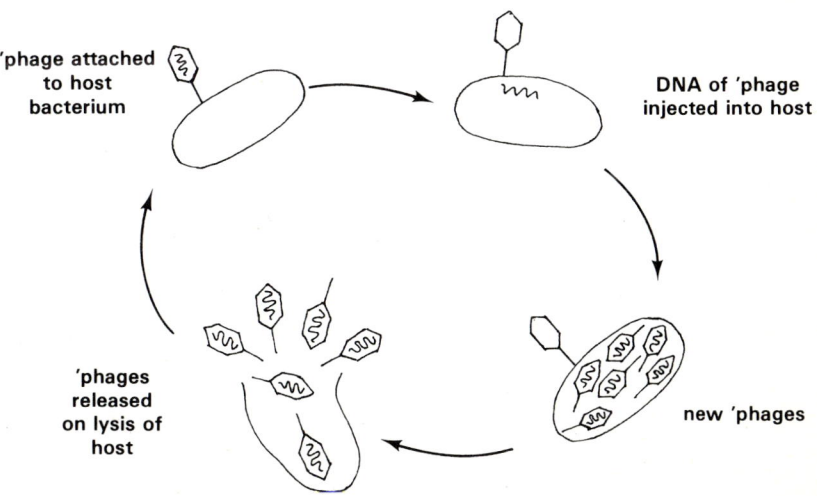

'phage attached to host bacterium

DNA of 'phage injected into host

new 'phages

'phages released on lysis of host

Measurements of the amount of radioactivity in the pellet and supernatant were then made separately for both radioisotopes. The experimental method and the results are summarised in Figure 6.11. Without doubt DNA controlled the production of the next generation of 'phages. DNA is therefore the hereditary substance.

FIG. 6.11 *The Hershey–Chase experiment*

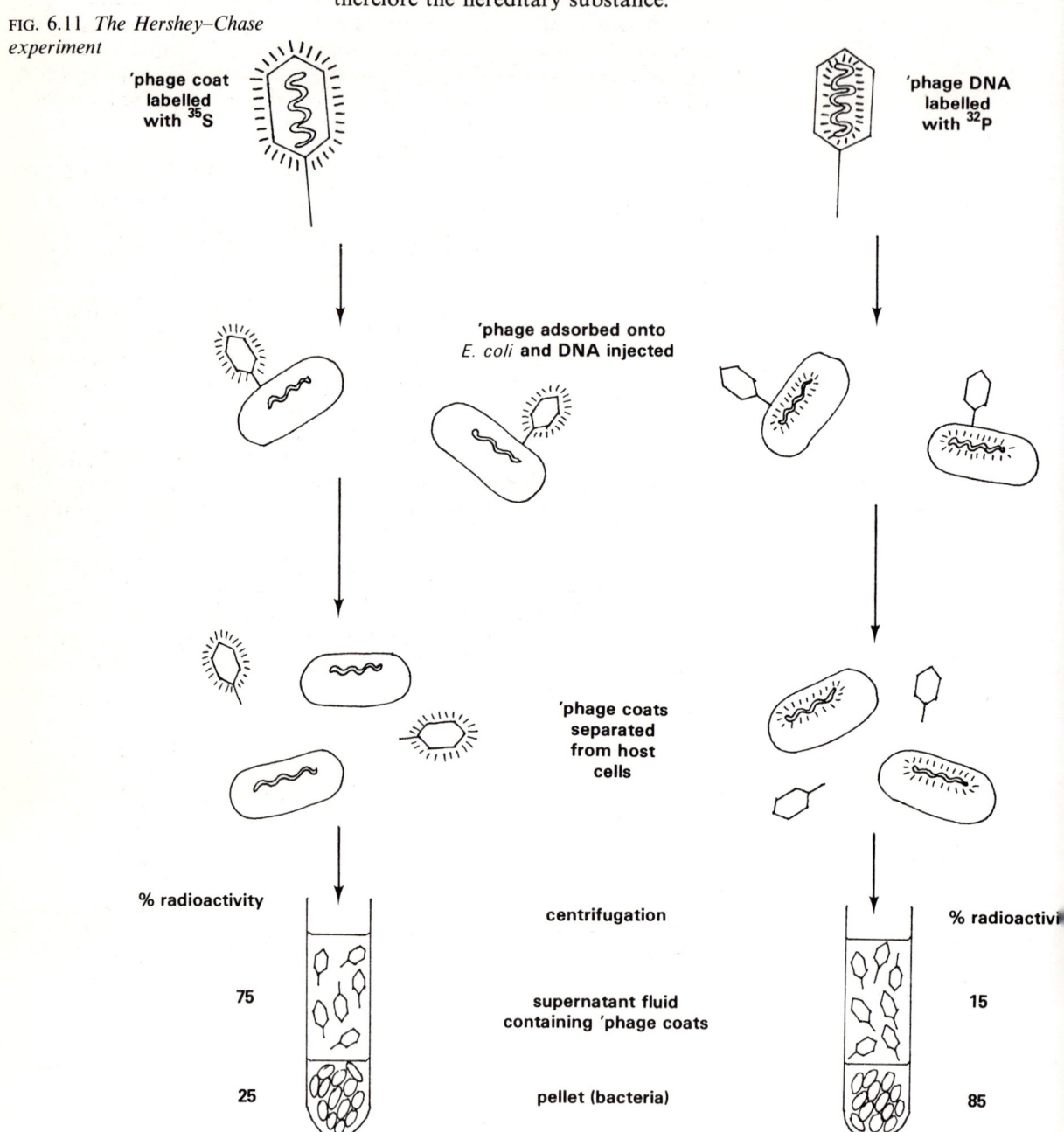

'phage coat labelled with ^{35}S

'phage DNA labelled with ^{32}P

'phage adsorbed onto *E. coli* and DNA injected

'phage coats separated from host cells

% radioactivity

centrifugation

% radioactivi

75

supernatant fluid containing 'phage coats

15

25

pellet (bacteria)

85

Experiments of these kinds riveted the attention of biologists to the nucleic acids at the beginning of the 1950s. The major question then being asked was how was it possible for DNA to be reproduced exactly so that it could be handed on to subsequent generations?

6.1.4 Replication of DNA

Watson and Crick's proposed double helix provided a strong hint as to how exact copies of DNA are normally produced but it was a few years later before there was any experimental proof.

The hydrogen bonds between the base pairs break and the two poly-nucleotide strands unwind. Each of the strands acts as a molecular mould or **template** on which a new polynucleotide strand is made. Synthesis of DNA is controlled by an enzyme called **DNA synthetase**. The new strands are complementary to the original strands (Fig. 6.12). This mode of DNA replication is described as **semi-conservative**. Compare it with conservative replication as shown in Figure 6.13.

FIG. 6.12 *Replication of DNA*

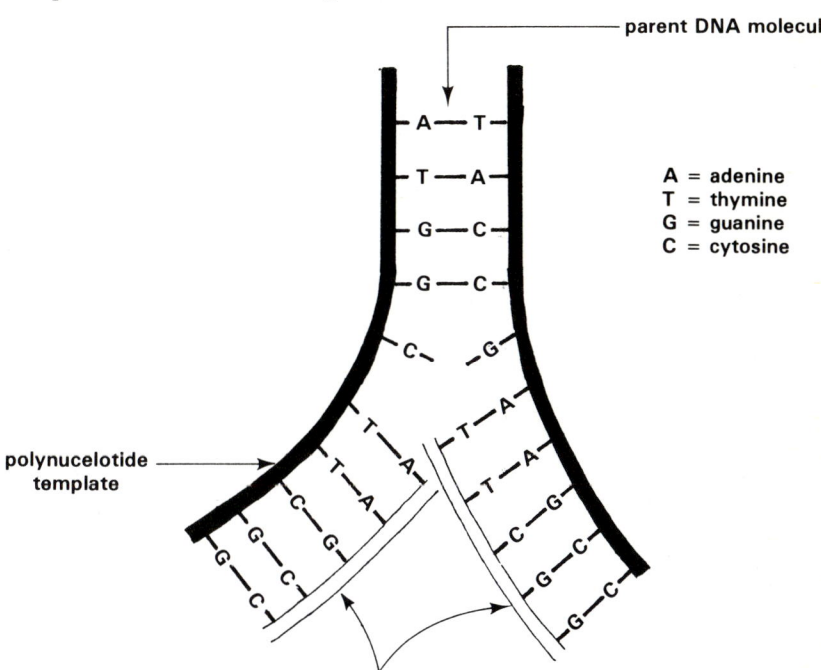

FIG. 6.13 *Semi-conservative and conservative replication of DNA*

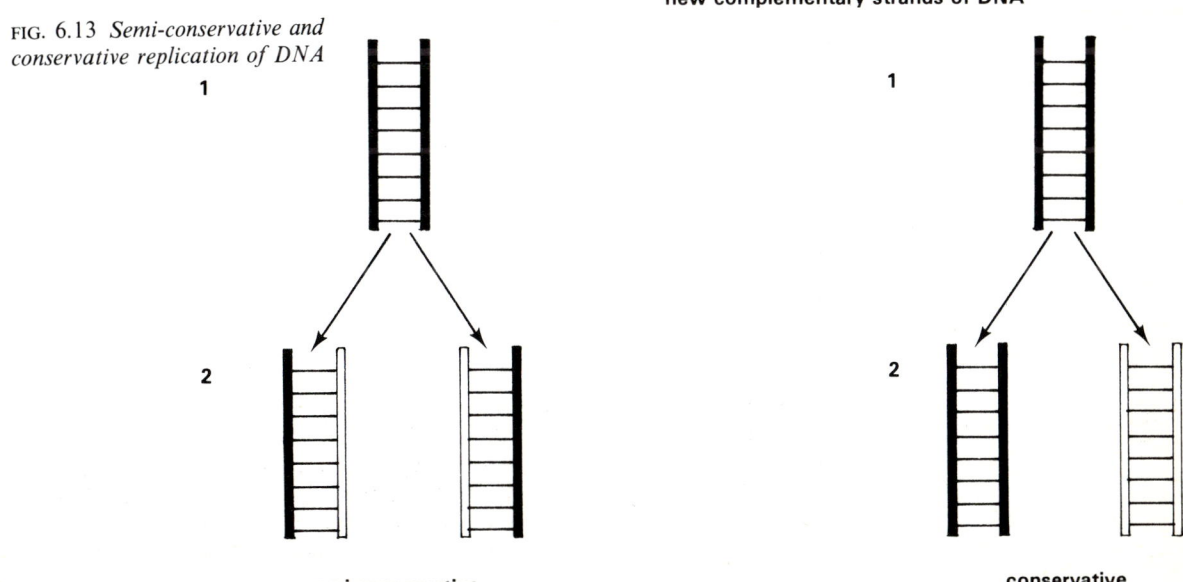

semi-conservative   conservative

Make drawings to show the result of replication of the DNA molecules in 2.

Experimental evidence that DNA replication is semi-conservative was obtained in 1957 by Meselson and Stahl. They grew the bacterium *Escherichia coli* for several generations in a medium containing $^{15}NH_4Cl$ as the only source of nitrogen. The DNA of the bacteria was thus labelled with ^{15}N, the heavy isotope of nitrogen. The bacteria were then transferred to a medium in which $^{14}NH_4Cl$ was the only source of nitrogen. As *E. coli* multiplied the density of its DNA was measured. After one generation the density of the DNA was intermediate between DNA of the bacterium made with the heavy isotope ^{15}N and DNA made using ^{14}N, the normal isotope of nitrogen (Fig. 6.14). The result was possible only if the DNA of *E. coli* had replicated in a semi-conservative way.

FIG. 6.14 *Meselson's and Stahl's experiment. The relative density of the DNA was determined by centrifugation What result would you have expected in B if replication of DNA was conservative?*

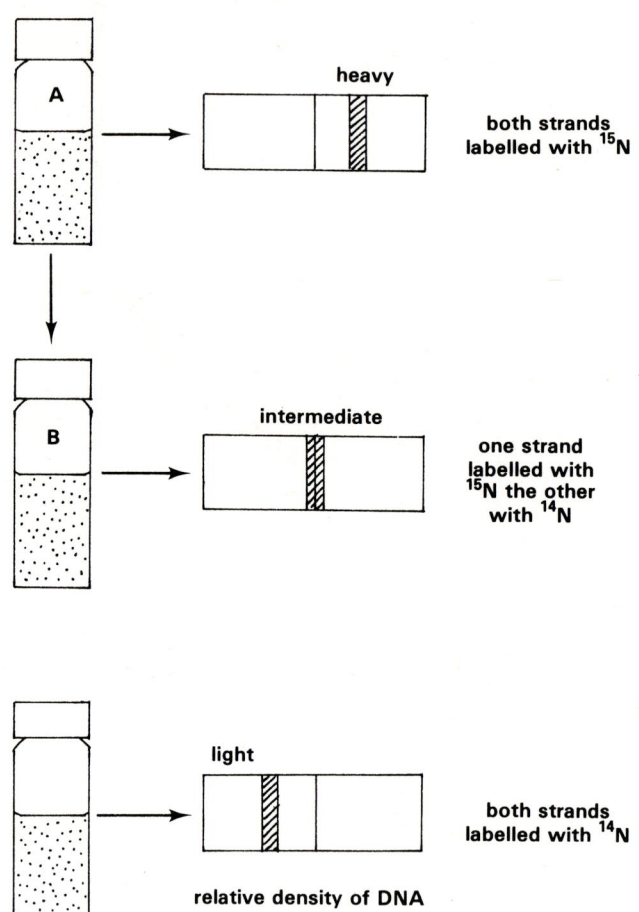

E.*coli* grown in ^{15}N medium

A heavy — both strands labelled with ^{15}N

E.*coli* from A transferred to ^{14}N medium for one generation

B intermediate — one strand labelled with ^{15}N the other with ^{14}N

E.*coli* grown in ^{14}N medium

light — both strands labelled with ^{14}N

relative density of DNA

Sometimes mistakes arise when DNA replicates. There are several ways in which errors can occur. One pair of bases may be replaced by another, for example C·G may be replaced by A·T. A pair of nucleotides may be lost or an extra pair may be added (Fig. 6.15). Whatever the cause the new DNA is not an exact copy of the original. Such changes are called **gene mutations**. The effects of erroneous DNA replication should become apparent after you have read the next section which describes the way in which DNA controls the synthesis of proteins.

FIG. 6.15 *Gene mutations*

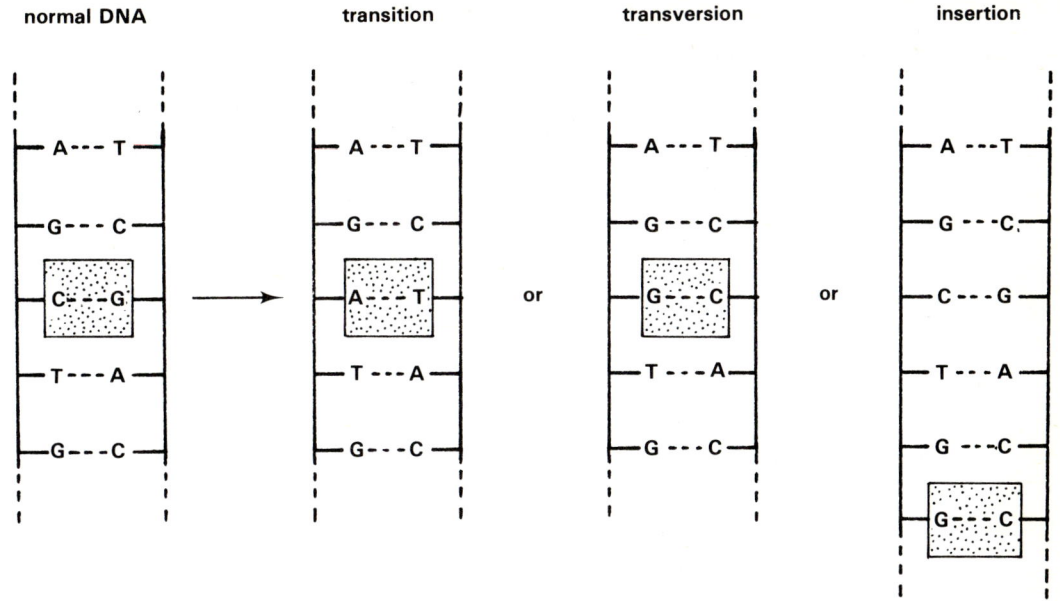

normal DNA   **transition**   **transversion**   **insertion**

If a base pair is lost the condition is called a deletion

6.2 Protein synthesis

If labelled amino acids, the building blocks of proteins, are fed to rats the ribosomes of the rats' liver cells soon become radioactive. Furthermore, peptide links are formed between the amino and carboxyl groups of amino acid molecules incubated with a suspension of ribosomes, RNA and ATP. Evidence of this sort indicates that the ribosomes are the sites of protein synthesis. But what parts do RNA and ATP play in the synthesis of proteins?

6.2.1 Role of the nucleic acids

Proteins are polymers consisting of up to twenty different types of amino acids joined together in a specific three-dimensional way (Chapter 3). Many different kinds of proteins can be made inside a living cell. The accuracy with which ribosomes, time after time, assemble such complex molecules suggests that some sort of control system is at work inside cells coding the production of proteins. The code, called the **genetic code**, is based on the sequence of nucleotides in DNA. How is information contained in the nucleus conveyed to and interpreted by the ribosomes? There is now a lot of evidence to indicate that protein synthesis takes place in a number of distinct stages.

1. TRANSCRIPTION

Just before proteins are synthesised, mRNA is made inside the nucleus. The double helix of DNA unwinds and single-stranded mRNA molecules are produced, one of the DNA strands acting as a template (Fig. 6.16). Remember that the mRNA is complementary to its DNA template. Exactly how one of the DNA strands is chosen to act as a template for transcription is unknown. Synthesis of mRNA is controlled by the enzyme **RNA synthetase**. The mRNA now passes through the pores of the nuclear membrane into the cytoplasm. Here one end of the mRNA molecule becomes attached to a ribosome.

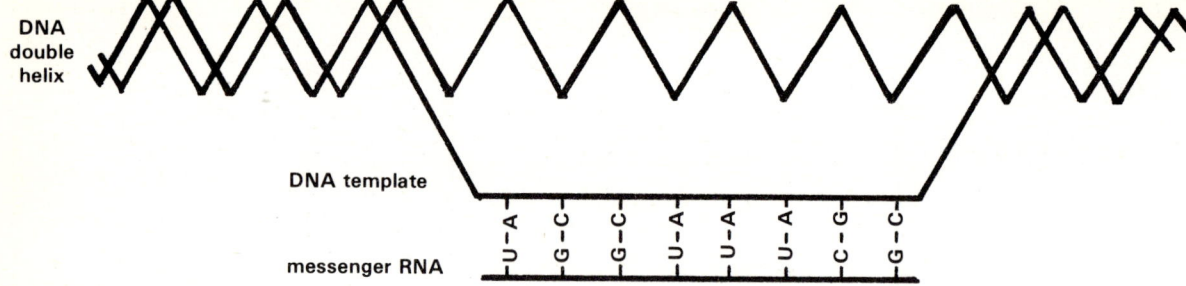

FIG. 6.16 *Transcription*

DNA
double
helix

DNA template

messenger RNA

| U – A | G – C | G – C | U – A | U – A | U – A | C – G | G – C |

2. ACTIVATION

During **activation**, energy from ATP is used to combine tRNA molecules with amino acid molecules. There are at least twenty different kinds of tRNA, the important difference between them being the sequence of nucleotides in their anticodons (section 6.1.2). Each type of tRNA binds with a specific amino acid. The amino acid molecules join to the free ends of the tRNA molecules where the A ← C ← C ← triplet of nitrogenous bases is found (Fig. 6.17). This rules out the possibility that the anticodon determines the amino acid with which the tRNA combines. It is still not known what determines the specific tRNA-amino acid combination. The tRNA-amino acid complexes now move to the ribosomes.

FIG. 6.17 *Activation*

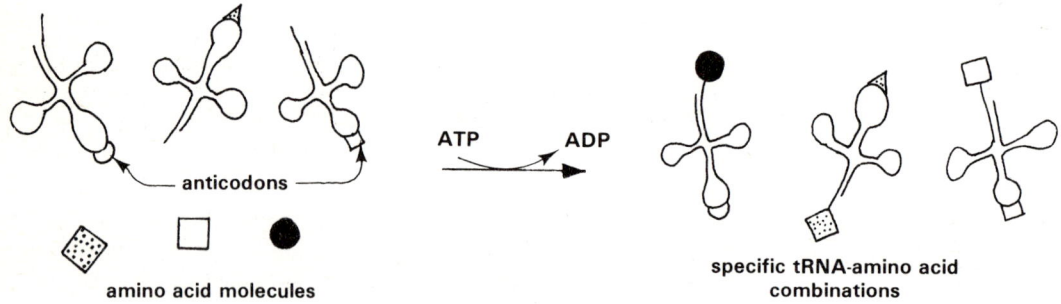

anticodons

amino acid molecules

ATP   ADP

specific tRNA-amino acid
combinations

3. TRANSLATION

A knowledge of protein structure (Chapter 3) is assumed in the following description of **translation**. Starting at one end of an mRNA molecule, a ribosome works its way along positioning the anticodon of each tRNA molecule on to a complementary codon of the mRNA strand (Fig. 6.18). In this way each of the amino acid molecules is brought to a specific, predetermined site. A polypeptide chain usually begins with methionine, the carboxyl group of which forms a peptide link with the amino group of the next amino acid molecule brought into position. The first amino acid built into a protein thus has a free amino group, while the last amino acid has a free carboxyl group. As soon as an amino acid is linked to its neighbour its tRNA partner is released back into the cytoplasm to pick up another molecule of the same amino acid.

Polysomes, groups of ribosomes connected by a common strand of mRNA, are often seen in cells (Chapter 1). This arrangement may mean that several proteins are made at the same time on one mRNA molecule

FIG. 6.18 *Anticodon binding*

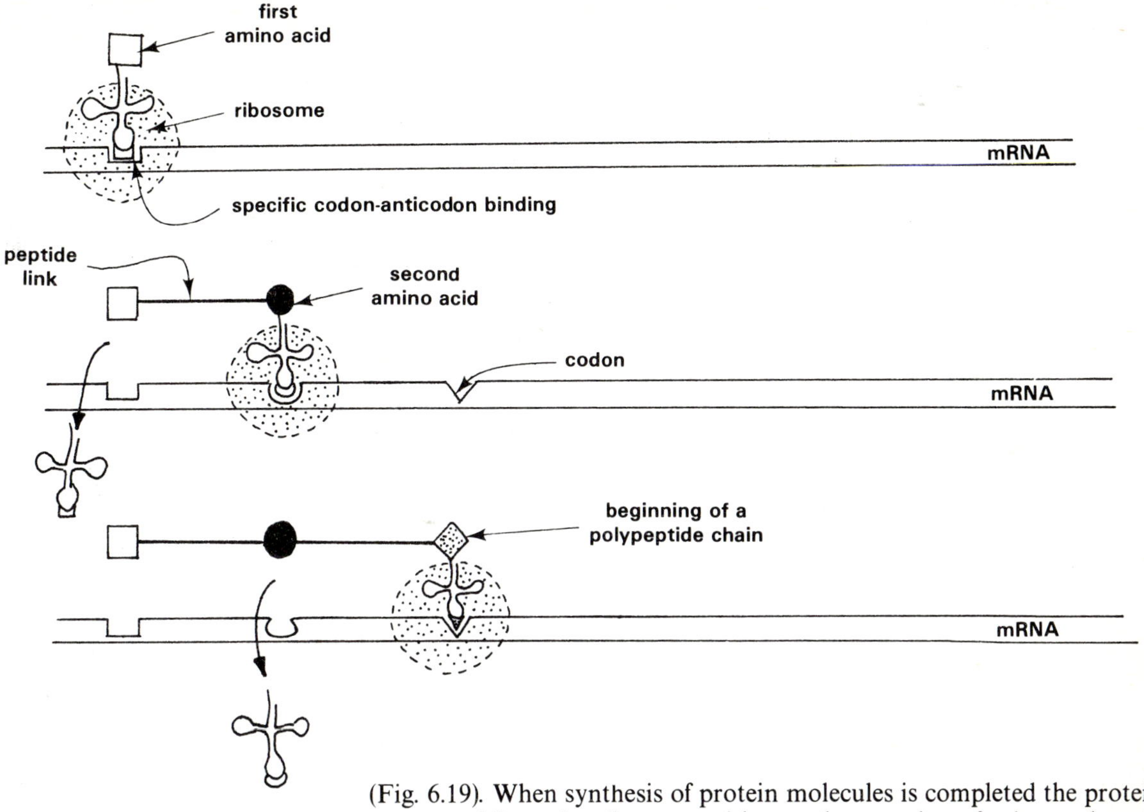

first amino acid

ribosome

mRNA

specific codon-anticodon binding

peptide link

second amino acid

codon

mRNA

beginning of a polypeptide chain

mRNA

(Fig. 6.19). When synthesis of protein molecules is completed the proteins are moved from the ribosomes either to the cytoplasm for internal use by the cell or to the endoplasmic reticulum ready for external secretion.

Not all proteins are synthesised at the ribosomes. Mitochondria and chloroplasts contain DNA and are able to code the production of their own proteins. The proteins are used by these organelles when they divide during the interphase of cell division (Chapter 7). Protein synthesis in mitochondria is very similar to that shown by many species of bacteria.

FIG. 6.19 (a) *Protein synthesis by a polysome*

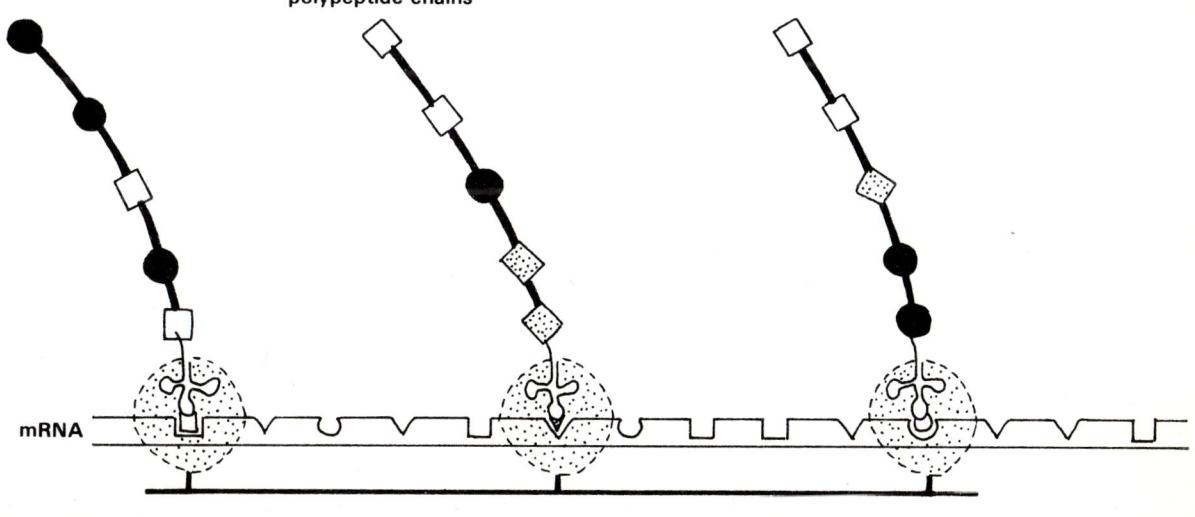

polypeptide chains

mRNA

linked ribosomes

FIG. 6.19 (b)
Electronmicrograph of polysomes, × 47 000

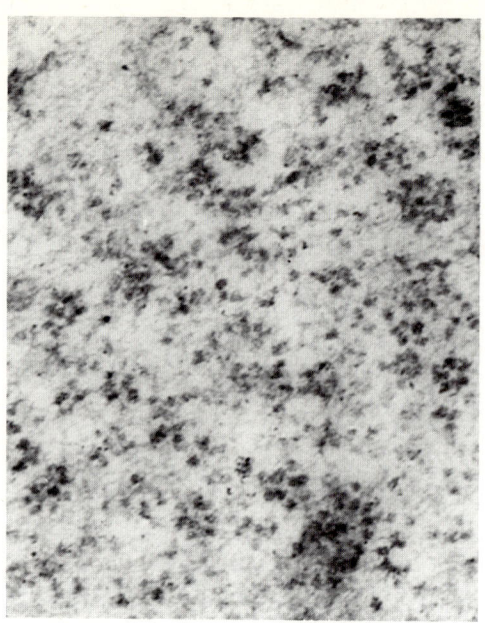

Observations of this sort have led some cell biologists to think that mitochondria are degenerate bacteria living symbiotically in the cytoplasm of plant and animal cells.

6.2.2 The genetic code

The description of the synthesis of proteins given above is based on the evidence provided by many, often highly intricate experiments. There is also a logical theoretical basis to what has been written.

In interpreting the mechanism of protein synthesis molecular biologists were faced with the problem of explaining how the nucleotides found in DNA could provide a code for the synthesis of many different kinds of proteins. Single nucleotides can hardly be the basis of the code as there are just four different nucleotides but at least twenty different amino acids. Neither does it seem logical that pairs of nucleotides are responsible as this would give only $4^2 = 16$ possible combinations of two nucleotides. It was therefore deduced that combinations of at least three nucleotides are required. The **triplet code** provides $4^3 = 64$ different permutations from the four nucleotides (Fig. 6.20).

In the early 1960s Nirenberg and his colleagues began to experiment with the production of artificial polypeptides using ribosomes extracted

FIG. 6.20 *The triplet code*

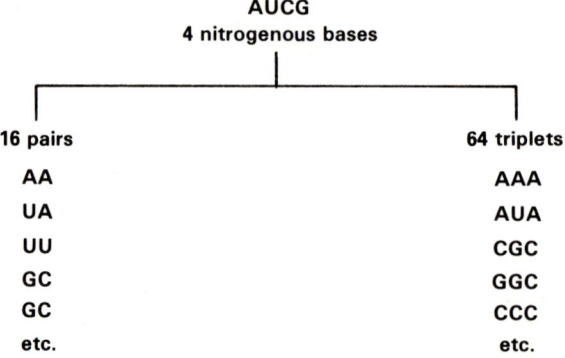

from *Escherichia coli*. When incubated with a mixture of amino acids, ATP and synthetically-made mRNA containing only the nitrogenous base uracil, the ribosomes synthesised a polypeptide made solely from the amino acid called phenylalanine. The triplet code for phenylalanine is therefore UUU. Nirenberg also discovered that AAA is the triplet code for lysine and CCC for proline. Since then the triplet codes for all twenty amino acids have been worked out. Methionine and tryptophan each have a single triplet code, the other eighteen amino acids have more than one triplet permutation. Some have just two alternative triplets, others have up to six. For this reason the genetic code is described as **degenerate**. A few of the sixty-four triplets, called **non-sense triplets**, do not code for any amino acid.

The way in which DNA transcribes the sequence of nucleotides in mRNA was outlined in section 6.2.1. But how is the sequence translated by the tRNA-amino acid complexes? The triplet code for methionine, the first amino acid built into a polypeptide chain, is AUG. The tRNA-methionine complex having the anticodon UAC is brought to the end of the mRNA molecule where the codon AUG is located. The ribosome binds the codon and anticodon then moves on to the next triplet of the mRNA strand. Here **codon-anticodon binding** again takes place and the next amino acid molecule is brought into position. A peptide link is formed with methionine and the polypeptide chain begins to take shape (Fig. 6.21). The ribosome continues to work its way along the mRNA strand until it meets a non-sense triplet. This is the signal that the protein molecule is complete. The part of a DNA molecule which codes the synthesis of a protein is usually called a **gene**. Molecular biologists call it a **cistron**.

FIG. 6.21 *Translation. What triplet of bases is found in the anticodon of the tRNA molecule which brings the next amino acid to the ribosome?*

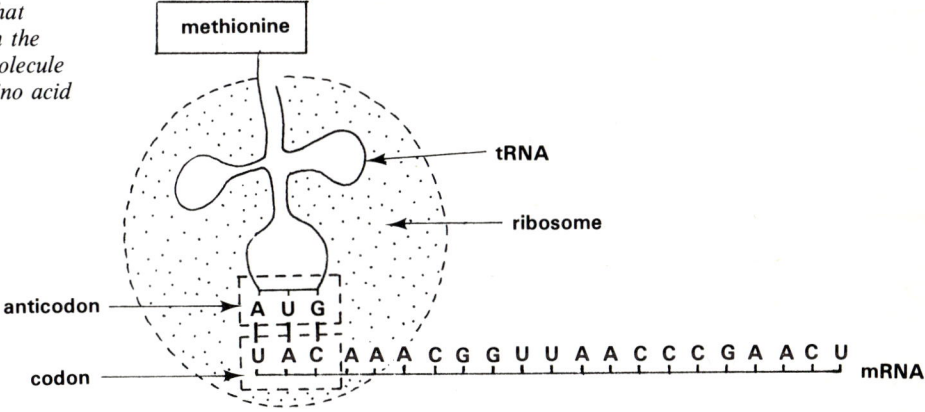

If a mistake is made during transcription then non-sense triplets appear elsewhere in the mRNA strand. When this happens the protein made is abnormal and may be unable to carry out its usual function. For example, haemoglobin S made by humans suffering from sickle cell anaemia (Chapter 8) differs from normal haemoglobin in having valine instead of glutamic acid in each of its two β-chains (Fig. 6.22). This is one example of the ways in which gene mutations are expressed. Gene mutations are probably also responsible for the ageing of cells. Mistakes in the genetic code responsible for synthesising DNA synthetase are more common in old cells than in young cells. This suggests that DNA replication gradually becomes less accurate as cells age. Cumulative mistakes in the genetic code may therefore be the cause of senility.

FIG. 6.22 *Two-dimensional analysis of haemoglobin. The haemoglobins were first broken down into peptide fragments using the enzyme trypsin. The peptides were then separated on paper by electrophoresis, after which the paper was turned at right angles and the peptides further separated by chromatography*

normal haemoglobin

haemoglobin S

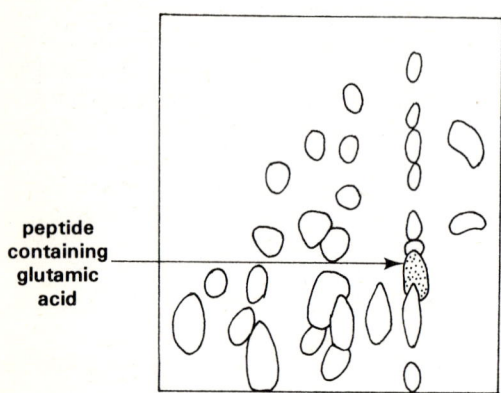

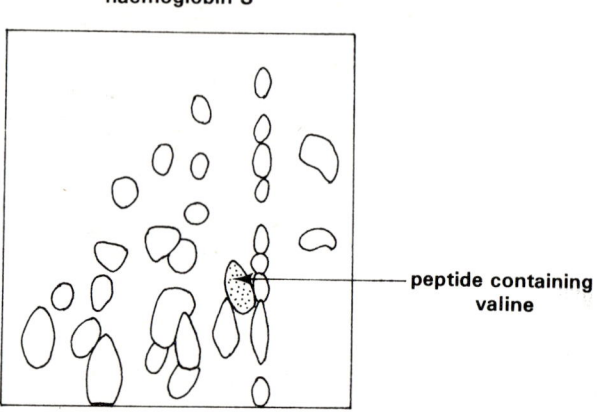

peptide containing glutamic acid

peptide containing valine

6.2.3 Regulating the synthesis of proteins

The ability of cells to regulate the amounts and types of proteins they make is important because it allows enzymes and metabolites to be produced as and when required. In this way the efficiency of metabolism is improved and a steady state can be maintained. One device which helps in **cellular homeostasis**, the maintenance of a steady state in living cells, is the temporary inhibition of an enzyme when the concentration of a substrate becomes too high (Chapter 4). Another is to regulate the synthesis of proteins.

In the late 1950s two French scientists, François Jacob and Jacques Monod, reported that the bacterium *Escherichia coli* was able to produce several enzymes 'on demand'. *E. coli* can synthesise all the metabolites and enzymes it requires from a medium consisting of glucose, mineral salts and water. Jacob and Monod observed that if lactose was substituted for glucose then *E. coli* makes an additional enzyme called β-D-galactosidase which hydrolyses lactose to glucose and galactose. When returned to the glucose medium the bacterium stops making β-D-galactosidase. The presence of lactose in the medium induces *E. coli* to make an enzyme capable of hydrolysing lactose. This is an example of **enzyme induction**.

E. coli can make all the amino acids it requires from a medium containing NH_4^+ as the only source of nitrogen. However, if the amino acid arginine is added to the medium the bacterium stops making the enzymes needed for the synthesis of arginine. In this case the presence of arginine in the medium inhibits the production of an enzyme, an example of **enzyme repression**.

Jacob and Monod produced a hypothesis to explain enzyme induction and repression. They proposed that the sequence of amino acids in an enzyme is determined by a **structural gene** which can be switched on or off by a **regulator gene**. The regulator gene transcribes the production of mRNA which has the codons to make a **repressor protein**. The repressor moves from the ribosomes where it is made to the nucleus. Here it binds with the structural gene stopping it from transcribing the manufacture of mRNA used in the synthesis of an enzyme. As a consequence, production of the enzyme is repressed. However, if an inducer substance is present it combines with the repressor. The inducer-repressor complex is unable to prevent transcription of the structural gene. Synthesis of the enzyme can then proceed. The formation of the inducer-repressor complex must be reversible because enzyme repression occurs when the inducer is removed. Induction and repression are thus linked. Induction is freedom from repression. The Jacob–Monod hypothesis is summarised in Figure 6.23.

FIG. 6.23 *The Jacob–Monod hypothesis*

(i) inducer absent

repressor gene

structural gene

R   S   DNA

transcription of mRNA blocked; enzyme synthesis prevented

mRNA

repressor protein

(ii) inducer present

R   S   DNA

mRNA

repressor now unable to block transcription by structural gene

mRNA

repressor   +   inducer   ⇌   inducer-repressor complex

enzyme   +   inducer (substrate)   ⇌   enzyme-substrate complex

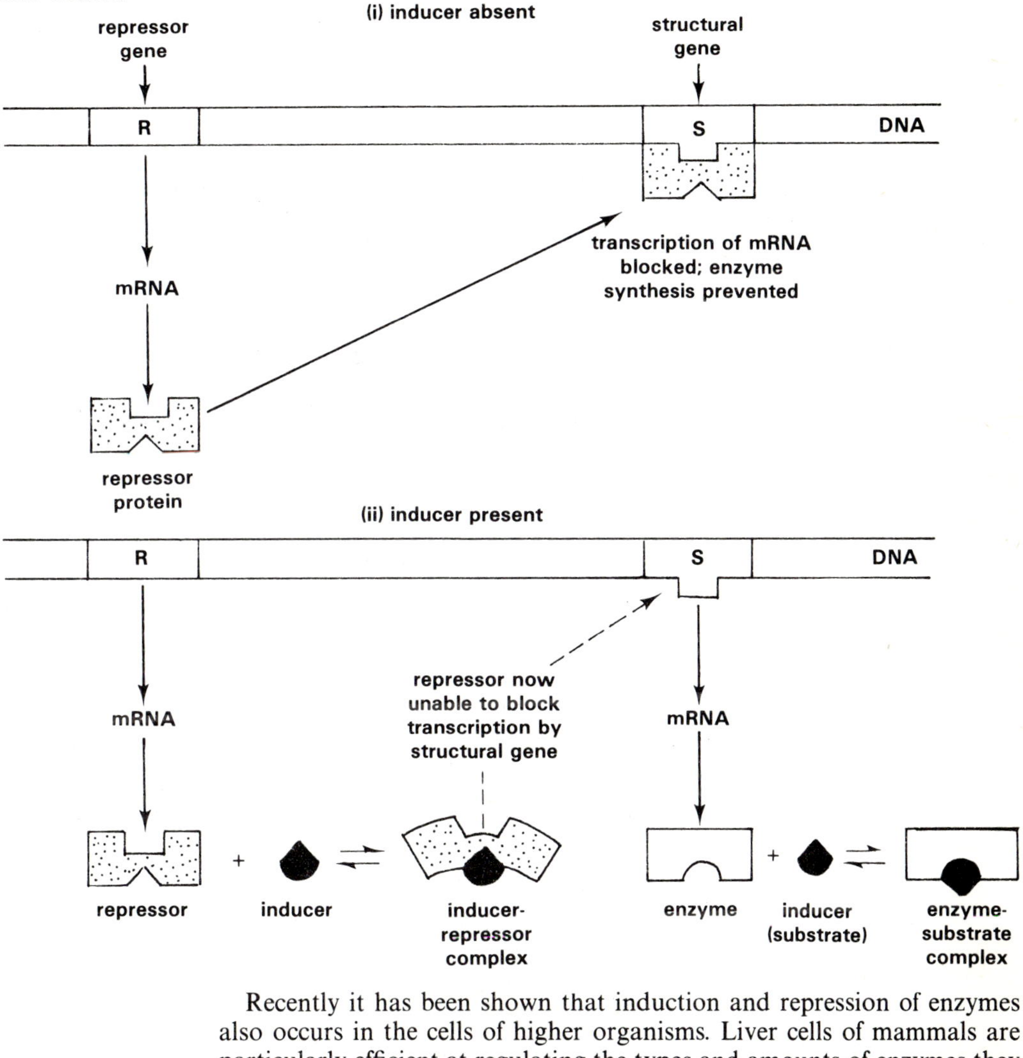

Recently it has been shown that induction and repression of enzymes also occurs in the cells of higher organisms. Liver cells of mammals are particularly efficient at regulating the types and amounts of enzymes they produce. In this way the liver can adjust to the fluctuating levels of substances it receives from the gut (Chapter 14).

Enzyme induction and repression are also important in the differentiation of living organisms. How it is possible for a one-celled fertilised egg to grow into a multicellular body made up of a number of different types of tissues is a process about which a lot has yet to be learned. However, induction and repression of enzymes probably enables the various types of tissues to differentiate by allowing different combinations of genes to be expressed at different stages during growth and development.

6.3 Genetic engineering

Avery in 1943 showed that it was possible to alter the genetic make-up **(genome)** of the R-strain of *Pneumococcus* by adding to its culture medium DNA from the S-strain (section 6.1.3). Since then biologists have devised sophisticated procedures whereby the genomes of experimental organisms such as bacteria can be altered with great precision. The experimental manipulation of the genomes of organisms is called **genetic engineering**. One of the methods used involves the transfer of genes from the cells of one organism to those of another. The DNA of the recipient is called **recombinant DNA**.

FIG. 6.24 (a) *Root system of a broad bean plant,* ×0·75

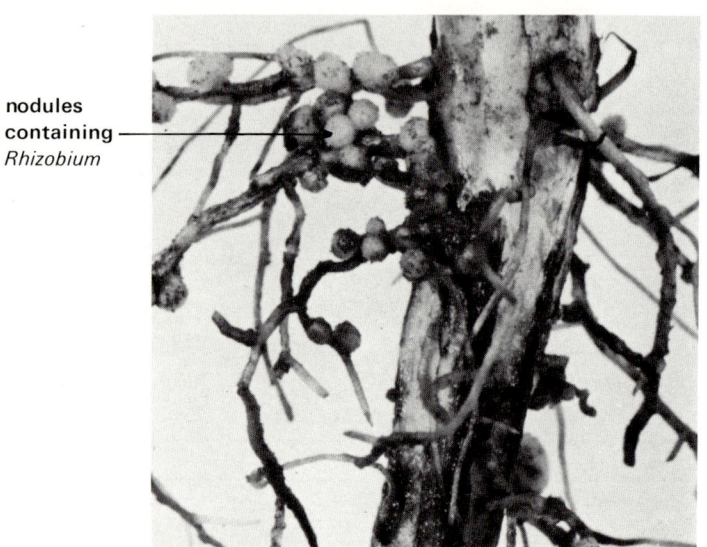

nodules containing *Rhizobium*

FIG. 6.24 (b) *Section through a root nodule,* ×30

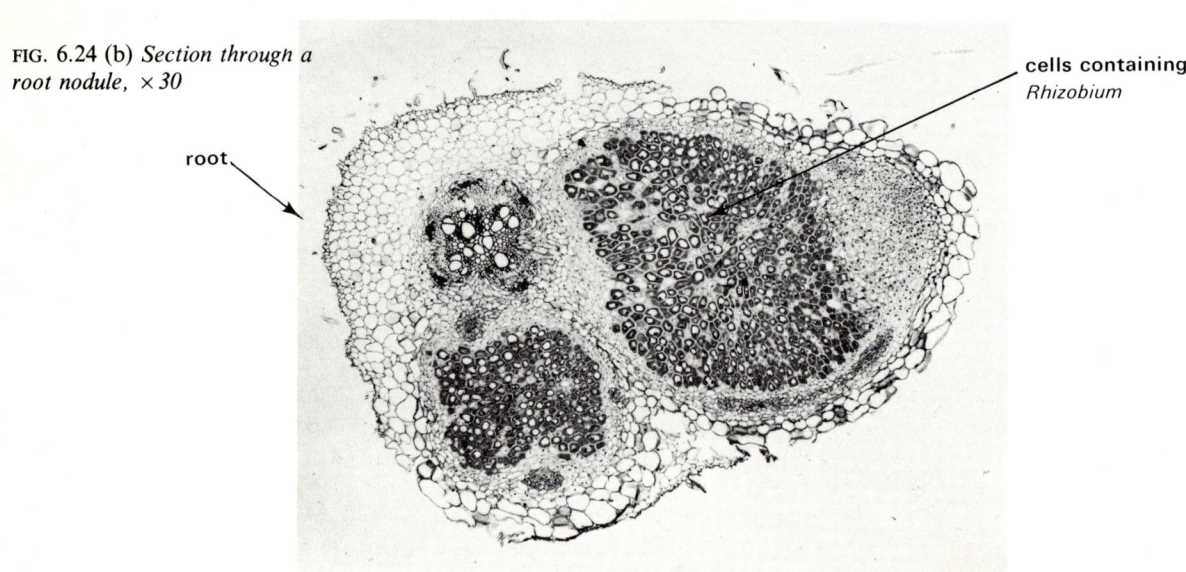

cells containing *Rhizobium*

root

Advantages of immense importance to the human race could arise from genetic engineering. For example, the roots of leguminous plants form a symbiotic relationship with the bacterium *Rhizobium leguminosarum* (Fig. 6.24). *Rhizobium* can fix nitrogen gas and some of the organic nitrogenous substances it makes are used by the leguminous host plants. The hosts as a result grow more profusely. Because many extensively grown crop plants such as peas, beans and clover are legumes the symbiotic relationship is of great importance to food production. But what of the many other types of crop plants which do not have this facility? The soil in which they are grown has to be treated with expensive artificial fertilisers or with manure (Chapter 13). Clearly it would be advantageous if non-legumes too could fix gaseous nitrogen which is plentiful in the atmosphere. For this reason attempts have been made to transplant the genes for nitrogen fixation into the cells of cereals and other non-leguminous plants.

There is, however, another side to genetic engineering. Concern has been expressed about the danger of producing organisms whose new genomes pose a threat to the well-being of man. One of the organisms widely used in such experiments is the gut bacterium *Escherichia coli* which lives commensally in the bowels of humans and other vertebrates. What would be the consequence if a strain of genetically-engineered *E. coli* which was highly pathogenic became resident in a human population? As Dr. Erwin Chargaff, an eminent molecular geneticist has put it:

'You cannot recall a new form of life. Once you have constructed a viable *E. coli* cell into which a piece of eukaryotic DNA has been spliced, it will survive you and your children and your children's children. An irreversible attack on the biosphere is something so unheard of, so unthinkable to previous generations, that I could only wish that mine had not been guilty of it.'

The potential hazards of such experiments have led to a call for a halt to certain types of genetic engineering.

Chapter 7

Cell division

There are two types of cell division, **mitosis** and **meiosis**. Mitosis takes place when new cells are added to multicellular organisms as they grow (Chapter 21) and when tissues are repaired or replaced. Asexual reproduction can also occur as a result of mitotic division. Meiosis occurs in the production of gametes by organisms which reproduce sexually (Chapters 19 and 22).

During both types of division the DNA and histone proteins of the cell nucleus can be seen as threads called **chromosomes**. Chromosomes can be stained and are visible using a conventional compound microscope (chromo = coloured; soma = body). The number of chromosomes in the nucleus is fixed for each species of living organism. What is more, the chromosomes can be arranged in **homologous pairs**. In humans for example, the body cells contain 46 chromosomes, 23 homologous pairs (Fig. 7.1). The characteristics of living organisms are determined by many pairs of genes called alleles. One of a pair of allelic genes is found on a chromosome. The other is found on its homologous partner. Each chromosome carries a number of genes.

Mitosis normally ensures that the cells produced contain exactly the same number of chromosomes as the cells from which they were formed. The cells of the bone marrow of humans for example, constantly give rise to blood cells which have 46 chromosomes. On the other hand, cells produced by meiosis have half the chromosome number. For example, human sperm and eggs each have 23 chromosomes. However, there is much more to it than this. Let us take a close look at the two types of cell division to see in what other ways they differ.

FIG. 7.1 (a) *Chromosomes of a normal human,* × 1000

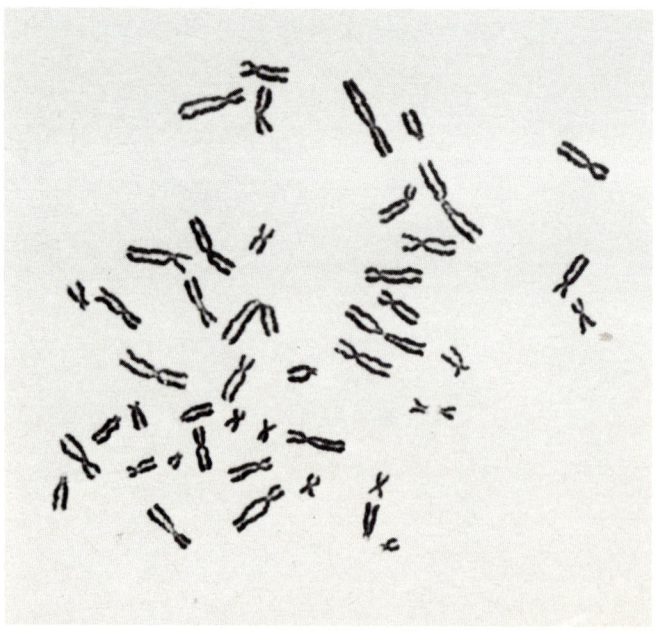

FIG. 7.1 (b) *Homologous pairs of chromosomes from a normal human male* (courtesy Philip Harris Biological Ltd.)

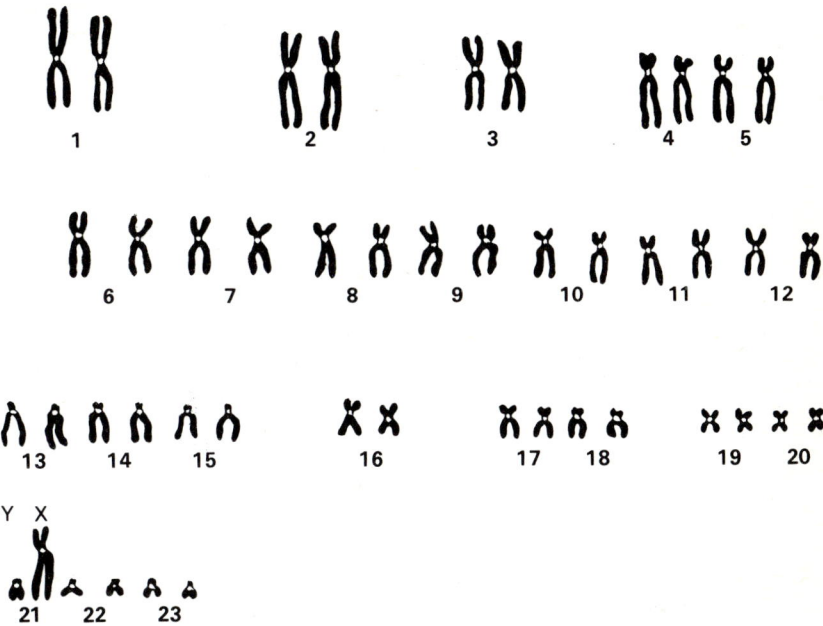

7.1 Mitosis

Before starting to divide a cell is at the **interphase** stage. The nucleus appears as a granular body. Inside the nuclear membrane are one or more dense nucleoli (Fig. 7.2). The absence of any visible signs of activity disguises the fact that intense metabolism is taking place. It is at this stage that replication of DNA (Chapter 6) and synthesis of nuclear proteins occurs. New ribosomes are also made at interphase. Mitochondria and centrosomes divide at this stage, so do proplastids from which chloroplasts originate. The proteins which later make up the microtubules of the spindle are also made at interphase although the spindle is not yet visible. The energy for the various forms of activity comes from respiration. It is therefore not surprising that the respiratory rate of cells at interphase is very rapid.

During both mitosis and meiosis the nucleus divides first, followed by cleavage of the cytoplasm.

FIG. 7.2 *Interphase*

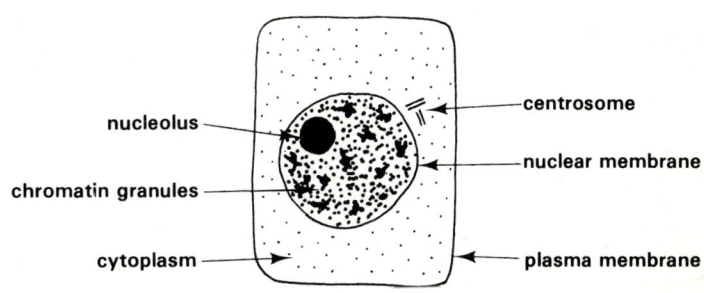

7.1.1 Mitotic nuclear division

FIG. 7.3 (a) *Cells dividing near the tip of an onion root* × *400*

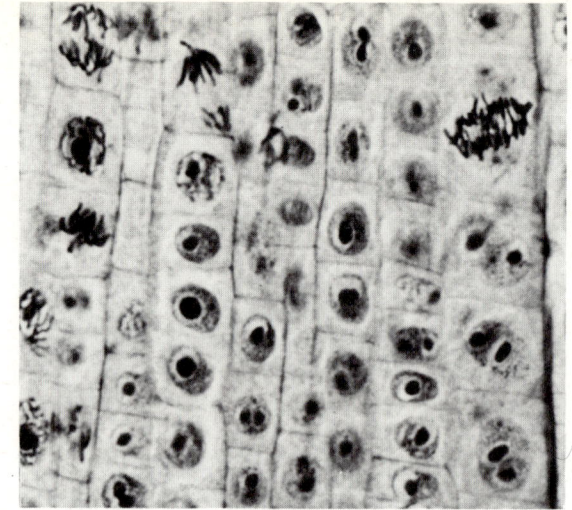

FIG. 7.3 (b) *Mitotic nuclear division (only four chromosomes shown for clarity)*

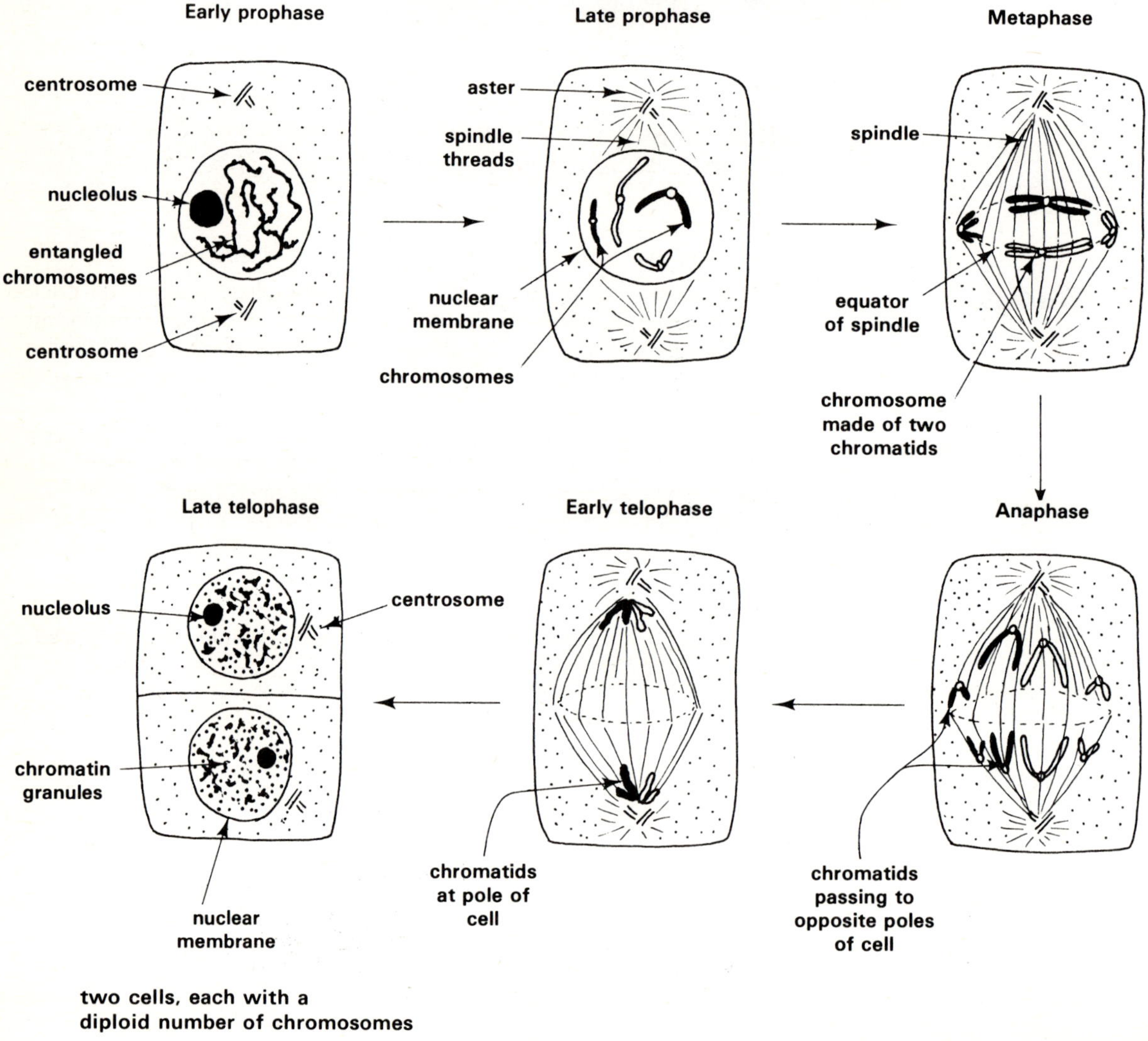

Early prophase

centrosome

nucleolus

entangled chromosomes

centrosome

Late prophase

aster

spindle threads

nuclear membrane

chromosomes

Metaphase

spindle

equator of spindle

chromosome made of two chromatids

Anaphase

chromatids passing to opposite poles of cell

Early telophase

chromatids at pole of cell

Late telophase

nucleolus

centrosome

chromatin granules

nuclear membrane

two cells, each with a diploid number of chromosomes

For convenience of description it is usual to separate mitotic division of the nucleus into four main stages. It is possible to see the various stages by examining stained sections and squashes of root tips or other suitable material under the microscope (Fig. 7.3). In this way what appears to be a series of static events can be observed. However, mitosis is an active process, the different stages merging into each other. Time-lapse photomicrography provides a means of seeing the dynamic nature of mitosis. Dividing cells are placed in a nutrient solution on a microscope slide and photographed through the microscope every 30 seconds or so over a period of several hours or even days. The exposed film is developed and run through a cine projector. It then becomes apparent that during division cells are particularly active.

1. PROPHASE

At the start of **prophase** chromosomes become visible inside the nuclear membrane. At first they are long, thin entangled threads. As time goes on the threads become shorter and thicker. The chromosomes disentangle and can be seen as separate structures. As the chromosomes become visible the nucleoli gradually disappear.

The centrosomes which duplicated at interphase begin to migrate to opposite ends (poles) of the cell. As they move apart the centrosomes lay down a large number of microtubules which extend from one pole of the cell to the other. The microtubules are the **spindle**, a fibrous structure which is widest at the centre (equator) of the cell. A mass of microtubules called an **aster** also radiates from the centrosomes at each of the poles of animal cells (Fig. 7.3(b)).

2. METAPHASE

The nuclear membrane breaks down. How and why this occurs is still a matter for debate. Mitochondria often congregate near the nuclear membrane at this stage. They may provide energy for some of the spindle microtubules to pull the nuclear membrane apart.

It is now possible to see that each chromosome consists of two threads called **chromatids** joined at a **centromere**. Unlike the rest of the chromosome the centromere is not easily stained. Independently of each other the chromosomes become attached by their centromeres to the equator of the spindle (Fig. 7.3(b)).

3. ANAPHASE

The centromere of each chromosome splits and the chromatids move to opposite poles of the cell. Separation of the chromatids appears to be caused by a shortening of the spindle microtubules to which the centromeres are attached. Consequently the chromatids are dragged, centromere first, away from the equator of the spindle. During their passage the chromatids slide over the other spindle microtubules which extend from pole to pole (Fig. 7.3(b)).

4. TELOPHASE

On reaching the poles of the cell the two groups of chromatids come together. Each group becomes surrounded by a newly formed nuclear membrane. It is not clear whether the new membranes are put together from fragments of the nuclear membrane destroyed in prophase or are made anew. Whatever their origin they are assembled from pieces of

membrane. Inside the nuclear membranes the chromatids become un-
coiled, nucleoli reform and the nucleus takes on the granular appearance it
had at interphase (Fig. 7.3(b)).

Study the drawings of the various stages of mitotic nuclear division. Try
to recognise the stages shown in the photographs in Figure 7.4.

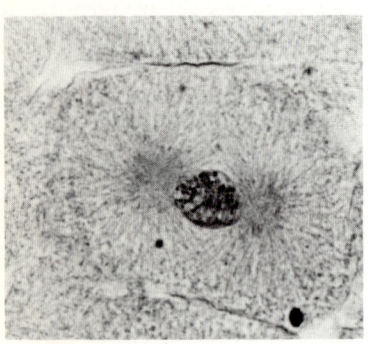
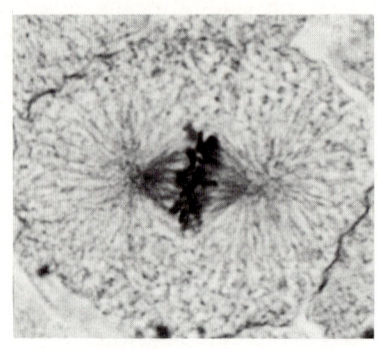
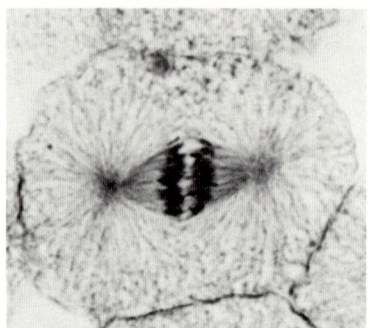
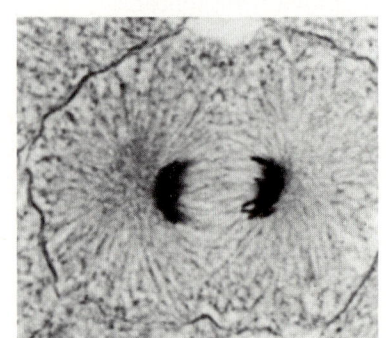
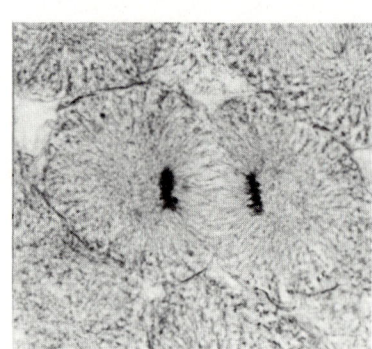

FIG. 7.4 *Mitotic division in a
blastula cell from a fish, × 1000.
Which stage of division is shown
in each photograph?*

7.1.2 Cytoplasmic cleavage

Soon after nuclear division the cytoplasm is separated into two more or
less equal parts, each part enclosing one of the newly formed nuclei. The
way in which **cytoplasmic cleavage** occurs differs in plant and animal cells.

In the cells of higher plants a series of flat vacuoles surrounded by unit
membrane, probably made by the Golgi body, appear at the middle of the
cell. The vacuoles extend across an area of dense cytoplasm called the
phragmoplast which contains numerous microtubules, the remains of the
spindle. The membranes around the vacuoles fuse to form the **cell plate**, a
double unit membrane which grows outwards and joins with the plasma
membrane of the dividing cell. Cell wall materials are laid down between
the two membranes of the cell plate and cleavage of the cytoplasm is thus
achieved. By this time the phragmoplast has disappeared (Fig. 7.5(a)).

A ring of microfilaments appears around the middle of animal cells just
inside the plasma membrane. A shallow **furrow** develops in the membrane,
due it is thought to contraction of the filaments. Further shortening of the
filaments ultimately pinches the cytoplasm into two more or less equal
parts, each part surrounding a nucleus (Fig. 7.5(b)).

What has been described so far is sometimes called a **cell cycle**. The time
taken by each of the stages varies considerably from species to species and
depends on the conditions in which the cells are maintained. Mammalian
skin cells kept at 36°C in an artificial culture medium divide every 24 hours
or so. Average figures for the various stages are: prophase 35 minutes,

metaphase 20 minutes, anaphase 10 minutes, telophase 40 minutes and cytoplasmic cleavage 25 minutes. This means that by far the longest period of the cell cycle is occupied by the interphase stage (Fig. 7.6).

FIG. 7.5. (a) *Cytoplasmic cleavage in a* Petunia *cell × 500*

FIG. 7.5 (b) *Cytoplasmic cleavage in a cell from the bone marrow of a mammal, ×1800*

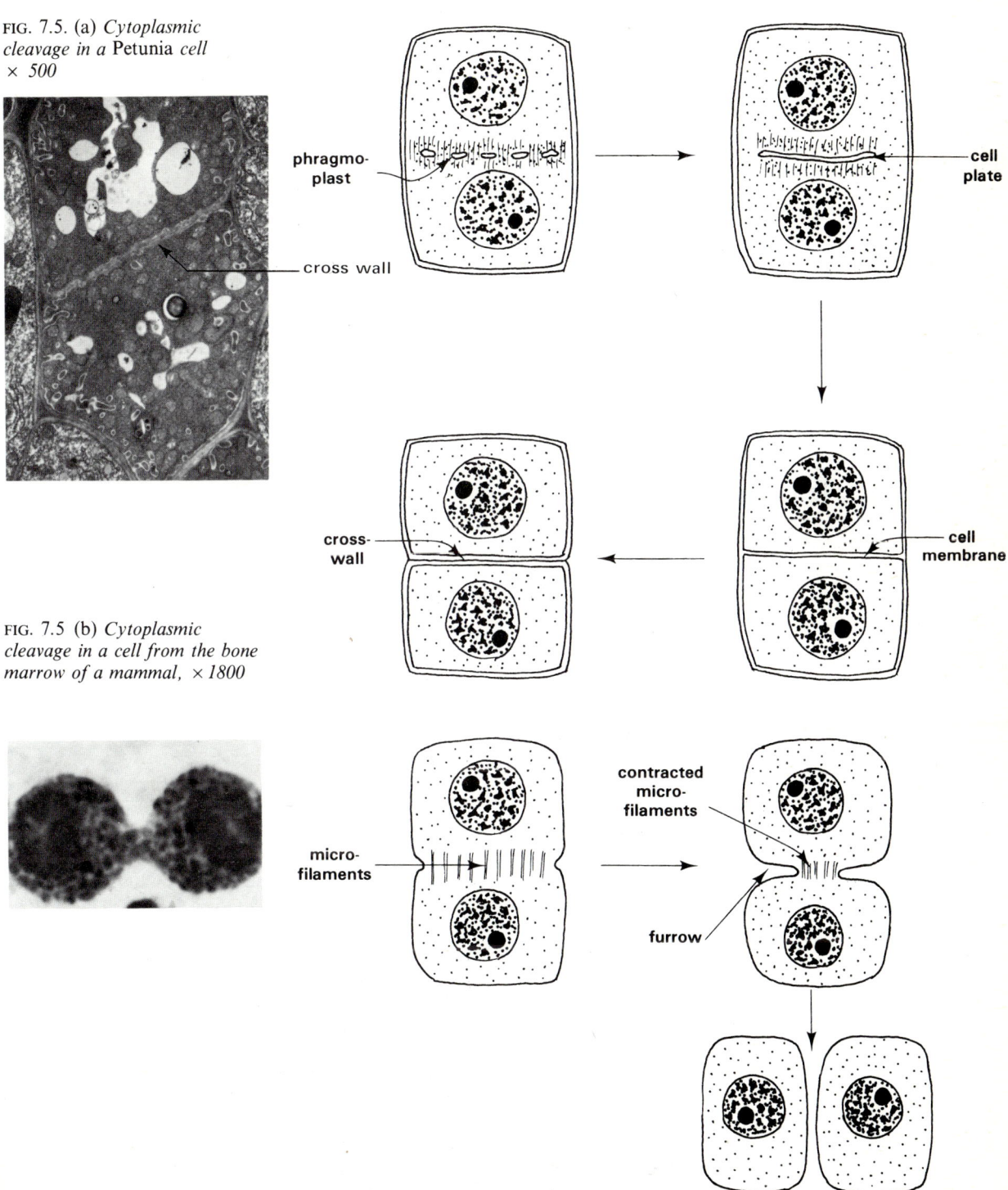

FIG. 7.6 *The cell cycle*

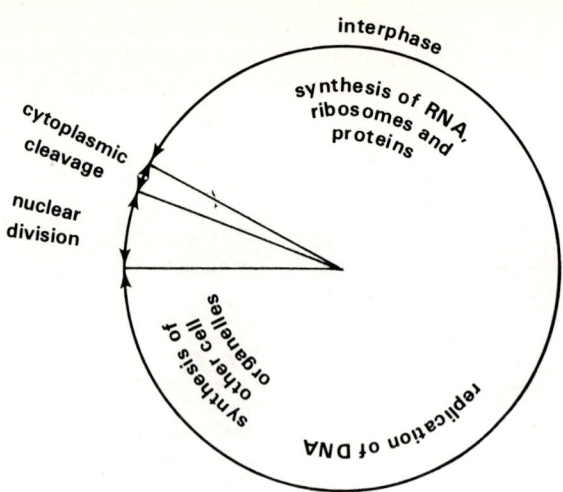

7.1.3 The significance of mitosis

The essential feature of mitosis is that it provides a means of distributing the hereditary material DNA equally between two cells. This does not mean that the DNA content is halved at each mitotic division. The hereditary material is normally reproduced in each cell exactly as it was in the parent cell.

FIG. 7.7 *Changes in the amount of DNA in a nucleus during mitotic division*

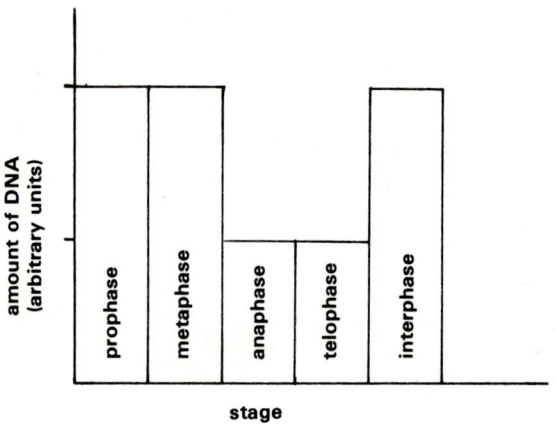

The way in which DNA is replicated is described in Chapter 6. What evidence is there that DNA replication occurs in the cell cycle? One way of finding out is to measure the amount of DNA in a cell at different times during mitosis. Figure 7.7 shows the results of such an investigation. The amount of DNA exactly doubles during interphase and is halved at anaphase. In this way each of the new cells receives the same amount of DNA as in the parent cell. It does not, however, prove that an exact copy of DNA is distributed to the new cells. Another approach is to use radioisotopes to label the DNA. Cells are grown in a nutrient solution containing thymine labelled with tritium (^{3}H) a radioisotope of hydrogen. As the cells divide the labelled thymine is incorporated into their DNA and after a while nearly all the DNA in the cells is radioactive. The cells are then placed in a non-radioactive medium. At each subsequent division some of the cells are removed and placed in photographic emulsion. Radiations given off by the labelled DNA cause the emulsion to develop giving labelled chromosomes a dark colour when viewed under a microscope. The results of such an experiment are shown in Figure 7.8. They prove conclusively that replication of DNA does occur during the cell cycle. Compare these results with the findings of Meselson and Stahl using the bacterium *Escherichia coli* (Chapter 6). There is thus no doubt that the same amount and type of DNA is normally distributed to each of the cells produced by mitosis.

FIG. 7.8 *Chromosomes of a Chinese hamster after labelling with ^3H thymidine (based on autoradiographs)*

(a) *metaphase immediately after labelling*

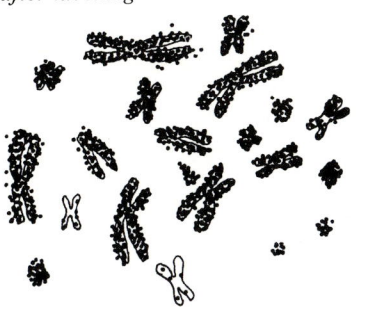

Both chromatids of most of the chromosomes are very radioactive. A few of the smaller chromosomes are only lightly labelled.

(b) *metaphase of next division*

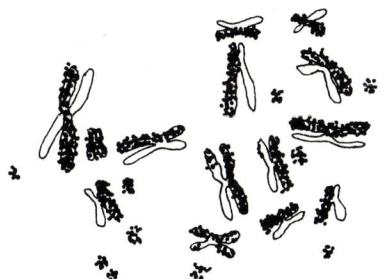

Only one chromatid of most of the chromosomes is very radioactive.

The significance of what has just been described is that mitosis normally gives rise to cells with the same combination of genes (genome). Successful genomes can be perpetuated generation after generation in organisms which multiply asexually. In this way breeders can maintain pure strains of many useful plants which can be propagated vegetatively (Chapter 19). Some of these are important crop plants such as the potato.

Even so, variations do arise in organisms which do not reproduce sexually. The variation is due in some instances to gene mutations (Chapter 6). In others it is caused by chromosome mutations (section 7.3).

7.2 Meiosis

Meiosis occurs in the formation of gametes in organisms which reproduce sexually. Cell squashes made from the testes of insects and the young anthers of flowering plants are suitable specimens for observing the stages of meiotic division. In meiosis, nuclear division is followed by cytoplasmic cleavage but in contrast with mitosis there are two nuclear divisions not one. Thus four cells are formed from a cell which undergoes meiosis, not two as in mitosis. However, there are other differences which are just as important.

The interphase stage of meiosis is the same as in mitosis.

7.2.1 The first meiotic division

FIG. 7.9 *The first meiotic division (only four chromosomes shown for clarity)*

Once more, for convenience of description, division of the nucleus is separated into a number of stages. The Roman numeral I is placed after each of the stages in the first meiotic division of the nucleus to distinguish them from the stages of the second meiotic division.

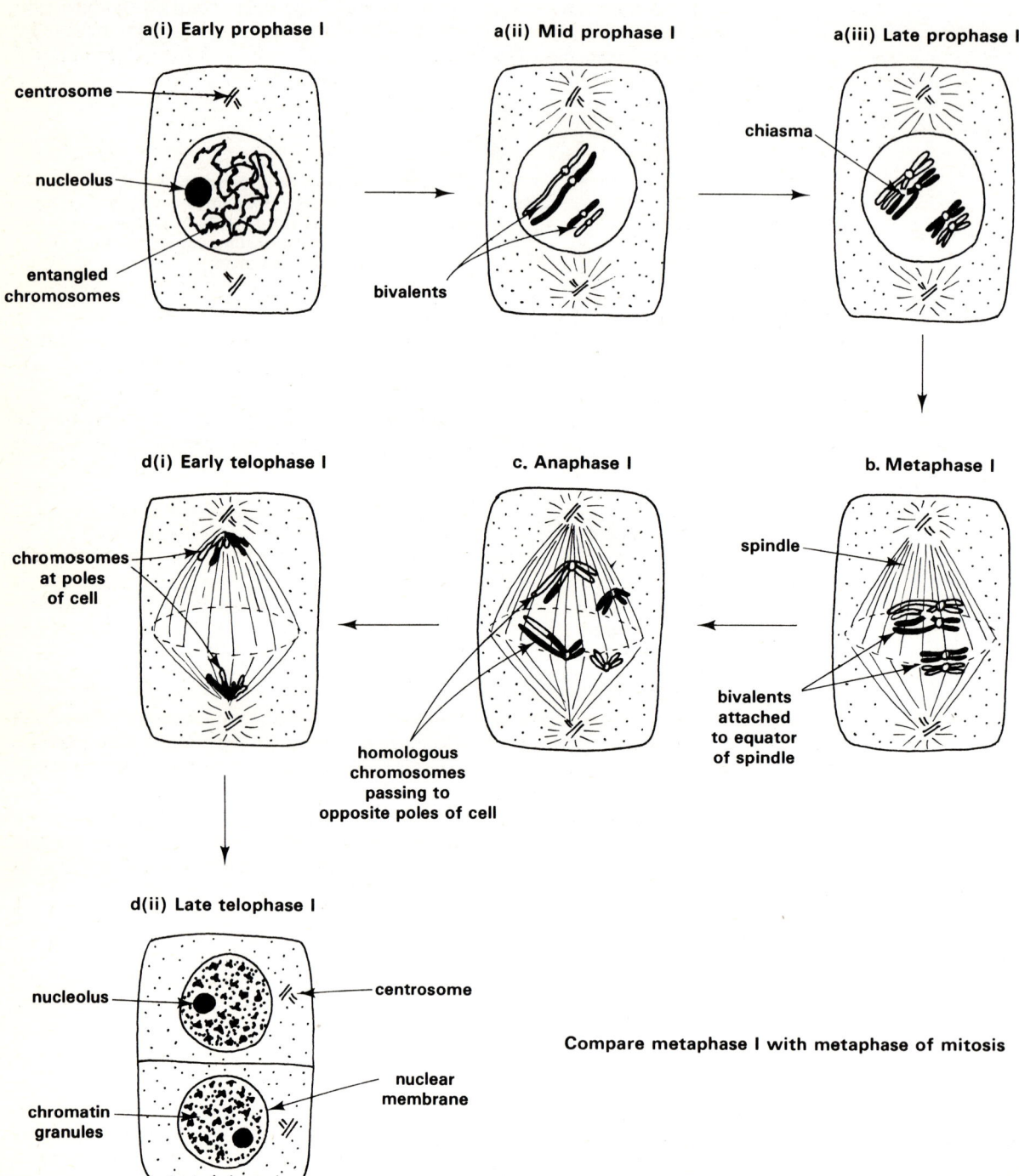

a(i) Early prophase I

centrosome

nucleolus

entangled chromosomes

a(ii) Mid prophase I

bivalents

a(iii) Late prophase I

chiasma

d(i) Early telophase I

chromosomes at poles of cell

c. Anaphase I

homologous chromosomes passing to opposite poles of cell

b. Metaphase I

spindle

bivalents attached to equator of spindle

d(ii) Late telophase I

nucleolus

centrosome

chromatin granules

nuclear membrane

Compare metaphase I with metaphase of mitosis

1. PROPHASE I

The events of prophase I are much more complex than those which occur in prophase of mitosis. At the start of prophase I the chromosomes appear as long, thin entangled threads inside the nuclear membrane. They then come together in pairs called **bivalents**. Each bivalent is a homologous pair of chromosomes. The homologous chromosomes are positioned so that their centromeres and allelic genes are adjacent. Gradually the chromosomes shorten and thicken. At this stage the two chromatids making up each chromosome can be clearly seen. Cross-links called **chiasmata** frequently develop between the chromosomes and their homologous partners. Bivalents of long chromosomes often display several chiasmata (Fig. 7.9(a)).

Pairing of homologous chromosomes and the formation of chiasmata do not occur during prophase of mitosis. However, two other events which take place in mitotic prophase also occur in prophase I. The nucleoli disappear and a spindle is laid down in the cytoplasm by the centrosome which divided at interphase.

2. METAPHASE I

The nuclear membrane breaks down and the bivalents become attached by their centromeres to the microtubules at the equator of the spindle. One chromosome of each bivalent is directed towards one of the poles of the cell, its homologous partner towards the other pole. It is important to realise that each bivalent is orientated at random with respect to each of the other bivalents (Fig. 7.9(b)).

3. ANAPHASE I

Shortening of the spindle microtubules drags the homologous chromosomes of each bivalent apart, pulling them to opposite poles of the cell. As the chromosomes begin to separate the chiasmata become very evident (Fig. 7.9(c)).

4. TELOPHASE I

The two groups of chromosomes come together at opposite poles. Each group becomes surrounded by a new nuclear membrane. The chromosomes uncoil, the nucleoli reappear and the nuclei take on a granular appearance. Cleavage of the cytoplasm occurs as in mitosis (Fig. 7.9(d)).

The two cells possess half the number of chromosomes as the cell from which they were derived. For this reason meiosis is sometimes called **reduction division**. A short interphase may follow but there is no replication of DNA as happens during interphase of mitosis. Often there is no interphase and the two cells proceed directly to the second division of meiosis.

7.2.2 The second meiotic division

The Roman numeral II is used after each of the stages to distinguish them from the stages of the first meiotic division.

1. PROPHASE II

Chromosomes appear in both of the nuclei formed in the first division of meiosis. The centrosome of each cell divides and moves to opposite poles of the cells laying down the microtubules of the spindle. There is no pairing of chromosomes and chiasmata do not develop as in prophase I (Fig. 7.10(a)).

2. METAPHASE II

The nuclear membranes disappear and the chromosomes become attached by their centromeres to the microtubules at the equators of the spindles. The two chromatids of each chromosome are now easily seen. What is not so obvious is that the chromosomes are orientated at random with respect to one another (Fig. 7.10(b)).

FIG. 7.10 *The second meiotic division*

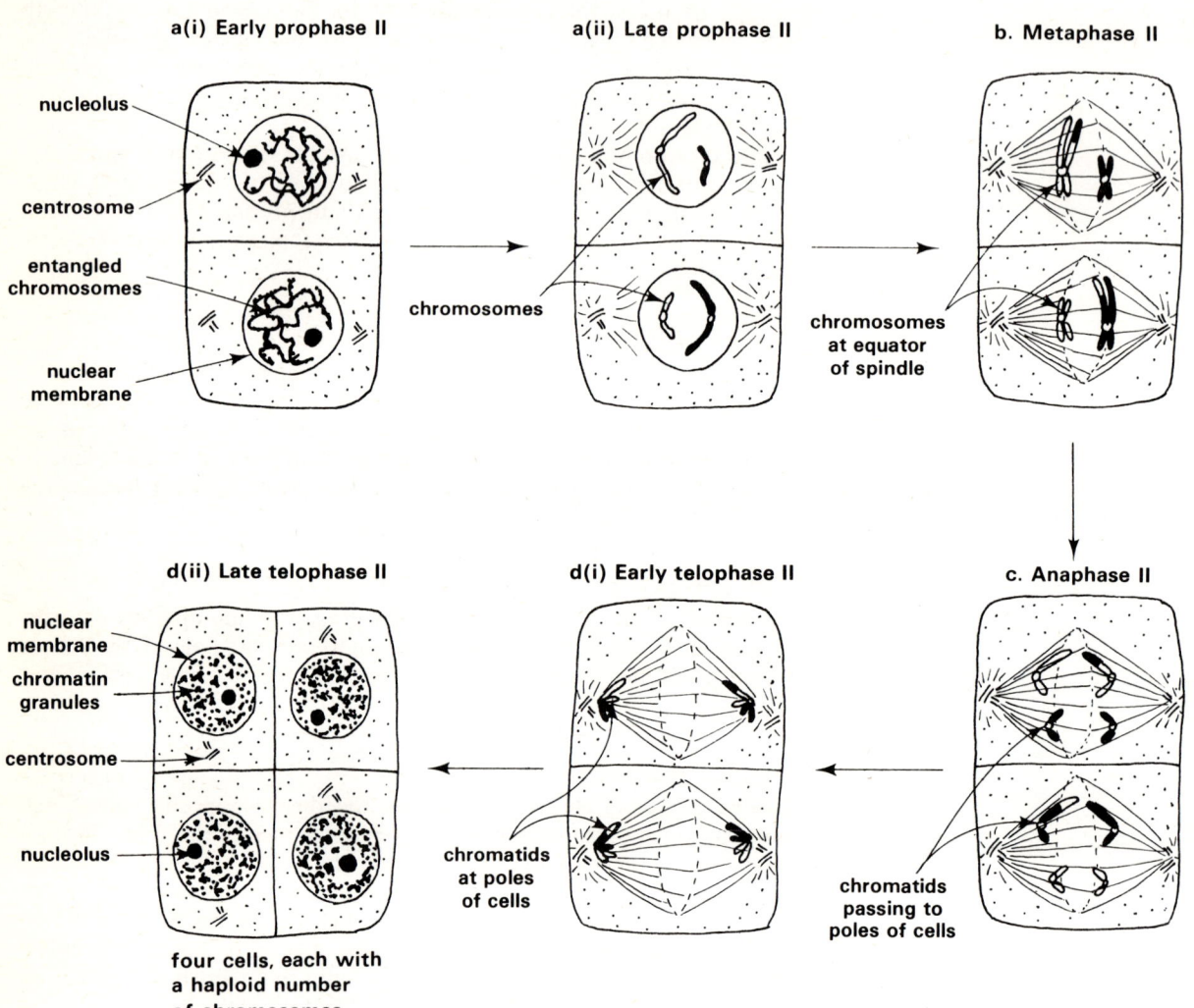

a(i) Early prophase II

nucleolus

centrosome

entangled chromosomes

nuclear membrane

chromosomes

a(ii) Late prophase II

b. Metaphase II

chromosomes at equator of spindle

d(ii) Late telophase II

nuclear membrane

chromatin granules

centrosome

nucleolus

four cells, each with a haploid number of chromosomes

d(i) Early telophase II

chromatids at poles of cells

c. Anaphase II

chromatids passing to poles of cells

3. ANAPHASE II

The centromeres of the chromosomes break in two and the chromatids are pulled, centromere first, towards opposite poles of the cell (Fig. 7.10(c)).

4. TELOPHASE II

The chromatids come together at opposite poles of the cells where they become surrounded by nuclear membranes. The chromatids uncoil and nucleoli appear. The spindle disappears and cleavage of the cytoplasm follows (Fig. 7.10(d)). Altogether four cells, each with half the number of chromosomes, are produced from each cell which divides by meiosis.

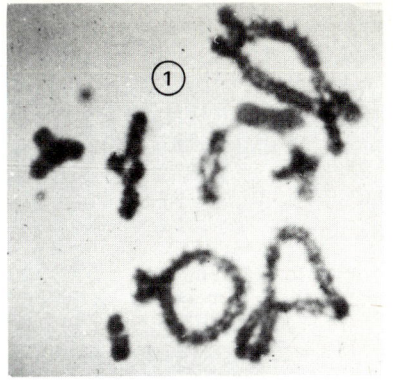

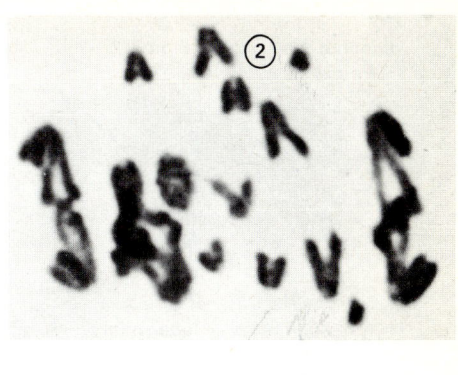

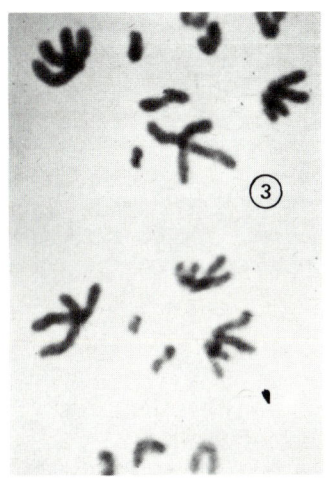

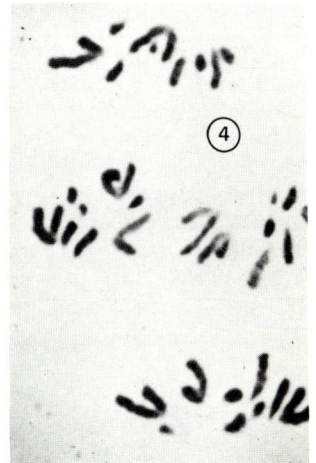

Examine the photographs in Figure 7.11. Which stage of meiosis is shown in each picture?

7.2.3 The significance of meiosis

Whereas mitosis normally produces cells with an exact replica of the genetic material found in the parent cell, meiosis gives rise to cells with half the amount of genetic material. The body cells of higher organisms have a **diploid** number of chromosomes. Gametes made by meiosis have a **haploid** number of chromosomes. At fertilisation the diploid number is restored. In the life cycles of organisms which reproduce sexually, meiosis ensures that the chromosome number is normally kept constant in each generation (Fig. 7.12 and Chapters 19 and 22).

Another significant feature of meiosis is that it produces gametes with varied combinations of genes. There are two important events in meiosis which create new genomes.

1. CROSSING OVER

Crossing over takes place when bivalents appear in prophase I. Chiasmata are formed and homologous chromosomes exchange genes (Fig. 7.13(a)). The homologous chromosomes later separate and end up in different gametes. As a result of crossing over linked genes are parted and gametes with new genomes are produced.

FIG. 7.12 *Life cycles of a mammal and flowering plant (simplified)*

(a) *Mammal*

(b) *Flowering plant*

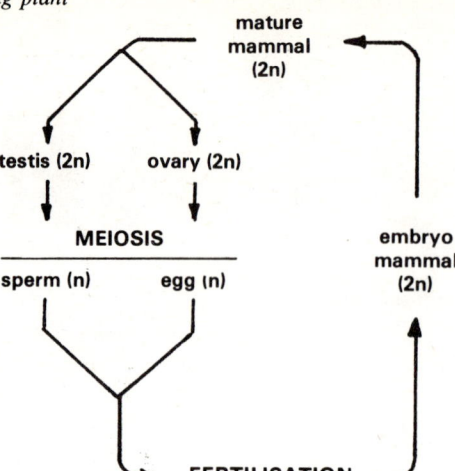

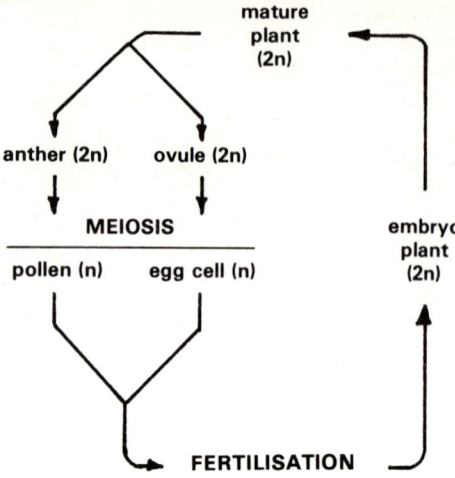

FIG. 7.13 (a) (i) *Crossing over. The photograph* (courtesy Philip Harris Biological Ltd.) *shows the chromosomes of a grasshopper*

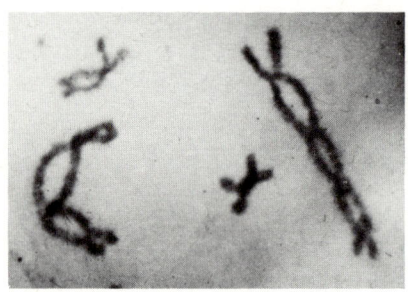

FIG. 7.13 (a) (ii) *Crossing over. For simplicity two pairs of alleles are shown. As a consequence of crossing over, the genes carried on the homologous chromosomes (linked genes) become separated. What combination of genes would result if there was no crossing over?*

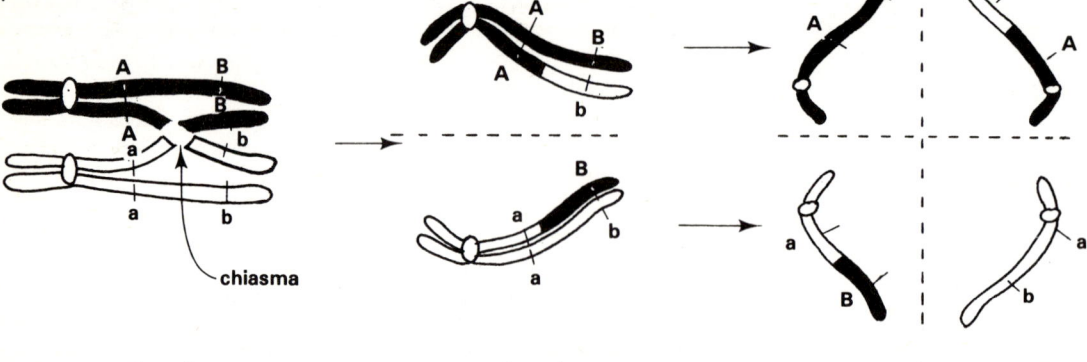

prophase I

anaphase I

anaphase II

FIG. 7.13 (b) *The consequences
of random orientation of
chromosomes. The genes on one
pair of homologous chromosomes
separate independently of those
on the other pair. For clarity,
crossing over has not been shown*

2. RANDOM ORIENTATION OF CHROMOSOMES

The separation of a pair of homologous chromosomes at anaphase I is independent of the separation of other pairs. As the chromosomes are orientated at random the alleles on one pair of homologous chromosomes separate independently of the alleles on others (Fig. 7.13(b)). Because of random assortment a vast permutation of genes is possible in the gametes.

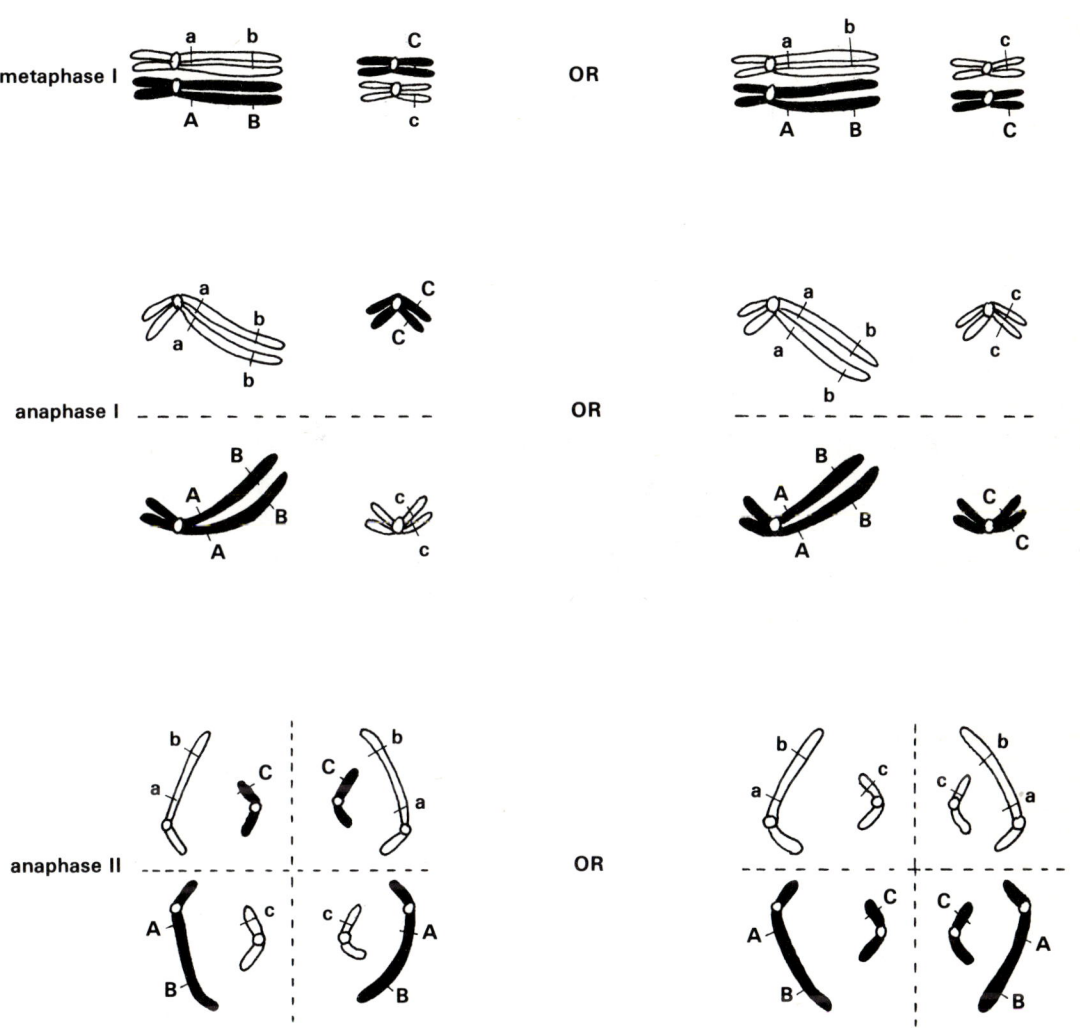

Crossing over and random assortment ensure that progeny resulting from sexual reproduction are genetically different from their parents. This is why no two humans for example, unless they are identical twins, have the same genome. Clearly meiosis is important in producing genetic variation in a species. There are however other ways in which genetic variation can arise (section 7.3).

7.3 Genetic variation

Do not forget that crossing over and random orientation of chromosomes in meiosis are important sources of genetic variation. Mutations are another important source of genetic variation. **Gene mutations** are described in Chapter 6. A second type of mutation which frequently occurs is called a **chromosome mutation**. There are several ways in which chromosome mutations can arise during either mitosis or meiosis.

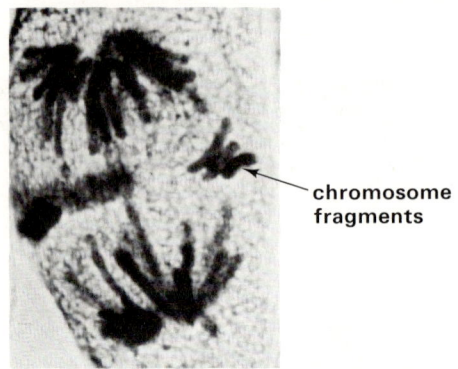

FIG. 7.14 (a) *Effect of radiation on chromosomes* (courtesy Dr. C. R. Davies and Philip Harris Biological Ltd.). *The chromosome fragments do not pass to the poles of the cell as do normal chromosomes*

chromosome fragments

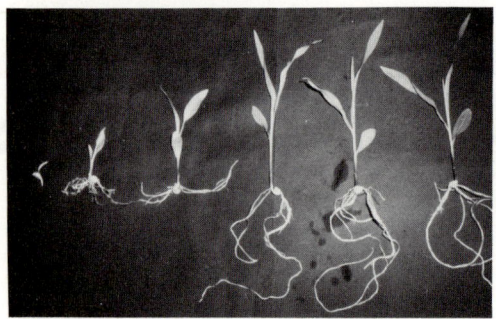

FIG. 7.14 (b) *Effect of γ radiation on the growth of maize seedlings* (courtesy Philip Harris Biological Ltd.). *The three seedlings on the right were grown from untreated seeds. Those on the left were grown from seed exposed to increasing doses of γ radiation*

Fragmentation of chromosomes often occurs when dividing cells are exposed to various forms of radiation such as X-rays (Fig. 7.14). The chromosome fragments pass at random into newly formed cells so producing cells with new genomes.

Failure of chromosomes to separate at anaphase gives rise to other types of chromosome mutation. If only one chromosome behaves in this way half of the new cells end up with an extra chromosome, the other half with one chromosome less than normal. Either condition is called **polysomy**. Down's syndrome (mongolism) in humans is a well-known example of polysomy (Fig. 7.15). Occasionally whole sets of chromosomes fail to separate. Half of the cells then receive twice the normal number of chromosomes, a condition called **polyploidy**. Polyploidy is more common in plants than in animals. Many modern varieties of cultivated plants are polyploids. For example, bread wheat has a diploid number of 42 chromosomes. Its ancestors have diploid numbers of 14 and 28.

Genetic variation is necessary for evolutionary change. If a new combination of genes enables its possessor to cope successfully with the pressures of its environment then the organism will survive. Should the organism reproduce asexually the new genome may be perpetuated unchanged for many generations. Equally well, the possessors of less effective genomes may not be able to compete successfully and will become extinct. This is the concept of **survival of the fittest**.

Charles Darwin was the first to draw attention to variation and survival of the fittest as the basis of evolution in his book *The Origin of Species by Natural Selection*. What has been outlined above is the genetic basis of evolution.

FIG. 7.15 (a) *Chromosomes of a male mongol* (courtesy Philip Harris Biological Ltd.)

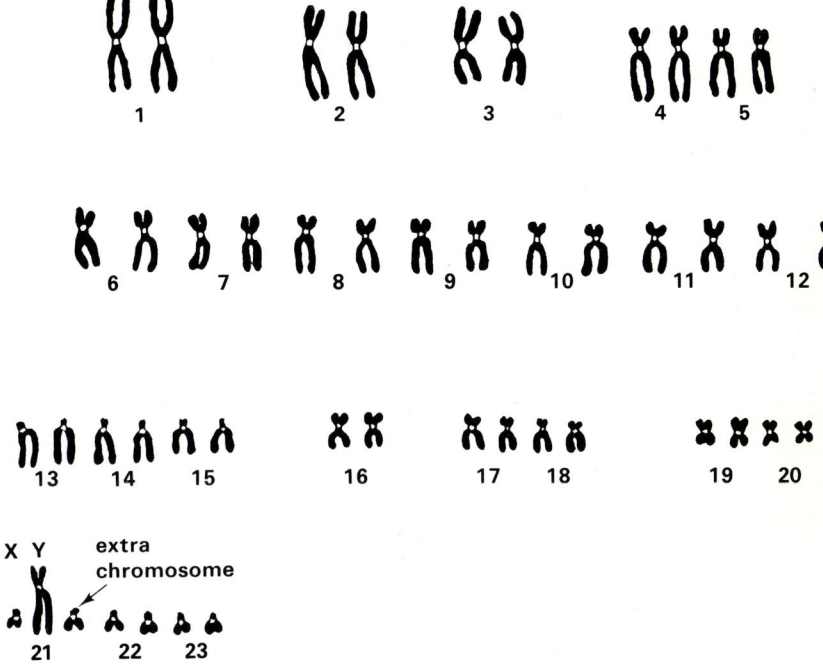

FIG. 7.15 (b) *A mongol child* (courtesy A. Bezear and Heinemann Medical Ltd.)

Chapter 8

Gas exchange in mammals

Mammals respire aerobically so they need a constant supply of oxygen. The role of oxygen in cellular respiration is described in Chapter 5. Oxygen is consumed by mammals at a fairly rapid rate. This happens because respiration provides heat which helps to maintain a stable and relatively high body temperature (Chapter 17).

8.1 Gas exchange and evaporation

Mammals obtain their oxygen from the atmosphere. Oxygen makes up about 21 per cent of the atmosphere's volume, the rest being nearly all nitrogen. Before oxygen gas can diffuse through the outer surface of an organism and into its tissues, oxygen must first dissolve in water. Consequently the surface through which oxygen is absorbed must be moist. The moisture on any surface exposed to the air will usually evaporate. Some 500 cm^3 water are lost daily from the lungs of a man out of a total water loss of about 2500 cm^3. The rest is lost through the skin and in the urine and faeces. Water loss by **evaporation** may thus be a major physiological problem during gas exchange (Table 8.1).

Table 8.1 Loss of water vapour from several mammals

SPECIES	EVAPORATIVE WATER LOSS AS A PERCENTAGE OF TOTAL WATER LOSS
kangaroo rat	68
rat	65
camel	51
man	42
donkey	36

No biological system is perfect and physiological adaptations are usually compromises between advantage and disadvantage. To obtain enough oxygen to meet their metabolic needs, air-breathing animals such as mammals bring the air into direct contact with a relatively large area of moist lung surface. In doing so, water evaporates from the lungs resulting in substantial water loss. The rate of evaporation from a moist surface is affected by environmental factors such as temperature, atmospheric humidity and air movement. The ways in which such factors affect the loss of water vapour from terrestrial plants are described in Chapter 11. Climate similarly affects the rate of water loss from terrestrial animals.

Because mammals can regulate their breathing rate there is some control over water loss from their lungs. There is no such control in many terrestrial invertebrates such as earthworms and lower vertebrates such as amphibians which rely on a moist skin for gas exchange. Evaporative water loss is one of the main factors responsible for restricting such animals to moist habitats. Many mammals however live in deserts where evaporation is very rapid. Desert-living mammals possess adaptations which enable them to succeed where others would dry up and die within hours.

FIG. 8.1 *Kangaroo rat*

One of these mammals is the kangaroo rat, *Dipodomys*, an inhabitant of the deserts of the south-west United States of America (Fig. 8.1). The kangaroo rat has such fine regulation of its water content that although it lives in the desert it does not drink water. Since it has to match its water loss with a limited water intake the control of evaporation from its lungs is critical. Kangaroo rats have an efficient heat-exchange mechanism in their nasal passages which reduces the rate of water vapour loss during breathing. When breathing in, moisture on the inner surfaces of its nasal cavities evaporates, lowering the temperature of its nasal membranes. When breathing out, the moisture in the exhaled air condenses on the cooled nasal linings. Thus oxygen is taken in from the atmosphere with less of the accompanying water loss than occurs in most mammals (Fig. 8.2).

FIG. 8.2 *Moisture retention in the kangaroo rat during breathing*

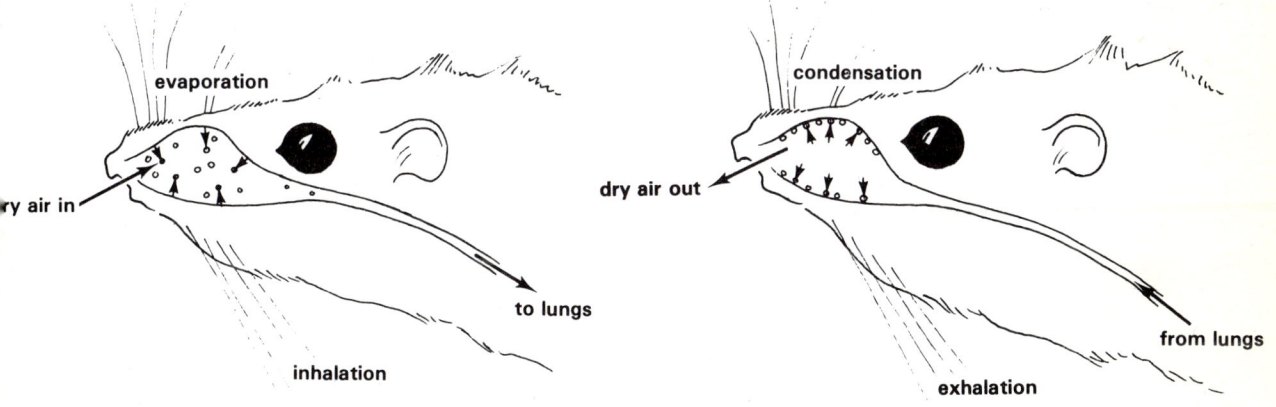

8.2 Breathing

8.2.1 The breathing mechanism

Mammalian **lungs** are compact organs. They are situated in the **thorax** which is separated from the abdominal cavity by a muscular **diaphragm** (Fig. 8.3). The lungs consist of millions of microscopic air sacs called **alveoli**. Each alveolus is surrounded by a network of fine capillary blood vessels (Fig. 8.4). It is across the delicate membranes separating blood from the alveolar air that diffusion of gases occurs. Oxygen is taken in from the alveolar air, carbon dioxide is released into it. Mammalian lungs have a considerable surface area for diffusion. For each gram of body mass there are about 7 cm^2 of lung surface in humans, 13 cm^2 in seals and 100 cm^2 in bats. This is a distinct improvement on the lungs of lower vertebrates which are little more than simple sacs (Fig. 8.5). In these animals the area of lung surface per unit mass of body is much less than in mammals.

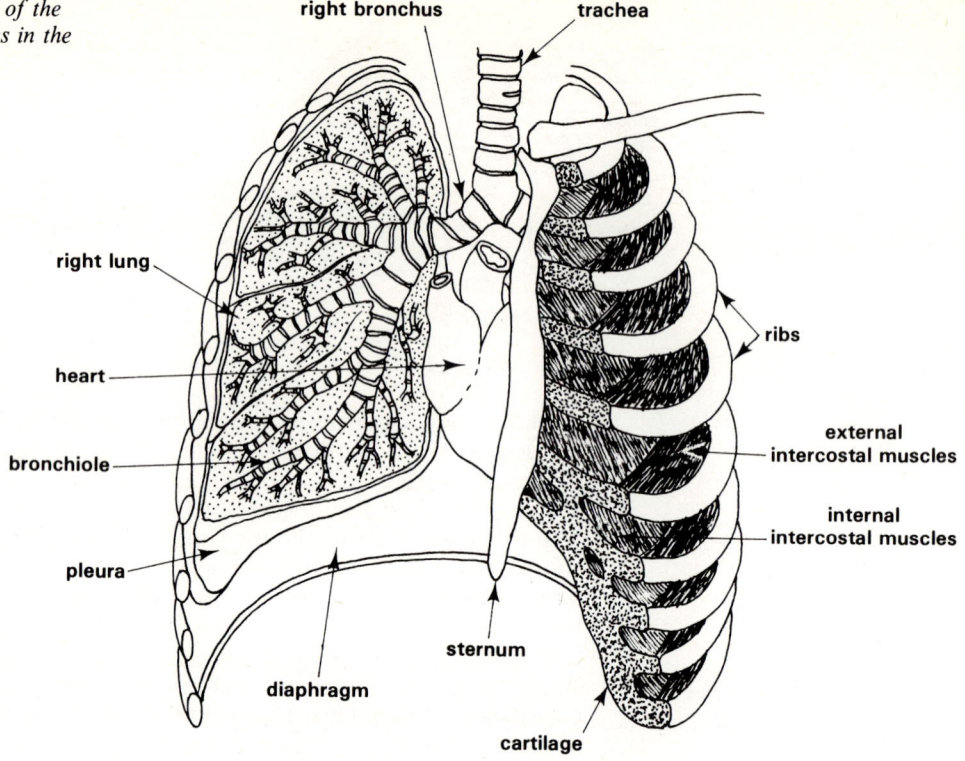

FIG. 8.3 *The position of the lungs and air passages in the human thorax*

right bronchus

trachea

right lung

heart

bronchiole

pleura

ribs

external intercostal muscles

internal intercostal muscles

diaphragm

sternum

cartilage

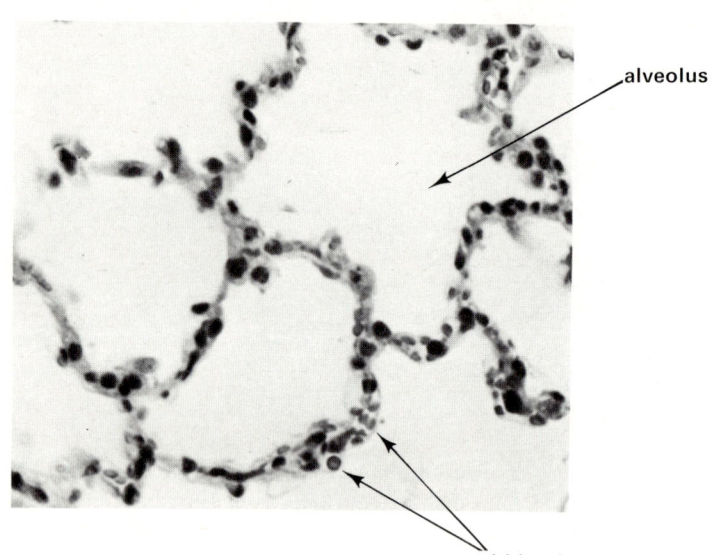

FIG. 8.4 *Photomicrograph of a thin section of cat lung,* × 200

alveolus

red blood cells in pulmonary capillaries

Breathing is a complex muscle-controlled activity in which the volume of the thorax is rhythmically increased and decreased. Air is sucked into and pushed out of the lungs when a mammal breathes. Breathing is brought about by movements of the ribs and diaphragm. Two sets of muscles, the **external and internal intercostal muscles** which work antagonistically, move the ribs. When the external intercostals contract, the internal intercostals relax and the ribs are forced obliquely out and up. Because they adhere to the inner wall of the thorax the lungs are stretched. At the same time the diaphragm contracts and pulls the lungs downwards towards the abdomen. These movements increase the volume of the lungs, thereby lowering the pressure of the air in them. As a result air is drawn into the lungs through the **trachea**, **bronchi** and **bronchioles**. Relaxation of the diaphragm and external intercostals together with contraction of the internal intercostals returns the thorax to its previous state. The pressure now applied to the lungs forces air from the lungs to the outside through the air passages (Fig. 8.6).

FIG. 8.5 *Longitudinal sections through the lungs of an amphibian and a reptile*

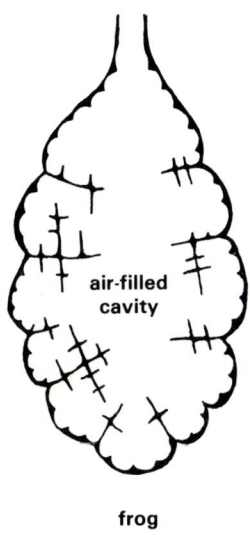

air-filled cavity

frog

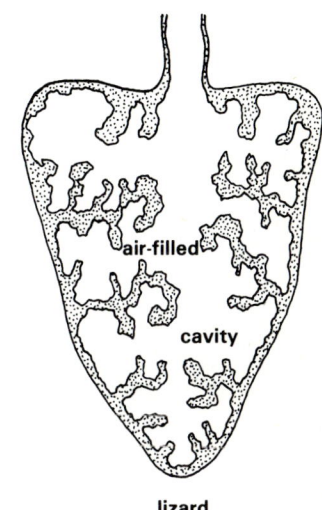

air-filled cavity

lizard

FIG. 8.6 *The process of breathing*

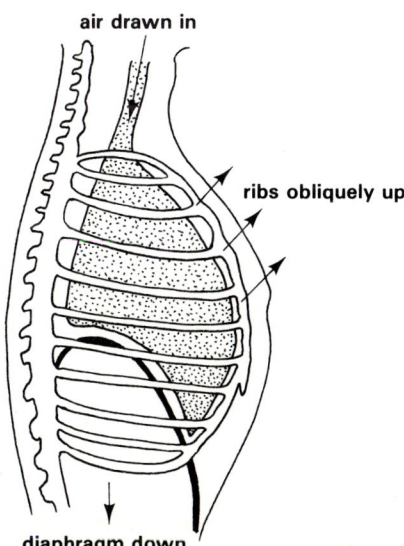

air drawn in

ribs obliquely up

diaphragm down

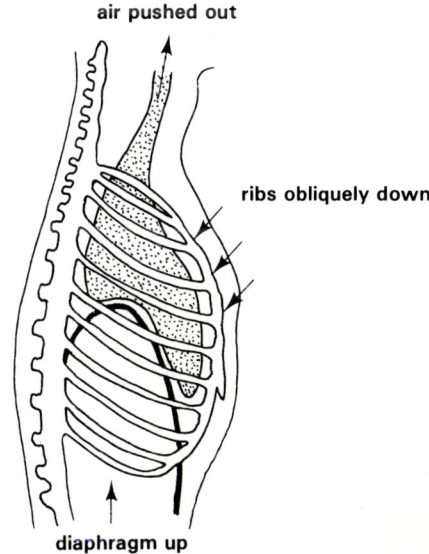

air pushed out

ribs obliquely down

diaphragm up

135

During each ventilation only about 10 per cent of the lungs' total air capacity is changed in most resting mammals. The volume of air exchanged with the atmosphere at each breathing cycle is called the **tidal volume**. A greater volume of air can be exchanged by forced breathing. The volume of air breathed out after breathing in to the fullest extent is called the **vital capacity**. About 70 per cent of the air breathed in enters the alveoli. The rest occupies **dead space** mainly in the bronchi and bronchioles. The total capacity of the lungs of an adult human is about 6 dm³. Consequently during each ventilation of the lungs, only a small proportion of the alveolar air is replaced.

FIG. 8.7 *A spirometer being used to measure lung volumes* (courtesy Philip Harris Biological Ltd.)

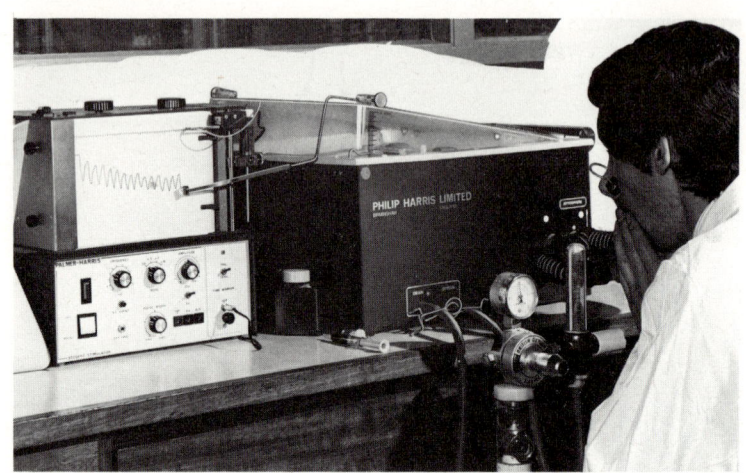

FIG. 8.8 *A spirometer tracing typical of a normal human adult. The inspiratory reserve volume is the maximum volume of air which can be inhaled during forced breathing. Similarly the expiratory reserve volume is the maximum volume of air exhaled. The residual volume is the minimum volume of air always present in the lungs*

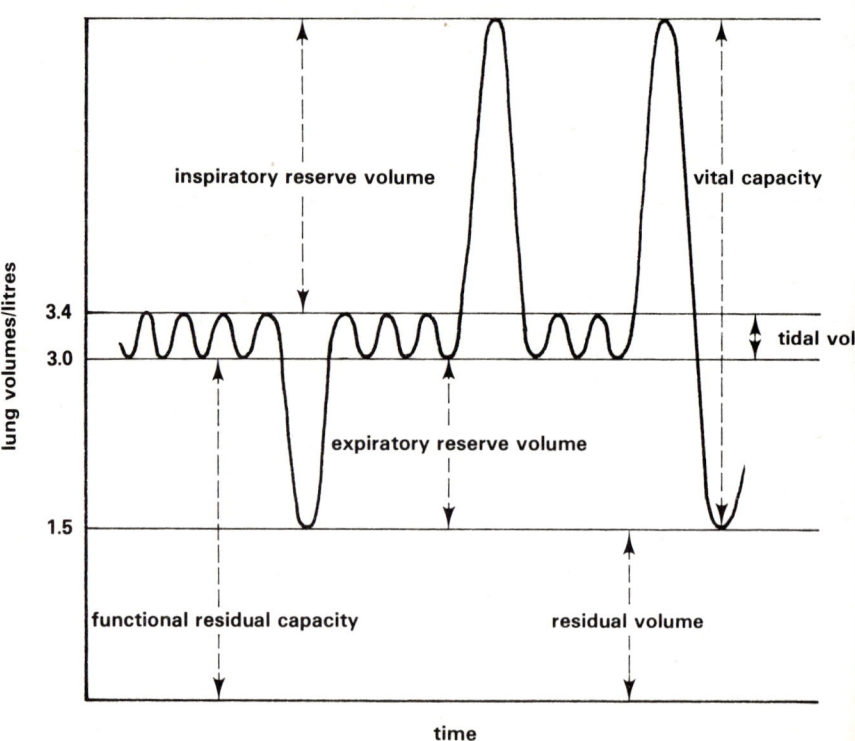

136

Tidal volume, vital capacity and other lung capacities can be measured using a **spirometer** (Fig. 8.7). A spirometer is an air-filled chamber which rises and falls when a subject breathes into an attached mouthpiece. Movement of the chamber is recorded as a trace on a moving paper (Fig. 8.8). Spirometers are used in hospitals to measure lung capacities of patients suffering from respiratory ailments.

The tidal ventilation of lungs is very different from the continuous flow ventilation of the gills of many aquatic animals. In fish for example, oxygenated water is continuously moved over the surface of the gills and carbon dioxide is taken away in the same movement (Fig. 8.9).

FIG. 8.9 *A comparison of continuous flow and tidal ventilation*

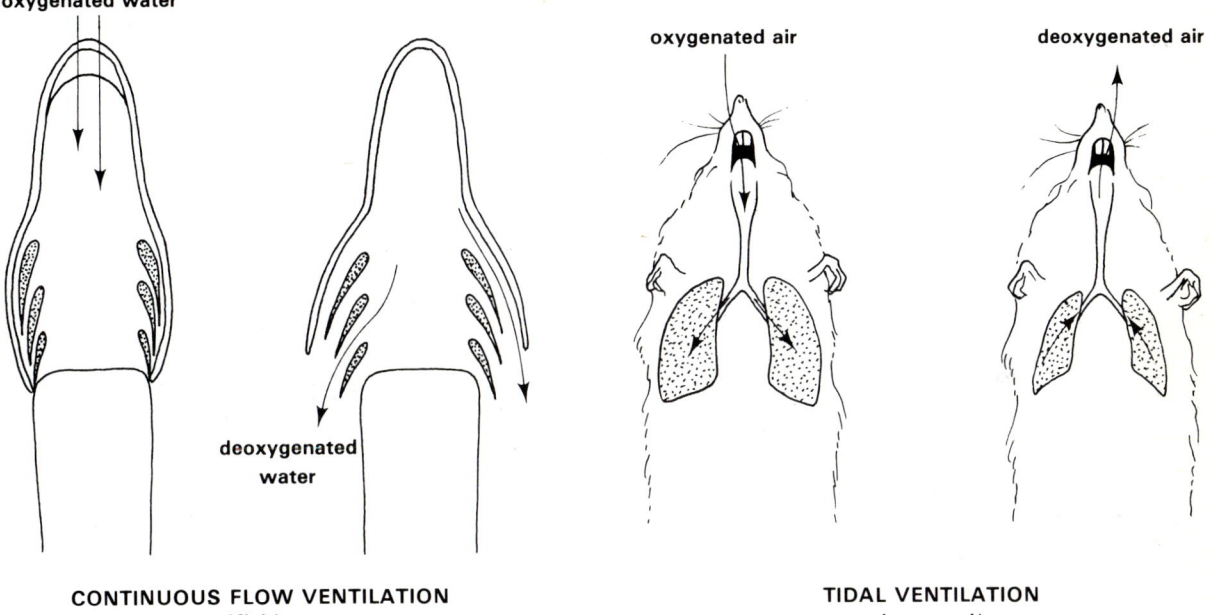

oxygenated water

deoxygenated water

oxygenated air

deoxygenated air

CONTINUOUS FLOW VENTILATION
(fish)

TIDAL VENTILATION
(mammal)

8.2.2 Control of breathing

The oxygen demand of an organism depends on the metabolic rate of its tissues. During periods of increased activity a mammal ventilates its lungs rapidly, so meeting the increased oxygen demand. Faster breathing also eliminates the extra carbon dioxide produced by increased respiration. Conversely the breathing rate slows down when the demand for oxygen drops during periods of rest. For example, a young woman resting on a bicycle may take in about 300 cm³ oxygen per minute. When pedalling rapidly, oxygen consumption may rise to about 1500 cm³ per minute. The fivefold increase in oxygen uptake is brought about by nearly trebling the breathing rate and nearly doubling the tidal volume. In other words she breathes faster and deeper. Changes in the rate and depth of breathing in response to exercise and rest are also accompanied by changes in the heart rate (Chapter 9).

The breathing rate is controlled by the **respiratory centre**, part of the medulla oblongata of the hindbrain (Chapter 15). Impulses from the respiratory centre travel along motor nerves to the intercostal and diaphragm muscles. The muscles respond by bringing about the thoracic movements necessary for lung ventilation. Whereas breathing is largely automatic, the responses of the intercostal and diaphragm muscles can be overridden by impulses from the higher centres of the brain. The pattern of breathing can thus be modified voluntarily. Activity of the respiratory centre is modified by several factors which indicate the need for a change in the breathing rate (Fig. 8.10).

FIG. 8.10 *Factors affecting the respiratory centre in the brain*

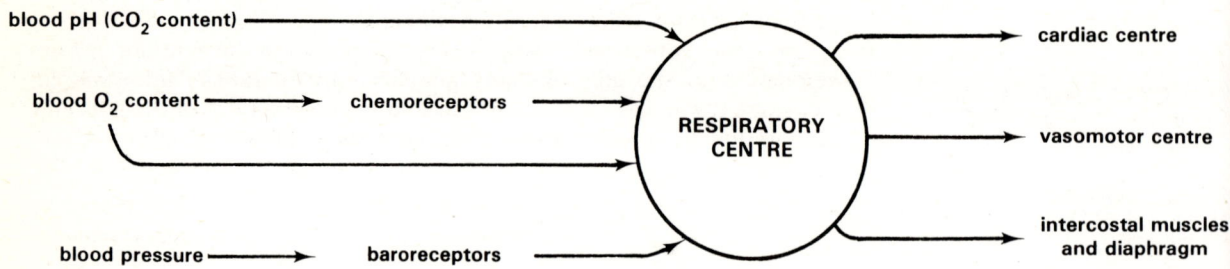

The respiratory centre is sensitive to the acidity of the blood flowing through it. Since the pH of blood is largely governed by the amount of carbon dioxide in it, increased breathing rate is an appropriate response as it leads to elimination of carbon dioxide from the blood. The carbon dioxide concentration of the blood is thought to be the main controlling factor in the automatic regulation of breathing.

The respiratory centre also responds to the oxygen tension of the blood. **Chemoreceptors**, cells sensitive to oxygen tension, are situated in small sinuses found in the aorta and carotid arteries. When the blood oxygen tension falls, the chemoreceptors stimulate the respiratory centre with impulses sent along sensory nerves. The respiratory centre then triggers off responses which bring about increased ventilation of the lungs.

FIG. 8.11 *Position of the baroreceptors (stippled) in the cat*

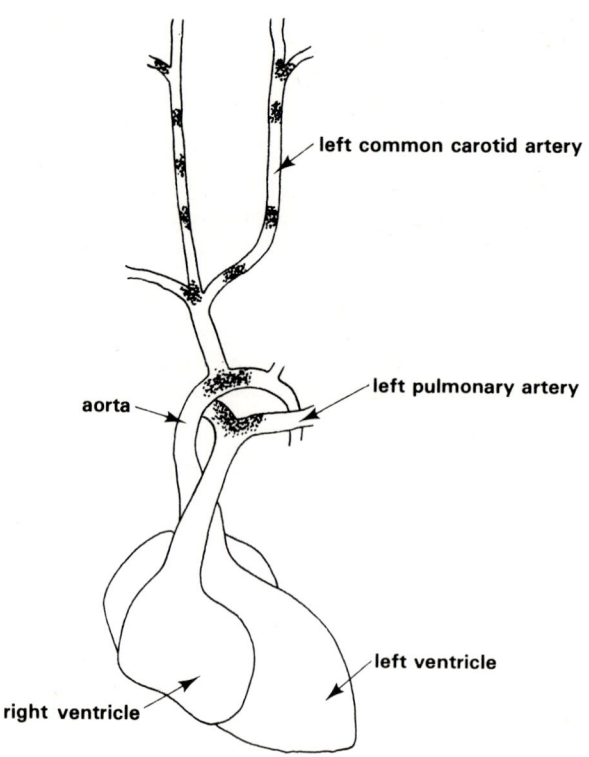

Cells sensitive to blood pressure, **baroreceptors**, are also found in the walls of the aorta, carotid arteries and pulmonary arteries (Fig. 8.11). Stimulation of the baroreceptors by a rise of blood pressure is monitored by the respiratory centre which reacts by triggering off responses which result in a decreased rate of lung ventilation.

Ventilation of the lungs in mammals is thus controlled by the net effect of the concentrations of carbon dioxide and oxygen in the blood, as well as blood pressure on the respiratory centre. In this way breathing and gas exchange are adjusted to meet the body's immediate needs. As with other homeostatic mechanisms the regulation of breathing involves **feedback** controls. The rate at which oxygen is taken in is regulated by factors which reflect the need for oxygen (Fig. 8.12).

FIG. 8.12 (a) *The main components of an automatic control system*

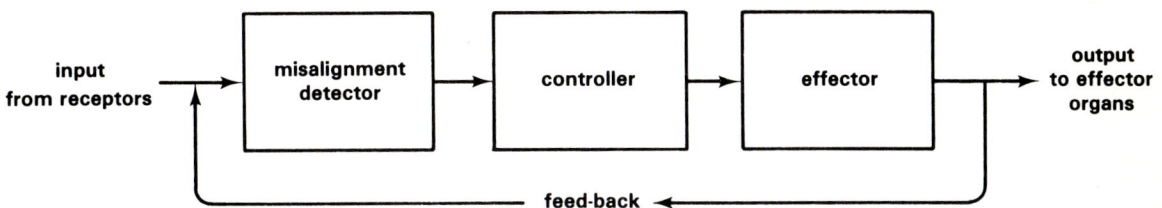

FIG. 8.12 (b) *The respiratory control system*

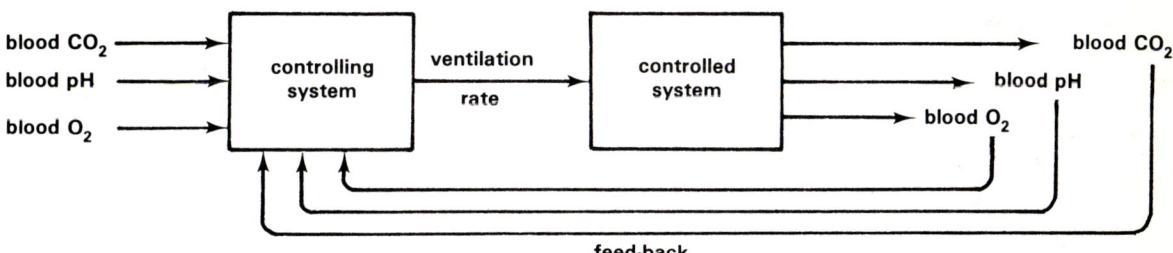

8.2.3 Oxygen debt

For short periods at a time the mammalian body can engage in severe exercise which requires more oxygen than can be provided by ventilating the lungs, whatever the rate of breathing. During intense exercise the reserves of free energy such as ATP and phosphocreatine are depleted. Anaerobic respiration occurs in the cells and lactic acid is produced (Chapter 5). During short periods of intense exercise an **oxygen debt** is built up. The debt is repaid by continued rapid and deep breathing when the period of exercise ends. The extra oxygen then absorbed by the blood corresponds to the oxygen debt and is used to oxidise the lactic acid produced in anaerobic respiration.

8.3 Oxygen uptake and carriage

One of the main functions of the lungs is to allow atmospheric oxygen to come into close proximity with the blood which can then absorb oxygen and transport it to the body organs and tissues.

8.3.1 Oxygenation of the blood

The red blood cells (**erythrocytes**) make up about 45 per cent of the total blood volume in man. Erythrocytes contain a respiratory pigment called **haemoglobin**. The main component of blood is plasma, a mixture of organic and inorganic materials dissolved in water. Oxygen is not very soluble in water. At 35°C, 100 cm³ of water contains about 0·5 cm³ oxygen. The blood of some mammals, such as the seal, can carry as much as 29 cm³ oxygen per 100 cm³ of blood. The ability of the blood to carry so much oxygen is due to haemoglobin (Table 8.2).

Table 8.2 Oxygen-carrying capacity of the blood of several mammals

SPECIES	cm³ oxygen per 100 cm³ blood
cat	15·0
rabbit	15·6
sheep	15·9
horse	16·7
kangaroo rat	17·5
rat	18·6
sea lion	19·8
man	20·0
porpoise	20·7
fox	21·7
llama	23·4
seal	29·3

FIG. 8.13 *Haemoglobin structure. Each molecule consists of four sub-units. Each sub-unit contains an iron-centred haem group attached to the protein globin. The total relative molar mass is about 68 000*

Haemoglobin is a complex molecule made of four almost identical sub-units. Each sub-unit consists of a molecule of the protein **globin** attached to a central porphyrin group called **haem** which contains an iron(II) atom (Fig. 8.13).

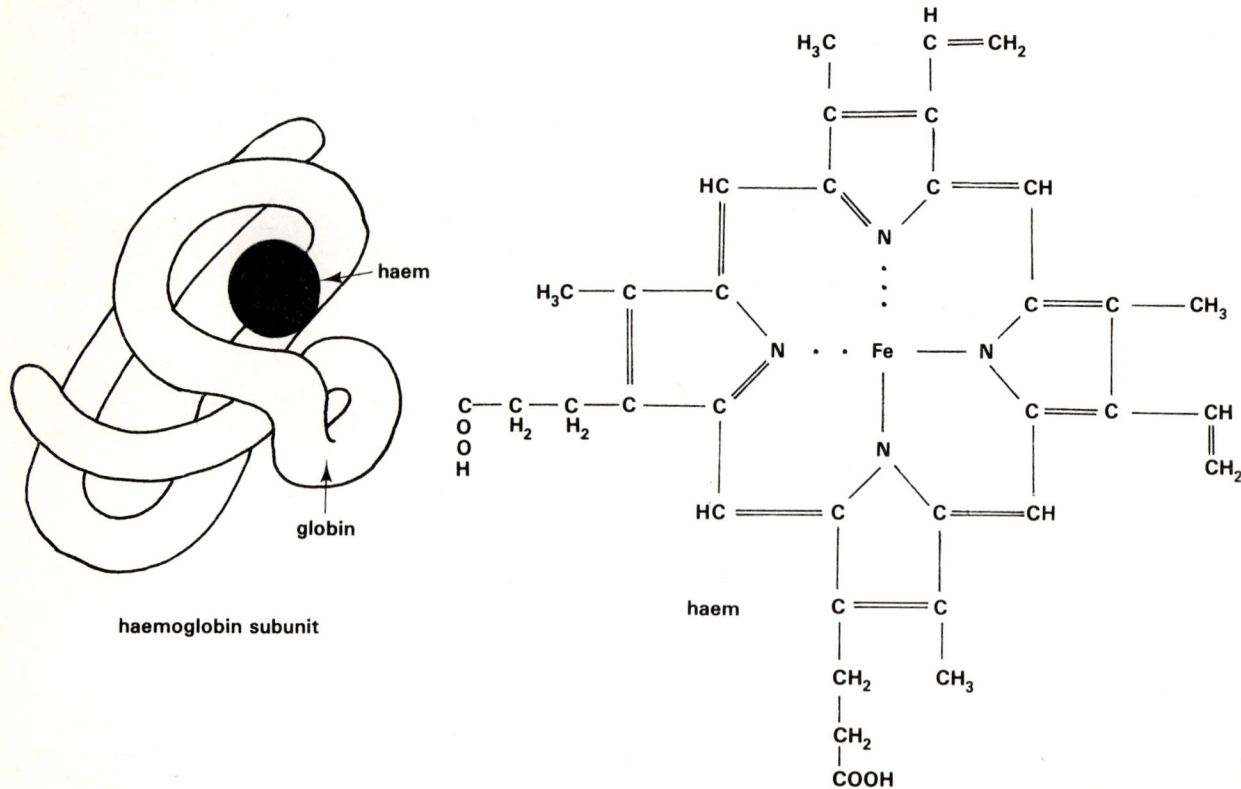

haem

globin

haemoglobin subunit

haem

FIG. 8.14 *Oxygenation of haemoglobin*

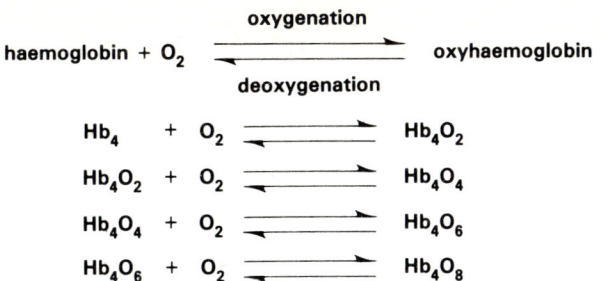

In the capillaries of the lungs oxygen combines loosely with haemoglobin to form **oxyhaemoglobin**. The iron of the haem groups is not oxidised and remains as iron(II) (Fig. 8.14). Certain chemicals can oxidise the iron(II) to iron(III) producing derivatives of haemoglobin which cannot carry oxygen. Carbon monoxide poisoning, for example, is caused by the formation of carbon monoxide haemoglobin containing iron(III). Traces of carbon monoxide haemoglobin occur in the blood of non-smokers whereas up to 10 per cent is present in the blood of smokers. Oxygenation of haemoglobin takes place in the blood capillaries surrounding the alveoli. Here the haemoglobin becomes saturated with oxygen. The relationship between oxygen tension and haemoglobin saturation is usually shown as an **oxygen dissociation curve** (Fig. 8.15).

FIG. 8.15 *Oxygen dissociation curve of human haemoglobin*

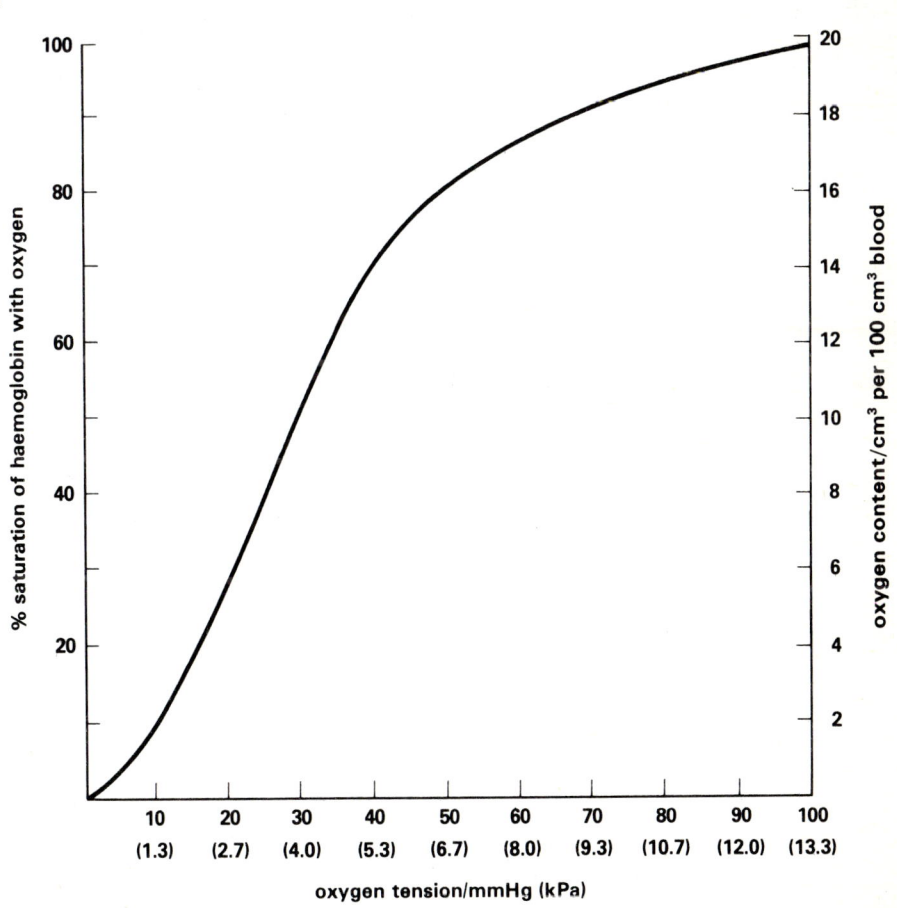

In air with a pressure of about 100 kPa (1 atm) the partial pressure exerted by the 21 per cent oxygen present is about 21 kPa. In the lungs, air breathed in is diluted with air in the dead space. Consequently the oxygen tension in the alveoli is lowered to about 13·3 kPa. Even so, blood circulating in the lung capillaries normally becomes fully saturated with oxygen from the alveoli.

8.3.2 Gas exchange in the tissues

Carbon dioxide is produced in body organs as a result of tissue respiration. The carbon dioxide diffuses through the walls of blood capillaries into the blood. In the red cells an enzyme called **carbonic anhydrase** greatly accelerates the combination of carbon dioxide with water to form carbonic acid:

$$\underset{\text{carbon dioxide}}{CO_2} + \underset{\text{water}}{H_2O} \overset{\text{carbonic anhydrase}}{\rightleftharpoons} \underset{\text{carbonic acid}}{H_2CO_3}$$

The carbonic acid partially dissociates into hydrogen (H^+) and hydrogencarbonate (HCO_3^-) ions:

$$\underset{\text{carbonic acid}}{H_2CO_3} \rightleftharpoons \underset{\text{hydrogen ion}}{H^+} + \underset{\text{hydrogencarbonate ion}}{HCO_3^-}$$

The HCO_3^- ions readily diffuse out of the red cells into the plasma. It is as HCO_3^- ions that most of the carbon dioxide is carried to the lungs (Table 8.3). In the lungs the reactions described above are reversed and the carbon dioxide gas so formed diffuses into the alveoli to be breathed out.

Table 8.3 Transport of carbon dioxide in the blood

	PERCENTAGE OF TOTAL CARBON DIOXIDE IN BLOOD
CO_2 in solution	7
carbon dioxide haemoglobin	7
hydrogencarbonate	86

FIG. 8.16 *Summary of red cell chemistry related to the carriage of respiratory gases*

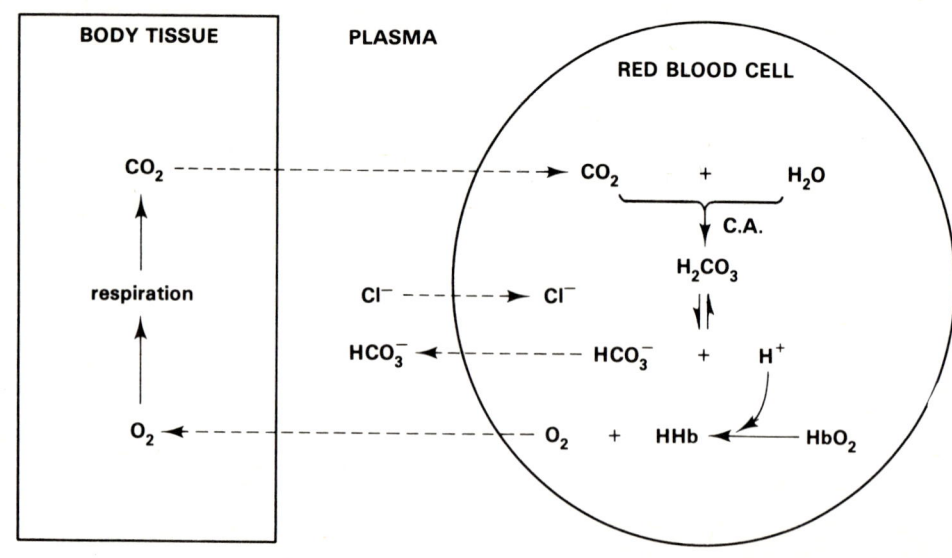

In response to the rapid loss of HCO_3^- ions from the red cells, an equal number of chloride ions (Cl^-) diffuse in the opposite direction. The **chloride shift** prevents electrochemical imbalances, particularly pH changes, which could adversely affect the red cells.

The H^+ ions which stay in the red cells cause the release of oxygen from oxyhaemoglobin:

$$H^+ + \underset{\text{oxyhaemoglobin}}{HbO_2} \rightleftharpoons \underset{\text{haemoglobinic acid}}{HHb} + \underset{\text{oxygen}}{O_2}$$

Gas exchange in the tissues takes place very rapidly. The reactions involved are summarised in Figure 8.16.

The main factor controlling the release of oxygen to the tissues is the volume of carbon dioxide produced by tissue respiration. As the carbon dioxide tension of the tissues and capillary blood rises the affinity of oxyhaemoglobin for oxygen is lowered. The phenomenon is called the **Bohr effect**. The oxygen dissociation curve moves to the right (Fig. 8.17).

Thus increased carbon dioxide production by respiring tissues causes the blood to release more oxygen. For example, assume that in a tissue somewhere in the body the oxygen tension is $4 \cdot 0$ kPa. If the carbon dioxide tension increases from $5 \cdot 3$ kPa to $9 \cdot 3$ kPa then, from the oxygen dissociation curves, it can be seen that the saturation of haemoglobin is lowered from 55 to 35 per cent. As a direct result of such an increase in

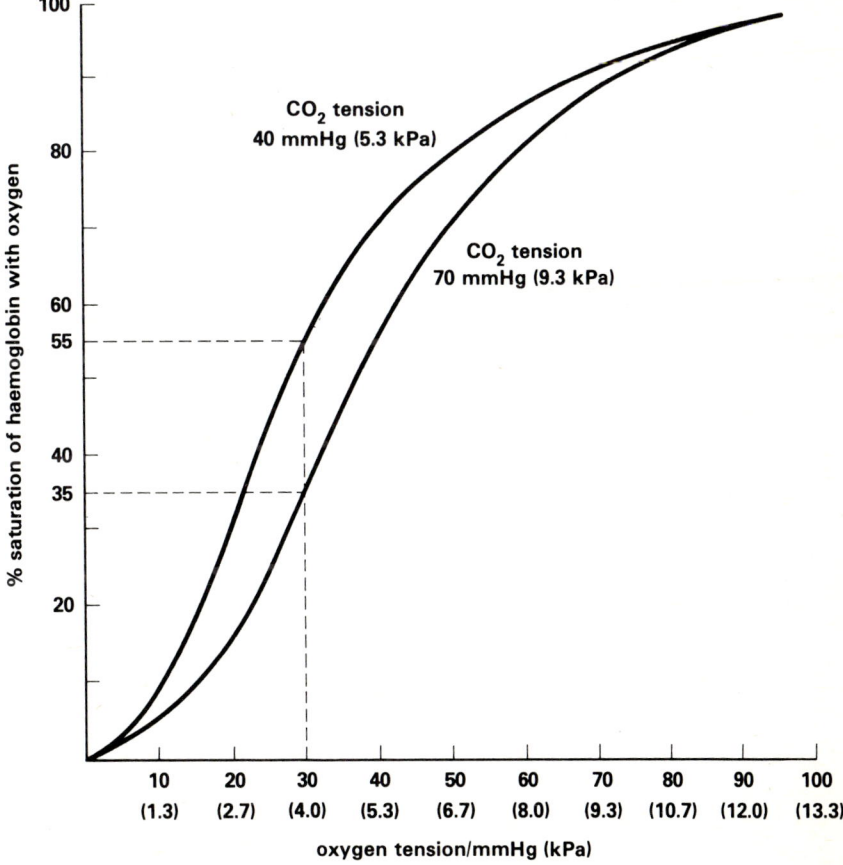

FIG. 8.17 *The effect of carbon dioxide tension on the oxygen dissociation curve of haemoglobin. The carbon dioxide tension in human arteries is about 40 mmHg (5·3 kPa). Tissue respiration increases this to about 46 mmHg (6·1 kPa). The oxygen tension of arterial blood is about 100 mmHg (13·3 kPa). Tissue respiration lowers this to about 40 mmHg (5·3 kPa).*

143

carbon dioxide production some 20 per cent of the haemoglobin releases its oxygen. The more carbon dioxide a tissue produces the greater is this effect. Consequently the tissues and organs such as muscles and the liver which have a relatively high respiration rate cause a rapid release of oxygen from the blood supplying them. Conversely those tissues with a low oxygen demand cause oxygen to be released less rapidly from the blood.

The carriage and exchange of respiratory gases also contributes to the regulation of the pH of blood and the body tissues (Chapter 2).

8.3.3 Adaptations to high altitude

As altitude is gained atmospheric pressure drops. The drop in atmospheric pressure is significant for organisms living at high altitudes because the partial pressure of oxygen in the atmosphere at high altitudes is less (Table 8.4).

Table 8.4 Approximate partial pressures of oxygen and nitrogen in the atmosphere at sea level and at 4848 metres

	PERCENTAGE	PARTIAL PRESSURE AT SEA LEVEL/kPa	PARTIAL PRESSURE AT 4848 METRES/kPa
oxygen	21	21	11
nitrogen	79	79	42
barometric pressure (approx)		100	53

The amount of haemoglobin and the number of red cells in the blood of mammals increase when they live at high altitudes (Table 8.5). This is a direct response to the smaller partial pressure of oxygen in the air. Red cell production occurs in the red bone marrow and is controlled by a hormone

FIG. 8.18 *Hormone control of red cell production (erythropoiesis). Erythropoiesis is a continuous process and red cells have a limited life in the circulation, about 4 months in man*

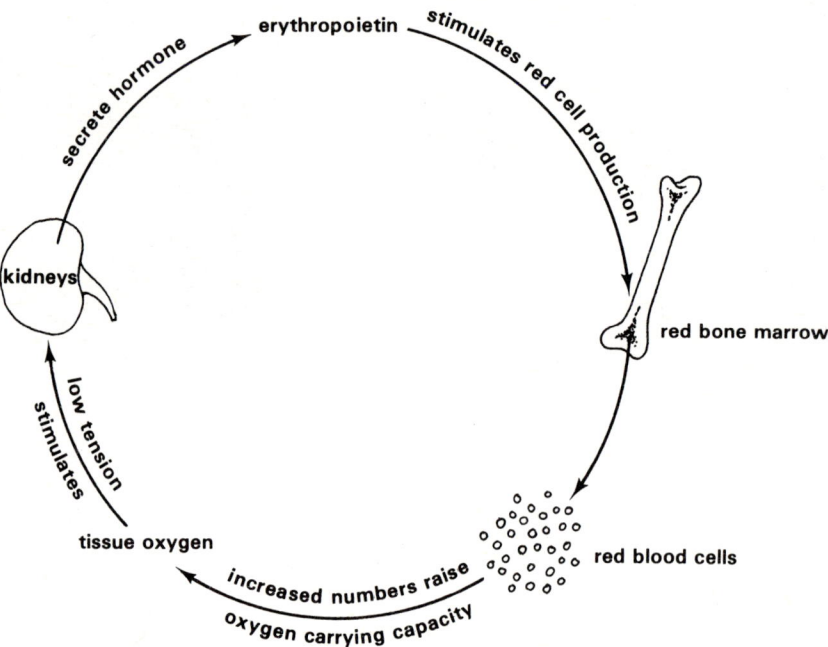

called **erythropoietin** which is made in the kidneys. Secretion of erythropoietin is stimulated by low tensions of oxygen (Fig. 8.18). However, when a mammal moves from low to high altitude time is needed for an increase of haemoglobin and red cells. The problem was highlighted at the 1968 Olympic Games in Mexico City, 2460 metres above sea level. Athletes

Table 8.5 Numbers of red cells in the blood of several mammals at sea level and at high altitude

SPECIES	ALTITUDE	AVERAGE RED CELL NUMBERS $\times 10^{12}$ per dm^3
man	sea level	5·00
	5333 m	7·37 (residents)
		5·95 (transients)
sheep	sea level	10·50
	4673 m	12·05
rabbit	sea level	4·55
	5303 m	7·00
llama	sea level	11·40
	2800 m	12·30

FIG. 8.19 (a) *Oxygen dissociation curves of llama and horse haemoglobins*

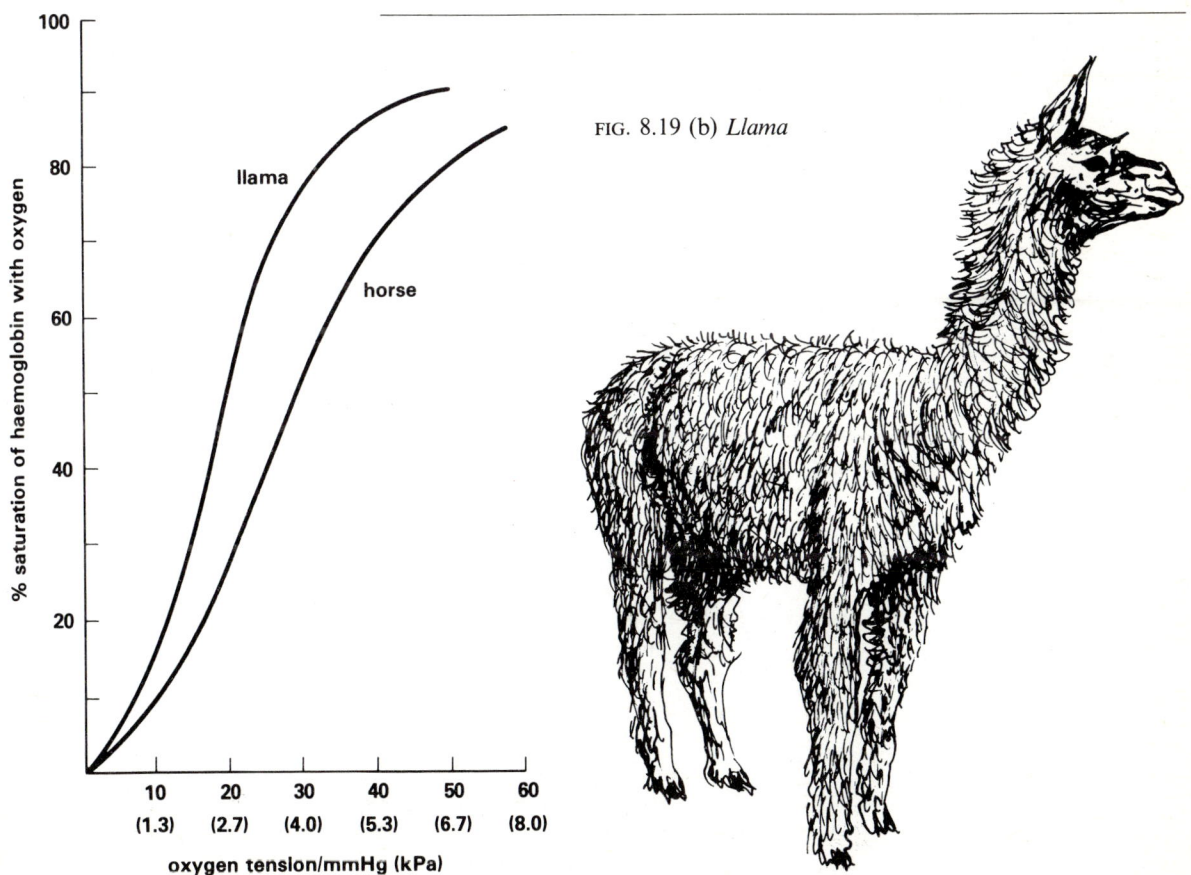

FIG. 8.19 (b) *Llama*

145

who were natives of high altitude countries or who had spent several months acclimatising to high altitude prior to the Games performed quite a lot better than those who had not.

Another adaptation which helps to overcome low oxygen tensions is seen in wild mammals living at high altitude. These animals possess haemoglobin which loads more readily with oxygen in the lungs. Haemoglobin of this sort has a dissociation curve to the left of normal haemoglobin (Fig. 8.19).

8.3.4 Pregnancy and foetal haemoglobin

During pregnancy several physiological changes take place inside the mother's body concerned with the provision of oxygen for the developing embryo. The changes occur partly as a result of the increased output of certain hormones during the gestation period (Chapter 22). For instance, the developing foetus in the uterus progressively displaces the mother's abdominal organs. The displaced organs push against the mother's diaphragm, as well as elsewhere, thus lowering the ventilating capacity of her lungs. Nevertheless, her rate of oxygen consumption increases by about 20 per cent as a result of widening of the thorax and an average 40 per cent increase in breathing rate.

Transport of the extra oxygen, partly to the foetus, is helped by an increase of about 1 dm^3 in the mother's total blood volume. The extra blood is circulated more rapidly than normal by increased cardiac output (Chapter 9) which rises from an average of 4·9 dm^3 per minute to about 6 dm^3 per minute. As it develops in the uterus, the embryo is attached to its

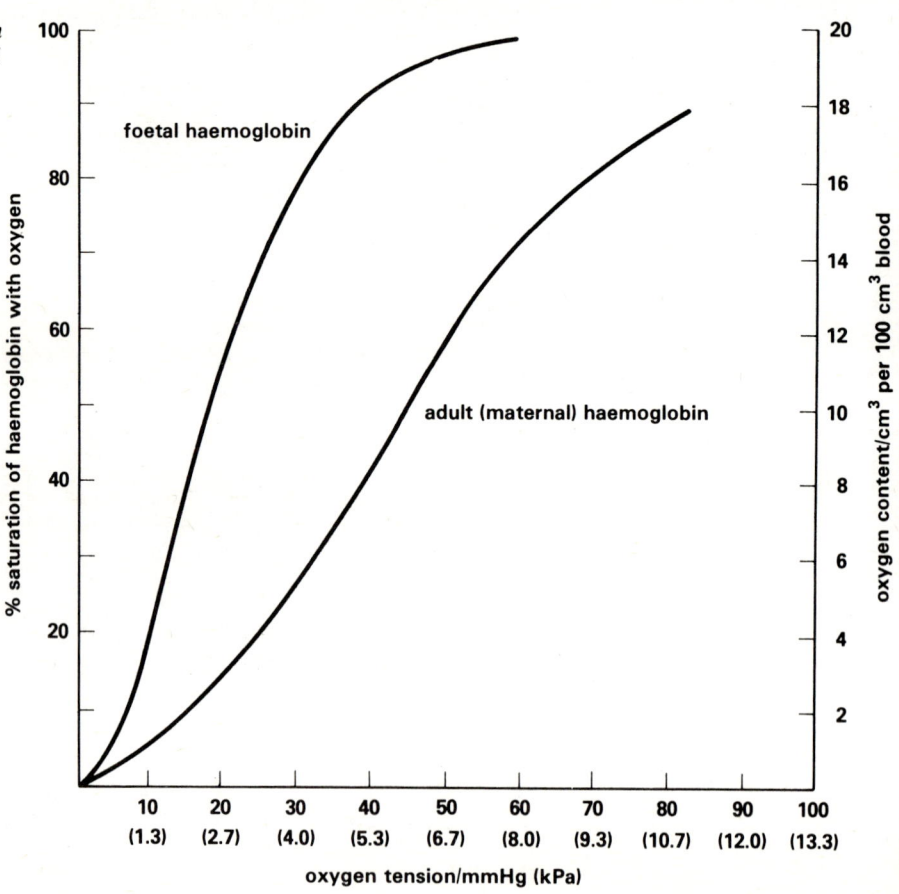

FIG. 8.20 *Oxygen dissociation curves of the adult and foetal haemoglobins of sheep*

mother by a placenta (Chapter 22). Inside the placenta the maternal blood vessels lie very close to the foetal blood vessels. A two-way exchange of nutrients, wastes and respiratory gases occurs across the membranes separating the two circulations. Because of its rapid growth and development a mammalian foetus has a very high oxygen demand. Unloading of oxygen from the mother's blood to the foetal circulation is brought about because the foetal red cells contain a variant of haemoglobin which has a greater affinity for oxygen than the mother's haemoglobin (Fig. 8.20). This important functional difference between foetal and adult haemoglobin results from very small differences in the amino acid sequences of their globin proteins. **Foetal haemoglobin** is replaced by adult haemoglobin at birth. Can you think what advantage results from the replacement?

Not all changes in the haemoglobin molecule are advantageous. A serious disease of man called **sickle cell anaemia** results from the production of a haemoglobin variant differing from normal haemoglobin by only a single amino acid out of several hundred in the whole molecule (Chapter 6). When deoxygenated, the variant, called **haemoglobin S**, is a 100 times less soluble than the normal adult haemoglobin. At low oxygen tensions, haemoglobin S crystallises in the red cells distorting them into a sickle shape (Fig. 8.21). Many sickle cells are destroyed in the circulation causing an anaemia, which considerably lowers the oxygen-carrying capacity of the blood. Some sickle cells block the capillaries in vital organs and the disease may thus be fatal.

The incidence of haemoglobin S in certain parts of Africa and Asia is very high compared with other areas. People who inherit the gene for haemoglobin S from only one parent display the heterozygous condition called **sickle cell trait**. Sufferers of sickle cell trait are resistant to some forms of malaria and are therefore more likely to survive to adulthood. The resistance explains the high frequency of haemoglobin S among natives of regions where malaria is prevalent.

FIG. 8.21 *Photomicrograph of a smear of human blood showing sickle cells, (× 750)*

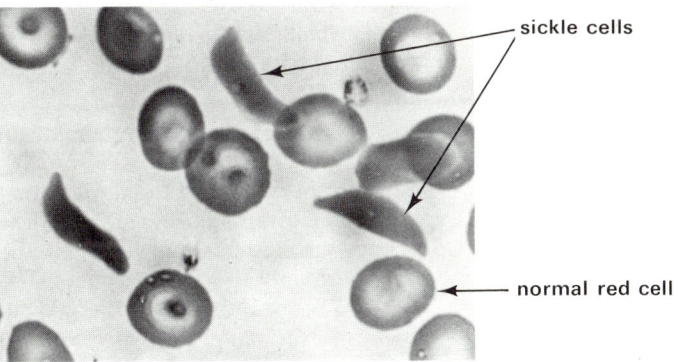

sickle cells

normal red cell

8.3.5 Myoglobin

The muscles of mammals contain a red pigment called **myoglobin** which is structurally similar to one of the four sub-units of haemoglobin (Fig. 3.27). Myoglobin has a higher affinity for oxygen than haemoglobin has and unloads its oxygen only when the blood is almost fully deoxygenated. This property is reflected in the oxygen dissociation curve for myoglobin which is further to the left than the curve for haemoglobin (Fig. 8.22).

Oxymyoglobin acts as an oxygen reserve in skeletal muscles which have a high oxygen demand during exercise. In this way the muscles can be provided with oxygen even though the blood flowing through them has all

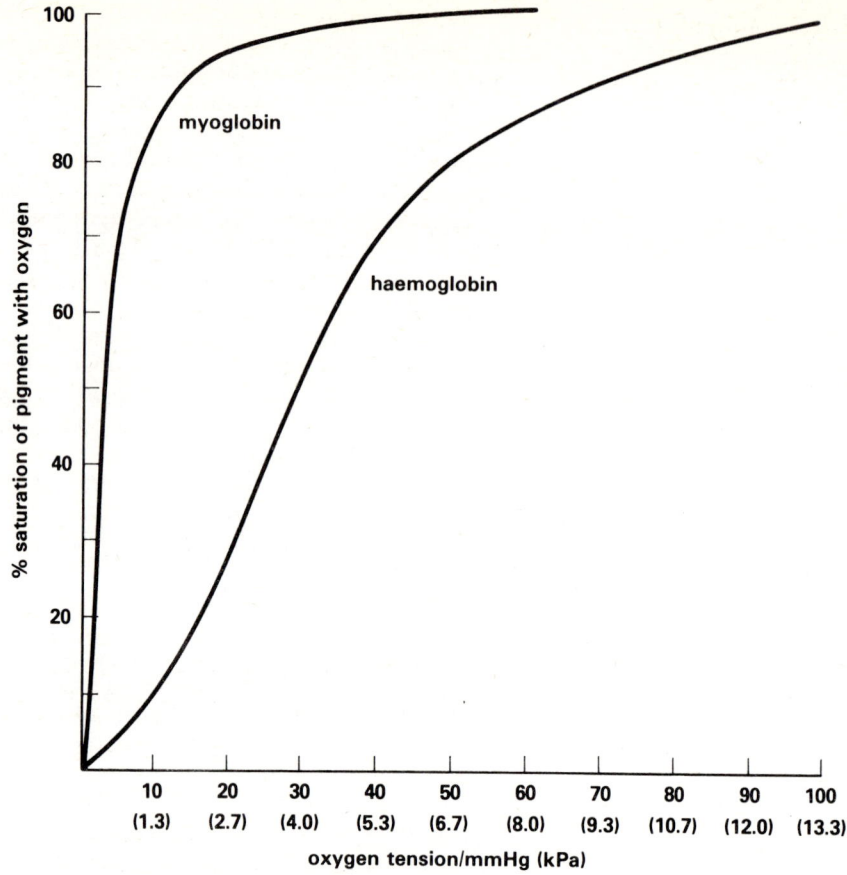

FIG. 8.22 *Oxygen dissociation curve of myoglobin. That of haemoglobin is shown for comparison*

oxygen removed from it. Mammalian muscles display several other adaptations which enable them to sustain intense activity for short periods of time (Chapter 15).

8.4 Respiratory physiology of diving mammals

Many marine mammals can remain submerged in water for long periods, often at great depths (Table 8.6). Diving involves several adaptations of the respiratory system and tissue physiology. For example, the tidal volume, the volume of air exchanged with the atmosphere during each breathing cycle, is some 80 per cent of the total lung capacity in the porpoise compared with only 10 per cent in man. This means that any oxygen debt built up in a porpoise during prolonged submergence in water can be repaid quickly on surfacing. There are, however, other adaptations which are just as important.

Table 8.6 Diving times of several mammals

SPECIES	AVERAGE DURATION OF DIVE/min
man	2·5
seal	15
finback whale	30
sperm whale	60–90
bottlenose whale	120+

8.4.1 Aquatic mammals

P. F. Scholander was one of the first to investigate the physiology of diving mammals, mainly seals. The blood of the seal has a very high oxygen-carrying capacity, 29·3 cm³ oxygen per 100 cm³ blood (Table 8.2). Even though there is a lot of oxygen in the seal's blood when it submerges, much of its tissue respiration during long dives is anaerobic. This is because breathing may stop for up to fifteen minutes when the seal is under water. Lactic acid and carbon dioxide are produced as a result of anaerobic respiration (Chapter 5). The lactic acid can increase to seven times its normal concentration in the blood of seals without ill-effect. The respiratory centre of seals must be far less sensitive than that of non-diving mammals to high concentrations of carbon dioxide in the blood.

Another feature of seals when diving is a remarkable drop in heart rate from about 150 beats per minute to about 10 beats per minute. Blood is also redirected in the vascular system so that the supply to the brain is maintained while the amount of blood pumped to other parts of the body is lowered. In this way the oxygen goes to organs which cannot do without it. Seals also expel air from their lungs as they dive, lowering the volume of gases which are compressed on submerging. The danger of bubbles of nitrogen forming in the blood and tissues when the seal surfaces is thus minimised.

The mechanisms described above ensure the efficient use of oxygen during submergence and allow anaerobic respiration without damage to the tissues. Less is known of such mechanisms in whales. However, the muscles of whales contain high concentrations of myoglobin.

8.4.2 Diving and man

There has been much interest recently in the physiological problems of human diving. One reason for concern is the high rates of accidents and fatalities among divers engaged in oil exploration in the North Sea off the coast of Britain.

The main problems relate to the effects of pressure from the surrounding water as divers descend (Fig. 8.23). The pressure increases the volumes of gases entering the blood from the alveoli. At depths below about 60 m the increased oxygen content of the blood is such that the tissues receive more oxygen than they normally need. The increased supply of oxygen leads to abnormal metabolism and damage to brain cells can occur. Consequently the diver can quickly lose control of his actions. Oxygen poisoning may be overcome by adjusting the oxygen content of a diver's breathing mixture as he descends.

High levels of dissolved nitrogen in the body have an anaesthetic effect on the central nervous system. At depths below 60 m **nitrogen narcosis** can make a diver unconscious. Prior to this, a state similar to drunkenness can give a diver a false sense of security in which he cannot make rational decisions. In this state, he may endanger his life. Substitution of helium for nitrogen in the breathing mixture eliminates the danger of nitrogen narcosis. One amusing, if inconvenient, side effect of the use of helium for such purposes is that it affects the voice which becomes very high pitched and squeaky.

Rapid **decompression**, on sudden surfacing, can cause a further problem, that of gas bubble formation in the tissues and body fluids. At a depth of 60 m the nitrogen in a diver's body is at a pressure of about 517 kPa, more than five times atmospheric pressure at sea level. At the surface however, nitrogen is at a pressure of 81 kPa. Sudden decompression causes rapid conversion of dissolved nitrogen in the blood into bubbles of nitrogen. The bubbles cause severe pain which in turn gives rise to terrible contortions of the body. For this reason, the condition is called the **bends**. The effects on the central nervous system can be severe and permanent. The bends or **decompression sickness** is prevented by controlling decompression at a rate slow enough for the extra nitrogen in the body to be gradually eliminated in the normal way by the lungs.

FIG. 8.23 *Relationship between pressure and depth beneath the sea*

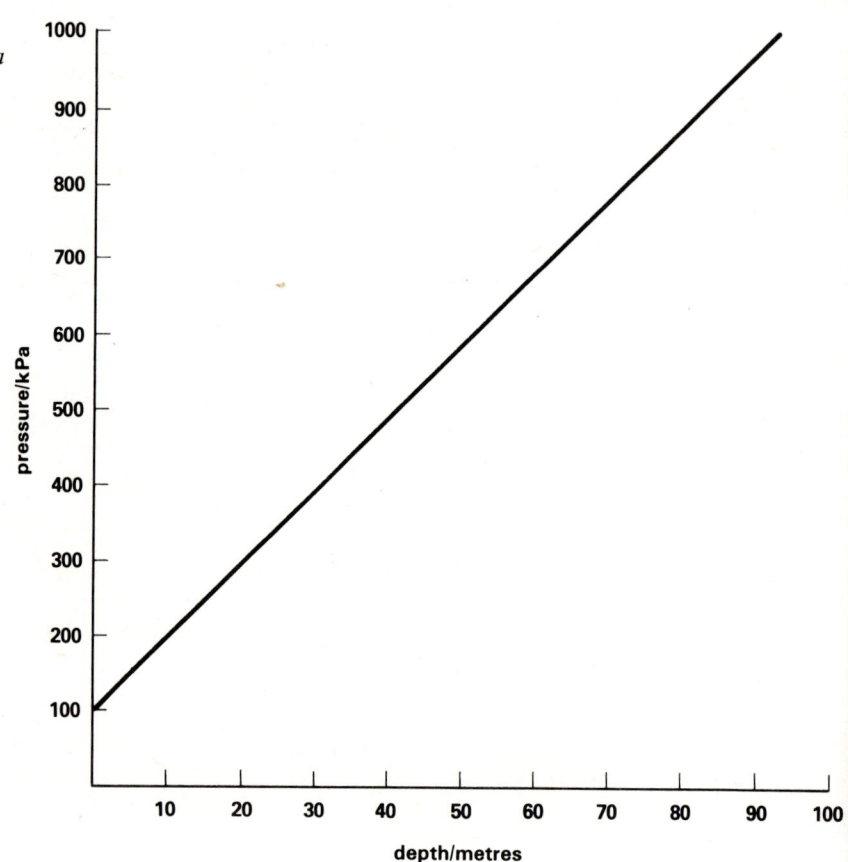

Chapter 9

Blood, lymph and circulation

Oxygen in the lungs and many nutrients in the gut enter the blood mainly by diffusion through the alveolar and intestinal surfaces respectively. Movement of dissolved metabolites around the body by diffusion alone, however, would be a very slow process. An inactive earthworm for example could only obtain about 10 per cent of its oxygen requirement if it had to rely on diffusion to distribute the gas throughout its body. Consequently all but the smallest of animals need a rapid means of internal transport. In mammals and other vertebrates transport is carried out mainly by the **blood**.

Blood distributes nutrients and respiratory gases to the tissues, transports hormones to target organs and carries metabolic wastes from the tissues to excretory organs. Blood also plays an important part in the defence against disease and in the repair of injured body tissues.

9.1 Circulation of blood and lymph

FIG. 9.1 *Diagram of the blood circulation of mammals*

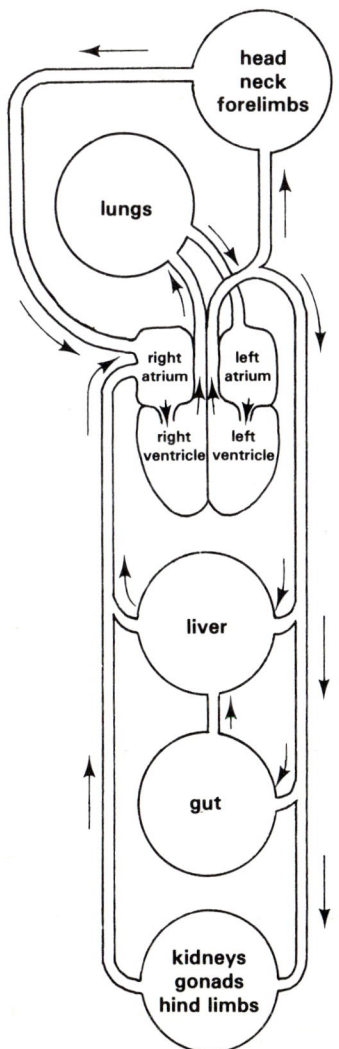

FIG. 9.2 *The main arteries and veins in man. The pulmonary and gastrointestinal vessels are omitted for clarity*

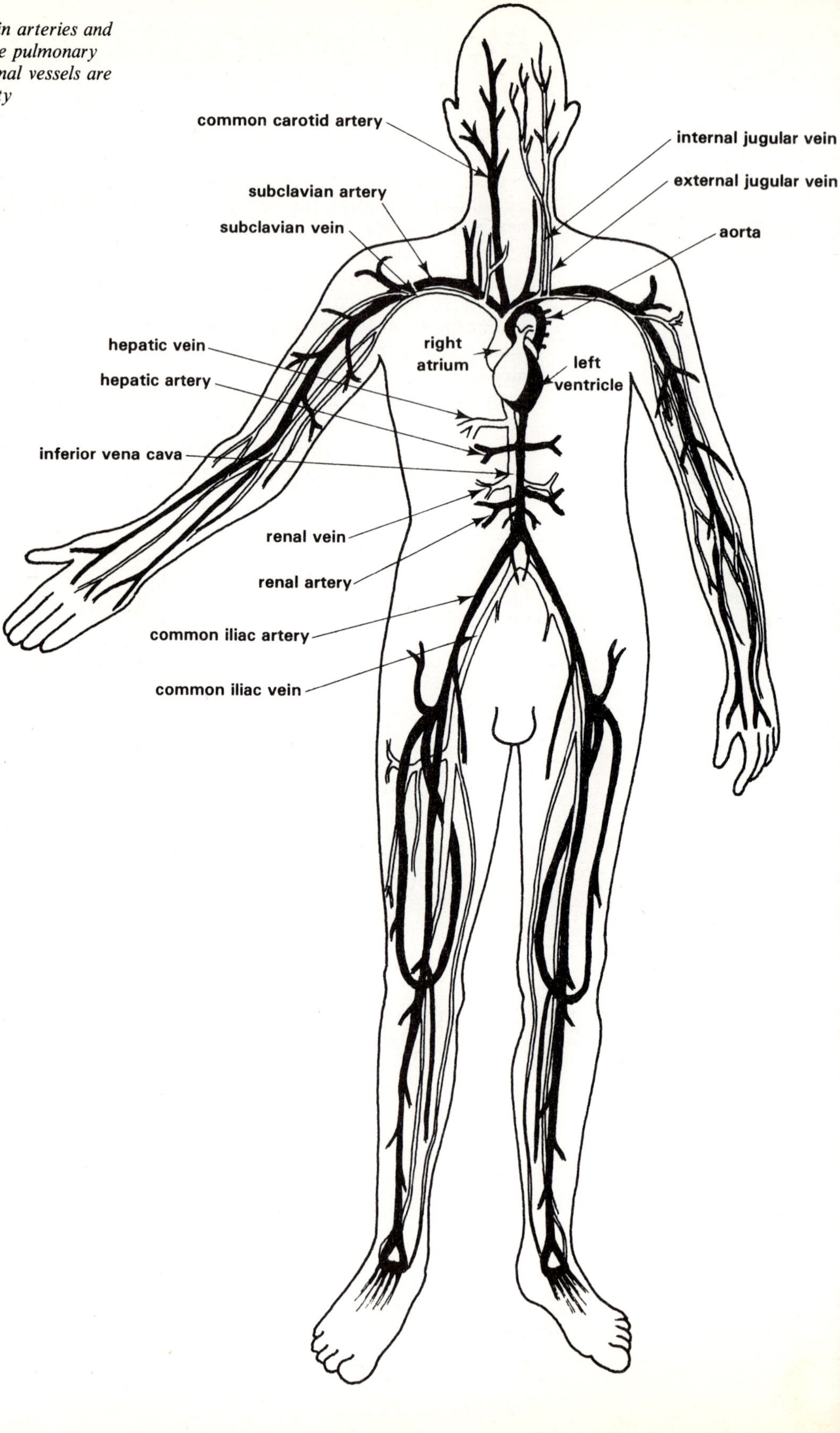

common carotid artery

subclavian artery

subclavian vein

hepatic vein

hepatic artery

inferior vena cava

renal vein

renal artery

common iliac artery

common iliac vein

internal jugular vein

external jugular vein

aorta

right atrium

left ventricle

The volume of blood in mammals is substantial (Table 9.1). To act as a transport system large volumes of fluid require a powerful pump, the **heart**. From the heart blood is pumped through an extensive system of **arteries** and **arterioles** to all the organs of the body. It then passes through microscopic blood **capillaries** where metabolites and wastes are exchanged with the tissue fluid and cells of the organs. Blood drains from the organs in **venules** and **veins** which carry it back to the heart. Figures 9.1 and 9.2 illustrate the extent and complexity of the mammalian blood vascular system.

Table 9.1 Blood volume as a percentage of body mass in several mammals

SPECIES	BLOOD VOLUME AS PER CENT BODY MASS
goat	6·1
rabbit	6·5
man	7·0
cow	7·5
guinea pig	7·5
pig	8·0
dog	9·4

9.1.1 The heart

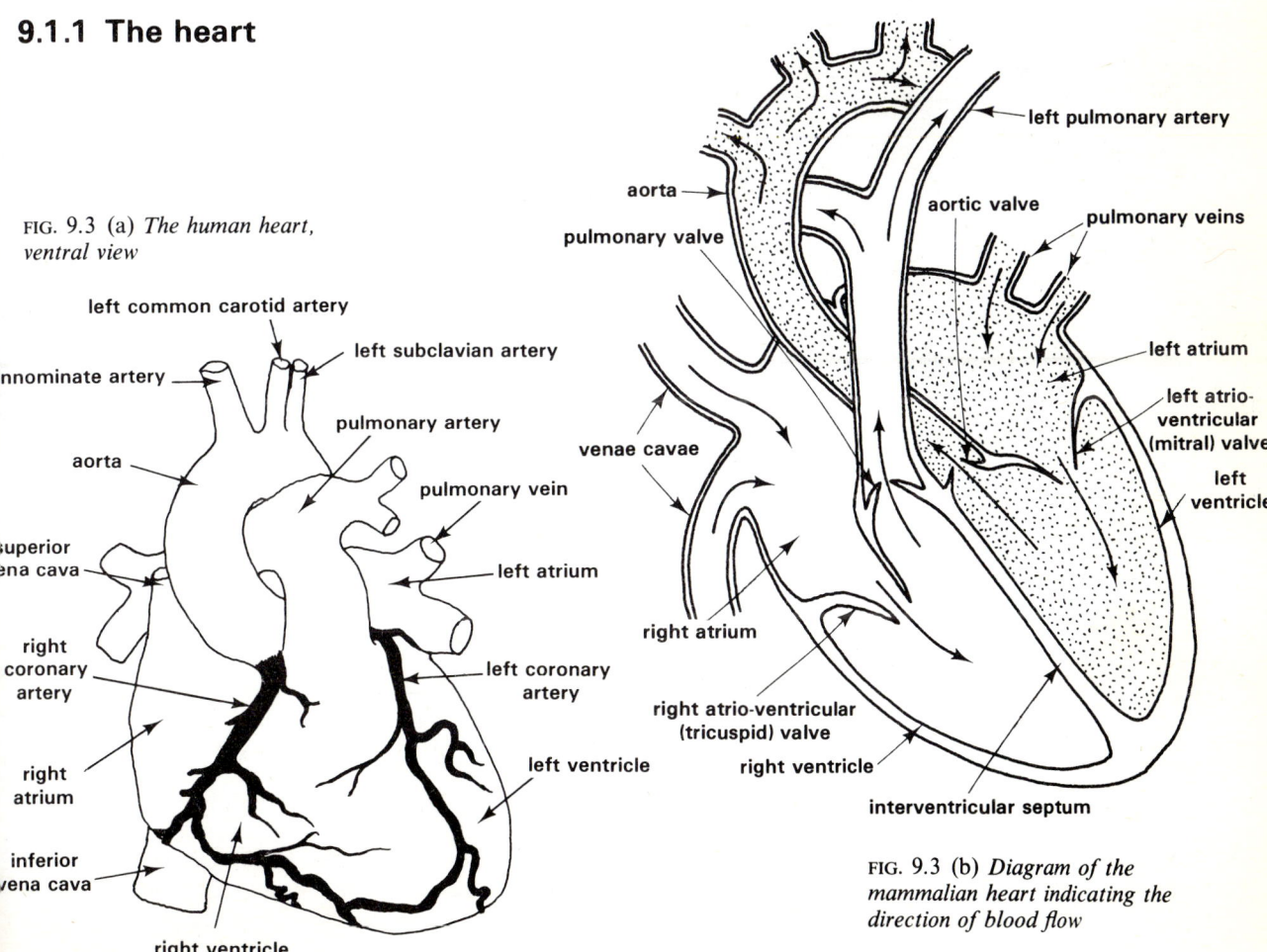

FIG. 9.3 (a) *The human heart, ventral view*

FIG. 9.3 (b) *Diagram of the mammalian heart indicating the direction of blood flow*

The mammalian heart is a remarkable organ. Functionally it is divided into right and left halves. The right half collects blood from the general body circulation and pumps it to the lungs where it is oxygenated (Chapter 8). The left half of the heart receives oxygenated blood from the lungs and pumps it into the general (systemic) circulation (Fig. 9.3).

1. THE CARDIAC CYCLE

The **cardiac output** is the volume of blood pumped from each ventricle in a minute. When the body is at rest the cardiac output of an adult human is about 4·9 dm³ per minute. During severe exercise the output can rise to 30 dm³ per minute. The pumping of blood requires considerable and sustained expenditure of energy. The mechanical work in pumping blood is performed by the **cardiac muscle** in the walls of the heart's four chambers. Cardiac muscle has inherent rhythmicity. Even if isolated from all its nervous supply it continues to contract and relax rhythmically (Fig. 9.4).

FIG. 9.3 (c) *Resin cast of the blood vessels of the human heart* (courtesy Professor McMinn, Department of Anatomy, The Royal College of Surgeons of England)

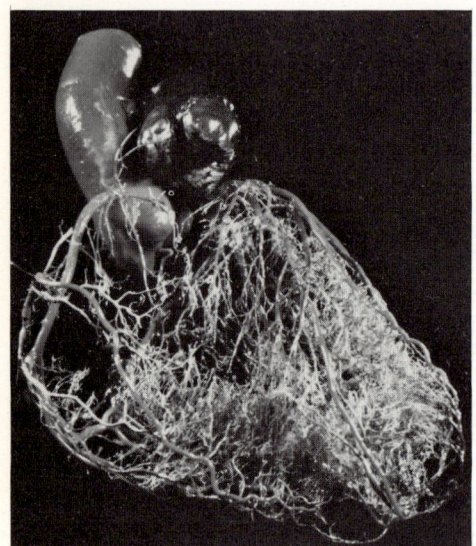

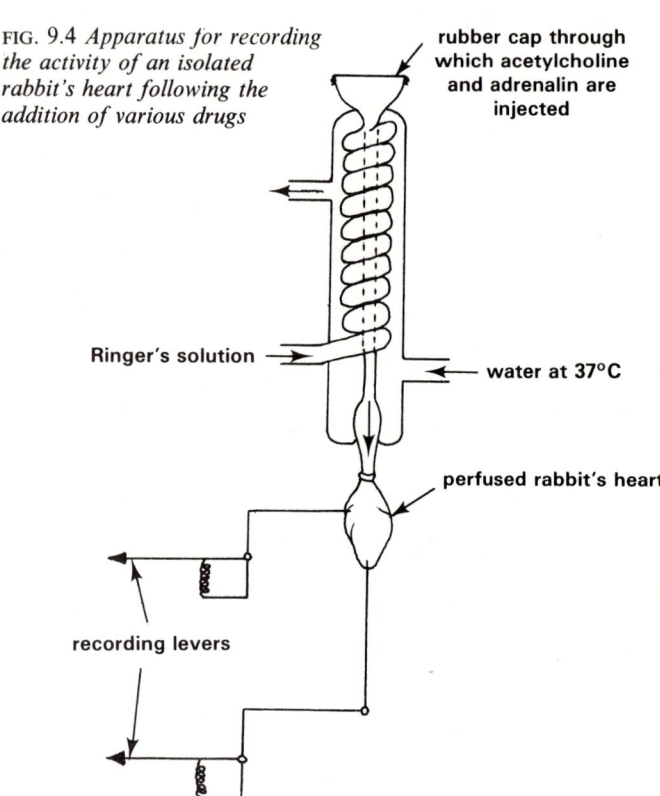

FIG. 9.4 *Apparatus for recording the activity of an isolated rabbit's heart following the addition of various drugs*

rubber cap through which acetylcholine and adrenalin are injected

Ringer's solution →

← water at 37°C

perfused rabbit's heart

recording levers

The **cardiac cycle** begins with atrial contraction. The electrochemical basis of muscle contraction is described in Chapter 15. The stimulus for the heart's rhythmic beat comes from a small part of the right atrium called the **sinu-atrial node (pacemaker)**. From the pacemaker waves of electrical activity similar to nerve impulses spread out very rapidly over both atria. Each wave contracts the atrial muscle, forcing blood in the **atria** through a pair of **atrio-ventricular valves** into the **ventricles**. **Semi-lunar valves** prevent a backflow of blood into the veins supplying blood to the heart when the atria contract (Fig. 9.5(a)). The non-conductive septum between the atria and ventricles prevents the cardiac impulse in the atrial muscles from spreading directly into the ventricles. However, a second node, the **atrio-ventricular (AV) node**, also in the wall of the right atrium, picks up the atrial impulse. The AV node transmits the cardiac impulse along the conductive **bundle of His** and its branches in the interventricular septum. When the impulse reaches the apex of the heart it spreads rapidly up the ventricular walls in a network of conductive **Purkinje fibres**. The arrival of the cardiac impulse in the ventricles stimulates contraction there. Blood is forced upwards from the ventricles into the pulmonary arteries leading to the lungs and the aorta leading to the rest of the body (Fig. 9.5(b)). The atrio-ventricular valves prevent blood flowing back into the atria when the ventricles contract. When the ventricles relax, pulmonary and aortic semi-lunar valves prevent blood flowing back into the ventricles. The valves of the heart are non-muscular structures. They respond passively to blood pressure changes brought about by contraction and relaxation of the heart's chambers (Fig. 9.6).

FIG. 9.5 (a) *The cardiac cycle begins with the cardiac impulse passing through the atria. Atrial constriction forces blood into the ventricles. The aorta and pulmonary arteries are omitted for clarity*

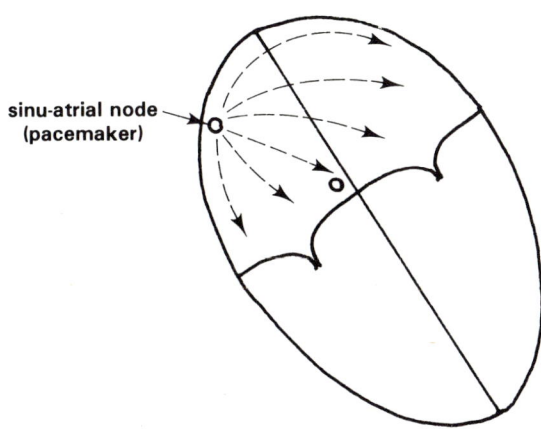

sinu-atrial node
(pacemaker)

ATRIAL SYSTOLE
(contraction of atrial walls)

blood forced into
ventricles

FIG. 9.5 (b) *The cardiac impulse reaches the ventricles along the conductive bundle of His. Ventricular contraction forces blood into the aorta and pulmonary arteries*

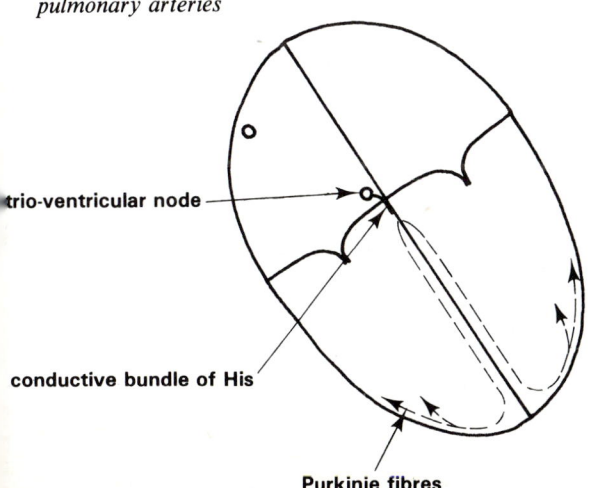

atrio-ventricular node

conductive bundle of His

Purkinje fibres

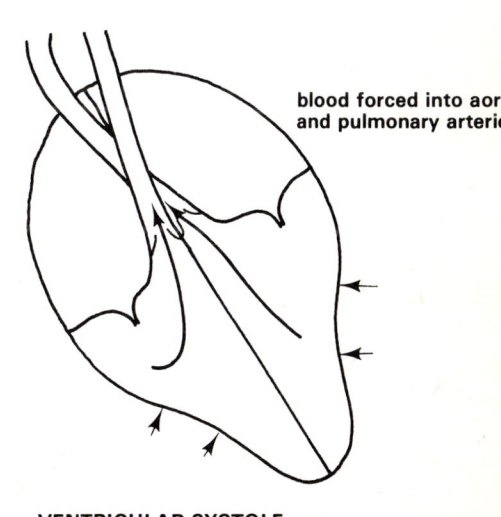

blood forced into aorta
and pulmonary arteries

VENTRICULAR SYSTOLE
(contraction of ventricular walls)

155

FIG. 9.6 *The passive action of* (a) *the mitral and* (b) *the aortic semi-lunar valves*

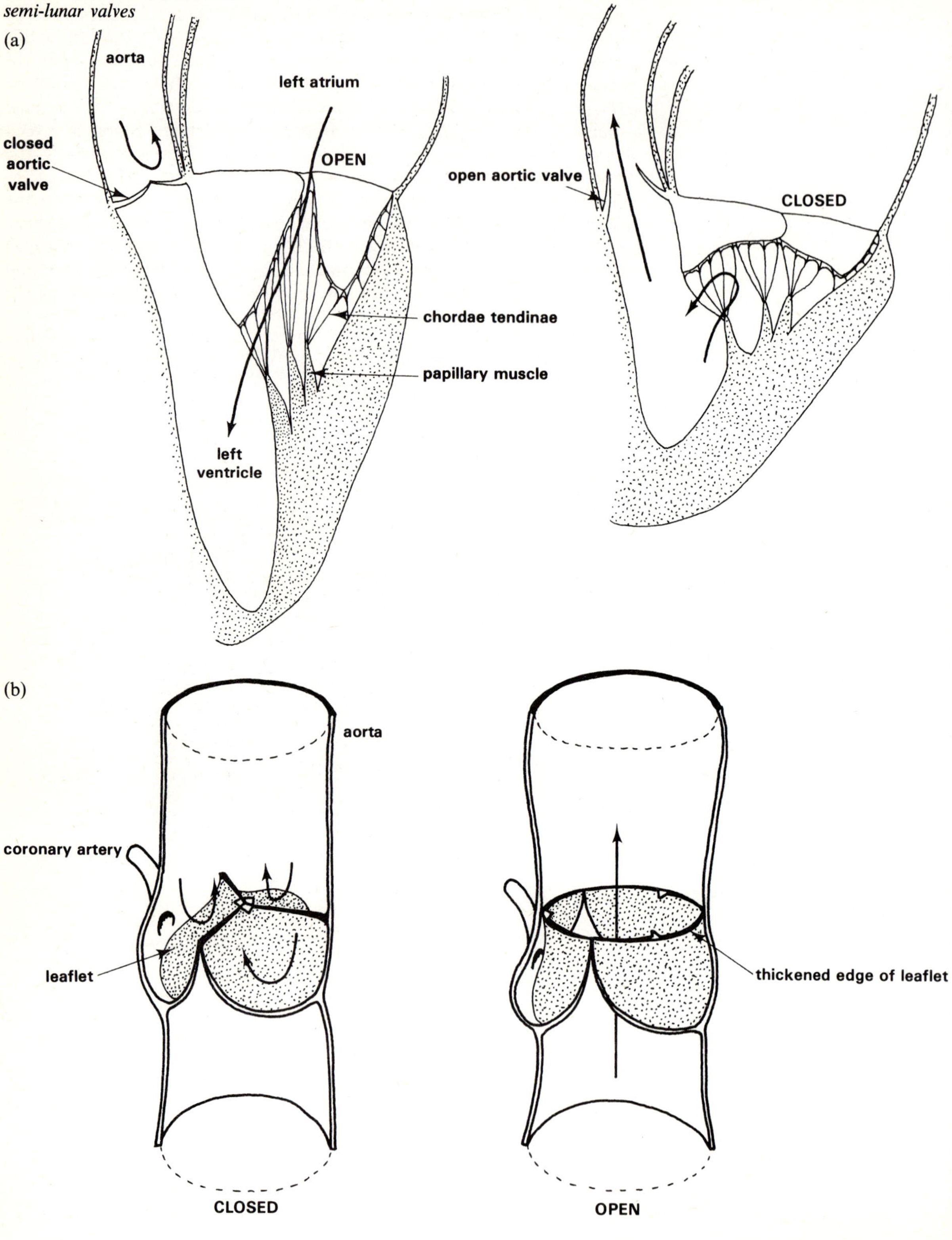

(a)

aorta

left atrium

closed aortic valve

OPEN

open aortic valve

CLOSED

chordae tendinae

papillary muscle

left ventricle

(b)

aorta

coronary artery

leaflet

thickened edge of leaflet

CLOSED

OPEN

The electrical events which occur during each cardiac cycle can be displayed as an **electrocardiogram (ECG)**. Taking an ECG involves placing metal electrodes at specific sites on the skin. The simplest system uses four electrodes, one on the right wrist, one on the left wrist and one on the left ankle. The fourth electrode is connected from the right ankle to earth so that any electromagnetic disturbances in the room which are picked up by the body do not interfere with the recording (Fig. 9.7(a)).

The cardiac impulse spreads through the heart starting at the pacemaker. Small voltages also spread through the body fluids which, because they contain electrolytes, are electrically conductive. Voltage changes therefore occur at the body surface and correspond to the intensity and direction of the cardiac impulse in the heart. The electrical changes are measured as voltage differences between pairs of electrodes called **leads**. The differences are recorded on moving paper or displayed on an oscilloscope screen (Fig. 9.7(b)). A typical ECG consists of **waves** and **complexes** which correspond to particular cardiac events. The *P*-wave is generated by the cardiac impulse passing over the atria. The *QRS* complex is generated by the passage of the impulse down the bundle of His into the Purkinje network and through the walls of the ventricles. The *T*-wave corresponds to the electrical recovery of the ventricles. Deviations from the normal ECG are readily seen. Certain heart complaints result in typical abnormal wave forms (Fig. 9.7(c)).

FIG. 9.7 (a) *Patient with electrodes in position for the recording of his ECG*

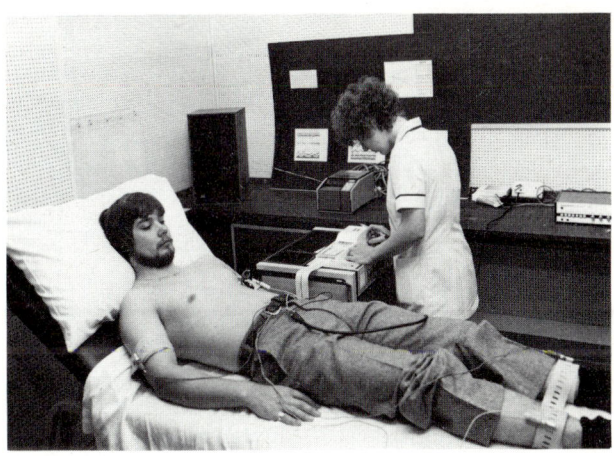

FIG. 9.7 (b) *A typical normal ECG pattern*

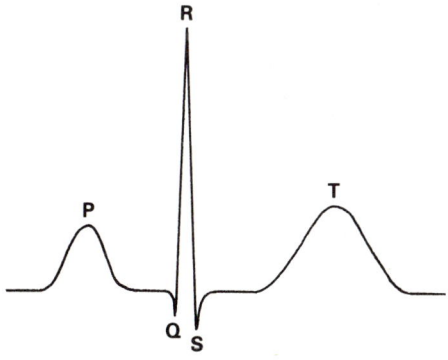

FIG. 9.7 (c) *Some abnormal ECG patterns*

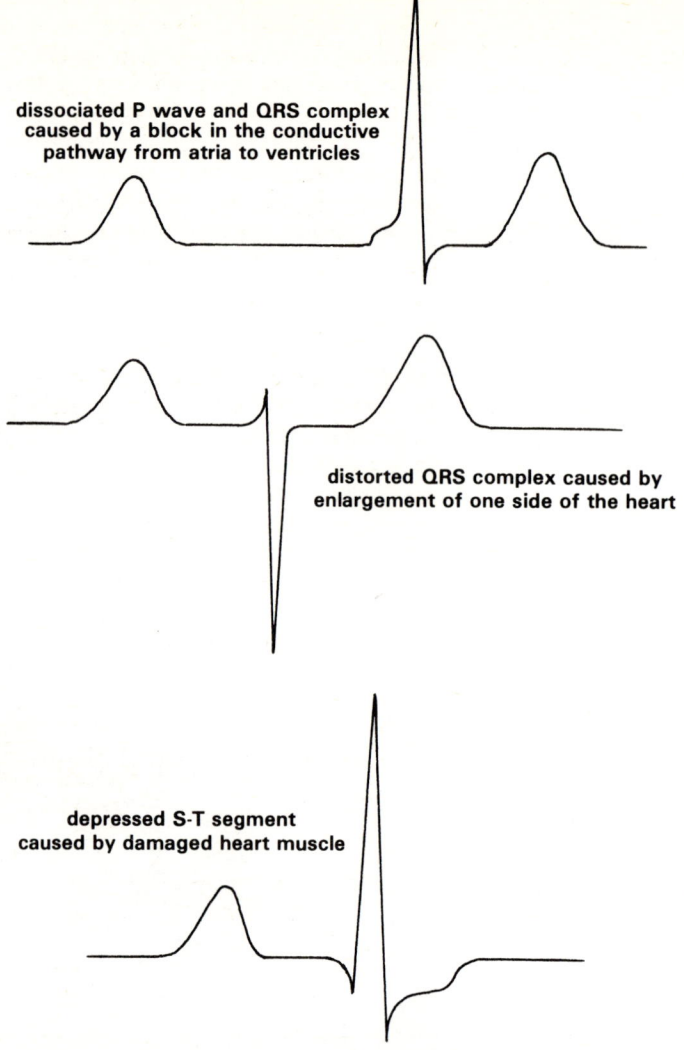

dissociated P wave and QRS complex
caused by a block in the conductive
pathway from atria to ventricles

distorted QRS complex caused by
enlargement of one side of the heart

depressed S-T segment
caused by damaged heart muscle

2. PRESSURE CHANGES IN THE HEART

Contraction of the atria, **atrial systole**, increases the pressure of blood in the atria. The rise in pressure forces the atrio-ventricular valves open and blood flows into the ventricles. Movement of blood from atria to ventricles is not entirely due to atrial systole. In a human heart about 70 per cent of the atrial blood is sucked into the ventricles as they relax, **ventricular diastole**, just before the atria contract. When the ventricles contract, **ventricular systole**, the pressure of the blood in the ventricles increases much more than atrial blood pressure. This is because the ventricles, especially the left one, have thicker and more muscular walls. During ventricular contraction blood pressure in man rises from zero to about 16 kPa (120 mm Hg) in the left ventricle and about 3·3 kPa (25 mm Hg) in the right ventricle. Blood enters the ventricles and stops for a moment before it is pumped into the arteries. The pumping action of the ventricles is so strong that it can be felt as the **pulse** in arteries far away from the heart. Although blood enters the arteries in spurts, its flow is continuous. How is this possible?

As the ventricles contract and the semi-lunar valves of the aorta and pulmonary arteries open, blood in the ventricles and arteries is continuous.

The blood pressure in the arteries is now the same as in the ventricles. Blood under pressure thus flows through the arteries causing the arterial walls to bulge. When the ventricles relax the main source of blood pressure is removed and the elastic arterial walls recoil keeping the blood in the arteries under pressure. In this way a continuous flow of blood is maintained. When the ventricular pressure falls to zero between contractions, aortic pressure drops to about 10·7 kPa (80 mm Hg) and pulmonary artery pressure to about 1·1 kPa (8 mm Hg). The term **blood pressure** as it is commonly used, refers to the pressure of blood in the aorta and the main arteries. Blood pressure in the arteries of the arm is often measured using a **sphygmomanometer** (Fig. 9.8).

FIG. 9.8 *A sphygmomanometer being used to take the blood pressure of a patient*

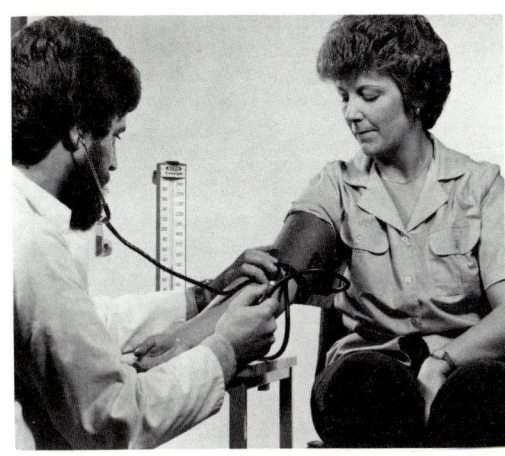

3. HEART SOUNDS

As the blood flows through the heart, **heart sounds** are produced as the valves open and close. The sounds can be heard with the aid of a **stethoscope** and are often described by the words **lub** and **dup**. Simultaneous closure of the atrio-ventricular valves when the ventricles contract makes the first sound, lub. The second sound, dup, 0·3 seconds later is caused by the simultaneous closure of the aortic and pulmonary valves. Extra sounds called **murmurs** can be heard if the blood rushes into the ventricles faster than normal, possibly because one of the valves is defective.

When the body is at rest the human heart beats about 70 times a minute and each cardiac cycle lasts for about 0·8 seconds. During the cycle the atria contract for about 0·3 seconds, the ventricles for 0·5 seconds. When we consider that the heart may beat more than 200 times per minute during severe exercise and that it beats without stopping throughout life it is evident that the heart is a remarkably durable and efficient pump.

4. REGULATION OF THE HEART

Cardiac activity is **myogenic**, the contractions initiated by the heart muscle itself. The heart is supplied with nerves of the autonomic system (Chapter 15) which regulate cardiac contraction. Regulation is important since it allows the heart rate and cardiac output to be varied according to the body's metabolic demands especially for oxygen. When the body is active the heart pumps oxygenated blood more rapidly into the circulatory system. Conversely, when the body rests the heart rate and cardiac output fall.

Two **autonomic nerves** are involved in cardiac regulation, the **accelerator nerve** of the sympathetic system and the **vagus nerve** of the parasympathetic

159

system. Each nerve affects the cardiac pacemaker in different ways. Impulses in the sympathetic nerve result in the secretion of small quantities of noradrenalin from the nerve endings onto the cardiac muscle. Noradrenalin is powerful in speeding up cardiac contractions and increases the volume of blood pumped by the heart at each beat. The **stroke volume** is usually about 70 cm^3 per ventricle but may rise to about 150 cm^3 during exercise. Impulses arriving at the parasympathetic nerve endings result in the release of small quantities of acetylcholine on to the cardiac muscle. Acetylcholine has the opposite effect to noradrenalin, reducing heart rate, stroke volume and hence the cardiac output (Fig. 9.9). It is the balance between sympathetic and parasympathetic nervous activity which adjusts the heart rate to meet prevailing physiological demands.

The questions now arise, what controls the autonomic nerves and how does the nervous system monitor the need to increase or decrease cardiac activity?

FIG. 9.9 (a) *The autonomic nerve supply to the cardiac pacemaker* (not drawn to scale) (after Green)

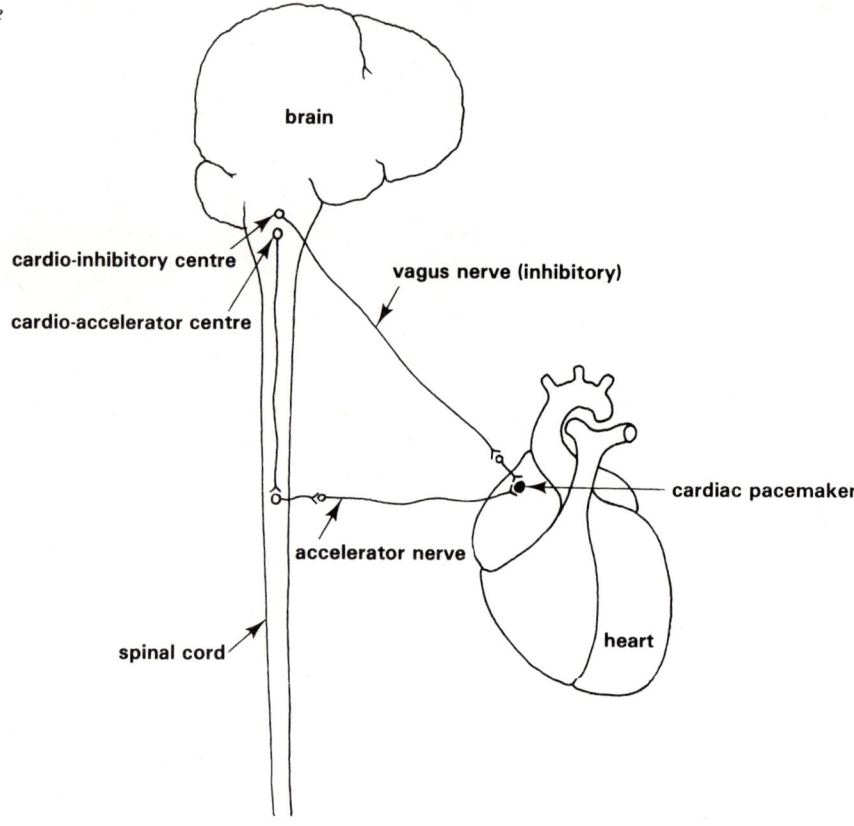

FIG. 9.9 (b) *Results of an experiment similar to that illustrated in Fig. 9.4, showing the effects on heart rate of the addition of adrenalin and acetylcholine*

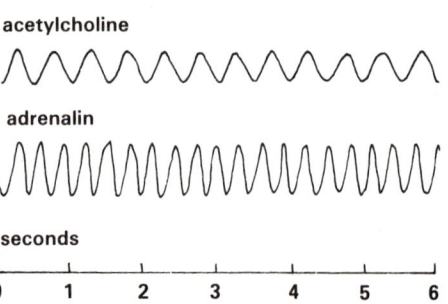

160

In the medulla oblongata of the brain there is a discrete region called the **cardiac centre** from which the nerves to the heart arise. The cardiac centre is sensitive to the tension of carbon dioxide in the blood which is high when the body is active and low when the body is at rest. In response to blood carbon dioxide tensions, impulses are directed from the cardiac centre along the autonomic nerves to the heart. The nerves regulate heart rate and cardiac output to meet the body's needs. A number of other sensory mechanisms monitor the extent of body activity. They include **chemoreceptors** in the walls of the carotid arteries which are sensitive to the oxygen tension of the blood. The oxygen tension is less when the body is active than when at rest. Near the chemoreceptors are patches of tissue called **baroreceptors** which are sensitive to increased blood pressure, a possible indication of an overactive heart. These sensory devices also help to regulate heart action via the cardiac centre in the brain (Fig. 9.10).

Reference is made in Chapter 8 to comparable mechanisms which control the breathing rate. Changes of breathing rate accompany changes in heart rate and enable efficient oxygenation of the blood as well as its distribution to body organs.

FIG. 9.10 *The main factors controlling the mammalian heart*

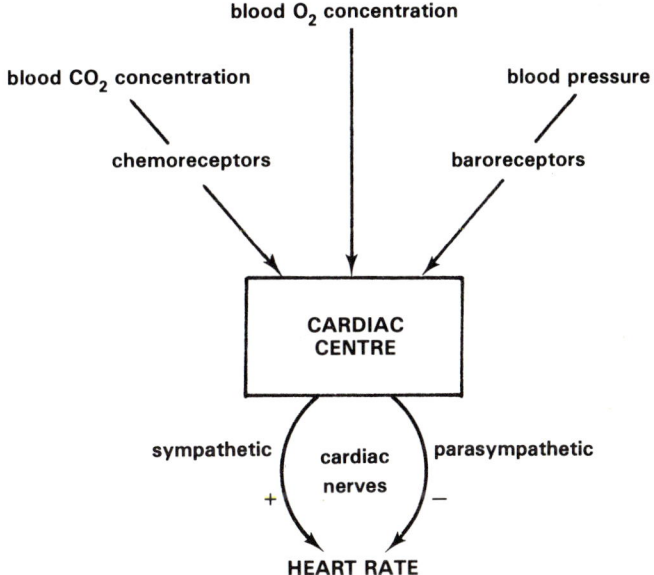

5. VASOMOTOR CONTROL

A further effect of stimulation of the baroreceptors is the transmission of nerve impulses to the **vasomotor centre**. Like the cardiac centre the vasomotor centre is found in the medulla oblongata of the brain. Sympathetic nerves originating in the centre lead to arterioles. By secreting noradrenalin from their endings the nerves bring about contraction of the smooth muscle in the walls of arterioles. As the muscle is arranged in a circular fashion its contraction results in narrowing of the arterioles called **vasoconstriction**. Conversely, decreased sympathetic activity from the vasomotor centre results in **vasodilation** (Fig. 9.11).

High blood pressure stimulates the baroreceptors which depress sympathetic output from the vasomotor centre. The resulting vasodilation lowers the blood pressure, an appropriate response to high blood pressure. Low blood pressure has the opposite effect (Fig. 9.12).

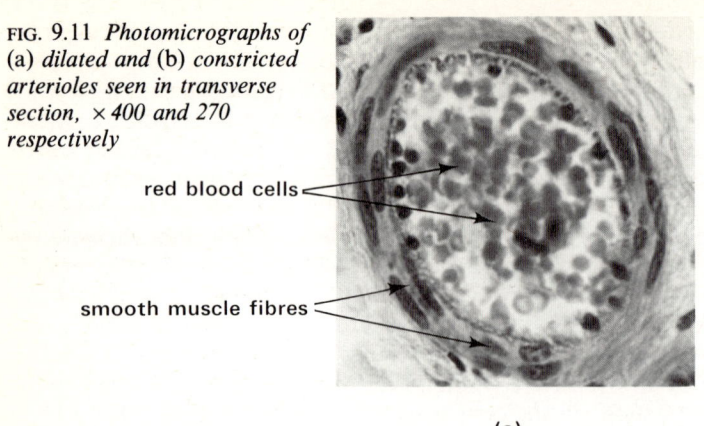

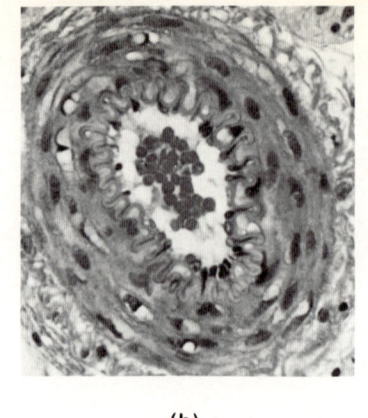

FIG. 9.11 *Photomicrographs of* (a) *dilated and* (b) *constricted arterioles seen in transverse section,* × 400 *and* 270 *respectively*

red blood cells

smooth muscle fibres

(a)

(b)

FIG. 9.12 *The main factors controlling mammalian blood pressure*

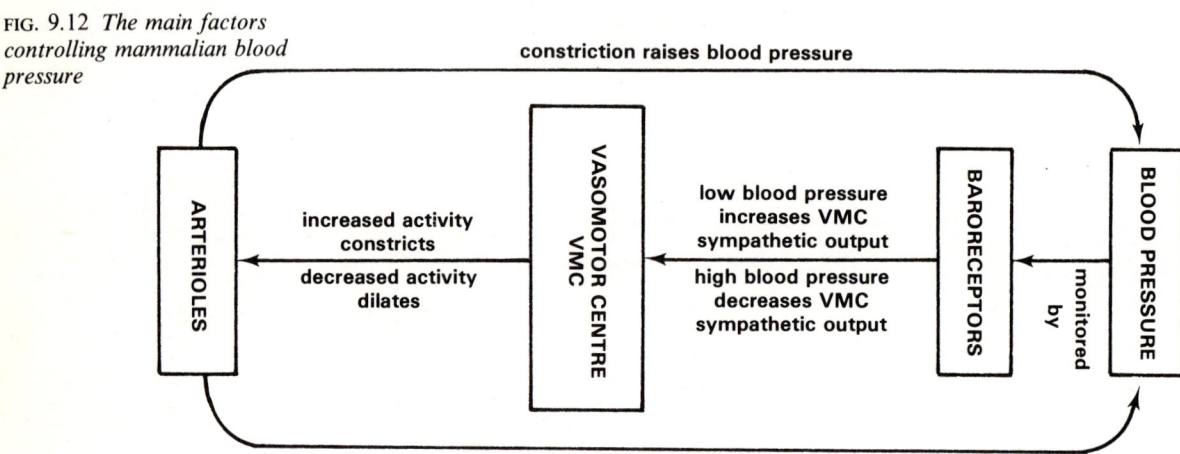

constriction raises blood pressure

ARTERIOLES

increased activity
constricts

decreased activity
dilates

VASOMOTOR CENTRE
VMC

low blood pressure
increases VMC
sympathetic output

high blood pressure
decreases VMC
sympathetic output

BARORECEPTORS

monitored
by

BLOOD PRESSURE

dilation lowers blood pressure

FIG. 9.13 *The control of breathing, heart rate and blood pressure in mammals*

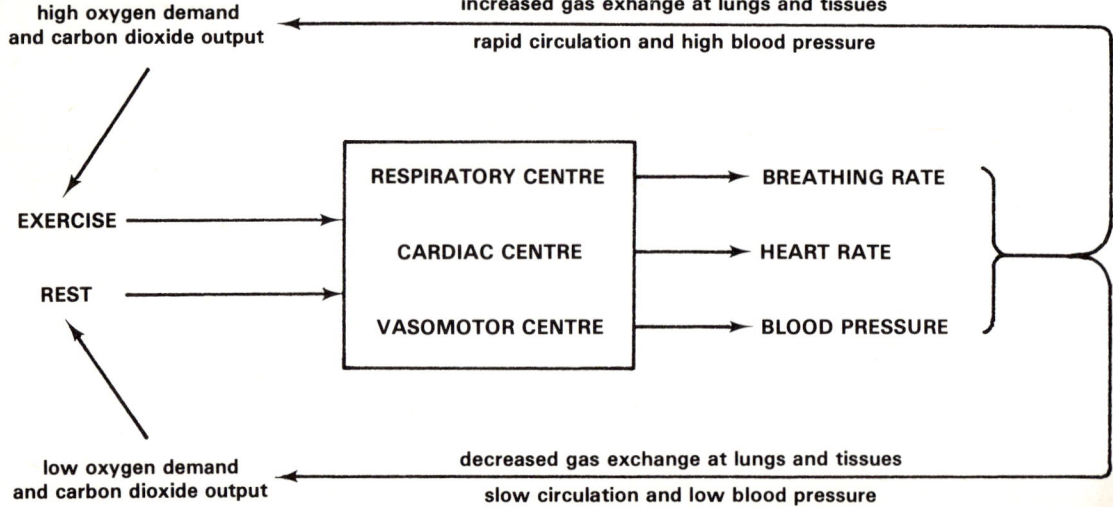

high oxygen demand
and carbon dioxide output

increased gas exhange at lungs and tissues

rapid circulation and high blood pressure

EXERCISE

REST

RESPIRATORY CENTRE

CARDIAC CENTRE

VASOMOTOR CENTRE

BREATHING RATE

HEART RATE

BLOOD PRESSURE

low oxygen demand
and carbon dioxide output

decreased gas exchange at lungs and tissues

slow circulation and low blood pressure

162

As with the cardiac and respiratory centres, the vasomotor centre is stimulated by changes in the tensions of oxygen and carbon dioxide in the blood. Generally all three centres respond to the same physiological stimuli (Fig. 9.13). Similar responses can be produced by secretions from the adrenal glands (Chapter 18).

9.1.2 Exchange of materials between blood and tissues

A major function of the circulatory system in mammals is to transport metabolites and wastes to and from the organs. Interchange of materials between blood and tissues occurs through the walls of numerous microscopic capillaries (Fig. 9.14). Each organ receives sufficient metabolites from the blood but not so much as to deprive other organs of vital materials. The means of controlling the release of oxygen from haemoglobin in the red blood cells is described in Chapter 8. Interchange of other substances is brought about by the interaction of two forces. The pumping action of the heart squeezes blood against the walls of the blood vessels, creating an outward-acting force called **hydrostatic pressure**. Counter to this is the attraction of certain blood constituents for water, the **colloid osmotic potential** of the blood. This is caused mainly by the plasma proteins which have a large relative molar mass.

FIG. 9.14 (a) *Some characteristics of different blood vessels. Note that the capillaries have the greatest cross-section area and the most permeable walls. Nearly all the interchange between the blood and the tissues occurs in the capillaries*

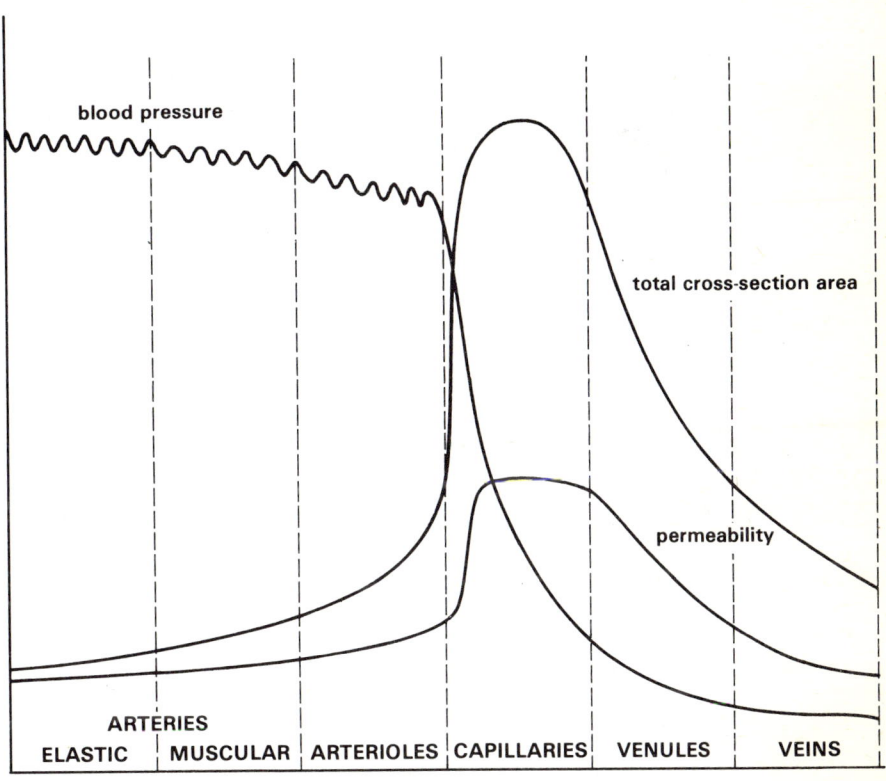

The hydrostatic pressure of the blood entering capillaries is normally about 4·4 kPa (32 mm Hg). Since the colloid osmotic potential is normally about −3·6 kPa (−27 mm Hg), a **filtration pressure** of −3·6 + 4·4 kPa = 0·8 kPa (5 mm Hg) exists. Consequently, any constituents of the blood to which the capillary walls are permeable leave the vessels. They include all substances of small relative molar mass such as water, oxygen, glucose, amino acids, fatty acids, hormones and inorganic ions. The fluid so formed is called **tissue fluid**. Proteins of high relative molar mass and blood cells remain in the capillaries. Because of the reduction in blood volume the hydrostatic pressure in the capillaries drops (Fig. 9.14). A point is eventually reached at which the filtration pressure controlling the movement of fluid between the blood and the tissues is negative. Consequently, water containing dissolved metabolic wastes such as carbon dioxide and metabolites in excess of the tissues' needs is absorbed back into the blood. The process is illustrated by Figure 9.15.

FIG. 9.14 (b) *Photomicrograph of an artery and vein seen in transverse section,* × 25

FIG. 9.14 (c) *Electronmicrograph showing a blood capillary in transverse section,* × 6000

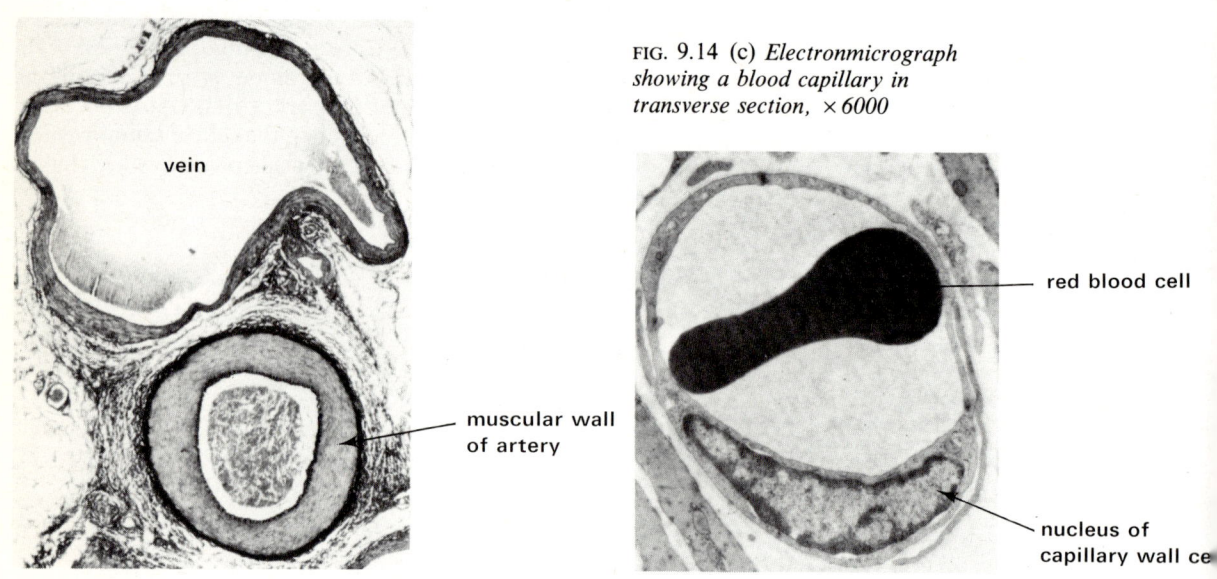

FIG. 9.15 *Diagrammatic interchange of metabolites between the capillary blood and tissue fluid*

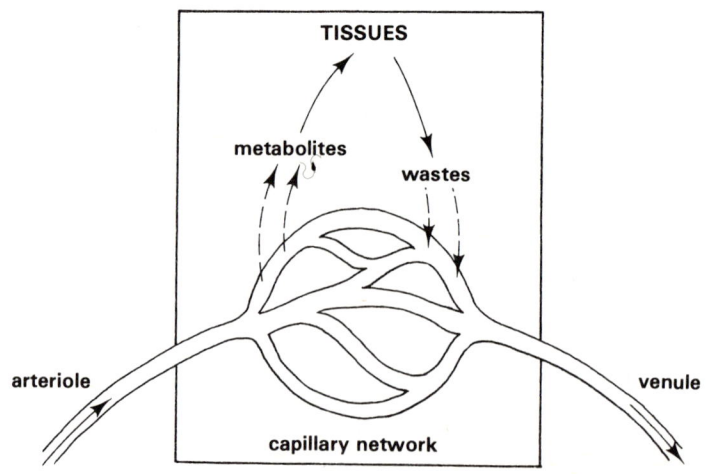

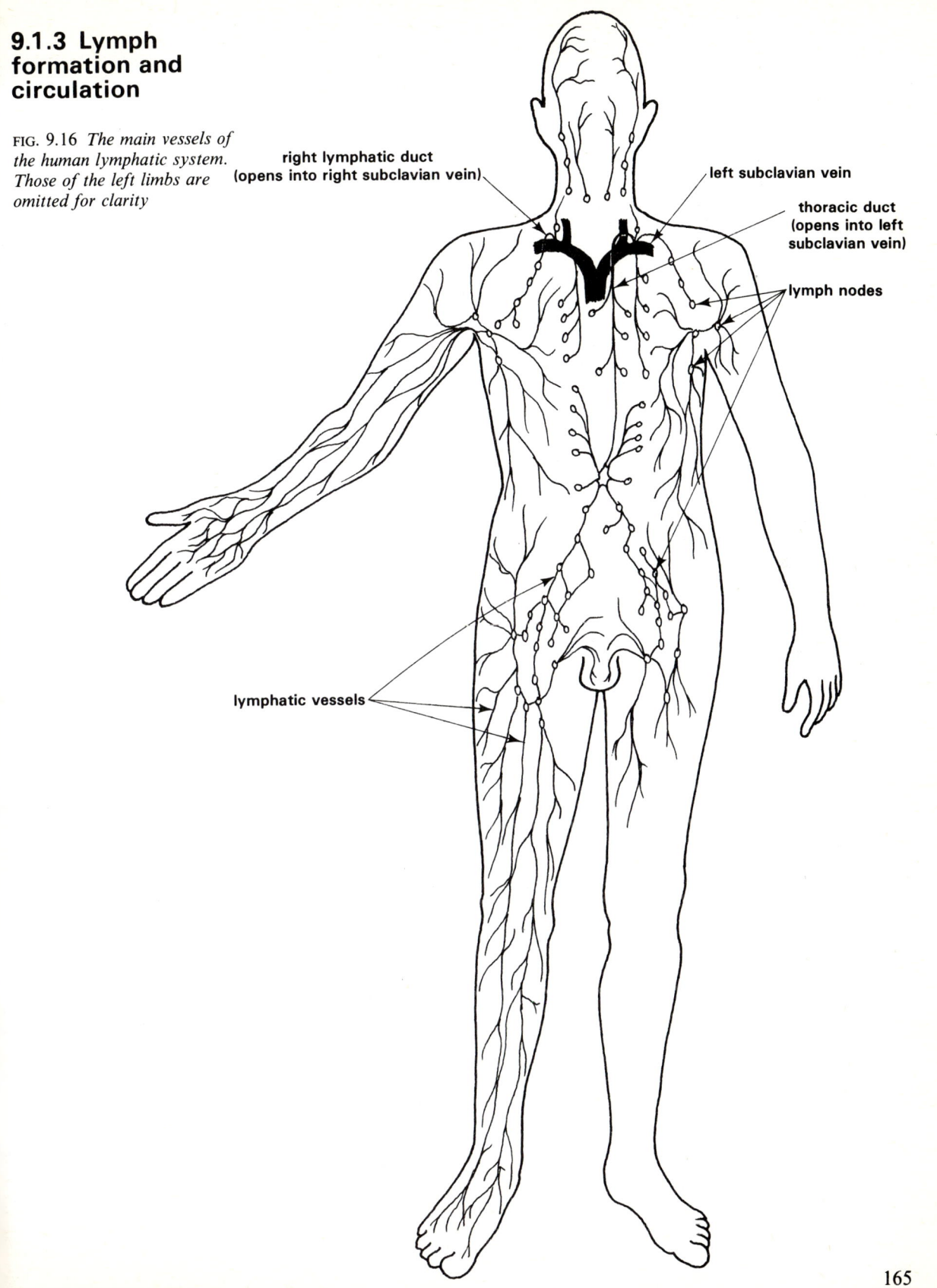

9.1.3 Lymph formation and circulation

FIG. 9.16 *The main vessels of the human lymphatic system. Those of the left limbs are omitted for clarity*

right lymphatic duct
(opens into right subclavian vein)

left subclavian vein

thoracic duct
(opens into left
subclavian vein)

lymph nodes

lymphatic vessels

During the formation of tissue fluid a small proportion of plasma proteins leaves the blood and enters the tissue fluid. These are plasma proteins of small relative molar mass. Any fluid not reabsorbed by the blood capillaries enters another set of vessels called the **lymphatic capillaries** which are similar to blood capillaries and have selectively permeable walls. They drain all the tissues of the body into an extensive system of larger **lymphatic vessels** (Fig. 9.16). One of the main functions of the lymphatic system is to remove materials that might otherwise accumulate in the tissues and impair normal function. For example, accumulation of water in the tissue fluid creates a swollen condition called **oedema**. The delivery of metabolites from the blood to oedematous tissues is impaired as the materials have to diffuse greater distances to reach the cells.

The fluid collected by the lymphatic system is called **lymph**. In the average healthy human body 100 cm^3 of lymph are formed in an hour. The lymphatic vessels return lymph to the blood through ducts joining the left and right subclavian veins (Figs. 9.16 and 9.17). The lymphatic system does more than this, however. On its passage through the lymphatic vessels, lymph enters numerous **lymph nodes**. The nodes contain stationary phagocytes which engulf small particles, micro-organisms and toxins in the lymph. The possible harmful effects of such materials are thus neutralised (Fig. 9.17). The roles of the lymph nodes in antibody production are described in section 9.2.4.

FIG. 9.17 *Diagrammatic relationship between the blood vascular and lymphatic systems*

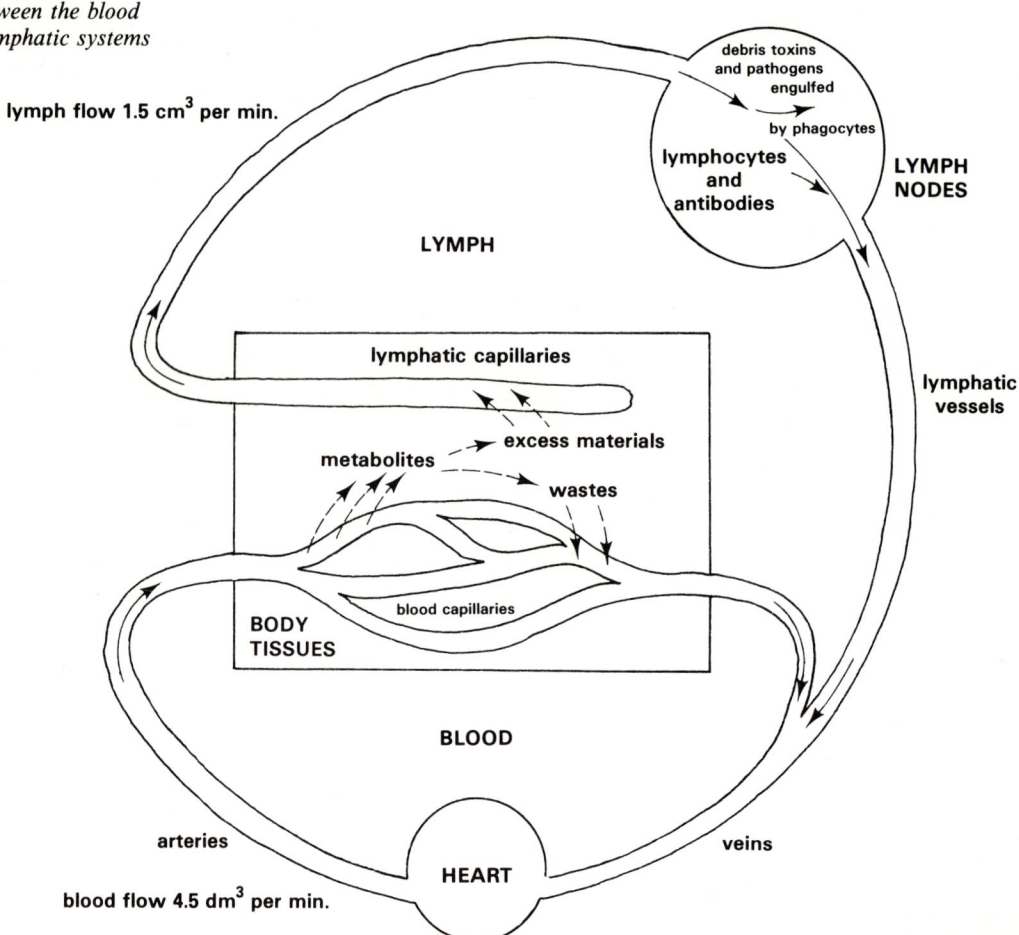

9.1.4 Circulation in the mammalian foetus

The mammalian foetus obtains its oxygen and nutrients from the **placenta** in the mother's uterus (Chapter 22). The placenta also removes wastes from the foetal blood. Maternal and foetal blood vessels come into close proximity to one another in the placenta (Fig. 9.18). The two-way exchange occurs mainly by diffusion. Thus the functions of the foetal lungs, kidneys and gut are performed by the placenta.

FIG. 9.18 *Human foetus and placenta inside the uterus*

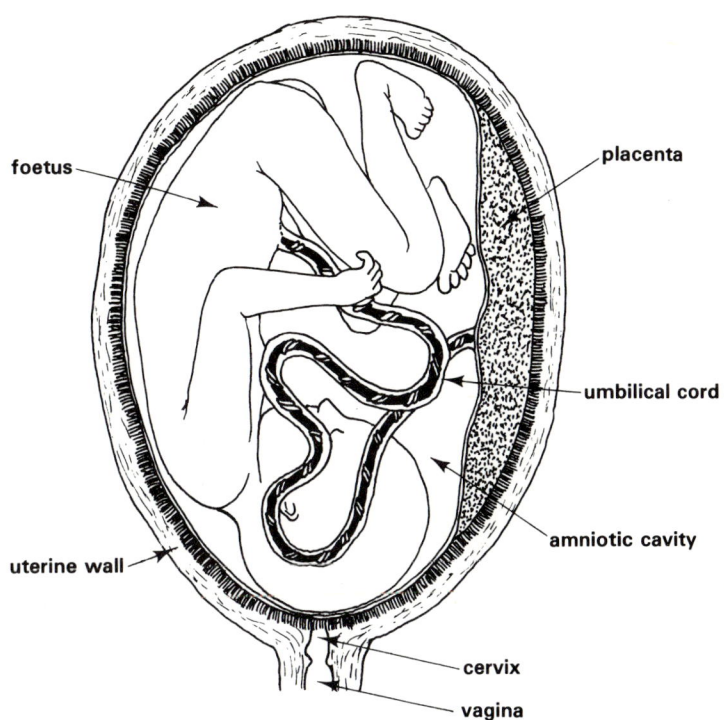

The pattern of circulation in the foetal body is different from that of an adult. Figure 9.19 illustrates the circulatory system of a human foetus. Compare it with Figure 9.1. An obvious difference between the foetal and adult circulation is the presence of **umbilical blood vessels** in the foetus which carry foetal blood to and from the placenta. The umbilical vessels are found in the umbilical cord. The other main differences are concerned with the transfer of the functions of the foetal lungs to the placenta. The pulmonary circulation in the foetus is almost completely by-passed due to a **ductus arteriosus** between the pulmonary arteries and the aorta, and a hole called the **foramen ovale** between the right and left atria. Such an arrangement involves mixing of oxygenated blood from the placenta and deoxygenated blood from the foetal tissues in the posterior vena cava and in the heart. The relative loss of efficiency in the transport of oxygen is compensated for by the presence of **foetal haemoglobin** which combines more readily with oxygen than does adult haemoglobin (Chapter 8). Further, the heart of the foetus does not pump blood to its unaerated lungs so energy is saved. The **ductus venosus** directs blood from the placenta and the foetal gut to the posterior vena cava, thus by-passing the foetal liver. Since most of the functions of the foetal liver are performed by the mother's liver, a supply of blood to the liver of the foetus is less important. The foetal liver is, however, the only source of foetal blood cells.

Shortly after birth the ductus venosus, ductus arteriosus and foramen ovale normally close. Failure to do so is the cause of severe circulatory problems some of which can be corrected by surgery.

FIG. 9.19 *Diagram of the blood circulation of the mammalian foetus*

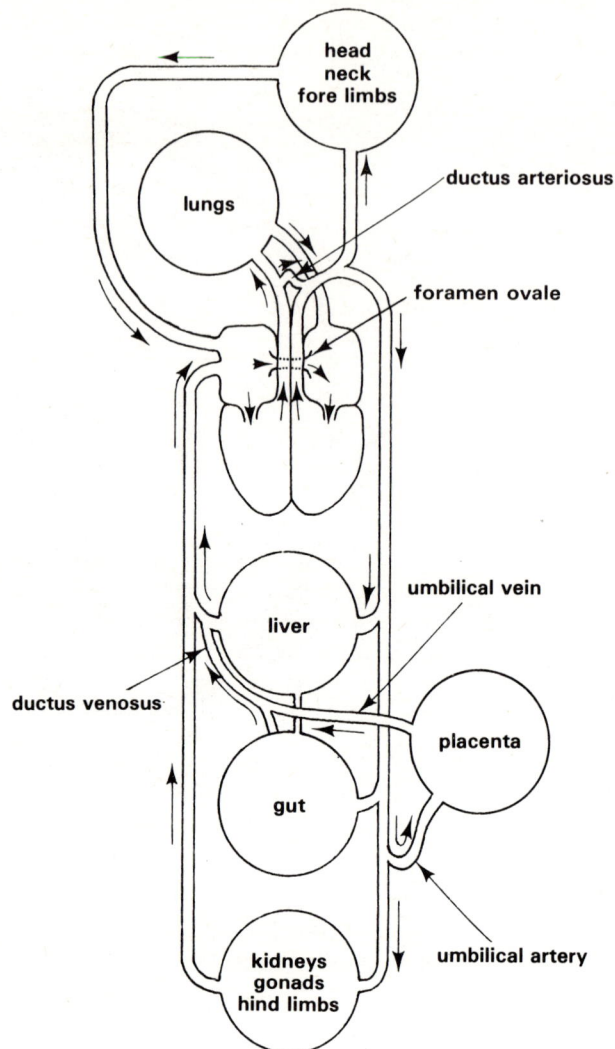

Table 9.2 Summary of the numbers and functions of human blood cells

CELL TYPE	AVERAGE NUMBERS × 10⁹ PER dm³ ADULT HUMAN BLOOD	FUNCTIONS
erythrocytes	3900–6500	oxygen carriage
neutrophils	2·50–7·50	phagocytosis
eosinophils	0·04–0·44	reduce inflammation; dissolve clots
basophils	0–0·10	unknown
monocytes	0·20–0·80	phagocytosis
lymphocytes	1·50–3·50	antibody production
thrombocytes	150–400	coagulation

9.2 Blood cells

Under the microscope **blood cells** are seen to be of several types. They make up about half the volume of mammalian blood (Table 9.2 and Fig. 9.20).

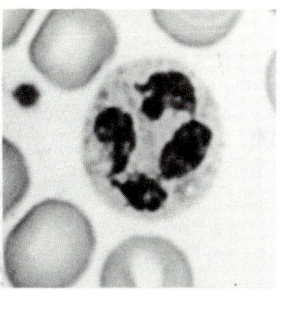
(a)

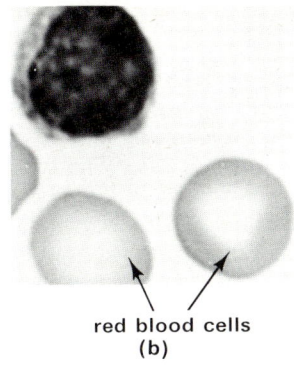
red blood cells
(b)

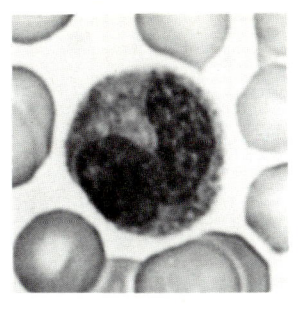
(c)

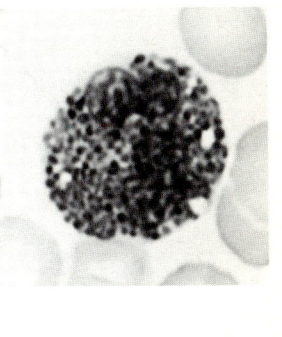
(d)

FIG. 9.20 *Photomicrograph of a smear of human blood showing* (a) *a neutrophil,* (b) *a lymphocyte,* (c) *a monocyte, and* (d) *an eosinophil, all* × *1500*

9.2.1 Erythrocytes

By far the most numerous of the blood cells are the red cells, **erythrocytes**. Their colour comes from the respiratory pigment **haemoglobin** which they contain. Human erythrocytes each contain about 30 pg of haemoglobin. Erythrocytes occupy between 40 and 60 per cent of the total blood volume in different mammals and their main function is to carry oxygen. Mammalian erythrocytes are biconcave discs and have a great surface area in relation to their volume. They are thus ideally shaped for the uptake of oxygen when in the pulmonary capillaries (Fig. 9.21). The function of haemoglobin in this respect is described in Chapter 8.

The absence of a nucleus in erythrocytes provides extra space for haemoglobin. The anucleate condition does have its disadvantages, however. In particular, erythrocytes have only a relatively short lifetime. Erythrocytes in man survive for about four months and have to be replaced continuously to keep a constant number in circulation. It is estimated that human erythrocytes are made at a rate of about 9000 million per hour. Erythrocyte production, **erythropoiesis**, occurs in the liver in the foetus but in adult life is restricted to the red marrow of certain bones. They include the cranium, vertebrae, ribs, sternum, parts of the pelvic girdle and some limb bones. Old erythrocytes are engulfed by phagocytic cells in the liver, spleen and red bone marrow. The haemoglobin is broken down and the iron it contains is retained for further haemoglobin synthesis. The rest of the pigment is excreted from the liver in bile (Chapter 14).

Impairment of the oxygen-carrying capacity of the blood is called **anaemia**. Anaemia can be caused by physical blood loss, **haemorrhage**, or functional blood loss, **haemolysis**. In haemolysis, erythrocytes are destroyed in the blood vessels and the haemoglobin is broken down in the blood plasma. Decreased blood production is another cause of anaemia. Anaemias in this category result from inadequate production of erythrocytes and haemoglobin in the bone marrow. **Impaired erythropoiesis** is often caused by deficiency of a vital nutrient in the diet. Iron deficiency is the commonest cause of human anaemia in the world today. Vitamin B_{12} is necessary for erythropoiesis. Absorption of vitamin B_{12} from the gut depends on the presence of a substance called intrinsic factor (IF) which is

FIG. 9.21 *Scanning electronmicrograph of some human red blood cells,* × *1700*

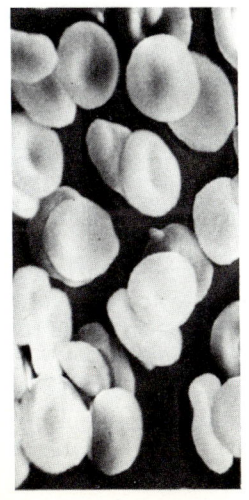

secreted in gastric juice (Chapter 14). Absence of IF leads to poor absorption of vitamin B_{12} and is the cause of **pernicious anaemia**. Production of abnormal haemoglobin is the cause of **sickle cell anaemia** (Chapters 6 and 8).

9.2.2 Leucocytes

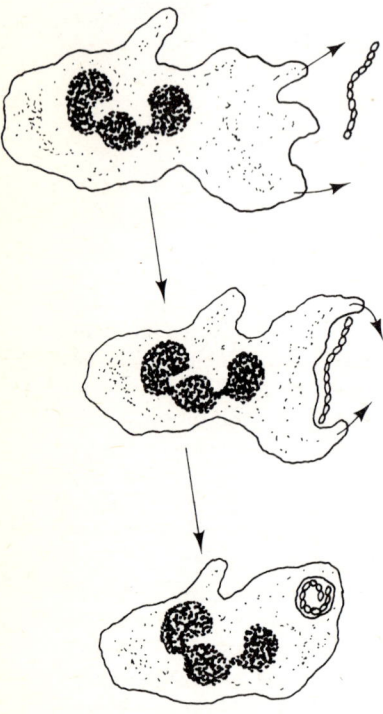

FIG. 9.22 (a) *A neutrophil engulfing a chain of bacteria by phagocytosis*

Far less numerous, but no less important than the erythrocytes are the white blood cells collectively called **leucocytes**. They include **neutrophils**, **lymphocytes**, **monocytes**, **eosinophils** and **basophils**. They are involved in defending the body against disease. The lymphocytes (section 9.2.4) are produced in lymphoid tissues, mainly the lymph nodes. Other leucocytes are made in the bone marrow. Leucocytes have a limited life time in the circulation and old leucocytes are disposed of by phagocytes in the liver and spleen.

Neutrophils and monocytes exhibit amoeboid movement. By extending pseudopodia they can encircle and engulf bacteria and other particles, a process called **phagocytosis** (Fig. 9.22). The phagocytic behaviour of living neutrophils can be observed microscopically and recorded by time-lapse photomicrography. Studies of this sort show that neutrophils can locate bacteria at a distance and actively seek them out for phagocytosis. This response is called **chemotaxis** and seems to be triggered off by chemicals released by the bacteria. Engulfed bacteria are taken into large food vacuoles where secretions of the leucocytes' lysosomes kill the bacteria after several minutes. When the bacteria are dead they are discharged from the trailing ends of leucocytes by reverse phagocytosis. Phagocytosis of some strains of virulent bacteria requires the presence of antibodies (section 9.2.4).

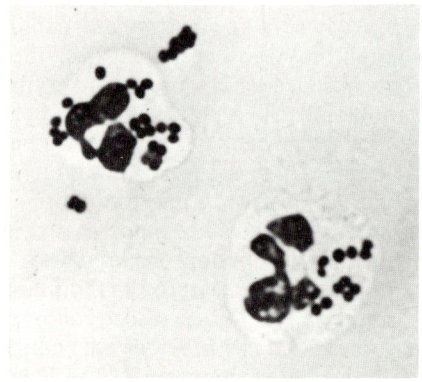

FIG. 9.22 (b) *Photomicrograph showing human neutrophils containing phagocytised bacteria, × 1400*

The numbers of phagocytes in the blood often increase in response to bacterial infection. Phagocytes can move in and out of blood capillaries by squeezing between the cells of capillary walls. They appear rapidly at sites of localised infection. As well as mobile phagocytes there are many fixed phagocytic cells in the mammalian body. Patches of fixed phagocytes are found in the lymph nodes, gut wall, alveoli, liver, spleen and red bone marrow. Fixed phagocytes form the **reticulo-endothelial system** which removes particulate debris including old blood cells from the blood and lymph.

Phagocytosis is thus a widespread and important means of neutralising infective agents and removing particulate debris generally from the body fluids.

9.2.3 Thrombocytes and coagulation

As well as erythrocytes and leucocytes, mammalian blood contains numerous anucleate cellular fragments called **thrombocytes** (platelets). Human blood contains between 150 and 400 thousand thrombocytes per mm³. They originate in the red bone marrow from large cells called **megakaryocytes** from which the thrombocytes are budded off as fragments of cytoplasm (Fig. 9.23).

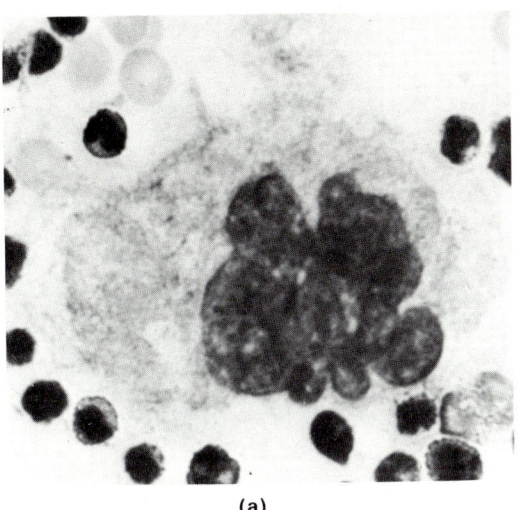

(a)

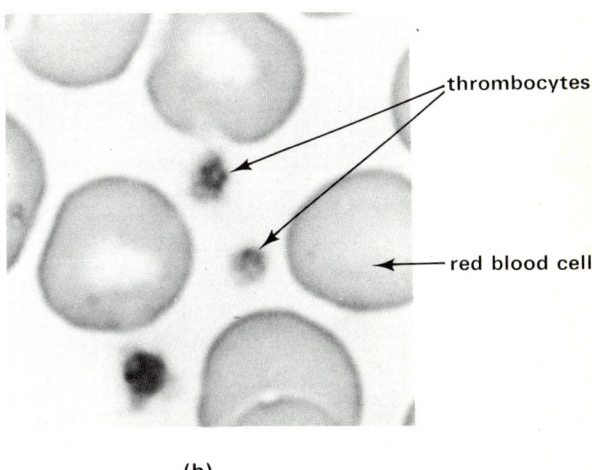

thrombocytes

red blood cell

(b)

FIG. 9.23 *Photomicrographs showing* (a) *a megakaryocyte in a bone marrow smear from a guinea-pig,* × 750, (b) *a smear of human blood with thrombocytes (platelets),* × 1500

Thrombocytes are involved in the **coagulation** of blood, a response to injury which prevents excessive blood loss. Coagulation is a complex sequence of chemical reactions which results in the deposition of insoluble fibrous proteins to form a **clot**. It is the clot which blocks damaged blood vessels and may stop bleeding. Thrombocytes stick to the surfaces of damaged vessels. Healthy blood vessels secrete into the blood a substance called **prostacyclin**. Prostacyclin prevents thrombocytes from sticking together and to the walls of blood vessels. In the absence of prostacyclin, however, agglutination of thrombocytes occurs. They then burst releasing thrombocyte factors into the plasma. The **thrombocyte factors** react with other **blood factors** in plasma to produce **thromboplastin**. Thromboplastin can also be produced when certain **tissue factors** react with the blood factors, a reaction which occurs for example when blood vessels are injured.

The next stage of coagulation is the conversion of **prothrombin**, a plasma protein, into an active enzyme called **thrombin**. Thromboplastin activates the conversion which requires the presence of calcium(II) ions, a normal constituent of plasma. The final stage is the conversion by thrombin of the soluble plasma protein **fibrinogen**, into insoluble **fibrin**. The entire clotting mechanism is summarised in Figure 9.24. Fibrin is a fibrous protein which precipitates as strands around the thrombocytes and damaged edges of the vessel. The meshwork of fibrin traps blood cells which increase the size of the clot, eventually sealing off the damaged vessel (Fig. 9.25). Burst thrombocytes also release a substance called **serotonin**. Serotonin causes vasoconstriction which may be enough to stop blood loss from damaged capillaries.

Important though coagulation is it is equally important that coagulation is restricted to sites of injury. A complex set of factors exists in mammalian blood to ensure that this normally happens. Fibrin inhibits excessive conversion of fibrinogen by negative feedback:

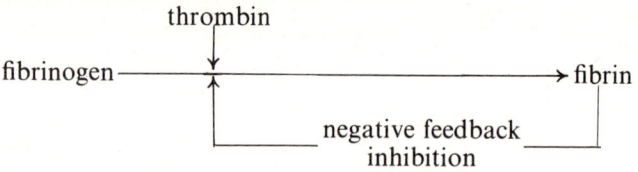

FIG. 9.24 *Summary of the coagulation mechanism*

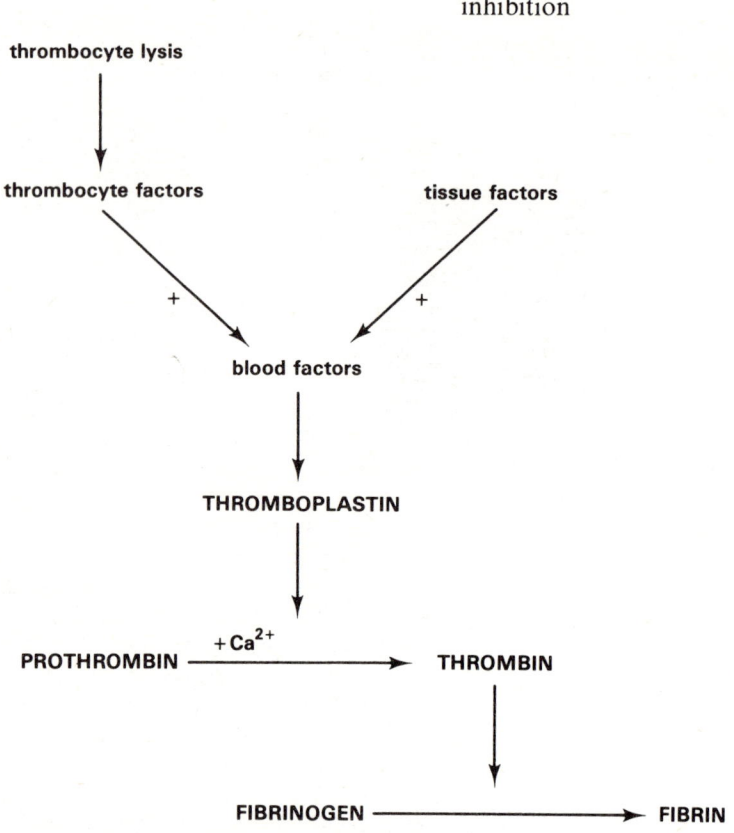

Enzymes capable of dissolving fibrin are carried in blood. The granules of eosinophils contain an inactive precursor called **profibrinolysin**. At sites of fibrin deposition profibrinolysin released by eosinophils is converted to **fibrinolysin** which helps to dissolve a clot after it has done its job.

The inner surfaces of blood vessels can become damaged, sometimes because of deposition of fatty substances particularly cholesterol. Deposits of this kind prevent secretion of prostacyclin and can trigger off coagulation, resulting in the formation of a clot inside the vessel. The clot is called an **intravascular thrombus**. Such a condition, commonly called **thrombosis**, can be very dangerous. For example, a thrombus in an artery feeding a vital organ such as the heart, can cause a sudden and fatal stoppage of blood to the organ. Coronary thrombosis accounts for a large number of deaths in industrialised countries. High-fat diets, sedentary jobs and tobacco smoking have been suggested as possible causes. An intravascular thrombus can occur in other vessels in the body such as a leg vein. If dislodged by the blood flow the thrombus is carried in the blood stream as an **embolus** which may eventually come to rest in the arteries of a vital organ. An embolus could for example be taken to the pulmonary arteries causing a **pulmonary embolism** which can cause death. The treatment of thrombosis and embolism involves administration of various anticoagulants such as warfarin which prevents coagulation by blocking the production of prothrombin in the liver. Other anticoagulants are used to dissolve the clot.

Evidently, normal coagulation depends on the delicate balance of many factors.

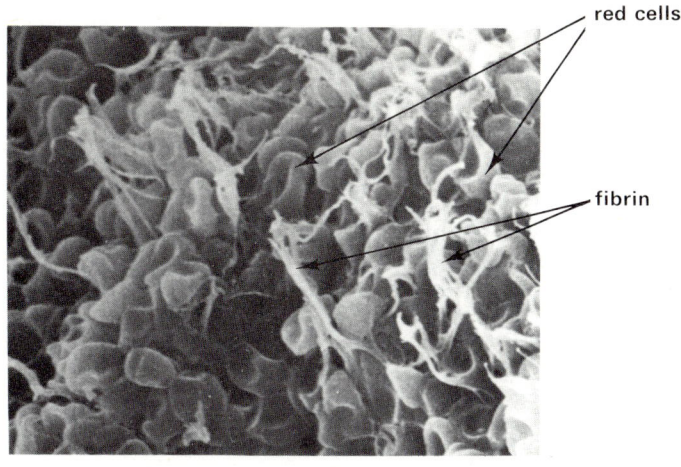

red cells

fibrin

9.2.4 Lymphocytes and antibodies

The term **immunity** generally means the ability to resist disease.

1. INNATE IMMUNITY
All mammals have **innate immunity** because of:
(i) The **skin's resistance** to penetration by micro-organisms.
(ii) **Enzymatic destruction** in the stomach of most micro-organisms ingested with food. **Lysozyme**, a powerful enzyme found in tears, can bring about lysis of bacteria.
(iii) The **acid environment** of the stomach and vagina inhibit the growth of most micro-organisms.
(iv) The **phagocytic activities** of neutrophils, monocytes and the reticulo-endothelial system.
(v) The action of certain chemicals produced in the body in response to microbial infection. For example, **interferon** is secreted by most mammalian tissues when they are attacked by viruses. Interferon is effective against many different viruses and it once looked like providing a major breakthrough in the medical treatment of viral diseases. Unfortunately interferon is species-specific. Mouse interferon works well in mice, but only human interferon is effective in man. Consequently it is difficult to obtain sufficient quantities of interferon for treating human diseases caused by viruses. Early experiments testing its use against diseases such as the common cold gave inconclusive results.
(vi) The **stickiness of mucus** which traps micro-organisms. Mucus is constantly produced by the linings of the respiratory and alimentary systems.

2. ADAPTIVE IMMUNITY
The means of protection outlined above are all general barriers to infection. Acquired or **adaptive immunity** is much more specific. It involves the production of specific **antibodies** directed against particular microbes or their toxins. Antibodies are produced by lymphoid tissue, as a direct

response to the presence in the body of substances which are recognised as foreign or **non-self**. Such substances are called **antigens**. Antigens are mostly proteins, a few are carbohydrates, and are present on micro-organisms, transfused blood cells and transplanted organs. Antibodies are proteins of a specific kind (Fig. 9.26) which can locate and neutralise or destroy the antigens which triggered their production. Normally the immune system can recognise the constituents of its own body. It does not normally produce antibodies against **self** (section 9.6). However, in **autoimmune diseases** self-recognition fails. The body then makes antibodies against its own tissues. Rheumatic fever and some forms of diabetes mellitus are diseases of this kind.

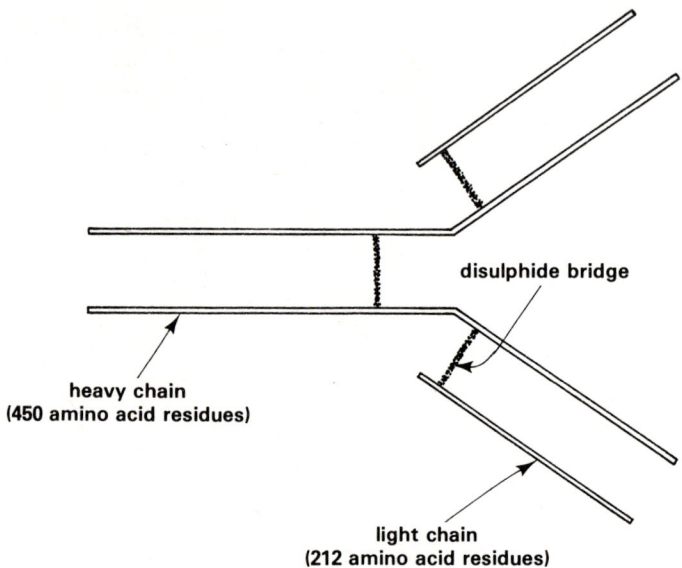

FIG. 9.26 *Diagrammatic structure of an immunoglobulin called IgG. It consists of two long (heavy) and two short (light) polypeptide chains joined by disulphide bridges. The sites for reaction with appropriate antigen are on the right where the molecule appears to gape open. The projection on the left may fix with complement*

disulphide bridge

heavy chain
(450 amino acid residues)

light chain
(212 amino acid residues)

Since the Second World War immunology has been an expanding science and is the focus of much research at the present time. Many theories of antibody action have been put forward and much is still not understood. Experiments indicate that as a result of stimulation by antigens, antibodies are made by lymphocytes in the lymph nodes. Antibodies enter the blood and are carried to sites of infection either in the plasma (**humoral immunity**) or on special lymphocytes (**cellular immunity**).

(i) **Humoral immunity**. On coming into contact with specific antigens, **B-lymphocytes** in the lymph nodes divide many times to produce a group of large **plasma cells**. The B-lymphocytes which react in this way are said to be **committed**. Plasma cells make the antibodies which are released into the lymph and eventually enter the blood. In the blood the antibodies form the γ-globulin (immunoglobulin) fraction of the plasma proteins (Fig. 9.27). Small numbers of committed B-lymphocytes remain in the lymph nodes for years, even after the antibodies have done their work and eliminated the original infection. This provides a very rapid response to any subsequent invasion by similar antigens.

(ii) **Cellular immunity**. In the lymph nodes other lymphocytes, called **T-lymphocytes**, may also be **sensitised** by antigens and become committed to produce specific antibodies. The antibodies remain attached to the T-lymphocytes which then enter the blood. About 80 per cent of the total lymphocyte population of human blood is of the T-variety. As in humoral immunity, a few committed T-lymphocytes remain in the lymph nodes and can respond to future invasion by similar antigens. Cellular immunity can be effective for the entire life-time of an organism.

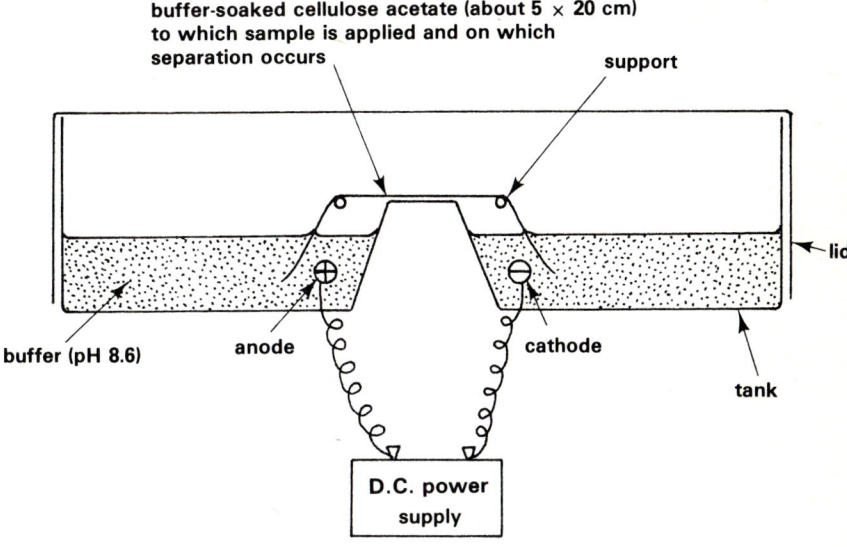

FIG. 9.27 (a) *The main fractions of plasma (or serum) proteins may be separated by* electrophoresis. *A sample of plasma (or serum) is subjected to an electric field in a special tank. At the pH used the different proteins in the sample carry different electric charges. Consequently the proteins migrate in different directions or at different speeds (some are heavier than others) when the current is switched on. Positive ions move towards the cathode, negative ions towards the anode*

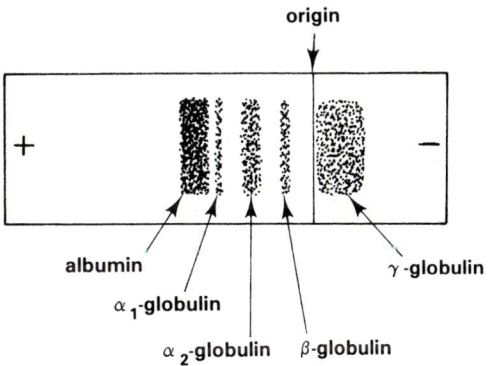

FIG. 9.27 (b) *When separation is complete the proteins are fixed and stained. The γ-globulin fraction contains the immunoglobulins*

Lymphocytes originate from **stem cells** in the bone marrow. The stem cells migrate to the lymph nodes before birth. On their way some of them enter the **thymus gland** lying near the heart. In the thymus gland the lymphocytes are exposed to self-components of the body. Those lymphocytes which react against self-components are destroyed. The remainder become T-cells ('T' for 'thymus-derived'). Removal of the thymus gland early on in life prevents T-cell processing and inhibits the development of the cellular immune system. Other stem cells are probably processed in the bone marrow. They produce B-cells ('B' for 'bone-derived'), the basis of the humoral immune system.

Some of the aspects of antibody production referred to above and the origins of lymphocytes are illustrated in Figure 9.28.

FIG. 9.28 *Diagrammatic origin of lymphocytes and their action in the immune response*

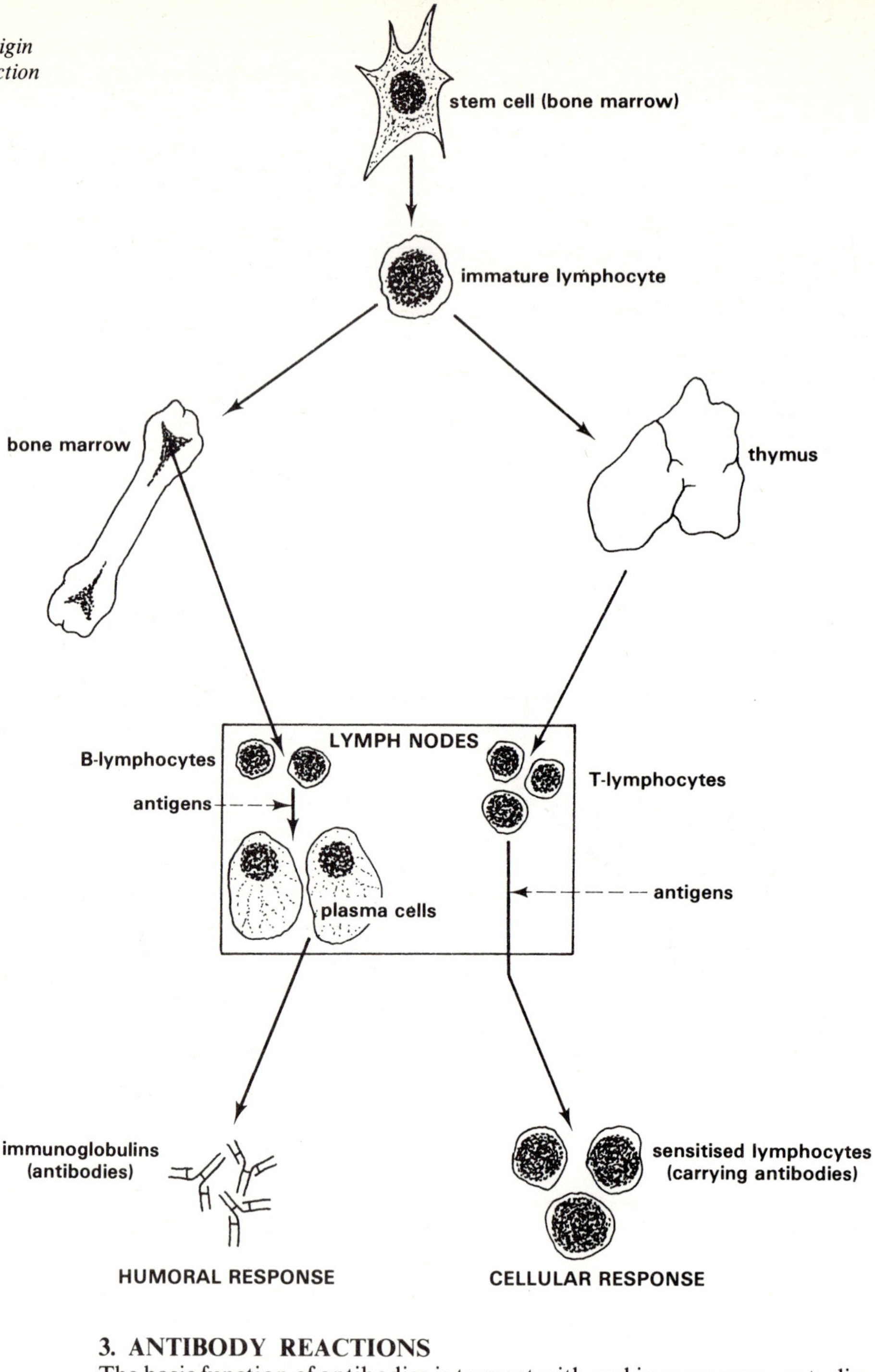

stem cell (bone marrow)

immature lymphocyte

bone marrow

thymus

LYMPH NODES

B-lymphocytes

antigens

plasma cells

T-lymphocytes

antigens

immunoglobulins
(antibodies)

sensitised lymphocytes
(carrying antibodies)

HUMORAL RESPONSE

CELLULAR RESPONSE

3. ANTIBODY REACTIONS

The basic function of antibodies is to react with and in some way neutralise antigens. In so doing, they eliminate or destroy the effect of the antigens. Since antigens usually are components of non-self cells, possibly pathogenic microbes, immunity results in resistance to disease. There are at least four types of antigen–antibody reactions, **neutralisation**, **precipitation**, **agglutination** and **lysis**. Some of the reactions make antigen-carrying cells susceptible to phagocytosis.

(i). **Neutralisation** is the reaction between antibodies and soluble antigens. Soluble antigens include toxins secreted by pathogenic microbes. The toxin produced by the bacterium *Clostridium tetani* which is the cause of tetanus is an example.

(ii). **Precipitation** involves the linkage of many antigen–antibody complexes to form a large insoluble lattice. Lattices are readily engulfed by phagocytes (Fig. 9.29).

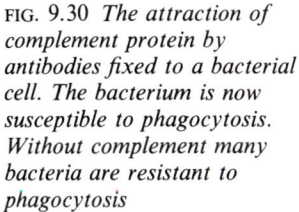

FIG. 9.29 *The ability of most antibodies each to react with two antigen molecules may result in the formation of a large* lattice *structure. Insoluble, the lattice precipitates and is readily engulfed by phagocytes*

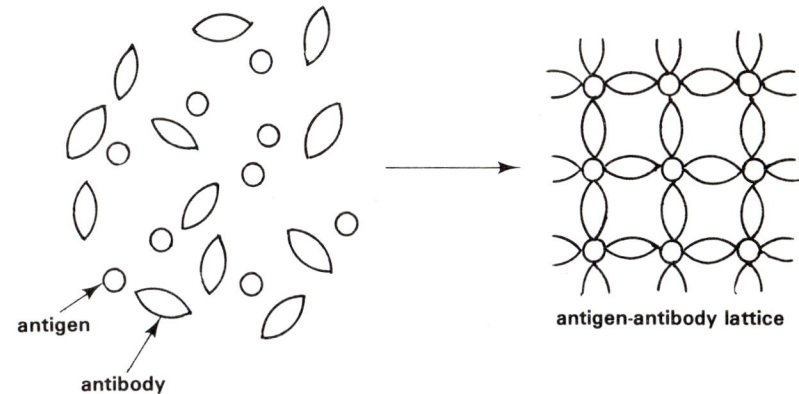

antigen

antibody

antigen-antibody lattice

(iii). **Agglutination** is the clumping together of antigen-carrying cells. Mammalian plasma contains a set of proteins called **complement** which helps antibodies to surround invading organisms such as bacteria (Fig. 9.30). Such action often brings about agglutination and promotes phagocytosis.

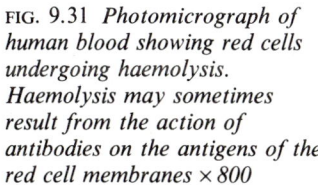

FIG. 9.30 *The attraction of complement protein by antibodies fixed to a bacterial cell. The bacterium is now susceptible to phagocytosis. Without complement many bacteria are resistant to phagocytosis*

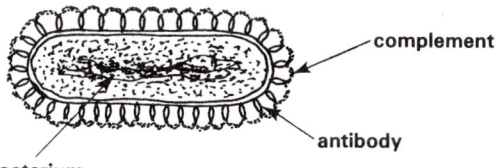

complement

antibody

bacterium

(iv). **Lysis** is the bursting of antigen-bearing cells. Complement, attracted to such cells by antibodies, contains digestive enzymes which dissolve the cells' outer covering (Fig. 9.31).

FIG. 9.31 *Photomicrograph of human blood showing red cells undergoing haemolysis. Haemolysis may sometimes result from the action of antibodies on the antigens of the red cell membranes × 800*

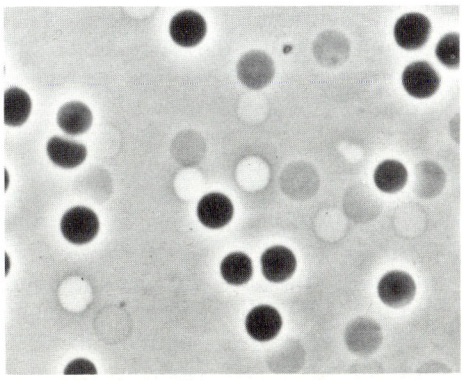

9.3 Vaccines

Among the main weapons in man's armoury against disease are **vaccines**. Vaccines are made from pathogenic microbes having the antigens which stimulate the body's immune system. The organisms are treated so that they do not give rise to disease when administered to a patient. The antibodies produced as a result of vaccination give immunity against disease following subsequent exposure to disease-causing microbes. Unfortunately it is not yet possible to make vaccines against all diseases of microbial origin.

In 1797 Edward Jenner, a Gloucestershire country doctor, was the first to use a vaccine successfully in preventing a human disease. He noticed that dairymaids who had milked cows suffering from cowpox were far less susceptible to the much more virulent and often fatal smallpox, although they often showed mild symptoms of the disease, such as hand sores. Jenner removed some of the liquid from a sore on the hand of a dairymaid and scratched it into the skin of a young boy, James Phipps. Later Jenner inoculated Phipps with material from the sores of someone suffering from smallpox. The boy was found to be immune to smallpox. Jenner had used an antigen to produce immunity against a dangerous pathogen. It is only relatively recently that mass vaccination programmes have virtually eliminated smallpox all over the world.

The reaction of the body to cowpox vaccine is an example of **active immunisation**. The body actively makes its own antibodies against the antigen. Protection against measles, whooping cough, poliomyelitis and tuberculosis is provided in this way. It is far too dangerous, however, to introduce some disease-causing organisms into the human body. Instead the pathogen is injected into another animal such as a horse. Antibodies soon appear in the horse's blood which is used to prepare an **antiserum**. The antiserum can then be injected into a human to provide **passive immunity**. It is in this way that humans are immunised against tetanus and diphtheria. Passive immunity is important in a foetus which absorbs antibodies from its mother across the placenta. The mother's milk also contains antibodies which provide passive immunity in the gut of new born mammals.

9.4 Allergy

About 10 per cent of the population in Britain has an **allergy**. Allergies include hay fever, asthma, childhood eczema and food allergies. An allergy is an immune response to an antigen called an **allergen**, to which most people show no reaction. Allergens are found on pollen grains, fungal spores, house dust, feathers, fur and in a variety of foods. Allergens on pollen for example, become attached to the mucus membranes in the breathing passages. The presence of allergens in people who suffer from an allergy stimulates the production of antibodies called **reagins**. It is thought that reagins are made in the lymph nodes. They circulate in the blood and become attached to cells throughout the body, particularly in the skin and mucus surfaces of the mouth, nose and breathing passages. The reagins can remain for years in these tissues which are said to be **hypersensitive**. Later, whenever there is exposure to allergens, an allergen–reagin reaction takes place. The reaction triggers a vigorous response involving the rupture of certain tissue cells and the release of **histamine**. Histamine causes inflammation of the affected tissues and constriction of the bronchi leading to breathing difficulties. Eosinophils increase in number during allergic responses. They are thought to have an anti-inflammatory effect by absorbing histamine from the tissues.

9.5 Blood groups

Since the pioneering work of Landsteiner in the 1900s it has been known that people can be categorized into one or more **blood groups**. A knowledge of blood groups is important because transfusion of blood from one person to another may cause an immune response in the recipient if the donor's blood is of a different group, that is if the donor's blood is **incompatible**. The response results in mass haemolysis of the donated red blood cells and oxygen carriage becomes affected. Before haemolysis some of the donor's red cells agglutinate and blockages occur in the capillaries of vital organs, usually with fatal results.

About twenty blood group systems have been described in man. They are based on the antigens which form part of the red cell membranes (Table 9.3). The best known blood group system is the **ABO system**. Human red cells possess one, both or neither of two antigens called **A** and **B**. Cells with A belong to blood **group A**, those with B belong to **group B**, those with both A and B belong to **group AB**. Cells with neither antigen belong to **group O**.

Table 9.3 The red cell antigens of the main blood group systems in man

SYSTEM	ANTIGENS
ABO	A_1 A_2 A_3 A_x B and others
Rhesus	D C c C^w C^x E D^u and others
MNSs	M N S s
P	P_1 P^k
Lutheran	Lu^a Lu^b
Kell	K k Kp^a Kp^b Js^a Js^b
Lewis	Le^a Le^b
Duffy	Fy^a Fy^b
Diego	Di^a Di^b
Yt	Yt^a Yt^b
I	I i
Xg	Xg^a
Kidd	Kj

In the plasma are two **isoantibodies** called **anti-A** and **anti-B**. Isoantibodies are not produced as a result of an immune response. Anti-A is found in the blood of group B individuals, anti-B in group A individuals. Both anti-A and anti-B are in the blood of group O individuals and neither antibody appears in the blood of group AB individuals. Table 9.4 shows the distribution of the antigens and antibodies of the ABO system. It can be seen that anti-A normally never encounters the A antigen, nor anti-B the B antigen. If they do, following incompatible transfusion for example, agglutination and haemolysis of the red cells occurs.

A person's blood group can be determined by mixing a small sample of blood cells with a variety of antisera. These are serum samples containing known blood group antibodies. Careful observation of which antisera agglutinate the red cells enables the selection of compatible blood for transfusion (Fig. 9.32).

The other well-known blood group system in man is the **rhesus system**. It was discovered from work on rhesus monkeys. The rhesus system involves several red cell antigens, mainly the **D-antigen**. About 75 per cent of the British population possess the D-antigen and are **rhesus-positive**. The rest without the D-antigen are **rhesus-negative**. There is no isoantibody to the D-antigen. However, if D-carrying red cells are transfused into a rhesus negative individual the recipient produces anti-D antibody by immune response. Anti-D antibody reacts with the D-antigen resulting in agglutination and haemolysis of the rhesus-positive cells. Consequently it is important to determine the rhesus group as well as the ABO group of blood before transfusion.

Table 9.4 The antigens and antibodies of the human ABO blood group system. There are sub-groups of A and AB called A_2 and A_2B but these have been omitted

Blood group	O	A	B	AB
red cell antigen	—	A	B	A + B
plasma antibody	anti-A + anti-B	anti-B	anti-A	—
British population/per cent	46·7	41·7	8·6	3

FIG. 9.32 *The results of blood group tests on four different samples. In each well blood cells to be tested are mixed with known blood group antibodies. What are the ABO and Rhesus groups of the four samples? What pattern of agglutination do you think cells of group O Rhesus negative would give in this test?*

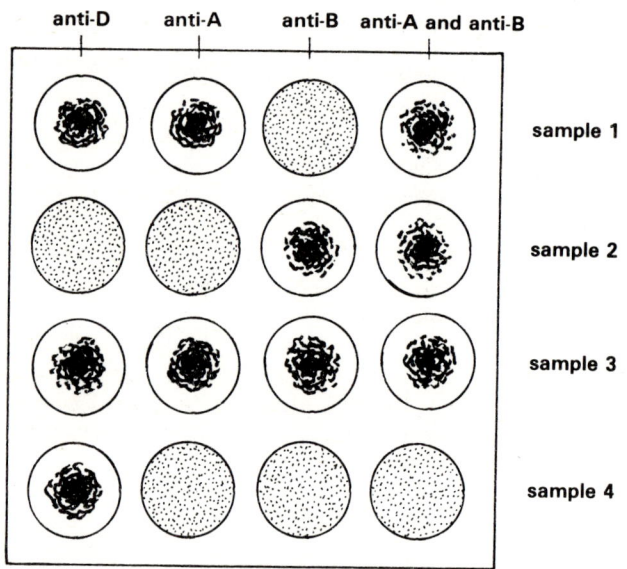

A complication concerning the rhesus system can arise during pregnancy if a rhesus-negative woman is carrying a rhesus-positive foetus. Normally the placenta keeps the D-carrying foetal red cells out of the mother's circulation. Nevertheless, the stresses caused by the muscular contractions of the uterus during birth can cause some foetal blood cells to enter the mother's blood. The mother's immune system recognises the D-antigen on the foetal cells as non-self and produces anti-D antibody. Since the anti-D antibody can pass across placental membranes, a rhesus-positive foetus of any future pregnancy is at risk. In about 10 per cent of such cases severe haemolysis of foetal red cells occurs during pregnancy.

The reaction is the cause of a dangerous condition called **haemolytic disease of the newborn**. Transfusion of fresh blood into the foetus while still in the uterus may be necessary several times during the pregnancy. If the reaction can be predicted from a knowledge of the mother's rhesus group, such drastic measures can be avoided. Anti-D antiserum is injected into her blood immediately after the first birth. Any D-carrying foetal cells which may have entered her system are destroyed before they stimulate her own anti-D production. A rhesus-positive foetus of a subsequent pregnancy should be safe from haemolytic disease.

9.6 Organ transplantation

Another area of medicine in which the immune system plays an important part is **organ transplantation**. The surgical difficulties of transplanting organs such as a kidney or a heart from one person to another are not great. However, the transplanted organs carry about forty antigens which stimulate the production of antibodies in the recipient. The antibodies eventually **reject** the transplanted tissues. To a certain extent it is possible to match donor and recipient in a way comparable to the selection of compatible blood before transfusion. The antigens of tissues which make up organs are similar to those of the white cells in the blood and they can be categorised into one of several groups. The system most often used is the **histocompatibility** or **HL-A system**. Even so, perfect matchings are virtually impossible, even in closely related individuals whose antigens are most likely to be similar.

One means of minimizing the rejection of transplanted organs is to administer **immunosuppressant drugs**. The drugs counteract the immune system and block antibody production in response to the antigens of the transplanted organ. The main problem arising from the use of immunosuppressant drugs is that they prevent an adequate immune response to other antigens such as those of pathogenic microbes. Mild infections of common diseases may thus become fatal in immunosuppressed patients. Much research is necessary in this field before the problem of rejection is conquered. Recent work has shown that the HL-A antigens are the means by which the body's immune system recognises self and non-self. When body cells are attacked by viruses the HL-A antigens on the affected cells are modified by the virus. T-lymphocytes recognise the change as non-self, attack the infected cell and the viruses inside. Autoimmune diseases may therefore be due to a change in the HL-A antigens of body cells. T-lymphocytes then misrecognise self cells as non-self cells and destroy them.

Chapter 10

The mammalian urinary system

The complex and numerous activities of the mammalian body require a stable environment in which to take place. Even slight chemical or physical changes can upset the smooth functioning of the body. Many of the physiological activities of mammals regulate the constitution of the body's internal environment, and do so to within relatively narrow limits. Maintaining a stable internal environment is called **homeostasis**.

The very functions and activities which need a stable environment change it continuously. Metabolism involves the production of wastes. If wastes accumulate in the body they can become toxic. Consequently mammals like other animals eliminate or **excrete** their wastes. Excretion is the elimination of any substances which are present in the body in concentrations exceeding normal levels, whether metabolic wastes or not. Among the organs concerned with excretion and homeostasis are the kidneys which form part of the **urinary system**.

10.1 The urinary system

The **kidneys** are paired organs found in the abdominal cavity. They are held firmly in position by the peritoneum, a thin layer of tissue lining the abdominal cavity. The kidneys are usually embedded in fat (Fig. 10.1).

Each kidney receives blood from the general circulation through a **renal artery** and is drained of blood by a **renal vein**. In kidneys the blood circulates in a capillary network which surrounds numerous microscopic urinary tubules called **nephrons**. The nephrons remove excess and unwanted materials from the blood. The excretory products collect as **urine** which passes from the kidneys through the **ureters**. Urine is stored temporarily in the **urinary bladder** before it is eliminated from the body. Emptying of the bladder is called **micturition** and is controlled by the autonomic nervous system (Chapter 15). The exit from the bladder into the **urethra** is closed by contraction of a ring of muscle called the **bladder sphincter**. As the bladder fills, cells in the bladder wall which are sensitive to stretching trigger off a reflex action which results in relaxation of the bladder sphincter. Simultaneous contraction of the smooth muscle in the bladder wall forces the urine out through the urethra. Micturition can be controlled by voluntary nervous activity which is learned by humans in early life (Fig. 10.2).

A human kidney is a compact organ about 7–10 cm long and 2·5–4 cm wide. Inside, the tissues are in distinct regions. There is an outer **cortex** and an inner **medulla** (Fig. 10.3). The internal appearance of a kidney is due to the arrangement of the blood vessels and nephrons which make up most of the organ. The nephrons are the functional units of the kidney and there are about a million nephrons in each human kidney.

Nephrons are very small thin-walled tubules between 2 and 4 cm long in man (Fig. 10.4). At one end is a cup-shaped **Bowman's capsule** which encloses a small group of capillaries called a **glomerulus**. A Bowman's capsule and the glomerulus together are called a **Malpighian body**. The capsule leads into a coiled structure called the **proximal convolution** which opens into the **loop of Henle**. The loop of Henle consists of a descending limb and an ascending limb and leads into a second coiled tube, the **distal convolution**. The distal convolutions of many nephrons join a common **collecting duct** and many collecting ducts lead through the medulla to the **renal pelvis**.

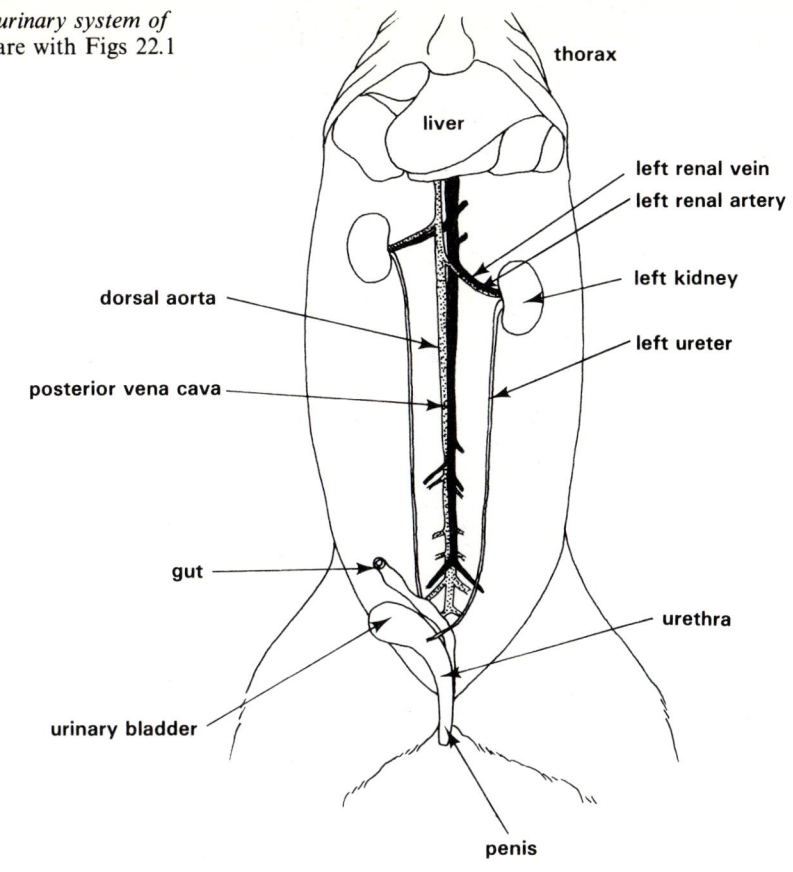

FIG. 10.1 *The urinary system of the rat.* Compare with Figs 22.1 and 22.2

thorax

liver

left renal vein

left renal artery

left kidney

left ureter

dorsal aorta

posterior vena cava

gut

urethra

urinary bladder

penis

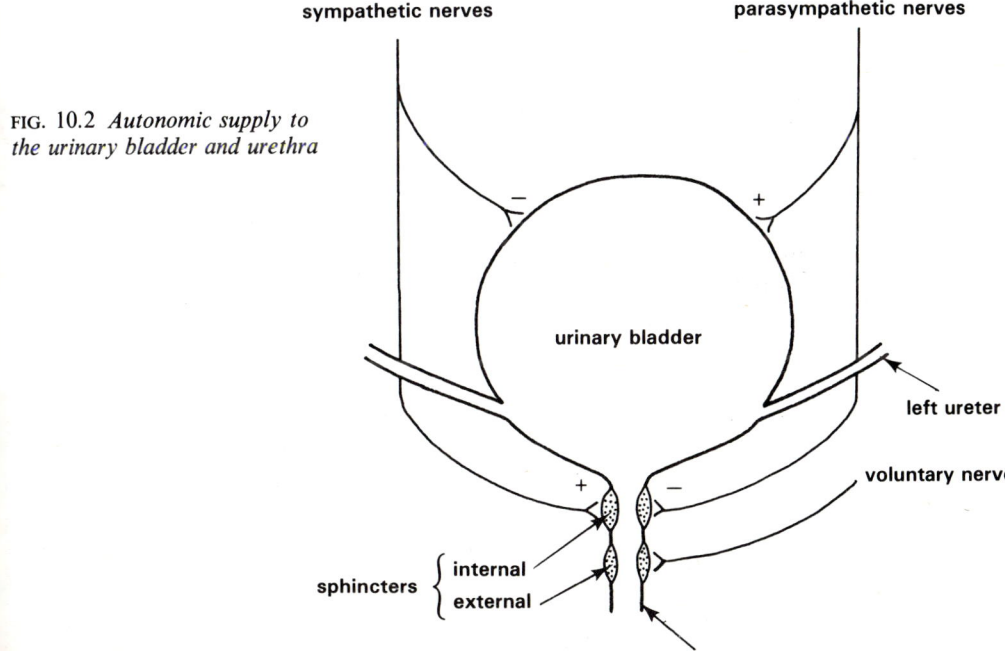

sympathetic nerves

parasympathetic nerves

FIG. 10.2 *Autonomic supply to the urinary bladder and urethra*

urinary bladder

left ureter

voluntary nerve

sphincters { internal
external

urethra

183

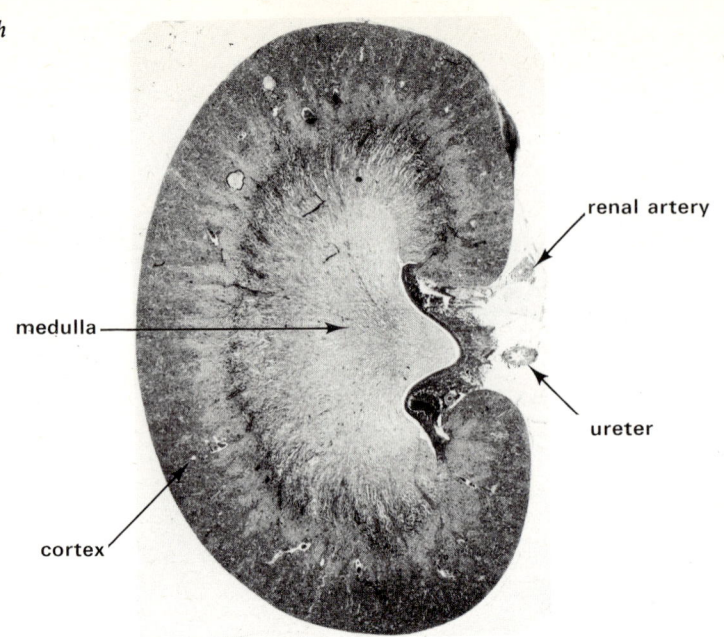

FIG. 10.3 (a) *Photomicrograph of a vertical section of a rat kidney,* ×5

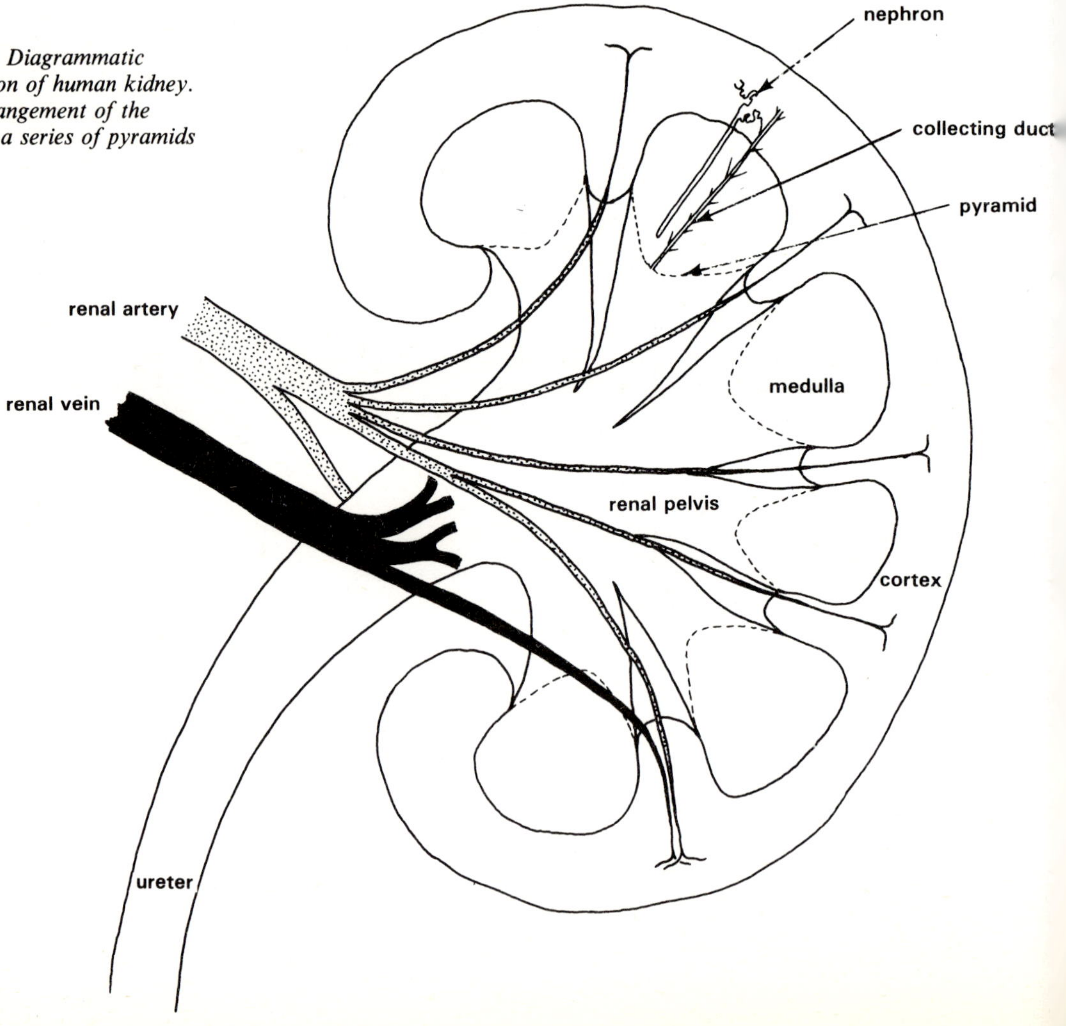

renal artery

medulla

ureter

cortex

FIG. 10.3 (b) *Diagrammatic vertical section of human kidney. Note the arrangement of the medulla into a series of pyramids*

nephron

collecting duct

pyramid

renal artery

renal vein

medulla

renal pelvis

cortex

ureter

FIG. 10.4 (a) *Photomicrograph of a thin section of monkey kidney in the cortical region, × 300*

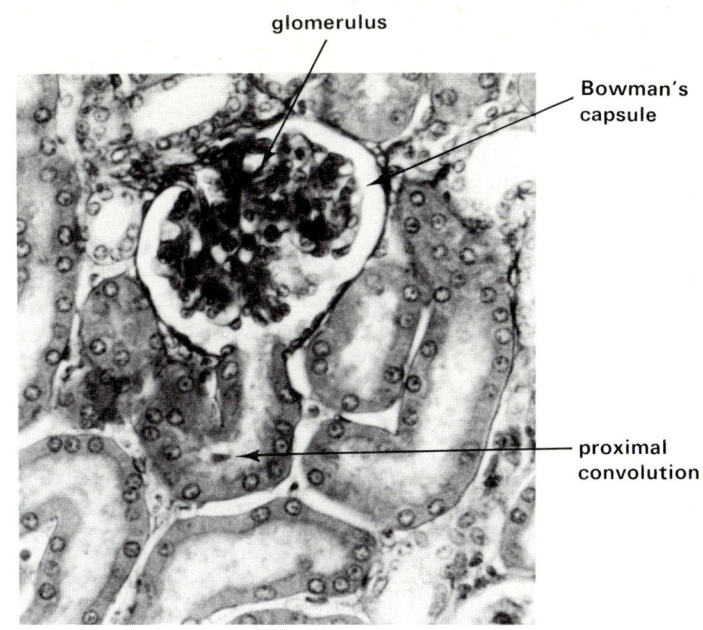

glomerulus

Bowman's capsule

proximal convolution

FIG. 10.4 (b) *A nephron and its blood supply*

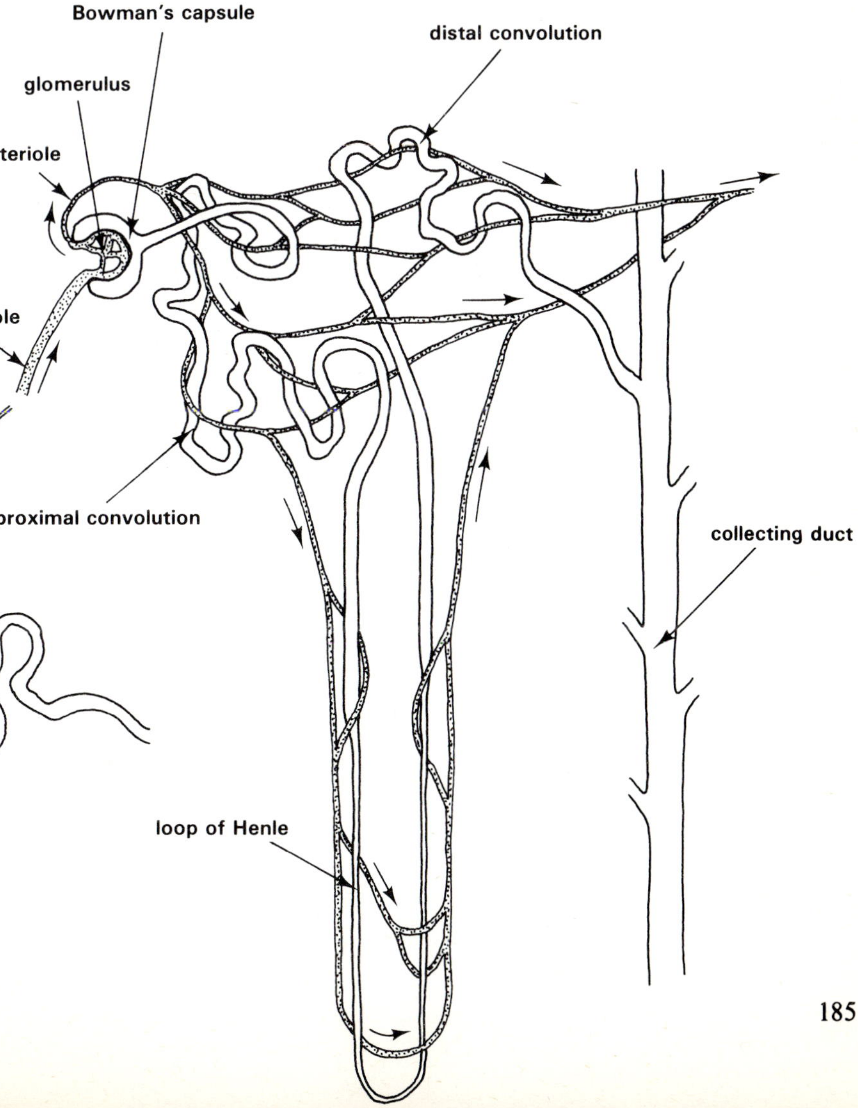

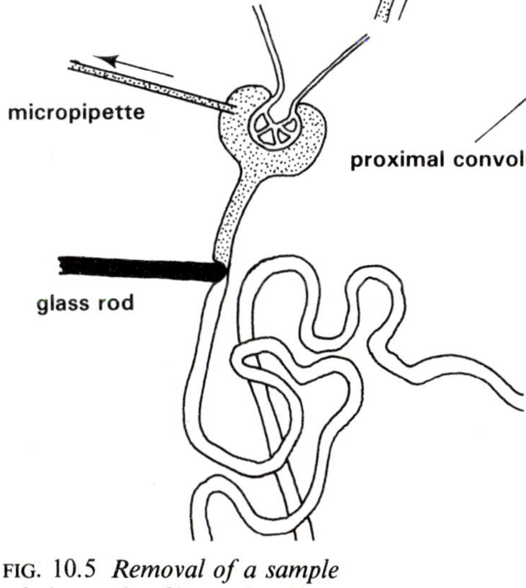

Bowman's capsule

distal convolution

glomerulus

efferent arteriole

afferent arteriole

micropipette

proximal convolution

collecting duct

glass rod

loop of Henle

FIG. 10.5 *Removal of a sample of glomerular filtrate for analysis*

185

10.2 The functions of nephrons

Much of what is known of the functions of kidneys comes from quantitative analyses of the fluid inside nephrons. This involves the insertion of delicate micropipettes into nephrons of experimental animals. The fluid is drawn off carefully and analysed to determine what changes have occurred during its passage through a nephron (Fig. 10.5).

10.2.1 Ultrafiltration

The first activity of a nephron is the **ultrafiltration** of blood brought to the Bowman's capsule by arterioles. The aterioles branch into the capillaries of the glomeruli, tightly nestled in the capsules. Electron microscope studies of the capsules show that the only effective barrier between the blood in the glomeruli and the cavity of the capsules is a thin porous **basement membrane** (Fig. 10.6).

FIG. 10.6 (a) *Electronmicrograph showing a section through a podocyte and a glomerular capillary,* × 12 000

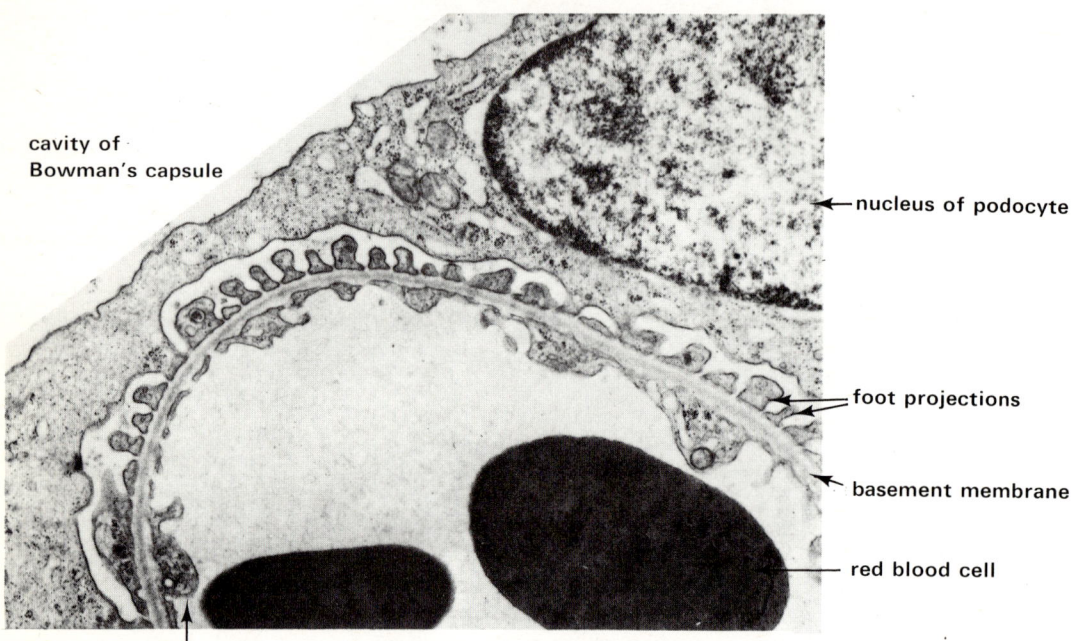

cavity of Bowman's capsule

nucleus of podocyte

foot projections

basement membrane

red blood cell

endothelium of glomerular capillary

The hydrostatic pressure of blood in the glomerular capillaries is relatively high because the diameter of the afferent arterioles is greater than that of the efferent arterioles. Blood pressure is normally maintained by the pumping action of the heart. If the blood pressure falls too low, for example when much blood is lost in an accident, temporary use of a kidney machine may be required until the patient's blood volume and pressure are restored (section 10.5).

The basement membrane of the capsule is permeable to some blood constituents but not to others. Because the blood in the glomerulus has a **high hydrostatic pressure** it is filtered through the basement membrane. Apart from the blood cells and substances of high relative molar mass such as the plasma proteins, the chemical composition of the glomerular filtrate is virtually the same as the plasma. There is almost total transfer of materials from the blood in the glomeruli to the Bowman's capsules. The composition of urine, however, is very different. The filtrate is therefore modified considerably while passing along the nephrons. The volume of urine excreted is also much less then the volume of filtrate produced in a given time (Table 10.1).

FIG. 10.6 (b) *Diagram of the barrier between the glomerular blood and the filtrate in Bowman's capsule.* Compare with the electronmicrograph, Fig. 10.6 (a)

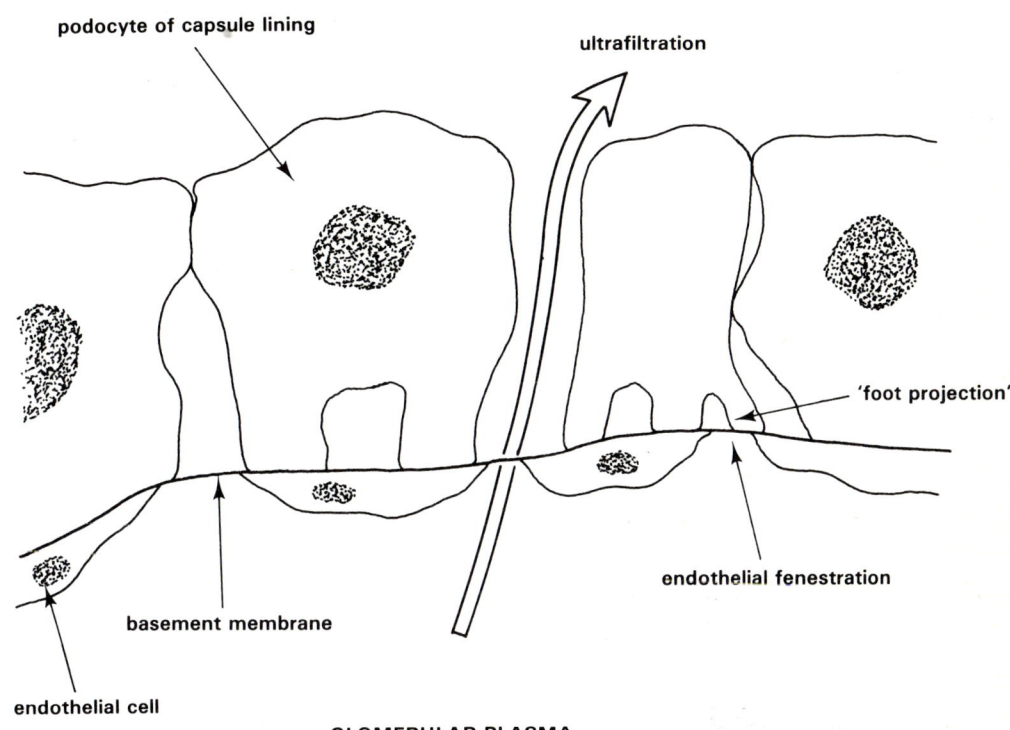

FILTRATE

podocyte of capsule lining

ultrafiltration

'foot projection'

endothelial fenestration

basement membrane

endothelial cell

GLOMERULAR PLASMA

Table 10.1 Some human blood constituents and the quantities filtered and reabsorbed by the kidneys in a day

CONSTITUENTS	AMOUNT IN FILTRATE/g	AMOUNT IN URINE/g	PERCENTAGE REABSORPTION
sodium ions	600	6	99
potassium ions	60	2	97
calcium ions	5	0·2	96
glucose	200	0	100
urea	60	35	42
water	180 dm³	1·5 dm³	99

10.2.2 Reabsorption

Changes in the composition of the filtrate begin in the proximal convolutions. Here the epithelial cells of the nephron wall reabsorb a large proportion of the filtrate, passing it back into the blood flowing in the surrounding vessels. **Reabsorption** of individual substances is at a rate just sufficient to maintain normal concentrations in the blood. Any excesses stay in the nephron and are later excreted.

187

Since the transfer of materials from the filtrate into the blood is against a concentration gradient, reabsorption is active (Fig. 10.7). The energy for **active uptake** is provided by respiration in the nephron's cells which contain many mitochondria. The efficiency of reabsorption is helped by the presence of numerous **microvilli** which greatly enlarge the surface area through which the materials pass (Fig. 10.8).

FIG. 10.7 *Active solute reabsorption from the proximal convolutions against a concentration gradient as the levels of reabsorbed materials exceed those remaining in the filtrate*

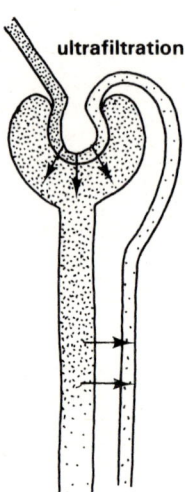

ultrafiltration

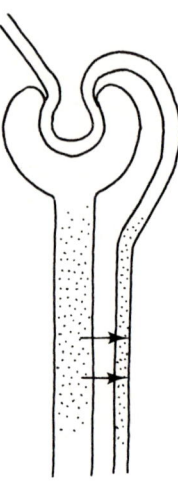

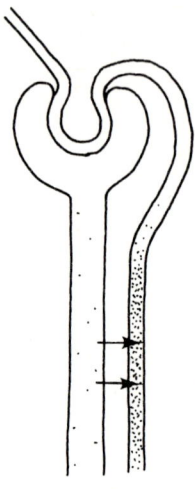

passive reabsorption (diffusion) along a concentration gradient

limit of passive reabsorption

active reabsorption against a concentration gradient

FIG. 10.8 *Electronmicrograph of a cell lining the proximal convolution of a nephron, ×5800*

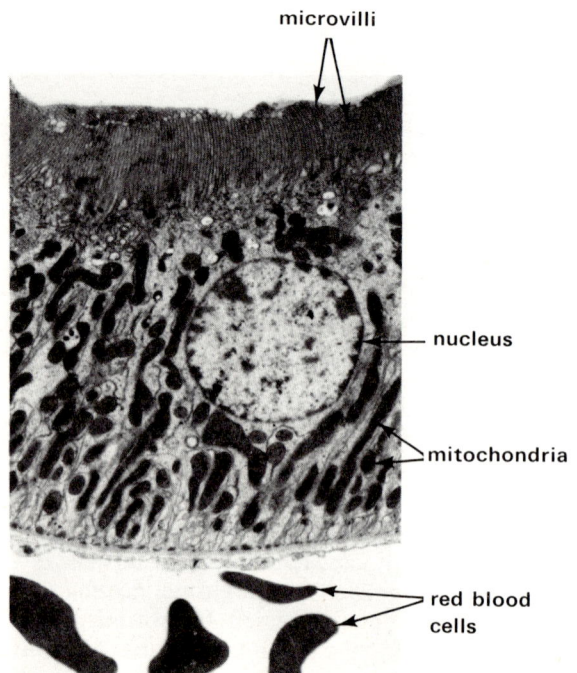

microvilli

nucleus

mitochondria

red blood cells

In humans about 120 cm³ of water pass into the nephrons every minute. Of this, about 100 cm³ per minute are reabsorbed passively from the proximal convolutions. The concentration gradient of solutes between the remainder of the filtrate in the proximal convolution and the blood in surrounding capillaries promotes **osmosis**. The osmotic reabsorption of water is further assisted by the colloid osmotic potential of the blood due to the plasma proteins (Chapter 9). Table 10.1 gives an indication of the extent of filtrate reabsorption from the proximal convolutions.

About 19 cm³ of every 20 cm³ of water left in the nephrons are reabsorbed every minute from the distal convolutions and collecting ducts. The extent to which water is reabsorbed from these parts of the nephrons depends on the body's state of hydration. If the body's water content is below normal the walls of the distal convolutions and collecting ducts become very permeable to water. If the body contains sufficient water, however, the walls of the distal convolutions become less permeable to water. Water is then reabsorbed slowly into the blood. Excess water then remains in the urine to be excreted (Fig. 10.9).

FIG. 10.9 *The fate of water in the distal parts of the nephron, depending on the body's state of hydration*

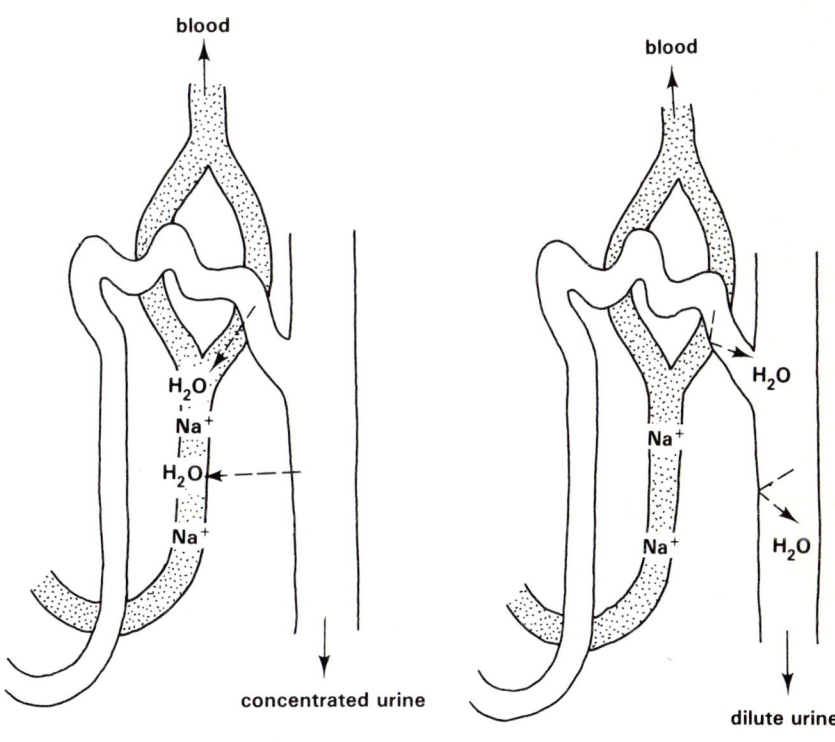

The arrangement of the loops of Henle and nearby blood vessels in the medulla of the kidney is important in efficiently maintaining the necessary water potential gradient for water reabsorption. The filtrate passing down the descending limb of the loop of Henle contains fewer sodium ions than the surrounding blood and tissue fluid which receive sodium ions pumped out of the ascending limb of the loop of Henle (Fig. 10.10). Consequently as the filtrate passes down the descending limb it absorbs sodium ions by diffusion. By the time the filtrate enters the ascending limb its sodium ion concentration is relatively high. The cells of the ascending limb expend only a little energy in pumping sodium ions out of the filtrate and into the blood and tissue fluid. This mechanism is called a **counter-current multiplication** system (Fig. 10.11).

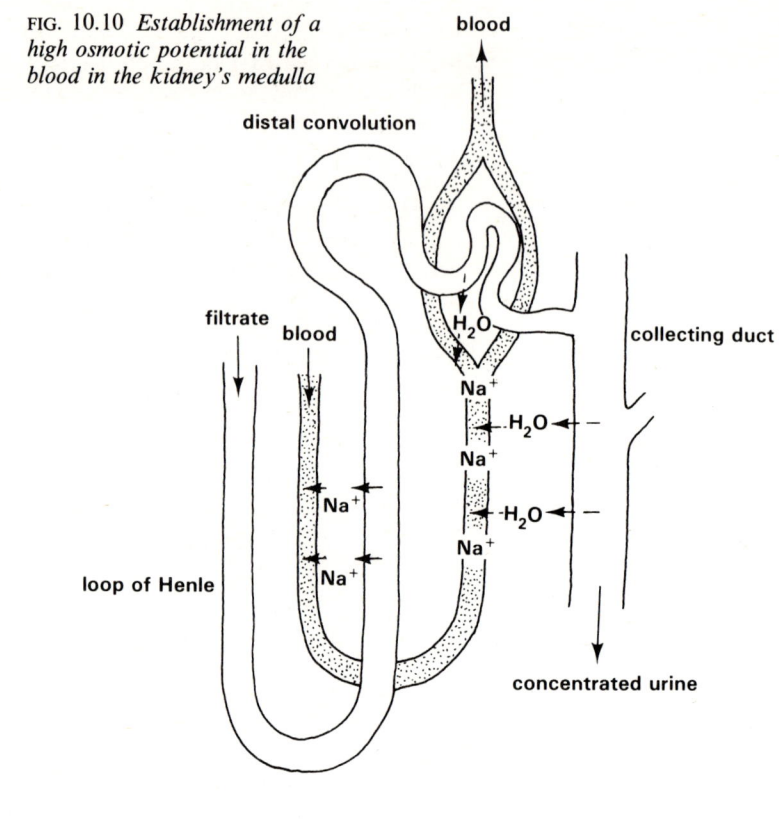

FIG. 10.10 *Establishment of a high osmotic potential in the blood in the kidney's medulla*

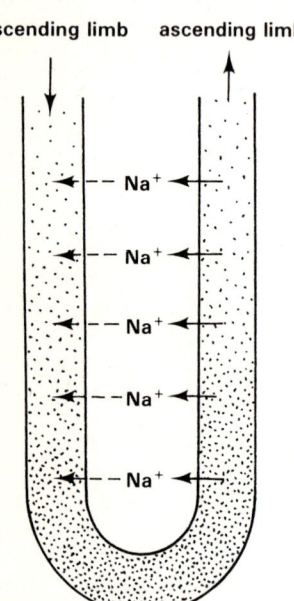

FIG. 10.11 *Counter-current multiplication system in the loop of Henle*

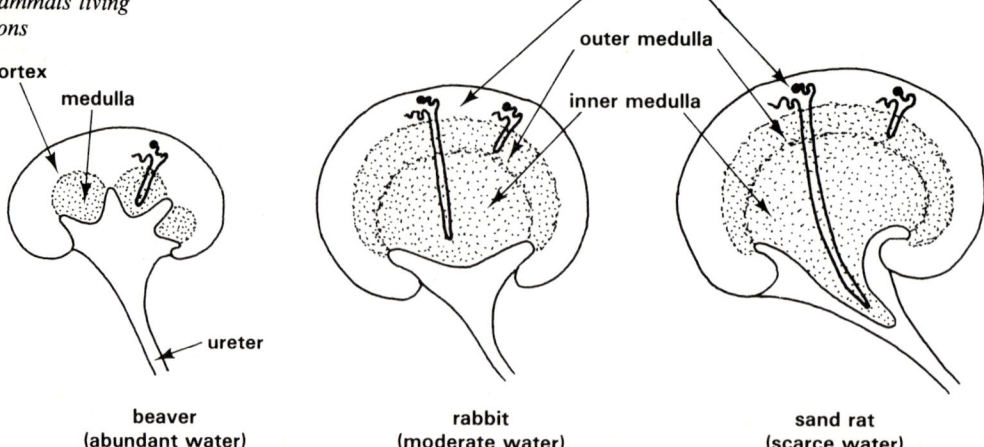

FIG. 10.12 *Relative proportions of the loops of Henle in the kidneys of three mammals living in different conditions*

beaver
(abundant water)

rabbit
(moderate water)

sand rat
(scarce water)

If the loops of Henle are long, more time is available for the filtrate passing down the descending limb to absorb sodium ions from the medulla. The counter-current effect is then increased and the ascending limb can pump more sodium ions into the blood and tissue fluid. The water potential gradient between blood and filtrate in the distal convolutions and collecting ducts is then much greater. The ability to concentrate urine is important to mammals living in arid conditions. It is therefore of interest that the nephrons of many desert mammals have unusually long loops of Henle (Fig. 10.12).

10.2.3 Regulation of nephron function

The active reabsorption of solutes from the proximal convolutions is regulated by hormones. Reabsorption of water from the proximal convolutions is passive by osmosis. Osmosis normally accounts for the reabsorption of over 80 per cent of the water in the filtrate from the glomeruli. As we saw in the previous section, water reabsorption from the distal parts of the nephrons is also by osmosis. The extent of water reabsorption depends on the body's state of hydration and the permeability of the walls of the distal convolutions and collecting ducts. How does the body detect and monitor its water content? How is this information translated into permeability changes of the walls of the distal convolutions and collecting ducts?

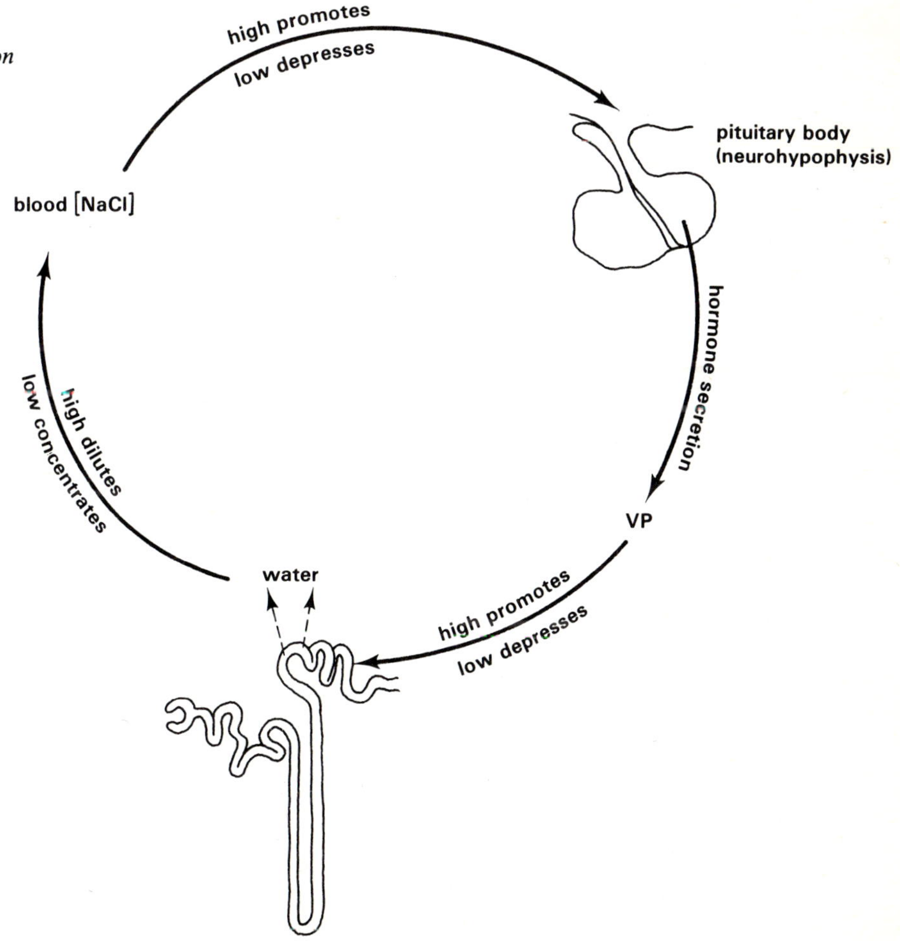

FIG. 10.13 *Mechanism controlling water reabsorption from the distal parts of the nephrons.* See Fig. 10.9

high promotes
low depresses

pituitary body (neurohypophysis)

blood [NaCl]

hormone secretion

low concentrates

high dilutes

VP

water

high promotes
low depresses

Beneath the **hypothalamus** in the brain and projecting downward from it is an endocrine gland called the **pituitary body** (Chapter 18). The hypothalamus is sensitive to the concentration of sodium chloride in the blood flowing through it. Since the concentration of sodium chloride is an indirect indication of the volume of water in the body, the hypothalamus is sensitive to the water content of the blood. If the water content is low the sodium chloride concentration is high. The pituitary body, stimulated by the hypothalamus, releases **vasopressin** (VP) sometimes called **antidiuretic hormone**. Vasopressin increases the permeability of the walls of the distal convolutions and collecting ducts to water, encouraging reabsorption of water into the blood. Consequently the small volume of urine eventually eliminated is relatively concentrated, a condition called **antidiuresis**. Conversely, if the body's water content is high VP output diminishes. The rate of water reabsorption from the distal ends of the nephrons then slows down. Urine flow increases and the urine becomes diluted, a condition called **diuresis** (Fig. 10.13).

Pituitary malfunction can lead to reduced VP output resulting in a daily urine production more than ten times the average 1·5 dm^3. This condition is called **diabetes insipidus** and is quite distinct from diabetes mellitus (Chapter 18).

10.3 Extreme cases of water economy

The camel is an animal which seems ideally fitted to life in the desert. Indeed, the camel can survive for many days without water. The exact length of time depends mainly on temperature and the availability of food. Camels have been reported to travel more than 900 km between watering points without drinking, taking three weeks to complete the trip.

For a long time it was assumed that the camel's hump, or even its stomach acted as a kind of water store. Investigations of the stomach ruled out such a function and the hump is filled with fatty tissue. Certainly the fat can be respired with the production of water (Chapter 5). However, in breathing, to obtain the oxygen needed for respiration, even more water is lost from the lungs by evaporation (Chapter 8). Schmidt-Nielsen and others have investigated the water relations of desert animals including camels. Their findings indicate that a camel's nephrons are particularly efficient in reabsorbing water from the glomerular filtrate. After allowing for differences in body size, the daily urinary water loss of a camel is only about 25 per cent that of a human.

More significant, however, is the camel's very great **tolerance of dehydration**. A camel can survive water loss of up to one third of its total body mass with no apparent ill-effects. This remarkable tolerance of dehydration cannot be matched by any other mammal. It is estimated that a water loss of 15 to 20 per cent of total body mass would be fatal to a human. Recent work has revealed the presence of a special protein in the membranes of the red blood cells of camels. The protein strengthens the red cells and protects them against collapse when the surrounding plasma becomes concentrated because of water loss. Most mammals sweat and pant both of which help maintain a constant body temperature. By sweating and panting very little camels preserve water, but in doing so camels have to tolerate wide internal temperature fluctuations. A camel's temperature can vary from about 34·5°C to 40·5°C, a 6°C variation which would be very harmful to other mammals (Chapter 17).

Another mammal well fitted for survival in desert conditions is the kangaroo rat, *Dipodomys* (Chapter 8). This rodent has such a well-balanced water economy that it does not normally drink at all. Its total water requirement is provided by the small amount of moisture in the rather dry vegetation it eats and more important, the moisture produced by respiration of its food.

$$C_6H_{12}O_6 + 6O_2 \longrightarrow 6CO_2 + 6H_2O$$

Survival on **oxidation water** depends on reducing water loss to an absolute minimum. *Dipodomys* has highly efficient kidneys which produce very concentrated urine. The kangaroo rat also keeps to a minimum the amount of water vapour it breathes out (Chapter 8).

In most terrestrial animals, drinking sea water raises the osmotic potential of the blood plasma, resulting in dehydration of the body tissues by osmosis. Schmidt-Nielsen predicted that kangaroo rats concentrate their urine so much that they should be able to drink sea water without ill-effect. To test the prediction he fed experimental kangaroo rats with soya beans, the high protein content of which resulted in a high concentration of blood urea (Chapter 14). This led to excessive water loss during urination since urea is toxic and is eliminated by the kidneys. Kangaroo rats which were given sea water to drink after this treatment recovered just as well as a control group provided with fresh water. The experiment demonstrates the ability of kangaroo rats to eliminate excess salts, such as sodium chloride, in forming highly concentrated urine.

10.4 The role of the kidneys in acid-base balance

Buffers such as the carbonic acid-hydrogencarbonate acid–base pair (Chapter 2) are limited in the extent to which they can cope with excess hydrogen ions in the body. The final regulation of the pH of body fluids is carried out by the lungs and kidneys. Gas exchange in the lungs disposes of carbon dioxide (Chapter 8). This prevents the build up of carbonic acid in the tissues with accompanying high concentration of hydrogen ions.

Relatively small but significant quantities of acids enter the body daily in food. These and other acids are disposed of by the kidneys in such a way as to keep basic ions like Na^+ (sodium ions) in the body. Acid excretion in the kidneys takes place in the distal convolutions of nephrons. Here, inside the tubule cells, hydrogen (H^+) and hydrogencarbonate (HCO_3^-) ions are formed from carbonic acid. The carbonic acid comes from the carbon dioxide produced by respiration in the tubule cells (Fig. 10.14). The hydrogen ions are secreted into the tubule where they meet and react with disodium hydrogenphosphate in the glomerular filtrate. The hydrogen ions replace some of the sodium ions producing sodium dihydrogenphosphate. Sodium ions are then actively reabsorbed into the blood where they combine with the hydrogencarbonate ions from the tubule cells (Fig. 10.15).

FIG. 10.14 *Production of hydrogencarbonate and hydrogen ions from respiration*

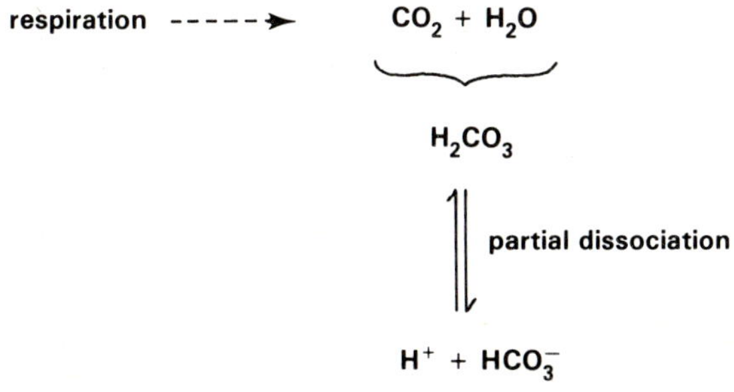

respiration $-----\blacktriangleright$ $CO_2 + H_2O$

H_2CO_3

partial dissociation

$H^+ + HCO_3^-$

FIG. 10.15

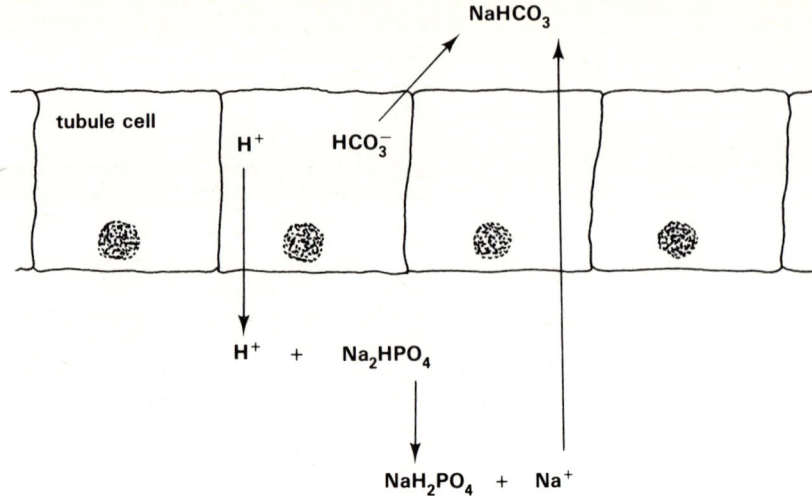

If the acid content of the blood is high, the resulting acidity of the glomerular filtrate stimulates the production of ammonia in the tubule cells. The ammonia completely replaces sodium ions from disodium hydrogenphosphate resulting in increased acid excretion and retention of Na^+ ions (Fig. 10.16).

FIG. 10.16

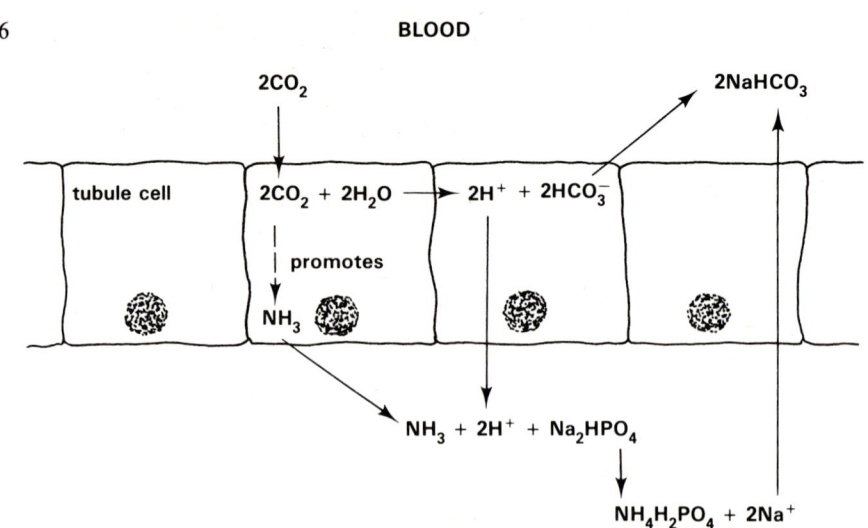

The pH of human blood lies normally between pH 7·35 and 7·45. A drop in pH below 7·30, a condition called **acidosis**, or a rise above 7·50 called **alkalosis**, results in metabolic malfunction. So sensitive are the body's physiological processes that failure to regulate the pH of the blood between 7·0 and 8·0 is usually fatal.

10.5 Kidney machines

Human kidney failure can be treated in two main ways. **Transplantation** of a healthy kidney from another person called a donor involves immunological reactions which have not as yet been completely overcome (Chapter 9). Alternatively, **kidney machines** can be used. A kidney machine is a mechanical device through which a patient's blood passes. The blood leaves the body usually from an artery in the forearm and returns to a nearby vein (Fig. 10.17). Inside the machine the blood flows over or between membranes which separate it from an aqueous **dialysing fluid** containing dissolved sugars and salts in concentrations normally found in blood. Soluble constituents in the blood in excess of normal concentrations diffuse across the membrane into the dialysing fluid. In this way wastes like urea which accumulate in the body are extracted. Blood cells and proteins remain in the blood. The process is called **renal dialysis** (Fig. 10.18).

FIG. 10.17 (a) *Patient connected to a kidney machine*

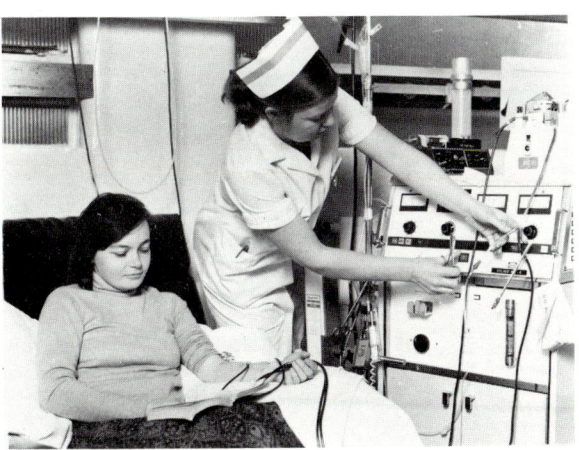

FIG. 10.17 (b) *Diagrammatic arrangement of the apparatus causing dialysis of a patient's blood in a kidney machine*

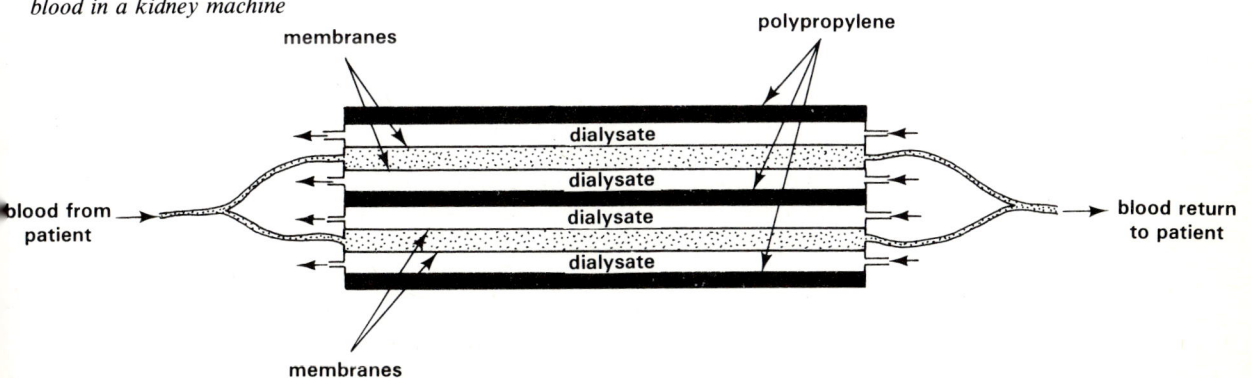

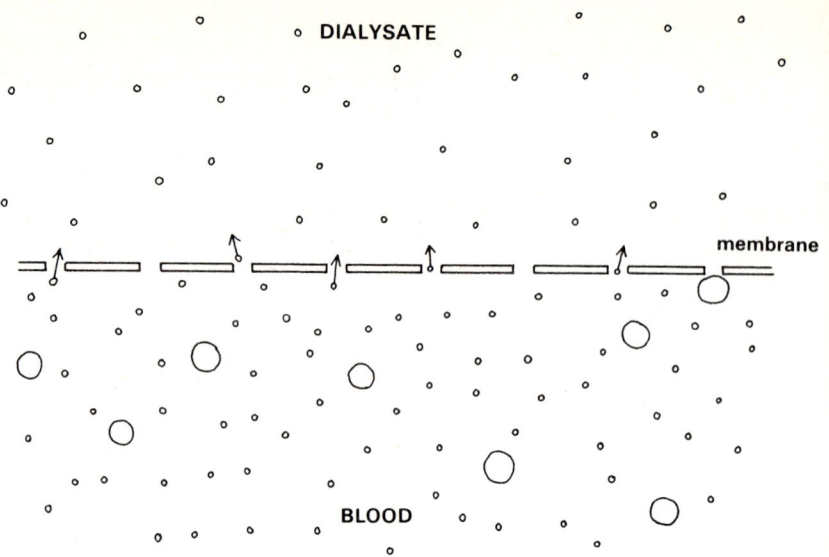

FIG. 10.18 *The mechanism of dialysis. Normal (for the blood) concentrations of dialysable substances (small circular symbols) in the dialysing fluid promote the diffusion of excesses and water from the blood across the membrane. The large circular symbols represent non-dialysable blood constituents such as proteins and cells. Dialysis continues until the concentration of dialysable substances on either side of the membrane are equal*

By using a clip to partially obstruct the tube carrying blood back to the patient's vein the blood pressure in the machine is increased. This makes it possible to remove water from the blood by **ultrafiltration**. Water is removed from the blood in a comparable way in a real kidney.

The pH of the blood falls from 7·4 to about 7·3 between two dialysis treatments. This is because the body's store of hydrogencarbonate ions is used to buffer the blood and in so doing is lowered to about 75 per cent of its normal value. The kidney machine replaces lost hydrogencarbonate. Earlier machines had sodium hydrogencarbonate in the dialysing fluid but sodium ethanoate is used now. The ethanoate diffuses into the blood and is metabolised in the body into hydrogencarbonate.

Usually a patient spends several hours about twice a week connected to a kidney machine during which time the dialysing fluid drains the blood of excess and toxic constituents. This is the basic function of healthy kidneys. However, man-made machines are not perfect. As we saw earlier in the chapter the activities of kidneys are regulated in response to changes in the body's internal environment. As yet kidney machines are not sufficiently sophisticated to be self-regulating. Furthermore, the machines are not designed to perform functions other than excretion. Kidneys functioning abnormally usually stop secreting the hormone **erythropoietin**. The hormone is necessary for normal red blood cell production (Chapter 9). Because of this, patients requiring dialysis are often anaemic. Nevertheless despite their shortcomings kidney machines very definitely have prolonged the lives of a large number of people.

In summary, the main functions of mammalian kidneys are to eliminate wastes and excesses from the blood. The kidneys thus help to maintain a stable internal body environment. Without the kidneys, changes in the composition of the body fluids soon occur. The body is then unable to maintain the optimum conditions for life.

Chapter 11

Uptake and transport of water in flowering plants

Water is essential to all forms of life. In plants, as in animals, water performs important functions as a solvent, as a thermal buffer and as a raw ingredient in hydrolytic reactions catalysed by enzymes (Chapter 2). There are some uses to which water is put in green plants alone. Water provides hydrogen to reduce the products of carbon dioxide fixation in photosynthesis (Chapter 5). In land plants the turgidity of water-containing cells helps to hold the shoot system erect in the atmosphere.

Most of the water required by a land plant is absorbed by its roots from the soil. The rate of water uptake is largely determined by the rate at which water is lost to the atmosphere from the shoot system. The loss of water vapour, called **transpiration**, also affects the rate at which water moves through the plant.

11.1 Transpiration

A major problem confronting all land-dwelling organisms is how to avoid desiccation. The chief advantages of a terrestrial as opposed to an aquatic existence for plants are direct access to sunlight, increased availability of carbon dioxide for photosynthesis and oxygen for respiration. However, in making the most of these advantages land plants continually lose water vapour to the atmosphere. Indeed, as much as 99 per cent of the water absorbed by a terrestrial plant can be lost by evaporation from the shoot system. Transpiration occurs most rapidly when the stomata are open and atmospheric gases come into direct contact with the internal tissues of stems and leaves (Chapter 12). Well over 90 per cent of the total water loss is due to **stomatal transpiration**. Water vapour may also pass directly through the epidermis of the shoot system, but this route is normally sealed by the waxy cuticle. **Cuticular transpiration** usually amounts to less than 5 per cent of the total water vapour loss. Woody stems lose small amounts of water vapour through lenticels by **lenticular transpiration**.

FIG. 11.1 *Venation of a leaf of elder, × 1*

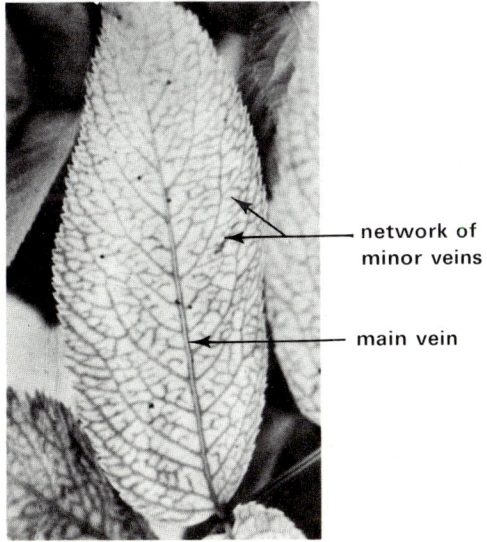

network of
minor veins

main vein

FIG. 11.2 (a) *Transverse section through part of the blade of a leaf of a dicotyledonous plant, × 100*

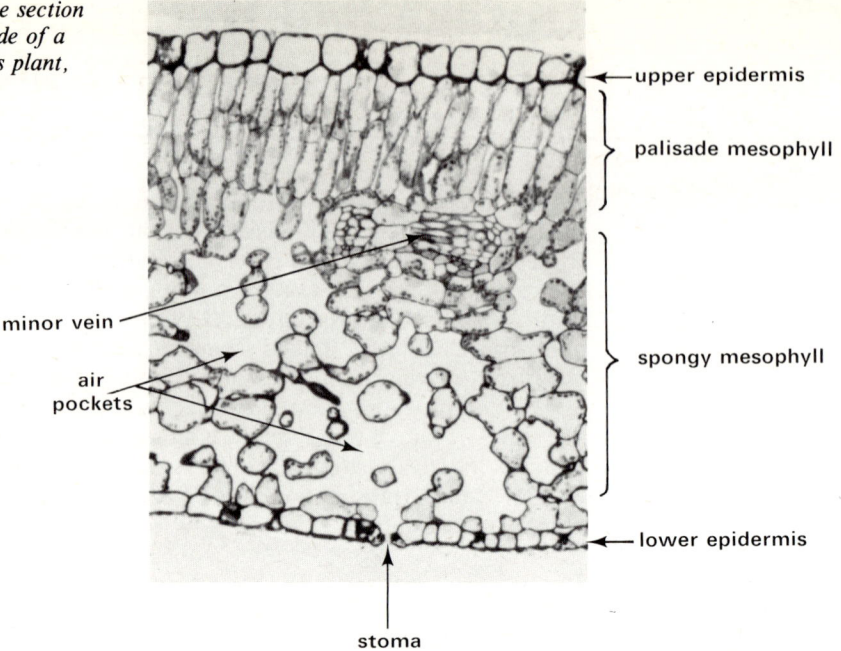

— upper epidermis

palisade mesophyll

minor vein

air pockets

spongy mesophyll

— lower epidermis

stoma

FIG. 11.2 (b) *Air pockets in the spongy mesophyll*

cuticle

upper epidermis

palisade mesoph

air pockets

water vapour

spongy mesophy

lower epidermis

cuticle

guard cell

water vapour

Because there are more stomata on the leaves than elsewhere on the shoot system it is evident that most of the water vapour is lost from the leaves. In many dicotyledonous species the leaves are broad, flattened organs with a vast network of minor veins (Fig. 11.1). Even the smallest vein contains transporting cells which supply the mesophyll tissue with water and dissolved minerals and remove soluble photosynthetic products. A conspicuous feature of the spongy mesophyll is the presence of numerous intercommunicating air pockets (Fig. 11.2). Water constantly evaporates into the air pockets from the walls of the mesophyll cells.

Since the mesophyll air pockets are enclosed spaces, the air in them is usually saturated or nearly so with water vapour. The air inside leaves therefore has a relatively high water potential (Chapter 2). The water potential of the atmosphere surrounding the shoot system fluctuates according to prevailing climatic conditions. In hot, damp greenhouses and in tropical jungles the atmosphere is extremely humid with a fairly high water potential. However, in most terrestrial habitats the atmosphere is generally much less humid and therefore has a relatively low water potential. Consequently when the stomata are open an area of high water potential is brought into contact with an area of low water potential and water vapour rapidly **diffuses** into the surrounding air down a **water potential gradient**. Clearly any factor which alters the size of the gradient will influence the rate of transpiration.

11.2 Factors affecting the transpiration rate

Climate is the product of a large number of interacting factors, notably the intensity and duration of sunlight, wind movement and precipitation, especially rainfall. The water potential of the atmosphere is determined by such interactions. As we have just seen, this affects the size of the water potential gradient between leaves and the surrounding atmosphere. However, the situation is complex because climatic factors may not affect leaves and the atmosphere in the same way or to the same extent. Although climate profoundly influences transpiration rates, other factors such as leaf structure, shape and physiology also determine the rate at which leaves lose water vapour to the atmosphere.

11.2.1 Climate

1. SOLAR RADIATION

Energy as heat and as light comes from the sun. About 10 per cent of **infra-red radiation**, the main source of heat from the sun, falling on leaves is absorbed and is therefore potentially available to raise the temperature of leaves. However, most of this energy provides latent heat for the evaporation of water from the leaf cells. A drastic increase in leaf temperature is thus averted. Nevertheless, leaf temperatures of between 5–8°C above atmospheric temperature are common even in Britain. The effect of such temperature differences is described more fully in the next section.

Light intensity affects transpiration by controlling the degree and to some extent the pattern of stomatal opening (Chapter 12). The stomata of most plants open fully in daylight hours, bringing the saturated air enclosed in a leaf into direct contact with the surrounding atmosphere. This is why there is usually a close correlation between the rate of transpiration and light intensity (Fig. 11.3).

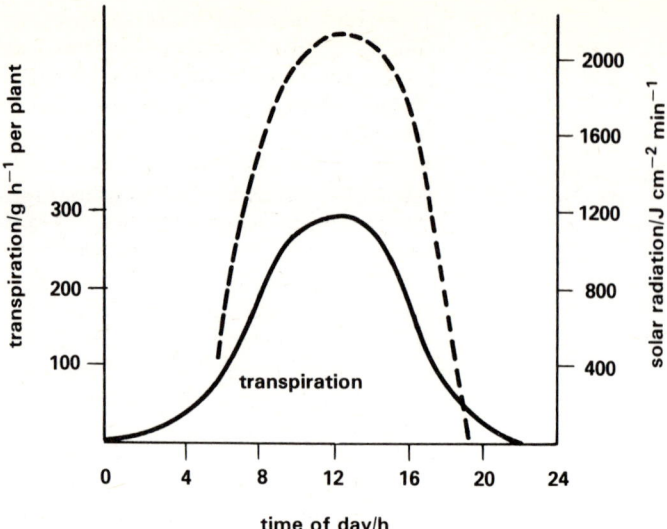

FIG. 11.3 *Correlation between rate of transpiration of oats and light intensity (after Sutcliffe 1968)*

2. TEMPERATURE

A rise in temperature provides additional **kinetic energy** for the movement of water molecules. The effect is to accelerate the rate of evaporation of water from the walls of mesophyll cells and, if the stomata are open, to speed up the rate of diffusion of water vapour into the surrounding atmosphere.

The water potential of the atmosphere generally becomes lower as its temperature is raised and it can then hold more moisture. Table 11.1 compares the water potential at two temperatures of a fully saturated atmosphere, as exists inside a leaf, with one which is 60 per cent saturated, as may occur outside a leaf. It is evident that even a 5°C rise in temperature considerably increases the water potential gradient between a leaf and the air surrounding it.

Table 11.1 Comparison of the water potential, expressed as water vapour pressure, of air at 100 and 60 %RH at two temperatures

TEMP °C	WATER VAPOUR PRESSURE/kPa		WATER VAPOUR PRESSURE GRADIENT/kPa
	100 %RH	60	
15	1·70	1·02	0·68
20	2·33	1·40	0·93

The air inside a leaf normally remains saturated with water vapour when its temperature is raised because of increased evaporation of water from the walls of the mesophyll cells. Thus, as the leaf temperature rises so does the water potential of the air in the leaf. It is not likely that a similar situation exists in the surrounding atmosphere. The water potential here will increase only if there is a nearby source of liquid water from which evaporation can take place. Transpiration from plants and evaporation from the soil and from lakes and oceans are the main sources of water vapour in the atmosphere. If the upper layers of the soil are dry following a long period without rainfall, a rise in temperature has little or no effect on the water potential of the atmosphere in a terrestrial environment. In hot, dry weather, therefore, the water potential gradient between leaf and atmosphere is great and transpiration is very rapid.

The effect of infra-red radiation in raising leaf temperatures by several degrees above ambient (section 11.2.1(1)) accelerates the rate of transpiration. However, when a leaf is hotter than the air around it, convection currents develop bringing cooler air in contact with the leaf. To some extent the cooling effect counteracts the expected increase in the transpiration rate by lowering the leaf temperature.

3. AIR MOVEMENT

Transpiration in still air results in the accumulation of a **boundary layer**, several millimetres thick, of saturated air at the surfaces of leaves. The boundary offers considerable resistance to the diffusion of water vapour through stomata and cuticle, thereby reducing the transpiration rate. Movement of the surrounding air reduces the thickness of the boundary layer and results in increased transpiration (Fig. 11.4). Air movements can occur as a result of convection (section 11.2.1(2)) or wind currents. As still air rarely exists in nature for any length of time, air movement is important in increasing the loss of water vapour from terrestrial plants. However, air movement generally brings cool air into contact with leaves. As we have already seen, this may lower the temperature of leaves and help to reduce transpiration. Continuous exposure to wind of high velocity generally causes wilting of leaves. This is because the rapid loss of water vapour by transpiration is not balanced by absorption of water from the soil.

FIG. 11.4 *Effects of air movement on thickness of boundary layer*

still air   moving air

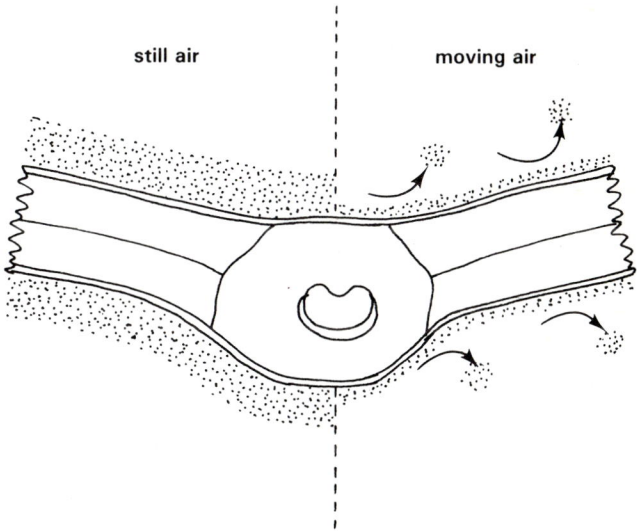

4. ATMOSPHERIC HUMIDITY

The water potential of air is usually expressed as **percentage relative humidity (%RH)**:

$$\%RH = \frac{\text{water vapour pressure of air at } t°C}{\text{water vapour pressure of saturated air at } t°C} \times 100$$

We have noted that the air inside a leaf is saturated with water vapour, that is at 100 %RH. Conversely the humidity of the atmosphere surrounding a leaf is changeable. Values exceeding 70 %RH are rare outdoors in Britain while figures as low as 30 %RH can occur following a long spell of hot, dry weather. In either case the water potential gradient between leaf and atmosphere is substantial and when the stomata are open water vapour rapidly diffuses from the leaf. As the %RH of the atmosphere falls the water potential gradient increases. There is thus a direct relationship between atmospheric humidity and the rate of transpiration (Table 11.2 and Fig. 11.5).

Table 11.2 Water potential, expressed as water vapour pressure, of air at four different humidities (readings taken at 20°C)

%RH	WATER VAPOUR PRESSURE/kPa	WATER VAPOUR PRESSURE GRADIENT/kPa
100	2·33	2·33 − 2·33 = 0·00
70	1·63	2·33 − 1·63 = 0·70
60	1·40	2·33 − 1·40 = 0·93
50	1·16	2·33 − 1·16 = 1·17

FIG. 11.5 *Effect of atmospheric humidity on the rate of transpiration*

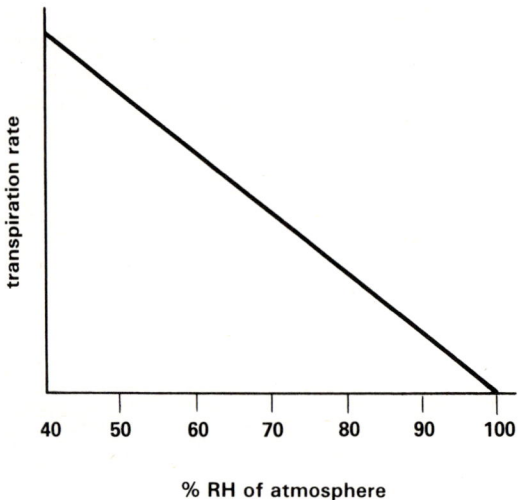

11.2.2 Leaf structure

Brown and Escombe in 1900 investigated the rates of diffusion of gases through membranes perforated by pores varying in size and number per unit area. Their findings of the movement of carbon dioxide and its implications in the absorption of the gas by leaves for photosynthesis are described in Chapter 12. Part of their work involved a study of the diffusion of water vapour. They observed that, providing the pores were not too close to one another, the rate of diffusion of water vapour through a perforated membrane is as much as 50 per cent of the total water loss from a body of water of surface area similar to the membrane. This occurred despite the fact that the pores occupied less than 3 per cent of the area of the membrane. Stomata usually account for about 5 per cent of the total area of a leaf. It may therefore be deduced that the epidermis is only a limited barrier to water loss when the stomata are open. The rate of diffusion of water vapour through stomata results, however, from the interaction of a number of factors.

1. PORE SIZE

As molecules of a gas such as water vapour escape through a pore into still air they spread out into concentric, hemispherical zones called **diffusion shells** (Fig. 11.6). The water potential of each shell diminishes the further it is from the pore. Water vapour molecules continually diffusing from the outer shell into the atmosphere are replaced by molecules diffusing from an inner shell. A continuous water potential gradient thus exists across the diffusion shells.

FIG. 11.6 *Diffusion of water vapour through a pore*

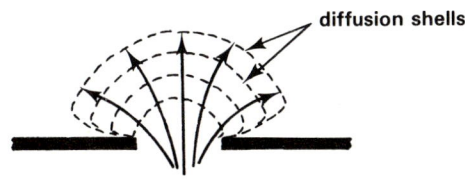

diffusion shells

The rate of diffusion **per unit area of pore** is greater through small pores than through large pores (Chapter 12). Furthermore the diffusion rate is proportional to the **diameter** of the pore (Table 12.1). The effectiveness with which stomata control transpiration depends on the humidity of the surrounding air and on the extent to which the stomatal apertures are open. In still air when outgoing water vapour accumulates above stomata a reduction in pore size is less effective in lowering the transpiration rate than in moving air (Fig. 11.7). Why do you think this is so?

FIG. 11.7 *Effect of pore size in controlling transpiration rate in* Zebrina *(after Bange, 1953)*

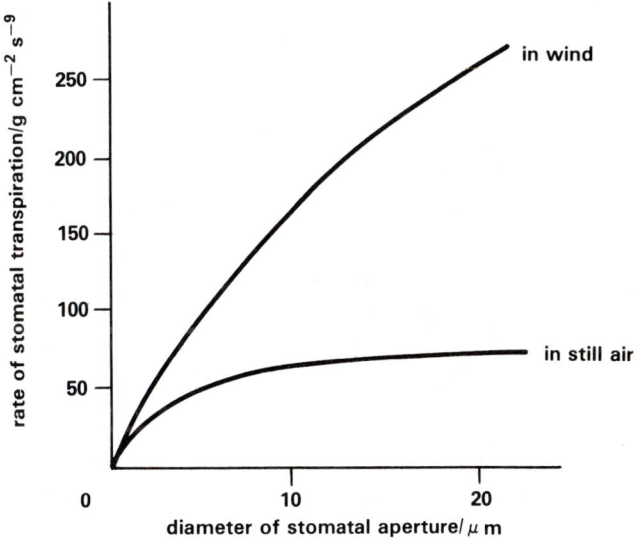

2. PORE DENSITY

Where stomata are spaced wide enough apart there is no interference between the diffusion shells of water vapour above each pore. However, where stomata are close to one another, as occurs in most land plants, there may be considerable **overlap** between the diffusion shells above adjacent stomata (Fig. 11.8). The effect is to form one large area from which water vapour diffuses into the atmosphere. Because the rate of diffusion per unit area from a large area is less than that from a number of smaller areas of equivalent area the transpiration rate is reduced. Other ways in which leaf structure affect transpiration are mentioned in the next section.

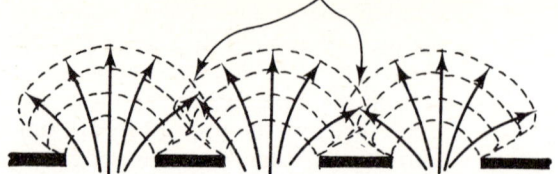

FIG. 11.8 *Diffusion of water vapour through adjacent pores*

overlap between diffusion shells

11.2.3 Xerophytic plants

There is a considerable variety of structure, stomatal distribution and physiology among higher plants. The differences often play a key part in determining the rate of transpiration. Species which inhabit areas where water is scarce are of particular interest in this respect. They are generally called **xerophytes** as opposed to most land plants known as **mesophytes** which inhabit areas where ample water is available and **hydrophytes** which live totally or partially submerged in water.

FIG. 11.9 (a) *Transverse section through a stoma of* Hakea, ×*375*

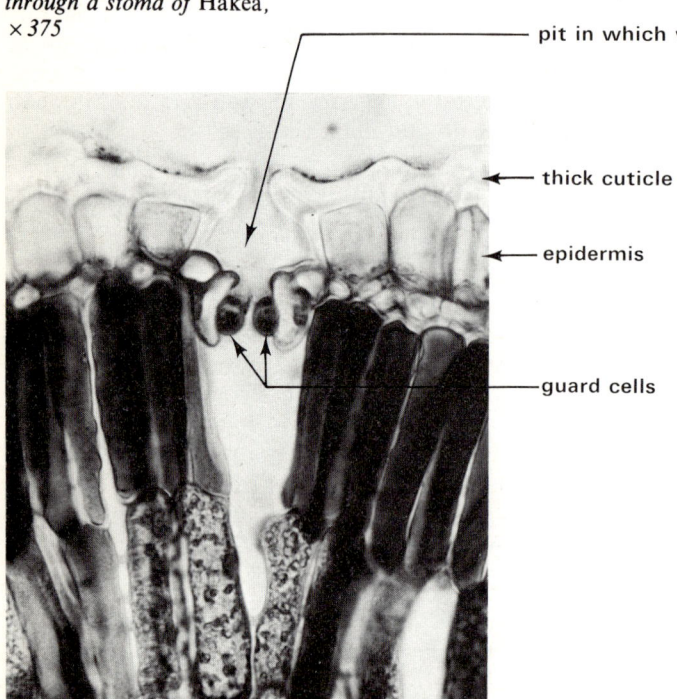

pit in which water vapour collects

thick cuticle

epidermis

guard cells

FIG. 11.9 (b) *Transverse section through part of a leaf of* Nerium, ×*100*

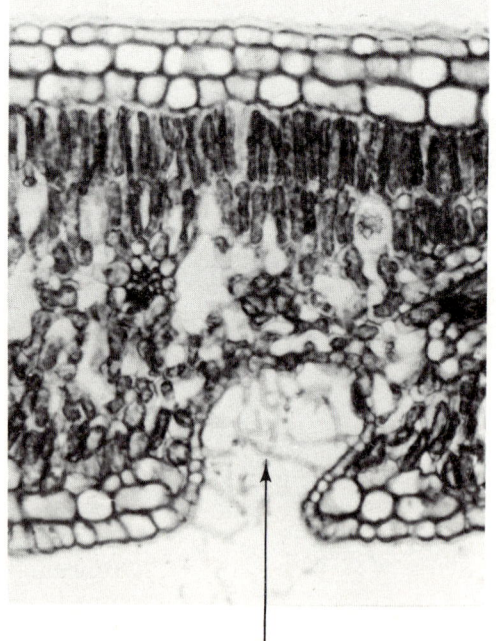

hairy groove in which water vapour collects

FIG. 11.9 (c) *Transverse section of a rolled leaf of marram grass, × 40. When the hinge cells lose water vapour excessively they become flaccid and the leaf rolls up as shown*

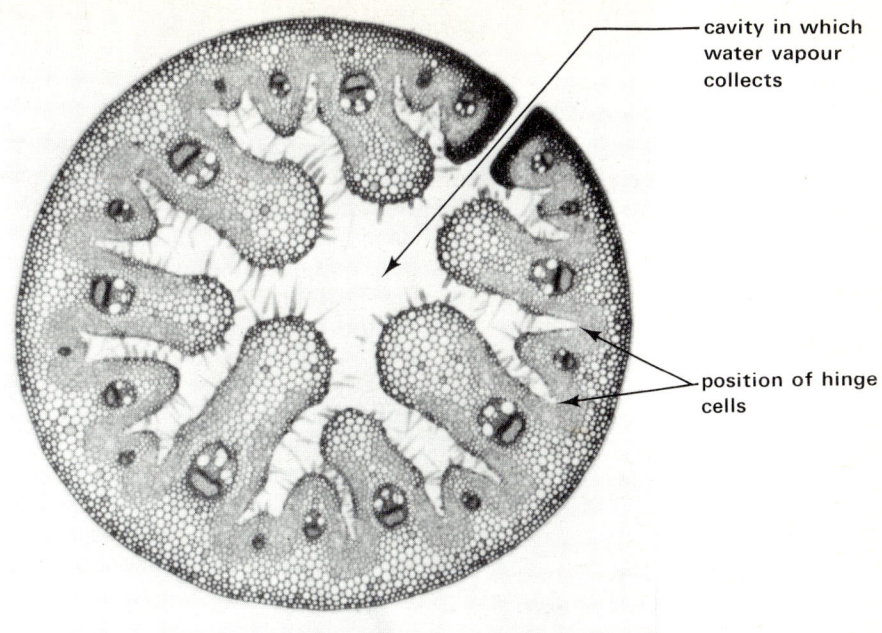

cavity in which water vapour collects

position of hinge cells

Frequently the guard cells of xerophytes are sunk in pits or in grooves well below the surface of the leaf or stem (Fig. 11.9(a)). Water vapour escaping through stomata accumulates immediately above the guard cells to form a **thick boundary layer** which reduces the transpiration rate even in moving air. Other modifications, such as **hairiness** of the shoot system and a relatively **thick cuticle**, also assist in minimising the rate at which water vapour is lost (Fig. 11.9(b)). In very dry weather the leaves of some xerophytes such as marram grass can even be **rolled** into a cylinder with the stomata on the inside. Water vapour is trapped inside the cylinder, so building up the humidity of the air into which transpiration takes place. Ultimately the water potential inside the cylinder becomes sufficiently high to stop transpiration (Fig. 11.9(c)).

Cacti are often thought of as plants ideally suited to conditions of drought. They are **succulent** plants and store water in a special tissue inside their bodies. The cylindrical shape of many cacti means that for each unit volume of tissue there is a small unit of surface area through which water vapour can be lost. Not all xerophytes are succulent. Some such as the creosote bush are highly lignified and can **tolerate** extremes of desiccation without wilting.

Perhaps even more significant is the rate at which some xerophytes fix carbon dioxide compared with mesophytes. Cacti and many other xerophytes are **C4 plants** (Chapter 5) and can make photosynthetic end-products more efficiently than C3 plants from the same volume of carbon dioxide. This means that the stomata of xerophytes need remain open for only short periods of time each day to acquire sufficient carbon dioxide. Cacti take in carbon dioxide at night when the air is cooler and conditions are less likely to encourage transpiration. Such adaptations considerably lessen the volume of water vapour transpired.

11.3 The effects of transpiration

11.3.1 Water movement through the shoot system

One of the effects of transpiration has already been mentioned: the cooling effect which prevents drastic rises in leaf temperature. There is also little doubt that transpiration is the main factor affecting the rate of water movement through plants.

A **potometer** (Fig. 11.10(a)) can be used to investigate the rate of absorption of water by a plant shoot. As the shoot transpires, the water vapour it has lost is replaced by liquid water drawn in from the potometer. What is more, the rate of absorption is affected by the same factors which influence the transpiration rate. If the leafy shoot is replaced with a water-filled porous pot (Fig. 11.10(b)) comparable results are obtained. As water evaporates from the tiny holes of the porous pot it is replaced by liquid water entering the pores by **capillarity**. The capillary movement of water is due partly to forces of **adhesion** between water molecules and the walls of the pores and of **cohesion**, the forces of attraction holding the water molecules to each other. It is therefore logical to deduce that the forces responsible for water movement in the pot are also at work in the shoot system of a plant. Water evaporating from the porous walls of leaf mesophyll cells is replaced by capillarity. But where does the water come from and what effect does the refilling of the pores in the mesophyll cell walls have on the rest of the plant?

FIG. 11.10 (a) *A potometer*

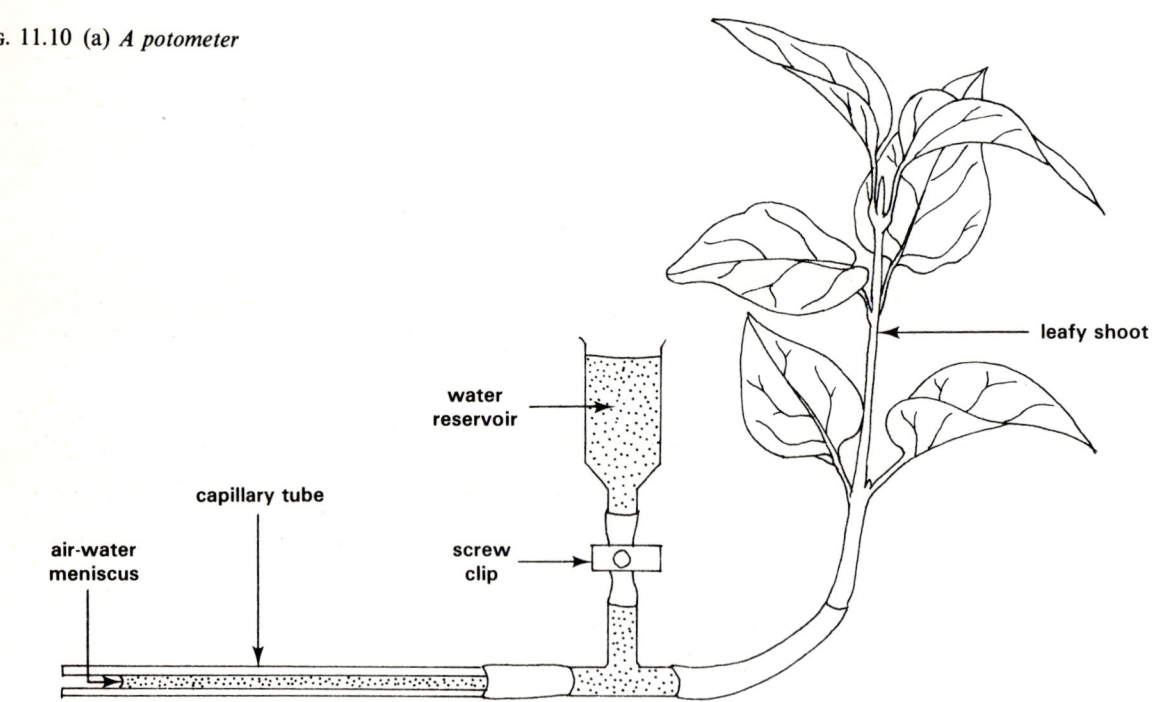

With the screw clip closed, the rate at which the meniscus moves along the capillary tube indicates the rate at which the leafy shoot is absorbing water. When all the water has been absorbed from the capillary tube it can be refilled by opening the screw clip.

FIG. 11.10 (b) *An atmometer.*

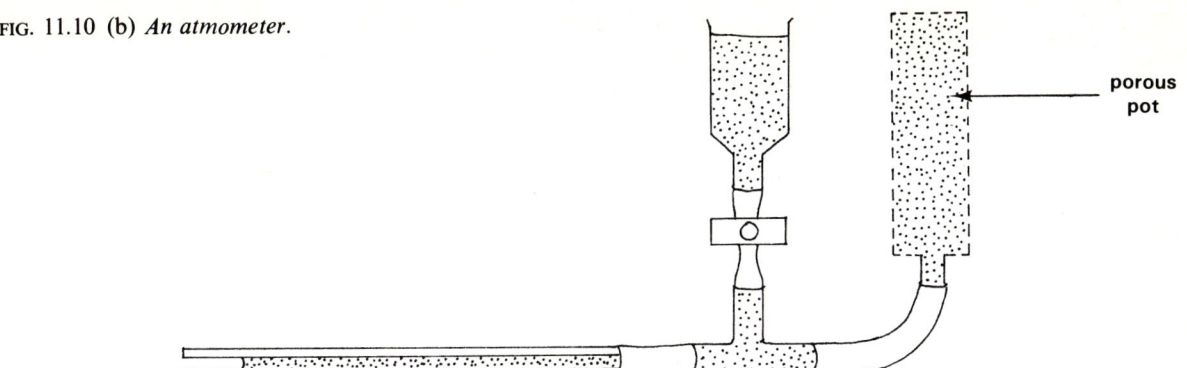

porous pot

Readings are taken as with the potometer

Typical results

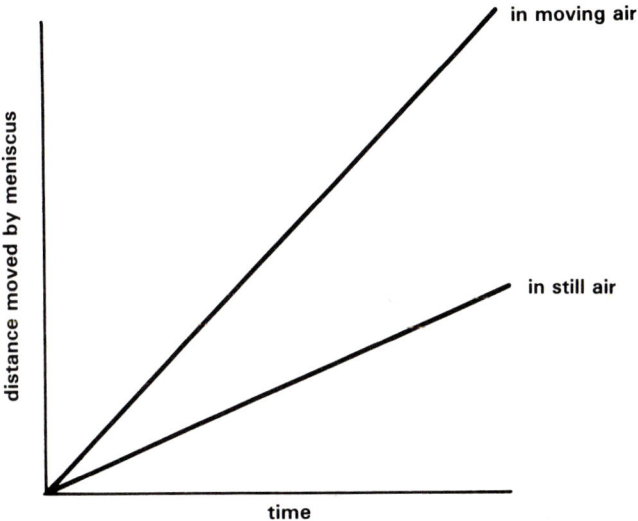

in moving air

in still air

distance moved by meniscus

time

The nearest reservoir of water for capillarity is the protoplasts of the mesophyll cells. The effect of losing water in this way is to lower the water potential of mesophyll cells. If a source of water of higher potential is available nearby the protoplasts of the mesophyll cells could absorb water through their plasma membranes by **osmosis**. Alternatively water could be drawn by **capillarity** from the source directly into the mesophyll cell walls where evaporation is taking place. Whichever route is taken, the water comes from a minor vein containing water-conducting xylem elements. One effect of transpiration therefore is to create **leaf suction** which removes water from the veins of leaves. In turn the leaf veins become replenished by absorbing water from the xylem elements of the stem. Again the movement of water from stem to leaf is along a water potential gradient.

The cut shoot in the potometer has direct access to water in the apparatus and can draw freely on it. Does a similar situation exist in a whole plant? Some idea can be obtained by comparing the rates of transpiration and water absorption of an intact plant over a period of time (Fig. 11.11). Absorption lags behind transpiration, indicating that the roots are a barrier to a continuous flow of water triggered by transpiration. The effect of leaf suction and root resistance working in opposite directions places the

207

water columns in the vessels and tracheids under strain or **tension**. The columns are sufficiently elastic to take the strain so they do not normally break and the flow of water, **the transpiration stream**, continues. What prevents breakage of the water columns? The answer to this question is the **cohesive forces** which exist between water molecules (Chapter 2).

FIG. 11.11 *Comparison of the rates of transpiration and water absorption by sunflower plants* (after Kramer, 1937)

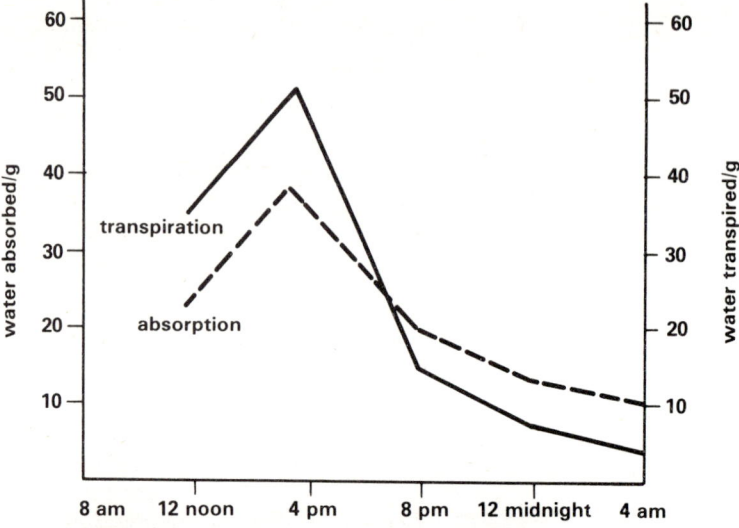

FIG. 11.12 *A giant California redwood* (Sequoia *tree*). (Photo Paul Conklin, permission Colorific)

The account of water transport given above is a summary of the **transpiration-cohesion-tension theory** described in 1894 by Dixon and Joly. What is of particular interest is that the proposed mechanism can account for the movement of water to the tops of extremely tall trees such as the redwood, *Sequoia*, which frequently grows to a height of well over 100 m (Fig. 11.12). Atmospheric pressure at sea level is 101·3 kPa (1 atm) which can raise a column of water to a height of only 10 m, and is quite inadequate to move water to the top of a tall tree. Water potentials of up to −3000 kPa (−30 atm) have been demonstrated in the mesophyll tissue of the leaves of tall trees. This means that leaves can generate a suction capable of pulling water to a height of 300 m. The tension required to stretch to breaking point columns of sap from the xylem of trees has been shown to be about 30 000 kPa. This is well in excess of the tension normally generated in plants.

What we have seen so far is that transpiration can lead to an upward movement of water through the shoot system. But what about water transport in roots?

11.3.2 Water absorption by roots

Water is absorbed by the outermost layer of the root which is in direct contact with the soil. Between 4 and 10 mm behind the root tip the **epidermal cells** have **root hairs** which increase the surface area of the root several fold (Fig. 11.13). Because of the unceasing apical growth of the root tip, a root hair zone has a functional life of only a few weeks. It is continually replaced by a new zone in the freshly extended part of a root. In older parts of the root the epidermis is replaced by a suberised **exodermis**. If secondary thickening occurs a **periderm** becomes the outermost layer (Chapter 21). It is commonly thought that water uptake by a root is confined to the root hair zone. Yet it has been shown that water absorption can occur rapidly through the surface of older parts of the root even where an exodermis or a periderm is present (Fig. 11.14).

FIG. 11.13 (a) *Transverse section through the root hair zone of a broad bean root,* ×40

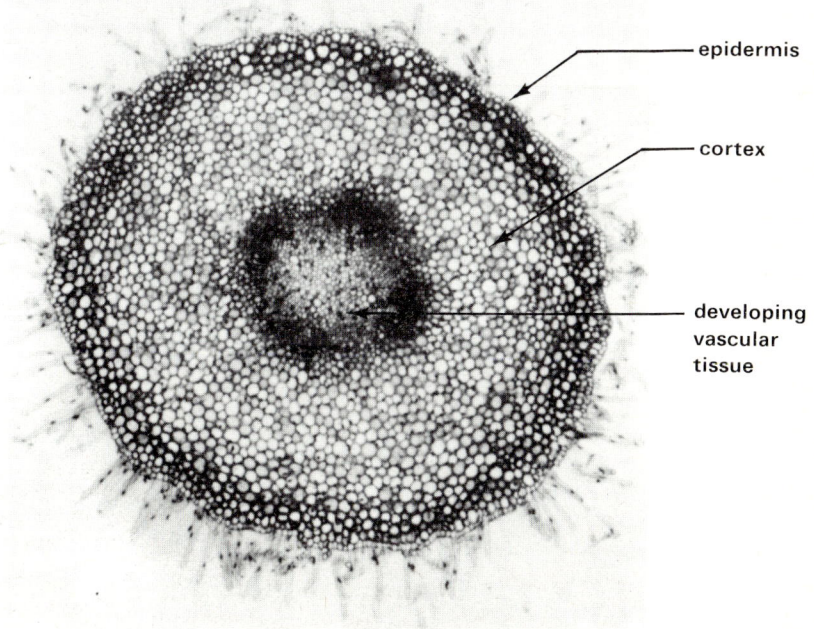

epidermis

cortex

developing vascular tissue

FIG. 11.13 (b) *Close-up view of
some root hairs,* × *100*

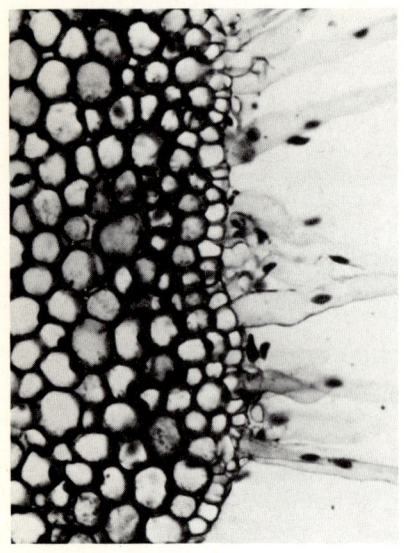

Inside the outermost root layer is the **cortex** of large, thin-walled, va-
cuolated parenchyma cells with conspicuous intercellular spaces. The innermost cortical layer is the **endodermis**, the cells of which, early in their
development, have a suberised Casparian band on their radial walls. Later
the suberisation becomes more extensive and covers the radial and inner
tangential walls (Fig. 11.15). Immediately inside the endodermis is the
pericycle, a region one to several cells wide of small, thin-walled parenchyma cells. The pericycle is the site of origin of lateral roots which may
add greatly to the absorptive surface area of the root system. At the core of
the root is the **vascular tissue**. In the root hair zone of the root of a
dicotyledonous plant there are between two and five groups of **protoxylem**
elements. At a later stage the central **metaxylem** elements differentiate and
the primary xylem looks star-shaped in cross section (Fig. 11.16).

The transpiration stream drags columns of water upwards through the
network of xylem elements which extend throughout the plant. The stream
eventually lowers the water potential in the root xylem sufficiently to
establish a gradient of water potential down which water is ultimately
absorbed from the soil. There are forces which lower the water potential of
the soil solution, thereby opposing absorption. The opposing forces include the osmotic potential of the soil solution, which is usually significant
only in certain habitats such as salt marshes. Capillary and imbibitional
forces also hold the soil solution in and between mineral particles and the
humus fraction of soil. Nevertheless the transpiration stream can generate
the necessary water potential gradient for water to pass from the soil
solution across the root cortex and into the root xylem elements.

FIG. 11.14 *Water uptake by
different parts of broad bean
roots* (after Brouwer, 1965)

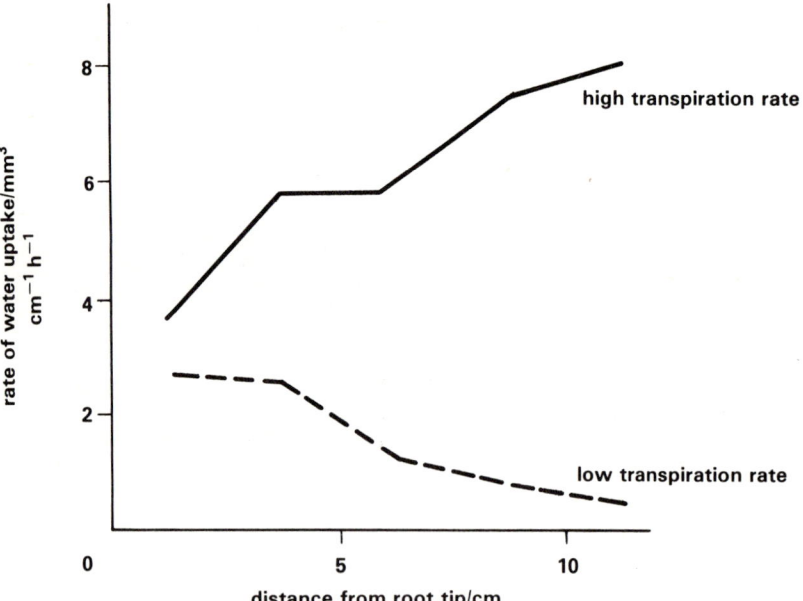

The classical view is that water moves through the living tissues of the
root cortex from vacuole to vacuole by **osmosis**. However, as in the leaves,
it is possible that cell to cell transport by **capillarity** along the cell walls is
also important. The wall route cannot be taken at the endodermis where
the waxy Casparian band is impervious to water. At the endodermis water
probably flows through the plasma membrane. The significance of this

FIG. 11.15 (a) *The endodermis* (i) t.s. young endodermis

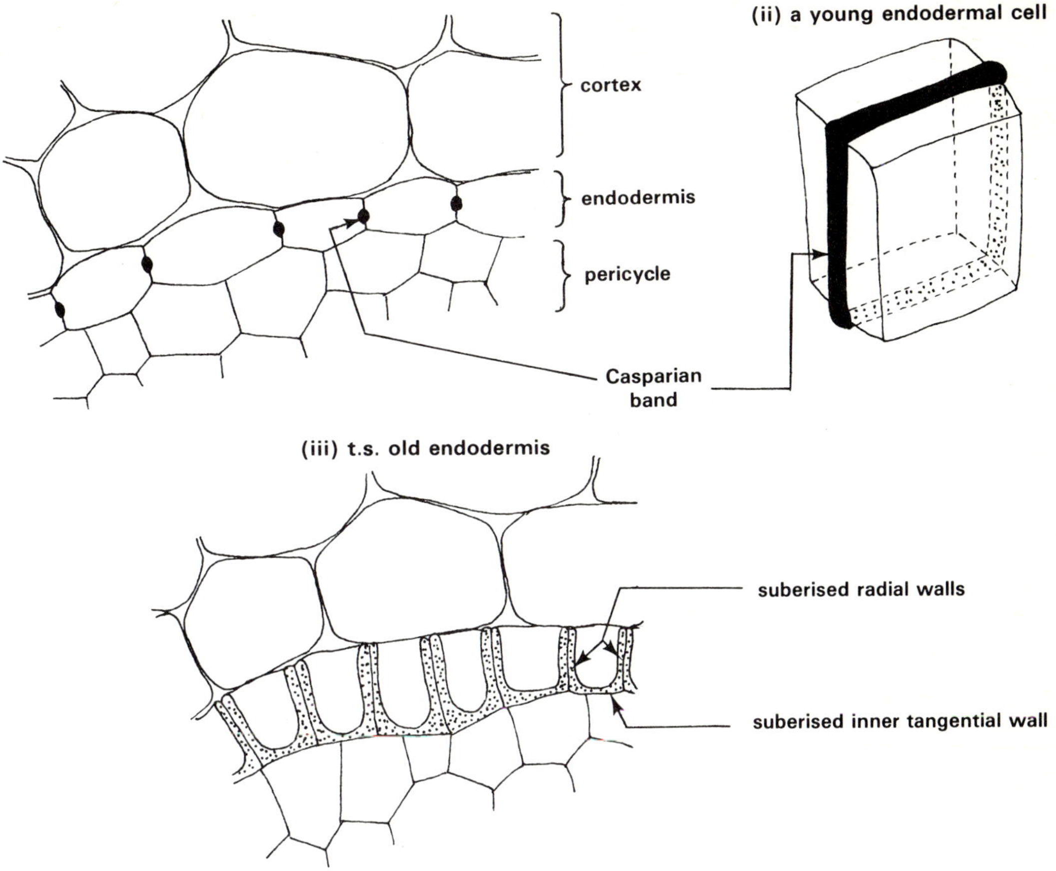

cortex

endodermis

pericycle

Casparian
band

(ii) a young endodermal cell

Casparian
band

(iii) t.s. old endodermis

suberised radial walls

suberised inner tangential wall

FIG. 11.15 (b) *T.s. young
endodermis,* × 400

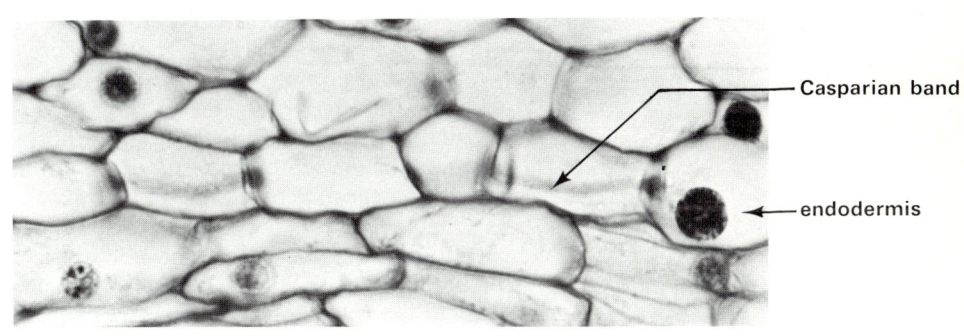

Casparian band

endodermis

FIG. 11.15 (c) *T.s. old
endodermis,* × 250

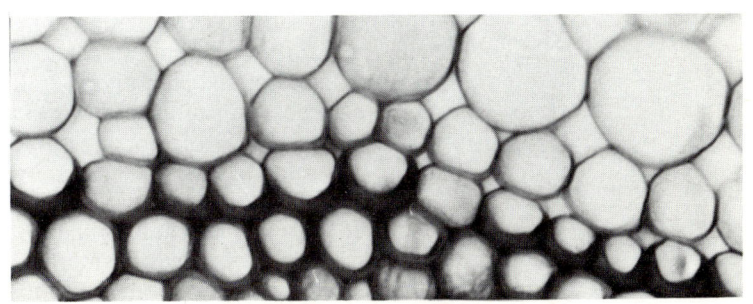

211

pathway in controlling the entry of dissolved mineral ions is discussed in Chapter 13. Eventually the absorbed water reaches the xylem elements of the root from where it moves upwards into the veins of the shoot system.

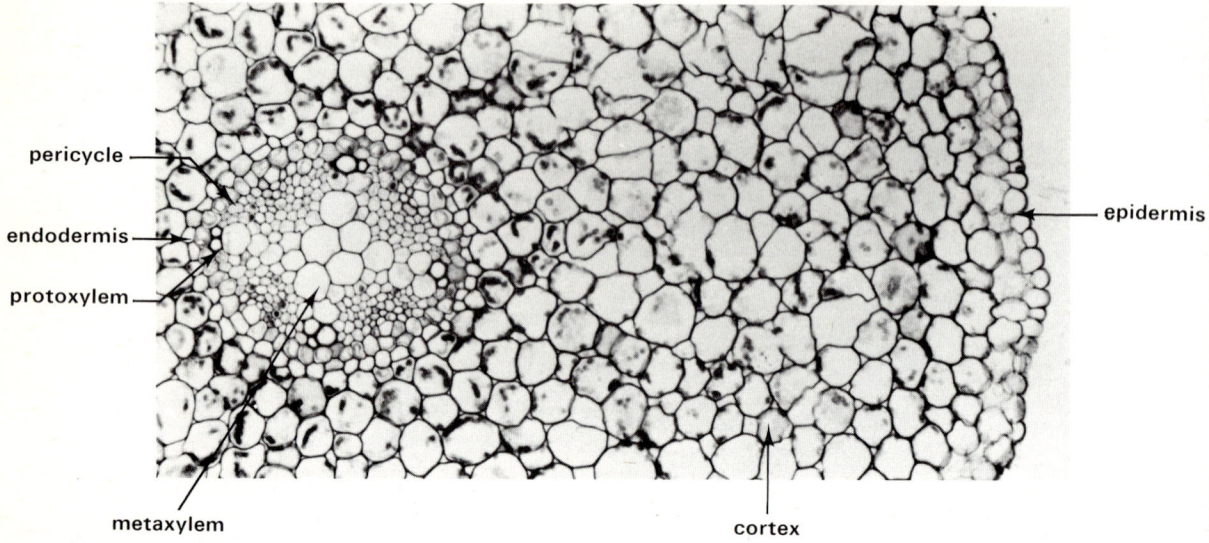

pericycle

endodermis

protoxylem

epidermis

metaxylem

cortex

11.4 Root pressure

Few plant physiologists would challenge the role of transpiration in the movement of water through terrestrial plants. Yet there are other forces at work in plants which can take part in the flow of water. If, for example, a glass tube is attached to the stump of a freshly decapitated plant such as a vine, sap is pushed up the tube to a height of several metres (Fig. 11.17). Clearly the flow of sap from the stump cannot be linked with transpiration. The force responsible is called **root pressure**. In some species, notably low-growing plants such as grasses, root pressure is sufficiently strong at times to cause exudation of liquid water from special pores called **hydathodes** located at the edges of the leaves (Fig. 11.18). This phenomenon is called **guttation**. It is therefore possible that the flow of water through plants may sometimes be the combined result of the pulling effect of transpiration and the pushing effect of root pressure. However, the available evidence suggests that the forces generated by transpiration are usually more important.

Root pressure is most evident in conditions favouring rapid water absorption but limiting transpiration. For example, it is responsible for the rising of sap in the trunks of deciduous trees during spring before the leaf buds have opened. But even in these circumstances the pressure generated is rarely more than 200 kPa (2 atm), a figure of minor importance compared with the transpiration pull. Furthermore the volume of sap exuded by the stumps of decapitated plants is generally little more than five per cent of the volume lost by transpiration from intact plants.

The development of root pressure seems to depend on the active secretion of mineral ions into the root xylem elements from surrounding parenchymatous transfer cells (Fig. 11.19). The effect of mineral secretion is to lower the water potential of the xylem sap leading to the absorption of water from the soil down a water potential gradient. Respiration by the transfer cells provides the energy required for movement of the ions against

FIG. 11.17 *A root pressure manometer*

a concentration gradient. It is not surprising therefore that root pressure disappears when plant roots are treated with respiratory inhibitors such as potassium cyanide.

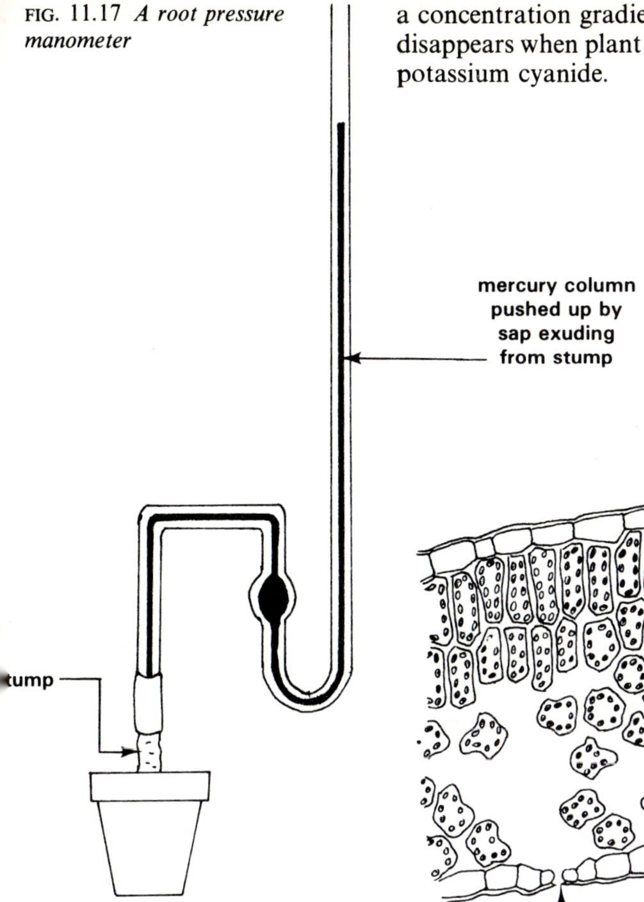

mercury column
pushed up by
sap exuding
from stump

FIG. 11.18 *A hydathode in t.s. from the edge of a cabbage leaf*

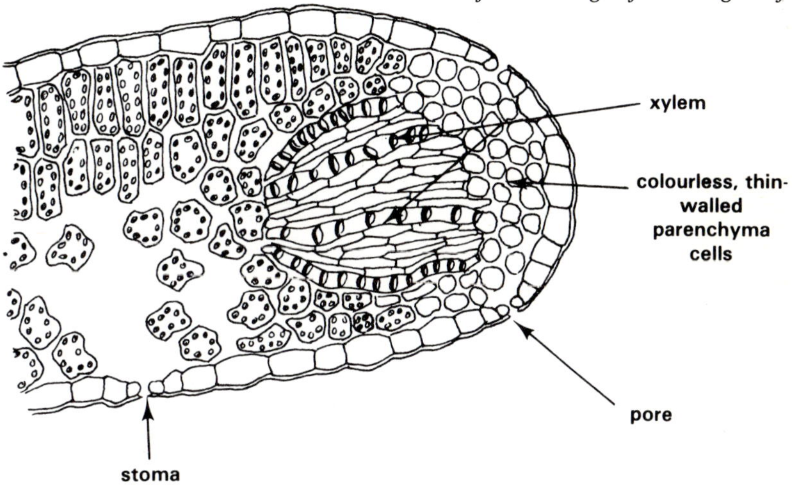

xylem

colourless, thin-walled parenchyma cells

pore

stump

stoma

FIG. 11.19 *Electronmicrograph of a section through a transfer cell from the pericycle of a root of clover (courtesy Dr. G. Briarty).* The cell wall ingrowths increase the surface area of the plasma membrane for the transfer of solutes. The mitochondria provide the free energy for active secretion of solutes

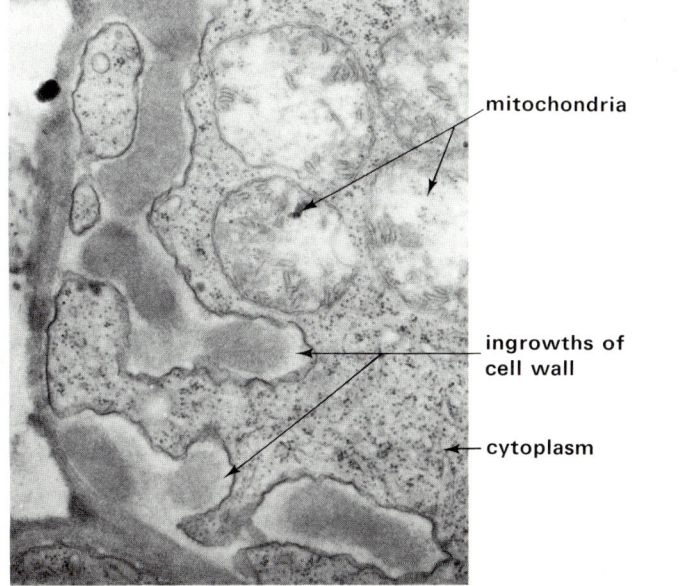

mitochondria

ingrowths of cell wall

cytoplasm

11.5 Tracheids, vessels and capillarity

FIG. 11.20 (a) *Tracheids and vessels*

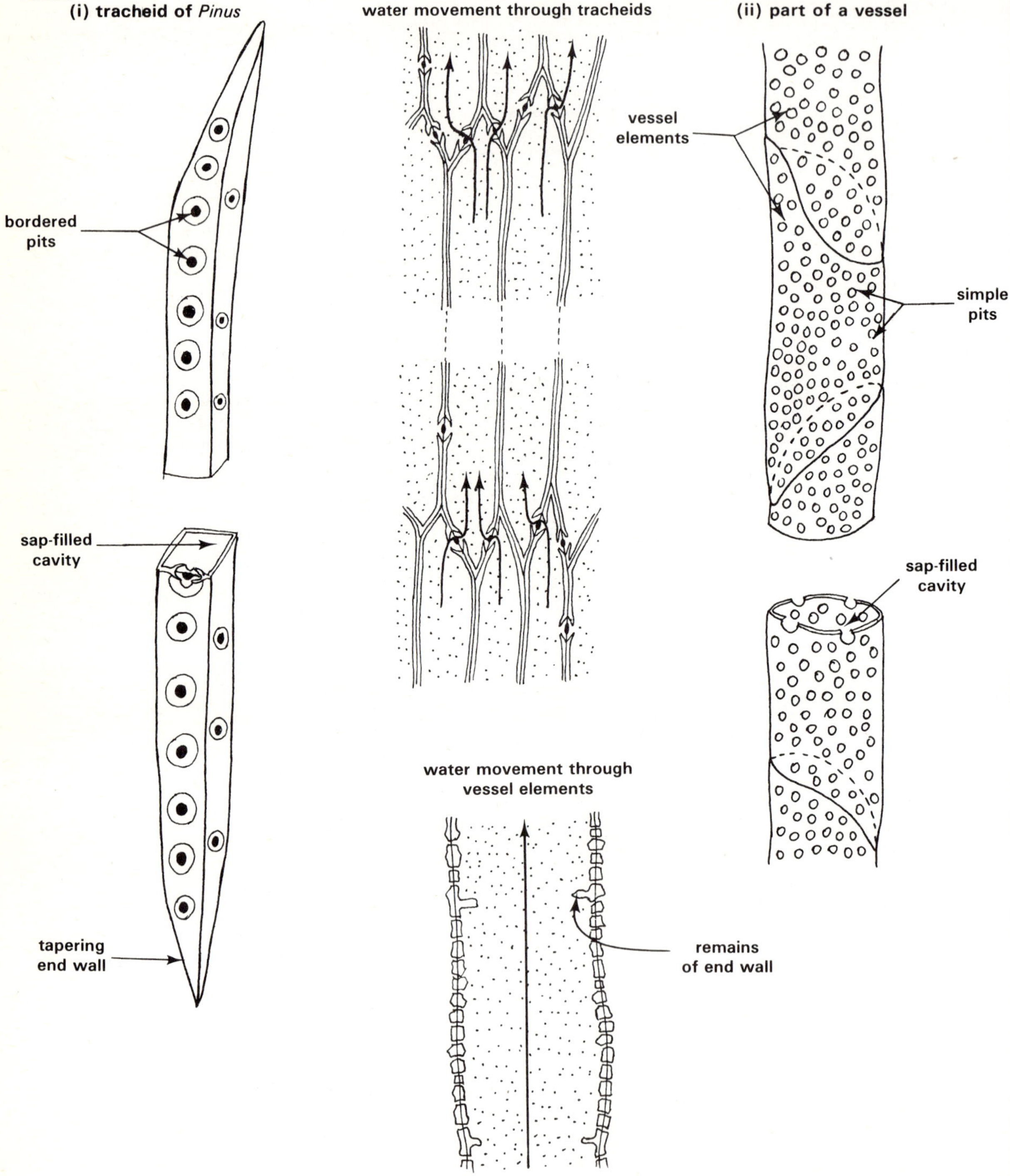

(i) tracheid of *Pinus*

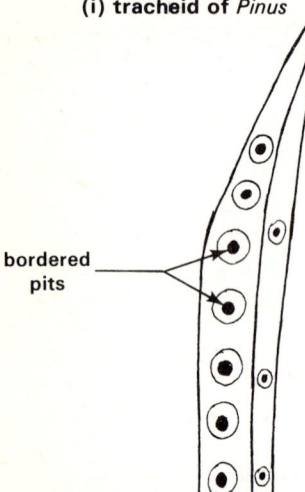

bordered pits

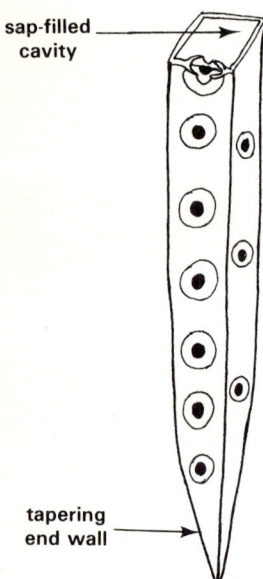

sap-filled cavity

tapering end wall

There is little doubt that xylem is the main water-carrying tissue of higher terrestrial plants. Proof of this comes from experiments in which the xylem elements of leafy shoots are blocked with wax and then given access to water. The leaves soon wilt compared with untreated shoots or where tissues other than xylem are blocked. As wax does not penetrate the

water movement through tracheids

(ii) part of a vessel

vessel elements

simple pits

sap-filled cavity

water movement through vessel elements

remains of end wall

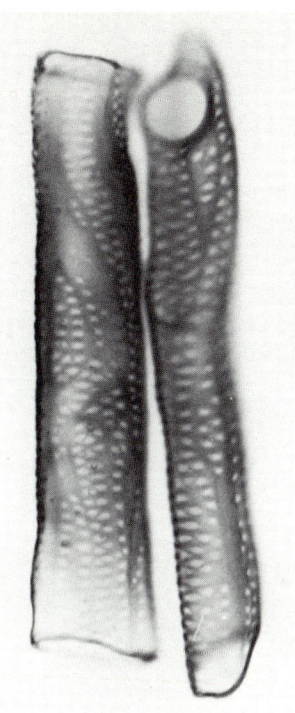

lignified walls of xylem elements it is clear that most of the water passes through the cavities of the xylem cells and not up the walls in a wick-like manner. These facts have recently been confirmed using isotopes of water.

The chief water-transporting cells found in flowering plants are **tracheids** and **vessels**. Tracheids were the water-carrying cells of the earliest land plants where they fulfilled the dual role of support and water conduction. They are still the main water-carrying cells in pteridophytes and gymnosperms. In conifers such as the pine, *Pinus*, mature tracheids are very long, narrow, angular cells measuring up to 5 mm long and 30 μm in diameter (Fig. 11.20(a)). Stiffening of the walls with lignin and interlocking of the tapered end walls are significant features in the role of tracheids as supporting cells. Water movement in the cavities of tracheids is helped by the lack of protoplasts. The presence of large, bordered pits allows the passage of aqueous solutions between adjacent tracheids.

Vessels are found in flowering plants. Unlike tracheids which differentiate from single cells derived from meristems, vessels originate from a vertical row of cells (Chapter 21). During differentiation the row of cells increases considerably in diameter, protoplasts are absorbed and the end walls between adjacent cells are partially or completely lost. Consequently, a long, open tube is formed, often several metres in length and averaging 250 μm in diameter (Fig. 11.20(b)). Because less resistance is offered, aqueous solutions can move three to six times more quickly through vessels compared with tracheids. As in tracheids, the walls of vessels are thickened with cellulose and hardened with lignin. The extent of wall thickening and lignification varies according to the stage of development of a plant organ when differentiation takes place. Protoxylem vessels which differentiate close to the apices of roots and stems have bands or spirals of strengthening material. The unthickened parts of the cell walls can be enlarged as the organs grow in length. Metaxylem vessels differentiate further back from root and shoot apices where growth in length has ceased. The walls of metaxylem vessels are more extensively thickened and cannot be enlarged.

Water rises in the bore of a fine tube by **capillarity** and plant physiologists have wondered whether capillarity is significant in the ascent of sap through the narrow cavities of xylem elements. Experiments have shown that water can rise to a height of 5–7 m in glass tubes with a bore comparable to the lumen of a tracheid. However, in wider tubes with bores similar to the lumen of a vessel the capillary rise is only a few centimetres. Even allowing for the fact that the many crevices in the walls of the transporting elements add to the capillary force it is generally considered that capillarity does not play an important role in the upward movement of sap through the xylem elements of land plants.

We have seen that the absorption of water and its transport through the body of a flowering plant is the consequence of several phenomena which are summarised in Figure 11.21. It is important to remember that water moves from one part of a plant to another and out into the atmosphere down **water potential gradients**. Heat from the sun is responsible for the largest water potential gradient between the shoot system and the atmosphere. Whereas emphasis in this chapter has been placed on the movement of water it should be mentioned that substances dissolved in the water are also moved simultaneously. The transport of mineral ions the main category of dissolved substances carried in xylem elements is described in Chapter 13.

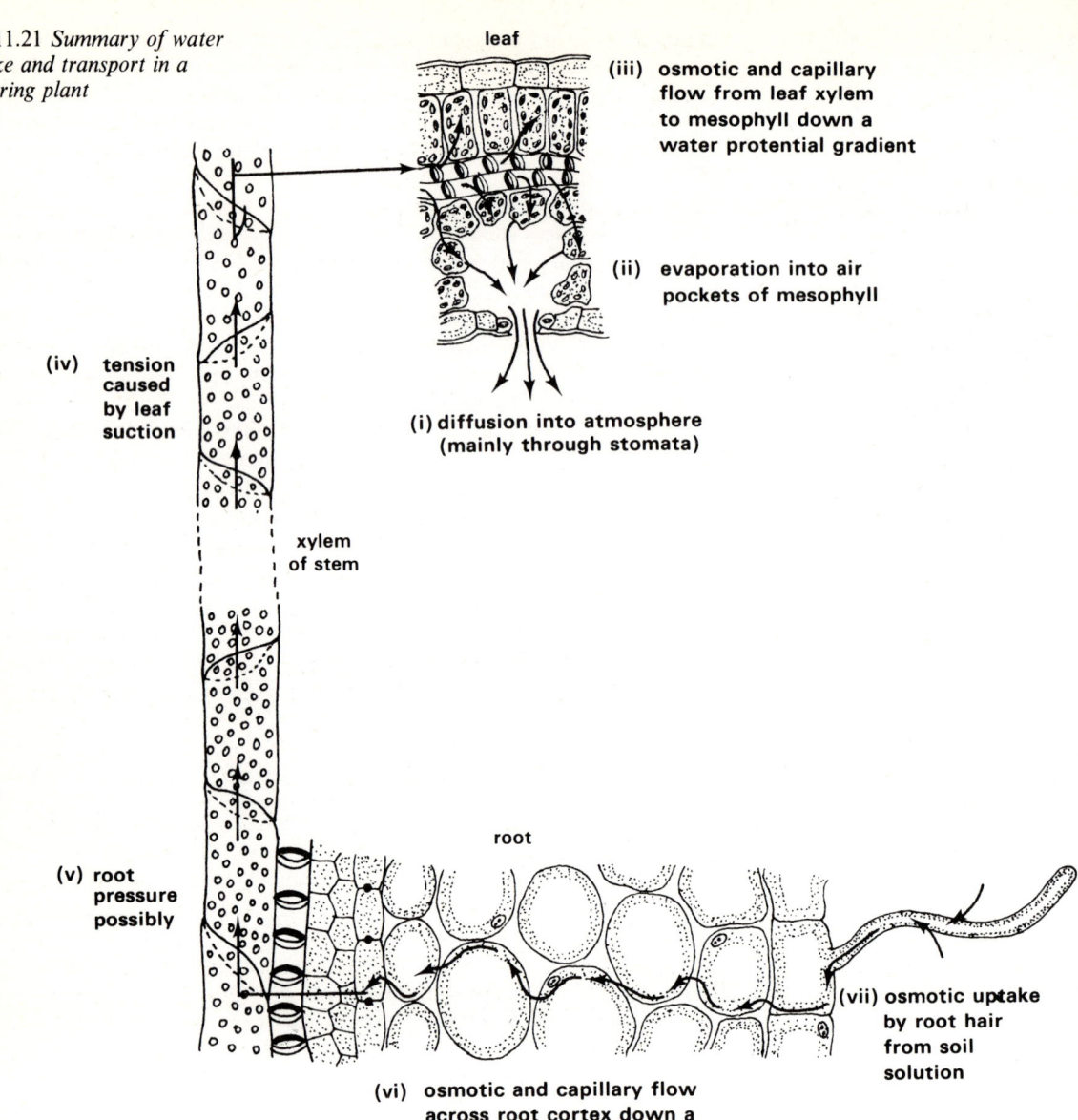

FIG. 11.21 *Summary of water uptake and transport in a flowering plant*

leaf

(iii) osmotic and capillary flow from leaf xylem to mesophyll down a water protential gradient

(ii) evaporation into air pockets of mesophyll

(i) diffusion into atmosphere (mainly through stomata)

(iv) tension caused by leaf suction

xylem of stem

root

(v) root pressure possibly

(vii) osmotic uptake by root hair from soil solution

(vi) osmotic and capillary flow across root cortex down a water potential gradient

Chapter 12

Photosynthesis and translocation of photosynthetic products

Animals and green plants need a supply of raw materials and a source of energy to synthesise the many complex organic substances from which protoplasm is made. Animals get their energy from organic substances in their diets. Green plants obtain their energy from sunlight. The process whereby green plants use sunlight to produce organic molecules is called **photosynthesis**. Carbon dioxide and water are the raw ingredients for the process, which can be summarised simply in the following equation:

$$\text{energy} + CO_2 + H_2O \longrightarrow (CH_2O) + O_2$$

energy (from sunlight) + CO₂ (carbon dioxide) + H₂O (water) ⟶ (CH₂O) (complex organic molecules) + O₂ (oxygen)

Photosynthesis is of the greatest importance to all living organisms. The organic substances made in photosynthesis can later be used by animals and other non-photosynthetic organisms. Thus in all ecosystems green plants are the **producers** on which the consumer organisms depend for energy and raw ingredients. On a global scale carbon dioxide is used in photosynthesis at a rate which more or less balances its output from respiration and the burning of fossil fuels such as coal and oil. What is more, the release into the environment of oxygen made in photosynthesis compensates for the uptake of oxygen by living organisms for respiration. Photosynthesis therefore helps to maintain an equilibrium or **steady state** in the environment.

Details of the way in which sunlight is converted into chemical bond energy in green plants are given in Chapter 5. In this chapter we shall examine how the structure of terrestrial plants enables them to obtain the raw materials and sunlight for photosynthesis. We shall also look at the ways in which various factors affect the rate of photosynthesis and what happens to photosynthetic products.

12.1 The leaf as a photosynthetic organ

The main organs of photosynthesis in terrestrial plants are the **leaves**. It is in the leaves that **chloroplasts** are present in abundance. If they are to function efficiently chloroplasts must have an adequate supply of **water** and **carbon dioxide**. They must also receive **light** of suitable wavelength and intensity. The structure of the leaves of mesophytes (Chapter 11) enables the photosynthetic requirements to reach the chloroplasts efficiently.

12.1.1 Penetration of light

The leaves of mesophytic plants are broad, thin, flattened structures. They thus have a relatively large surface area through which light can pass. The thinness of leaves means that light reaching the leaf surface has to pass only a short distance before reaching the mesophyll tissue where the chloroplasts are mainly located. Penetration of light to the mesophyll tissue is helped by the **transparency** of the leaf epidermis (Fig. 12.1(a)).

The leaves of many plants grow so that the leaf blade is usually at 90° to the sun's rays. Shading by leaves on the same shoot is often avoided because the leaves are normally arranged in a **mosaic** (Fig. 12.1(b)). These adaptations help to ensure that the maximum amount of available sunlight is received by each unit area of leaf surface. Leaves which are shaded, **shade leaves**, have a larger surface area than those exposed to full sunlight, **sun leaves**. Shade leaves are also thinner and have more chloroplasts. Because of this shade leaves make efficient use of the dim light they receive.

Inside chloroplasts the photosynthetic pigments are spread out in thin layers (Chapter 1). This arrangement presents a relatively extensive surface area for the absorption of sunlight. Chloroplasts can also change their positions to make the best use of available light (Fig. 12.1(c)).

FIG. 12.1 (a) *Passage of light through the outer surface of a leaf*

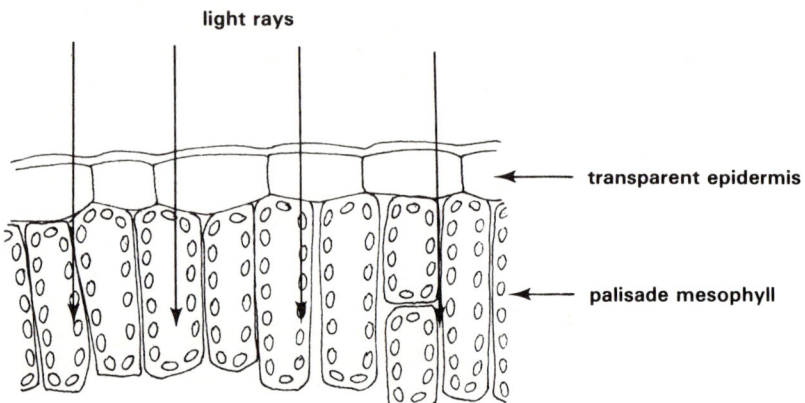

light rays

← transparent epidermis

← palisade mesophyll

FIG. 12.1 (b) *Leaf mosaic of sycamore*

FIG. 12.1 (c) *Movement of chloroplasts in response to light intensity*

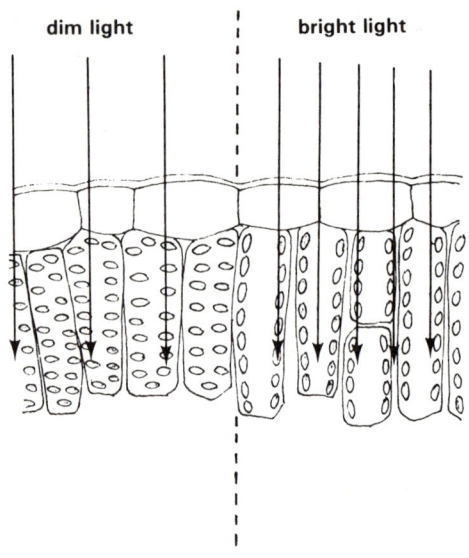

dim light   bright light

12.1.2 Supply of water

Water is carried into a leaf through a main vein in the midrib. An extensive network of minor veins arising from the main vein penetrates the mesophyll tissue. In an oak leaf for example, each mm² of leaf area contains a total length of about 10 mm of vein. Under normal circumstances the **xylem elements** of the veins maintain a continuous flow of water to the leaf tissues (Chapter 11). Another important function of the leaf veins is to carry away the organic products of photosynthesis to other parts of the plant. Phloem elements in the veins are responsible for this task (section 12.4).

12.1.3 Uptake of carbon dioxide

The waxy cuticle covering the epidermis of leaves is permeable to carbon dioxide in some plants. In other species carbon dioxide reaches the mesophyll tissue mainly through the **stomata**. There are three main patterns of stomatal distribution in leaves. In many mesophytic plants stomata are confined to the lower epidermis. The leaves of some mesophytes have a few stomata on the upper epidermis. Equal numbers of stomata occur on both surfaces in many monocotyledonous species. Each stoma consists of a pair of **guard cells** between which a **pore** is formed when the stoma is open. The walls of the guard cells are unevenly thickened. The inner walls where the two cells face each other is much thicker than the outer walls (Fig. 12.2). The guard cells are nucleate and several small chloroplasts are often present in the dense cytoplasm.

FIG. 12.2 (a) (i) *Surface view of stoma of privet*

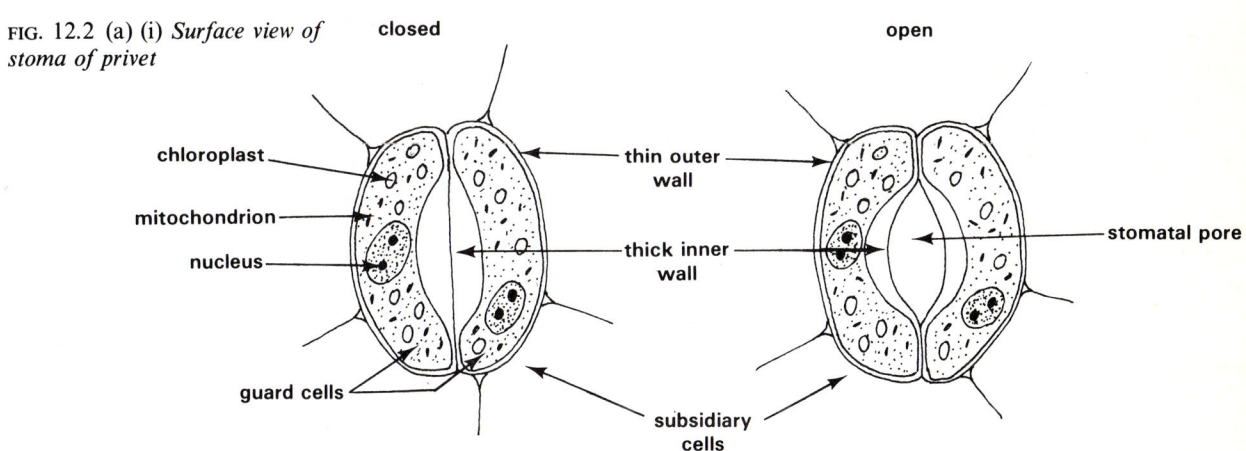

FIG. 12.2 (a) (ii) *T.s. stoma of privet*

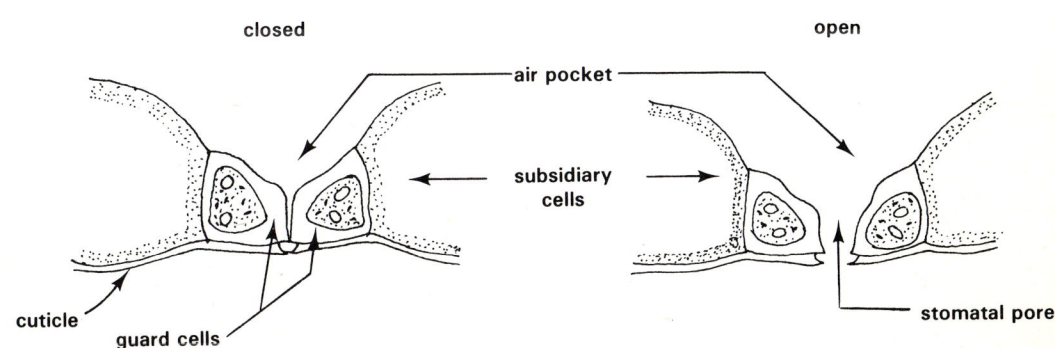

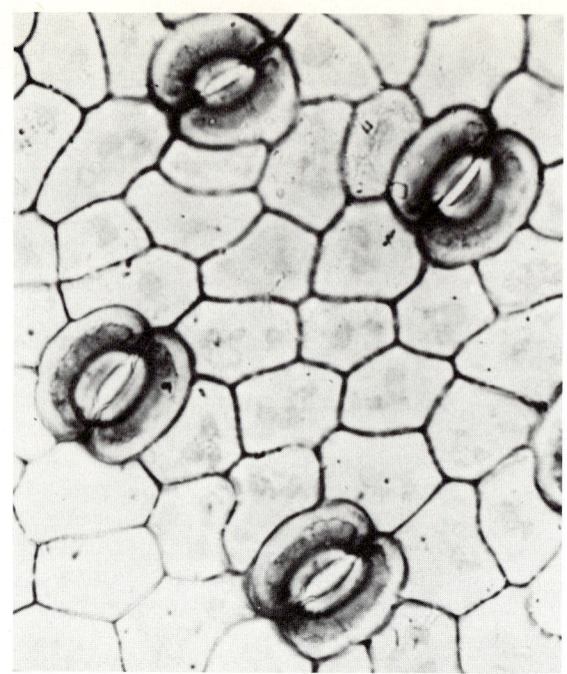

FIG. 12.2 (b) *Surface view of open stomata on the leaf epidermis of a dicotyledonous plant*, × 300

It might be expected that the tiny stomatal pores should offer considerable resistance to the passage of carbon dioxide from the surrounding air into the mesophyll tissue. However, experiments carried out by Brown and Escombe at the end of the last century demonstrate that gases can pass quickly through small pores (Table 12.1). The data shows that for a given area of pore, carbon dioxide flows more rapidly through small pores than through large pores. Stomatal pores are on average 10–30 μm long and 5–10 μm wide when fully open. The rate of gas flow per unit area through stomatal pores is much more rapid than that shown for the smallest pores used in the experiments.

Table 12.1 Diffusion of carbon dioxide through pores of different diameter (after Brown and Escombe, 1900)

DIAMETER OF PORE (mm)	cm^3 OF CO$_2$ DIFFUSING h^{-1}	cm^3 DIFFUSING h^{-1} cm^{-2} OF PORE
22·70	0·24	5·9 × 10^{-4}
12·06	0·10	8·8 × 10^{-4}
6·03	0·06	21·0 × 10^{-4}
3·23	0·04	49·0 × 10^{-4}
2·00	0·02	64·0 × 10^{-4}

Plot the data given in Table 12.1 as a graph. Use the graph to calculate the rate at which carbon dioxide passes through pores 0·2 mm in diameter. Pores of this size are at least ten times larger than stomatal pores. Further experimental investigations by Brown and Escombe showed that the rate of diffusion of gases through porous membranes is also affected by the closeness of the pores (Table 12.2). It may therefore be anticipated that the uptake of carbon dioxide by a leaf to some extent depends on the number of stomata per unit area of leaf. Stomatal densities of between 50 and 200 per mm^2 of leaf area are commonly found on the leaves of mesophytes. The combined area of open pores is usually between 0·5 and 1·5 per cent of the total leaf area. Experiments and observations such as these suggest that atmospheric carbon dioxide enters the air spaces of leaves when the stomata are open almost as though there is no epidermis present at all.

Table 12.2 Diffusion of carbon dioxide through membranes with different densities of pores (after Brown and Escombe, 1900)

NO. OF PORES cm^{-2} OF MEMBRANE	% AREA PERFORATED	DIFFUSION AS % OF VALUE WITHOUT MEMBRANE
100·00	11·3	87·6
25·00	2·8	63·7
11·11	1·25	44·0

The mesophyll tissue is permeated by air-filled spaces which amount to 30–40 per cent of the total leaf volume in many species. The internal **air spaces** are very important because they allow carbon dioxide taken in from the atmosphere to diffuse rapidly to photosynthesising cells. If the carbon dioxide had to diffuse through water instead it would move 10 000 times less quickly. However, once inside the mesophyll cells carbon dioxide has to move through the aqueous cell solution to reach the chloroplasts. As the chloroplasts are situated just inside the plasma membrane the distance that carbon dioxide passes through water is minimal. Uptake of carbon dioxide by the mesophyll cells is enhanced by the enzyme **carbonic anhydrase** which catalyses the formation of carbonic acid:

$$CO_2 + H_2O \xrightleftharpoons[\text{}]{\text{carbonic anhydrase}} H_2CO_3$$

carbon dioxide   water   carbonic acid

In the air spaces of a photosynthesising leaf the concentration of carbon dioxide is lowered to 0·01 per cent compared with 0·03 per cent in the external air. The result is that atmospheric carbon dioxide diffuses down a concentration gradient through stomatal pores into the air spaces of the mesophyll tissue.

12.1.4 Opening and closing of stomata

The stomata of most terrestrial plants open in light and close in darkness. Water loss due to transpiration is thus confined to periods when the plant is obtaining carbon dioxide from the surrounding air. Conservation of water is essential in a terrestrial environment because of the dehydrating effect of the atmosphere. Some xerophytic plants open their stomata only at night when the drying power of the air is less intense (Chapter 11).

Changes in turgidity (Chapter 2) of the guard cells are responsible for the opening and closing of stomatal pores. When the guard cells are turgid the pores are open. The guard cells of closed stomata are flaccid. Stomata close in wilted plants because the guard cells lose turgidity following excessive water loss by evaporation (Chapter 11). The reasons for changes in the turgidity of guard cells of non-wilted plants, however, are still a matter for debate. The factors which are thought to be mainly responsible include photosynthesis by the guard cells, carbon dioxide concentration of the air inside leaves and the concentration of mineral ions in the guard cells.

1. PHOTOSYNTHESIS BY THE GUARD CELLS

The earliest theories state that the formation of sugars by photosynthesis in the guard cells is the key factor causing stomata to open. The water-soluble sugars lower the water potential of the guard cells causing water to enter by osmosis from surrounding epidermal cells. When exposed to light therefore the guard cells gradually increase in turgidity and the stomata open. However, not all guard cells contain chloroplasts. Think of those on the non-green parts of variegated leaves. Even guard cells with chloroplasts are unlikely to make sugars quickly enough to bring about the rapid increase in pressure potential necessary for opening stomata.

2. CARBON DIOXIDE CONCENTRATION INSIDE LEAVES

Guard cells are sensitive to changes in the concentration of carbon dioxide in the air around them. High concentrations of carbon dioxide cause stomata to close, low concentrations cause them to open. What is more, there are fluctuations in the concentration of carbon dioxide in and around leaves each day. In darkness, when photosynthesis stops, respiratory carbon dioxide accumulates. During the daytime, however, carbon dioxide is used for photosynthesis. The concentration of carbon dioxide inside the leaf is then much lower than at night. But how can changes in carbon dioxide concentration bring about changes in turgidity of the guard cells? After a period of darkness most guard cells contain starch grains which are changed to sugar on exposure to light. Could it be that the conversion of starch to sugar is affected by carbon dioxide concentration, and if so how?

Leaf mesophyll cells contain the enzyme carbonic anhydrase which catalyses the reaction between carbon dioxide and water to form carbonic acid. The acid dissociates weakly into hydrogencarbonate (HCO_3^-) and hydrogen (H^+) ions:

$$\underset{\text{carbonic acid}}{H_2CO_3} \rightleftharpoons \underset{\text{hydrogencarbonate ion}}{HCO_3^-} + \underset{\text{hydrogen ion}}{H^+}$$

High concentrations of hydrogen ions therefore accumulate in leaf cells, guard cells included, in darkness. When light is available the carbonic acid is used for photosynthesis thereby lowering the hydrogen ion concentration $[H^+]$ in leaf cells during the daytime. But how are $[H^+]$ changes in leaf cells linked with changes in turgidity of the guard cells?

The starch-sugar conversion in guard cells is catalysed by a group of enzymes. Like most enzyme-catalysed reactions, the conversion is affected by pH. In acid conditions, high $[H^+]$, the enzymes convert glucose phosphate to starch. However at a lower $[H^+]$ the direction of the reaction is reversed and sugar accumulates in the guard cells:

$$\text{starch} \underset{\text{high } [H^+]}{\overset{\text{low } [H^+]}{\rightleftharpoons}} \text{glucose phosphate}$$

The daytime accumulation of sugar lowers the water potential of the guard cells. The guard cells take in water from nearby epidermal cells and become turgid. In darkness the sugar is converted to starch which, being insoluble in water, causes the water potential of the guard cells to be raised. Water now passes from the guard cells into surrounding epidermal cells down a water potential gradient. The resulting flaccidity of the guard cells closes the stomatal pores.

Plausible as this explanation is, it does not account for recent observations showing that there are fluctuations in the concentrations of certain mineral ions in guard cells during the opening and closing of stomata.

3. MINERAL ION CONCENTRATION OF GUARD CELLS

Increases in mineral ion content of the guard cells, especially of potassium ions (K^+), accompany stomatal opening (Fig. 12.3). ATP is produced rapidly in guard cells exposed to light. It is thought that energy from ATP may be used to work an **ion pump** which draws in K^+ ions from surrounding epidermal cells. The accumulation of K^+ ions could lead to the increase in turgidity of the guard cells which brings about stomatal opening.

However, there are simultaneous increases in concentration of other substances in the guard cells of open stomata. One of the substances is malic acid, a product of the breakdown of starch. Some plant physiologists think that an increase of hydrogen ions (H^+) in the guard cells due to accumulation of malic acid is prevented by exchanging the H^+ ions for K^+ ions from surrounding epidermal cells. The water-soluble malate ions help to lower the water potential of the guard cells leading to stomatal opening.

Although stomata have been studied for a very long time it is evident that there is still much to be learned about the mechanisms which control stomatal opening and closing. It may even be that different mechanisms operate in different species or that a combination of mechanisms is involved.

FIG. 12.3 *Relative amounts of potassium (continuous line) and phosphorus (dotted line) in guard cells of open and closed stomata* (after Humble and Raschke, 1971)

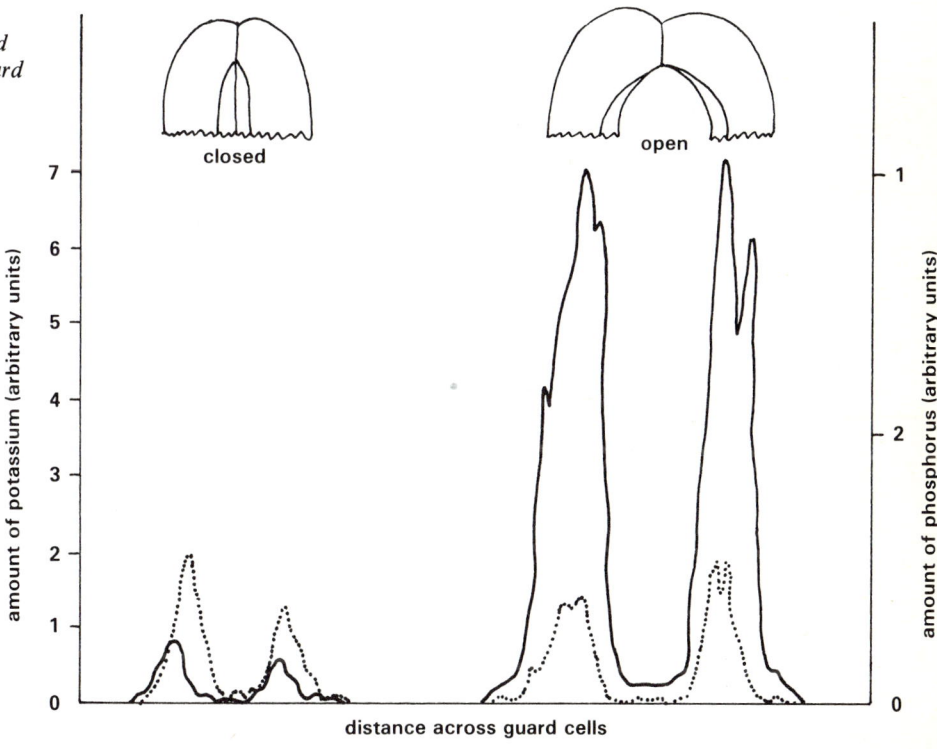

12.2 Factors affecting the rate of photosynthesis

The rate at which oxygen is given off by green plants can be used to measure the rate of photosynthesis. Aquatic plants such as *Elodea* are particularly suitable for this purpose. Bubbles of oxygen are given off from the cut end of the stem of *Elodea* when it is immersed in illuminated pond water or a dilute solution of sodium hydrogencarbonate (Fig. 12.4). Using this method in the 1930s the British plant physiologist F. F. Blackmann investigated the effect of light intensity on the rate of photosynthesis. The intensity of light was varied by increasing or decreasing the distance (D) between the lamp and the plant. Light intensity (I) is inversely proportional to the square of the distance ($I \propto \dfrac{1}{D^2}$). The results of the experiment are shown in Figure 12.5.

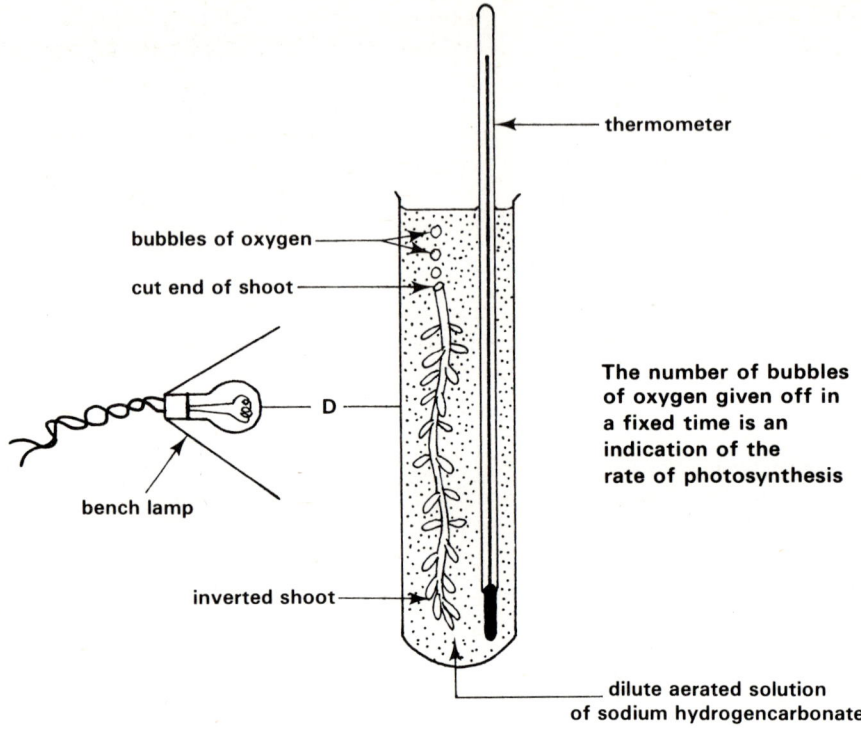

FIG. 12.4 *Measuring the rate of photosynthesis of the pondweed* Elodea

thermometer

bubbles of oxygen

cut end of shoot

The number of bubbles of oxygen given off in a fixed time is an indication of the rate of photosynthesis

bench lamp

inverted shoot

dilute aerated solution of sodium hydrogencarbonate

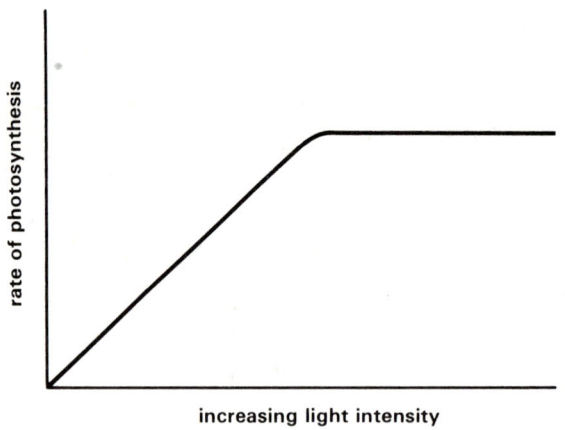

FIG. 12.5 *Effect of light intensity on the rate of photosynthesis*

rate of photosynthesis

increasing light intensity

At low intensities of light Blackmann observed a straight line relationship between the rate of photosynthesis and light intensity. However, once a critical light intensity is reached the rate of photosynthesis remains constant. The results surprised Blackmann who reasoned that if photosynthesis is driven by light energy then it should go faster if more intense light is provided. He concluded that factors other than the availability of light energy limit the rate of photosynthesis at high light intensities.

Knowing that carbon dioxide is an essential raw ingredient for photo-synthesis and that the process was probably catalysed at least in part by enzymes, Blackmann went on to investigate the effects of carbon dioxide concentration and temperature on the rate of photosynthesis. The results of these investigations are shown in Figures 12.6 and 12.7. Clearly photo-synthesis is controlled by a combination of factors and it is apparent that the photosynthetic rate is limited by whichever factor is nearest its minimum value. This is a good example of the **law of limiting factors**.

FIG. 12.6 *Effect of carbon dioxide concentration on the rate of photosynthesis*

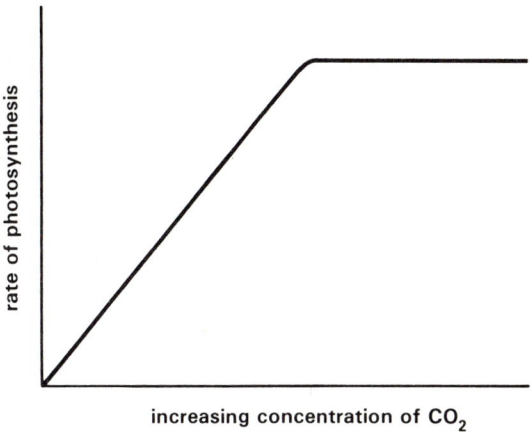

FIG. 12.7 *Effect of temperature on the rate of photosynthesis*

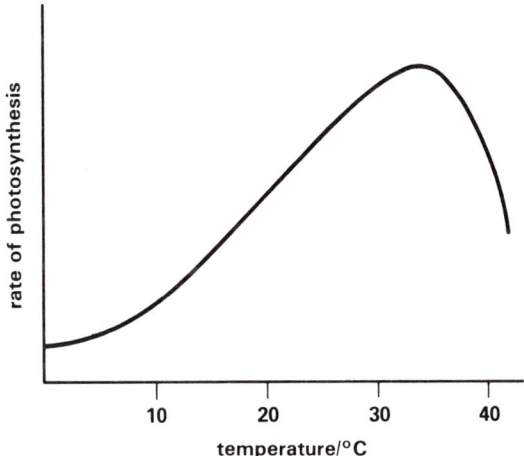

Blackmann did not attempt to find out whether the availability of water affects the rate of photosynthesis. In non-wilted plants there is always sufficient water for photosynthesis. The closure of stomata of wilted leaves may however limit the rate of photosynthesis by preventing the diffusion of carbon dioxide to the mesophyll tissue. In normal circumstances tempera-ture, light intensity and carbon dioxide concentration are the factors most likely to limit the rate of photosynthesis.

12.2.1 Temperature

Photosynthesis consists of light-dependent and light-independent stages (Chapter 5). It is the enzyme catalysed reactions of the **light-independent stage** which are affected by temperature. The temperature coefficient, Q_{10} (Chapter 4), for the light-independent stage is between 2 and 3. Figure 12.8 shows the effect of temperature on photosynthesis at high and low light intensities. Notice that increasing the temperature has no effect on the rate of photosynthesis at low light intensity. Why is this so? For C4 plants the optimum temperature for photosynthesis is between 35–40°C and for C3 plants it is 20–25°C.

FIG. 12.8 *Effect of temperature on the rate of photosynthesis at two intensities of light*

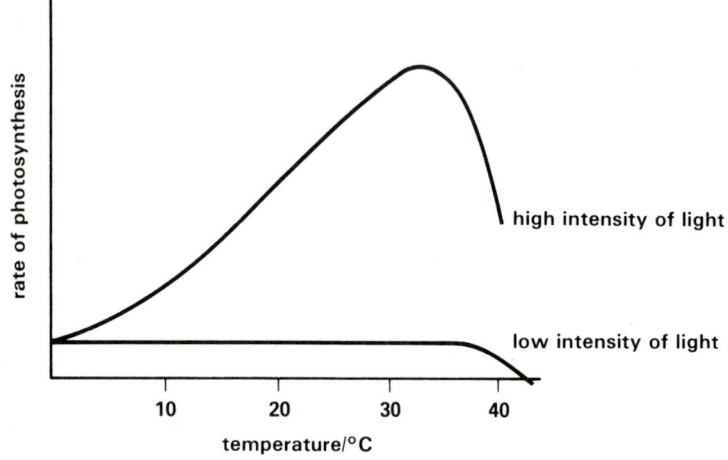

12.2.2 Light intensity

Light reaching the shaded leaves of a plant is less intense than light received by leaves exposed to full sunlight. Shade leaves compensate for this by structural adaptations which enable them to make the most of dim light (section 12.1.1). In some habitats such as dense beech forests and coniferous woodland the shade of the trees prevents the growth of herbaceous plants which normally appear as a field layer in open woodland where light can penetrate to the ground (Fig. 12.9).

FIG. 12.9 *Effect of shading on the field layer of a beech wood. The deep shade cast by beech trees prevents the development of a field layer, so the ground is bare*

In darkness photosynthesis cannot take place, although other metabolic processes such as respiration go on just as they do in daylight. In the absence of light, green plants give off carbon dioxide made in respiration. When light is available respiratory carbon dioxide provides a proportion of a plant's photosynthetic needs. The rest of the carbon dioxide is absorbed from the surrounding air. For a plant to grow it has to synthesise organic materials more rapidly than it oxidises them in respiration. When photosynthesis and respiration occur at rates such that there is no gain or loss of organic matter a plant is at its **compensation point**. The time taken for a plant which has been in darkness to reach the compensation point is called the **compensation period**. The most important factor governing the length of the compensation period in natural conditions is light intensity. Shade leaves have shorter compensation periods than sun leaves (Fig. 12.10). This is a physiological adaptation which enables shade leaves to make efficient use of light of low intensity.

FIG. 12.10 *Effect of light intensity on CO_2 exchange by sun and shade leaves*

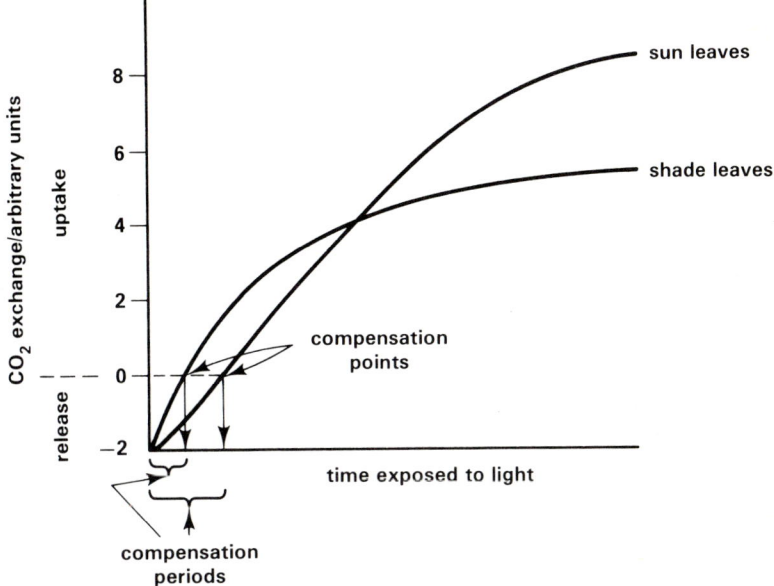

The stomata of most mesophytic plants open in light and close in darkness (section 12.1.4). In intense light stomatal pores are at their widest, enabling carbon dioxide to diffuse rapidly into the mesophyll tissue. Not all wavelengths of light are equally effective in stimulating photosynthesis. Blue and red light are more effective than others in this respect (Chapter 5).

12.2.3 Concentration of carbon dioxide

On a warm sunny day the concentration of carbon dioxide in the air is probably the factor which limits photosynthesis more than any other. Enriching air with carbon dioxide has a significant effect on crop plants grown in greenhouses (Table 12.3). It is now common practice for commercial growers of salad plants to raise the concentration of carbon dioxide of the air in greenhouses during daylight. The gas is either pumped in

Table 12.3 Yield of lettuces and tomatoes grown in normal air and in air enriched with carbon dioxide

CROP	WITHOUT ADDED CO_2	WITH ADDED CO_2	YIELD
lettuces	0·9	1·1	Fresh mass/ kg per 10 heads
tomatoes	4·4	6·4	Fresh mass/kg per plant

directly or is released by the burning of fuels such as paraffin, propane and natural gas which are used to heat greenhouses. Concentrations of up to 0·2 per cent carbon dioxide produce increases in crop yield which are economically worthwhile.

Of course, increasing the concentration of atmospheric carbon dioxide for crops grown in the open air is impracticable. On average the amount of carbon dioxide in the atmosphere outdoors is 0·03 per cent. It is controlled mainly by the rate at which living organisms give off carbon dioxide from respiration compared with the rate at which it is used by green plants for photosynthesis. An old-fashioned way of improving crop production in greenhouses was to add large quantities of manure to the soil. Carbon dioxide, released by soil micro-organisms which decomposed the manure, accumulated in the air and stimulated photosynthesis resulting in bigger crops.

12.3 Productivity of plants and plant communities

So far we have concentrated on the leaves of terrestrial plants as photosynthetic organs. It is just as important to consider photosynthesis in whole plants. After all, the entire plant is potential food for consumer organisms. Plants normally grow with others of the same species or of different species in **plant communities**. A field of wheat is an example of a man-made plant community. Natural forest and woodland are plant communities not made by man. The efficiency with which whole plants and plant communities produce dry matter determines how much food is available for the higher trophic levels in an ecosystem.

It is only recently that biologists have begun to appreciate which factors cause some plants and plant communities to produce dry matter more efficiently than others. Studies on crop plants show that two factors of fundamental importance to crop yield are **leaf area index** and **unit leaf rate**.

12.3.1 Leaf area index

Plants with a large surface area of leaves and other parts which can photosynthesise may be expected to produce more dry matter than plants having shoot systems with a small surface area. The area of leaves available for photosynthesis can be expressed as the **leaf area index (LAI)**:

$$LAI = \frac{\text{total leaf area of plant}}{\text{area of ground covered by plant}}$$

It indicates the amount of light intercepted by the shoot system of a plant. During the early stages of growth, crop plants have small LAI values because each plant has only a few small leaves and is surrounded by a patch of bare ground. As growth proceeds and the shoot system enlarges, the LAI increases. Maize grown in the U.S.A. achieves a maximum LAI of 4. Some crop plants have larger values than this. Sugar cane for example has a maximum LAI of about 7.

The shape of the shoot system is particularly important in determining the leaf area index of a plant. Plants which can be grown close to each other and which have leaves held vertically have higher LAI values than those with horizontally-held or drooping leaves. With this in mind plant breeders have produced new varieties of wheat and rice with erect leaves. The new varieties grow so that there is little mutual shading of their leaves and they yield extremely good crops of grain.

12.3.2 Unit leaf rate

Whatever the LAI value, increases in organic matter occur efficiently only if most of the photosynthetic products are converted to plant tissue or storage materials. If most of the products of photosynthesis are respired, dry matter accumulates slowly.

The term **unit leaf rate** (**ULR**) expresses the efficiency of dry matter accumulation by green plants. ULR is the mean rate of increase in dry mass of a whole plant per unit area of leaf. The ULR of a plant can be calculated from measurements of the leaf area and dry mass of a representative sample of plants at different stages of growth. Most species of crop plants grown in temperate areas have mean ULR values of between 4 and 6 g m^{-2} day^{-1}. In ideal growing conditions, when there are no factors limiting the rate of photosynthesis, the ULR can reach 15 g m^{-2} day^{-1}. Even so, some species have higher ULR values than others. Because they do not photorespire and have very short compensation periods C4 plants such as sugar cane and maize have a much greater unit leaf rate than most C3 plants (Chapter 5).

12.3.3 Net primary productivity

Synthesis of dry matter by green plants is called **primary production**. The total amount of dry matter produced per unit area of ground per year is called **gross primary productivity**. Some of the dry matter is used by green plants in respiration. What is left is called **net primary productivity** (**NPP**) and it is this which is available for consumer organisms including man.

$$NPP = LAI \times ULR$$

Clearly plants which quickly achieve high LAI values and which sustain an efficient ULR over a long growing period are highly productive. In attempting to raise the amount of food for human consumption plant breeders now have such factors foremost in mind.

12.4 Translocation of photosynthetic products

Leaves can be thought of as a **source** of photosynthetic products. Although leaves can store organic compounds such as starch, the products of photosynthesis are generally exported fairly quickly to tissues which actively use them or where they can be stored. Such places are called **sinks**. The meristems of root and shoot apices and the vascular and cork cambium (Chapter 21) are among the more important sinks for photosynthetic products. It is there that the organic materials made by leaves are used to produce new cells and provide the energy for growth. Most green plants have some means of storing food which can be used at a later stage. The cotyledons and endosperm of seeds are the only sites of food storage in many species (Chapter 20). In others, an enlarged part of the vegetative plant body acts as a storage organ (Fig. 12.11).

FIG. 12.11 *Storage organs of flowering plants*

bulb of onion

tap root of carrot

stem tuber of potato

The long distance movement of metabolites from sources to sinks is called **translocation**. The term translocation is often used to describe the movement of all water-soluble substances in the vascular tissue of plants. Here we shall confine our account mainly to the movement of organic substances made in photosynthesis.

12.4.1 The pathway of translocation

There is a lot of experimental evidence which suggests that the **phloem** is the tissue mainly concerned with the translocation of organic substances. An elegant technique applied to translocation studies makes use of the fact that phloem elements are penetrated by the mouthparts of **aphids** such as greenfly and blackfly when they are feeding on plant juices (Fig. 12.12). If a feeding aphid is anaesthetised and the mouthparts carefully cut across, the stylets left in the plant act as a tiny sampling tube at the end of which droplets of phloem sap soon appear. Analyses show that the phloem sap, unlike xylem sap, is rich in sugars, mainly sucrose, amino acids and potassium and phosphate ions (Table 12.4). Furthermore, the composition of the sap varies according to the photosynthetic activity of the plant. Sucrose for example is at its highest concentration a few hours after sunrise.

Table 12.4 Composition of phloem sap of the castor oil plant *Ricinus communis* (after Hall and Baker 1972)

COMPONENT	CONCENTRATION/mg cm^{-3}
sucrose	80–106
reducing sugars	0
protein	1·45–2·20
amino acids	5·2 (as glutamic acid)
carboxylic acids	2·0–3·2 (as malic acid)
phosphate(V) ions	0·35–0·55
sulphate(VI) ions	0·024–0·048
chloride ions	0·355–0·675
nitrate(V) ions	0
hydrogencarbonate ions	0·010
potassium ions	2·3–4·4
sodium ions	0·046–0·276
calcium ions	0·020–0·092
magnesium ions	0·109–0·122
ammonia	0·029
ATP	0·24–0·36
auxin	$10·5 \times 10^{-6}$
gibberellin	$2·3 \times 10^{-6}$
cytokinin	$10·8 \times 10^{-6}$

FIG. 12.12 (a) (i) *Aphids feeding on a rose shoot*

FIG. 12.12 (a) (ii) *Ventral view of aphid × 50*

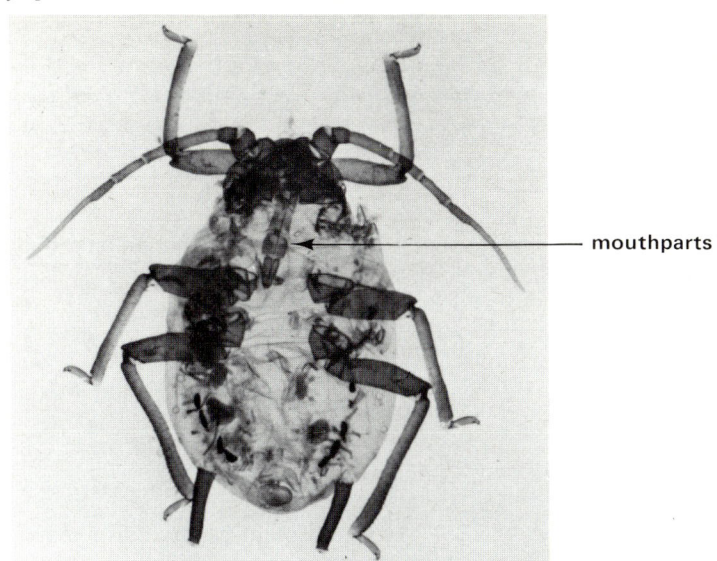

FIG. 12.12 (b) (i) *Aphid feeding on phloem sap* (modified after Dixon, 1973)

FIG. 12.12 (b) (ii)

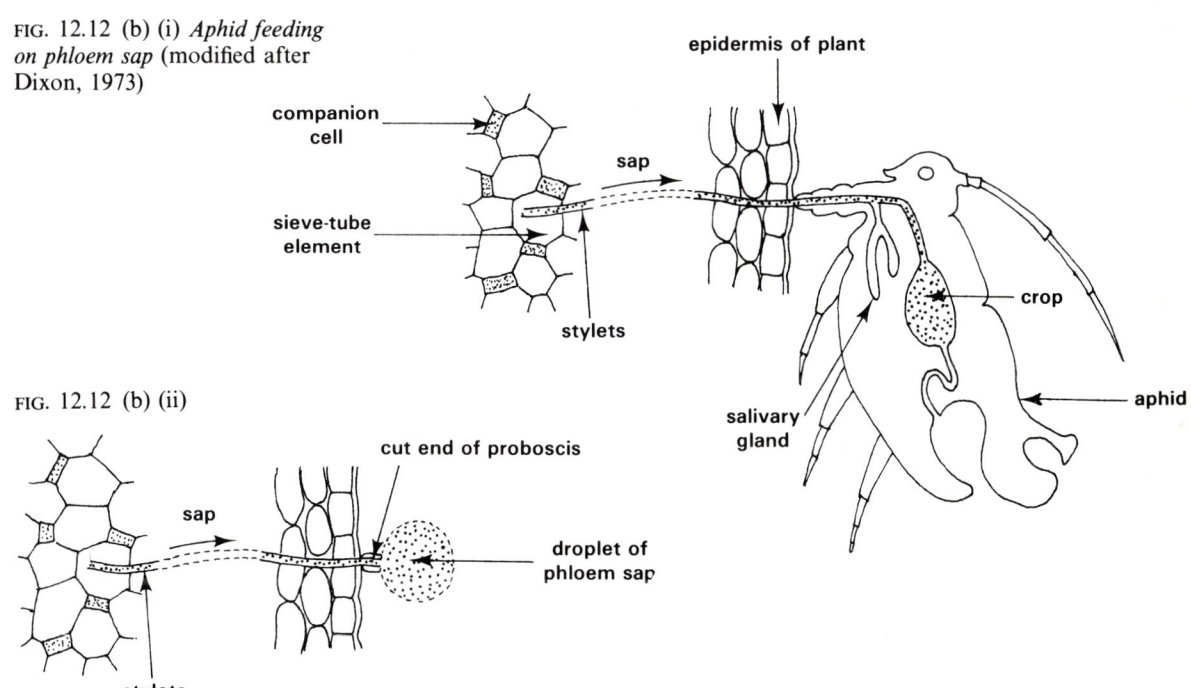

Another technique which has been used to great advantage in recent years involves supplying carbon dioxide containing the **radioisotope** ^{14}C to illuminated plant leaves. The isotope is fixed in the organic products of photosynthesis which are then translocated to other parts of the plant. The spread of radioactivity can be followed and is found almost entirely in the phloem (Fig. 12.13).

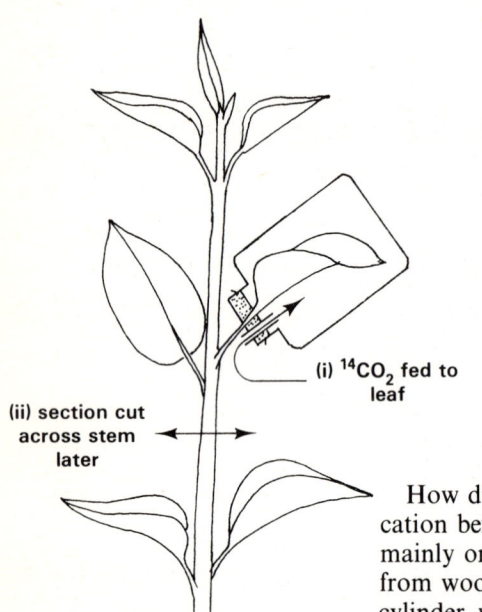

FIG. 12.13 *Investigating the spread of a carbon radioisotope in a plant*

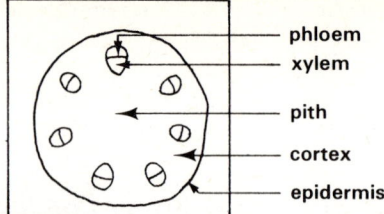

(iii) Section placed on photographic film

phloem
xylem
pith
cortex
epidermis

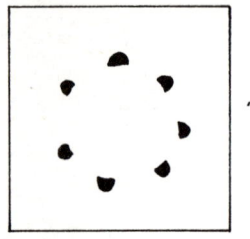

(iv) Section later removed

'fogging' only where phloem was in contact with film

(i) $^{14}CO_2$ fed to leaf

(ii) section cut across stem later

How did plant physiologists manage to learn anything about translocation before such sophisticated techniques were developed? They relied mainly on **ringing experiments** in which cylinders of bark were removed from woody stems and the contents of the phloem above and below the cylinder were later analysed (Fig. 12.14). Although crude, the ringing method provided some evidence that phloem is the main tissue in which organic compounds are translocated.

In flowering plants there are two conspicuous types of phloem cell concerned with translocation, **sieve-tubes** and companion cells. They lie alongside one another, derived from the same meristematic cell by longitudinal division (Chapter 21).

FIG. 12.14 *A ringing experiment*

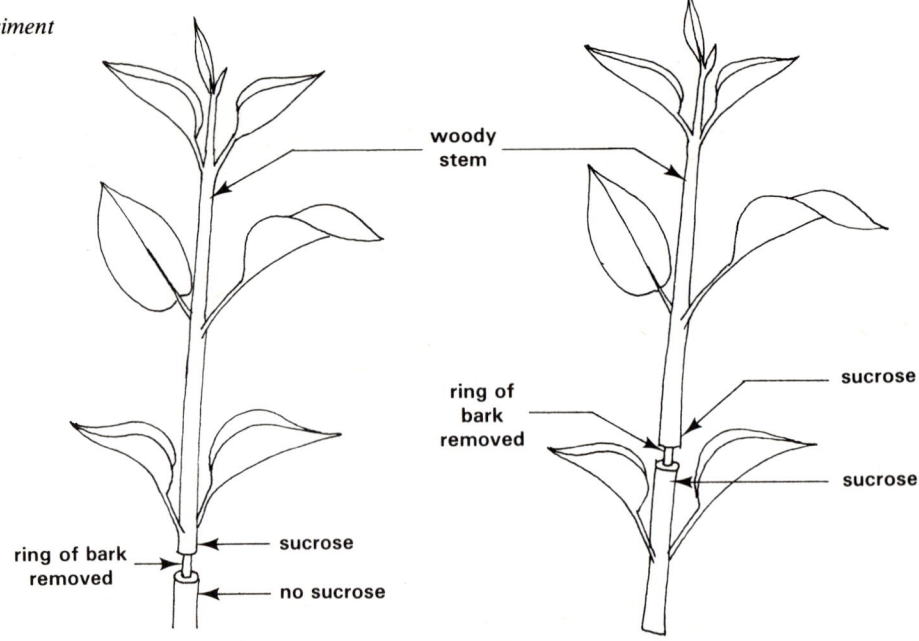

woody stem

ring of bark removed

sucrose

sucrose

ring of bark removed

sucrose

no sucrose

The bark includes the secondary phloem. What does the experiment tell us about the movement of organic materials like sucrose in the stem?

1. SIEVE-TUBES

Sieve-tubes are composed of a series of **sieve-tube elements** joined end to end. Each element is 10–50 μm in diameter and 150–1000 μm long. When first produced from a meristematic cell by mitosis, sieve-tube elements have a large nucleus and the usual cytoplasmic organelles. However, at maturity, when translocating organic materials rapidly in solution, the nucleus, ribosomes, Golgi body and tonoplast degenerate and the side walls are thickened mainly with cellulose. Small mitochondria remain, as does part of the endoplasmic reticulum which exists as parallel layers near the side walls. Plastids containing small starch grains are also present and the lumen is filled with slimy sap containing fibrils of a material called **phloem protein** (Fig. 12.15).

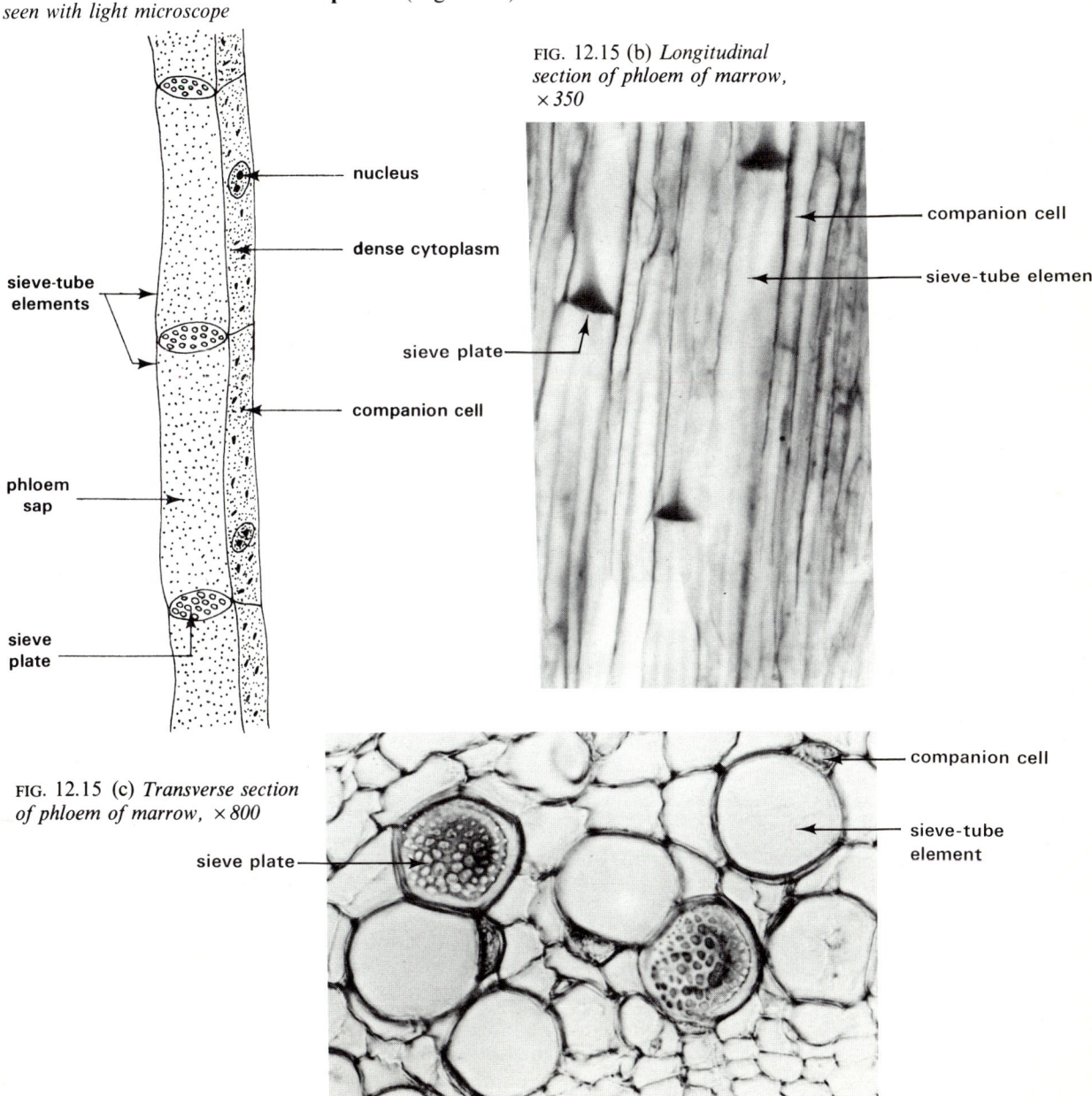

FIG. 12.15 (a) *L.s. of phloem as seen with light microscope*

nucleus

dense cytoplasm

sieve-tube elements

companion cell

phloem sap

sieve plate

FIG. 12.15 (b) *Longitudinal section of phloem of marrow,* ×350

companion cell

sieve-tube element

sieve plate

FIG. 12.15 (c) *Transverse section of phloem of marrow,* ×800

sieve plate

companion cell

sieve-tube element

During differentiation the end walls between adjacent sieve-tube elements are modified to form **sieve plates** perforated by pores. The pores are lined with deposits, which vary in thickness, of a glucose polymer called **callose**. The mean diameter of the pores is 2–6 μm but where thick deposits of callose occur the pores appear to be blocked.

FIG. 12.15 (d) *Ultrastructure of phloem (based on electronmicrographs)*

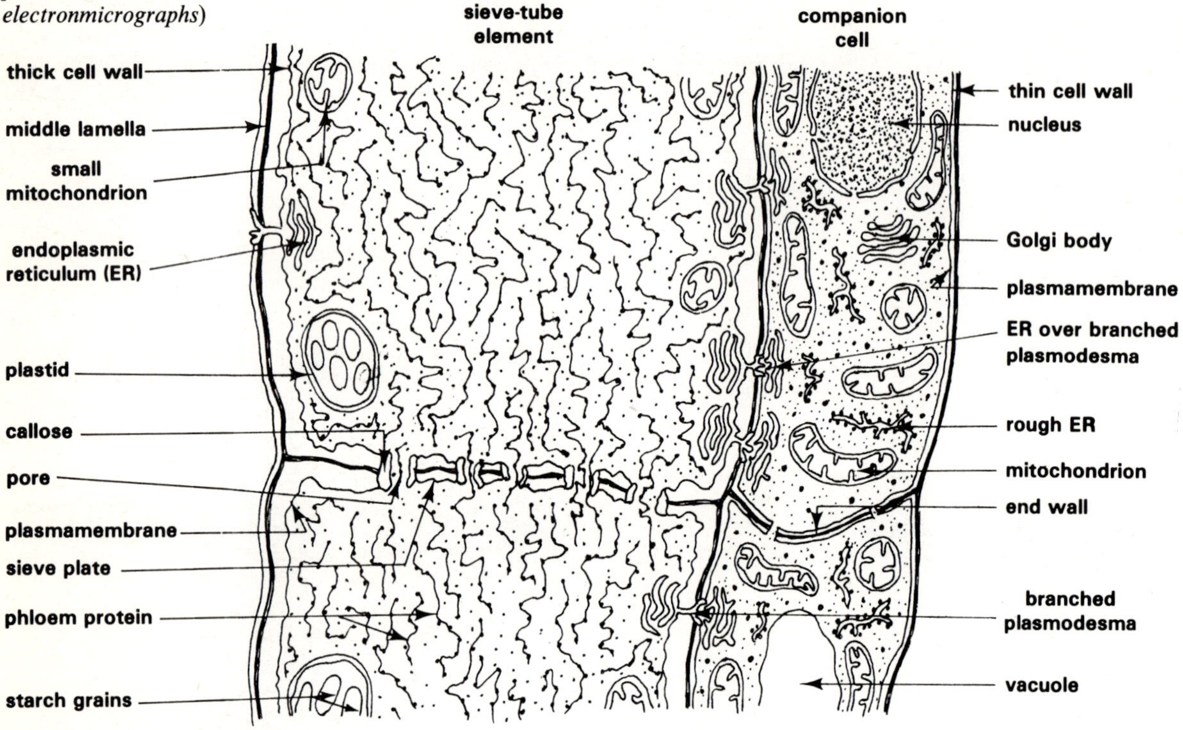

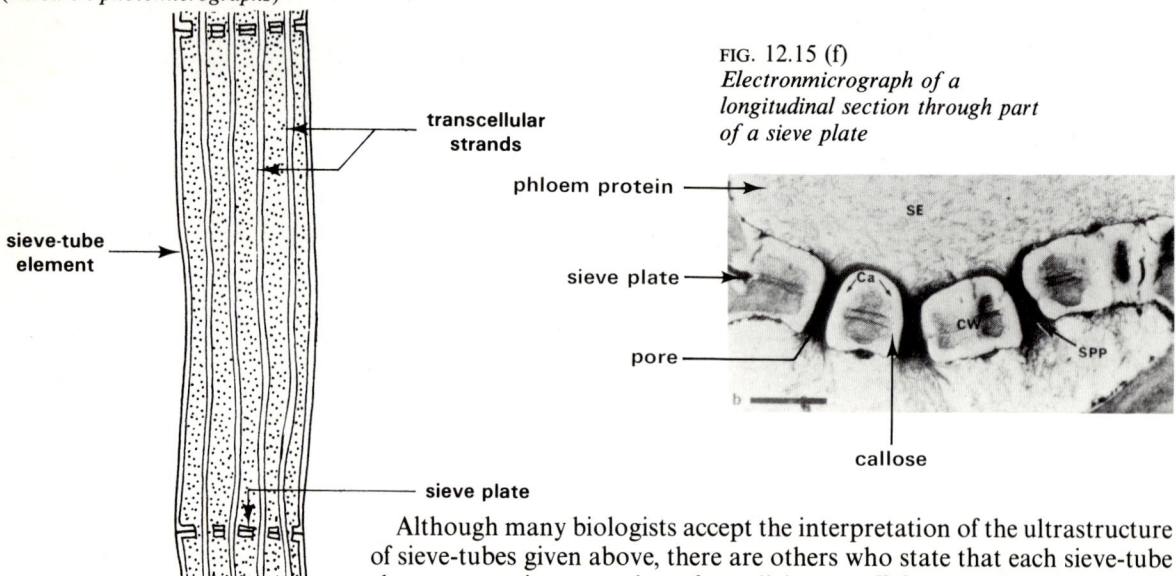

sieve-tube element

companion cell

thick cell wall
middle lamella
small mitochondrion
endoplasmic reticulum (ER)
plastid
callose
pore
plasmamembrane
sieve plate
phloem protein
starch grains

thin cell wall
nucleus
Golgi body
plasmamembrane
ER over branched plasmodesma
rough ER
mitochondrion
end wall
branched plasmodesma
vacuole

FIG. 12.15 (e) *Longitudinal section of part of sieve-tube showing transcellular strands (based on photomicrographs)*

transcellular strands

sieve-tube element

sieve plate

FIG. 12.15 (f) *Electronmicrograph of a longitudinal section through part of a sieve plate*

phloem protein
sieve plate
pore
callose

Although many biologists accept the interpretation of the ultrastructure of sieve-tubes given above, there are others who state that each sieve-tube element contains a number of parallel **transcellular strands** 1–7 μm wide.

The strands are thought to be hollow, membranous tubes which pass through the pores of sieve plates and extend through a series of sieve-tube elements (Fig. 12.15). Such a divergence of opinion may be due to differences in the species used for investigation. It may also result from differences in the techniques used to examine sieve-tubes.

2. COMPANION CELLS

A **companion cell** has a thin cellulose wall enclosing a protoplast which contains a prominent nucleus, numerous mitochondria, a well-developed endoplasmic reticulum with attached ribosomes, a Golgi body, small vacuoles and colourless plastids. The protoplasts of companion cells and adjacent sieve-tube elements are connected by many plasmodesmata (Fig. 12.15) suggesting that transfer of substances between the two types of cell can occur. At the ends of minor leaf veins the companion cells are much bigger compared with the sieve-tube elements than elsewhere in the plant. These companion cells also have ingrowths of their cell walls which greatly increase the surface area of the plasma membrane for the uptake of substances from surrounding mesophyll cells. The name **transfer cell** is given to specialised companion cells of this sort (Fig. 12.16).

FIG. 12.16
Electronmicrograph of a section of part of a transfer cell from lentil, × 20 000. Note the ingrowths of the cell wall which increase the surface area of the plasma membrane for the transfer of solutes. What functions may the mitochondria have?

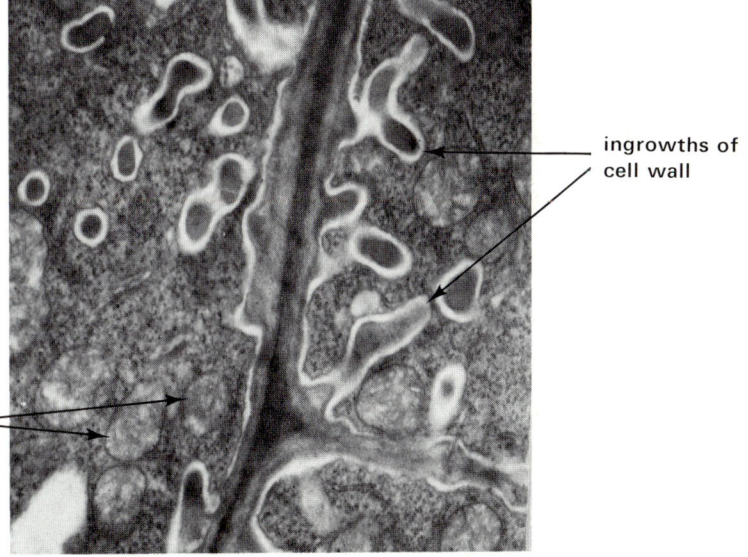

ingrowths of cell wall

mitochondria

12.4.2 The mechanism of translocation

A common method used to determine the rate of translocation in phloem is to measure the rate at which growing organs increase in dry mass. If the area of cross-section of phloem supplying the organ is also measured it is possible to calculate the rate at which organic matter is transferred per unit area of phloem in a known period of time (Table 12.5). Another technique is to follow the spread of radioactivity in a plant after a labelled compound such as $^{14}CO_2$ has been supplied to a photosynthesising leaf. This procedure has been particularly rewarding when used in conjunction with aphids (Fig. 12.17). The results of such experiments clearly demonstrate that translocation is a rapid process, much too rapid to be explained by diffusion. Sugars such as sucrose are translocated at a rate of 25–200 cm h^{-1} in sieve-tubes compared with a meagre 0.2 mm day^{-1} possible by diffusion. How much faster does sucrose move when translocated compared with its movement by diffusion?

Table 12.5 Rates of translocation measured by transfer of dry mass

SPECIES	ORGAN	MASS TRANSFER/ g DRY MASS cm^{-2} PHLOEM h^{-1}
potato	tuber	2·1–4·5
marrow	fruit	3·3–4·8
grasses	leaves	4·4–14·9

Having dismissed diffusion as a possible mechanism of translocation it is not easy to provide a really convincing alternative. There are many theories but relatively few facts to support them. One of the more favoured explanations of translocation is the mass-flow hypothesis.

FIG. 12.17 *Use of aphids and radioactive carbon to measure the rate of translocation. Phloem sap at A and B is analysed for radioactivity at frequent intervals. The time taken for the radioactivity to spread from A to B is a measure of the rate of translocation*

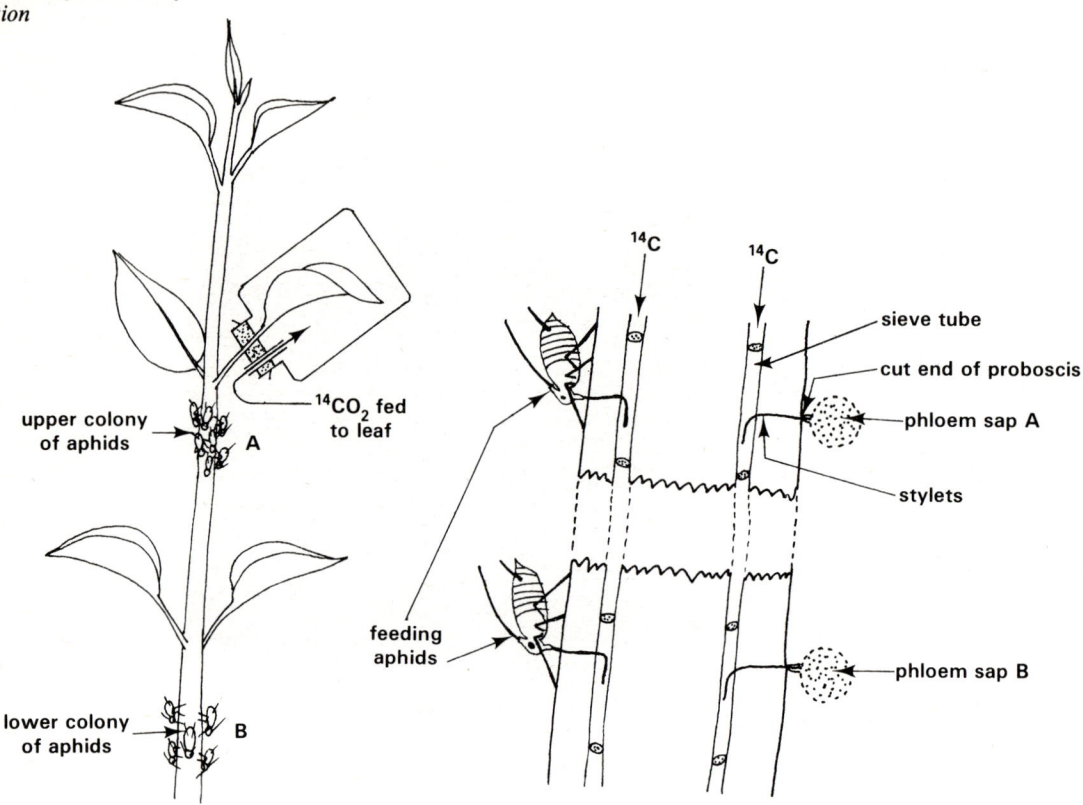

1. MASS-FLOW HYPOTHESIS

Figure 12.18 shows the principle of the **mass-flow hypothesis**. The sap in the sieve-tube elements next to leaf mesophyll cells, the **source**, is rich in organic solutes and thus has a low water potential. Water is absorbed into the sieve-tube elements by osmosis from surrounding tissues. A high pressure potential is thus created near the source and the solution inside the sieve-tube elements is pushed towards a sink along a pressure potential gradient. In this way organic solutes can be carried downwards to the roots of the plant or upwards to developing flowers, fruits and young leaves all of which are **sinks** (Fig. 12.19).

236

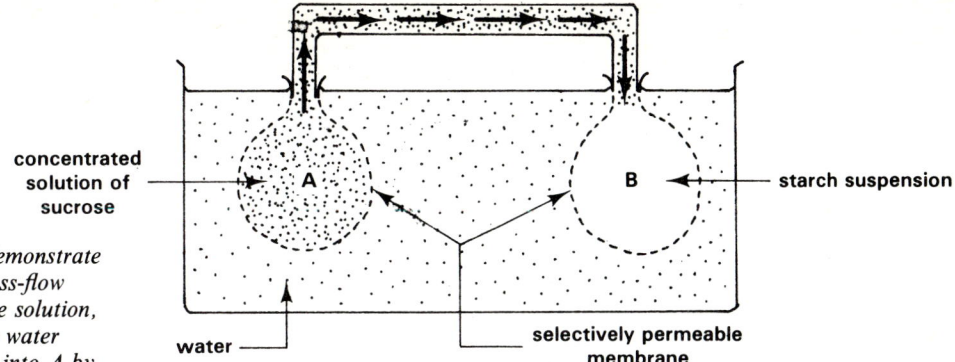

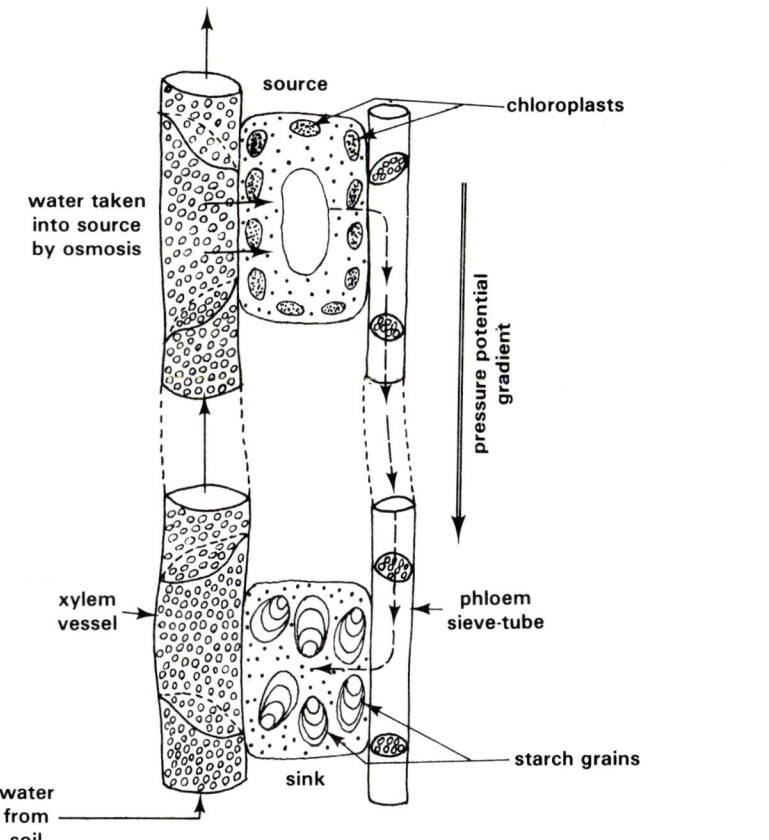

FIG. 12.18 *Model to demonstrate the principle of the mass-flow hypothesis. The sucrose solution, having a relatively low water potential, draws water into A by osmosis. The resultant rise in pressure in A causes the sucrose solution to enter the connecting tube and pass into B. Starch, being osmotically inactive, does not draw water into B from the container. The model could be made into a continuous-flow system if sugar could be continually added to A and removed from B*

FIG. 12.19 *Mass-flow of organic solutes in a plant*

The rapid and continued exudation of phloem sap from the severed mouthparts of feeding aphids shows that the contents of the sieve-tube elements have a high pressure potential. Furthermore, suitable concentration gradients of translocated substances exist between sources and sinks. Even so, the hypothesis does not account for all that is known of the physiology of phloem. There are doubts too that the structure of sieve-tubes in all plants is suitable for mass-flow. The main points of criticism of mass-flow are as follows:

(i) Phloem tissue has a relatively high rate of oxygen consumption (Table 12.6) and translocation is slowed down or stopped altogether if respiratory poisons such as potassium cyanide enter the phloem. Observations of this sort suggest that metabolic energy is used in translocation. The companion cells produce energy because they contain numerous mitochondria. Yet the mass-flow hypothesis is based on a passive, physical phenomenon and does not suggest a role for companion cells.

Table 12.6 Rates of oxygen uptake by various plant tissues (after Coult 1971)

TISSUE	mm^3 O_2 CONSUMED g^{-1} h^{-1}
vascular tissue	800
whole leaves	400
petioles	200
carrot taproot	30–40

(ii) It has recently been shown that plant hormones such as indole-3-acetic acid (IAA) assist the loading of sugars into sieve-tubes and their unloading into sinks. The mass-flow hypothesis makes no mention of hormones in translocation.

(iii) The sieve plates may offer a resistance which is greater than could be overcome by the pressure potential of the phloem sap. Higher pressure potentials than have been recorded would be necessary to squeeze sap through partially blocked pores in sieve plates. However, in view of the controversy centred around sieve plate structure this criticism may not be valid for all plants.

2. ALTERNATIVE HYPOTHESES

There has been no shortage of ideas about alternative mechanisms of phloem transport. One suggestion is that respiratory energy produced by companion cells maintains an electrical potential difference across sieve plates. The potential difference is achieved by the active removal of potassium ions (K^+) from one side of the plate by the companion cells and their secretion on the other side. The movement of K^+ ions through the pores of the sieve plate rapidly draws polar molecules of water and dissolved solutes through the pores. This phenomenon is called **electro-osmosis**. Experimental evidence to support the theory is sparse. Even so it is known that potassium ions stimulate the loading of phloem with sugars. Furthermore, potassium uptake is promoted by IAA.

Circulation of protoplasm has been observed in sieve-tubes and some physiologists have thought that this may be involved in translocation. It is unlikely though that **protoplasmic streaming** on its own would account for the speed at which substances are normally translocated.

Finally, supporters of the existence of transcellular strands have proposed that solutions may be pumped through the strands by **peristaltic waves** of contraction passing along the strands!

The various ideas embodied in the alternative hypotheses are summarised in Figure 12.20. There is obviously much debate surrounding the structure and physiology of phloem. At present, plant physiologists do not have a complete understanding of what appears to be a sophisticated and subtle mechanism of translocation.

FIG. 12.20 *Alternative mechanisms of translocation*

(a) electro-osmosis

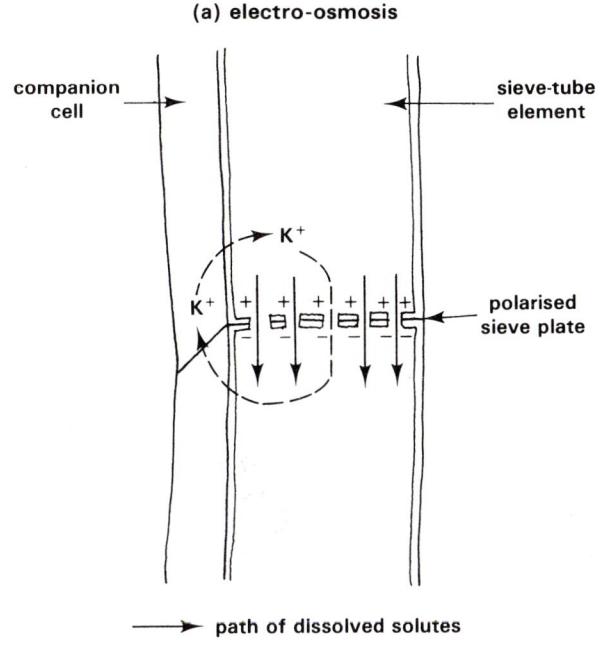

companion cell

sieve-tube element

K⁺

K⁺

polarised sieve plate

path of dissolved solutes

(b) protoplasmic streaming

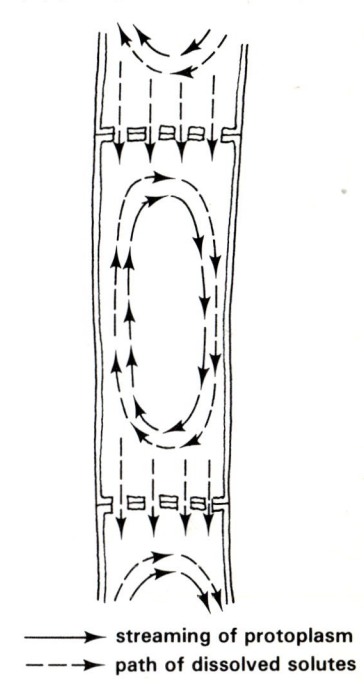

streaming of protoplasm

path of dissolved solutes

(c) peristalsis in transcellular strands

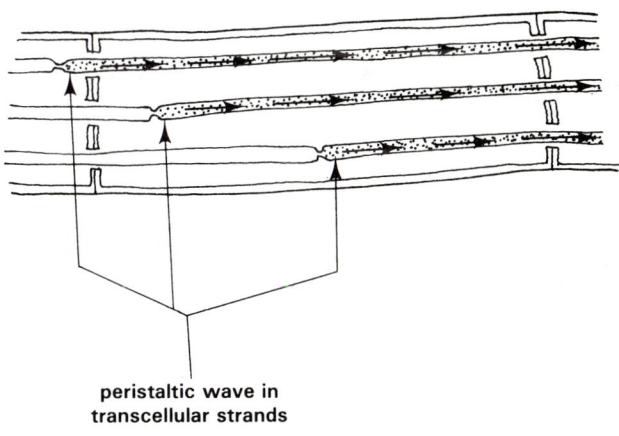

peristaltic wave in transcellular strands

path of dissolved solutes

Chapter 13

Mineral nutrition of flowering plants

The great improvement in crop yields which has come from a combination of mechanisation on the farm, the application of fertilisers and pesticides to the land and the breeding of high-yielding varieties of crop plants has been called the **green revolution**. Fertilisers improve the fertility of soils, that is the ability of soils to sustain vigorous, healthy plant growth. **Minerals**, inorganic chemical compounds, are used extensively as fertilisers in agriculture and horticulture. Mixtures of minerals called compound fertilisers are applied to the soil before and at various times during the growth of a crop (Fig. 13.1). Few arable farmers in developed countries now grow crops without treating their land with fertilisers, especially those containing nitrogen, phosphorus and potassium.

FIG. 13.1 *Artificial fertiliser being applied to the land*

It is therefore evident that minerals perform a crucial role in the growth of flowering plants. But what minerals are required and what part do they play in the functioning of plants?

13.1 Mineral requirements of flowering plants

Seventeenth-century records indicate that scientists were then aware that plant growth depends on a supply of minerals. Yet it was two centuries later before any real progress was made in identifying precisely which minerals are required by higher plants.

13.1.1 Water culture experiments

In 1860 the German plant physiologist Julius von Sachs described a system for growing flowering plants without soil (Fig. 13.2). A few years later Knop published the formula of a liquid medium which was suitable for growing a wide variety of plants (Table 13.1). The main advantage of using water culture techniques to investigate a plant's mineral requirements is that the chemical composition of the liquid medium can be exactly fixed. Such techniques make it possible to find out how plants respond to any combination of minerals.

Table 13.1 Knop's culture solution (1865)

MINERAL	g dm^{-3} WATER
KNO_3	0·2
$Ca(NO_3)_2$	0·8
KH_2PO_4	0·2
$MgSO_4.7H_2O$	0·2
$FePO_4$	0·1

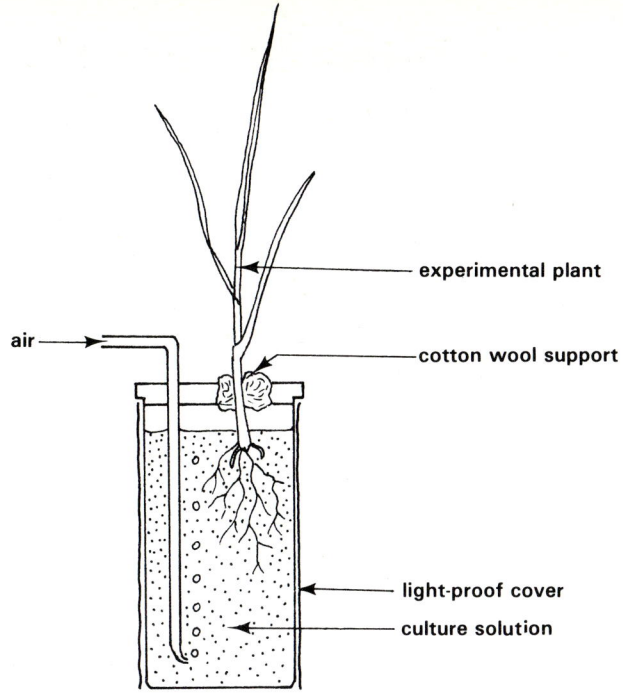

FIG. 13.2 *Equipment for water culture investigation using grass plants. The solution is aerated to encourage root growth. The light-proof cover discourages the growth of algae in the culture solution*

experimental plant

air

cotton wool support

light-proof cover

culture solution

Sachs, Knop and others discovered that given a supply of carbon dioxide, water and light, flowering plants require seven chemical elements in relatively large amounts if they are to show healthy and vigorous growth. The seven **macronutrients** as they have since been called are **nitrogen (N)**, **phosphorus (P)**, **potassium (K)**, **calcium (Ca)**, **magnesium (Mg)**, **sulphur (S)** and **iron (Fe)**. It was later found that relatively small amounts of a number of other elements are also needed for normal plant growth. These are called **micronutrients** or trace elements. **Manganese (Mn)**, **zinc (Zn)**, **copper (Cu)**, **boron (B)**, **molybdenum (Mo)** and **chlorine (Cl)** are the most important micronutrients.

The reason why Sachs and his contemporaries failed to discover the importance of micronutrients was because the chemicals they used, although the purest available at the time, probably contained small amounts of trace elements as impurities. Liquid media used in present-day water culture experiments are prepared from pure chemicals dissolved in de-ionised water (Table 13.2). Glassware is cleaned with hot 50 per cent hydrochloric acid and then thoroughly washed with de-ionised water. Otherwise the vessels may be a source of minerals which the experimental plants can absorb.

Table 13.2 A modern culture solution (modified from Hewitt 1974)

MINERAL	g dm^{-3} WATER	g dm^{-3} WATER		
KNO_3	0·404	K	0·156	
		N	0·057	
$Ca(NO_3)_2$	0·656	Ca	0·160	Macronutrients
		N	0·113	
$MgSO_4.7H_2O$	0·368	Mg	0·036	
		S	0·048	
$NaH_2PO_4.2H_2O$	0·208	P	0·041	
Fe citrate.$5H_2O$	0·03350	Fe	0·005	

$MnSO_4.4H_2O$	0·00223		Mn	0·00055	
$ZnSO_4.7H_2O$	0·00029		Zn	0·00006	
$CuSO_4.5H_2O$	0·00025		Cu	0·00006	
H_3BO_3	0·00310		B	0·00054	Micronutrients
$Na_2MoO_4.2H_2O$	0·00012		Mo	0·00005	
NaCl	0·00580		Cl	0·00350	

The effects of mineral deficiencies on plants grown in water culture can be assessed in a variety of ways. One of the simplest is to compare the fresh and dry mass of whole plants or the organs of plants kept in a medium deficient in one or more elements with similar plants grown in a complete medium (Table 13.3). Differences in growth pattern and leaf colour can also be noted.

Table 13.3 Effect of mineral deficiency on growth of lettuce (after Strafford 1963)

ELEMENTS IN SOLUTION	AVERAGE FRESH MASS/g	
	SHOOTS	ROOTS
macronutrients only	71·4	14·5
macronutrients plus Zn, Cu, Mn, B, Cl	105·7	22·0

Good subjects for water culture work are seeds with small food reserves because they contain only small amounts of stored minerals. Many grasses produce seeds of this kind. Large seeds and cuttings are less suitable as they often have a substantial reserve of one or more minerals.

13.2 Functions of minerals in flowering plants

Valuable as the results of water culture experiments are, they rarely give a complete picture of the roles of minerals in the functioning of a plant. The results often need to be supplemented with data from metabolic studies in which the rates of processes such as respiration, photosynthesis and protein synthesis are measured. Analysis of the chemical composition of experimental and control plants may also be required.

13.2.1 Macronutrients

With the exception of iron, **macronutrients** are normally required at concentrations of 0·04–0·2 g dm^{-3} in mineral culture solutions to sustain vigorous, healthy plant growth (Table 13.2). About 0·005 g dm^{-3} of iron gives satisfactory results. For this reason iron is sometimes described as a micronutrient.

1. NITROGEN
Nitrogen is absorbed by plant roots as nitrate(V) ions (NO_3^-) or as ammonium (NH_4^+) ions. Insufficient NO_3^- or NH_4^+ in the soil is usually one of the factors which most limits the growth of crops on cultivated land. Some important crop plants such as clover, beans and peas have symbiotic nitrogen-fixing bacteria living in nodules in their roots. The bacteria can use nitrogen gas to form organic nitrogenous compounds, some of which are used by the higher plant partner.

FIG. 13.3 *A cabbage showing effect of nitrogen deficiency* (Crown copyright). *The young leaves at the top of the plant are green but the older, larger leaves are yellow*

In flowering plants nitrogen is found in amino acids, proteins, nucleic acids and chlorophyll. The functions of these important compounds are described in Chapters 3, 4, 5 and 6. Nitrate ions absorbed by roots are first reduced to ammonia which combines with carboxylic acids to form amino acids. Ammonium ions are used directly for amino acid synthesis.

Typical symptoms of nitrogen deficiency are reduced growth of all organs and chlorosis, a yellowing of the leaves due to inadequate production of chlorophyll (Fig. 13.3). Chlorosis appears first in older leaves.

2. PHOSPHORUS

Phosphorus is absorbed from the soil as dihydrogenphosphate(V) ions ($H_2PO_4^-$). The quantity of $H_2PO_4^-$ probably limits plant growth more than the amounts of NO_3^- or NH_4^+ in uncultivated soils (Table 13.4). Nucleic acids, phospholipids and ATP are among the phosphorus-containing compounds found in plants. The roles of these substances are described in Chapters 3, 5 and 6.

Table 13.4 Effect of N and P on dry mass of shoots of *Lolium perenne* **(after Goodman 1969)**

PHOSPHORUS/g kg^{-1}	NITROGEN/g kg^{-1}				MEAN DRY MASS/g
	0·000	0·040	0·100	0·200	
0·000	343	323	344	332	335
0·120	455	466	459	416	470

Severe lack of phosphorus obviously affects processes which use free energy provided by ATP. Poor growth, especially of roots (Fig. 13.4), and a reduction in uptake of all minerals are the usual symptoms of phosphorus deficiency.

3. POTASSIUM

Potassium is absorbed from the soil as potassium ions (K^+). Whereas nitrogen and phosphorus appear in complex organic compounds in flowering plants potassium does not. Nevertheless potassium is found in green plants in large amounts. An important function of potassium is to activate enzymes (Chapter 4). Over forty enzymes depend on potassium for optimum activity. Potassium is also probably essential for translocation in phloem sieve-tubes (Chapter 12).

A feature of plants grown in a potassium-deficient environment is the mottled appearance of the older leaves. The leaves may also display chlorosis (Fig. 13.5).

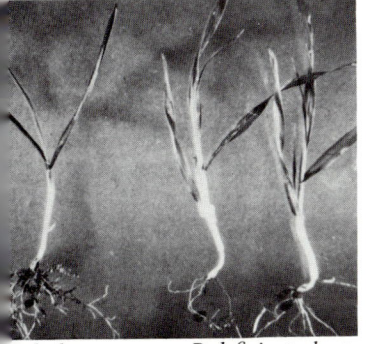

FIG. 13.4 *Spring barley showing effect of phosphorus deficiency* (Crown copyright)

rol plants   P-deficient plants

FIG. 13.5 *Apple leaves showing effect of potassium deficiency* (Crown copyright)

4. CALCIUM

Calcium is taken up by plant roots as calcium ions (Ca^{2+}). The main function of calcium in flowering plants is to form salt linkages between pectic acid molecules in the middle lamella which binds adjacent cells to each other (Fig. 13.6). Some enzymes are also activated by calcium.

Typical deficiency symptoms for calcium are chlorosis of young leaves, die-back of shoots due to death of apical buds and poor root growth.

FIG. 13.6 *The role of calcium in the middle lamella*

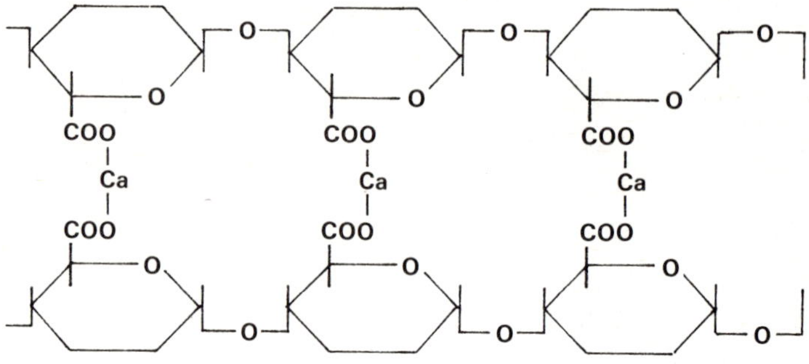

5. MAGNESIUM

Magnesium is absorbed from the soil as magnesium ions (Mg^{2+}). In green plants magnesium forms part of the chlorophyll molecule where it donates electrons in the light-dependent reactions of photosynthesis (Chapter 5). Magnesium also activates certain enzymes. Pronounced chlorosis beginning between the veins of older leaves is the main symptom of magnesium deficiency.

6. SULPHUR

Sulphur is taken up by plant roots as sulphate(VI) ions (SO_4^{2-}). It is needed for the formation of amino acids such as methionine and cysteine which contain thiol ($-SH$) groups. Sulphur-containing amino acids are present in most proteins. For this reason the symptoms of sulphur deficiency are very similar to those of nitrogen deficiency.

7. IRON

Iron is absorbed mainly as iron(II) ions (Fe^{2+}). Iron is present in cytochromes which act as electron carriers in photosynthesis and respiration and in the reduction of nitrate(V) ions. Iron is also required for the synthesis of chlorophyll and to activate certain enzymes.

Characteristic signs of iron deficiency are chlorosis, especially in young leaves, and inhibition of photosynthesis and respiration.

13.2.2 Micronutrients

A large number of chemical elements given to plants in very small quantities stimulate plant growth. However, not all of these elements are essential for plant metabolism. The **micronutrients** essential for plant growth are manganese, zinc, copper, boron, molybdenum and chlorine. Water culture experiments show that they are required at much lower concentrations than the macronutrients (Table 13.2).

The main functions of the micronutrients are summarised in Table 13.5. Chlorosis can be a symptom of insufficient Zn, Cu, Mo or Cl. Even so, the deficiency symptoms for most of the micronutrients are so characteristic that certain plants may be used as indicator species in diagnosing soils which are deficient in trace elements (Fig. 13.7).

Table 13.5 Roles of micronutrients in plant metabolism

ELEMENT	ROLES
Mn	Activates carboxylase enzymes. May act as an electron donor for excited chlorophyll b.
Zn	Present in the enzyme carbonic anhydrase and also in various dehydrogenase enzymes.
Cu	Present in a number of oxidase enzymes. Also found in plastocyanin which acts as an electron carrier for chlorophyll a.
B	Not fully known. Essential for meristem activity and for growth of pollen tubes.
Mo	Required for nitrogen fixation. Activates the enzyme nitrate reductase.
Cl	Essential for oxygen evolution in photosynthesis.

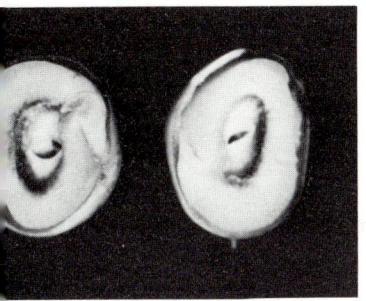

13.3 Toxic minerals

In recent years large-scale reclamation of derelict industrial sites has taken place in Britain and in other parts of the world. One of the difficulties in work of this sort is to get vegetation to grow in such places where the soil, mainly derived from spoil heaps, is sometimes contaminated with heavy metals such as copper, lead, zinc and nickel. In spoil heaps the concentrations of these elements are toxic to most green plants.

Old spoil heaps from heavy metal workings are colonised by a select group of plants called **metallophytes** because of their tolerance to high concentrations of heavy metals. It is said that prospectors often located surface deposits of ores rich in heavy metals by the type of vegetation growing on the sites. One approach to the problem of getting vegetation to grow on spoil heaps has been to plant metallophytes. Varieties of grasses such as *Festuca ovina* and *Agrostis canina* tolerant of high concentrations of heavy metals have been used successfully in projects of this kind.

13.4 Soil as a source of minerals

Soil is much more complex than solutions used in water culture experiments. A fertile soil consists of a **mineral skeleton** derived from weathered rock, **organic matter** mainly as dead and decaying plant material, **water**, **air** and **living organisms** notably bacteria, fungi and soil-dwelling animals. The interaction of the various components mainly determines the availability of minerals to the roots of higher plants.

In natural soils nearly all the nitrogen comes from the breakdown of organic matter by **decomposer organisms** such as bacteria and fungi. Most of the sulphur, phosphorus, calcium, magnesium and potassium are also of organic origin. The rest, together with iron and the trace elements, are obtained from weathering of the mineral skeleton.

Cations such as Ca^{2+}, Mg^{2+} and K^+ are attracted to and held by the negative charges of clay and humus particles. However, the attraction is not permanent and the elements can be released by **ion exchange**. Anions such as NO_3^-, SO_4^{2-} and Cl^- are not usually held in this way and are found dissolved in soil water. Phosphorus is normally found in soil as insoluble phosphates of calcium, iron and aluminium.

In an undisturbed terrestrial ecosystem the soil is in a state of equilibrium. Minerals absorbed by plants are returned to the soil when the plants die and are decomposed. On the contrary, substantial quantities of mineral elements are removed annually from cultivated soils. Most of the loss is due to the high mineral content of the crop at harvesting (Table 13.6). Thus to maintain the fertility of land used for growing crops it is necessary to replenish the soil with minerals. This is done by applying **manures, composts** or **mineral fertilisers.**

Table 13.6 Mineral content of leaves of sugar beet (after Wallace 1961)

ELEMENT	Ca	Mg	K	P	Fe	Mn	B
% dry mass	2·64	0·55	4·21	0·35	0·0125	0·0046	0·0029

13.5 Mineral absorption by plants

Roots are the main organs concerned with mineral uptake by terrestrial plants although minerals can be absorbed through intact leaves. Foliar sprays containing fertilisers are now quite widely used to feed crops grown on a large scale.

13.5.1 Roots as organs of mineral absorption

The absorption of water by roots is described in Chapter 11. Minerals are taken into roots in aqueous solution yet there is no direct link between the absorption of minerals and water. Water uptake is controlled mainly by the rate at which a plant transpires. Mineral uptake can be rapid even when transpiration has stopped. Indeed, submerged aquatic plants absorb minerals even though they do not transpire. Most of a terrestrial plant's mineral requirements are absorbed by its roots from the top six inches of soil. It is here that roots grow profusely and minerals are most abundant.

FIG. 13.8 *Mineral accumulation by different parts of barley roots* (after Kramer, 1969)

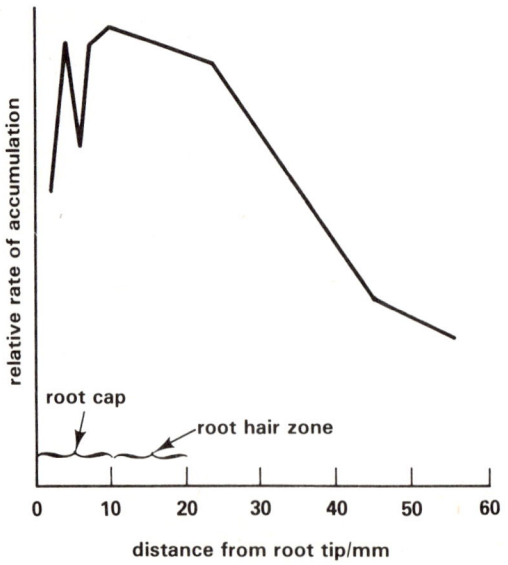

Roots which penetrate deeper into the soil absorb mainly water. By growing constantly in length, roots continually occupy new areas of soil from which minerals can be absorbed. All parts of the root system do not absorb minerals with equal efficiency. Mineral uptake takes place most rapidly in the zone of elongation just behind the apices of roots where root hairs are present (Fig. 13.8). The root hairs increase the area of contact between the soil and the root surface.

The roots of many terrestrial plants form a symbiotic association with the hyphae of certain species of soil-dwelling fungi. The association is called a **mycorrhiza**. Mycorrhizal roots absorb minerals more efficiently than ordinary roots (Table 13.7). What do you think are the reasons for this? In addition there are thriving populations of free-living saprophytic bacteria and fungi in the immediate vicinity of living roots. The micro-organisms feed on sugars and amino acids released into the soil by the root cells and on dead cells sloughed off as roots grow through the soil. The term **rhizosphere** is used to describe the soil around roots affected in this way. What effect do you think the rhizosphere micro-organisms have on mineral uptake by plant roots?

Table 13.7 Uptake of N, P and K by *Pinus strobus* **(after Hatch 1937)**

TYPE OF ROOT	NUTRIENTS ABSORBED/% DRY MASS		
	N	P	K
mycorrhizal	1·24	0·196	0·744
non-mycorrhizal	0·85	0·074	0·425

13.5.2 Mechanisms of mineral absorption

Minerals are usually absorbed against concentration gradients. Not all minerals are absorbed at the same rate and uptake is most rapid by tissues respiring aerobically. Let us examine two of the mechanisms which could account for mineral absorption.

1. DIFFUSION

FIG. 13.9 *A comparison of the concentrations of various ions in the sap of the alga* Nitella *and the pond water in which it lives* (after Hoagland, 1944)

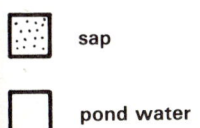

sap

pond water

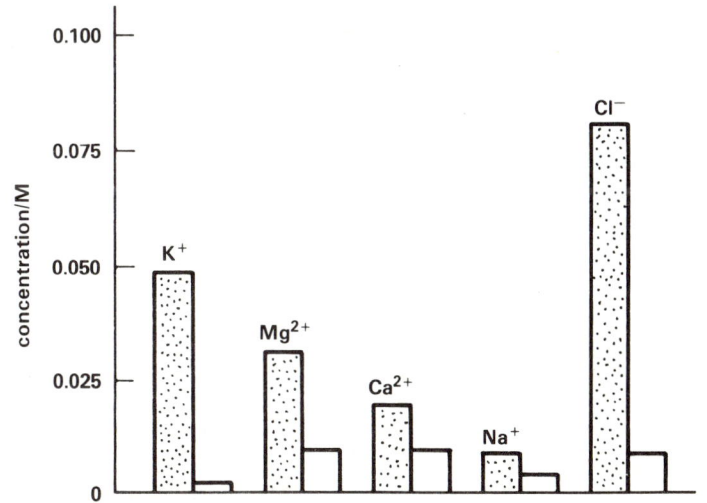

Diffusion is the random movement of ions or molecules down a concentration gradient. Most of the available evidence indicates that mineral absorption does not occur by diffusion. Minerals entering plants usually move against concentration gradients (Fig. 13.9). It is also difficult to explain why respiring cells absorb minerals more rapidly than non-respiring cells if diffusion is the mechanism of mineral absorption.

2. ACTIVE UPTAKE

When roots are transferred from distilled water to a dilute solution of a mineral salt the respiratory rate of the roots increases. Furthermore the increase in respiration is proportional to the rate of ion absorption (Fig. 13.10). It is thought that respiratory energy is used to absorb ions against a concentration gradient.

FIG. 13.10 *Relationship between the rate of respiration and bromide uptake by carrot root discs* (after Steward, Berry and Broyer, 1936)

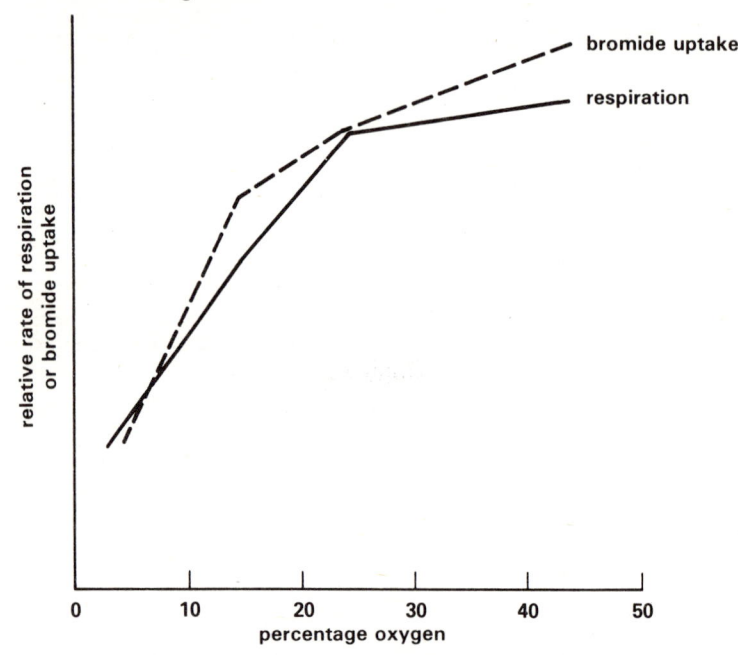

FIG. 13.11 *Two ways in which carrier molecules may be used to carry ions into cells*

(a) active transport (against a concentration gradient)

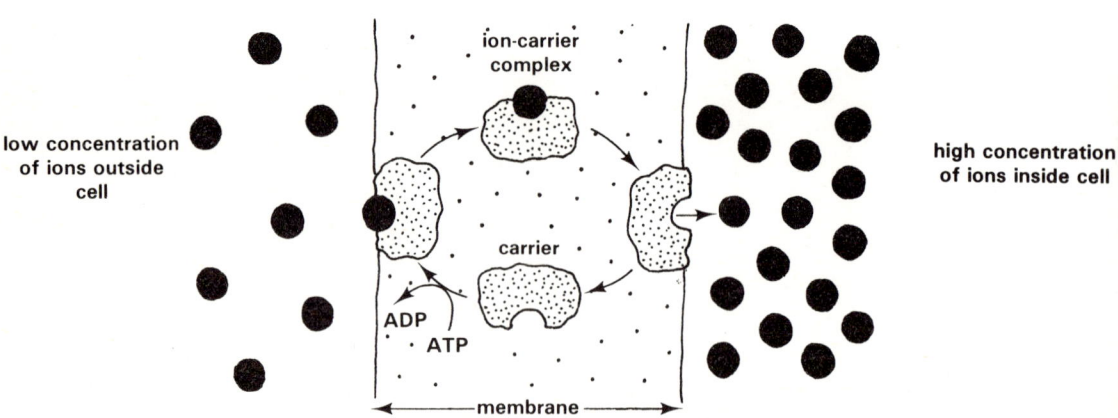

Energy from ATP is used to form the ion-carrier complex.

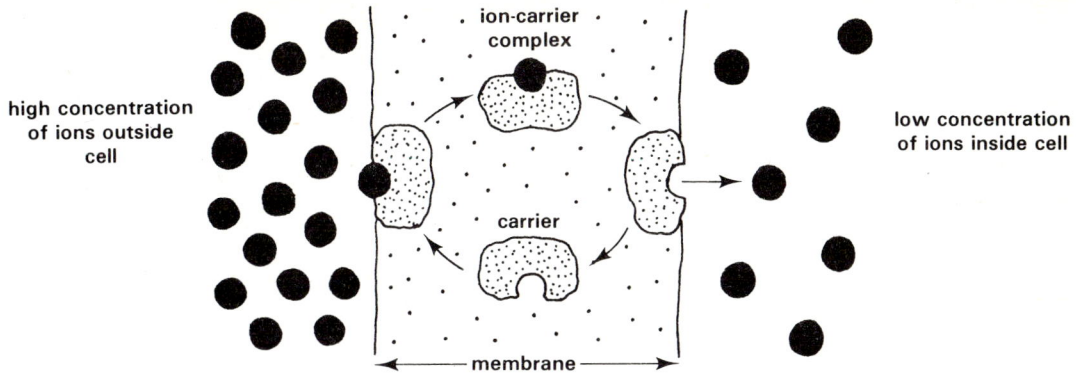

(b) facilitated transport (down a concentration gradient)

high concentration of ions outside cell

ion-carrier complex

carrier

low concentration of ions inside cell

← membrane →

The carrier conveys the ions across the membrane without using energy.

Carrier molecules are probably used to transport mineral ions into cells (Fig. 13.11). Evidence that carriers are involved has come from experiments in which the rates of ion uptake are measured in the presence of increasing concentrations of mineral ions (Fig. 13.12). The results are similar to the relationship between the rate of an enzyme-catalysed reaction and substrate concentration (Chapter 4). The similarity in the relationships suggests that ions combine with carriers in much the same way as enzyme-substrate complexes are formed. In Chapter 4 it is explained how inhibitors compete for active sites on enzyme molecules. A comparable phenomenon occurs in mineral uptake when the presence of one mineral sometimes inhibits the uptake of another. Why do you think this occurs?

FIG. 13.12 *Effect of concentration of ions on ion uptake*

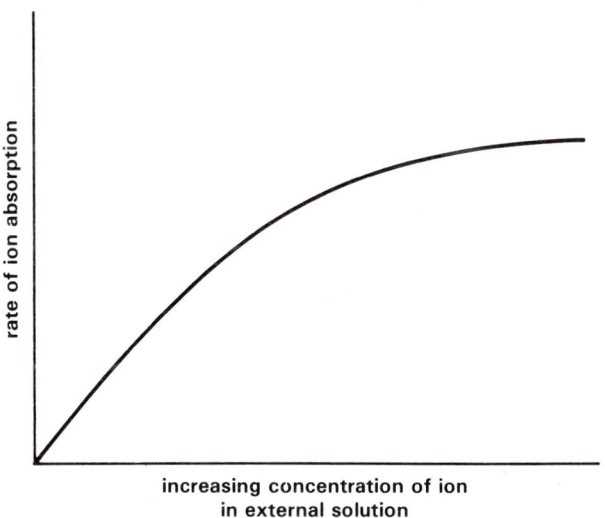

rate of ion absorption

increasing concentration of ion in external solution

13.6 Transport of minerals in plants

Because they absorb minerals, roots can be thought of as a **source** of minerals for the rest of the plant. Actively growing tissues which have a demand for minerals can be thought of as **sinks**. Let us see how minerals move from a source to sinks.

13.6.1 Passage across the root tissues

After absorption by the epidermal cells of roots, minerals eventually enter the root xylem elements. Plant physiologists are still uncertain as to how the minerals cross the intervening tissues. The path of least resistance across the root cortex is probably through the cytoplasm, movement from cell to cell taking place along plasmodesmata. Alternatively mineral transport may take place along the walls of the cortical cells. At the endodermis the wall route is blocked by the suberised **Casparian band**. It is likely that minerals pass through the selectively permeable plasma membrane of the endodermal cells. Entry of minerals into the pericycle cells is controlled in this way. Movement of minerals from the pericycle to the xylem elements is thought to occur by **active transport**. The pumping of ions into the vascular tissue uses respiratory energy and is probably responsible for root pressure (Chapter 11).

Not all of the minerals absorbed by roots end up in the plant's vascular tissue. Some minerals are retained by the root cells for metabolic purposes. Metallophytes often store large deposits of heavy metals in the vacuoles and cell walls of the root cortex.

13.6.2 Movement into the shoot system

The ascent of solutes in the xylem elements of the stem can be demonstrated by placing the stalk of a leafy transpiring shoot in a solution of a dye such as eosin. Microscopic examination of thin sections of the stem cut a few hours later reveals the presence of dye in the vessels and tracheids. Some of the dye is found in the cell walls but the results of recent experiments with radioisotopes indicate that the cavities of tracheids and vessels are the most important route for mineral transport. The ascent of minerals is a **passive** process dependent on transpiration. A correlation between the rate of transpiration and the upward movement of solutes has been demonstrated with indicators such as dyes and radioactive isotopes. Although many minerals ascend the xylem as inorganic ions, some are carried as organic compounds. While most of the sulphur is transported as sulphate(VI) ions, a small proportion is moved as sulphur-containing amino acids such as cysteine and methionine. Nitrate(V) ions are generally converted in the roots to amino acids.

On reaching a sink in the **transpiration stream**, minerals are moved into the metabolising tissue. **Transfer cells** which have a large surface area to volume ratio are found in the minor veins of many species of flowering plants and are thought to transport minerals from the xylem into metabolising tissue (Fig. 13.13). Because transfer cells have large numbers of mitochondria it is probable that **active transport** is involved. Some lateral transport also takes place between xylem elements and adjacent tissues such as the vascular cambium and phloem.

13.6.3 Circulation of minerals in plants

What happens to individual mineral nutrients after they have been taken into a sink varies considerably. Calcium and iron accumulate in young leaves and display little movement into new organs formed later. The current needs of a plant for Ca and Fe are met by absorption from the soil. Thus deficiency symptoms of these elements often appear in the young leaves at shoot apices.

Most minerals are relatively mobile and after entering initial sinks they then pass into developing sinks such as young leaves, flowers and fruits.

For this reason deficiency symptoms of N, K, Mg, P and S are often seen in mature leaves. It has been shown that phosphate(V) ions are re-exported in phloem sieve-tubes and it is likely that other mobile elements follow the same pathway. The internal recycling of minerals enables plants to withdraw nutrients from old leaves before they fall off. The yellowing of older leaves for example is accompanied by the breakdown of nitrogenous compounds including chlorophyll and withdrawal of the end products into other parts of the plant. A small proportion of mineral nutrients, potassium especially, is leached from leaves by rainfall but the natural recirculation of the elements largely depends on the activities of micro-organisms in decomposing dead plant organs returned to the soil.

The main processes and pathways thought to be involved in mineral uptake and transport are summarised in Figure 13.13.

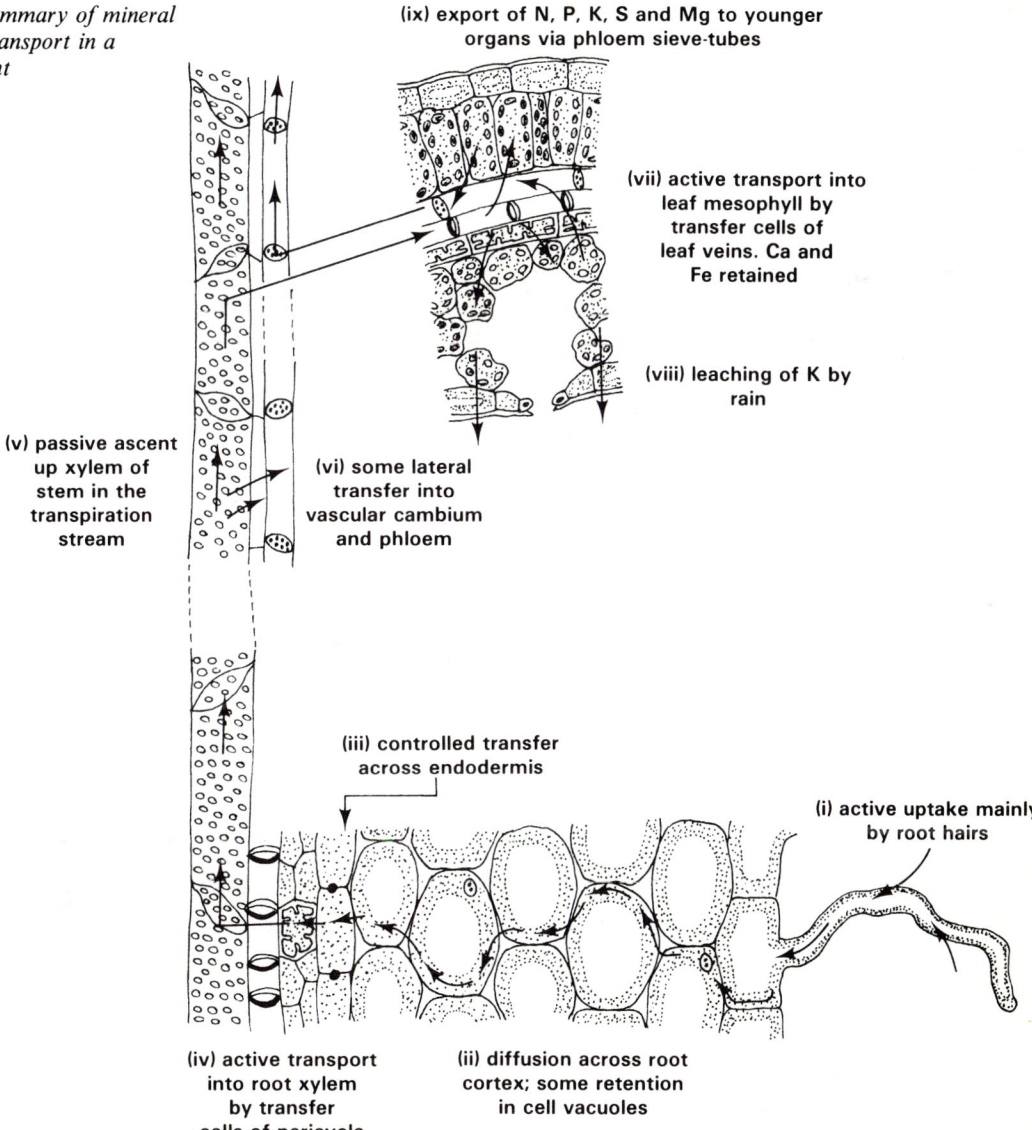

FIG. 13.13 *Summary of mineral uptake and transport in a flowering plant*

(ix) export of N, P, K, S and Mg to younger organs via phloem sieve-tubes

(vii) active transport into leaf mesophyll by transfer cells of leaf veins. Ca and Fe retained

(viii) leaching of K by rain

(v) passive ascent up xylem of stem in the transpiration stream

(vi) some lateral transfer into vascular cambium and phloem

(iii) controlled transfer across endodermis

(i) active uptake mainly by root hairs

(iv) active transport into root xylem by transfer cells of pericycle

(ii) diffusion across root cortex; some retention in cell vacuoles

252

13.7 The nutrient-film technique

This chapter ends as it began with water culture techniques. In recent years a system has been developed at the Glasshouse Crops Research Institute in Britain for the large-scale cultivation of crops without soil. The system is called the **nutrient-film technique**. The plants are grown in plastic troughs through which circulates a film of water containing balanced amounts of macro- and micronutrients (Fig. 13.14). So successful is the system that commercial production of tomatoes, cucumbers and strawberries has now begun using the nutrient-film technique. One of the important advantages of this form of farming is that it dramatically increases crop production. One grower in Australia has obtained ten lettuce crops in a year using the technique. Another very useful benefit is that it can be used in arid areas because little water is lost from the troughs by evaporation.

Chapter 14

The mammalian alimentary system

A constant supply of organic and inorganic nutrients is required to sustain the many and complex activities which take place in the mammalian body.

Organic nutrients such as proteins and lipids are the raw materials for the synthesis of cellular components in growth and repair. The energy needed to drive synthetic processes comes mainly from the oxidation of sugars in cellular respiration (Chapter 5). Mammals also require energy for movement and to help maintain a constant body temperature (Chapter 17). Vitamins are organic nutrients which participate in a wide range of body functions.

Inorganic nutrients (minerals) are needed to create and maintain electrical potentials across the membranes of nerve and muscle cells (Chapter 15). The coagulation of blood (Chapter 9), activation of enzymes (Chapter 4), growth of teeth and bones and maintenance of the osmotic potential of the body fluids are among the many other essential activities which depend on a supply of minerals.

Mammals obtain nutrients from their diet which consists of food and water. Food is a mixture of organic and inorganic nutrients. Water contains dissolved minerals. It is the alimentary system which first deals with food and water taken into the body.

14.1 Mammalian nutritional requirements

The nutritional requirements of mammals can be divided into three categories, **organic nutrients**, **vitamins** and **inorganic nutrients**.

14.1.1 Organic nutrients

The chemical nature of the main organic constituents of the diet is described in Chapter 3.

1. CARBOHYDRATES

About 50 per cent of the energy from the average human diet in Britain comes from **carbohydrates**. Glucose is the main respiratory substrate. Reserves of carbohydrate are stored as glycogen in the liver and muscles (section 14.5.2).

2. PROTEINS

Proteins are necessary in the diet as a source of **amino acids**. Amino acids are respired in emergencies but only about five per cent of the human body's energy requirement normally comes from amino acids. The main function of amino acids is to provide the raw materials for the synthesis of proteins. A constant supply of amino acids is needed for the production of enzymes and cell membranes. Metabolism, growth and repair thus depend on the presence of proteins in the diet. Proteins are also made in secretory cells for export to other parts of the body. Plasma proteins in the blood for example are made in the liver. Antibodies are proteins made in the lymph nodes. Digestive enzymes and certain hormones are proteins too. **Transamination** in the liver enables some amino acids called **non-essential amino acids** to be made from others (section 14.5.2). A few amino acids cannot be made in this way and these **essential amino acids** are a vital constituent of the diet.

3. LIPIDS

Lipids can be used as respiratory substrates. Fats and oils account for about 45 per cent of the energy obtained from the average human diet in Britain. The synthesis of cell membranes requires lipids as well as proteins. The fatty membranes of the myelin sheath around certain nerve cells is important in impulse transmission. Surplus fats are stored in mammals beneath the skin and around internal body organs such as the kidneys. As well as being a foodstore fat acts as a heat insulator which helps to keep a constant body temperature.

14.1.2 Vitamins

Vitamins are a range of substances necessary for good health. They affect a variety of body functions. Vitamins are varied chemically and are often required only in trace quantities. Some vitamins are soluble in fats and are absorbed from the gut dissolved in lipids. Others are water-soluble and are absorbed into the blood dissolved in water.

1. FAT-SOLUBLE VITAMINS

(i) VITAMIN A (retinol) is made from the plant pigment β-carotene and is needed for synthesis of the visual pigments rhodopsin and iodopsin. Deficiency of vitamin A can lead to **night blindness** (Chapter 16). Vitamin A is also essential for maintaining epithelial tissues in good condition. Infection of the epithelium of the respiratory tract often results from serious lack of vitamin A in the body.

(ii) VITAMIN D is a group of several sterols including ergocalciferol (vitamin D_2). Ergocalciferol is produced by the action of ultraviolet radiation from the sun on ergesterol in the skin. Vitamin D_2 stimulates the absorption of calcium from the gut. Deficiency of vitamin D_2 can lead to the release of calcium from the bones under the influence of parathyrin (Chapter 18). Malformation of the skeleton due to vitamin D_2 deficiency is called **rickets**.

(iii) VITAMIN E. The function of vitamin E in the human body is not fully understood. In rats it is essential for normal **spermatogenesis** (Chapter 22).

(iv) VITAMIN K is necessary for the production of **prothrombin** in the liver. Deficiency of vitamin K leads to impaired coagulation of the blood (Chapter 9).

2. WATER-SOLUBLE VITAMINS

(i) VITAMIN B is a complex of several vitamins most of which are components of **co-enzymes** (Chapter 4). **Vitamin B_1** (thiamin) is required to make the co-enzyme of carboxylases (Chapter 4). Deficiency of vitamin B_1 leads to **beri-beri** with malfunction of the nervous and cardiovascular systems.

Vitamin B_2 (riboflavin) is part of the co-enzyme FAD used as a hydrogen acceptor in Krebs' tricarboxylic acid cycle (Chapters 4 and 5). **Vitamin B_6** is required for **protein metabolism** including transamination (section 14.5.2 and Chapter 6). **Vitamin B_{12}** (cyanocob(III)alamin) is necessary for the production of red blood cells. Deficiency of vitamin B_{12} causes **pernicious anaemia** (section 14.4 and Chapter 9). Nicotinamide and nicotinic acid are other B-group vitamins. They are required to make the hydrogen acceptor co-enzymes NAD and NADP (Chapters 4 and 5).

(ii) VITAMIN C (ascorbic acid) is thought to act as an electron carrier in respiration. Recent evidence suggests that a high daily intake of vitamin C in the diet may have many beneficial effects including reduction of blood cholesterol (Chapter 9). Deficiency of vitamin C causes **scurvy**, symptoms of which include breakdown of connective tissues and blood vessels. Scurvy was once common among sailors who were deprived of fresh fruit containing vitamin C while on long voyages. When it was found that drinking the juice of citrus fruits prevented scurvy the Royal Navy provided lime juice on its ships for consumption by the crew.

14.1.3 Inorganic nutrients

Minerals are inorganic nutrients which participate in a wide variety of body functions. Some minerals called **macronutrients** are needed in relatively large quantities. Others called **trace elements** are needed in small amounts.

1. MACRONUTRIENTS

Sodium ions are more abundant than any other cation in the body fluids. Sodium ions play an important role in the electrochemical activities of nerves and muscles. Sodium chloride helps to maintain the osmotic potential of tissue fluid and blood plasma. **Potassium** ions are the most abundant of cations in cells. Potassium ions are involved, like sodium, in nerve and muscle impulse transmission. **Calcium** and **phosphate(V)** ions are the main mineral constituents of bone and teeth. Calcium ions are also required for coagulation of blood and for muscle contraction.

2. TRACE ELEMENTS

Trace elements must be provided in the diet to maintain health. Large quantities of trace elements can be poisonous.

    Cobalt is a constituent of vitamin B_{12} (cyanocob(III)alamin) and is vital for the production of haemoglobin and red blood cells. Providing cobalt(II) ions are present, vitamin B_{12} can be synthesised by bacteria in the guts of ruminant mammals (section 14.3.3). Non-ruminants must have vitamin B_{12} in their diets. **Iodine** is a constituent of the hormone thyroxin and is absorbed from the blood by the thyroid gland. Lack of iodine in the diet leads to decreased thyroxin production and enlargement of the thyroid gland called a goitre (Chapter 18). **Copper** is vital for the activation of a variety of enzymes and for the production of haemoglobin. **Iron** is a constituent of haemoglobin and myoglobin (Chapter 8) and cytochromes (Chapter 5). Reserves of iron are stored in the liver (section 14.5.2). **Zinc** is an activator of several enzymes, including carbonic anhydrase (Chapter 8) and is required for synthesis of the hormone insulin (Chapter 18).

14.2 Mammalian dentition

Ingested food first enters the **mouth** where it is **chewed**. Chewing makes it easier to swallow food and helps the action of digestive enzymes secreted by the various glands which open into the gut. The cutting and grinding of food which takes place when food is chewed is the function of the **teeth**.

14.2.1 Structure and development of teeth

Tooth buds in the developing jaw give rise to the teeth. The buds each contain an enamel organ in which **ameloblast cells** produce **enamel** made of calcium salts. Enamel is the hardest material in the mammalian body and coats the protruding surfaces called the **crowns** of the teeth. Inside the cup-shaped enamel organ, **odontoblast cells** of the dental papilla produce **dentine** (Fig. 14.1). During tooth development, sockets of bone grow around the **roots** of the teeth (Fig. 14.2).

    Many mammals including man, have two sets of teeth during their lives. The first set called **milk teeth** grow in early life and replaced by the **permanent teeth**. In children, tooth replacement begins usually at about five years of age and takes several years to complete. The final molars are called the wisdom teeth and usually emerge in early adult life. Permanent teeth develop from separate rows of tooth buds which arise under the buds from which the milk teeth grow (Figs. 14.1 and 14.3).

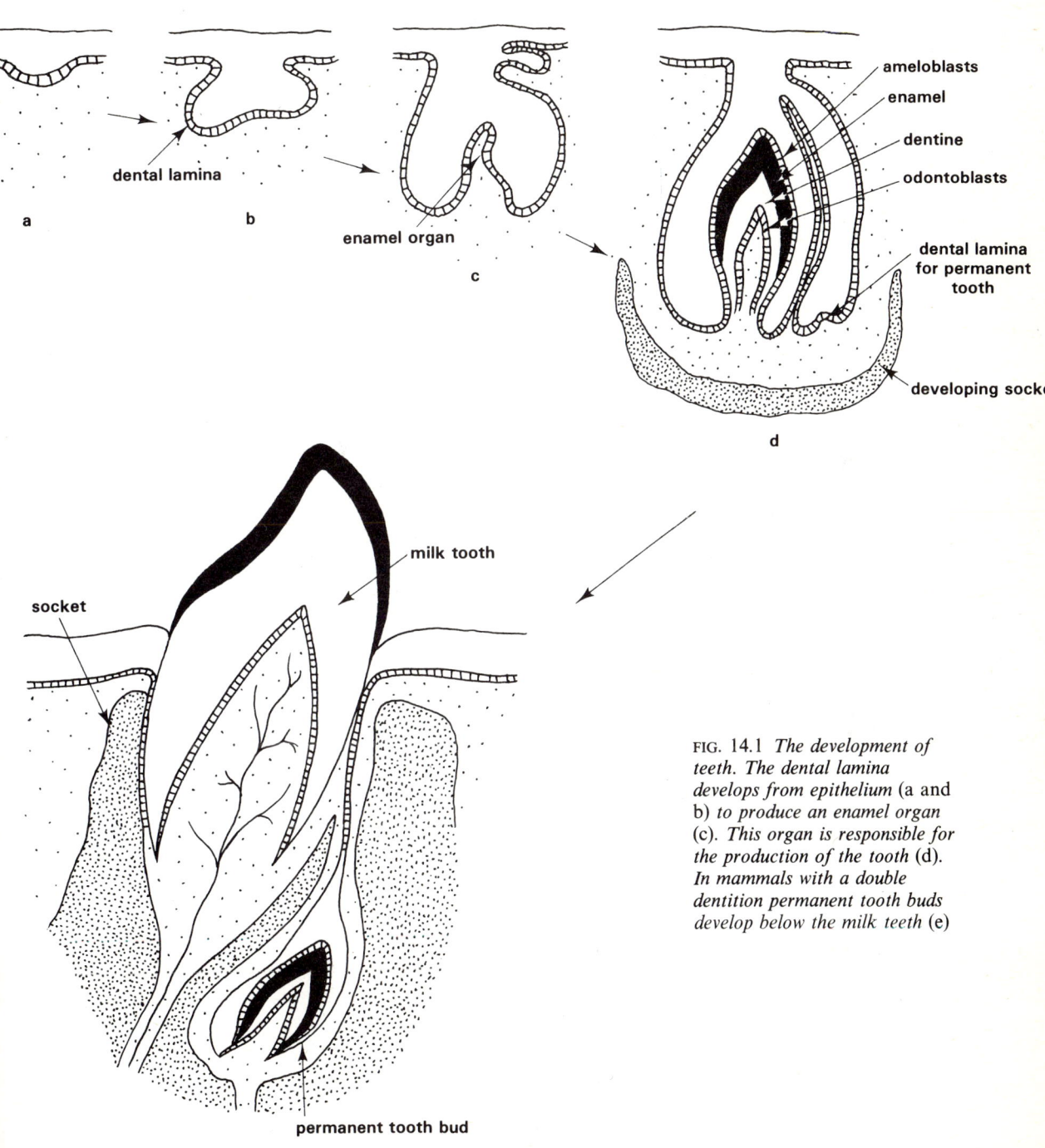

ameloblasts

enamel

dentine

odontoblasts

dental lamina for permanent tooth

developing socket

dental lamina

enamel organ

a

b

c

d

milk tooth

socket

permanent tooth bud

e

FIG. 14.1 *The development of teeth. The dental lamina develops from epithelium* (a and b) *to produce an enamel organ* (c). *This organ is responsible for the production of the tooth* (d). *In mammals with a double dentition permanent tooth buds develop below the milk teeth* (e)

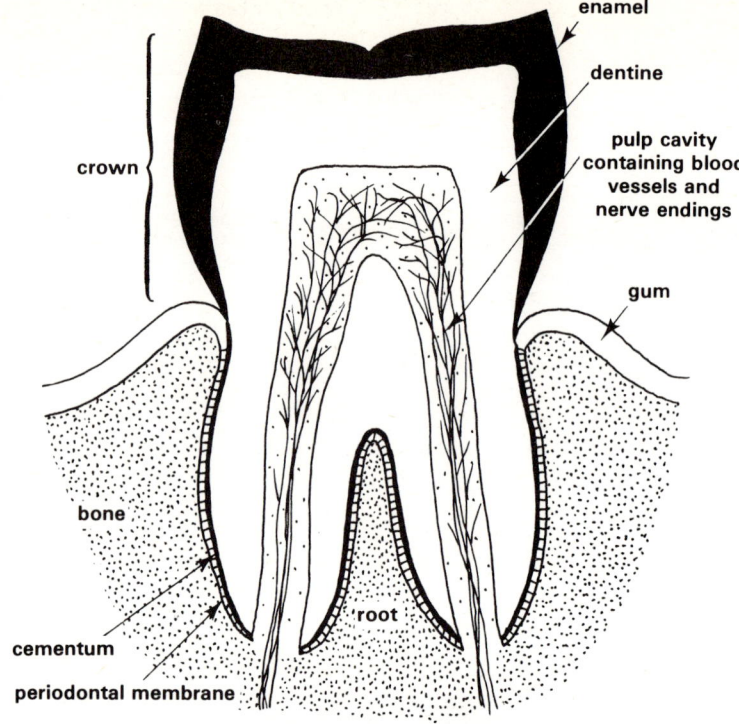

FIG. 14.2 *Diagrammatic vertical section through a human molar. The root is held in the bony socket by cementum joined to a fibrous connective layer called the periodontal membrane. The elasticity of the fibres in this membrane allows limited movement of the root in the socket, thus acting as a 'shock absorber' during chewing*

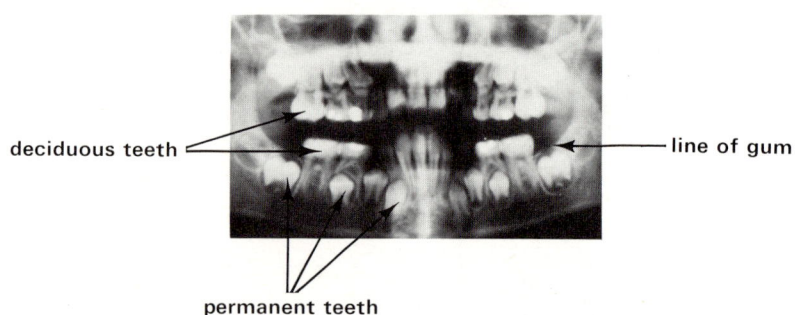

FIG. 14.3 *X-ray photograph showing the dentition of an 8-year-old child. Note the presence of both milk (deciduous) and permanent teeth* (courtesy Mr. Gould, Dental Department, General Hospital, Nottingham)

14.2.2 Variety of mammalian teeth

Mammals are **heterodont**. Their dentition includes several kinds of teeth. Moving from the front of the jaw to the rear there are **incisors**, **canines**, **premolars** and **molars**. The various kinds of teeth perform different functions during feeding.

Carnivores eat meat which is often captured alive. The teeth pierce and grip the prey to prevent it escaping and rip and cut it when chewing. In the dog these functions are performed mainly by the canine and **carnassial** teeth respectively. The latter are the fourth upper premolar and the first lower molar teeth (Fig. 14.4). Attachment of the lower jaw to deep grooves on the sides of the skull of the dog restricts the jaw to up and down movement, ideal for cutting flesh (Fig. 14.5). The molar and premolar teeth act like scissors, the upper teeth biting outside the lower ones.

Herbivores eat vegetation. The canine teeth are absent and a toothless space, the **diastema**, separates the incisors from the premolars (Fig. 14.6). In sheep the upper incisors are also missing. When grazing, the lower set of incisors bite against a horny pad in the upper jaw. The main chewing action in herbivores involves the premolars and molars. Because herbivores chew almost constantly the premolars and molars become very worn. The wearing is uneven, as the enamel on the outer surfaces of the teeth is harder than the dentine inside. Consequently the grinding surfaces of the premolars and molars develop a ridged pattern (Fig. 14.7).

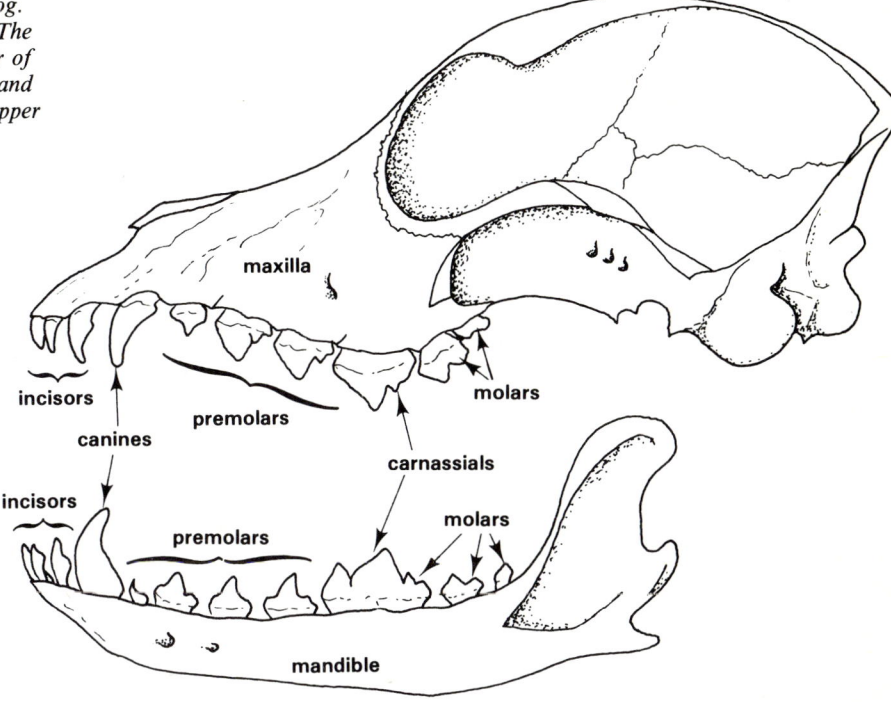

FIG. 14.4 *Dentition of the dog. The dental formula is $\frac{3142}{3143}$. The formula refers to the number of incisors, canines, premolars and molars on each side of the upper and lower jaws*

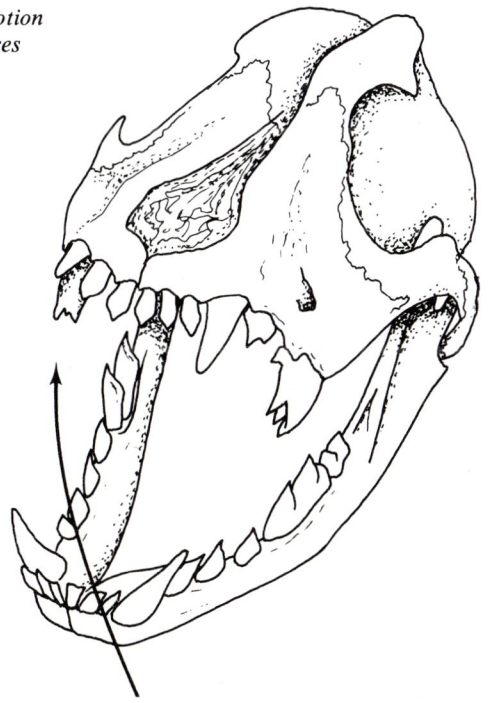

FIG. 14.5 *Vertical biting motion of the lower jaw in carnivores*

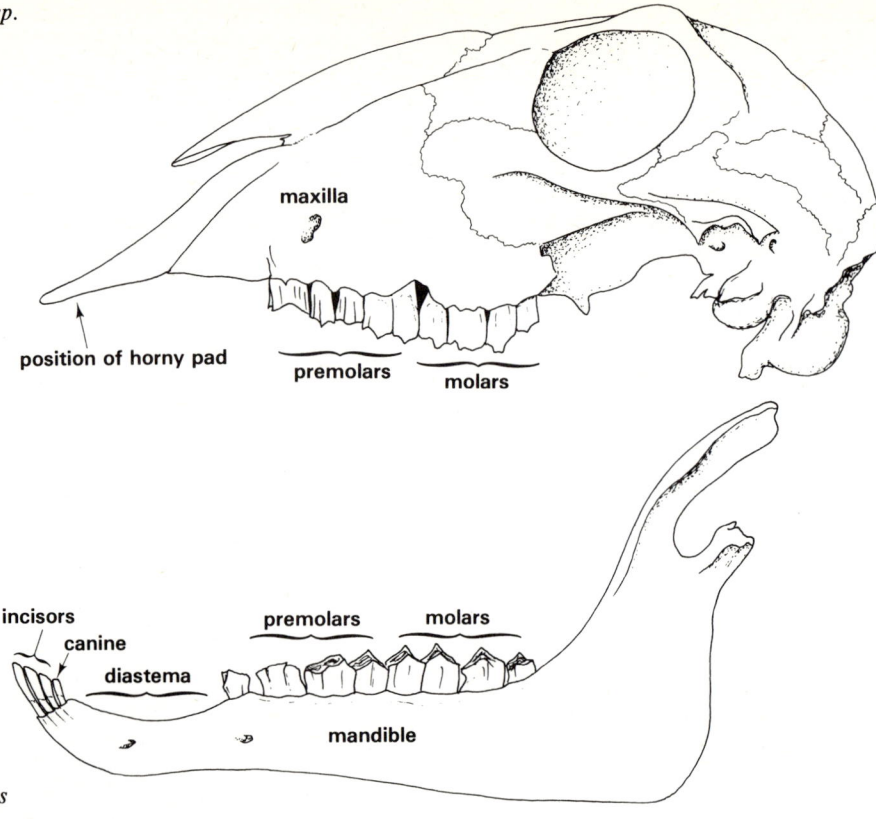

FIG. 14.6 *Dentition of the sheep. The dental formula is* $\frac{0033}{3133}$*. Compare with Fig. 14.4*

maxilla

position of horny pad

premolars

molars

incisors

canine

diastema

premolars

molars

mandible

FIG. 14.7 *The grinding surfaces of a sheep's molars*

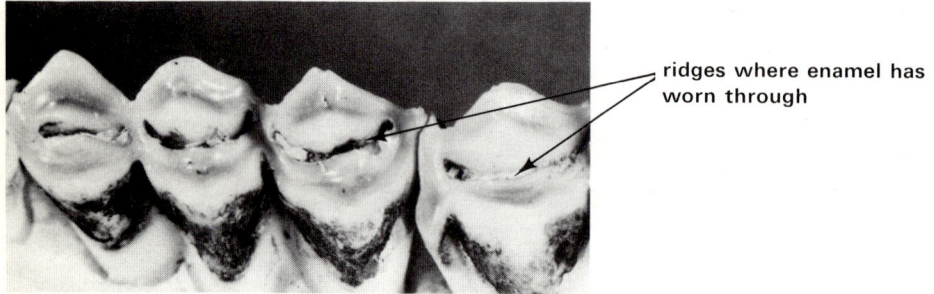

ridges where enamel has worn through

Attachment of the lower jaw to the skull allows a circular jaw motion in the horizontal plane. The ridges of the upper and lower premolars and molars slide over one another with the food in between (Fig. 14.8). In this way vegetation is ground to very fine particles before swallowing. Humans are called **omnivores** because we usually eat meat, fruits and vegetables. We have a relatively unspecialised dentition (Fig. 14.9).

The fossil record shows that primitive mammals had a **dental formula** of:

$$\frac{3 \quad 1 \quad 4 \quad 3}{3 \quad 1 \quad 4 \quad 3}$$

That is, three incisors, one canine, four premolars and three molars on each side of both upper and lower jaws, a total of 44 teeth. Present-day mammals have either this full dentition or, more usually, a reduced number of teeth. Some species of whales and a few other mammals have more teeth.

Modern mammals are thought to have evolved from early insect-eating species. Mammals belonging to the order *Insectivora* which includes shrews, hedgehogs and moles usually have a full dentition (Fig. 14.10).

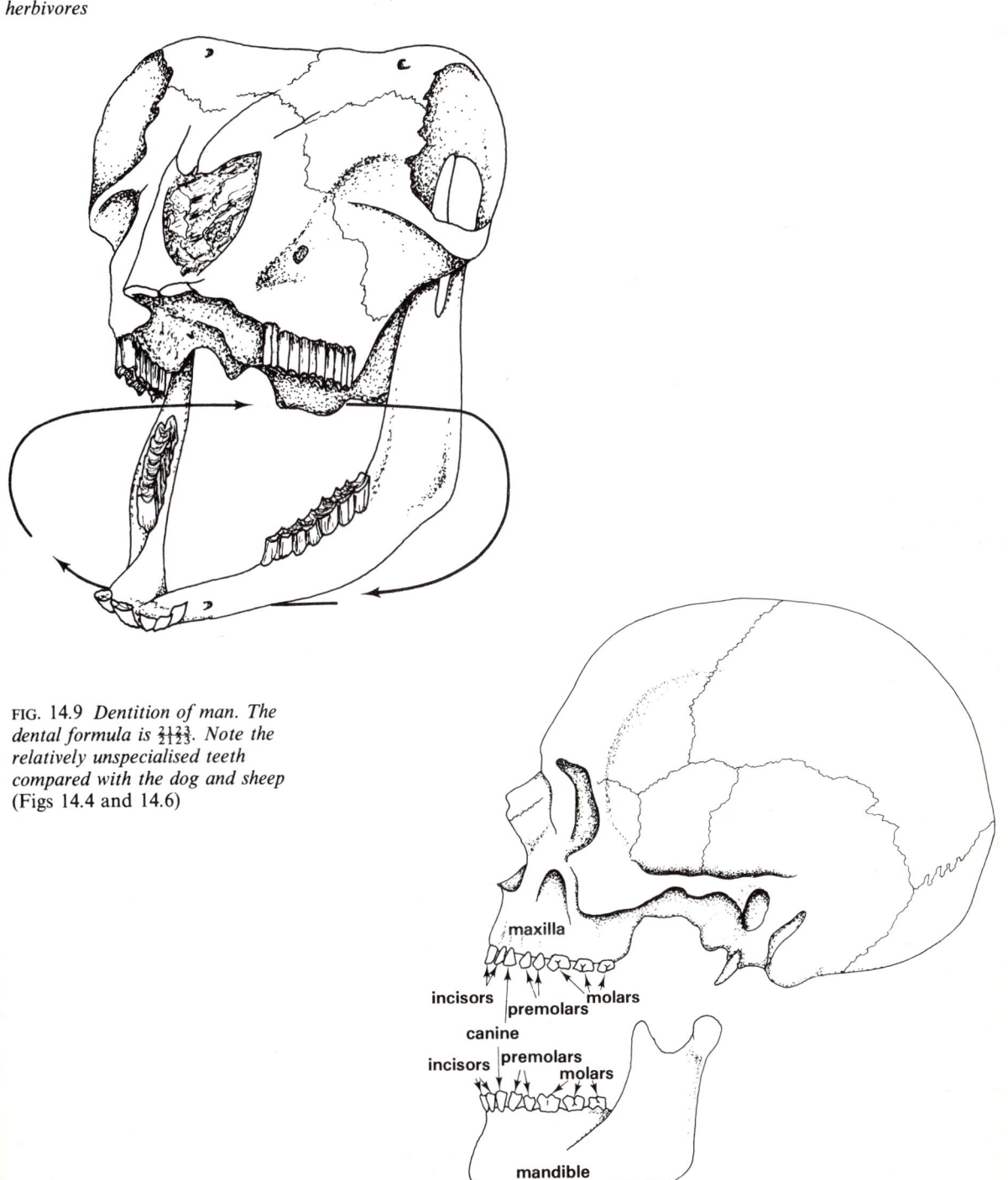

FIG. 14.8 *Circular grinding motion of the lower jaw in herbivores*

FIG. 14.9 *Dentition of man. The dental formula is* $\frac{2123}{2123}$. *Note the relatively unspecialised teeth compared with the dog and sheep (Figs 14.4 and 14.6)*

maxilla

incisors

premolars

molars

canine

incisors

premolars

molars

mandible

FIG. 14.10 *Dentition of the hedgehog*

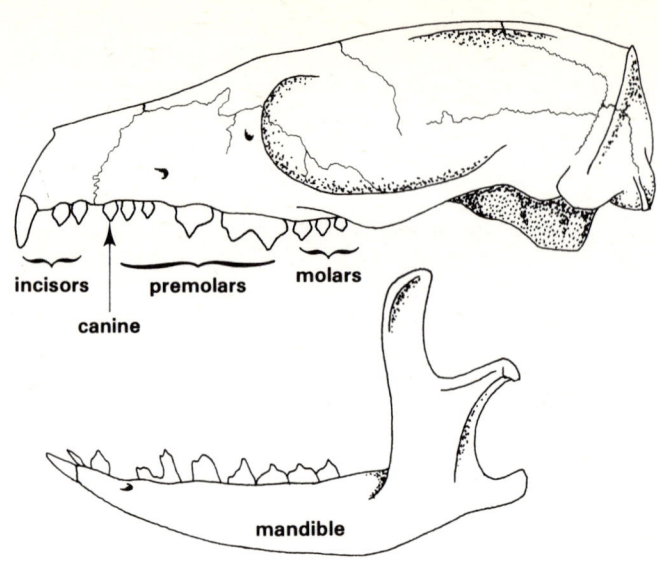

incisors
premolars
molars
canine
mandible

Rodents, such as rats, mice, squirrels, voles and beavers have incisors with **open roots** which allow the incisors to grow continually. Constant growth of the incisors is necessary to replace the tooth substance worn away by the gnawing action of the jaws. Enamel is present only on the outer surface of the incisors and a sharp sloping cutting edge is created as the teeth are worn. Rabbits and hares, which belong to the order *Lagomorpha*, have the same pattern of incisor growth. However, lagomorphs have a small second pair of incisors just behind the first pair (Fig. 14.11).

FIG. 14.11 (a) *Dentition of the rat. The dental formula is* $\frac{1003}{1003}$

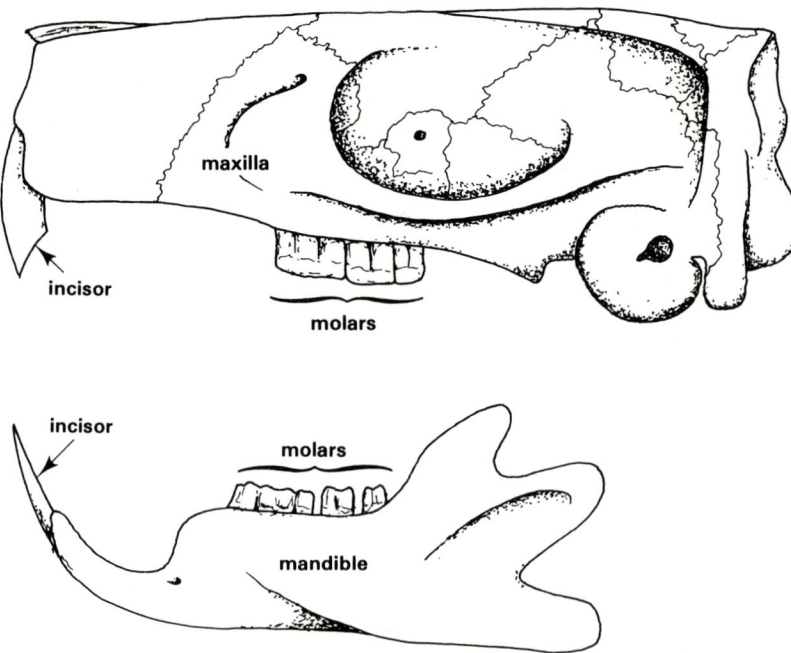

maxilla
incisor
molars
incisor
molars
mandible

FIG. 14.11 (b) *Dentition of the rabbit. The dental formula is* $\frac{2033}{1023}$

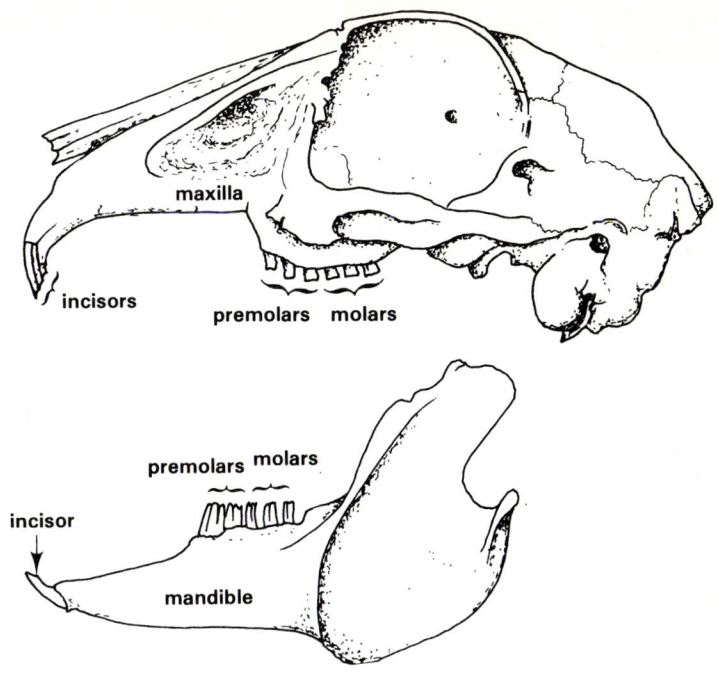

Toothed whales and **dolphins** are mostly flesh-eaters. They have very long jaws with rows of many peg-like teeth which are used to grip prey. **Whalebone whales** on the other hand feed on plankton. Only the foetus whalebone whale has teeth. In the adult, rows of large keratinous plates hanging transversely from the roof of the mouth trap plankton from sea water drawn into the mouth. The muscular tongue then forces sea water out of the mouth and the food is swallowed (Fig. 14.12).

Elephants have huge molar teeth with very large grinding surfaces. The large grinding area enables elephants to cope with the quantities of food necessary to support their massive bodies. The tusks of elephants are a single pair of continually growing upper incisors. They are composed of solid dentine except for a cap of enamel at their tips (Fig. 14.13).

FIG. 14.12 (a) *Dentition of a toothed whale*

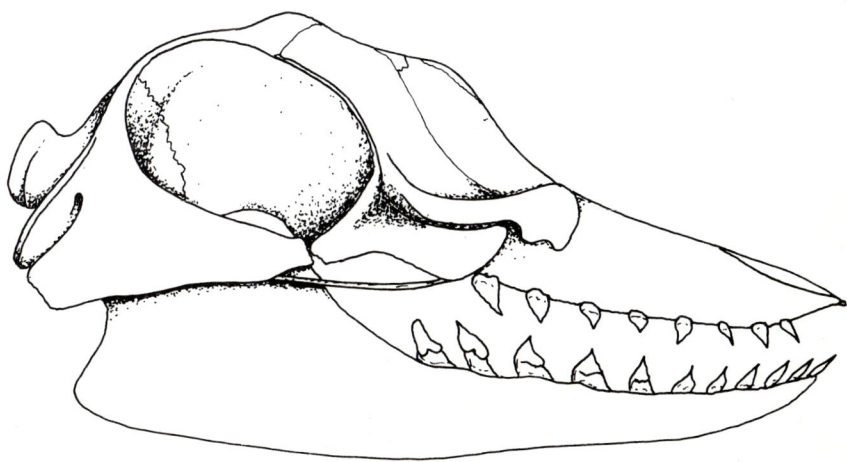

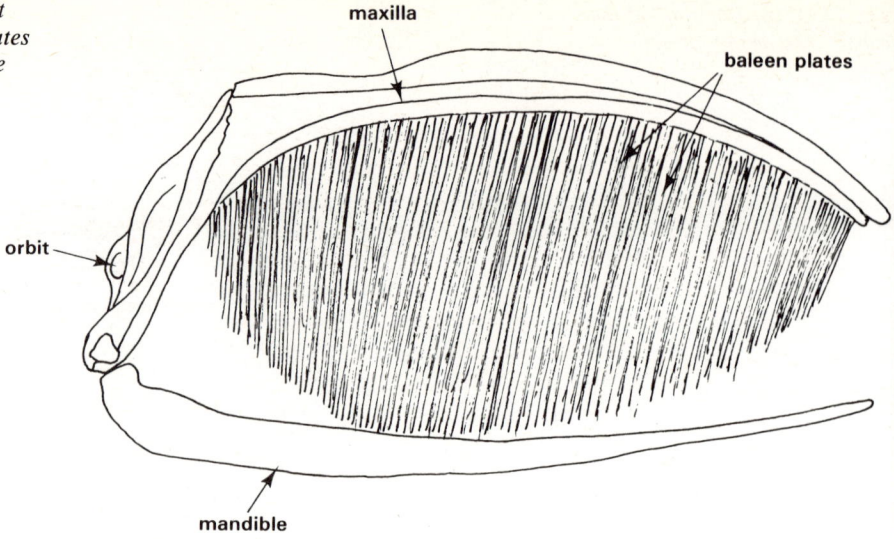

FIG. 14.12 (b) *Skull of a right whale, showing the baleen plates used for sieving food from the sea*

maxilla

baleen plates

orbit

mandible

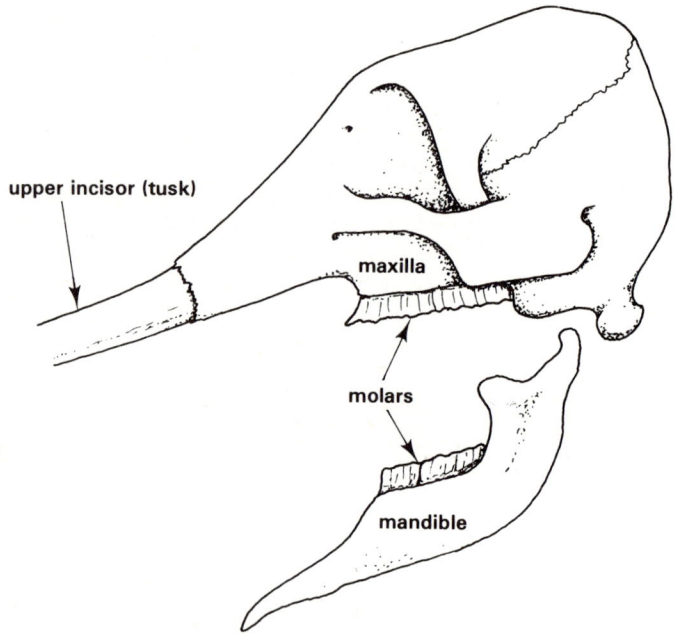

FIG. 14.13 *Dentition of the elephant*

upper incisor (tusk)

maxilla

molars

mandible

14.3 Digestion

Many of the organic nutrients in food such as proteins and polysaccharides are substances of high relative molar mass. They are insoluble in water and cannot pass through the membranes of body cells. It is necessary for such substances to be broken down into units small enough to be absorbed into the body fluids in which they can be transported to all parts of the body. The breaking down of food is called **digestion**. It takes place in the gut where food is acted on by hydrolytic enzymes (Chapter 4).

Except where otherwise stated the following account of digestion relates to man. The human digestive system is illustrated in Figure 14.14.

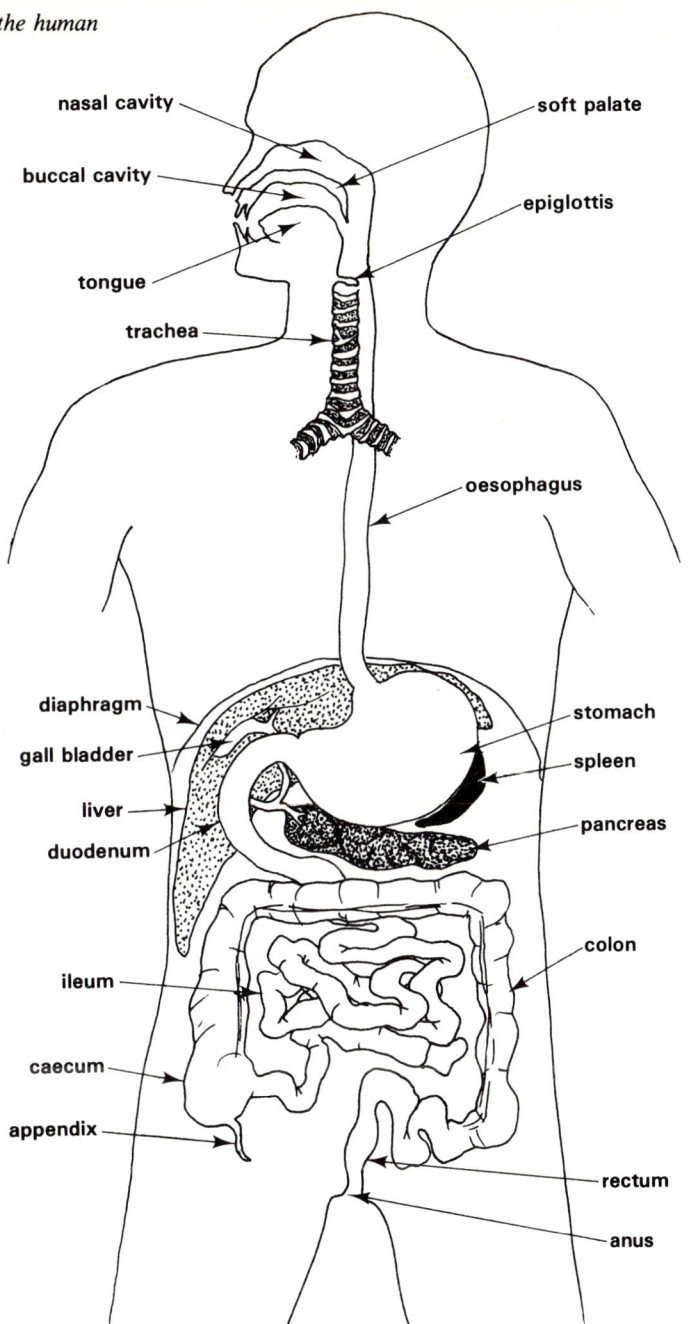

FIG. 14.14 *Diagram of the human digestive system*

nasal cavity

soft palate

buccal cavity

epiglottis

tongue

trachea

oesophagus

diaphragm

stomach

gall bladder

spleen

liver

pancreas

duodenum

colon

ileum

caecum

appendix

rectum

anus

14.3.1 The mouth

The smell and sight of food as well as the mechanical stimulation of food in the mouth triggers a reflex action which results in the secretion of **saliva**. The reflex ensures a flow of saliva when food is in the mouth. Secretion of saliva slows down when a mammal is not eating. Saliva enters the mouth from three pairs of salivary glands (Fig. 14.15).

FIG. 14.15 *Position of the salivary glands in man*

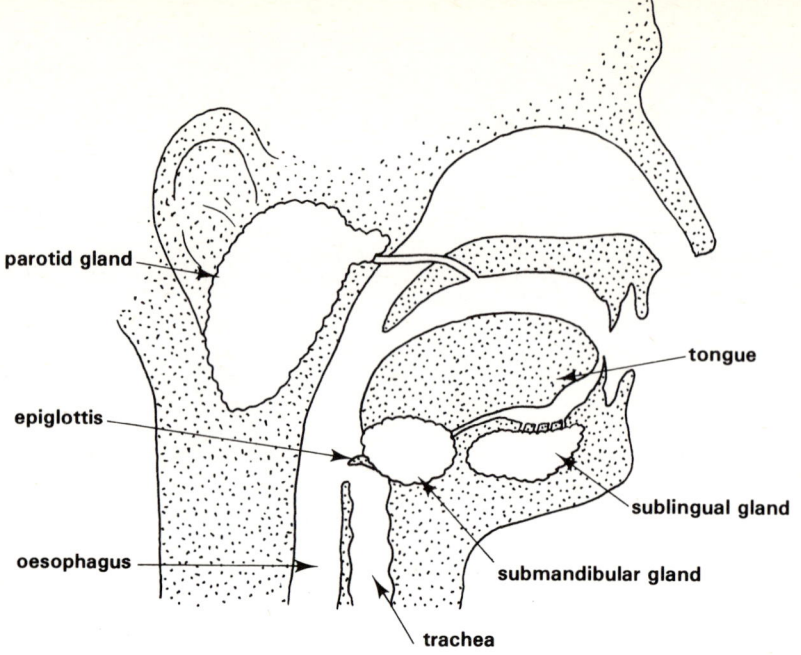

An enzyme called **salivary amylase** is usually found in saliva. Salivary amylase hydrolyses glycosidic linkages in **starch** breaking it down to the disaccharide sugar **maltose** (Chapter 3). Apart from in herbivores, salivary amylase probably contributes little to digestion. This is because most mammals hold food in the mouth for only a short time. Salivary amylase works best in neutral conditions and the pH of saliva is about 7. After swallowing, food passes down the oesophagus into the stomach where the pH is much lower. In acid conditions salivary amylase becomes inactive.

Saliva **lubricates** the pharynx and oesophagus making it easier for food to be swallowed. The food is moulded into a ball (bolus) by the tongue then pushed into the pharynx. Contraction of the pharyngeal wall pushes the bolus into the oesophagus. Rhythmical contractions of the oesophagus called **peristalsis** move the bolus towards the stomach. Peristalsis is a property of all parts of the gut and helps to move food along the alimentary canal. During swallowing the larynx is raised and the glottis is covered by the epiglottis so preventing food from entering the trachea (Fig. 14.16).

FIG. 14.16 *Peristalsis*

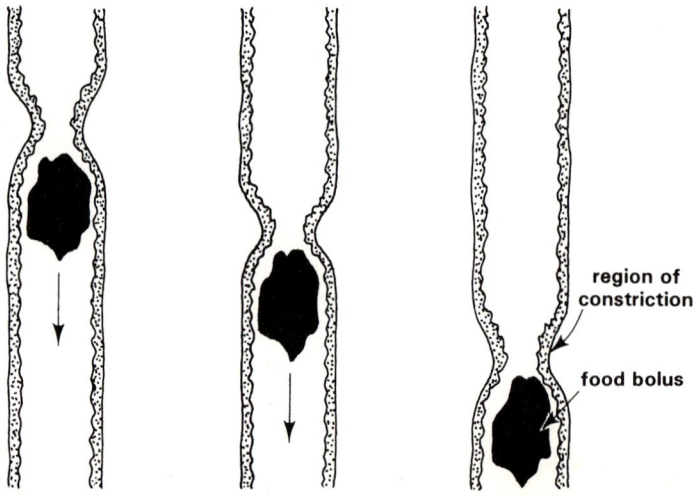

14.3.2 The stomach

The stomach is the first region of the gut where any significant digestion of food takes place. In man the stomach is a muscular bag with a volume of about two litres. It has a delicate inner folded membrane called the **gastric mucosa**. **Gastric juice** is secreted by numerous microscopic **gastric pits** embedded in the mucosa (Figs. 14.17 and 14.18). The juice is very acid with a pH of 2. The acidity is caused by **hydrochloric acid** secreted by **oxyntic cells**. **Peptic cells** secrete **pepsinogen** which on contact with the acid is converted to the peptidase enzyme **pepsin**. The gastric juice of many mammals contains a second peptidase enzyme called **rennin**. Rennin coagulates the milk protein casein and is particularly important in young mammals whose diet consists solely of milk. Pepsin hydrolyses peptide linkages in proteins producing polypeptides (Fig. 14.19).

FIG. 14.17 *The main regions of the human stomach. Three layers of muscles in the wall bring about stomach movements which mix the food with digestive gastric juice during digestion*

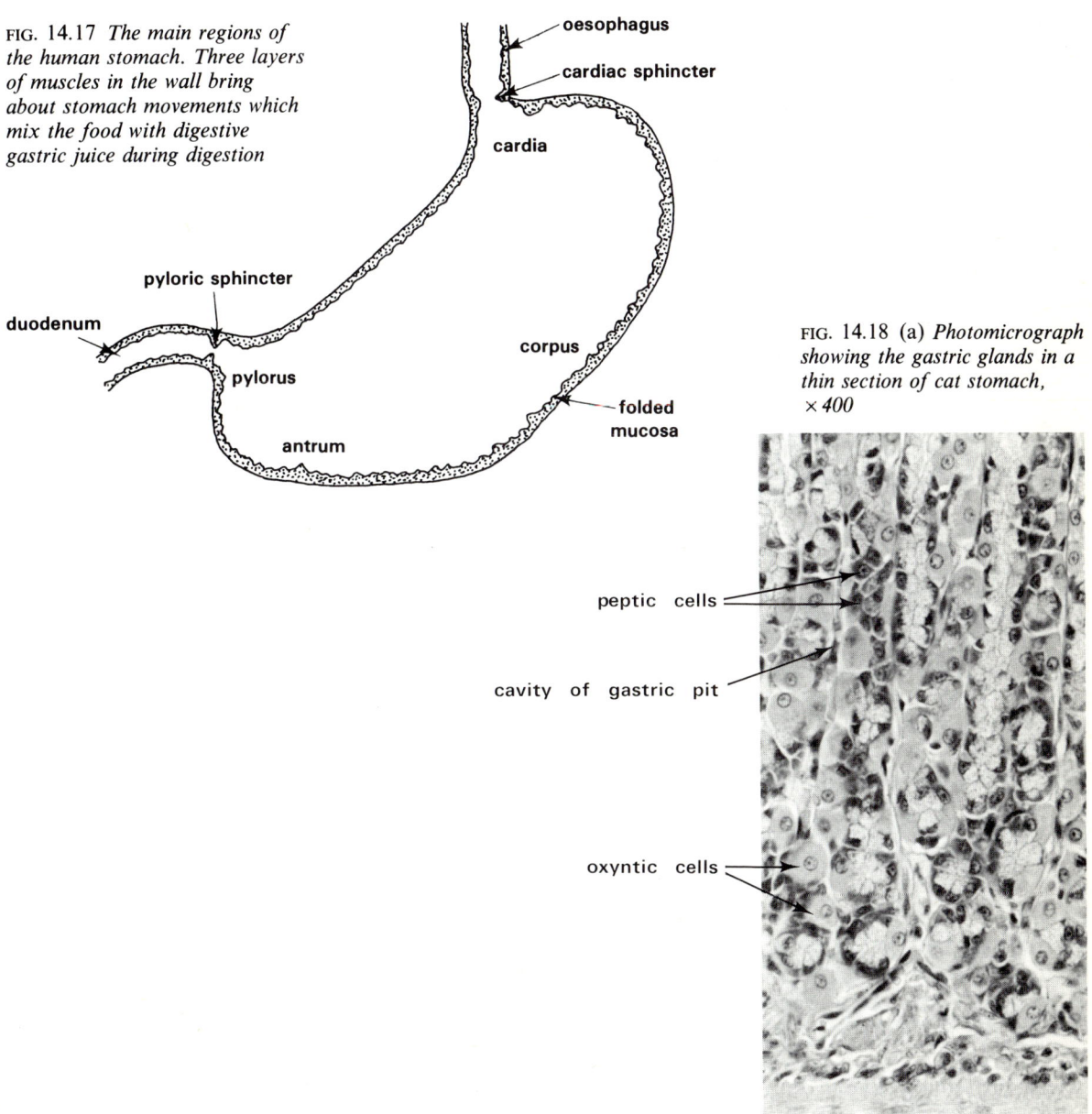

oesophagus

cardiac sphincter

cardia

pyloric sphincter

duodenum

pylorus

corpus

folded mucosa

antrum

FIG. 14.18 (a) *Photomicrograph showing the gastric glands in a thin section of cat stomach,* ×400

peptic cells

cavity of gastric pit

oxyntic cells

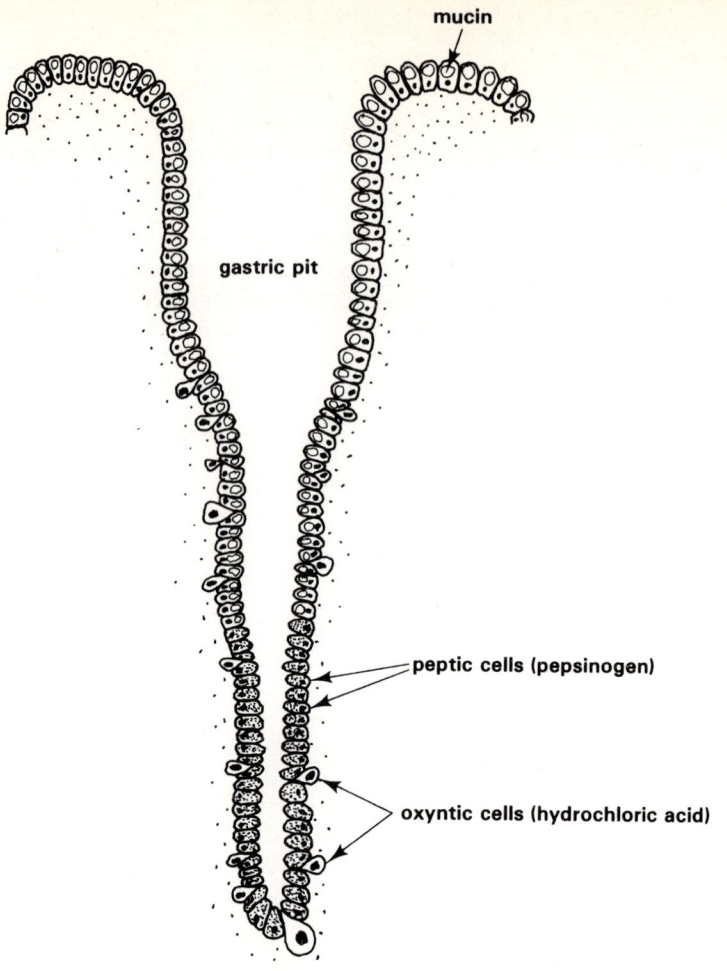

FIG. 14.18 (b) *Diagram of a gastric gland*

mucin

gastric pit

peptic cells (pepsinogen)

oxyntic cells (hydrochloric acid)

FIG. 14.19 *Action of rennin and pepsin. Gastric hydrochloric acid provides optimum pH for the action of these enzymes. Rennin is important in the stomach of many young mammals*

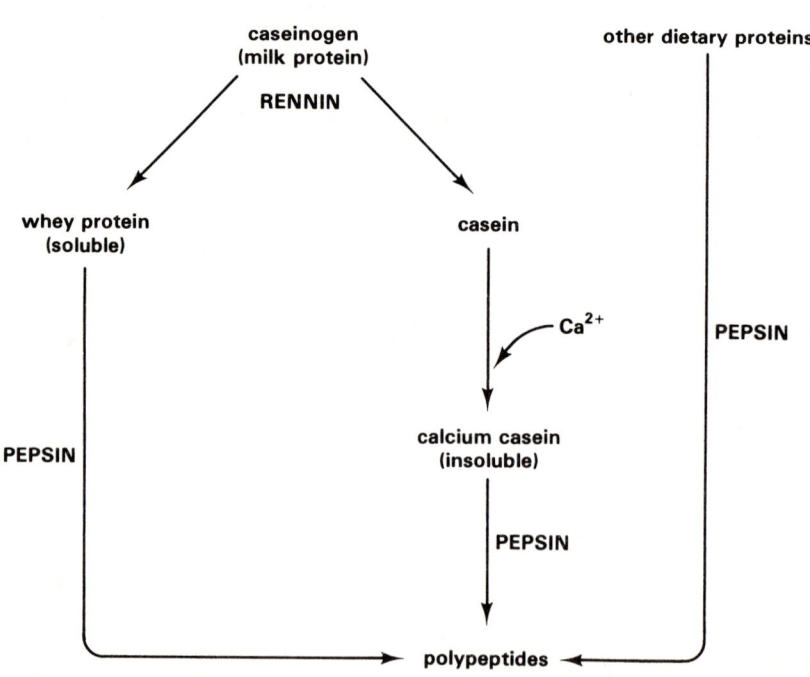

caseinogen
(milk protein)

other dietary proteins

RENNIN

whey protein
(soluble)

casein

Ca²⁺

PEPSIN

PEPSIN

calcium casein
(insoluble)

PEPSIN

polypeptides

The corrosive and digestive properties of hydrochloric acid and pepsin place the delicate gastric mucosa at risk. Protection is given by a slimy glycoprotein called **mucin**. Mucin is secreted by goblet cells in the mucosa and forms a layer of mucus over the stomach lining. Mucus protects against mechanical as well as chemical injury. If the protection is not effective the mucosa and stomach wall are attacked by the gastric juice, causing an ulcer to form.

Like the release of saliva, gastric secretion is controlled by autonomic reflexes. Secretion of gastric juice begins when food is in the mouth. The presence of food in the stomach triggers the stomach lining to produce a hormone called **gastrin** which is taken into the blood. Gastrin stimulates continued secretion of gastric juice after the faster nervous mechanism has started it off (Fig. 14.20).

FIG. 14.20 *The relative importance of nervous (vagus) and hormonal (gastrin) influence on gastric juice secretion*

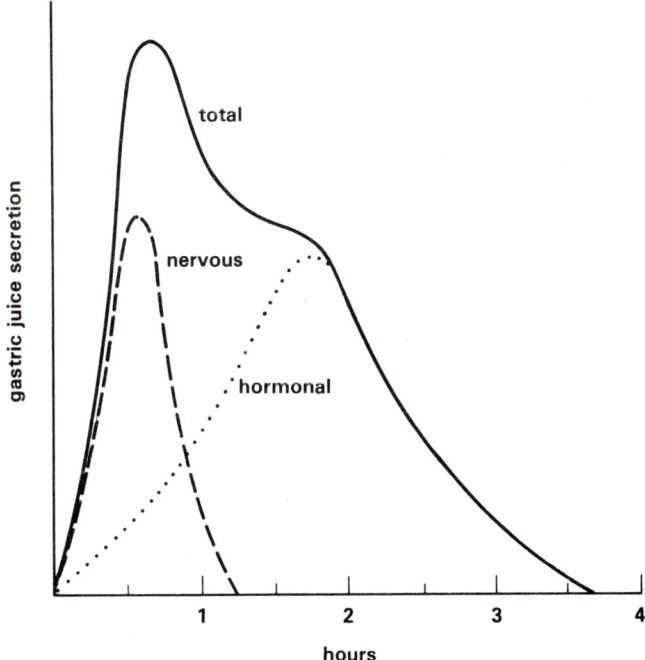

FIG. 14.21 *Churning action of the human stomach during digestion* (from X-ray photographs)

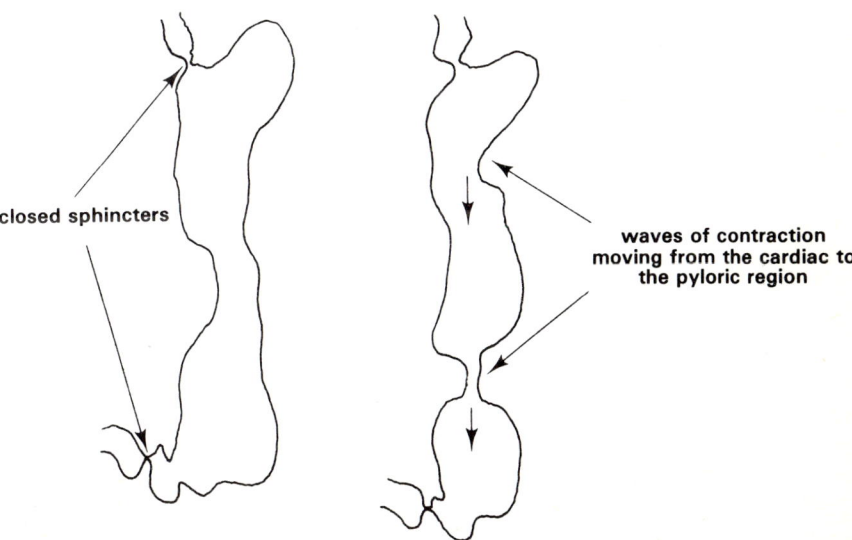

closed sphincters

waves of contraction moving from the cardiac to the pyloric region

269

During the three or four hours food is held in the human stomach, rhythmic muscular contractions of the stomach wall churn the food, mixing it with gastric juice, so promoting digestion. The food is gradually converted into a creamy, acid suspension called **chyme** (Fig. 14.21).

14.3.3 Rumination

In **ruminant mammals** such as cattle and sheep, the stomach is large and modified for the digestion of vegetation. Cellulose in plant cell walls is the main carbohydrate constituent of the food eaten by ruminants. The ruminant stomach consists of four chambers, the **rumen, reticulum, omasum** and the **abomasum** (Fig. 14.22).

FIG. 14.22 *Ruminant stomach. Continuous arrows indicate the course of food in the rumen and reticulum after swallowing. Following a return to the mouth for cudding, food enters the true stomach by a course shown by a broken arrow*

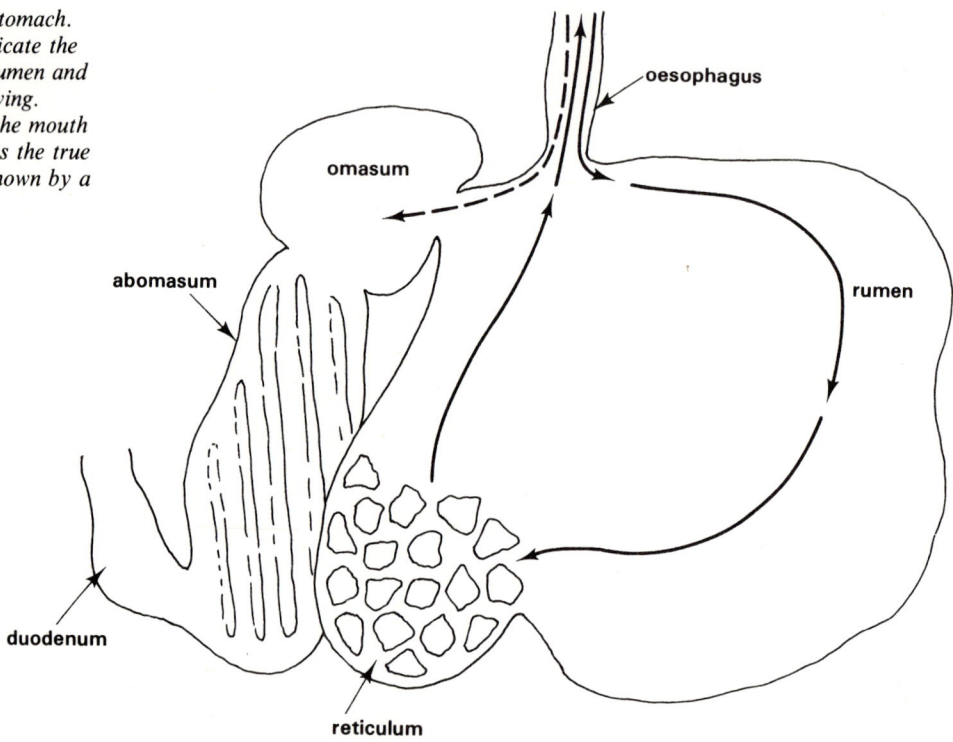

Food is passed first into the largest chamber, the rumen, where huge populations of **anaerobic cellulolytic bacteria** begin the fermentation of cellulose. Fermentation continues in the reticulum into which the contents of the rumen are later directed. The partially digested food is then forced back into the mouth for further grinding called **chewing the cud**. Thorough grinding of plant material is essential if the cellulolytic bacteria are to be effective. On reswallowing, the semi-liquid cud is directed by closure of a groove on the side of the reticulum into a small chamber, the omasum. Inside the omasum, water is pressed out of the cud and is absorbed. The solid material passes into the abomasum. This is the true stomach and secretes an acid peptic juice.

The bacterial fermentation which occurs in the rumen produces sugars, methane gas and carboxylic acids such as ethanoic acid. Bacteria entering the abomasum with the cud are digested and add to the material available for absorption.

14.3.4 The small intestine

During gastric digestion, food is kept in the stomach by the constriction of two **sphincters**. These are circular muscles which seal off the stomach where it is joined to the oesophagus and small intestine (Fig. 14.17). Relaxation of the pyloric sphincter allows the propulsion of chyme into the small intestine by peristalsis of the involuntary muscles in the gastric wall.

From the stomach, chyme passes first into the **duodenum** which in man is about 30 cm long. Beyond the duodenum is the **jejunum** and then the **ileum** making up the rest of the small intestine (Fig. 14.14). The total length of the small intestine is about 6 m in man and 45 m in cattle.

In most mammals, digestion occurs mainly in the small intestine where enzymes complete the chemical breakdown of food into a soluble state suitable for absorption. The enzymes come from two sources, the pancreas and the glands in the wall of the small intestine. Food is also exposed to a non-enzymic fluid called bile made in the liver (Fig. 14.23).

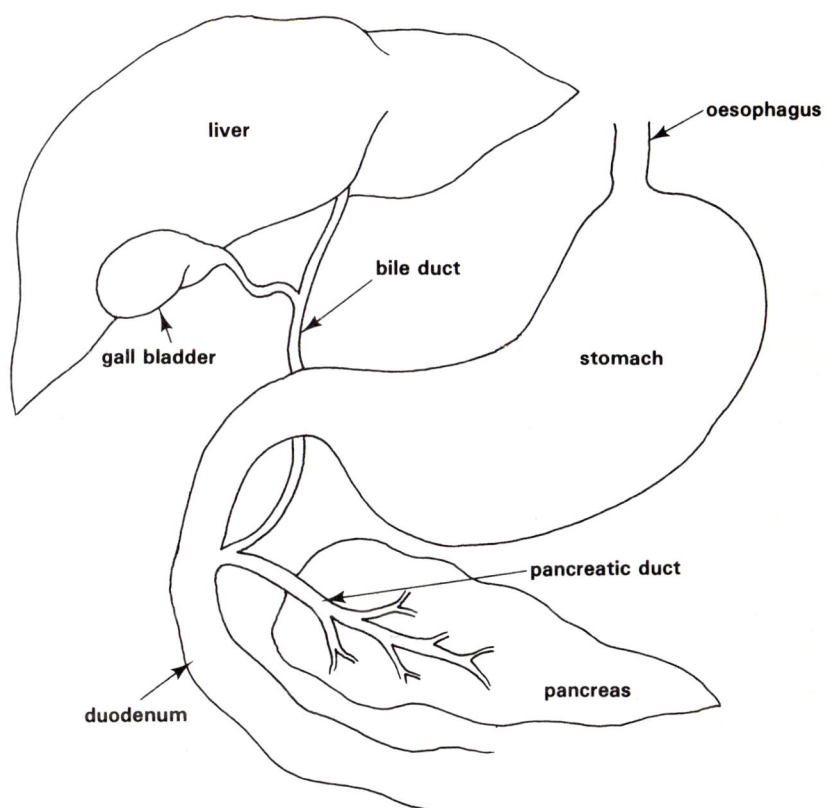

FIG. 14.23 *Relationship between the liver, gall bladder, pancreas, stomach and small intestine in man*

1. BILE

Bile is a greenish fluid containing bile pigments which are the excretory products of the breakdown of haemoglobin (section 14.5.2). Though containing no enzymes bile helps digestion in several important ways. First, it contains sodium hydrogencarbonate which gives it a pH of between 7 and 8. This is the optimum pH for the action of pancreatic and intestinal enzymes. Bile is therefore important in **neutralising** the acid chyme from the stomach. The second digestive function of bile is due to the bile salts, sodium and potassium glycocholate and taurocholate. They **emulsify lipids**, causing them to break down into numerous small droplets.

271

Emulsification provides a relatively large surface area of lipid for the action of lipase enzymes and hence speeds up digestion of fats and oils (Fig. 14.24).

FIG. 14.24 *Emulsification of fat droplets*

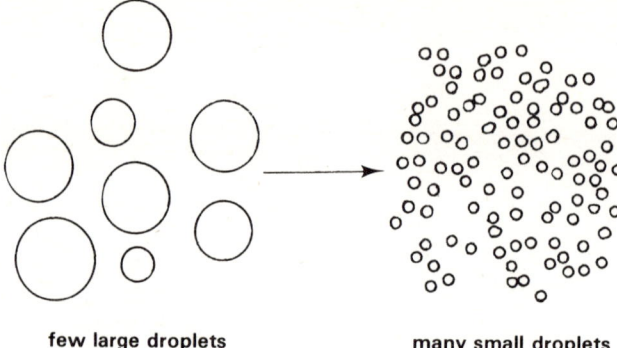

few large droplets　　　many small droplets

2. PANCREATIC JUICE

The **pancreas** is situated just beneath the stomach and is connected to the small intestine by a pancreatic duct through which **pancreatic juice** is discharged. The bile duct joins the pancreatic duct in man. In some mammals the bile and pancreatic ducts are joined separately to the small intestine. The endocrine component of the pancreas, the islets of Langerhans (Chapter 18), plays no part in digestion.

Pancreatic juice contains four enzymes. **Pancreatic amylase** hydrolyses glycosidic linkages in starch, converting it to maltose in the same way as salivary amylase. Because food remains in the duodenum for some time there is more opportunity for hydrolysis of starch in the small intestine than in the mouth. **Pancreatic lipase**, probably a group of several lipase enzymes, hydrolyses lipids to fatty acids and glycerol (Fig. 14.25).

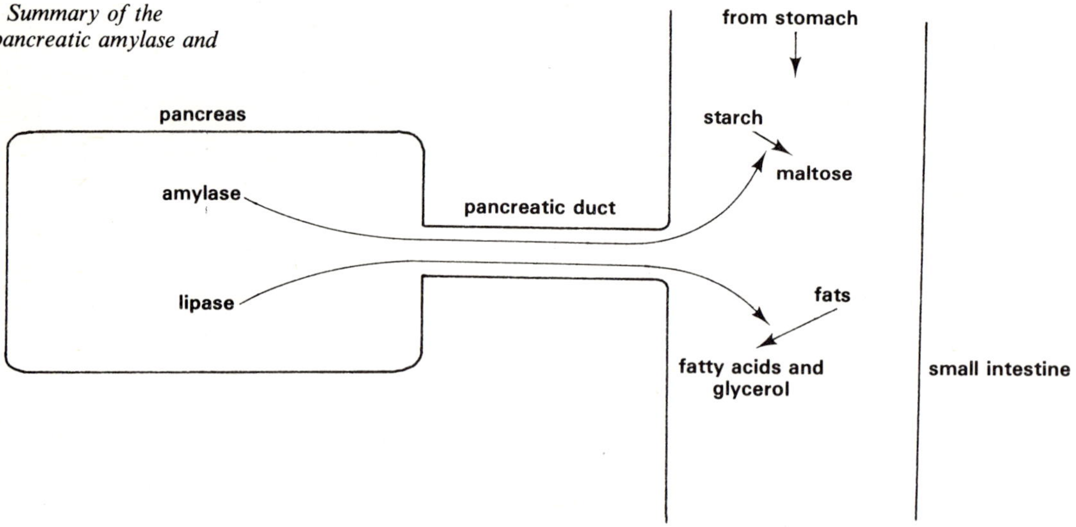

FIG. 14.25 *Summary of the action of pancreatic amylase and lipase*

from stomach

pancreas

amylase

pancreatic duct

starch

maltose

lipase

fats

fatty acids and glycerol

small intestine

Two inactive enzyme precursors are also found in pancreatic juice. They are **trypsinogen** and **chymotrypsinogen**. Trypsinogen is converted to the active enzyme **trypsin** by an activator called **enterokinase** made in the intestinal glands. Trypsin activates the conversion of chymotrypsinogen to

chymotrypsin. Trypsin and chymotrypsin are peptidase enzymes having an effect on proteins similar to that of gastric pepsin. They bring about a partial breakdown of proteins to polypeptides (Fig. 14.26).

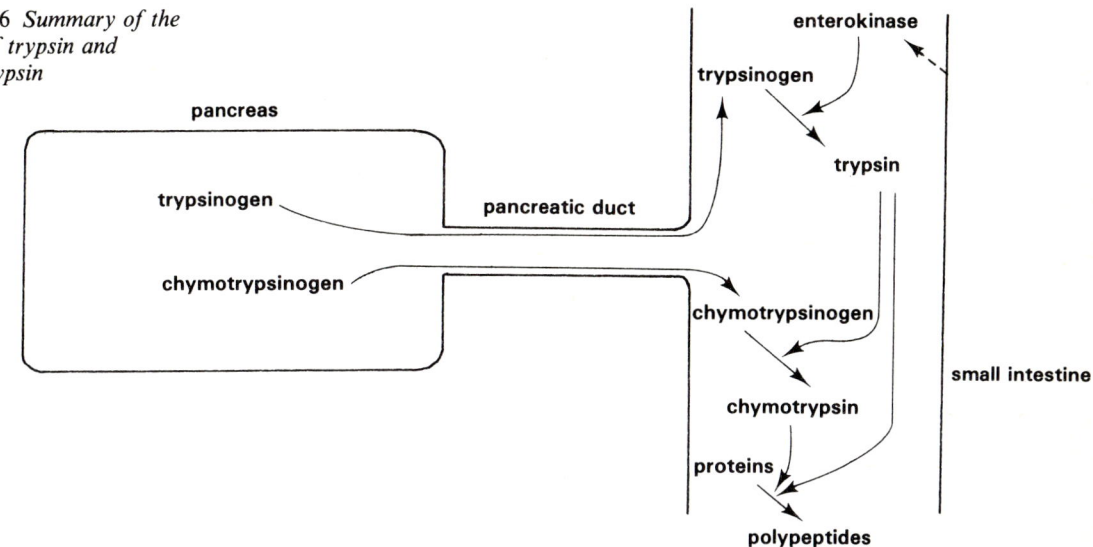

FIG. 14.26 *Summary of the action of trypsin and chymotrypsin*

3. INTESTINAL FLUID

Embedded in the intestinal wall is a great number of microscopic pits called the **crypts of Lieberkühn**. In the duodenum, coiled **Brunner's glands**, lying deep in the sub-mucosa, empty their products into the crypts. Enzymes from Brunner's glands complete the process of digestion started by the mouth, stomach and pancreas. Intestinal juice, called **succus entericus**, contains a large number of enzymes which convert complex carbohydrates to monosaccharide sugars, proteins and polypeptides to amino acids, and lipids to fatty acids and glycerol. The enzymes and their activities are summarised in Table 14.1. Some of the final stages in the digestion of carbohydrates and polypeptides occur in the cells lining the intestinal mucosa.

Table 14.1 Summary of the action of enzymes secreted by the small intestine

ENZYMES	SUBSTRATES	PRODUCTS
amylase	starch	maltose
maltase	maltose	glucose
sucrase	sucrose	glucose and fructose
lactase	lactose	glucose and galactose
peptidases	polypeptides	amino acids
lipases	fats	fatty acids and glycerol

The mechanisms controlling the release of bile, pancreatic juice and succus entericus are complex. They are triggered by the presence of chyme in the small intestine. Chyme stimulates the release of at least three hormones from the intestinal mucosa into the blood. **Pancreozymin** triggers

273

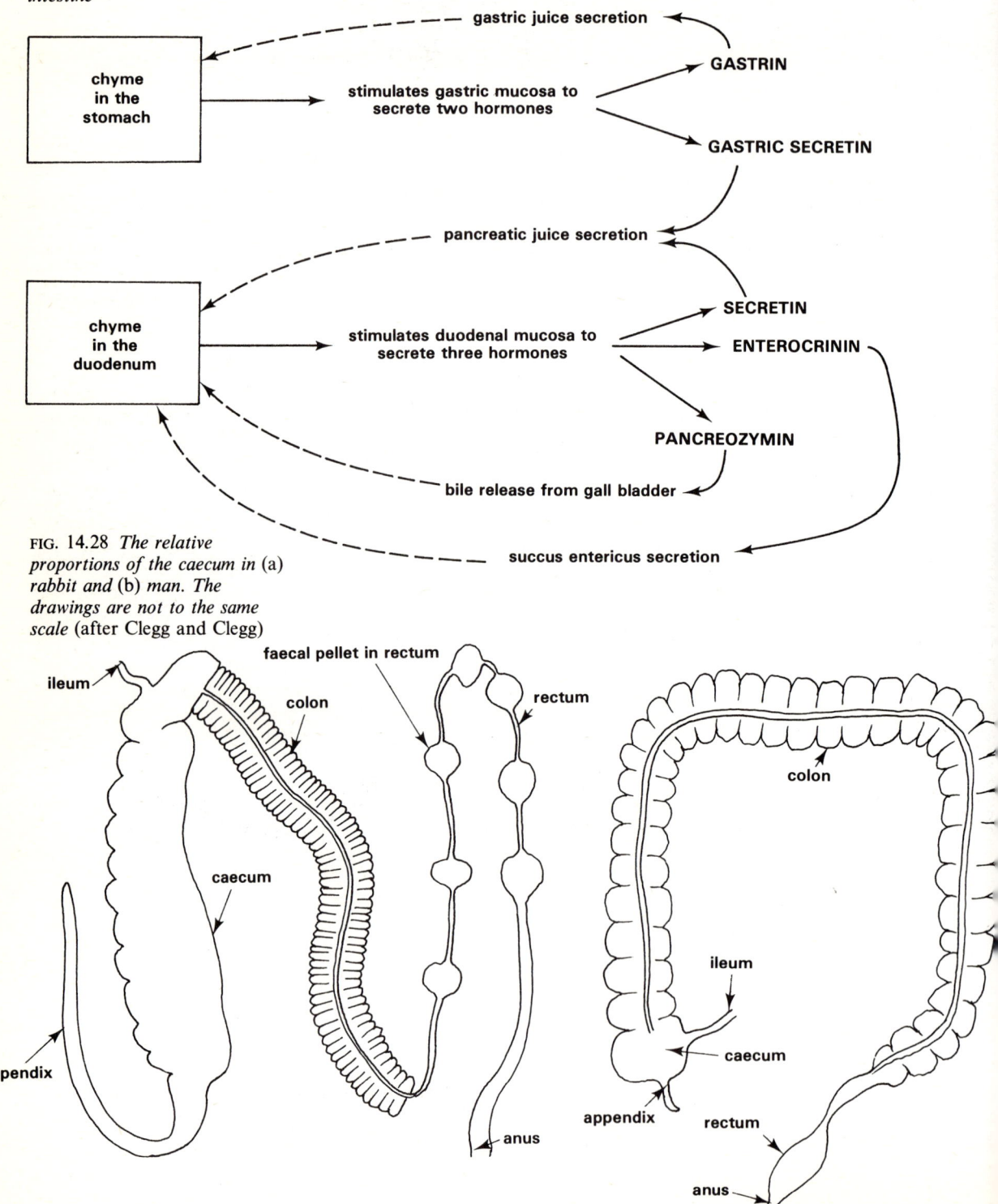

FIG. 14.27 *Summary of the gastrointestinal hormones involved in the release of bile and the digestive juices for the stomach, pancreas and small intestine*

the release of bile from its temporary store, the gall bladder. **Secretin** stimulates pancreatic secretion which may have started before chyme enters the duodenum under the influence of another hormone, **gastric secretin** produced by the stomach lining. Finally **enterocrinin** stimulates the intestinal glands to secrete their digestive juice (Fig. 14.27).

gastric juice secretion

chyme in the stomach

stimulates gastric mucosa to secrete two hormones

GASTRIN

GASTRIC SECRETIN

pancreatic juice secretion

chyme in the duodenum

stimulates duodenal mucosa to secrete three hormones

SECRETIN

ENTEROCRININ

PANCREOZYMIN

bile release from gall bladder

succus entericus secretion

FIG. 14.28 *The relative proportions of the caecum in* (a) *rabbit and* (b) *man. The drawings are not to the same scale* (after Clegg and Clegg)

faecal pellet in rectum

ileum

colon

rectum

caecum

colon

ileum

caecum

appendix

appendix

rectum

anus

anus

(a)

(b)

14.3.5 The caecum

In non-ruminant herbivores such as rabbits and horses fermentation of cellulose occurs in the caecum. The caecum is a bag-like structure and is the first part of the large intestine (Fig. 14.28). Inside the caecum **cellulolytic bacteria** perform the same function as those in the rumen of cattle and sheep (section 14.3.3). In many herbivores digestion and fermentation render the food soluble enough to be absorbed on its first passage through the gut. In others such as rabbits, the food must pass twice through the gut before it is sufficiently broken down for absorption to take place. For this reason rabbits eat the faeces formed from food which has passed once through the gut.

14.4 Absorption

Peristalsis of the small intestine helps to mix the digestive juices with chyme, so promoting hydrolysis of food by the enzymes. As a result, chyme is turned into a watery emulsion called **chyle**.

The **absorption** of nearly all the digestive products takes place through the extensive mucosa mainly in the duodenum and the first part of the jejunum. The internal surface area of the small intestine is greatly enlarged by its folded nature and even more so by numerous **villi**. Each villus is a microscopic tubular projection containing blood and lymph capillaries and covered with epithelium (Fig. 14.29).

Monosaccharide sugars and amino acids are absorbed into the **blood capillaries** of the villi. The capillaries drain into the **hepatic portal vein** which carries blood from the gut to the liver. While some fatty acids, glycerol and even lipid globules, are absorbed directly into the blood, they are mainly taken up by the lymphatic capillaries called **lacteals**. Fatty acids and glycerol are converted back to lipids in the epithelial cells of the villi. From here they are transferred into the lacteals. As a result, lymph in the lacteals appears white. Lymph eventually enters the blood through the thoracic ducts (Figs. 14.30 and 9.16).

FIG. 14.29 (a) *Photomicrograph of a transverse section of cat ileum. The blood vessels have been impregnated with a dye.* × 20

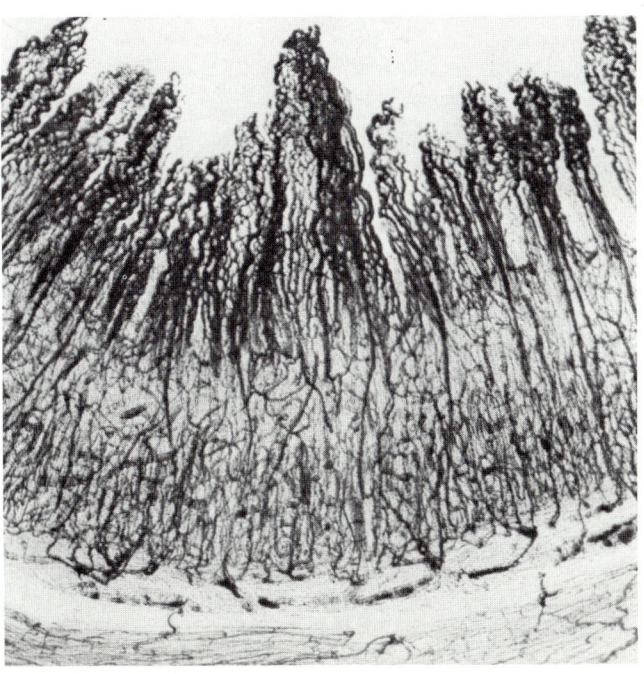

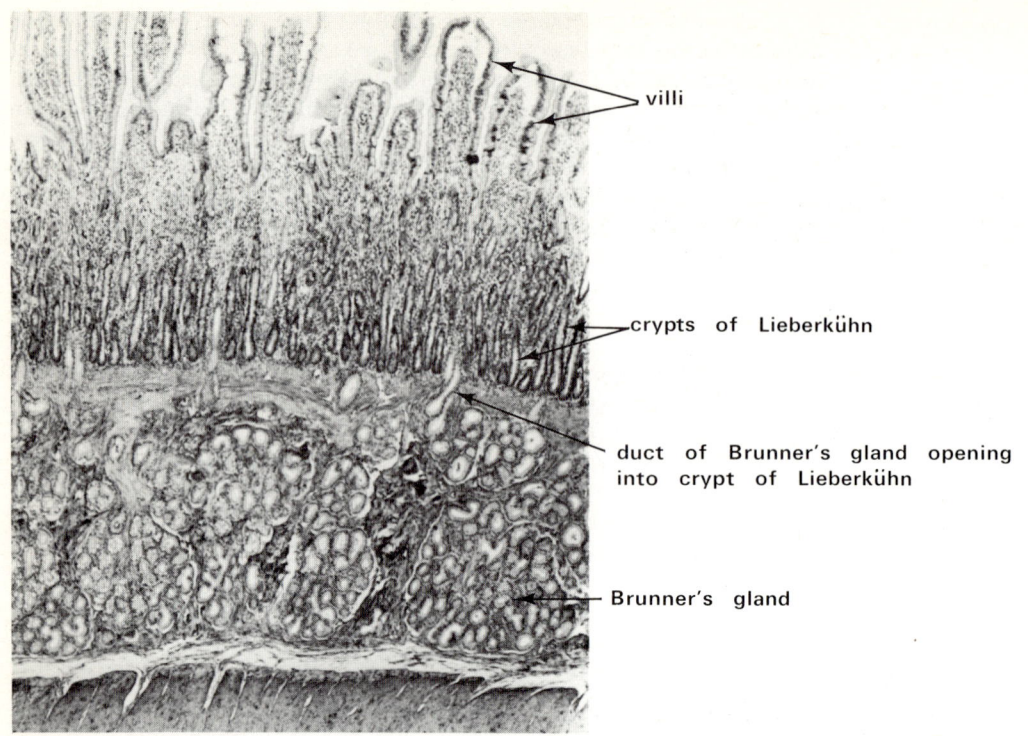

FIG. 14.29 (b) *Photomicrograph of a transverse section of cat duodenum,* × 45

Table 14.2 Factors affecting the absorption of nutrients

Glucose	Active absorption may be linked with the transport of sodium ions across the membranes of mucosal cells.
Amino acids, some peptides	Active absorption may be affected by absorption of sugars and may also be linked with sodium ion transport through the mucosa.
Vitamin B_{12}	Absorption dependent upon presence of intrinsic factor (IF) in gastric juice.
Vitamin K	Absorption promoted by bile.
Calcium ions	Active absorption promoted by vitamin D_2 and probably also by parathyrin.
Iron ions	Absorbed in the iron(II) state. Rate of absorption depends on degree of saturation of transferrin with iron in the blood.
Water	Passive absorption by osmosis depends upon solute absorption.

The absorption of the products of digestion as well as inorganic minerals and vitamins which have not been changed by the gut enzymes involves **active and passive mechanisms**. Many substances require the presence of certain factors before they are absorbed in any amount. For example, the absorption of vitamin B_{12} requires intrinsic factor (IF), a mucoprotein component of gastric juice. Vitamin B_{12} is vital to red blood cell production and inability to absorb vitamin B_{12}, say through lack of IF, causes pernicious anaemia (Chapter 9). Reference is made in Chapter 18 to the role of parathyrin in the absorption of calcium ions. Table 14.2 summarises some of the factors concerned with the active absorption of nutrients from the gut.

It is in the large intestine, expecially the **colon**, that the absorption of water takes place. The digestive secretions of the gut contain a lot of water. If most of the water were not reabsorbed the loss would seriously reduce the body's water content. Faecal water accounts for about four to eight per cent of the total water loss in man.

Unabsorbed matter including the excretory products of bile is passed to the exterior through the **anus** as **faeces**.

FIG. 14.30 *Diagram of a villus*

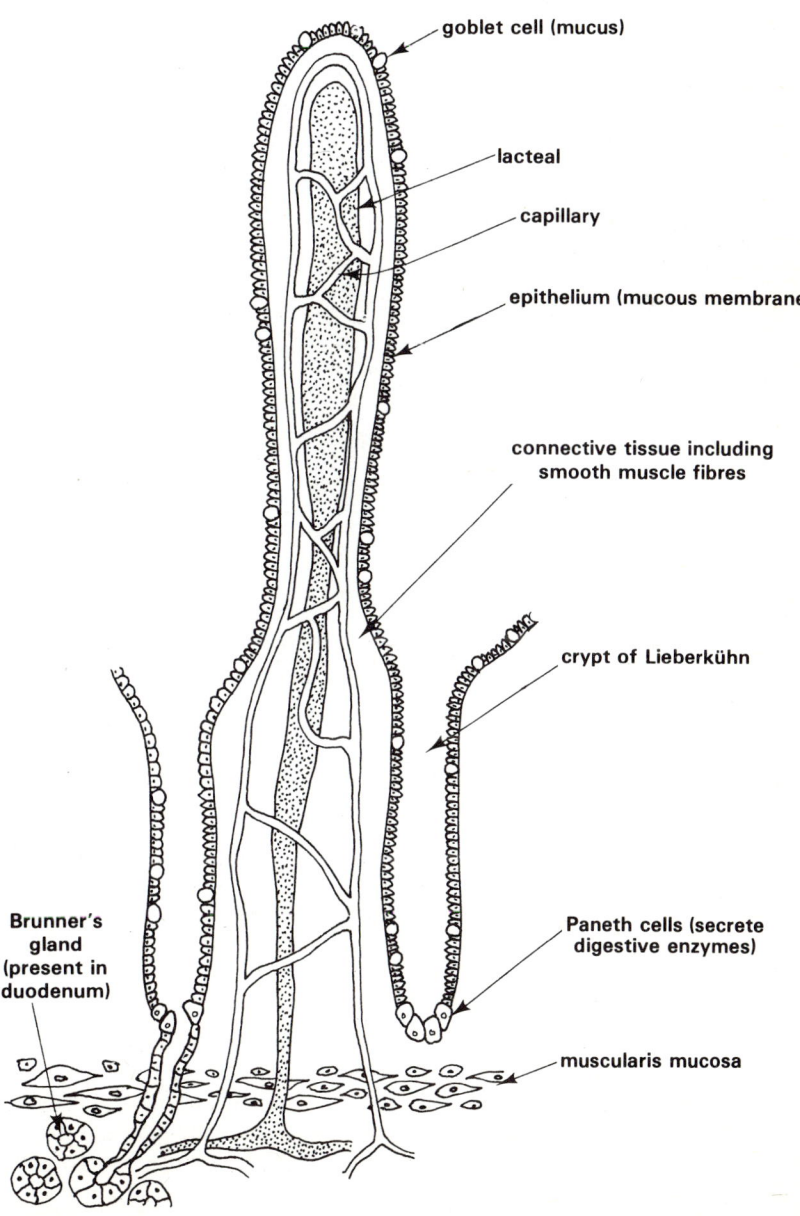

goblet cell (mucus)

lacteal

capillary

epithelium (mucous membrane)

connective tissue including smooth muscle fibres

crypt of Lieberkühn

Brunner's gland (present in duodenum)

Paneth cells (secrete digestive enzymes)

muscularis mucosa

14.5 The liver

Several chapters in this book refer to **homeostasis**, the maintenance of a steady internal environment. Regulating the composition of blood and tissue fluid is an important aspect of homeostasis. The composition of blood leaving the gut in the hepatic portal vein is largely determined by the diet and thus the substances absorbed from the small intestine. However, the blood which enters the general circulation is often very different in composition. The hepatic portal vein carries blood to the liver. One of the functions of the liver therefore is to regulate the composition of blood. The liver is the largest organ in a mammal. It performs many functions, not all of which are concerned with homeostasis.

14.5.1 Structure of the liver

The mammalian liver consists of a large number of **lobules** (Fig. 14.31). Each lobule contains many vertical **plates of liver cells**, arranged radially around a central blood vessel which is a branch of the **hepatic vein**. The hepatic vein drains blood from the lobules and releases it into the general circulation. The blood supply to each lobule is from two sources, the **hepatic artery** and the **hepatic portal vein**. Branches of these vessels are found between the liver lobules. The artery delivers oxygenated blood from the general circulation and the vein delivers food-laden blood from the gut. Blood from the hepatic artery and hepatic portal vein flows between the plates of liver cells inwards and into the central vein in channels called **sinusoids**.

FIG. 14.31 (a) *Diagram of a liver lobule*

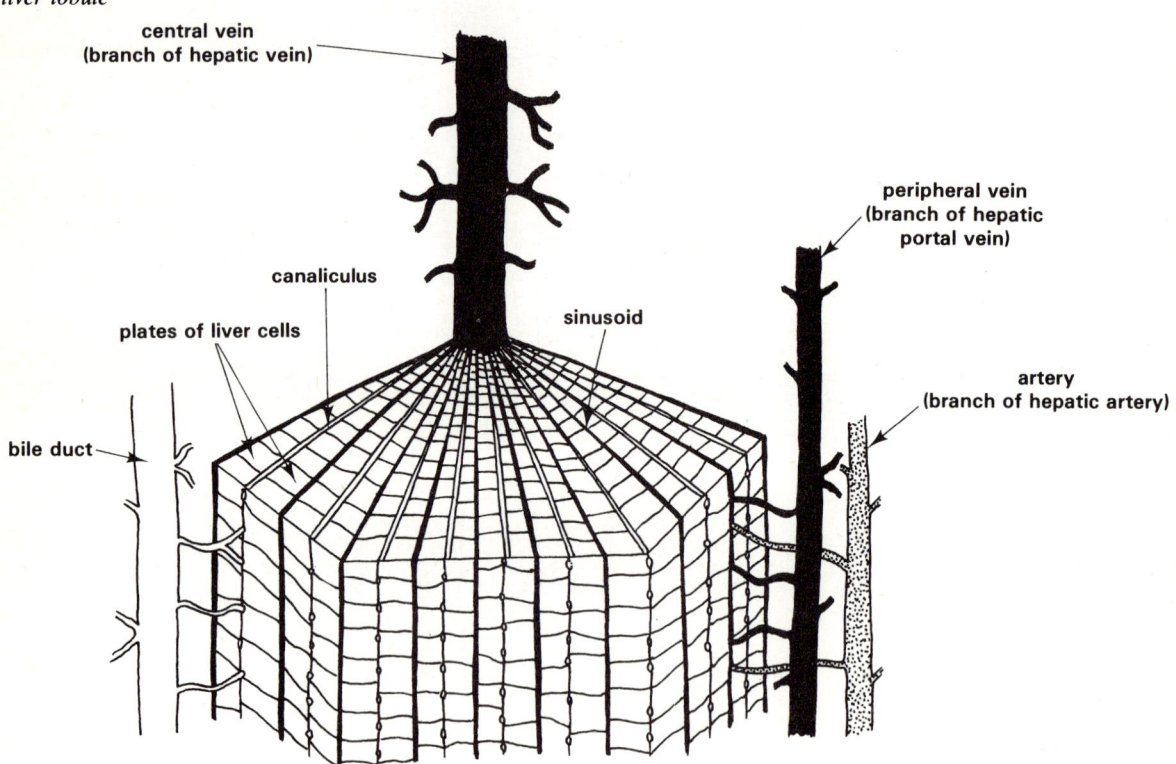

Between the plates of liver cells are channels called **canaliculi** which receive bile. The bile moves outwards to the periphery of the lobules where it collects into bile ducts. In most mammals bile is stored temporarily in a sac-like **gall bladder** before its periodic release into the small intestine.

FIG. 14.31 (b) *Photomicrograph of a transverse section of a lobule of pig's liver,* ×45

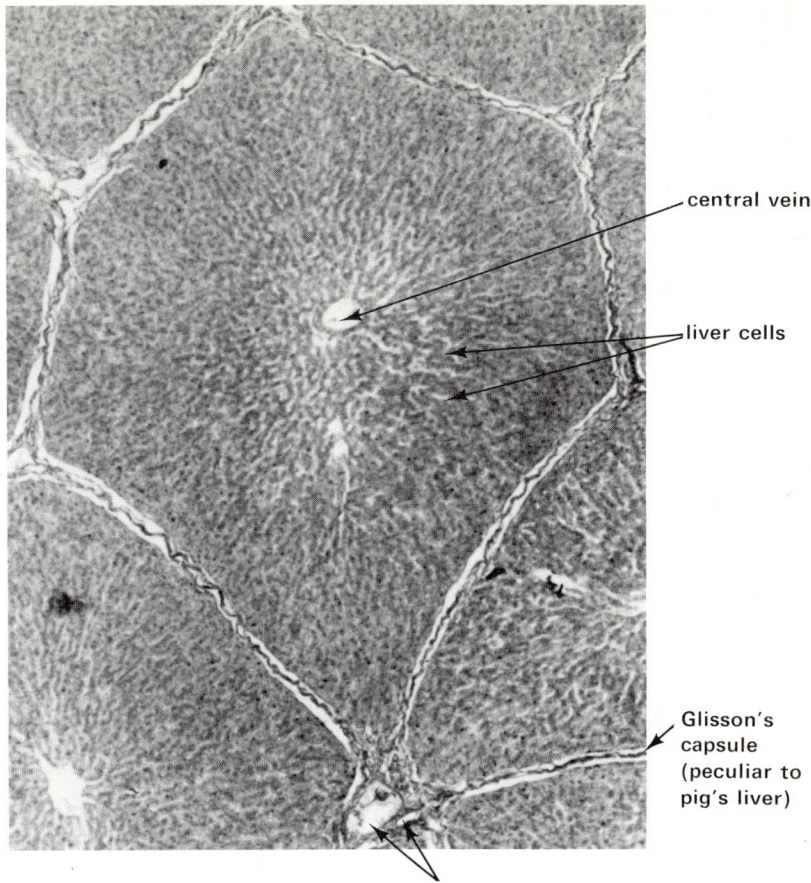

central vein

liver cells

Glisson's capsule (peculiar to pig's liver)

peripheral vessels

FIG. 14.31 (c) *Diagram showing the flow of blood and bile in a* sinusoid *and* canaliculus *of a liver lobule*

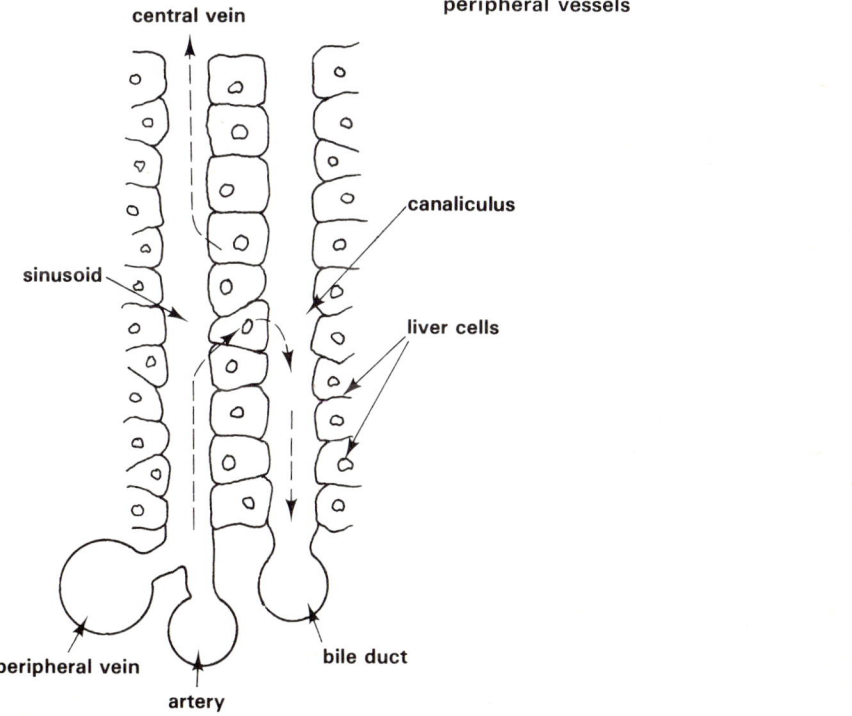

central vein

canaliculus

sinusoid

liver cells

peripheral vein

artery

bile duct

14.5.2 Functions of the liver

The functions of the liver are numerous and vital to life. For convenience they are described under three main headings.

1. METABOLISM OF ABSORBED FOOD

(i) CARBOHYDRATE METABOLISM. Soluble sugars, mainly **glucose**, are carried from the gut to the liver in the hepatic portal vein. Following a carbohydrate-rich meal the concentration of glucose in blood going to the liver is relatively high. When absorption is completed the glucose level drops. However, the concentration of glucose in the general blood circulation remains fairly stable. This is because excess glucose absorbed after a meal is converted in the liver cells to a storage polysaccharide called **glycogen**. The conversion is controlled by the hormone **insulin** from the islets of Langerhans in the pancreas (Chapter 18).

Glucose carried in the general blood circulation is used for tissue respiration throughout the body. It is replaced mainly from the liver's store of glycogen. The liver thus regulates the concentration of glucose in the general circulation irrespective of the rate of glucose absorption from the gut.

(ii) PROTEIN METABOLISM. Unlike carbohydrate, protein is not stored in the body. Amino acids from the gut pass through the liver and enter the **amino acid pool** in the general circulation. Amino acids are absorbed by the body cells and used for protein synthesis. The liver cells are important sites of **protein synthesis**, producing the plasma proteins (Chapter 9). A stable protein pool is established in the blood. Synthesis is balanced by an equal breakdown of plasma proteins which have a limited lifetime in the circulation.

Plasma proteins at the end of their lifetime together with excess amino acids absorbed from the gut are deaminated by liver cells. **Deamination** is the removal of amino groups and their conversion to ammonia. The remains of the acid molecules enter the Krebs' tricarboxylic acid cycle and are used to produce respiratory energy. Ammonia is very toxic and enters a sequence of reactions called the **ornithine cycle** which converts ammonia to the less toxic **urea** (Fig. 14.32). Urea is released into the general circulation and is excreted by the kidneys (Chapter 10).

Some of the twenty amino acids used for protein synthesis are **non-essential**. They can be produced from other amino acids in the liver by **transamination**. Liver cells contain a number of transaminase enzymes (Chapter 4) which transfer amino groups from amino acids to carboxylic acids, thus producing new amino acids. For example, glutamic-oxaloacetic transaminase (GOT) catalyses the following reaction;

FIG. 14.32 *The ornithine cycle*

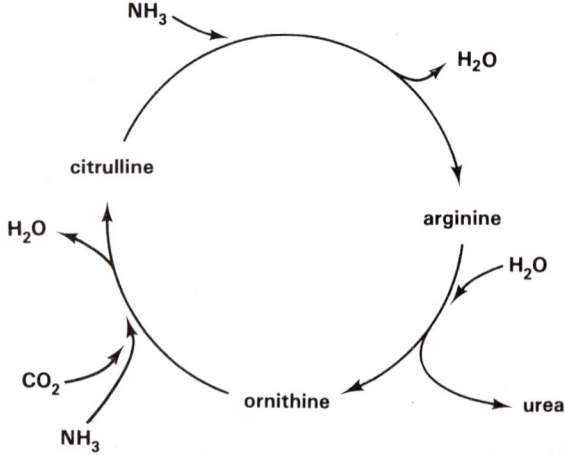

$$\text{glutamic} \underset{\text{acid}}{\text{}} + \underset{\text{acid}}{\text{oxaloacetic}} \xrightarrow{\text{GOT}} \underset{\text{acid}}{\text{aspartic}} + \underset{\text{acid}}{\text{oxaloglutaric}}$$

glutamic   oxaloacetic ——GOT——→ aspartic   oxaloglutaric
acid   acid   acid   acid
(amino group) (no amino group) (amino group) (no amino group)

Such enzymes enable the liver cells to make a wide range of amino acids even though the variety of amino acids in the diet is limited.

(iii) LIPID METABOLISM. Fats and oils absorbed from the gut can be converted to glycogen which is stored in the liver. Fat-soluble vitamins such as vitamins A and D are also stored in liver cells. **Cholesterol**, a sterol (Chapter 3) carried in the blood, is used in a variety of syntheses, particularly cell membrane production. The liver makes cholesterol when the dietary intake is inadequate. Excess cholesterol absorbed from the gut is eliminated in bile. Gross excesses of cholesterol however may be precipitated as gall stones in the bile ducts and gall bladder.

2. IRON METABOLISM

Some of the liver cells lining the sinusoids are part of the body's **reticulo-endothelial system**. The system removes particles of debris from the blood. The **Küpffer cells** are phagocytic liver cells which remove old red cells from the blood (Chapter 9). After old red cells have been engulfed the haemoglobin is split into two, an iron-globin complex and an iron-less haem group. The haem is then converted to **bilirubin**. Bilirubin combines with glucuronic acid to form the bile pigment **bilirubin diglucuronide** which is excreted in bile. In man about 7–8 g haemoglobin are removed from the circulation in this way each day.

FIG. 14.33 *Summary of haemoglobin breakdown in the liver and the fate of the products*

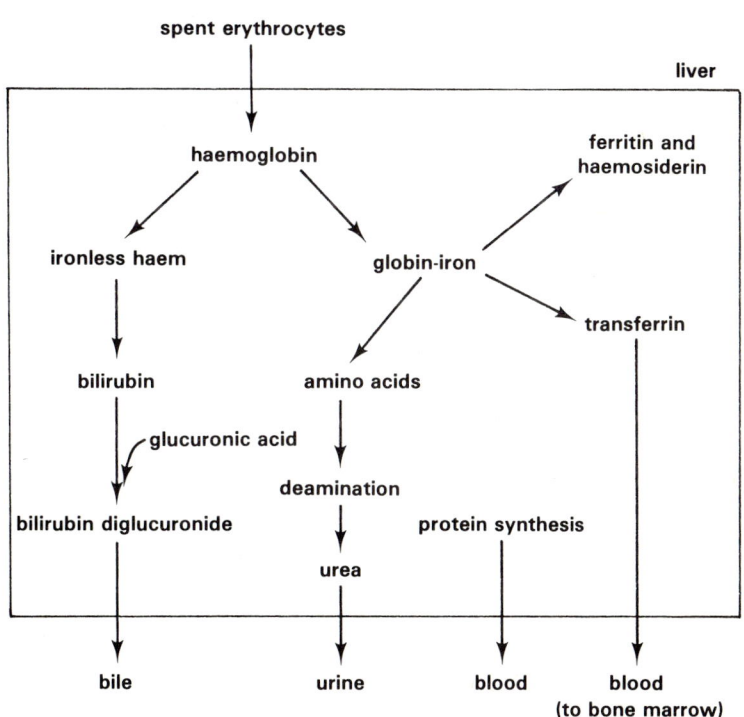

281

Any physical obstruction, such as **gall stones** in the bile ducts, or an increased breakdown of red cells results in an increase in the amount of bilirubin in blood. The skin takes on a yellow colour and the condition is called **jaundice**.

The iron-globin complex is metabolised further in the liver. The globin is broken down into amino acids which are used for protein synthesis. The iron is retained for the production of fresh haemoglobin. Haemoglobin synthesis and red cell production, erythropoiesis, occur in the liver of the foetal mammal. In the adult these processes are restricted to the red bone marrow (Chapter 9). Iron released from broken-down haemoglobin becomes attached to a plasma protein called transferrin and is transported in the blood to the bone marrow. Excess iron is stored in the liver cells as **ferritin** and **haemosiderin**.

The total iron content of the adult human body is normally between 3 and 5 g. Of this, about 1·5–5 g is in haemoglobin and about 1–1·5 g is stored in the liver. Pregnancy and lactation can lower a woman's total iron by as much as 20 per cent. This is why iron-containing tablets are given to pregnant women. Figure 14.33 summarises some of the factors involved in iron metabolism.

The liver helps erythropoiesis by producing a substance called the **haematinic principle**. Vitamin B_{12} is necessary for the production of the haematinic principle and deficiency of vitamin B_{12} causes pernicious anaemia (Chapter 9).

3. DETOXIFICATION

Many chemical substances which pass through the liver in the blood are modified by the liver cells. The substances include a variety of hormones. Much of the insulin from the pancreas is broken down by enzymes in the liver. Many sex hormones are also **inactivated** in the liver and **excreted** in the bile or released into the blood and excreted by the kidneys. Some chemicals are **destroyed** by liver cells, others are combined with various substances to render them less toxic. For example, benzenecarboxylic acid, a commonly used food preservative, is joined to the amino acid glycine to form N-benzoylglycine which is excreted in the urine. Certain chemicals such as tetrachloromethane, trichloromethane and ethanol damage the liver cells.

Within limits then, the liver acts as a filter, removing toxic substances from the blood, making them less harmful and preparing them for excretion. The liver performs a variety of other functions which are described elsewhere in the book. Notable among them is the contribution the liver makes to heat production in the body (Chapter 17).

Chapter 15

Nervous control and co-ordination in mammals

Changes constantly take place inside and outside the body. Mammals respond to the changes and their internal environment remains fairly constant. The ability to respond in this way is called **homeostasis**. An important aspect of homeostasis is the **co-ordination** of responses. A co-ordinating system contains several components. There are means of **detecting stimuli**, means of **transmitting information** about the stimuli and means of **responding to the stimuli**. The system also has a means of directing information along the most appropriate of many possible channels. Co-ordination also involves **feedback** which ensures that the degree of response is related to the intensity and direction of stimuli (Fig. 15.1). Note the similarity between Figure 15.1 and Figure 8.12 which summarises the mechanism controlling ventilation of the lungs. Other comparable examples are described elsewhere in the book such as control of the heart (Chapter 9) and body temperature (Chapter 17).

Mammals have two main co-ordinating systems, the **nervous system** and the **endocrine system**. Each system has different means of detecting and transmitting information. However, the two systems often work together (Fig. 15.2). The endocrine system is the subject of Chapter 18.

15.1 Electrochemical basis of nervous activity

The mammalian **nervous system** is an organised collection of numerous nerve cells called **neurons** (Fig. 15.3). Neurons consist of a bulbous cell body called the **soma** which contains a nucleus. Branched hair-like structures called **dendrites** which receive information from other neurons project from

FIG. 15.1 *Basic components of a feed-back control mechanism*

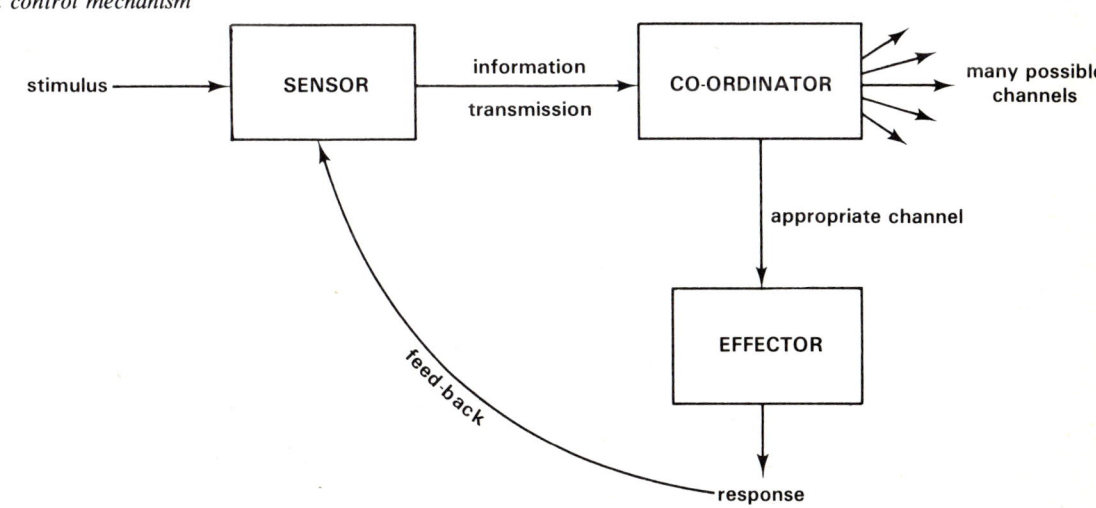

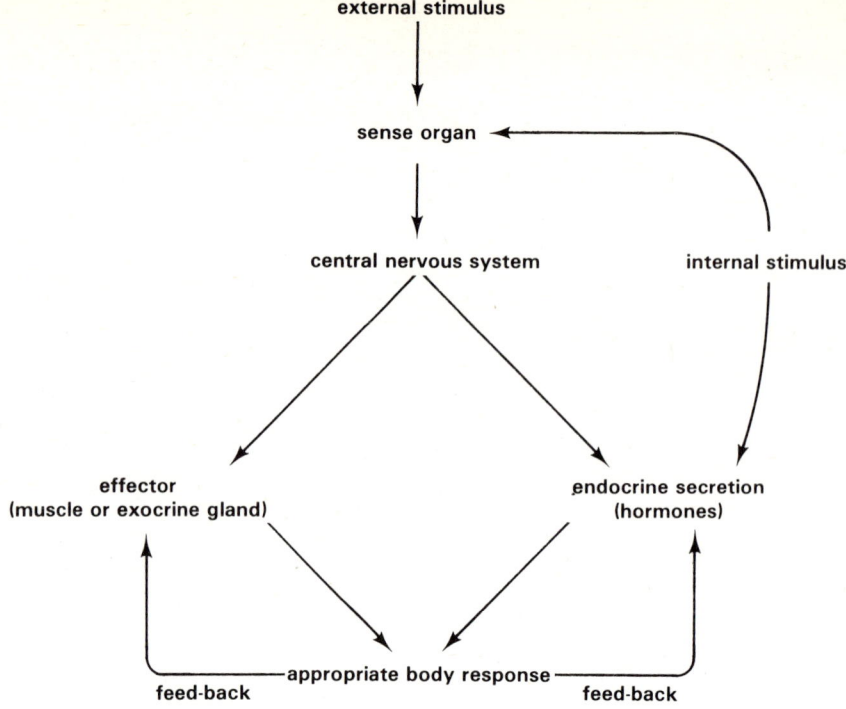

FIG. 15.2 *Interrelationships between the nervous and endocrine systems in mammals*

the soma. Some neurons have a long membrane-enclosed cytoplasmic thread called an **axon**. At their ends the axons divide into many branches which nearly touch the dendrites of adjacent neurons. Intricate pathways thus exist along which information can be transmitted very rapidly from one part of the body to another. The main function of neurons is to transmit nerve impulses over relatively long distances.

15.1.1 The resting potential

Analysis of the cytoplasm of the axon shows striking differences between its chemical composition compared with the surrounding tissue fluid (Table 15.1). The concentration of sodium ions $[Na^+]$ in the tissue fluid greatly exceeds that in the **axoplasm**. Conversely the concentration of potassium ions $[K^+]$ in the tissue fluid is far less than that in the axon. This state of affairs cannot be due to diffusion. Diffusion would result in equal concentrations of the ions on either side of the membrane. **Active transport** involving a sodium-potassium exchange pump is responsible for the non-random distribution of sodium and potassium ions. **Carrier molecules** in the axon membrane transport sodium ions out of, and potassium ions into, the axon against concentration gradients (Fig. 15.4). The pumping mechanism uses respiratory energy and is inhibited by metabolic poisons.

Table 15.1 Concentrations of various ions in cells and tissue fluids

ION	CELLULAR FLUID /m mol dm^{-3}	TISSUE FLUID /m mol dm^{-3}
sodium	16	140
potassium	100	4·4
chloride	4	103
hydrogencarbonate	8	27

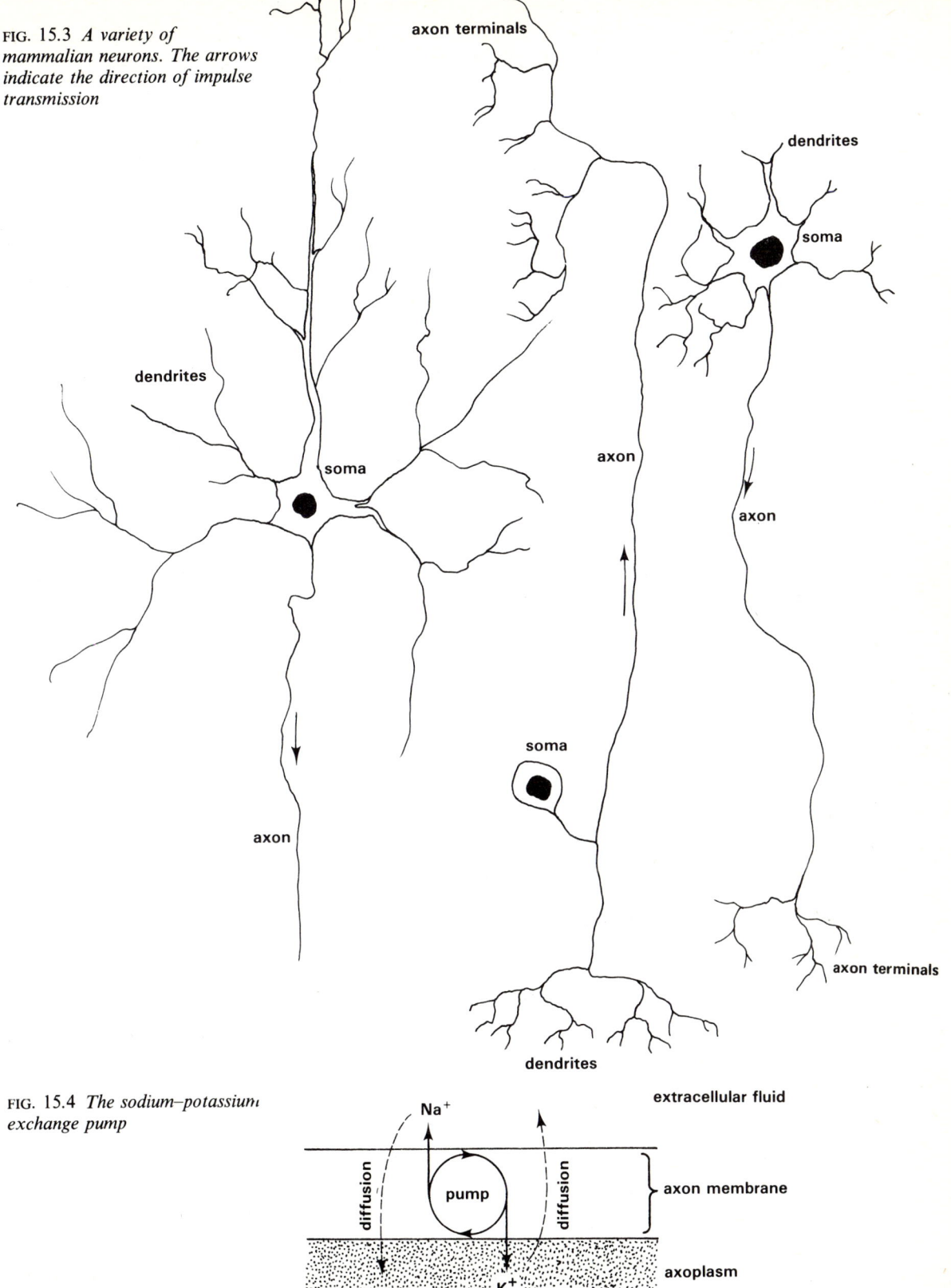

FIG. 15.3 *A variety of mammalian neurons. The arrows indicate the direction of impulse transmission*

axon terminals

dendrites

soma

dendrites

soma

axon

axon

axon

soma

axon

soma

axon terminals

dendrites

FIG. 15.4 *The sodium–potassium exchange pump*

extracellular fluid

Na$^+$

diffusion

pump

diffusion

axon membrane

K$^+$

axoplasm

The axon membrane is about twenty-five times more permeable to potassium ions than to sodium ions. Consequently potassium ions can leave the axoplasm more rapidly than sodium ions can enter by diffusion. The net effect is a greater total concentration of both ions outside the axon:

$$[Na^+] + [K^+] > [Na^+] + [K^+]$$
$$\text{out} \qquad \text{out} \qquad \text{in} \qquad \text{in}$$

Since both ions are cations the outer surface of the membrane has a net positive electric charge relative to the inner surface. In other words an electrical potential difference, the **transmembrane potential**, exists across the membrane. The membrane is **polarised**. Microelectrodes can be inserted into an axon and connected up with extracellular electrodes to a sensitive recording device such as a cathode ray oscilloscope (Fig. 15.5). A transmembrane potential of about -60 mV is recorded when neurons are not transmitting nerve impulses. This is called the **resting potential** and is the basis of the ability of neurons to transmit impulses. Even minor changes in the resting potential trigger off changes which may produce an electrochemical force. The force is propagated along the axon as a **nerve impulse**.

FIG. 15.5 *Measuring the transmembrane potential of an axon*

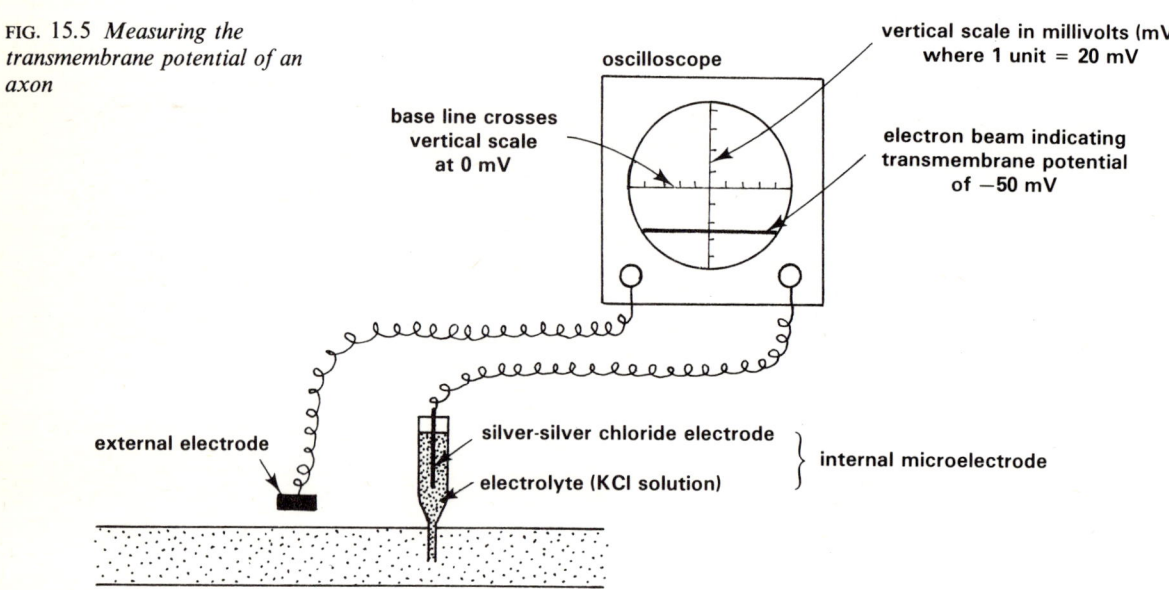

vertical scale in millivolts (mV) where 1 unit = 20 mV

oscilloscope

base line crosses vertical scale at 0 mV

electron beam indicating transmembrane potential of −50 mV

external electrode

silver-silver chloride electrode

electrolyte (KCl solution)

internal microelectrode

axon

15.1.2 The action potential

Resting neurons are affected by a variety of stimuli including heat, stretching the membrane, electric currents and chemicals. Such stimuli change the permeability of the membrane at the point of application resulting in an influx of sodium ions. The membrane becomes **depolarised** (Fig. 15.6) and the transmembrane potential becomes progressively less. A point may be reached when the cation content of the axoplasm equals that in the tissue fluid. The electrical potential difference across the membrane is then zero. If the stimulus is sufficiently intense the polarity of the membrane is reversed. When the change of charge reaches a threshold value a nerve impulse results and the new transmembrane potential is called an **action potential**. If on stimulation the electrical change is not great enough the threshold is not reached and an impulse is not produced. This is the **all-or-none effect** (Fig. 15.7).

FIG. 15.6 *Depolarisation of an axon membrane following stimulation*

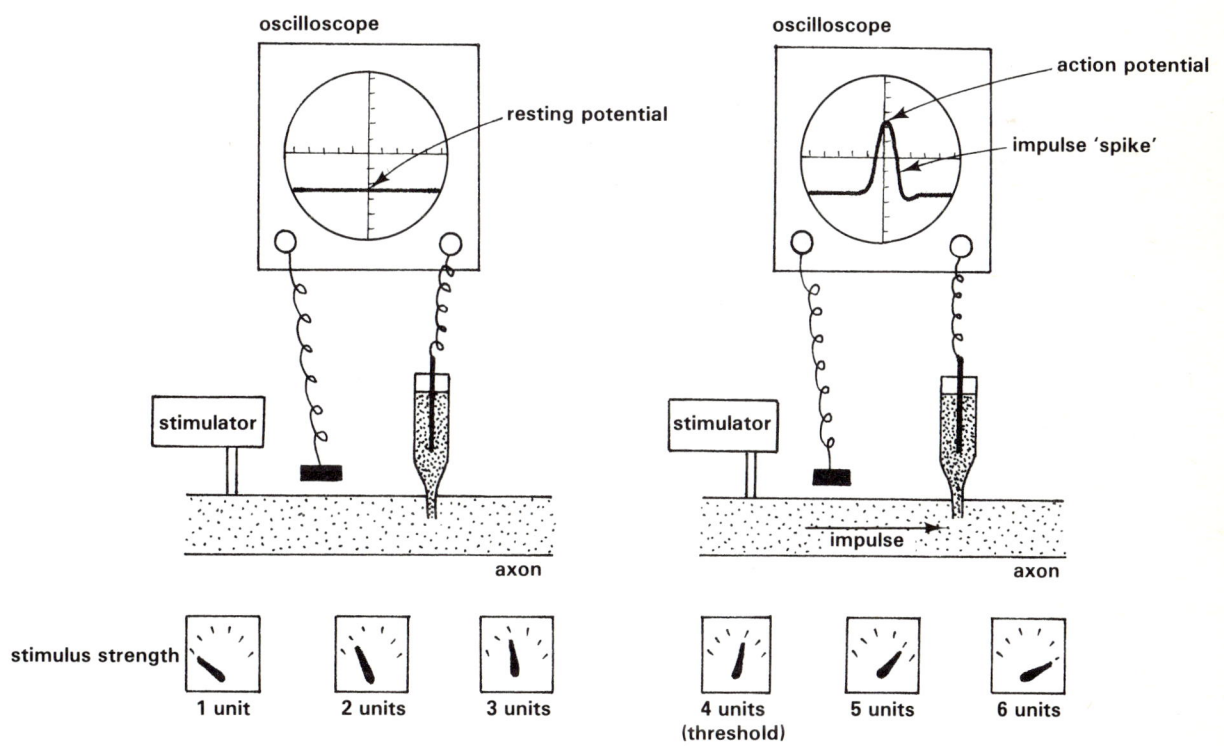

FIG. 15.7 *The all-or-none effect. Raising the stimulus strength from zero does not produce an impulse in the axon until a stimulus threshold is reached. Further increase in stimulus strength results in the same action potential across the axon membrane*

15.1.3 The nerve impulse

The production of an action potential causes the flow of a small electric current in the axoplasm and tissue fluid. The current crosses the axon membrane on either side of the point at which the stimulus is applied causing membrane depolarisation. In turn depolarisation spreads along the axon membrane and in this way a nerve impulse is **propagated** along the axon (Fig. 15.8). Recordings made using giant neurons of squids show that the membrane potential during passage of an impulse is reversed from a resting value of -60 mV to a peak of about $+45$ mV (Fig. 15.9).

(a)   The resting state

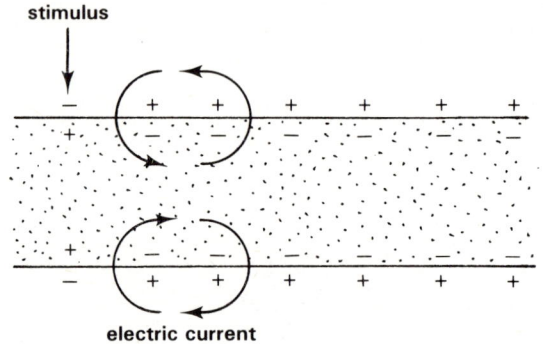

(b)   Depolarisation of the membrane causing the flow of electric current

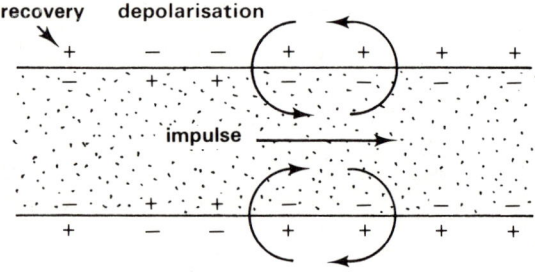

(c)   Electric current stimulates the membrane nearby causing further depolarisation

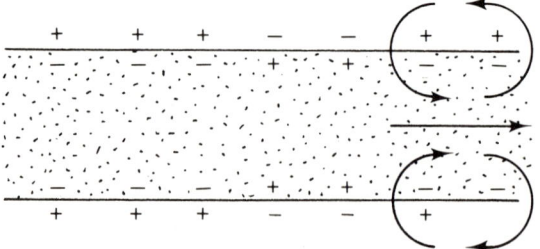

(d)   The process repeats itself along the axon

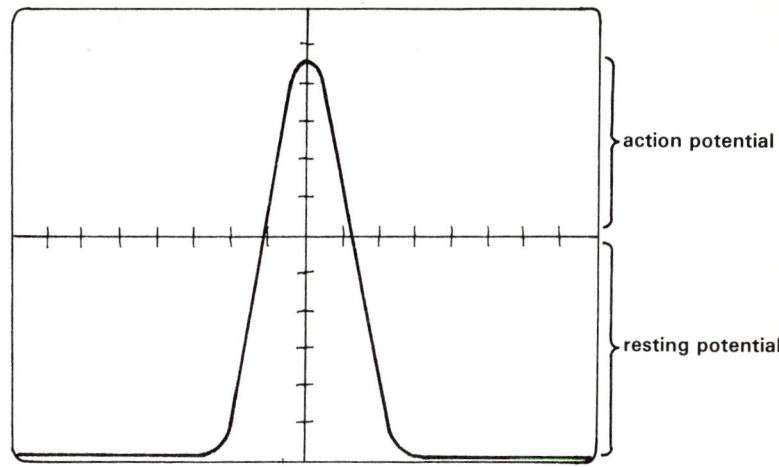

FIG. 15.9 *Oscilloscope display of the action potential produced during the propagation of an impulse along a giant axon from a squid. The vertical scale is graduated in millivolts (mV) where 1 unit = 10 mV. The horizontal scale is graduated in milliseconds (ms) where 1 unit = 0·25 ms*

15.1.4 Membrane recovery

During the brief period of membrane depolarisation, which is only one millisecond, an axon cannot conduct a second impulse since impulse transmission depends on disturbance of the resting potential. Consequently a period of non-conductability called the **refractory period** follows each impulse (Fig. 15.10). During this time the membrane recovers its resting potential. Depolarisation is accompanied by diffusion of sodium ions from the tissue fluid into the axoplasm. In the refractory period potassium ions diffuse from the axoplasm into the tissue fluid. The sodium–potassium exchange pump now operates to restore the original ionic state of the neuron and the resting potential is recovered (Fig. 15.11).

FIG. 15.10 *Oscilloscope traces showing refractory period*

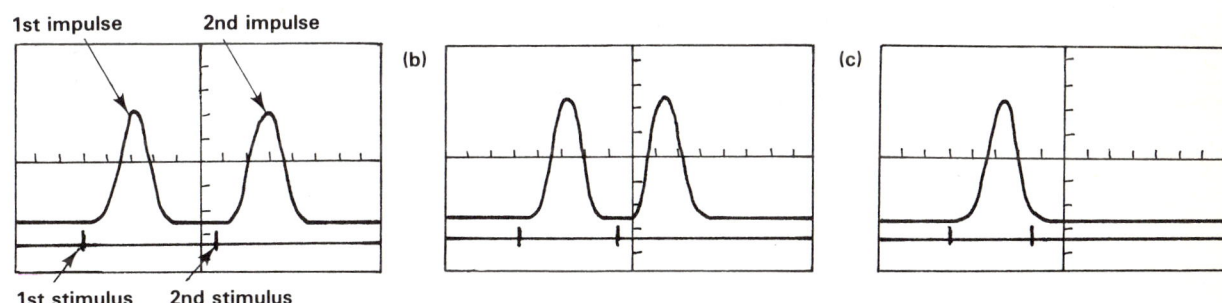

15.1.5 Rate of impulse conduction

Peripheral neurons are surrounded by **Schwann cells** which may respond to the proximity of neurons by growing around their axons. The axon is then wrapped in several membranous layers called the **myelin sheath**. The myelin sheath limits contact between the axon membrane and tissue fluid to only a few sites called **nodes of Ranvier** (Fig. 15.12). Because less of the axon membrane is used in impulse transmission the rate of impulse conduction in myelinated axons is greater than in unmyelinated axons (Fig. 15.13).

289

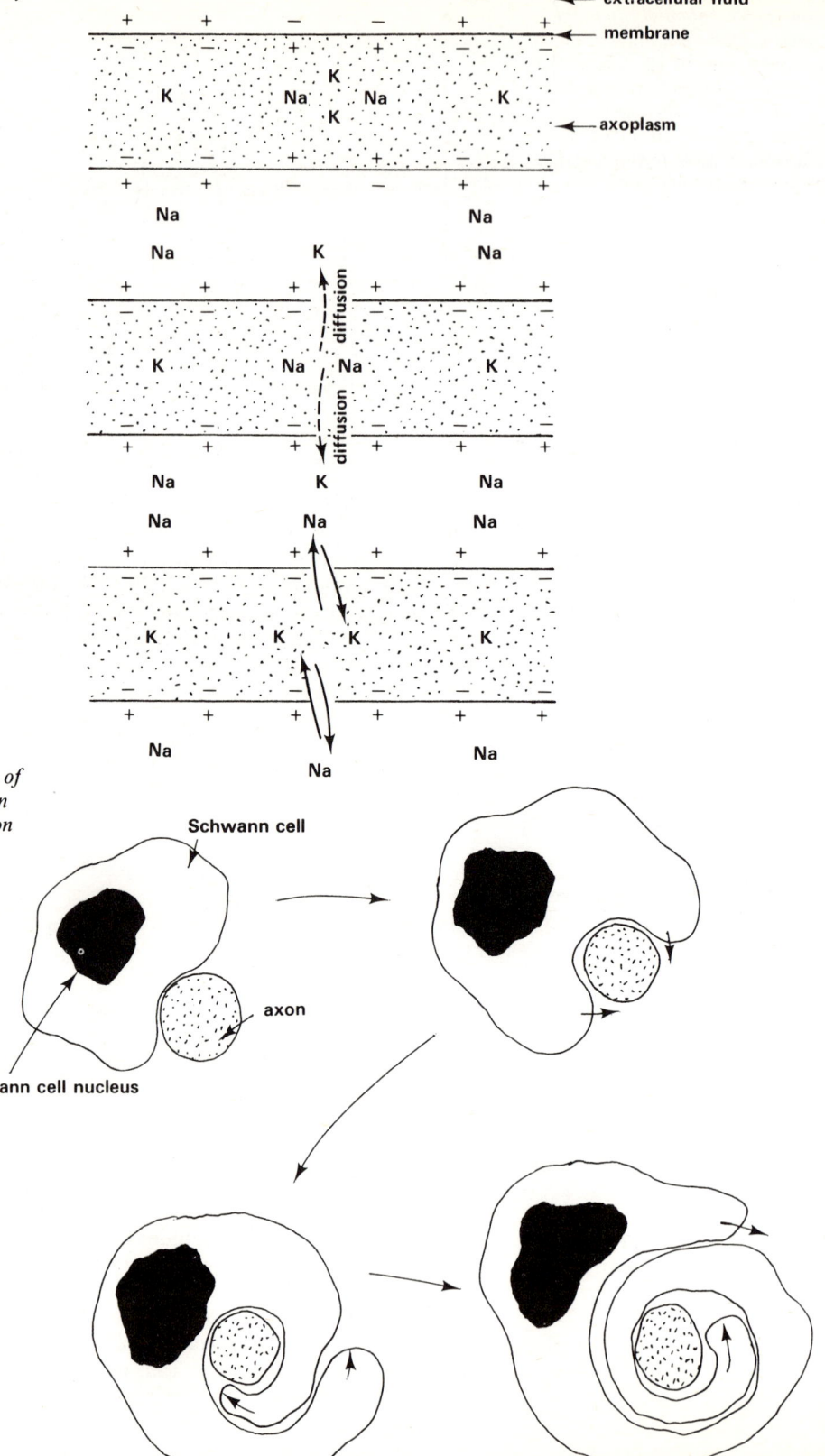

FIG. 15.11 *Membrane recovery*

FIG. 15.12 (a) *Development of myelin sheath from Schwann cell seen in transverse section*

FIG. 15.12 (b)
*Electronmicrograph of a
transverse section of a
myelinated axon, × 25 000*

nucleus of Schwann cell

myelin sheath

axon

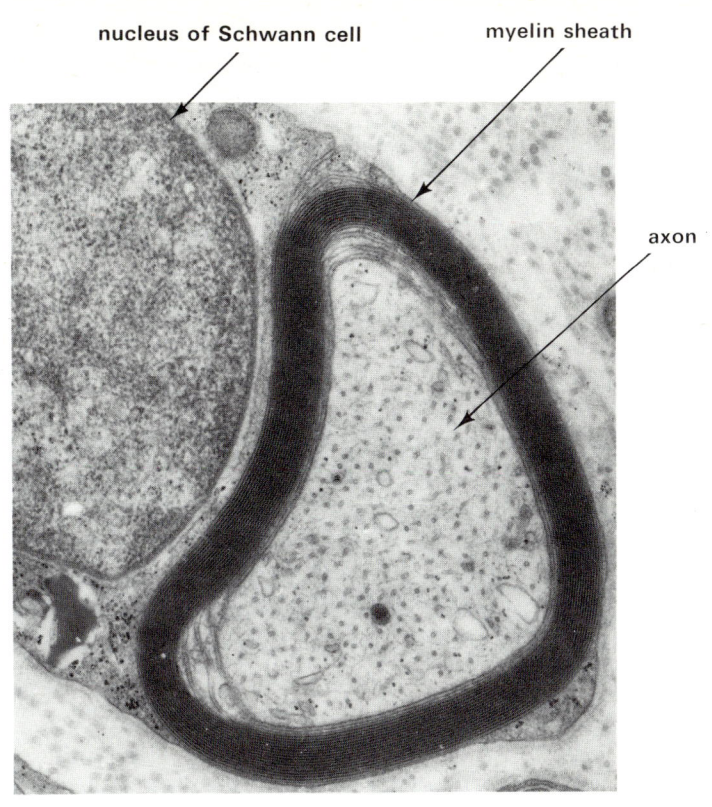

FIG. 15.12 (c) *Photomicrograph
of some human myelinated
neurons showing the nodes of
Ranvier, × 400*

node of Ranvier

myelin sheath

axon

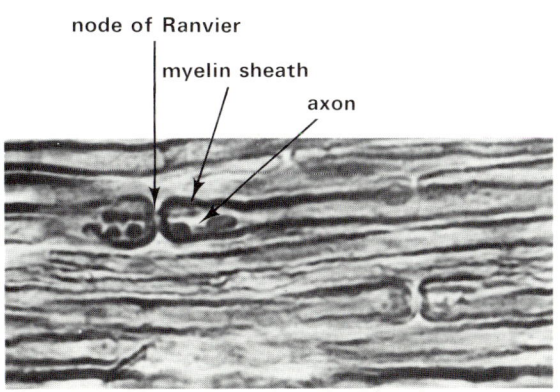

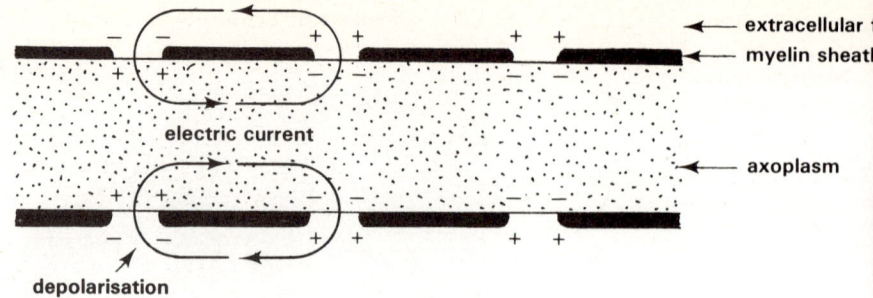

FIG. 15.13 *Electrochemical events at the nodes of Ranvier during the passage of an impulse along a myelinated axon. Compare with Fig. 15.8*

extracellular f▮
myelin sheath▮

electric current

axoplasm

depolarisation

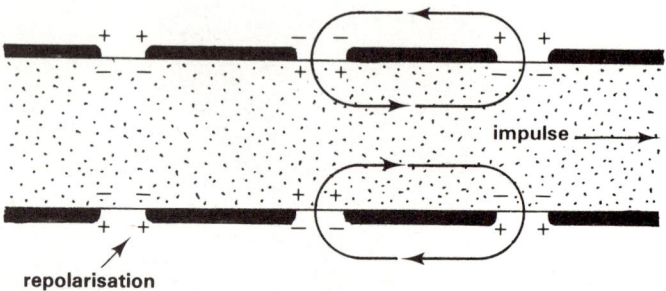

impulse ⟶

repolarisation

Another factor affecting the rate of impulse transmission is the diameter of the axon. The relationship for large myelinated mammalian nerve fibres is approximately:

$$v \propto 7 \times d$$

where v is the velocity of impulse conduction in m per sec and d is the fibre diameter in μm.

Thus, other factors excepted, a doubling of fibre diameter increases impulse velocity by about 14 times. Table 15.2 shows the conduction velocities of neurons from different animals. It was Helmholz in 1850 who first measured the velocity of a nerve impulse. He obtained a value of about 25 m per sec in a frog's nerve. Prior to his experiments impulses were thought to travel at about the speed of light!

Table 15.2 Speeds of impulse conduction in the nerves of various animals (from Bendal 1969)

TISSUE	TEMPERATURE /°C	FIBRE DIAMETER /μm	IMPULSE VELOCITY /m s^{-1}
crab nerve	20	30	5
squid giant axon	20	500	25
cat nerve (unmyelinated)	38	0·3–1·5	0·7–2·3
cat nerve (myelinated)	38	2–20	10–100
prawn nerve (myelinated)	20	35	20
frog nerve (myelinated)	24	3–16	6–32

15.1.6 Synaptic transmission

The junctions between neurons are called **synapses**. They are found where the branches of axons lie close to the dendrites of other neurons (Fig. 15.14). Electron microscopy has revealed a narrow gap about 20 nm across at the synapse. The gap has to be crossed when impulses are transmitted from neuron to neuron (Fig. 15.15). The transmission of impulses between neurons depends on **chemical transmitters** synthesised in the axon terminals and stored there in numerous small membranous vesicles. The arrival of a

FIG. 15.14 *Many axons terminate as synapses on the dendrites and soma of a neuron (light microscope view)*

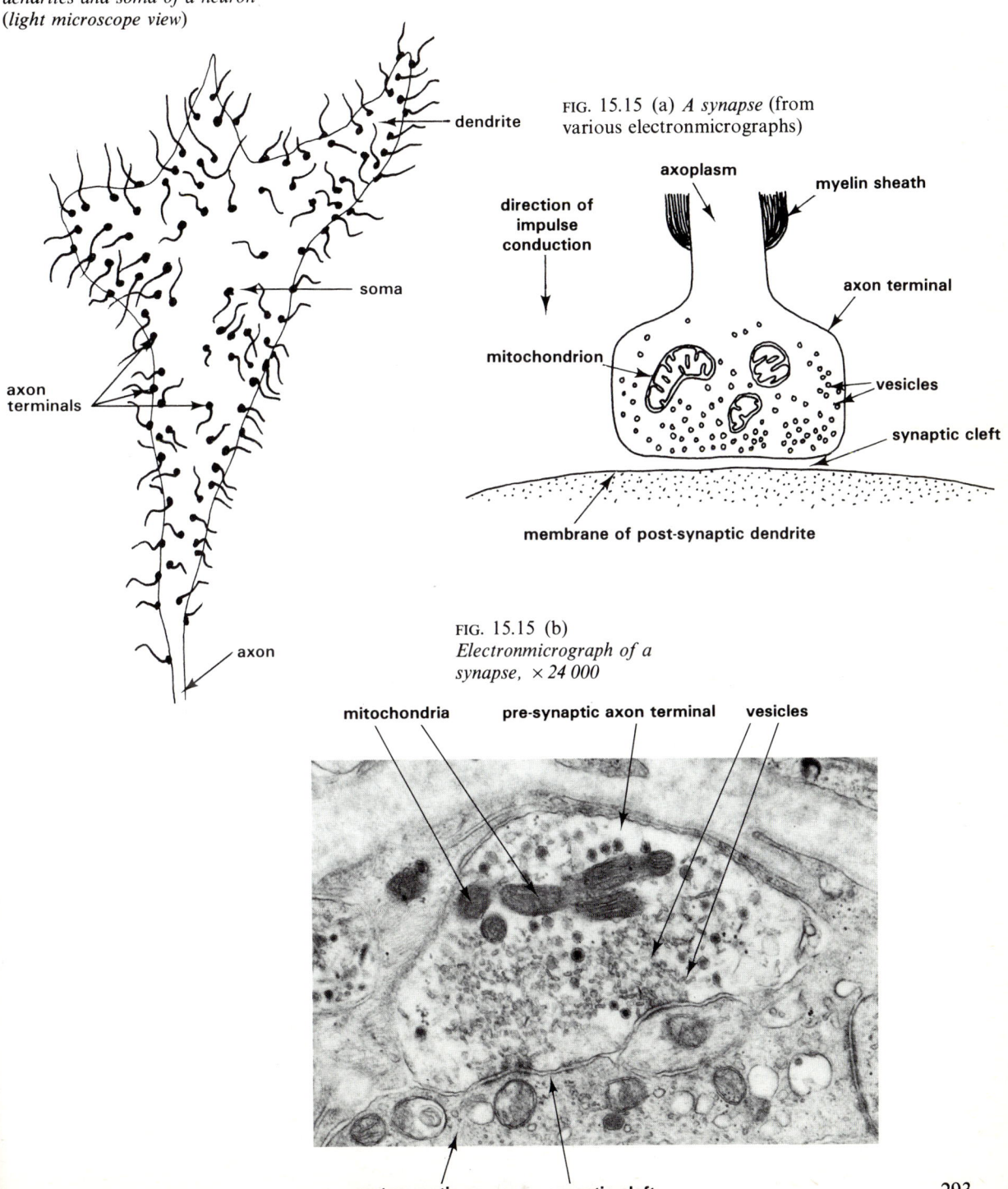

FIG. 15.15 (a) *A synapse* (from various electronmicrographs)

dendrite

soma

axon terminals

axon

direction of impulse conduction

axoplasm

myelin sheath

axon terminal

mitochondrion

vesicles

synaptic cleft

membrane of post-synaptic dendrite

FIG. 15.15 (b) *Electronmicrograph of a synapse, × 24 000*

mitochondria

pre-synaptic axon terminal

vesicles

post-synaptic neuron

synaptic cleft

nerve impulse at an axon terminal triggers the release of small quantities of transmitter into the **synaptic cleft.** The transmitter becomes attached to receptive sites on the **post-synaptic membrane** of the dendrite of the next neuron. As a result the post-synaptic membrane is depolarised and an impulse may be transmitted along the second neuron (Fig. 15.16).

FIG. 15.16 *Synaptic transmission*

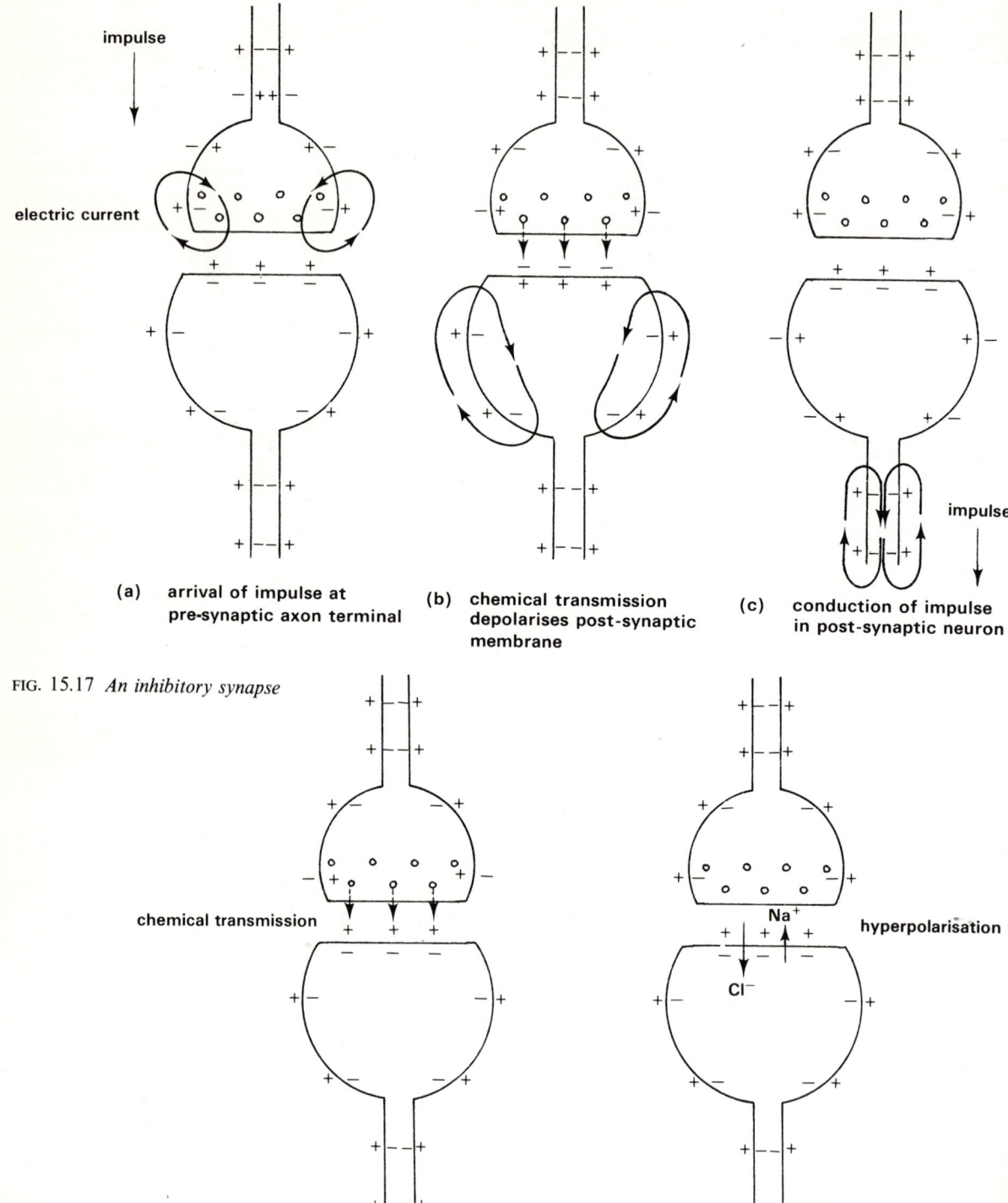

(a) arrival of impulse at pre-synaptic axon terminal

(b) chemical transmission depolarises post-synaptic membrane

(c) conduction of impulse in post-synaptic neuron

FIG. 15.17 *An inhibitory synapse*

Excitation of the post-synaptic dendrite does not always occur. Impulses can be inhibited by some transmitters (Fig. 15.17). **Inhibitory synapses** are important in the overall control of the nervous system (section 15.2.1).

The transmitters released from the pre-synaptic terminals are quickly destroyed by enzymes in the tissue fluid. Consequently their effect is only temporary. One transmitter widely found in the mammalian nervous system is **acetylcholine**. After its secretion acetylcholine is broken down by the enzyme **cholinesterase** into the inactive end products choline and ethanoic acid. The end products are reabsorbed by the axon terminals and remade into acetylcholine using respiratory energy provided by mitochondria (Fig. 15.18). A general effect of transmitter destruction is to prevent the merging of successive impulses passing from neuron to neuron. The frequency of impulses in a sequence of neurons is determined primarily by the intensity of stimulus at the beginning of the sequence. If the frequency was amplified by uncontrolled synaptic transmission the response might be inappropriately excessive.

FIG. 15.18 *Cycle of release, breakdown, reabsorption and resynthesis of synaptic transmitter*

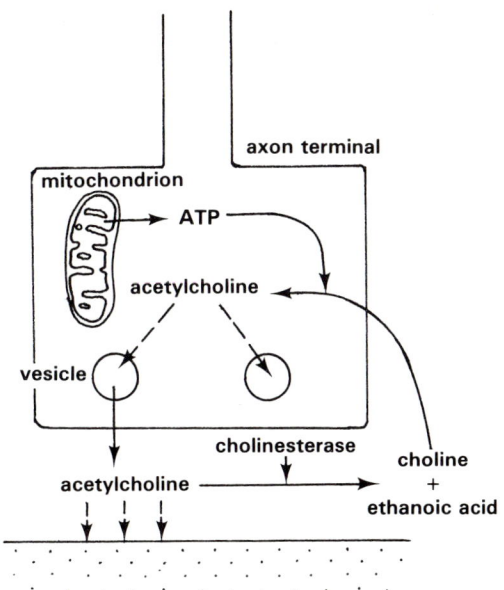

FIG. 15.19 (a) *Photomicrograph showing the junctions between motor nerve cell endings and striped muscle fibres,* × 450

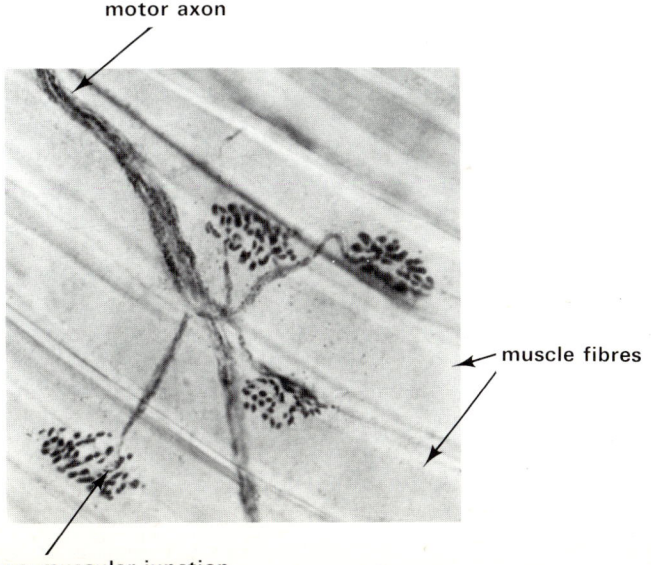

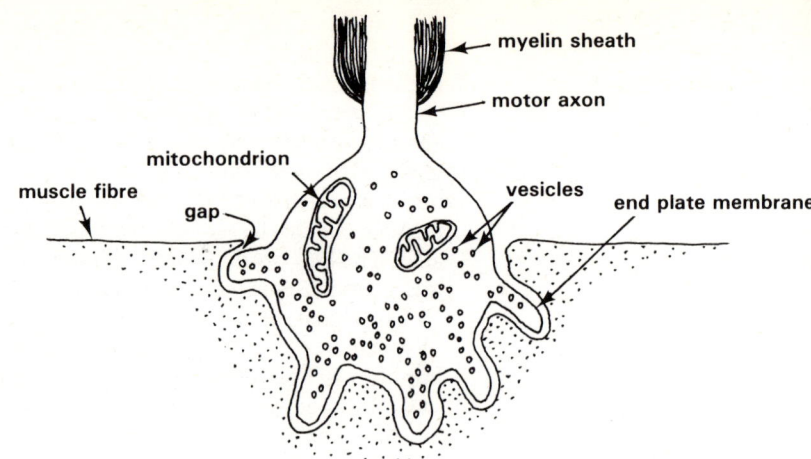

FIG. 15.19 (b) *Neuromuscular junction as seen with the aid of an electron microscope.* Compare with Fig. 15.15

myelin sheath

motor axon

mitochondrion

muscle fibre

gap

vesicles

end plate membrane

15.1.7 Neuroeffector junctions

The junctions between neurons and effectors such as muscles and glands are very similar to synapses (Fig. 15.19). The mechanism of impulse transmission from neurons to effectors also involves the secretion of transmitters. How does chemical transmission from a neuron bring about a response in an effector?

FIG. 15.20 (a) *The structure of striped muscle*

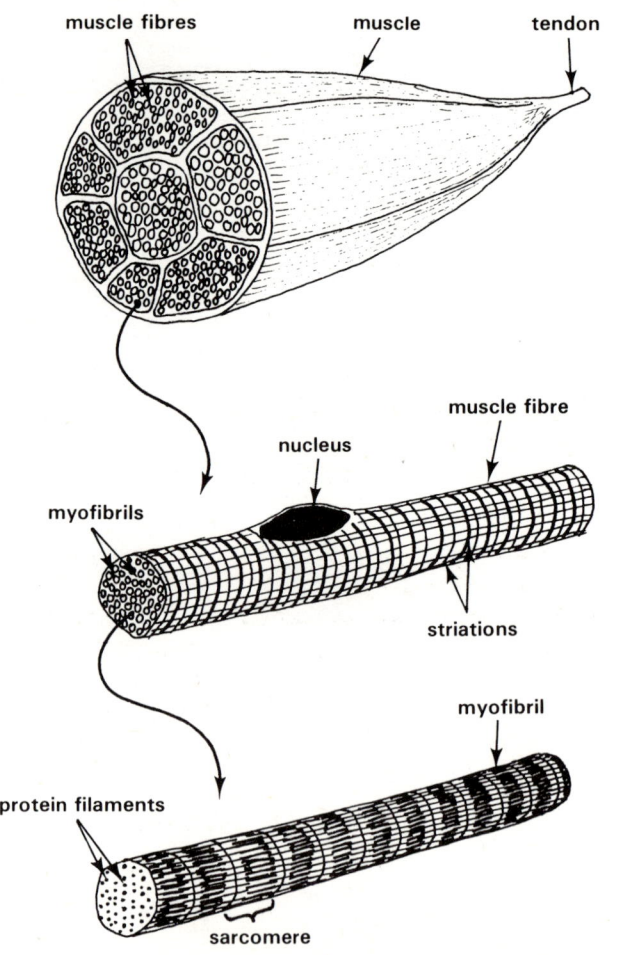

muscle fibres

muscle

tendon

nucleus

muscle fibre

myofibrils

striations

protein filaments

myofibril

sarcomere

FIG. 15.20 (b) *Photomicrograph of some striped muscle fibres teased apart,* ×600

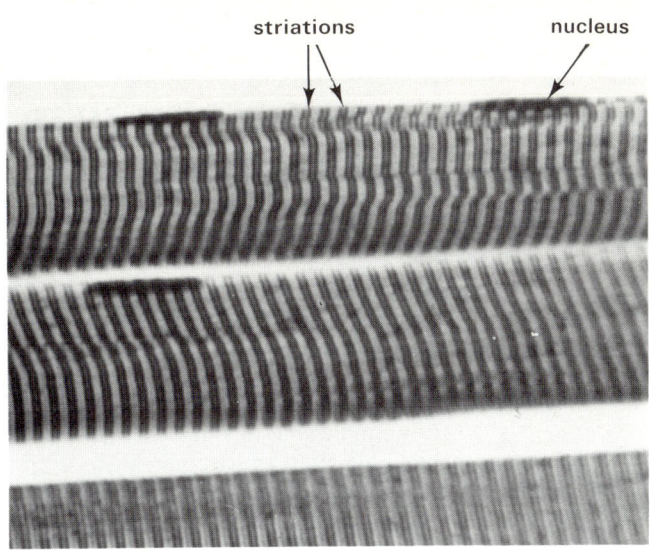

15.1.8 Muscle and contraction

Mammals have three kinds of muscle. The muscle used to move the skeleton is called **skeletal (striped) muscle**. The walls of the gut, urinogenital system and arteries contain **visceral (smooth) muscle**. Most of the heart is made of **cardiac muscle**.

Striped muscle consists of **fibres** which are made of many **myofibrils**. Each myofibril is divided by cross-partitions called **Z-lines** into numerous compartments called **sarcomeres** (Fig. 15.20). Chemical analysis and electron microscope studies show that the myofibrils have parallel strands. The arrangement of the strands, composed of the proteins **actin** and **myosin**, explains the striped appearance of the muscle when viewed with a light microscope. Compare Figures 15.20(a) and (b). Strands of actin project into the sarcomeres from the Z-lines. Between the actin strands, thicker filaments of myosin are held in position by molecular bonds.

FIG. 15.20 (c) *Electronmicrograph of a myofibril showing a single sarcomere,* ×65 000

mitochondria   glycogen granules

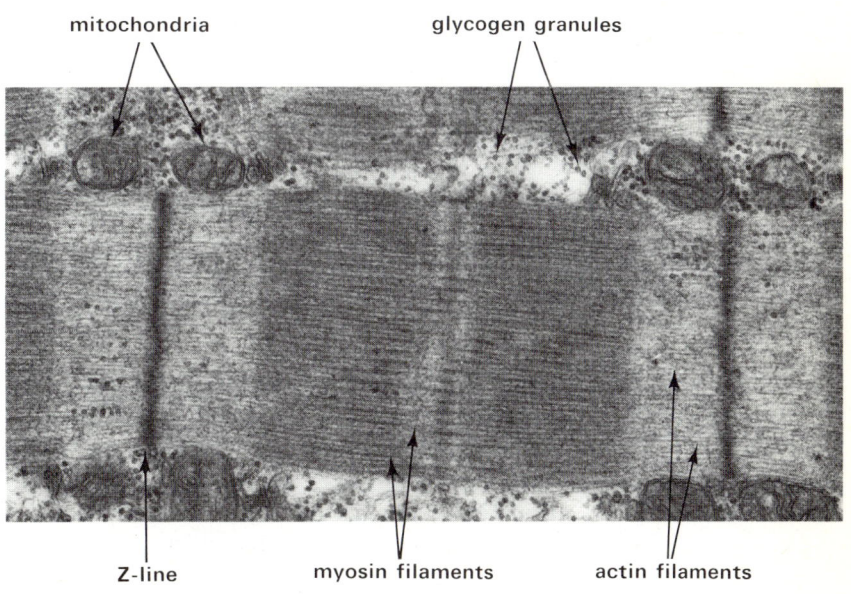

Z-line   myosin filaments   actin filaments

Isolated actin–myosin filaments contract when ATP is applied to them. As ATP is always present in living cells then muscles must contain an inhibitor, preventing continuous contraction. The inhibitor is neutralised by calcium ions. In relaxed muscle calcium ions and sodium ions are pumped out of the muscle fibres into the tissue fluid. The membrane of the fibres is thus polarised. Depolarisation of the membrane occurs when the muscle is stimulated and calcium ions enter the fibre. Here the calcium ions neutralise the inhibitor. This enables ATP to provide the energy for actin and myosin to interact resulting in muscle contraction.

Impulses spread rapidly all over the muscle in much the same way as nerve impulses are transmitted. Consequently stimulation of a muscle fibre by a nerve brings about impulse **propagation** in the muscle which causes contraction of the whole muscle. Using energy from ATP the bonds between actin and myosin break temporarily. The bonds reform further along the actin threads nearer each Z-line. The Z-lines of adjacent sarcomeres are thus pulled closer together, since actin and myosin do not stretch. In this way the sarcomeres shorten (Fig. 15.21). Because the stimulus to contraction spreads rapidly over the entire muscle all the myofibrils contract at once.

FIG. 15.21 *Mechanism of sarcomere action*

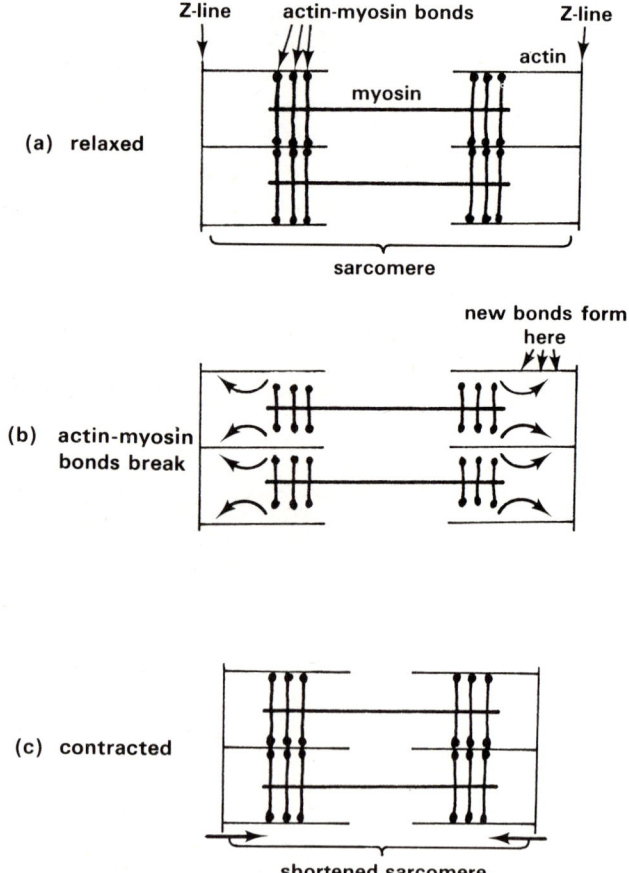

(a) relaxed

(b) actin-myosin bonds break

(c) contracted

Muscle contraction requires a lot of free energy. It is therefore not surprising to find that muscles are well equipped to produce energy. Between the muscle fibres are numerous **mitochondria**. **Glycogen** is stored in muscle as a source of glucose, the main substrate used to provide free energy in respiration (Chapter 5). Muscle also contains a red respiratory pigment, **myoglobin**, which reacts with oxygen in much the same way as does haemoglobin, the pigment in blood. Oxygenated myoglobin acts as a reserve of oxygen in muscle (Chapter 8). Also peculiar to muscle is a substance called **creatine** which can become phosphorylated by transfer of high-energy phosphate groups from ATP. The ADP so formed is converted back to ATP in respiration:

$$\text{respiratory substrate} \Big) \Big(\begin{array}{c} \text{ADP} \\ \text{ATP} \end{array} \Big) \Big(\begin{array}{c} \text{phosphocreatine} \\ \text{creatine} \end{array}$$

$$\text{oxidised respiratory substrate}$$

When ATP is used to provide free energy for muscular contraction, phosphocreatine replenishes the ATP supply in muscle fibres:

$$\text{phosphocreatine} \Big) \Big(\begin{array}{c} \text{ADP} \\ \text{ATP} \end{array} \Big) \Big(\begin{array}{c} \text{contracted actin-myosin} \\ \text{relaxed actin-myosin} \end{array}$$

$$\text{creatine}$$

Phosphocreatine therefore acts as an energy store in addition to ATP (Chapter 5).

FIG. 15.22 *Photomicrograph of some smooth muscle cells from the urinary bladder of a cat, × 1000*

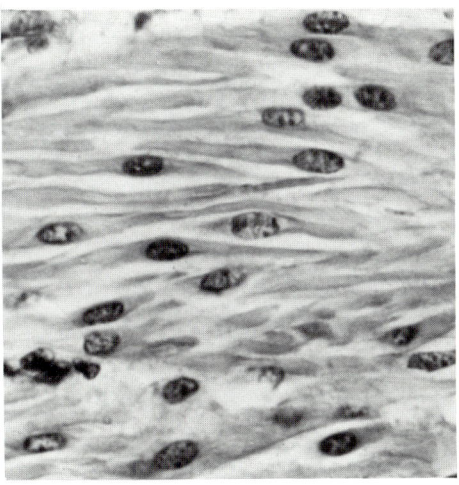

Smooth muscle is made of fibres capable of slow but sustained contraction (Fig. 15.22). The electrochemical properties of smooth muscle are different from those of striped muscle. Smooth muscle has a lower and less stable resting potential. Changes of the resting potential result in the rhythmic production of action potentials. Hence smooth muscle exhibits **rhythmic contractions** even without nervous stimulation. This is called **myogenic** contraction as opposed to **neurogenic** contraction which requires nervous stimulation. Nerve fibres are usually present in smooth muscle but they regulate activity rather than initiate it (Fig. 15.23).

Cardiac muscle is structurally very similar to striped muscle (Fig. 15.24).

FIG. 15.23 (a) *Experimental arrangement to measure the contraction of smooth muscle in mammalian gut. Change in length of the gut piece is recorded as a fluctuation on a pen recorder*

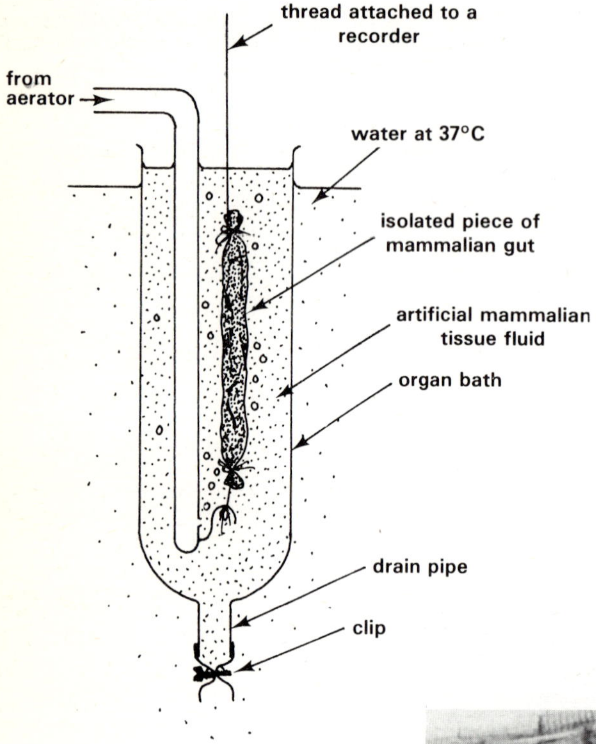

FIG. 15.23 (b) *Addition of adrenalin and acetylcholine to the isolated gut piece (Fig. 15.23 (a)) simulates the effects of the autonomic nerves on smooth muscle. Note the rhythmic myogenic contractions before the substances were added*

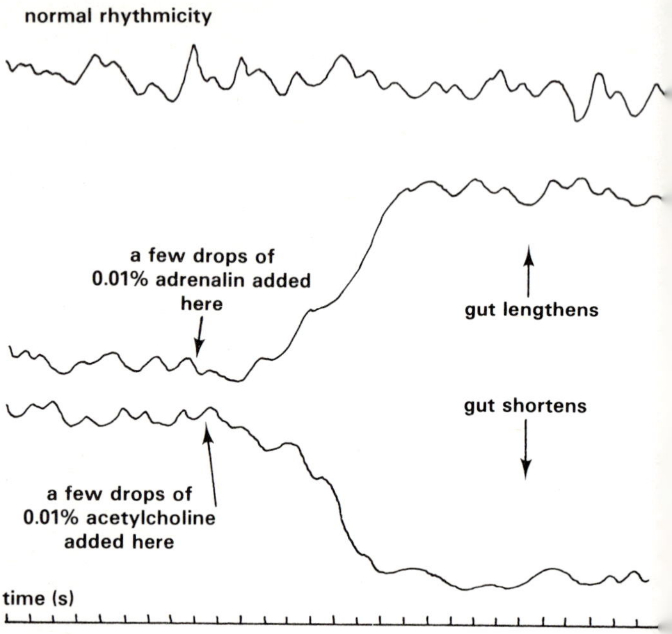

thread attached to a recorder

from aerator →

water at 37°C

isolated piece of mammalian gut

artificial mammalian tissue fluid

organ bath

drain pipe

clip

normal rhythmicity

a few drops of 0.01% adrenalin added here

gut lengthens

a few drops of 0.01% acetylcholine added here

gut shortens

time (s)

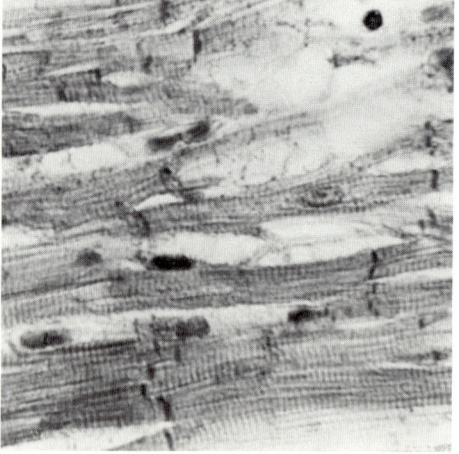

FIG. 15.24 *Photomicrograph of some human cardiac muscle fibres × 450*

However, like smooth muscle, cardiac muscle exhibits **rhythmic myogenic contractions**. The cross-connections between adjacent fibres of cardiac muscle enable contraction to spread through the heart wall. Cardiac muscle is also very conductive. These properties account for the ability of cardiac muscle to contract and relax smoothly and continuously when the heart beats (Chapter 9).

15.2 The mammalian nervous system

FIG. 15.25 *Diagram of the main divisions of the human nervous system*

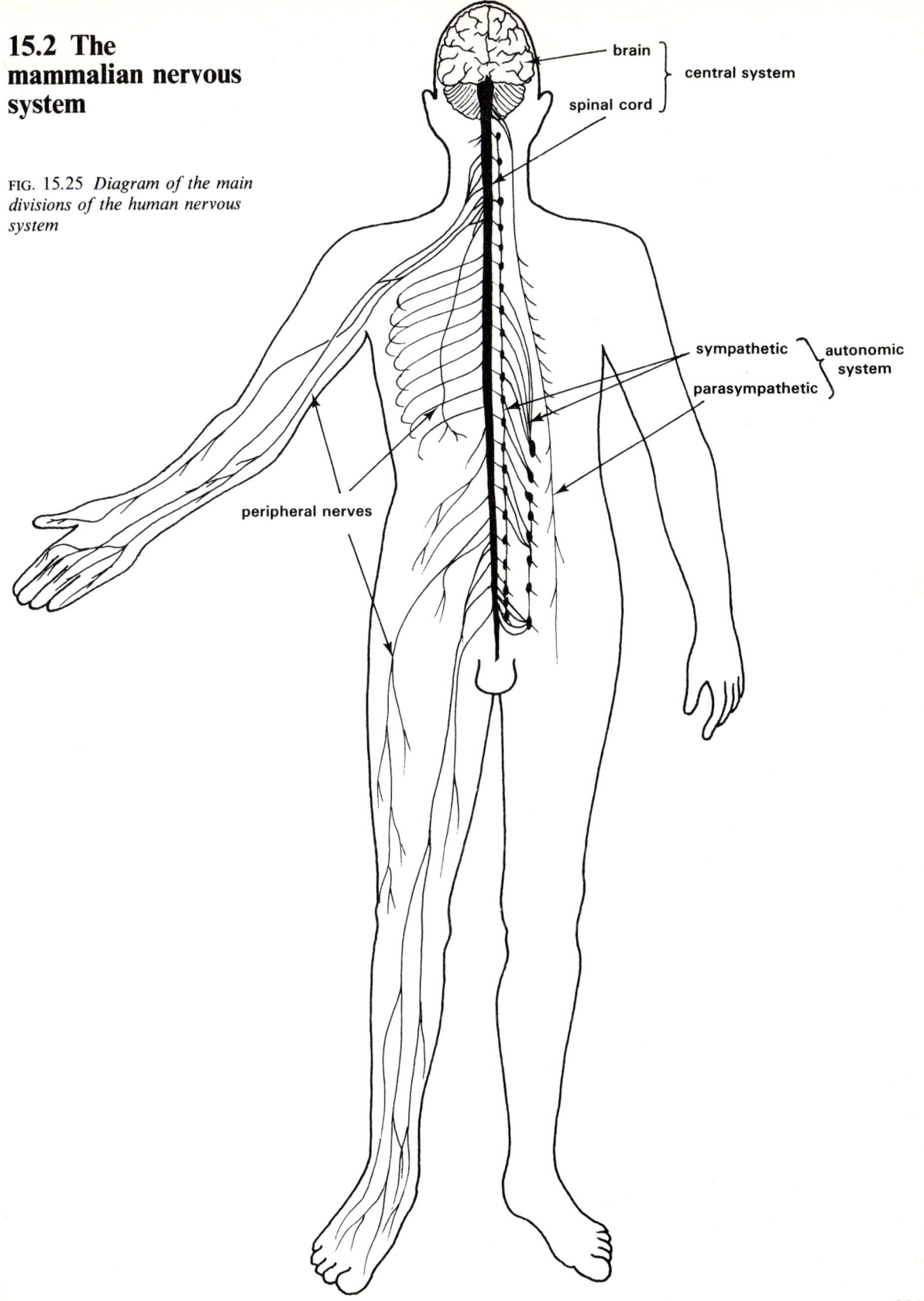

brain
central system
spinal cord
sympathetic
autonomic system
parasympathetic
peripheral nerves

While integrated into a functional unit the mammalian nervous system can be divided into several structural sections. The sections include the **peripheral, autonomic** and the **central nervous systems** (Fig. 15.25).

The **peripheral system** includes **afferent (sensory) neurons** which receive impulses from **receptors (sense organs)** and transmit them to the central nervous system. Impulses may then be transmitted from the central system along **efferent (motor) neurons** of the peripheral system to **effectors**. Effectors are muscles and glands which respond to the stimulus (Fig. 15.26).

15.2.1 The peripheral nervous system and reflexes

FIG. 15.26 *The main components of a nervous control mechanism.* Compare with Fig. 15.1

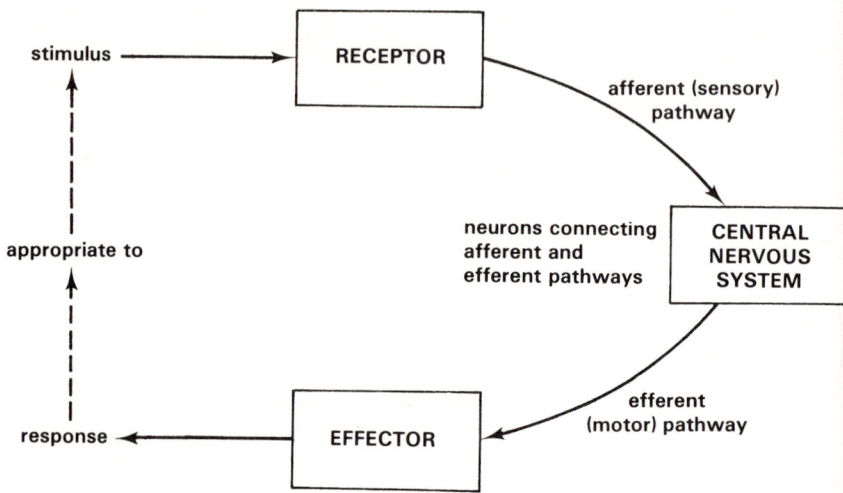

The type of activity in which the central system acts as a **relay** for incoming and outgoing impulses is the simplest form of nervous behaviour in mammals. Such **reflex activity** controls many aspects of mammalian behaviour. Reflex actions allow rapid responses to be made to a variety of stimuli without the conscious involvement of the brain. Most of the responses involve the control of posture and body movements under the subconscious control of the autonomic nervous system.

The involvement of the spinal cord in reflex actions is important since it connects afferent neurons to appropriate efferent neurons in different parts of the body. The mammalian **spinal cord** is a posterior extension of the brain and has a characteristic structure (Fig. 15.27). It contains **intermediate neurons** making up a central area of **grey matter** which relays information between afferent and efferent neurons. The grey matter is surrounded by **white matter**. White matter is composed mainly of the axons of longitudinal neurons connecting many parts of the body with the brain. Peripheral nerves contain afferent and efferent neurons lying side by side (Fig. 15.28). Near the spinal cord the afferent neurons enter the cord by pairs of **dorsal roots** and the efferent neurons leave in pairs of **ventral roots**.

An example of reflex activity is the body's response to changes of skin temperature. This is a convenient example to study in some detail as it illustrates how many co-ordinated responses can result from a single stimulus. Mammalian skin contains receptors sensitive to cold. When stimulated by low temperature, impulses are transmitted from the receptors along afferent neurons to the spinal cord. Here relays direct impulses along efferent neurons to a number of effectors. The effectors include skeletal muscles whose rapid contraction and relaxation result in shivering. Subcutaneous arterioles also constrict and the erector pili muscles at the base of the hair follicles contract causing erection of the hairs. Shivering produces heat. The other two responses minimise heat loss (Fig. 15.29). Control of body temperature in mammals is more complex than this and involves other reflexes described in Chapter 17.

FIG. 15.27 *Structure of the spinal cord. The central grey matter contains transverse fibres connecting afferent and efferent neurons in one spinal nerve. The outer white matter contains longitudinal fibres. Thick arrows indicate direction of impulses*

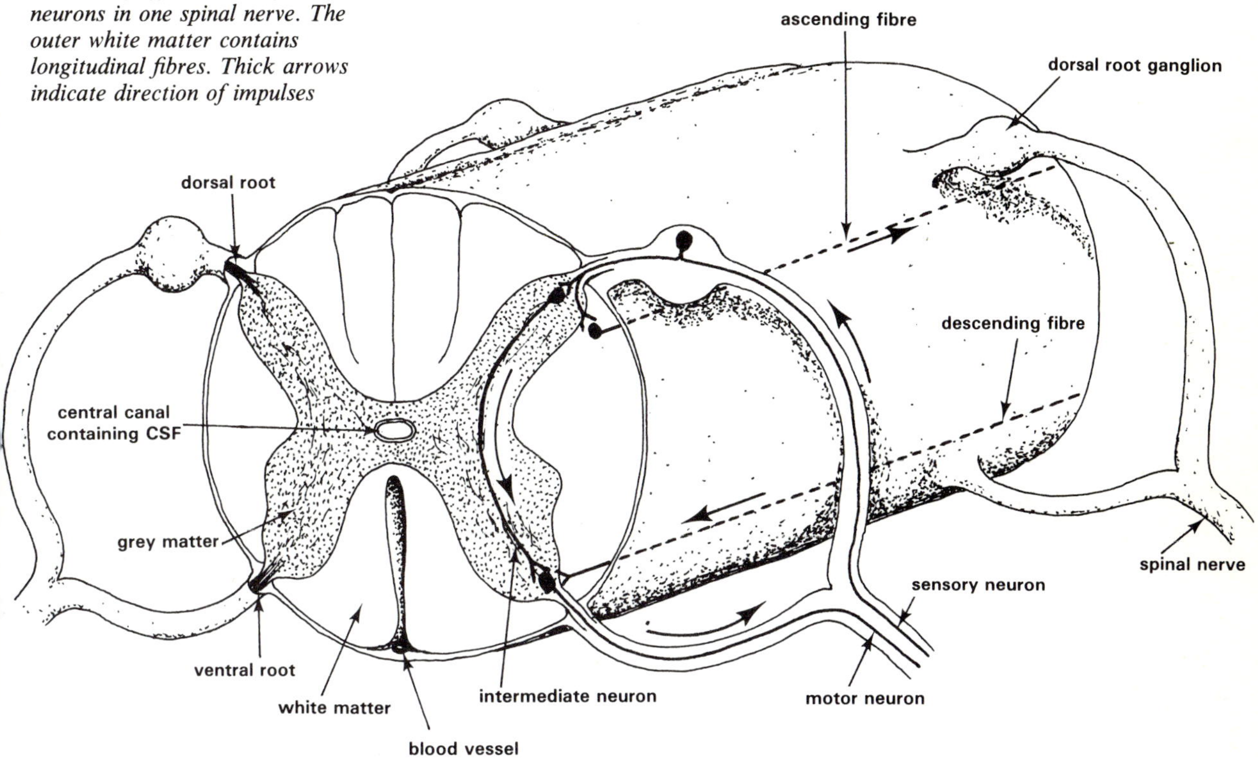

ascending fibre

dorsal root ganglion

dorsal root

descending fibre

central canal containing CSF

grey matter

spinal nerve

ventral root

sensory neuron

white matter

intermediate neuron

motor neuron

blood vessel

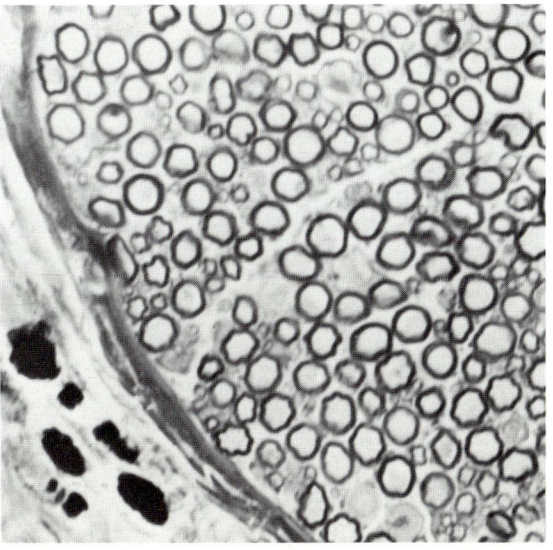

FIG. 15.28 *Photomicrograph of a transverse section of human median nerve. Note the variety of axon diameters.* ×850

FIG. 15.29 *Nervous pathways involved in the responses to cold.* Compare with Fig. 15.26

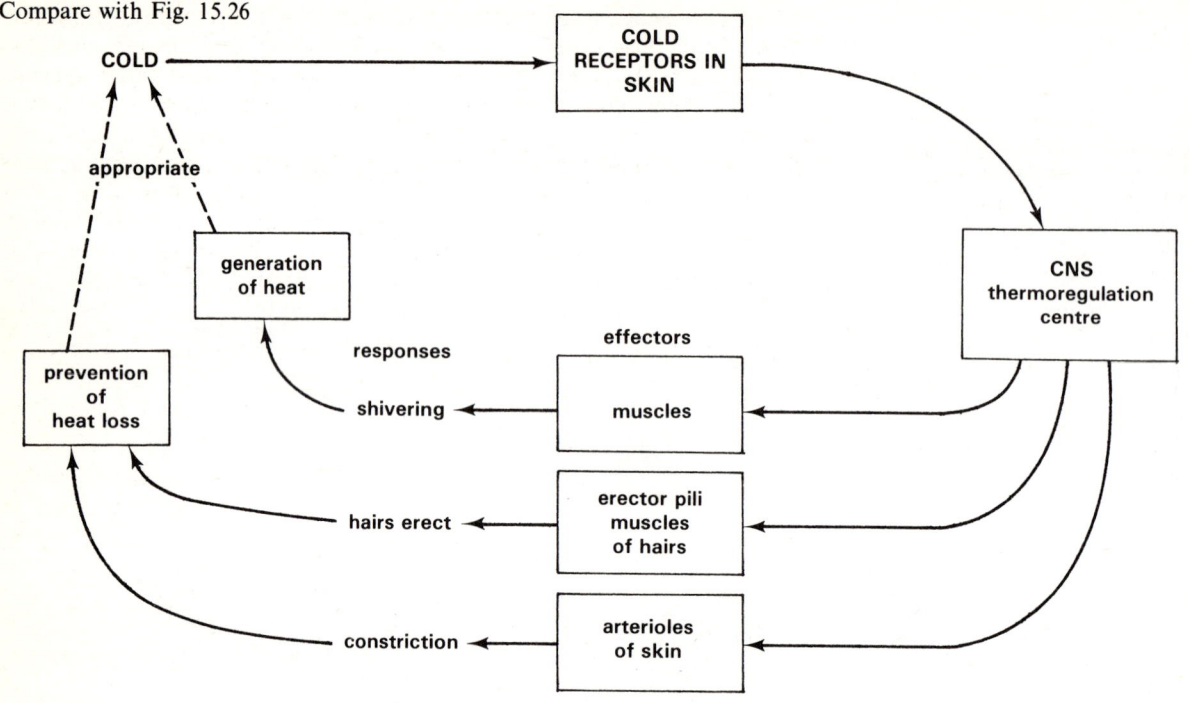

New-born mammals display a variety of **innate reflexes**. A baby for example will suck almost any object placed in its mouth. This is a vital reflex which activates the expulsion of milk from the mother's mammary glands during suckling (Chapter 22). Very young infants acquire additional reflexes and later override them at certain stages of their growth and development. Testing **acquired reflexes** is used to assess the developmental progress of premature babies (Fig. 15.30).

FIG. 15.30 *Some infantile reflexes* (after Williams and Wendell-Smith)

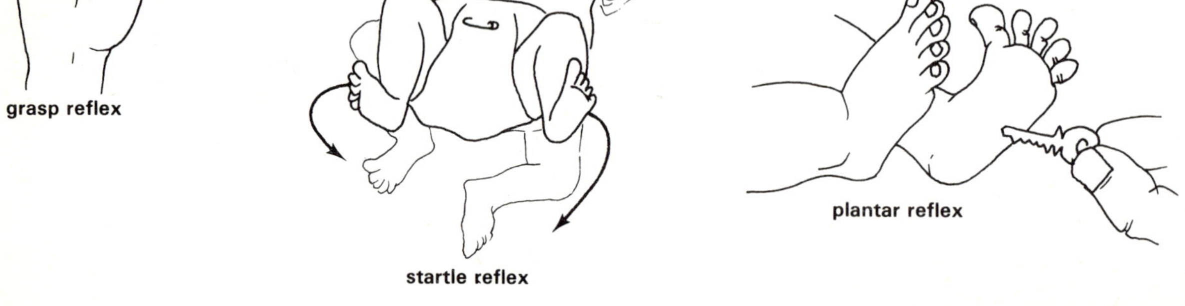

grasp reflex

startle reflex

plantar reflex

Pavlov in 1910 demonstrated another kind of reflex in mammals, the **conditioned reflex**. He noticed that hungry dogs salivate when presented with food, this being an innate reflex. He then observed that dogs salivate when exposed to a **secondary stimulus**, a ringing bell, at the same time as the **primary stimulus**, food. Pavlov discovered that in time the dogs learned to associate the secondary stimulus with food and salivated when a bell was rung even though food was not present (Fig. 15.31). This kind of conditioned reflex is common in mammalian behaviour. Knowledge of conditioning is very effectively used by the advertising industry. It is an interesting exercise to try to detect Pavlov's technique in television advertisements. How many describe the true nature of the product to be sold and how many associate it with something attractive but possibly quite unrelated to the product?

FIG. 15.31 *Pavlov's observations: (a) the hungry dog responds by salivating to the primary stimulus of food; (b) the dog is subjected to a period in which a second stimulus, a ringing bell, is presented with the primary stimulus to the salivation reflex; (c) eventually the dog salivates where only the secondary stimulus is given. The dog is said to be conditioned*

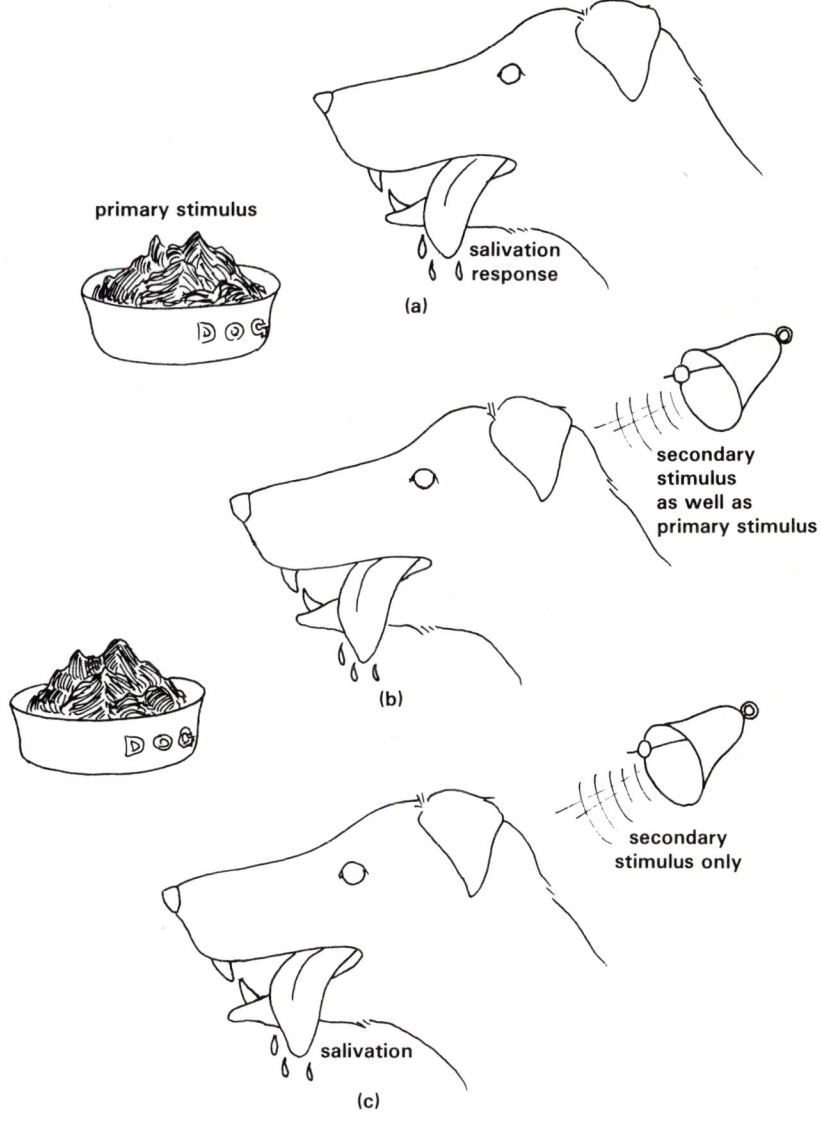

primary stimulus

salivation response

(a)

secondary stimulus as well as primary stimulus

(b)

secondary stimulus only

salivation

(c)

Body movements, often brought about by reflexes, involve the action of more than one muscle. Raising the forearm for example requires contraction of the **biceps (flexor) muscle** and relaxation of the **triceps (extensor) muscle** (Fig. 15.32). The nerve pathways in **antagonistic activity** of this sort excite one muscle and inhibit the other. The control mechanism is thought to involve excitatory and inhibitory synapses.

FIG. 15.32 *Reciprocal inhibition. During body movements involving the action of antagonistic muscles, one muscle is excited while the other is inhibited. Here, the flexing of the forearm is produced by contraction of the biceps and relaxation of the triceps. Nerve pathways involving inhibitory and excitatory synapses are thought to control this type of activity*

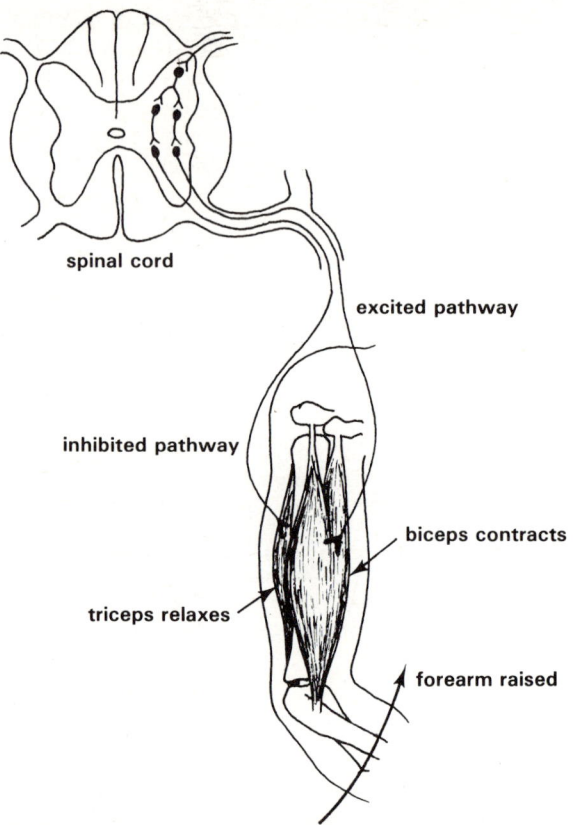

spinal cord

excited pathway

inhibited pathway

biceps contracts

triceps relaxes

forearm raised

15.2.2 The autonomic nervous system

The **autonomic nervous system** consists of two sets of nerves, **sympathetic** and **parasympathetic** (Fig. 15.33). The autonomic nerves carry impulses to many organs in the body and the two systems normally act **antagonistically** (Table 15.3). The balance between the activity of sympathetic and parasympathetic nerves results in co-ordinated regulation of organs and glands, usually by reflex action. As with peripheral reflexes, the central nervous system is involved in autonomic activity. Many of the receptors which regulate autonomic activity are situated in the brain. They include the cardiac centre, the vasomotor centre and the respiratory centre (Chapters 8 and 9). The receptors detect body changes which require adjustment of the cardiovascular and gas exchange systems and convey appropriate impulses to the systems along autonomic nerves. However, unlike the peripheral system, the autonomic system is less easily controlled by conscious activity of the brain. Although it is possible to train an individual to reduce the heart rate for instance, autonomic function relies almost entirely on reflexes. The ability to control breathing consciously is due to the fact that the muscles involved are innervated by peripheral as well as autonomic nerves.

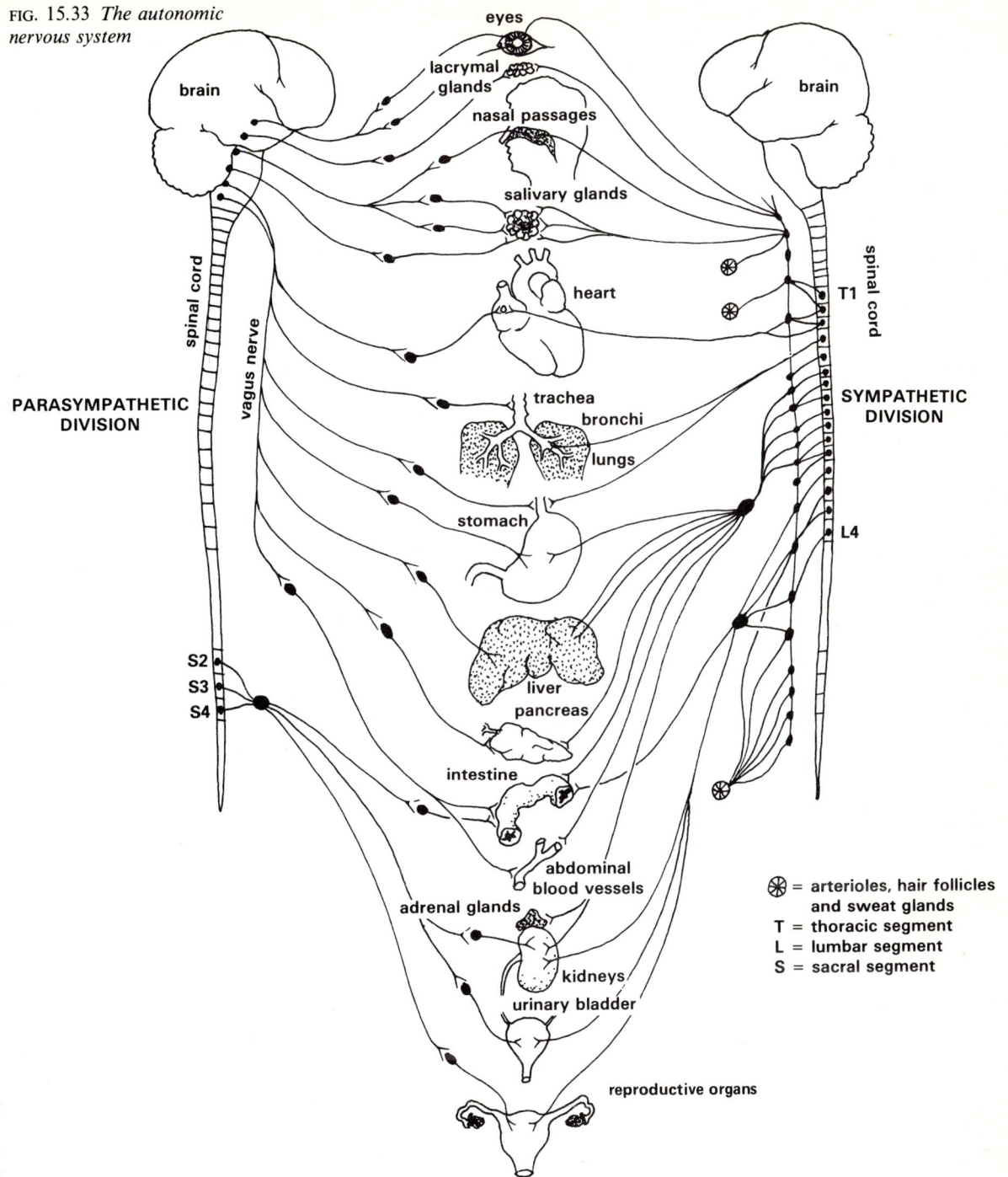

FIG. 15.33 *The autonomic nervous system*

PARASYMPATHETIC DIVISION

SYMPATHETIC DIVISION

brain

brain

spinal cord

spinal cord

vagus nerve

eyes

lacrymal glands

nasal passages

salivary glands

heart

trachea

bronchi

lungs

stomach

liver

pancreas

intestine

abdominal blood vessels

adrenal glands

kidneys

urinary bladder

reproductive organs

T1

L4

S2
S3
S4

⊛ = arterioles, hair follicles
    and sweat glands
T = thoracic segment
L = lumbar segment
S = sacral segment

    The antagonistic effects of sympathetic and parasympathetic nerves result from the different chemical transmitters secreted at their neuroeffector junctions. **Acetylcholine** is the transmitter secreted by parasympathetic neurons. For this reason parasympathetic neurons are **cholinergic**. Sympathetic neurons secrete **noradrenalin** and are **adrenergic** (Fig. 15.34).

Table 15.3 Summary of the effects of the autonomic nervous system

TARGET	SYMPATHETIC EFFECT	PARASYMPATHETIC EFFECT
pupil of eye	dilation	constriction
bronchi	dilation	constriction
heart	increase in rate	decrease in rate
sphincters of gut	constriction	dilation
urinary bladder	dilation	constriction
blood vessels	vasoconstriction	—
sweat glands	secretion of sweat	—
salivary glands	—	secretion of saliva
stomach	—	secretion of gastric juice
pancreas	—	secretion of pancreatic juice
genitalia	—	vasodilation of erectile tissue

FIG. 15.34 *Parasympathetic fibres are cholinergic; sympathetic fibres are adrenergic*

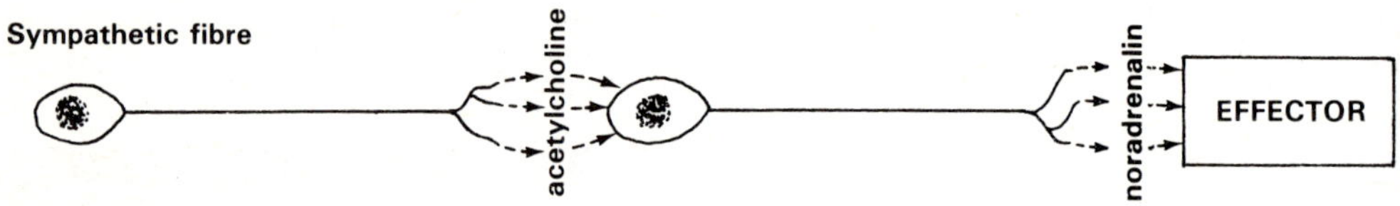

FIG. 15.35 *Relative development of the main portions of the brain in amphibians, reptiles, birds and mammals. Note the relative size of the cerebrum in the mammalian brain*

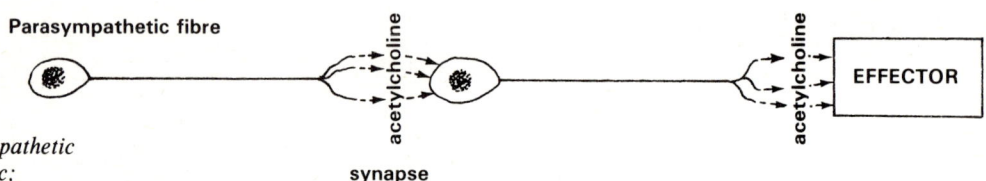

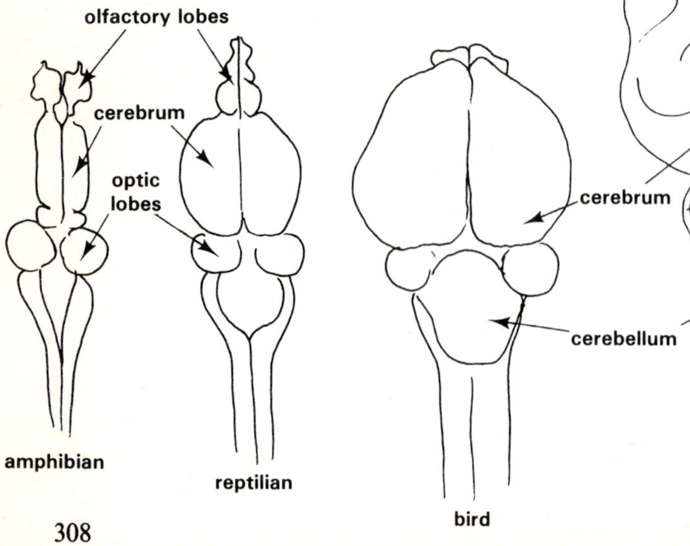

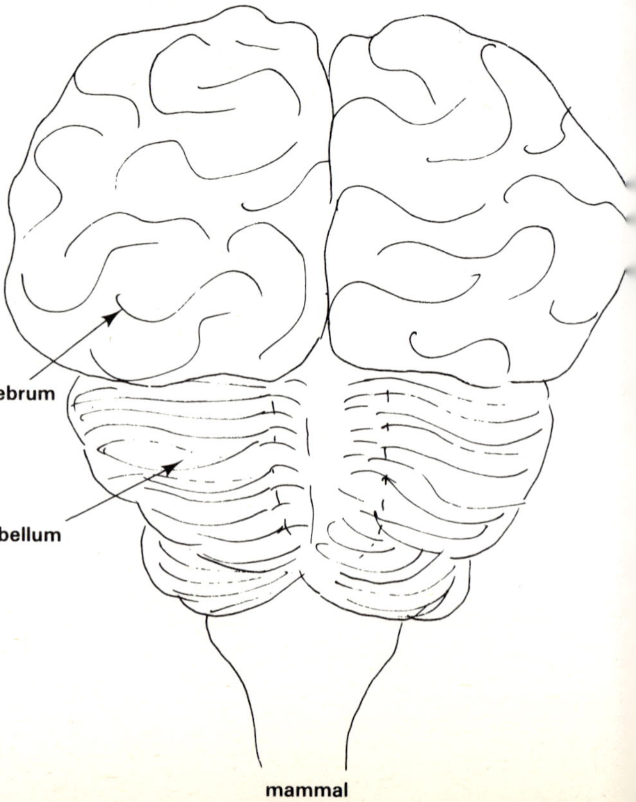

Adrenalin as well as noradrenalin is secreted into the blood from the medulla of the adrenal glands which are modified portions of the sympathetic system. In conditions of stress, secretion of adrenalin is increased. The significance of the response is discussed in Chapter 18.

15.2.3 The central nervous system

At the beginning of the chapter the basic components of a co-ordination system are listed. Stimuli are first detected by sensors called receptors. Information, conveyed in the nervous system as impulses, is then transmitted to a co-ordinator. The co-ordinator relays the impulses along appropriate channels to an effector which brings about the response (Fig. 15.1). The **central nervous system** comprising the **brain** and **spinal cord**, performs the role of the **co-ordinator** in the nervous system. We have already seen how the spinal cord relays impulses from afferent neurons to appropriate efferent neurons (section 15.2.1). However, the brain can act in much more complex ways.

1. STRUCTURE OF THE MAMMALIAN BRAIN

The mammalian brain is divisible into three main regions, **hindbrain**, **midbrain** and **forebrain** (Fig. 15.35). The hindbrain is continuous with the spinal cord which runs through the spinal canal of the vertebrae. The hindbrain consists of the **medulla oblongata** and a large dorsal structure called the **cerebellum**. The midbrain is relatively short in mammals, linking the hindbrain with the very large forebrain. The mammalian forebrain has two large dorsal structures called the **cerebral hemispheres** which make up the **cerebrum**. The hemispheres are so highly folded that only about a third of their surface area is visible on the outside of a human brain (Fig. 15.36). Projecting from beneath the forebrain is the hypophysis or **pituitary body** (Chapter 18).

FIG. 15.36 *The human brain viewed from above*

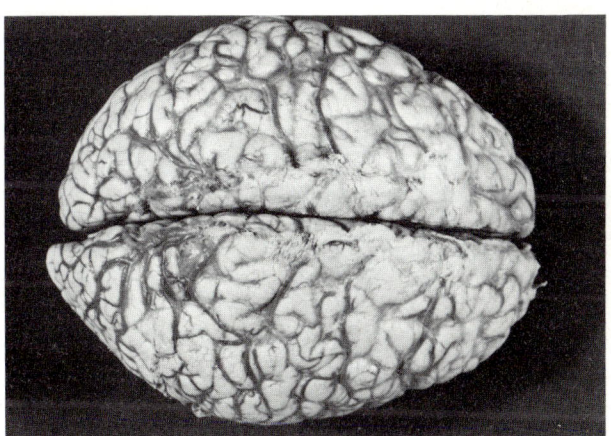

The entire central nervous system is surrounded by three protective membranes called the **meninges**. Between the inner two meninges is a fluid called the **cerebrospinal fluid (CSF)**. CSF is also found inside the brain, in four cavities called **ventricles** (Fig. 15.37). Ciliated epithelium lining the ventricles keeps the CSF in motion. Respiratory gases, nutrients and metabolic wastes pass between the CSF and the blood carried in capillaries which are plentiful in two of the ventricles. Disturbance of the blood supply can rapidly lead to permanent brain damage and may be fatal (Fig. 15.38). CSF also circulates in the central canal of the spinal cord. Impulses are transmitted to and from the brain along the spinal cord and twelve pairs of cranial nerves (Fig. 15.39).

The central nervous system is supported and protected by the **skeleton**. The **cranium** of the skull surrounds the brain and the **vertebral column** surrounds the spinal cord.

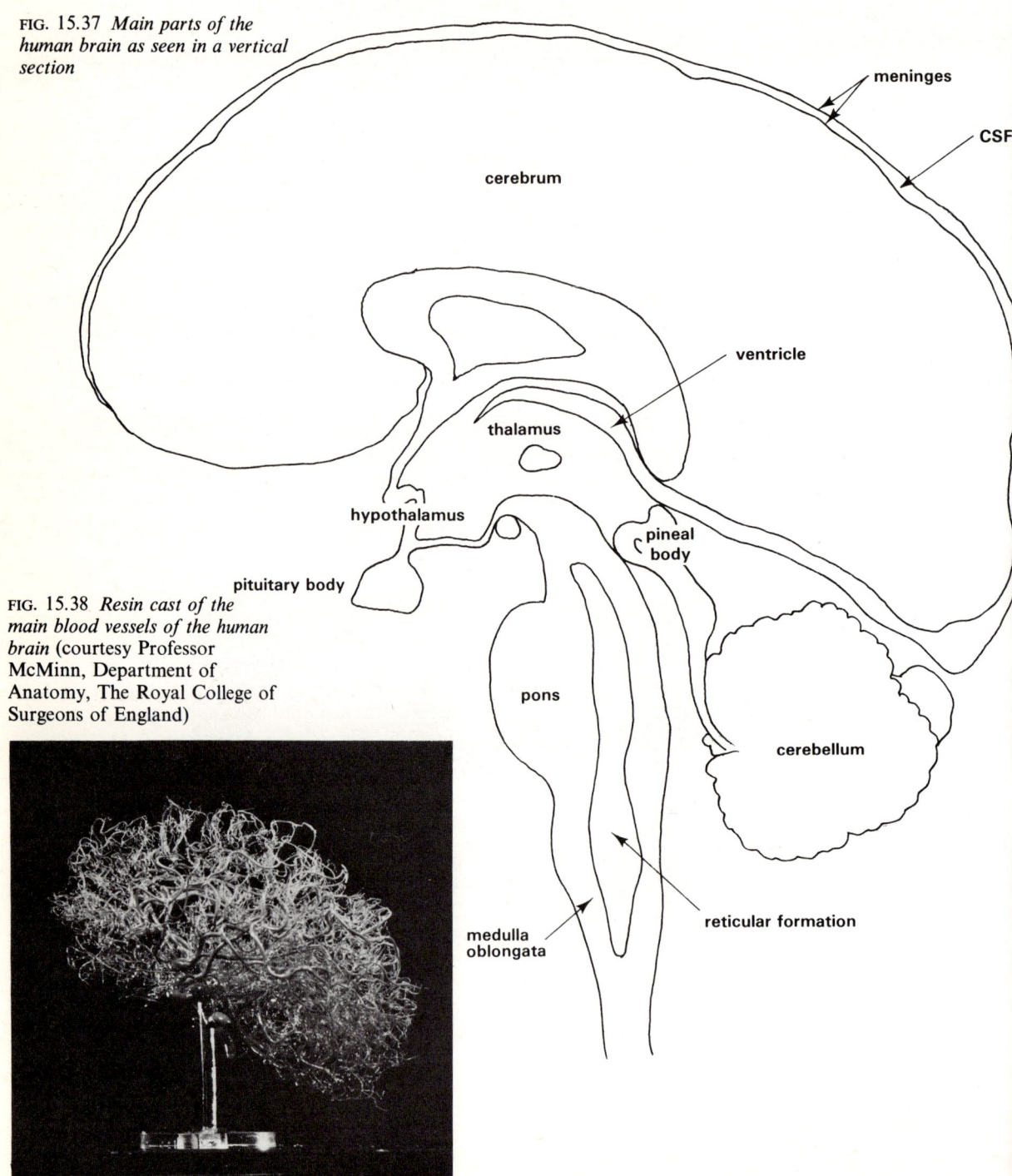

FIG. 15.37 *Main parts of the human brain as seen in a vertical section*

meninges

CSF

cerebrum

ventricle

thalamus

hypothalamus

pineal body

pituitary body

pons

cerebellum

medulla oblongata

reticular formation

FIG. 15.38 *Resin cast of the main blood vessels of the human brain* (courtesy Professor McMinn, Department of Anatomy, The Royal College of Surgeons of England)

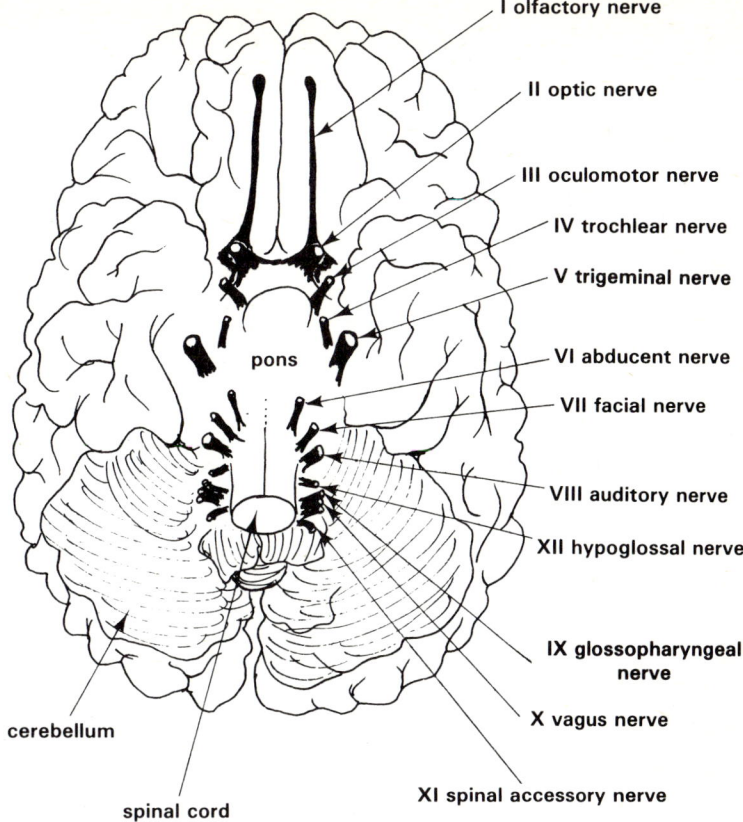

FIG. 15.39 *The cranial nerves of the human brain (viewed from the base)*

I olfactory nerve

II optic nerve

III oculomotor nerve

IV trochlear nerve

V trigeminal nerve

VI abducent nerve

VII facial nerve

VIII auditory nerve

XII hypoglossal nerve

IX glossopharyngeal nerve

X vagus nerve

XI spinal accessory nerve

pons

cerebellum

spinal cord

2. FUNCTIONS OF THE MAMMALIAN BRAIN

The human brain contains more than a thousand million neurons. It has been estimated that in the cerebral cortex there are something like $10^{2\ 783\ 000}$ synapses! Considering the numbers of neurons and synapses in an organ weighing only about 1·3 kg it is evident how complex the human brain is.

(i) HINDBRAIN. The **medulla oblongata** contains the **cardiac**, **respiratory** and **vasomotor centres**. The centres contain neurons which are directly sensitive to blood carbon dioxide or more precisely hydrogencarbonate (HCO_3^-) concentration. An increase in blood carbon dioxide triggers off reflex activity starting in the centres and results in modifications of heart rate, breathing rhythm and blood pressure (Chapters 8 and 9). Refer again to Figure 15.1 and describe the roles played by the central nervous system in regulating cardiac and respiratory function.

The **cerebellum** is concerned with **co-ordination** of body movements and the **maintenance of posture**. Signals arising in the motor region of the forebrain travel through the cerebellum on their way to the spinal cord and the rest of the body. The cerebellum also receives sensory information from a variety of receptors. In some way, the cerebellum modifies the motor signals in the light of the sensory information received. The cerebellum thus ensures that appropriate body movements are made as intended and are not weakened or exaggerated in any way (Fig. 15.40).

FIG. 15.40 *Some nerve pathways illustrating the role of the cerebellum in modifying motor signals from the motor cortex of the cerebrum in the light of the degree of response produced. The stretch receptors in the muscle respond to muscle movements. If these movements are too great or too little the pathways illustrated enable 'damping down' or further stimulation* (after Guyton)

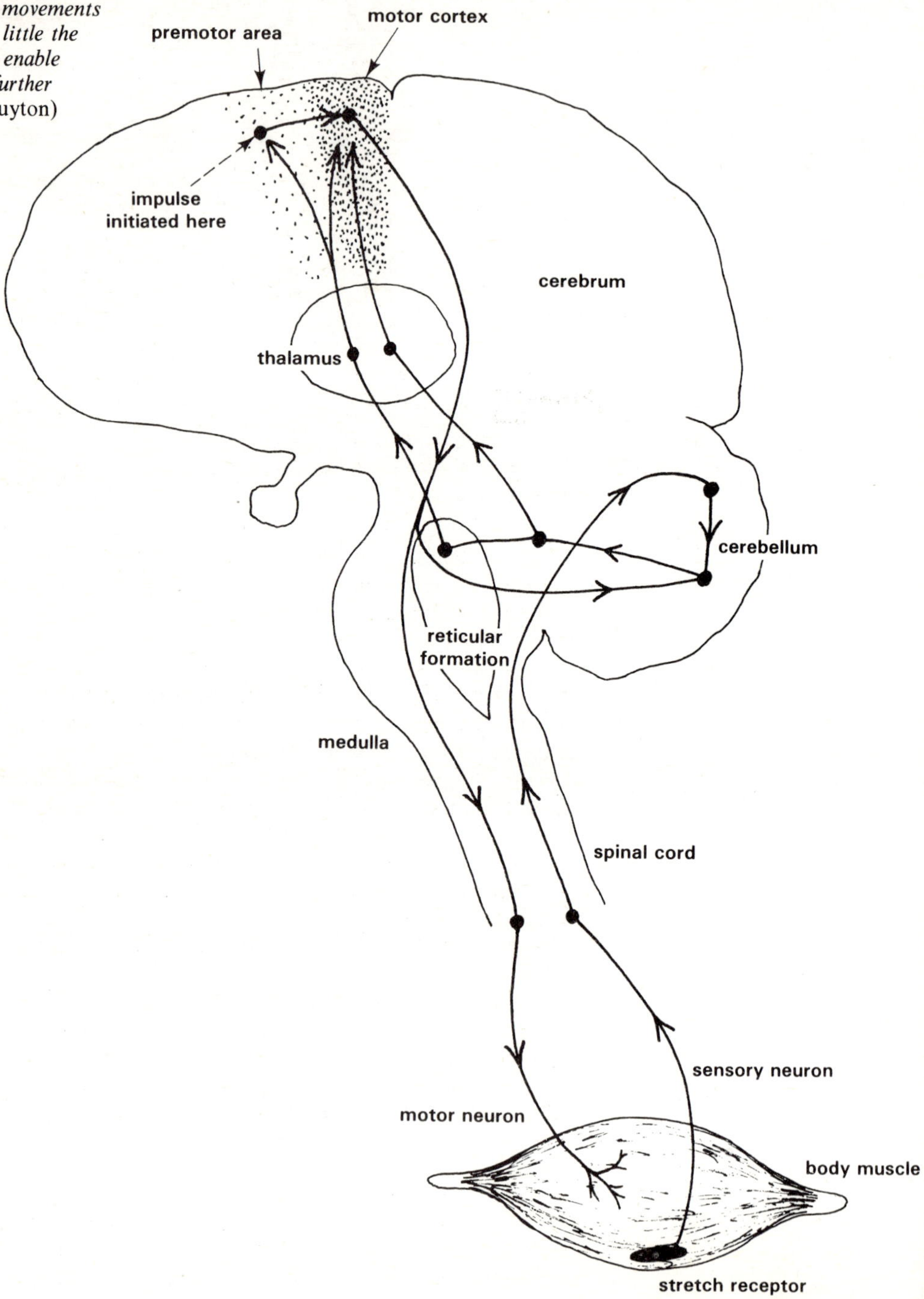

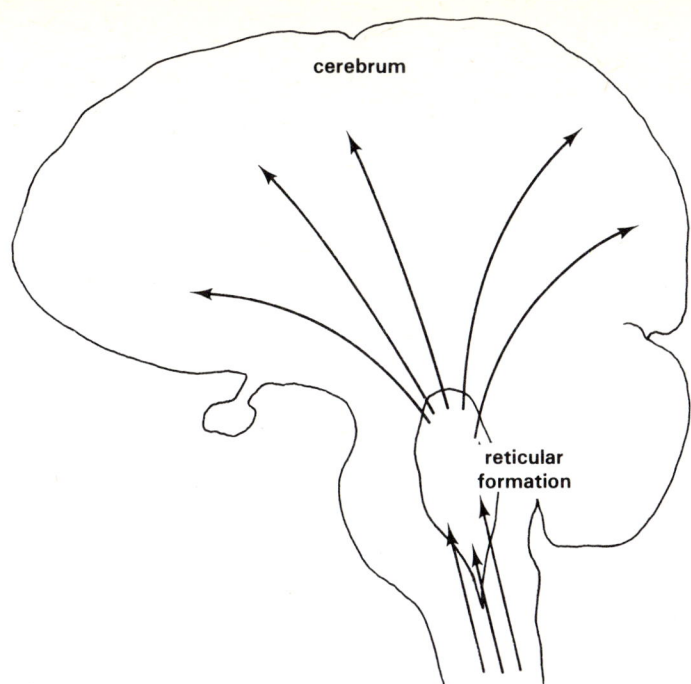

(ii) MIDBRAIN. As you sit reading this book you are bombarded by a variety of environmental stimuli. The more successfully you concentrate on your reading the less aware you are of outside distractions. When studying, unwanted stimuli may include the noise of traffic, people talking, a radio or television. You also constantly receive signals from your skin in response to the touch of your clothes or of a chair you may be sitting on. In other words, of the great number of environmental stimuli you receive at any one time, you are consciously aware of only a few whether you are studying or not. The **filtering effect** prevents overloading of your conscious awareness. Most people experience difficulty in concentrating on more than one thing at a time. The result of overloading the brain with many stimuli is lowered mental efficiency and often confusion.

One of the functions of the midbrain is to **activate the forebrain** with appropriate sensory signals. Lying in the midbrain is a group of nerve fibres called the **reticular formation**. Sensory pathways lead from the spinal cord to the reticular formation as well as directly to the forebrain. The forebrain does not respond to sensory signals unless it is activated by the reticular formation (Fig. 15.41).

Activation of the cerebral cortex by the reticular formation is necessary for **wakefulness**. The cortex is very active when you are awake and your nervous system constantly transmits impulses. Presumably as a result of neuronal or synaptic fatigue the system eventually needs rest. The reticular formation stops activating the cerebral cortex and a state of **sleep** follows. The full significance and physiology of sleep are not known at present. Sleep is nevertheless vital for the normal balance of nervous activity. Lack of sleep and excessive nervous fatigue can lead to serious mental disturbances. Sleep deprivation has been used to lower the mental resistance of prisoners prior to interrogation.

(iii) FOREBRAIN. The forebrain contains the dominant regions of the central system, notably the **cerebrum**. The cerebrum in mammals is very large relative to that of other vertebrates. It is in the cerebrum that higher

FIG. 15.42 *Neurons in the cerebral cortex of a rat, ×75*

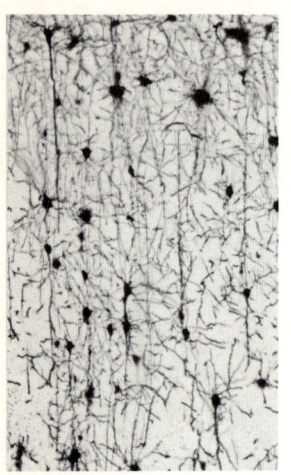

mental processes occur, including conscious **thought**, **memory** and the detailed **interpretation** of sensory information. The main pathways of motor neurons originate in the cerebrum. Less than one per cent of the enormous number of neurons in the cerebrum synapse with fibres which transmit impulses out of the forebrain. Little is known of the significance of the rest of the cerebral neurons. The outer cortex of the cerebrum is grey in colour and contains dense masses of nerve cell bodies (Fig. 15.42).

It is possible to map out distinct **sensory and motor regions** on the surface of the cerebrum. Localisation of function has been revealed by experiments with animals. Specific parts of the cerebral cortex can be stimulated with small electric currents and the response recorded. Deficiencies of sensation and motor activity have also been observed in people following injury to specific parts of the cerebrum. The sensory and motor regions are found on either side of a groove called the **central sulcus** which runs down the **parietal lobe** of each cerebral hemisphere (Fig. 15.43). Other regions of the cerebrum deal with **vision**, **hearing**, **speech** and **personality**.

FIG. 15.43 (a) *Location of the main motor and sensory regions in the human cerebral cortex*

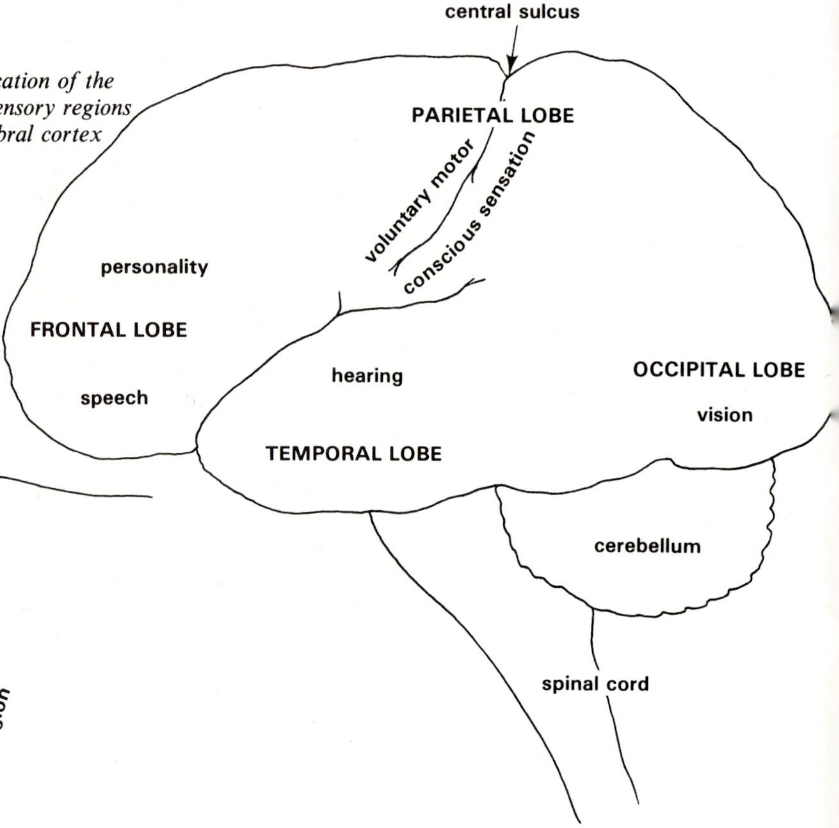

central sulcus

PARIETAL LOBE

voluntary motor

conscious sensation

personality

FRONTAL LOBE

speech

hearing

OCCIPITAL LOBE

vision

TEMPORAL LOBE

cerebellum

spinal cord

FIG. 15.43 (b) *Localised areas within the motor and sensory regions on either side of the central sulcus*

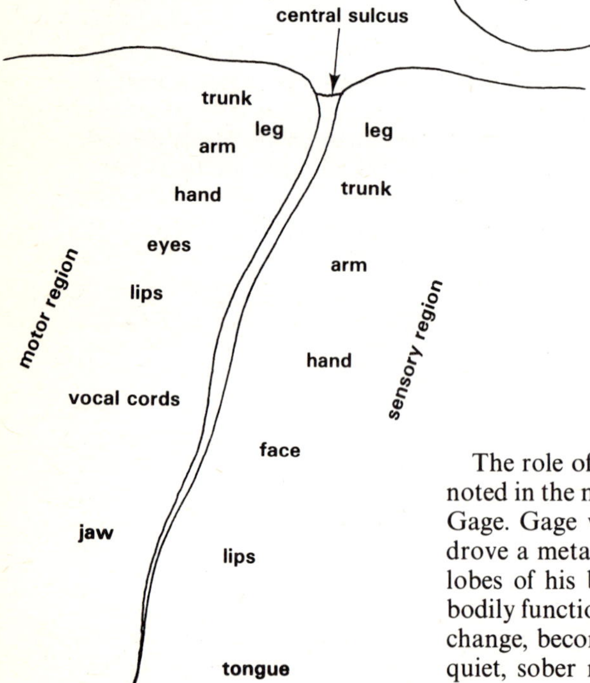

central sulcus

motor region

trunk

leg

leg

arm

hand

trunk

eyes

arm

lips

hand

vocal cords

sensory region

face

jaw

lips

tongue

The role of the **frontal lobes** in the maintenance of personality was first noted in the middle of the nineteenth century following the case of Phineas Gage. Gage was working in a quarry when an accident with explosives drove a metal bar through his head (Fig. 15.44). Even though the frontal lobes of his brain were severely damaged he survived with most of his bodily functions unaffected. However, Gage suffered a marked personality change, becoming irresponsible and vulgar. Before his accident he was a quiet, sober man. More recently, the frontal lobes have been surgically isolated in some mental patients in an attempt to cure severe personality problems. This practice of **prefrontal lobotomy** has now been replaced by the use of drugs.

FIG. 15.44 *Skull of Phineas Gage showing the damage caused by the metal rod which pierced it during a mining accident*

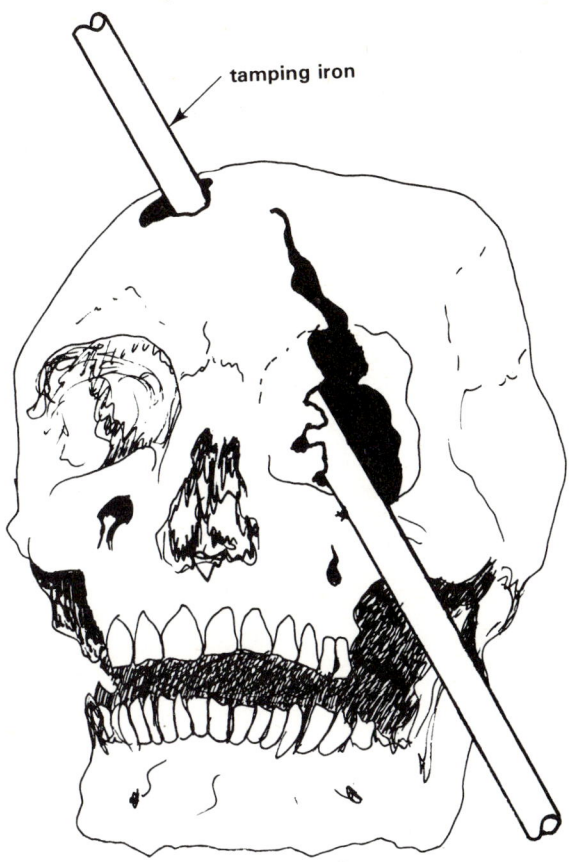

tamping iron

FIG. 15.45 (a) *Electrodes being placed on the scalp of a patient prior to taking an electroencephalogram* (courtesy Sally York, Leicester Royal Infirmary)

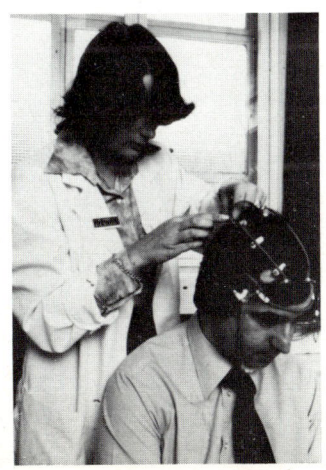

The enormous numbers of neurons in the cerebral cortex are constantly active. The electrical activity accompanying transmission of impulses in the brain can be measured by placing electrodes on the scalp. Rhythmic waves are produced and these are displayed as an **electroencephalogram (EEG)**. Mental disturbances may produce EEG patterns which are abnormal and are of value in the diagnosis of some mental diseases (Fig. 15.45).

Sensory information reaching the cerebral hemispheres first passes through a region in the centre of the forebrain called the **thalamus**. The thalamus contains nerve pathways which activate appropriate regions of the cerebral cortex when sensory signals come in from the body's receptors. The function of the cortex is to interpret the details of the sensory signals and to link these with the memory of previous similar stimulations. The thalamus controls this function by channelling signals to the appropriate cortical region.

Beneath the thalamus is a part of the forebrain called the **hypothalamus**. The hypothalamus contains the **osmoregulation** and **thermoregulation centres**, important in regulating the body's water content and temperature (Chapters 10 and 17). Several **hormones** are made in the hypothalamus including ocytocin and vasopressin. Also made in the hypothalamus are hormones which control the release of tropins from the anterior pituitary (Chapter 18).

FIG. 15.45 (b) *EEG traces. The upper trace was recorded from the frontal region of a normal 18-year-old male. The lower trace was recorded from an individual with 'petit mal' epilepsy*

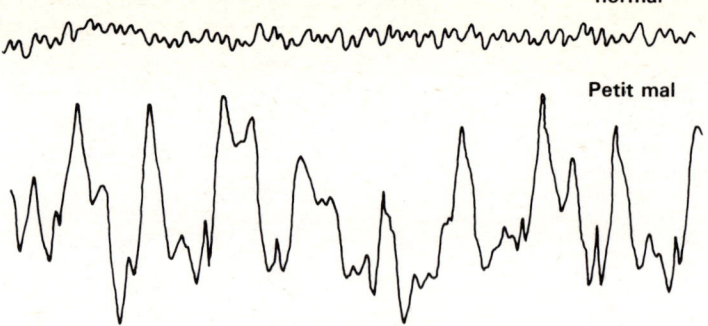

Petit mal

The brief survey of the brain given here serves to introduce some of the main functions of this highly complex organ. Many aspects of brain function are little known and still present one of the greatest challenges to science today.

Chapter 16

The mammalian eye and ear

The detection of stimuli depends on the conversion of the stimuli by **receptors** into impulses in the nervous system. The mammalian body contains a variety of receptors capable of feeding impulses into the nervous system about an equal variety of stimuli. Different receptors are sensitive to heat, light, sound, touch, stretching, spatial orientation and to chemicals.

The simplest receptors are single nerve cells which respond directly to a stimulus. Receptors in the skin are of this type (Fig. 17.11). Some receptors consist of groups of sensitive neurons such as the cardiac and respiratory centres in the brain (Chapter 15). Other receptors are grouped in complex **sense organs**. The structure of sense organs causes the stimulus to be channelled into a receptive region.

The way in which impulses are generated by receptor cells is important in discriminating between strong and weak stimuli. Measurement of action potentials (Chapter 15) in sensory neurons shows no significant difference between one impulse and another. However, a strong stimulus usually causes impulses to be generated rapidly in sensory neurons. The frequency of impulses generated by weak stimuli is lower. The appropriate sensory region of the cerebral cortex of the brain interprets the intensity of a stimulus according to the frequency with which it receives impulses.

The most obvious sense organs in mammals are the **eyes** and **ears**.

16.1 The mammalian eye

Most mammals depend greatly on vision to sense their environment. While reading this page your conscious awareness is almost totally stimulated by visual information.

16.1.1 Structure of the eye

Mammalian **eyes** are spherical structures (Fig. 16.1). The wall of the eye consists of three layers, the outer **sclera**, the middle **choroid** layer and the inner **retina**. The sclera is a tough, fibrous coating which protects the delicate inner layers. The sclera is white in appearance except at the front

FIG. 16.1 (a) *Photomicrograph of a longitudinal section of the eye from a monkey,* ×4

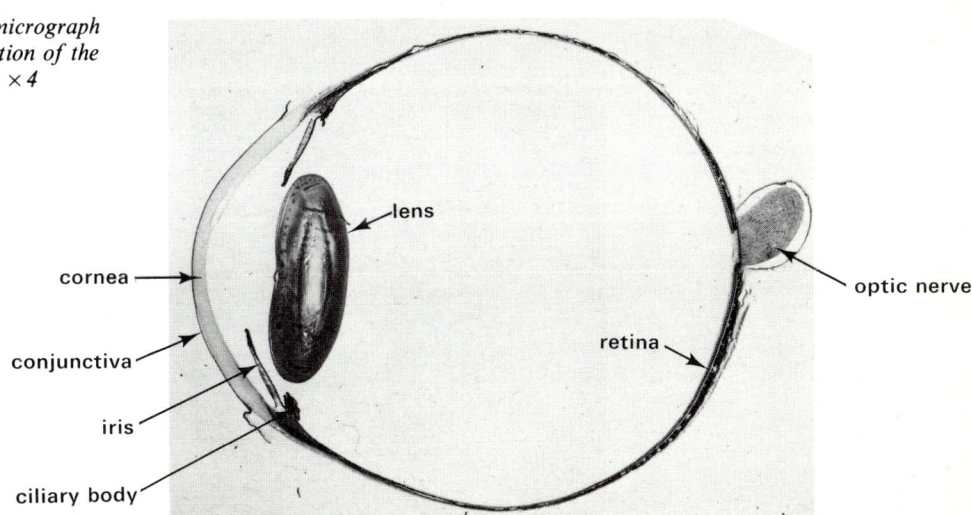

FIG. 16.1 (b) *Diagram of a*
vertical section through the eye
of a mammal

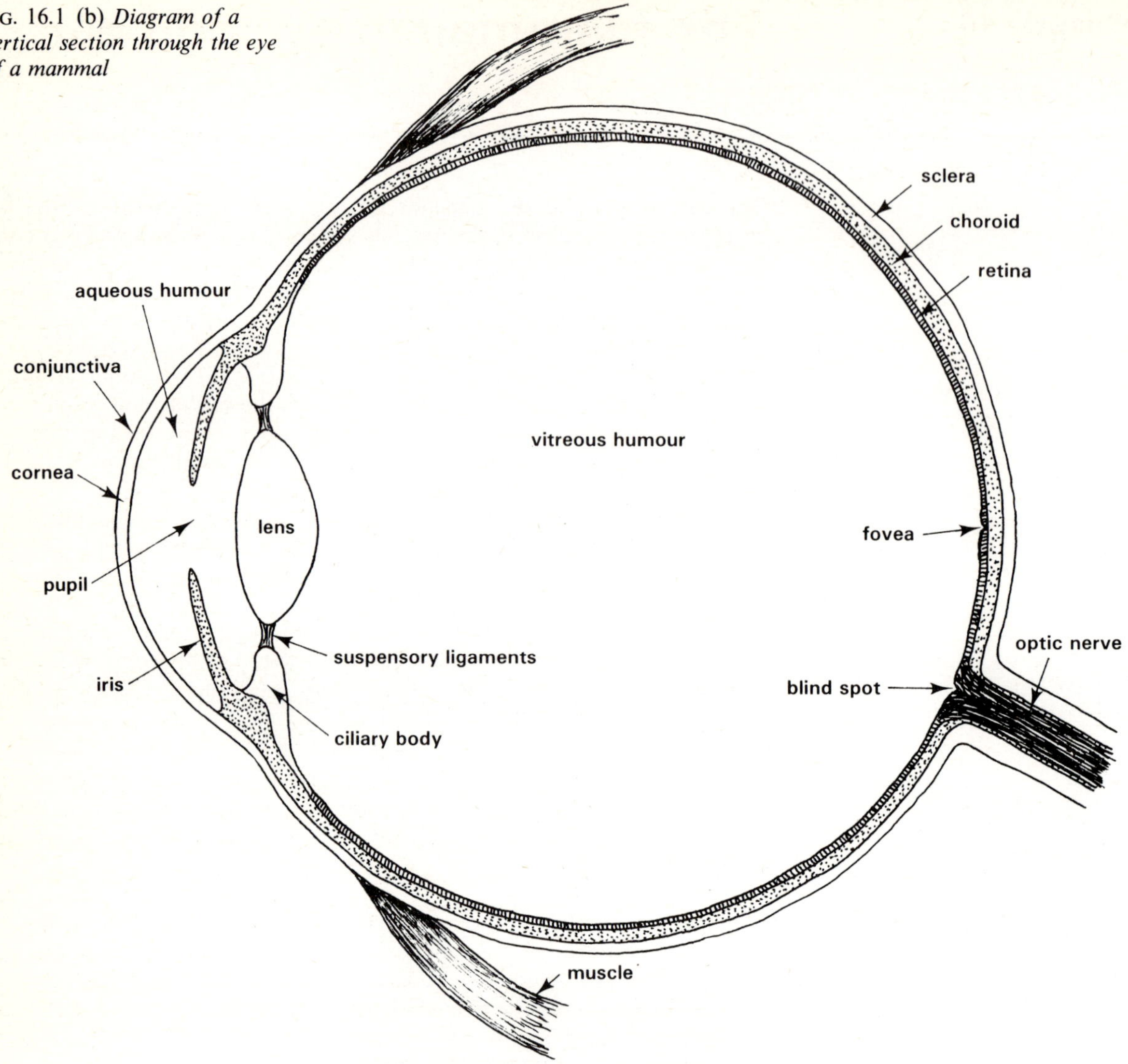

sclera

choroid

retina

aqueous humour

conjunctiva

vitreous humour

cornea

fovea

lens

pupil

optic nerve

suspensory ligaments

blind spot

iris

ciliary body

muscle

FIG. 16.2 *Angiogram showing*
the choroid blood vessels
(injected with fluorescein) of the
human eye

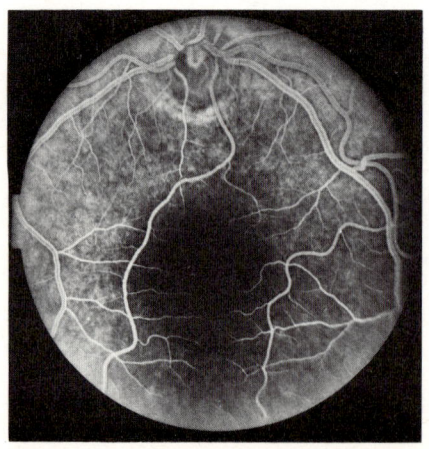

where there is a transparent area called the **cornea**. The cornea allows light into the eye. Tears secreted by the **lachrymal glands** lubricate the exposed surface of the eye, including the **conjuctiva** covering the cornea. The watery secretion helps to prevent abrasion of the eye's surface by dust particles and helps combat infection of the eye. Periodic closure of the eyelids, blinking, clears away debris.

Inside the sclera is the choroid layer which contains numerous blood vessels (Fig. 16.2). At the front of the eye the choroid layer is modified as the **iris** containing pigments which give the eye its colour. The iris also contains radial bands of smooth muscle. Contraction of the radial muscles causes dilation of an aperture called the **pupil**, in the centre of the iris. Constriction of the pupil occurs when the radial muscles of the iris relax. Variation in pupil size is controlled by autonomic reflexes and is usually a response to change in the intensity of light entering the eye. In bright light the pupil constricts and prevents excessive illumination of the interior of the eye. In dim light the pupil dilates allowing the maximum amount of light to reach the photoreceptor cells.

The photoreceptors are located in the retina situated immediately inside the choroid layer. Fibres of sensory neurons lead from the retina at the back of the eye as the **optic nerve**. The optic nerve transmits impulses generated in the retina to the brain. The retina develops in the embryo as an outgrowth of the brain, and can be regarded as a modified part of the central nervous system.

Suspended in the fluid inside the eye and just behind the pupil is a biconvex, crystalline **lens**. The lens is held in position by **suspensory ligaments** attached to a ring of smooth muscle called the **ciliary body**.

16.1.2 Focusing

Light rays entering the eye are bent or **refracted**. Refraction occurs at three surfaces of the eye before the light reaches the retina. The first of the refracting surfaces is the cornea, then the front surface of the lens and finally the rear surface of the lens (Fig. 16.3).

Between the cornea and the lens is a colourless, watery fluid called **aqueous humour**. At the back of the eye between the lens and the retina is another fluid called **vitreous humour**. Vitreous humour is also clear and is made of a gelatinous mucoprotein. The humours are transparent so that transmission of light through the cavities of the eye to the retina is not normally impeded.

FIG. 16.3 *The three refracting surfaces in the mammalian eye*

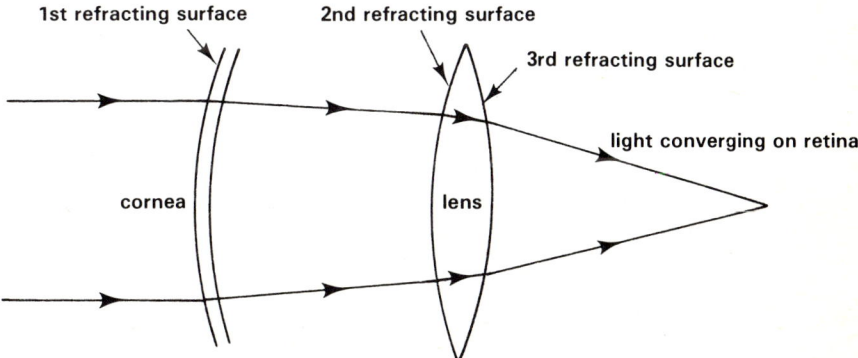

In the normal eye light rays are refracted sufficiently to be brought to a point on the retina. In this way the object viewed is brought into focus and a clear image is formed (Fig. 16.4). The **image** is **inverted** by the lens.

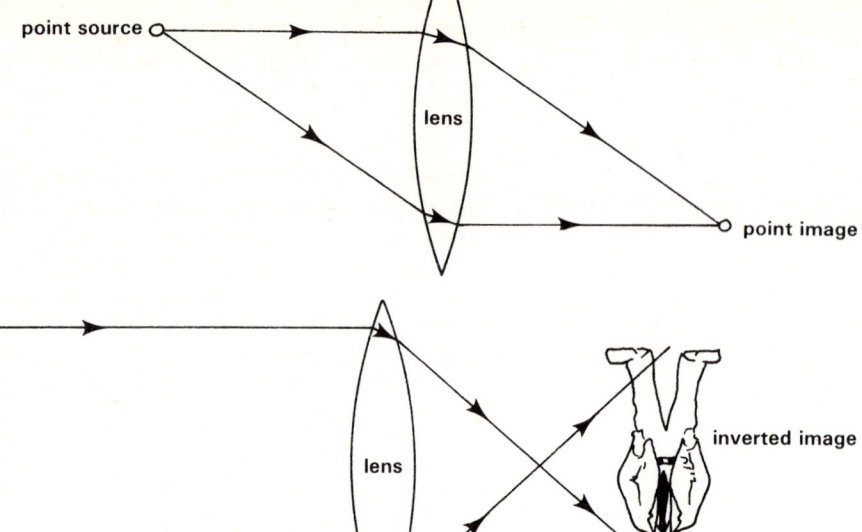

FIG. 16.4 (a) *Focusing of a light from a point source by the eye's lens. The cornea has been omitted for clarity*

FIG. 16.4 (b) *Light from all points on an object are focused in such a way that the image on the retina is inverted. For clarity, only two light rays are shown here*

However, objects are seen the right way up because of the way in which the brain interprets images. Experiments have been performed in which human volunteers wore special spectacles, the lenses of which produced upright images on the retina. For a while the subjects in the experiment were confronted with an upside down world. However, after a few days they became used to the situation and their perception became adjusted so that once more they perceived things the right way up. When the spectacles were removed they again experienced a period of seeing things upside down before normal perception returned. Thus the mechanism of image interpretation, situated in the brain's cerebral cortex, is somewhat flexible.

FIG. 16.5 *Focusing (accommodation)*

(a) *Light from a distant object is focused on the retina by a flattened lens*

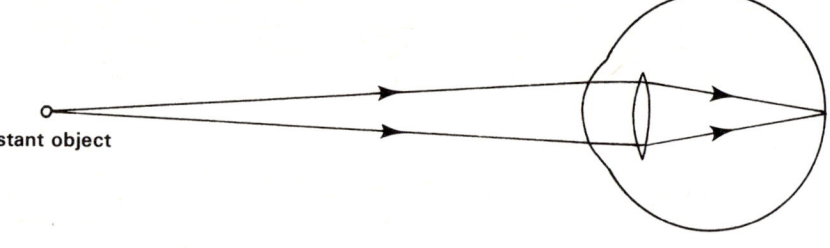

(b) *Light from a near object is focused on the retina by a near-spherical lens*

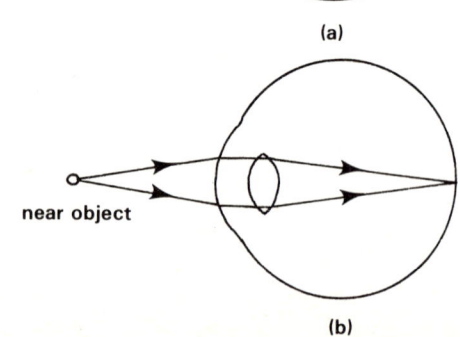

Light rays from an object near the eye strike the cornea and lens at an acute angle depending on the object's size. If the same object is moved further away from the eye the angle is less acute. Consequently the degree of refraction necessary to **focus** light rays on the retina is greater for close objects than for distant objects. Changes in the degree of refraction of light are achieved by altering the curvature of the lens' surfaces (Fig. 16.5). The ciliary body contains muscles the contraction or relaxation of which alters the tension of the suspensory ligaments which hold the lens in place. It is the tension of the ligaments applied to the lens which determines the shape of the lens. When the tension is increased the lens is pulled into a flattened shape suitable for focusing distant objects. When the tension is decreased, the lens becomes a more spherical shape suitable for focusing near objects.

16.1.3 Defects of the eye

A number of abnormalities of the eye's focusing mechanism exists in man and other mammals. The commonest are **myopia (short-sightedness)**, **hypermetropia (long-sightedness)**, and **astigmatism**, which is a mixture of the other two defects.

1. MYOPIA

Myopia results when the lens' curvature is too great (Fig. 16.6(a)). Light rays entering the eye are refracted more than is necessary. Consequently light is focused in front of the retina. By the time the light stimulates the retina it has diverged from the focal point of the lens. The image perceived is thus blurred. The condition is called short-sightedness because objects near the eye are less out of focus than those further away. This is because the light rays from near objects require greater refraction to be focused on the retina than rays from distant objects. Since the lens in a myopic eye refracts light excessively, distant objects appear more blurred than near ones. Myopia can be corrected by placing a **concave lens** in front of the eye. The surface of the concave lens refracts light rays in such a way that the rays diverge slightly from their original path. The lens of the myopic eye now refracts the diverged light rays into focus on the retina (Fig. 16.6(b)).

FIG. 16.6 *Myopia*

(a) *Light is focused on a point in front of the retina*

(b) *Light is focused on the retina by placing a concave lens in front of the eye*

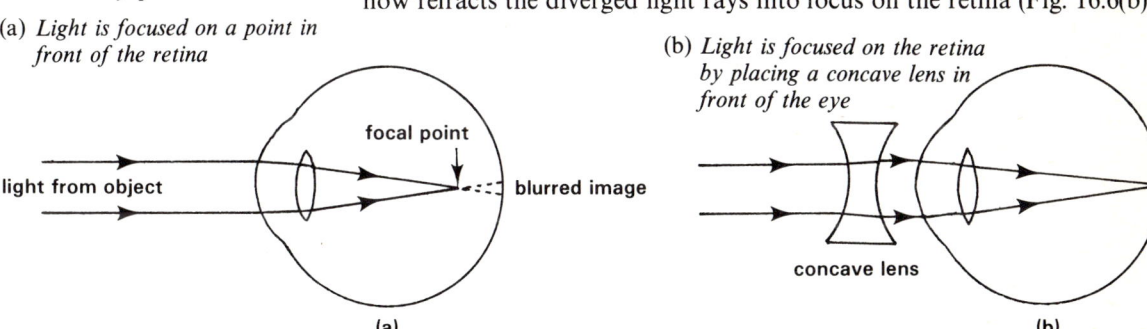

2. HYPERMETROPIA

Hypermetropia results when the curvature of the eye's lens is not great enough. Light rays are not refracted enough and are thus focused behind the retina (Fig. 16.7(a)). The condition is called long-sightedness because distant objects are less out of focus than near ones. This happens because light rays from distant objects require less refraction than rays from near objects. Correction of hypermetropia requires placing a lens in front of the eye. This time the lens is **convex**. The lens converges light rays before they enter the eye so that the eye's lens focuses the light correctly on the retina (Fig. 16.7(b)).

FIG. 16.7 *Hypermetropia*

(a) *Light is focused at a point behind the retina*

(b) *Light is focused on the retina by placing a convex lens in front of the eye*

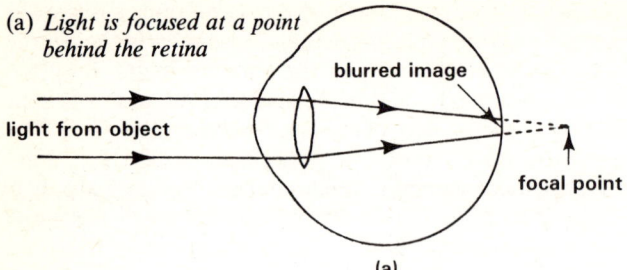

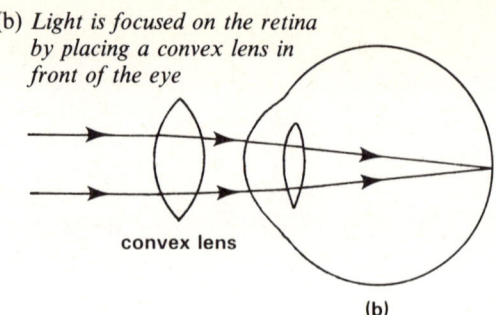

(a)

(b)

3. ASTIGMATISM

Astigmatism occurs if either the cornea or lens is distorted. One part of the focusing mechanism then refracts light rays too much, another not enough. Usually most of the image perceived is out of focus. Light rays from parts of the object are focused in front of the retina, as in myopia. Rays from other parts are focused behind the retina, as in hypermetropia. Astigmatism can be corrected by placing a lens in front of the eye. The curvature of this lens varies from one part to another to compensate for the eye's deficiencies.

The eye has often been compared with a camera. Look at Figures 16.1 and 16.8 and assess how valid the comparison is.

FIG. 16.8 *The main components of a simple camera*

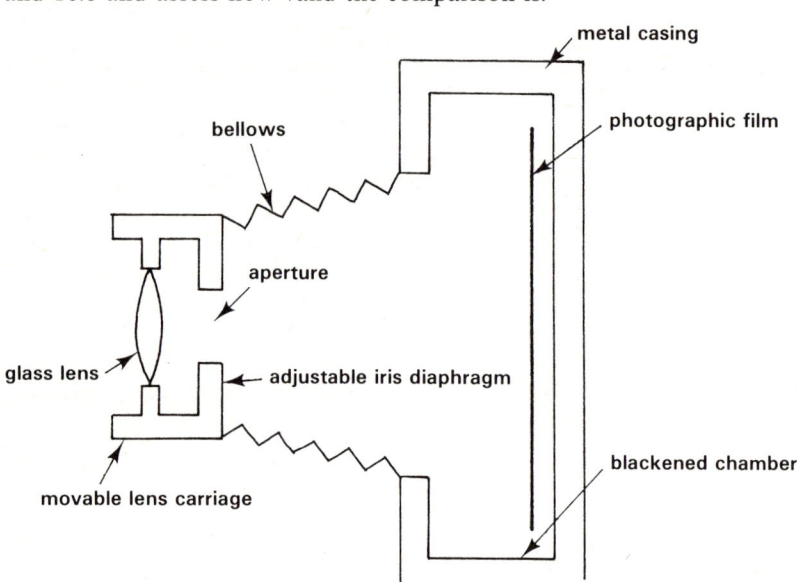

16.1.4 Photoreception

The transmission of nerve impulses to the brain in response to light is the function of the optic nerves and the **photoreceptors** in the retina. In human eyes and those of some other mammals there are two types of photoreceptors called **rods** and **cones**.

1. RODS AND CONES

Rods are sensitive to different intensities of light. Most mammals have only rods. **Cones** are sensitive to different wavelengths of light and enable some mammals to see things in colour. The assumption that an angry bull will charge a red object is somewhat misplaced, since the retina of a bull's eye has no cones. The bull is just as likely to charge an object of some other colour if it is annoyed.

2. DISTRIBUTION OF RODS AND CONES

The arrangement of photoreceptors in the mammalian retina is such that light has to travel through several layers of neurons which are not sensitive to light before reaching the rods and cones. The retina is **inverted** (Fig. 16.9).

FIG. 16.9 *The main cellular components of the retina*

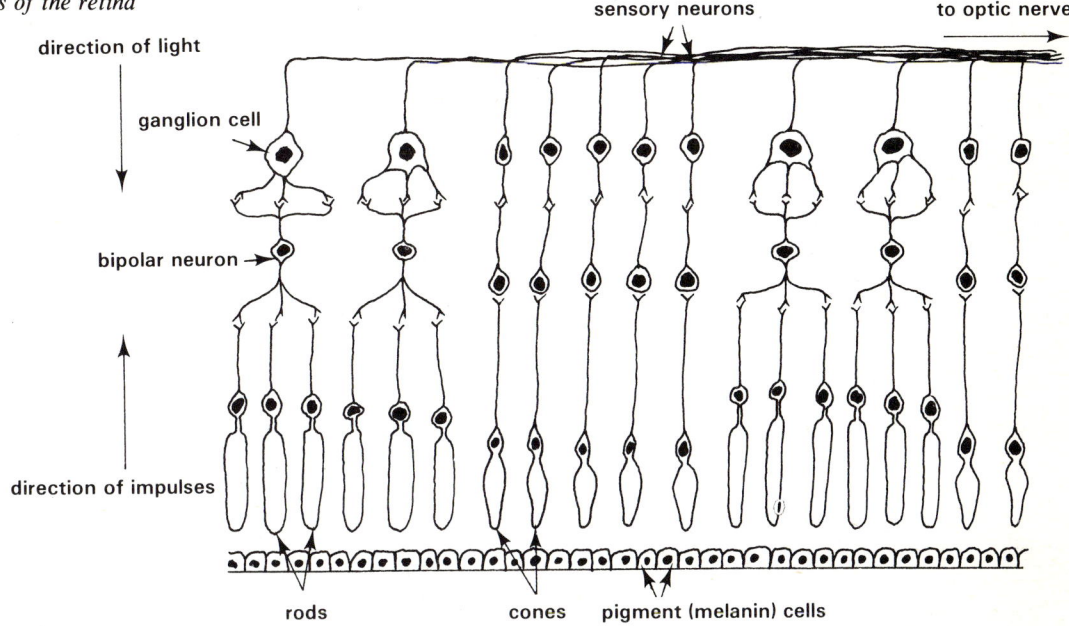

Beneath the photoreceptors is a layer of cells containing the black pigment melanin. Melanin is not sensitive to light but absorbs light rays which would otherwise pass through the retina. In this way the formation of hazy images caused by reflection of light to other parts of the retina is prevented. Some mammals have a reflective layer called the **tapetum** in the retina. The reflective 'cat's eyes' along the centre of roads create a similar effect to the reflective tapetum of real cat's eyes. A tapetum is common in nocturnal mammals and enables the maximum use of what little light is available at night.

Many synapses link the photoreceptors with sensory neurons. Impulses generated by the photoreceptors are first transmitted to a small region where the neurons project through the retina into the optic nerve. This region is called the **blind spot**, since no photoreceptors are located there (Fig. 16.1). It is possible to demonstrate that there are blind spots in your eyes by referring to the circle and cross illustrated below. Close your right eye and hold the page about 60 cm away from your open left eye. Keeping your left eye focused on the circle, the cross should be visible but slightly less clear. Now move the page slowly towards you and notice that at a distance of about 30 cm from your left eye the cross disappears. At this point light from the cross falls on the blind spot of your left eye. If you now move the page even nearer to your left eye the cross should reappear.

In this simple demonstration the circle appears clearer than anything around it in the field of vision. The reason for this is that when you direct your eye at the circle the light from it is focused on to a region of the retina

323

called the **fovea**. Only cones are found in the foveas of the eyes of animals which have colour vision. Each cone forms a synapse with very few, sometimes only one, sensory neurons. Consequently the signals sent to the brain from each cone come from a small area of the retina on which a small part of the image is focused. The cones can thus discriminate between two points of the image which are close to one another. For this reason cones are said to have high **visual acuity**. On the other hand, many rods synapse with each sensory neuron. The signals transmitted to the brain from the rods therefore come from a relatively large area of the image (Fig. 16.10). The rods are distributed throughout the retina but are absent from the fovea. The part of the image focused on the fovea therefore appears much clearer than the rest of the image.

FIG. 16.10 *Cones synapse with single sensory neurons, whereas many rods synapse with a single sensory neuron. Cones are therefore important in the discrimination of close points in the image*

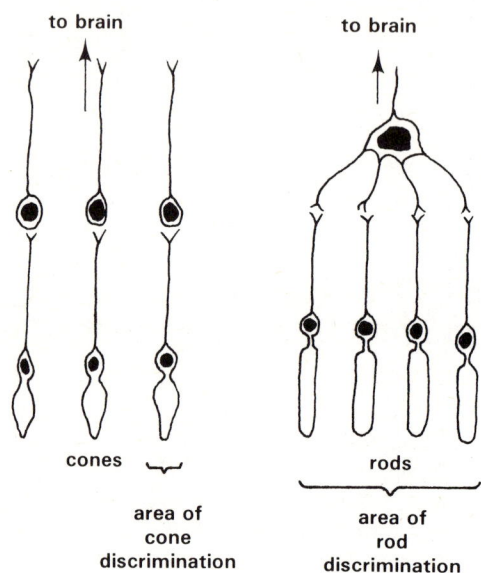

3. PHOTOSENSITIVE PIGMENTS

The functioning of rods and cones depends on **photosensitive pigments**. Electron microscopy has revealed the intricate subcellular structure of rods (Fig. 16.11). The outer segments of rods contain a great number of membranous, disc-like lamellae. The lamellae contain a photosensitive pigment called **rhodopsin**. Cones contain a pigment called **iodopsin**.

(i) RHODOPSIN consists of a protein called **opsin** attached to **retinal**, a derivative of vitamin A (Chapter 14). When exposed to light rhodopsin is split into opsin and retinal. The breakdown of rhodopsin produces a **generator potential** which is transmitted from the rod as an action potential to a sensory neuron. Rhodopsin, split in this way, has to be resynthesised to maintain the rod's ability to respond to light.

Resynthesis of rhodopsin requires energy and the rods have many mitochondria which make ATP for this purpose. However, rhodopsin resynthesis takes time. It is a common experience to suffer a brief period of poor vision after going from a well-lit room into darkness. When exposed to bright light rhodopsin is broken down rapidly and the reserve of rhodopsin in the rods is low. The eyes are **light-adapted**. If the retina is then exposed to dim light the rods show little response so vision is poor. The several minutes getting used to the dark is the time taken for enough rhodopsin to be resynthesised. When the retina is sufficiently sensitive for us to see in dim light the eyes are **dark-adapted**.

A brief period of poor vision is also experienced when you go from a very dark to a brightly lit room. When dark-adapted, the retina can work in dim light. Exposure of a dark-adapted retina to bright light overloads the photoreception mechanism. Light rays, even from the darker areas of an object in view, stimulate the rods which are now rich in rhodopsin. The eyes are light-adapted when excess rhodopsin is broken down and the retina once more adjusts to working in bright light. Prolonged exposure to very intense light, however, can reduce sensitivity of the retina too much. The rate of rhodopsin resynthesis may then be unable to keep pace with its breakdown. **Snow blindness** is caused by such an effect.

FIG. 16.11 (a)
Electronmicrograph of parts of several rod cells from the retina of a cat, ×5800

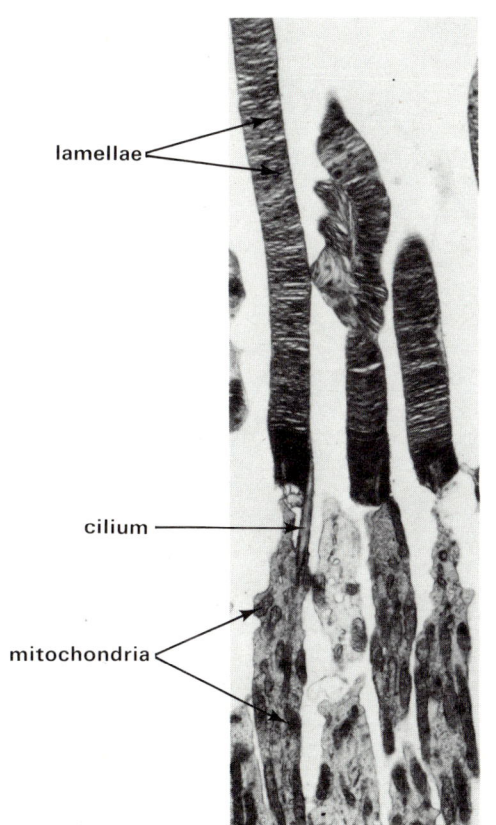

lamellae

cilium

mitochondria

FIG. 16.11 (b) *Structure of a rod cell, based on electronmicrographs*

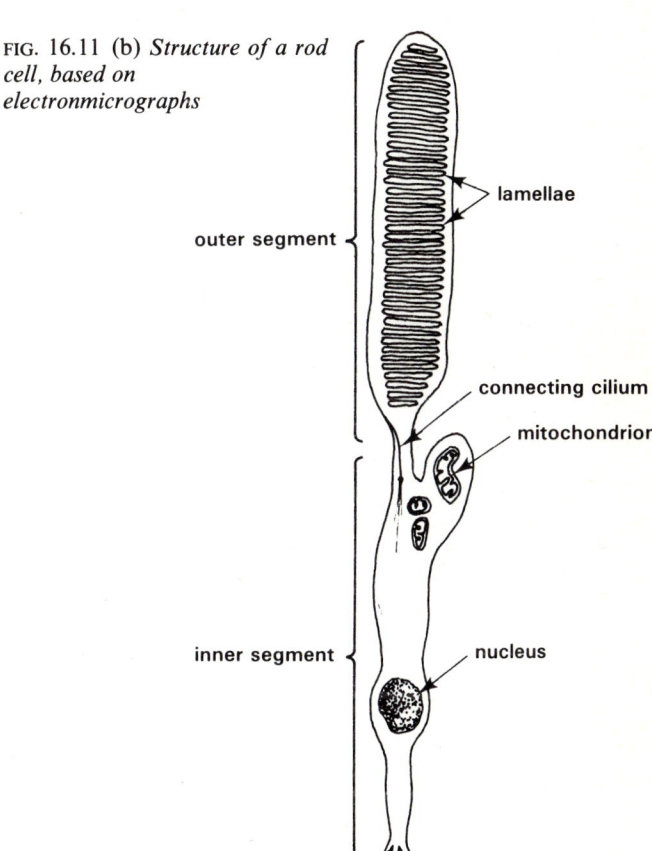

outer segment

lamellae

connecting cilium

mitochondrion

inner segment

nucleus

(ii) IODOPSIN is less sensitive to changes in light intensity than rhodopsin, so the cones are of little value in helping us to see in dim light. A popular theory of colour vision suggests that there are three variants of iodopsin each of which is sensitive to light of the primary colours, red, blue and green (Fig. 16.12(a)). For this reason the theory is called the **trichromatic theory** of colour vision. Each type of iodopsin is probably located in different cones. Light with a wavelength between those of the primary colours stimulates combinations of cones. Yellow light for instance, simultaneously stimulates cones which are sensitive to red and green light. The action potentials generated in the sensory neurons in response to generator

325

potentials in the cones are interpreted by the brain as the appropriate intermediate colour, yellow in this case (Fig. 16.12(b)).

Deficiency of one or more of the three primary colour cones results in **colour blindness**. A glance at Figure 16.12(a) shows that there is some overlap in colour sensitivity between the three types of cones. For example, a green cone is sensitive to red light but its sensitivity to red is far less than that of a red cone. Absence of red cones therefore, means that it is still possible to perceive green, yellow, orange and red. However, the brain cannot distinguish satisfactorily between these colours because there are no impulses from red cones with which to contrast impulses from green cones. A similar effect occurs when green cones are absent. The condition is called **red-green colour blindness**. Very rarely, blue cones are absent causing **blue weakness**. The exact mechanism of colour vision is not known. Many people challenge the trichromatic theory and several alternative theories have been proposed in recent years.

FIG. 16.12 (a) *Absorption spectra showing the wavelengths (colours) of light absorbed most by the iodopsin in the three main types of cones* (after Marks, Dobelle, MacNichol, Brown and Wald)

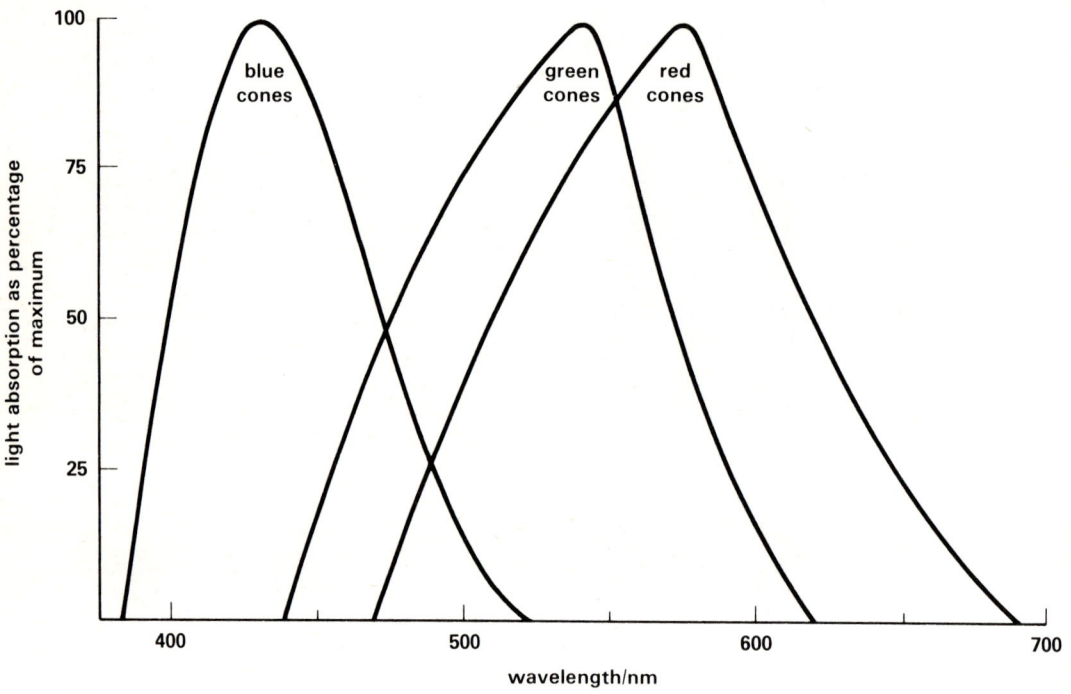

FIG. 16.12 (b) *Different colours are perceived in the brain from the sensory information received from the cones. Signals from combination of different cones produce the sensation of intermediate colours based on red, green and blue detected by the three primary cones*

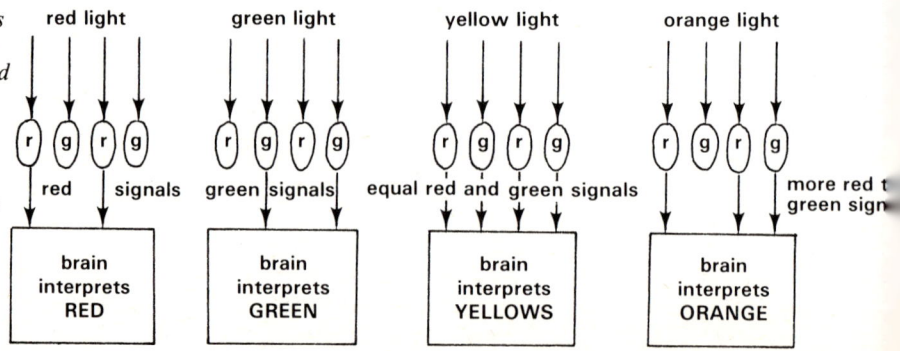

16.2 The mammalian ear

The ability to generate sounds, as well as receive and interpret them, is valuable as a means of communication between mammals. Such a form of communication is particularly well developed in human language. Sound waves trigger off the transmission of sensory nerve impulses from the ear to the cerebral cortex of the brain.

The mammalian ear also performs another, equally important function. It transmits to the brain information about the body's relative position in space.

16.2.1 Structure of the ear

The mammalian ear is divided into three main regions, the **outer ear**, **middle ear** and **inner ear** (Fig. 16.13). The outer ear channels sound waves from the surrounding air into the middle ear where the energy of sound waves is converted to mechanical vibrations. In the inner ear nerve impulses are generated in response to vibrations received from the middle ear and to changes in position of the head.

FIG. 16.13 *Structure of the human ear*

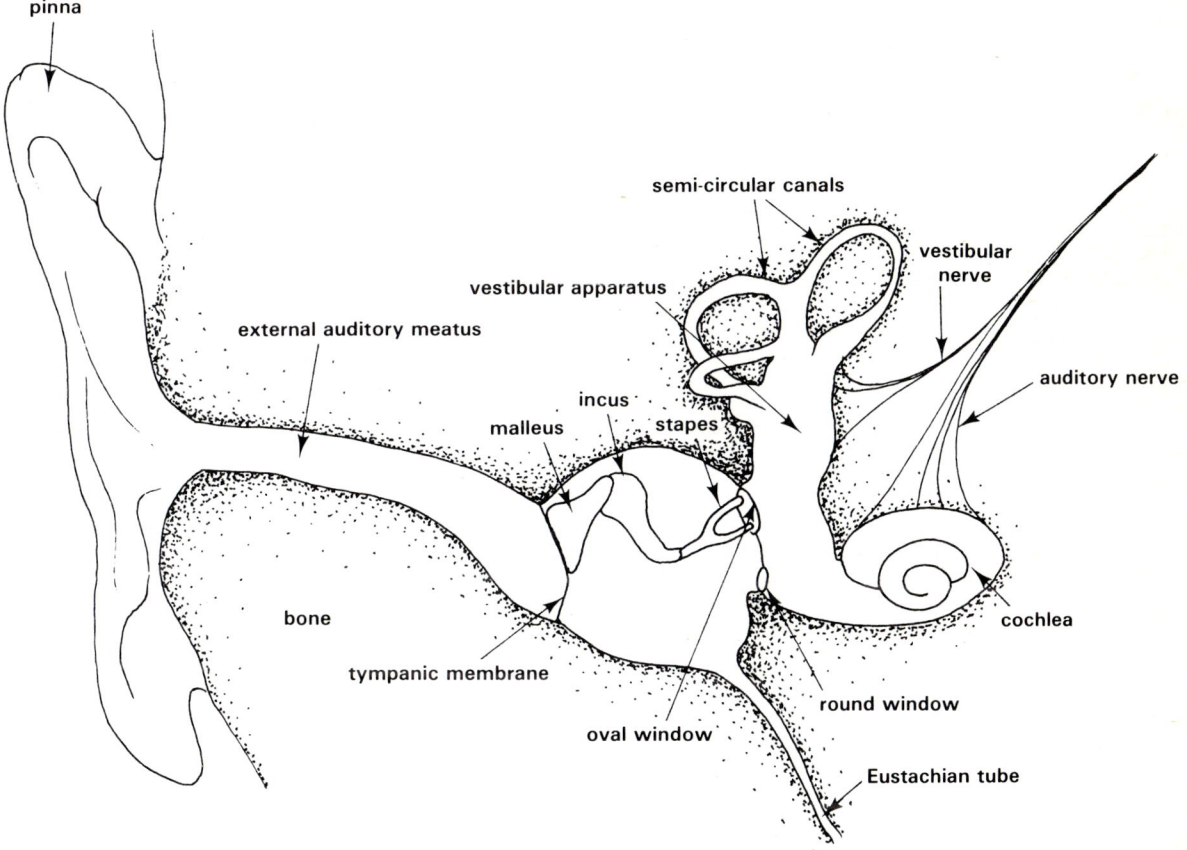

16.2.2 Hearing

Sound waves enter the ear along a short tube called the **external auditory meatus**. The external flap of skin, the **pinna**, helps in directing sound waves from the environment into the meatus. Some mammals have large and mobile pinnae which enable them to locate the source of sound without moving the head. At the inner end of the external auditory meatus is an elastic membrane called the **tympanic membrane** or eardrum.

Bridging the middle ear are three small bones called **ossicles** held in place by muscles and tendons. The ossicles are the **malleus**, **incus** and **stapes**. Sound waves entering the ear vibrate the tympanic membrane which in turn vibrates the ossicles. The innermost ossicle, the stapes, is connected to another membrane called the **oval window** which is part of the inner ear. The membrane of the oval window is less than five per cent of the area of the tympanic membrane. Consequently vibrations of the tympanic membrane are amplified about 20 times in the oval window. Amplification makes it easier for vibrations to pass through the dense liquid in the inner ear.

An air-filled canal, called the **Eustachian tube**, connects the middle ear with the pharynx. The air pressure in the external air and in the middle ear is normally the same. Should there be a sudden large increase in external air pressure there is a possibility that the eardrum will burst. However, this danger is usually prevented when air taken in from outside enters the Eustachian tube during swallowing. In this way the air pressure on either side of the eardrum is equalised.

The oval window transmits the vibration of the ossicles into a coiled, fluid-filled tube called the **cochlea** (Fig. 16.14). The cochlea contains three canals separated from each other by two flexible longitudinal membranes. The upper **vestibular canal** is connected to the oval window. Between the vestibular canal and the **median canal** is **Reissner's membrane**. The **basilar membrane** separates the median canal from the lower **tympanic canal**.

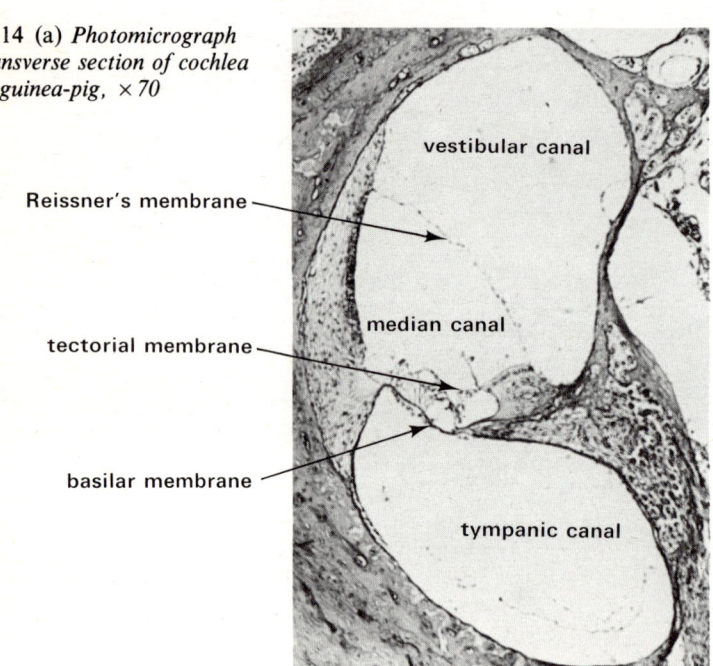

FIG. 16.14 (a) *Photomicrograph of a transverse section of cochlea from a guinea-pig,* × 70

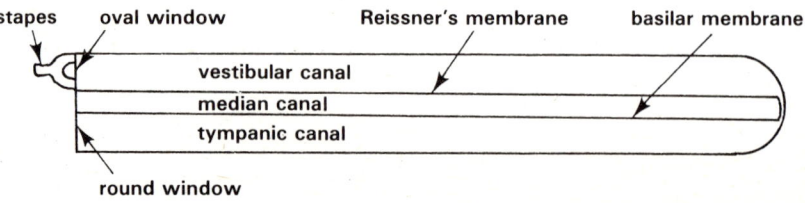

FIG. 16.14 (b) *Diagram of the cochlea 'straightened out' to show the three canals and their separating membranes*

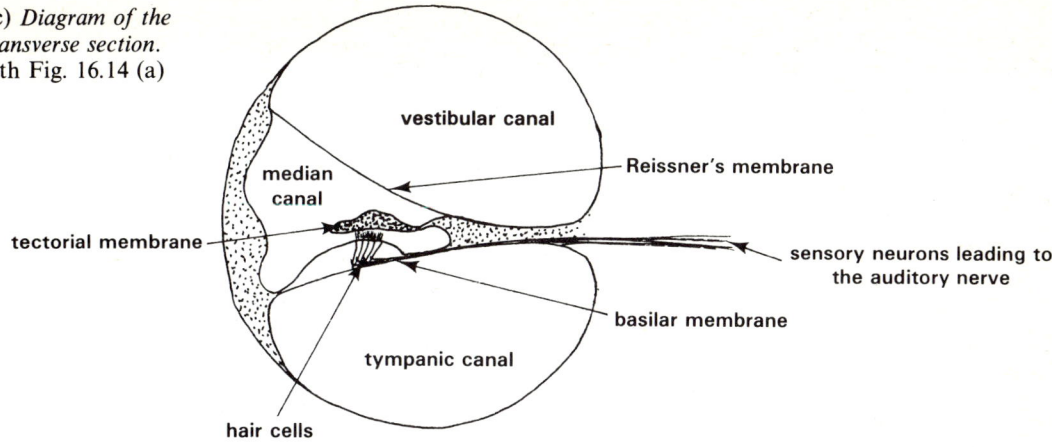

FIG. 16.14 (c) *Diagram of the cochlea in transverse section. Compare with Fig. 16.14 (a)*

vestibular canal

median canal

Reissner's membrane

tectorial membrane

sensory neurons leading to the auditory nerve

basilar membrane

tympanic canal

hair cells

Vibrations of the oval window generate pressure waves in the fluid filling the vestibular canal. The waves cause Reissner's membrane to vibrate which in turn causes pressure waves to pass through the fluid in the median canal. The pressure waves in the median canal eventually bring about vibration of the basilar membrane. The tympanic canal is connected to a circular membrane called the **round window** just beneath the oval window. Pressure waves in the fluid inside the tympanic canal cause vibration of the round window. This arrangement causes the basilar membrane to vibrate in response to vibrations transmitted to the cochlea by the ossicles of the middle ear (Fig. 16.15).

FIG. 16.15 *Vibrations of the stapes move the oval window back and forth. The oscillations of the oval window generate pressure waves in the cochlear fluids and vibrations in the cochlear membranes. Note the effects of the vibrations on the round window, (a) when the oval window is 'pushed' and (b) when the oval window is 'pulled'*

The sensory region of the cochlea is the **organ of Corti** in the median canal (Fig. 16.14(c)). The organ contains many hair cells which are rooted in the basilar membrane. Short sensory hairs project from the free ends of these cells and touch the **tectorial membrane**. Vibrations of the basilar membrane cause the hair cells to move towards the tectorial membrane, deflecting the sensory hairs. Consequently nerve impulses are generated in sensory neurons which synapse with the hair cells. The impulses are transmitted to the brain along the auditory nerve.

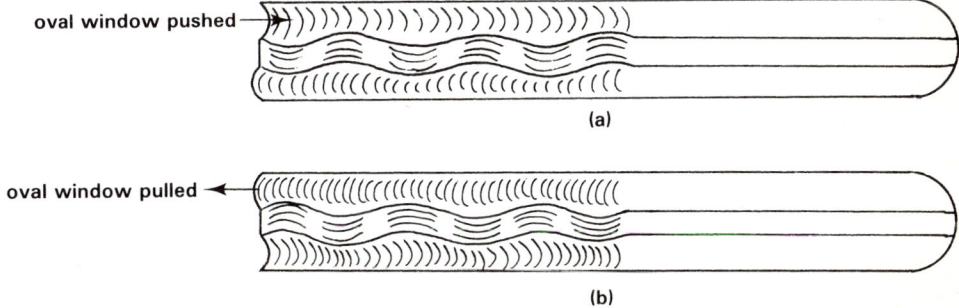

oval window pushed

(a)

oval window pulled

(b)

Vibrations of the basilar membrane are crucial to the conversion of sound into sensory nerve impulses. The basilar membrane is about 2·5 times wider at the apex of the cochlea than it is at its base, between the oval and round windows. Sound waves of short wavelength which have a relatively high frequency vibrate just a short portion of the basilar membrane. Only the hair cells near the oval window are thus stimulated. The

329

impulses arriving at the brain are interpreted as sounds of a high pitch. A greater length of the basilar membrane is vibrated by sound waves of long wavelength. Hair cells further along the membrane now generate impulses in the auditory nerve. The brain interprets these impulses as sounds of a low pitch. Sounds of intermediate wavelength stimulate the middle part of the basilar membrane (Fig. 16.16).

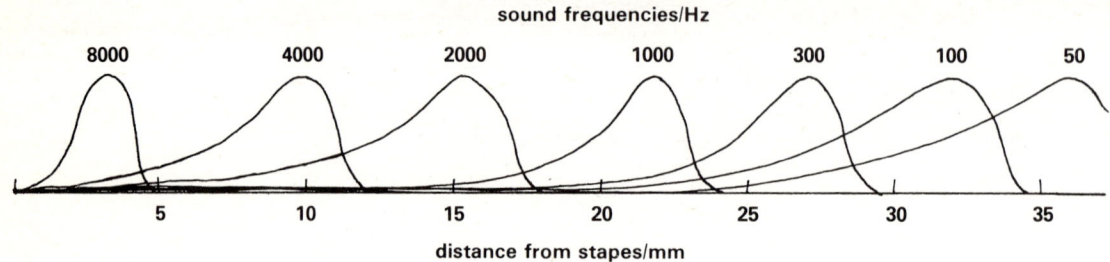

The intensity of sound is discerned according to the intensity with which appropriate sections of the basilar membrane are vibrated. Weak sounds generate fewer nerve impulses than loud sounds. The interpretation of sounds takes place in the cerebral cortex of the brain (Chapter 15).

16.2.3 Balance

Just above the cochlea and connected to it by a short tube is the **vestibular apparatus** (Fig. 16.17). The vestibular apparatus consists of two lymph-filled sacs called the **saccule** and the **utricle**. Projecting from the top of the utricle are three **semicircular canals**, also containing lymph.

FIG. 16.17 *The vestibular apparatus of the human ear*

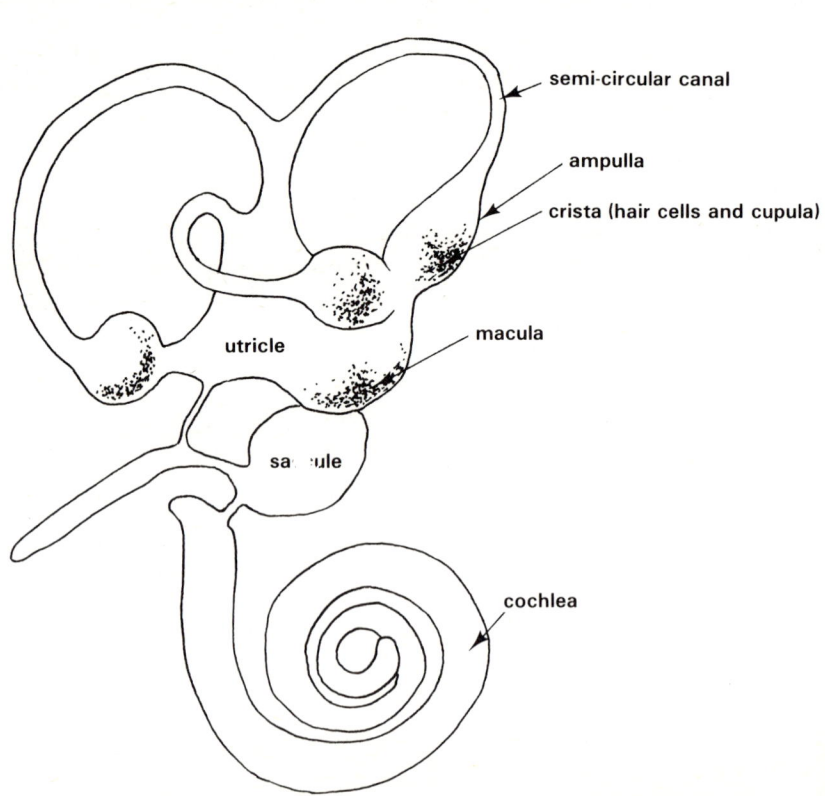

The saccule and utricle contain receptors called **maculae** which are sensitive to gravity. Small hairs project from the receptor cells into the lymph. The hairs are attached to calcium carbonate granules called **otoliths** (Fig. 16.18). Gravity causes the otoliths to distort the sensory hairs in a direction dictated by the position of the head. In response to the distortion, nerve impulses pass along the vestibular nerve to the brain. If the head is moved to a different position the otoliths distort the sensory hairs in a different direction. Information about the head's new position is interpreted in the brain.

FIG. 16.18 (a) *Structure of the macula from the utricle*

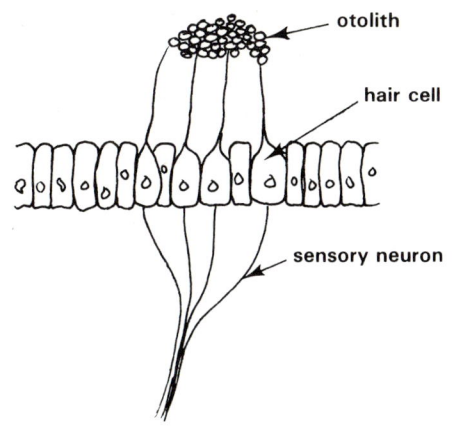

otolith

hair cell

sensory neuron

FIG. 16.18 (b) *The otolith deflects the sensory 'hairs' in a direction dictated by gravity and the position of the head in relation to the earth*

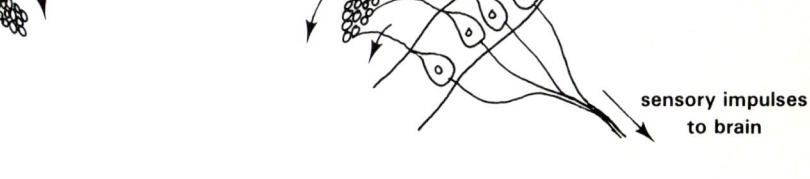

sensory impulses to brain

sensory impulses to brain

The semicircular canals provide information about head movements rather than the position of the head when it is stationary. The ends of the canals are enlarged to form an **ampulla**. Each ampulla contains a receptor, consisting of hair cells similar to those of the maculae. The ampullary hairs, however, project into a gelatinous mass called a **cupula** which is suspended in the lymph inside the ampulla (Fig. 16.19(a)).

Any movement of the head of course moves the semicircular canals in the same direction. The lymph inside the canals, however, lags behind and pushes the cupulae in the opposite direction (Fig. 16.19(b)). As a result, the hairs projecting into the cupulae are bent and nerve impulses are sent to the brain along the vestibular nerve. There are three semicircular canals and each is arranged at a right angle to the others. Consequently, at least one cupula is stimulated by lymph movements whatever direction the head is moved.

Information about the orientation and movement of the head is vital, especially when the whole body is moved. Visual information from the eyes also contributes to the body's awareness of spatial position. The information is used by the brain to co-ordinate movement and posture of the body.

16.2.4 Defects of the ear

The most common defects of the ear lead to hearing loss or **deafness**. Deafness can be caused by inability of the outer and middle ear to conduct vibrations to the cochlea. **Conductive deafness** is of this sort. One of the commonest causes of conductive deafness in man is blockage of the external auditory meatus with wax. Little is known of the function of wax which is secreted from the **ceruminous glands** in the skin inside the meatus. In some people the wax accumulates in the meatus and hardens, sometimes pressing against the eardrum. Normal hearing is usually restored after the hardened wax is removed with a special syringe.

FIG. 16.19 (a) *Structure of the receptors in the ampullae of the semi-circular canal*

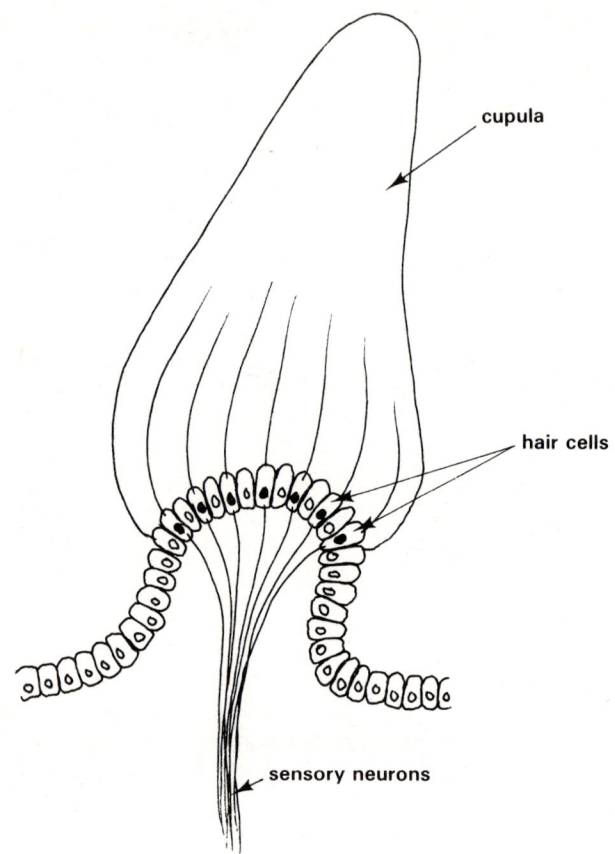

FIG. 16.19 (b) *Movements of the fluid inside the semi-circular canals, caused by movements of the head, displace the gelatinous cupula and sensory 'hairs'*

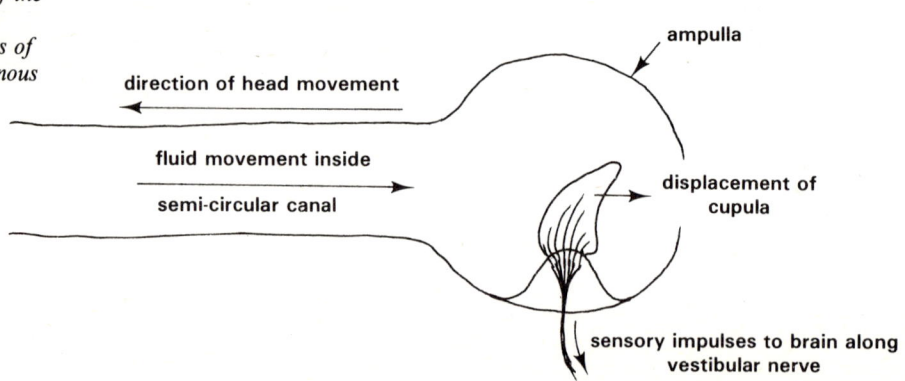

FIG. 16.20 *A 'behind-the-ear' hearing aid* (from DHSS booklet *General Guidance for Hearing Aid Users*)

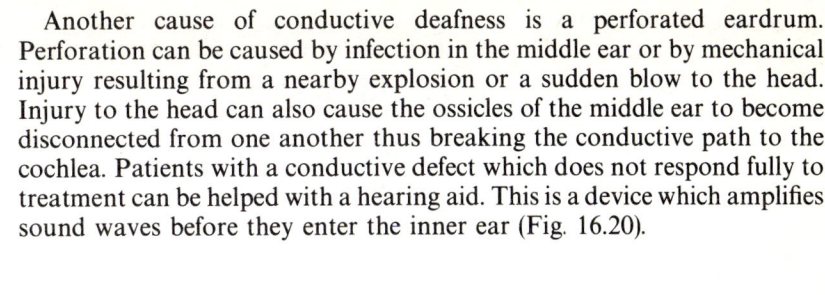

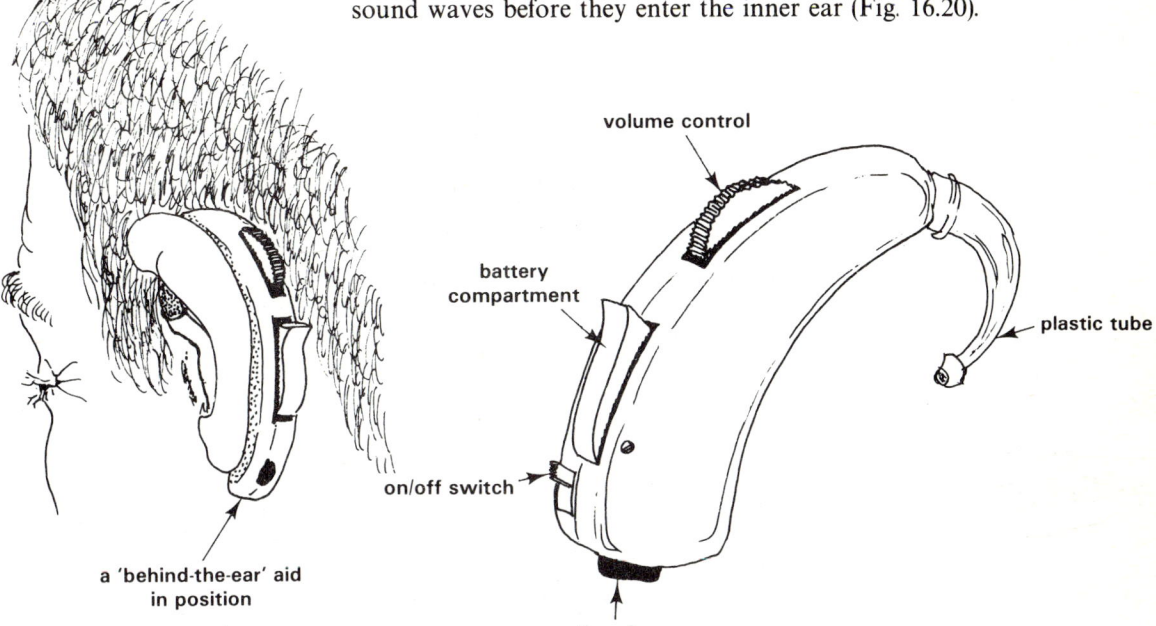

volume control

battery compartment

plastic tube

on/off switch

microphone

a 'behind-the-ear' aid in position

Another cause of conductive deafness is a perforated eardrum. Perforation can be caused by infection in the middle ear or by mechanical injury resulting from a nearby explosion or a sudden blow to the head. Injury to the head can also cause the ossicles of the middle ear to become disconnected from one another thus breaking the conductive path to the cochlea. Patients with a conductive defect which does not respond fully to treatment can be helped with a hearing aid. This is a device which amplifies sound waves before they enter the inner ear (Fig. 16.20).

Malfunction of the cochlea and auditory nerve can be the cause of deafness even though vibrations are conducted perfectly into the inner ear. Such deafness is called **sensorineural (perceptive) deafness**. Sensorineural deafness can be inherited though it is often acquired. Acquired forms of the condition can result from infection, head injury, blast from explosions or exposure to excessive noise. Much concern has been expressed in recent years about the possible harmful effects on hearing of the very high noise levels in discotheques, at airports and where noisy machinery is used. So serious is the problem that people employed in extremely noisy places are encouraged to wear ear muffs to protect their hearing from permanent damage.

Complex though sense organs such as the eye and ear are, it is the interpretation of sensory information which ultimately limits their use. Information processing occurs in the brain by mechanisms which are, as yet, poorly understood.

Chapter 17

Thermoregulation in mammals

One of the many aspects of homeostasis is the regulation of body temperature. It is a characteristic feature of birds and mammals. These animals are called **endotherms**. They maintain a relatively high and constant body temperature which enables them to lead active lives even when the temperature of their surroundings is low. All other animals are called **ectotherms**. Their body temperature varies with temperature fluctuation of the environment (Fig. 17.1). Because the metabolic rate slows down when the body temperature is low, ectotherms are sluggish in cold weather. In these conditions they are less likely to succeed in obtaining food and avoiding predators (Fig. 17.2). Endothermy is thus of obvious survival value.

The regulation of body temperature is called **thermoregulation** and is brought about by balancing heat production in the body with heat loss to the environment.

17.1 Heat production

The main source of heat in a mammal is tissue respiration. Over 50 per cent of the energy released in respiration is heat energy (Chapter 5). The rate at which heat is produced is proportional to the metabolic rate which varies with the intensity of exercise. Humans produce about 250 kJ h^{-1} of heat energy when resting. Heat production increases to 1000 kJ h^{-1} during moderate exercise while the rate may be the equivalent of 8000 kJ h^{-1} in a few minutes of intense exercise.

FIG. 17.1 *Effects of environmental temperature on the body temperatures of three mammals (endotherms) and the lizard (an ectotherm)*

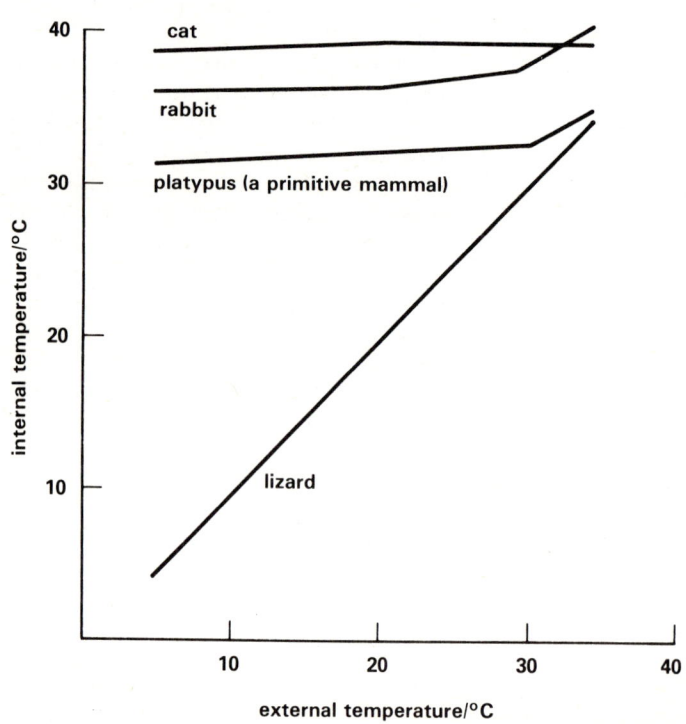

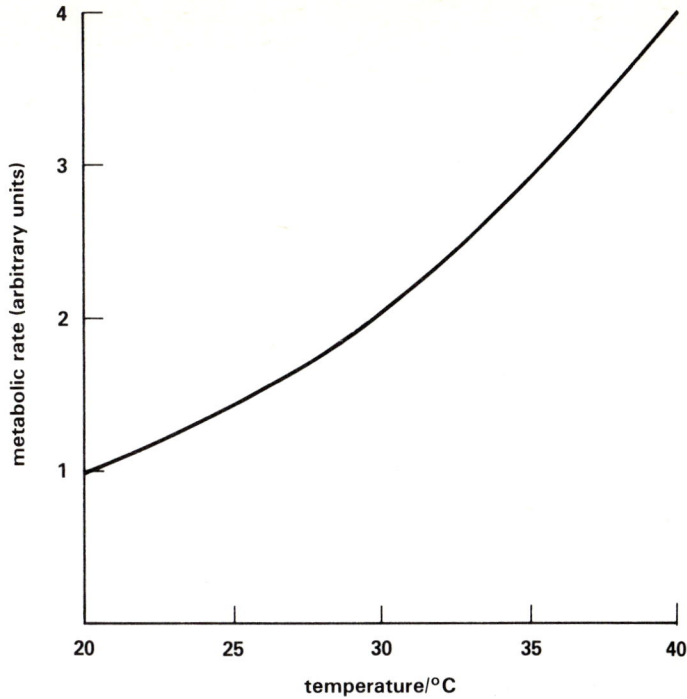

FIG. 17.2 *Effects of temperature on metabolic rate.* Compare with Fig. 4.7

17.1.1 Measuring metabolic rate

When an animal is fasting energy is released in respiration of food reserves such as fat and glycogen in the body. When at rest nearly all the energy released in respiration is eventually given off as heat.

Measurement of the heat given off by a resting, fasting individual gives an indication of the energy required to maintain the body's vital functions.

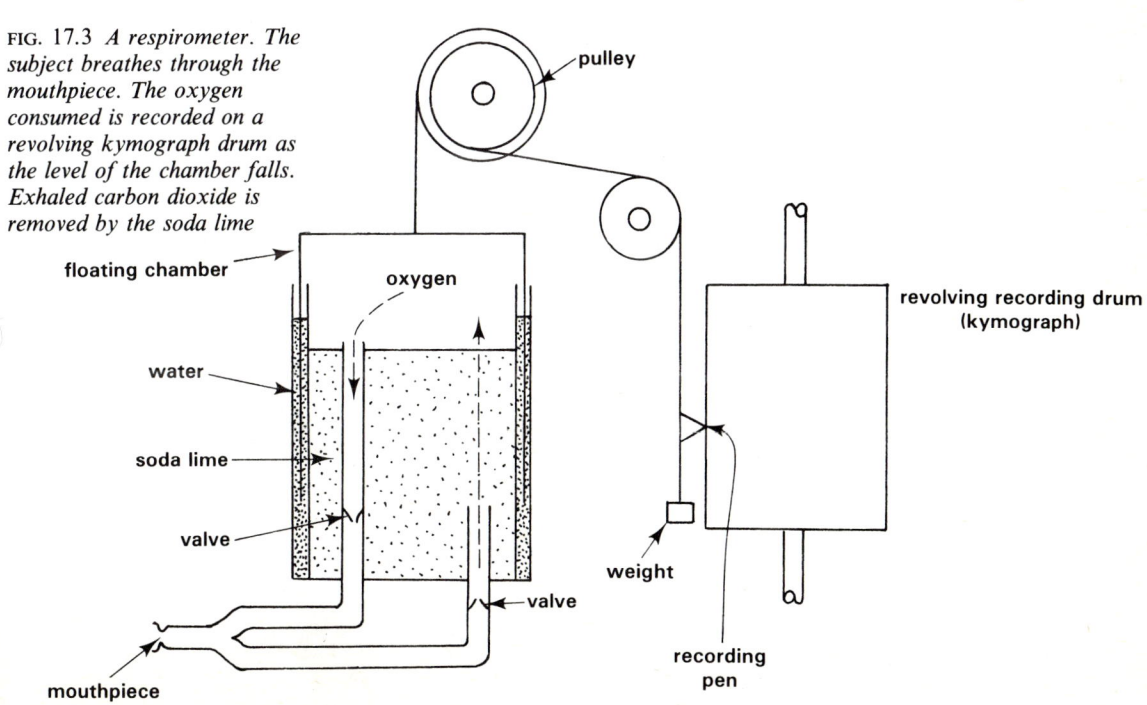

FIG. 17.3 *A respirometer. The subject breathes through the mouthpiece. The oxygen consumed is recorded on a revolving kymograph drum as the level of the chamber falls. Exhaled carbon dioxide is removed by the soda lime*

The rate at which the body respires to provide this amount of energy is called the **basal metabolic rate (BMR)**. The BMR of a man can be measured using a **human calorimeter** which is an insulated room containing water-filled pipes. The temperature of the water entering and leaving the room through the pipes is recorded. So is the temperature of the air in the room. Heat lost from the body over a period of time causes a rise in temperature of the water and air. From the data the BMR can be calculated.

A much quicker way of calculating the BMR involves using a **respirometer** (Fig. 17.3). A respirometer allows the volume of oxygen taken in by the subject in a given period of time to be measured. For every 1 dm^3 oxygen used in respiration, approximately 20·17 kJ of heat energy are released. This is the average of the amount of heat energy released from the oxidation of carbohydrates, lipids and proteins. If a subject takes in 1·5 dm^3 oxygen in 5 minutes, which is equivalent to 18 dm^3 h^{-1}, the rate at which heat energy is released is:

$$18 \times 20·17 \quad = 363·06 \text{ kJ h}^{-1}$$
$$\text{Thus the BMR} = 363·06 \text{ kJ h}^{-1}$$

The figure is generally adjusted to take into account differences in surface area of the body (section 17.2.4).

17.1.2 Distribution of body heat

Heat is produced unevenly in the body. Skeletal muscle releases a lot of heat during exercise. Another important source of heat is the liver (Chapter 14). Blood moving through the circulatory system has kinetic energy which is converted to heat energy when the blood meets resistance mainly in the arterioles. As blood flows through the skeletal muscles and liver it absorbs heat and distributes it to parts of the body where little heat is produced (Figs. 17.4 and 17.5). When blood flows near the body surface heat is lost through the skin.

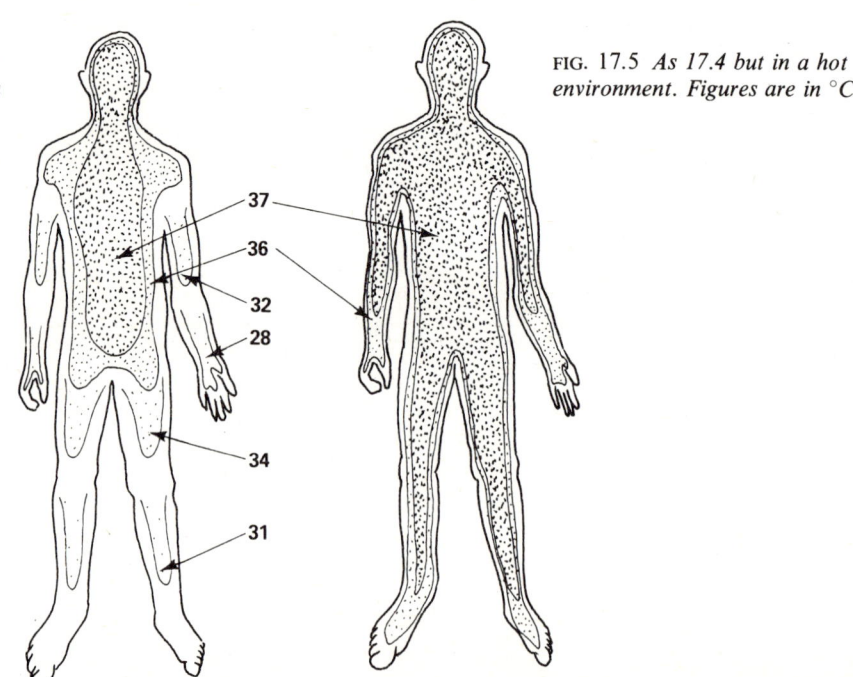

FIG. 17.4 *Distribution of heat on the surface of the human body in a cold environment*

FIG. 17.5 *As 17.4 but in a hot environment. Figures are in* °C

17.2 Heat loss

Mammals lose body heat in three main ways, **radiation**, **evaporation** and **conduction**.

17.2.1 Radiation

Radiation is the emission of heat from a body to its surroundings, mostly the air. Except in tropical regions, the body temperature of mammals is generally higher than the surroundings. Mammals therefore radiate heat to their surroundings more rapidly than they gain heat back by the same means. It has been estimated that 60 per cent of the total heat loss from a naked man sitting in a room kept at 33°C is by radiation (Fig. 17.6).

Measurements of heat coming from the body as infra-red radiation are useful in the detection of certain types of tumour. The patient is kept in a room of constant temperature, generally between 18° and 20°C, to minimise fluctuations in skin temperature. Photographs are taken with a special camera which is sensitive to infra-red radiation coming from the body. Hot areas of the skin show up as light patches on the picture. Cool areas show up as dark patches. Hot spots occur in areas where blood flow and metabolism are increased, which may be diseased areas of the body. The technique is called **thermography** and is often used in the detection of breast cancer.

FIG. 17.6 *Approximate relative heat loss in man by radiation (wavy line) evaporation (broken line) and conduction (straight line)*

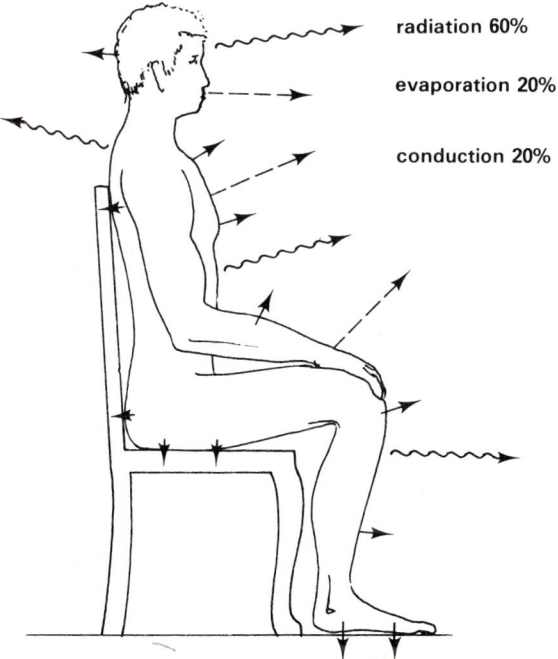

radiation 60%

evaporation 20%

conduction 20%

17.2.2 Evaporation

As water evaporates from a moist surface, energy is taken from the surface which consequently cools. The energy used is the **latent heat of vaporisation** (Chapter 2). The mammalian skin contains **sudorific glands** which secrete **sweat** on to the body surface when the body is overheated. Sweat is mainly water and when it evaporates the skin is cooled. Evaporation of body water also occurs from the moist linings of the nasal cavities, mouth, trachea and the extensive internal surface of the lungs. In very hot climates the surroundings often have a temperature higher than body temperature. In such conditions the body gains heat from the environment by radiation and conduction (section 17.2.3). Profuse sweating and subsequent evaporation of water from the skin's surface prevents overheating. A man can secrete up to 4 dm^3 of sweat per hour in very hot dry conditions. Dehydration of the body can then become a critical problem (Chapter 8).

Even in a moderately warm room when sweating is virtually nil, about 20 per cent of a man's total heat loss is due to evaporation. This is because even without sweating, small quantities of water vapour escape through the skin. Some of the heat loss is also due to evaporation from the mouth, nose, trachea and lungs (Fig. 17.6).

Many environmental factors affect the rate of evaporation and hence heat loss from the skin. Most notable are the water potential of the air in contact with the skin, the air temperature and air movement. They control the rate of evaporation from the mammalian body in much the same way that they influence transpiration from terrestrial plants (Chapter 11).

17.2.3 Conduction

Heat passes from a warm object to a cooler one in direct contact with it by **conduction**. The ground, if cooler than any part of the body in contact with it removes heat from the body by conduction. The greater the temperature difference between the body and other objects touching it, the greater is the rate of conductive heat loss. A cool breeze passing over the skin removes heat by conduction to the air. About 20 per cent of the total heat lost by the man illustrated in Figure 17.6 can be due to conduction.

17.2.4 Surface area and body volume

Heat is released in the body mainly in respiration. The amount of heat released depends on the **volume** of the body. Because most of the heat is lost through the skin, the amount of heat lost depends on the **surface area** of the skin.

As an animal grows, its volume increases in three dimensions whereas its surface area increases only in two dimensions. Consequently, relative to its volume its surface area increases at a slower rate. In terms of heat exchange, a bulky animal has a larger volume of tissues in which heat is released, but relative to this, a smaller surface through which heat is lost to the environment. The **surface area to volume (SA/V) ratio** is the area of skin per unit of body volume. A deer of volume 150 000 cm^3 and 19 000 cm^2 surface area has a SA/V ratio of:

$$\frac{19\ 000}{150\ 000} = 0.127\ \text{cm}^2\ \text{cm}^{-3}$$

In comparison, a squirrel of 625 cm^3 volume and 550 cm^2 surface area has a SA/V ratio of:

$$\frac{550}{625} = 0.88\ \text{cm}^2\ \text{cm}^{-3}$$

In other words, each cm^3 volume of the squirrel has about seven times the skin area available for heat loss compared with the bulkier deer (Fig. 17.7).

Such considerations highlight one of the factors which limits the extremes of size in terrestrial mammals. Bulky animals face overheating while small ones lose heat rapidly. Experimenting with small mammals, Pearson in 1957 obtained data for the metabolic rate of a variety of shrews. He related the metabolic rate to body mass (Fig. 17.8). The rate of metabolism controls the rate of heat production. It is, therefore, not surprising that the smaller shrews which have a greater relative surface area through which heat is lost show the highest metabolic rate. However, there is a limit to which increased metabolism can compensate for heat loss. It is estimated that a mammal smaller in size than the smallest species of shrew would be unable to make energy quickly enough from its food to make good the rate at which heat is lost through its body surface.

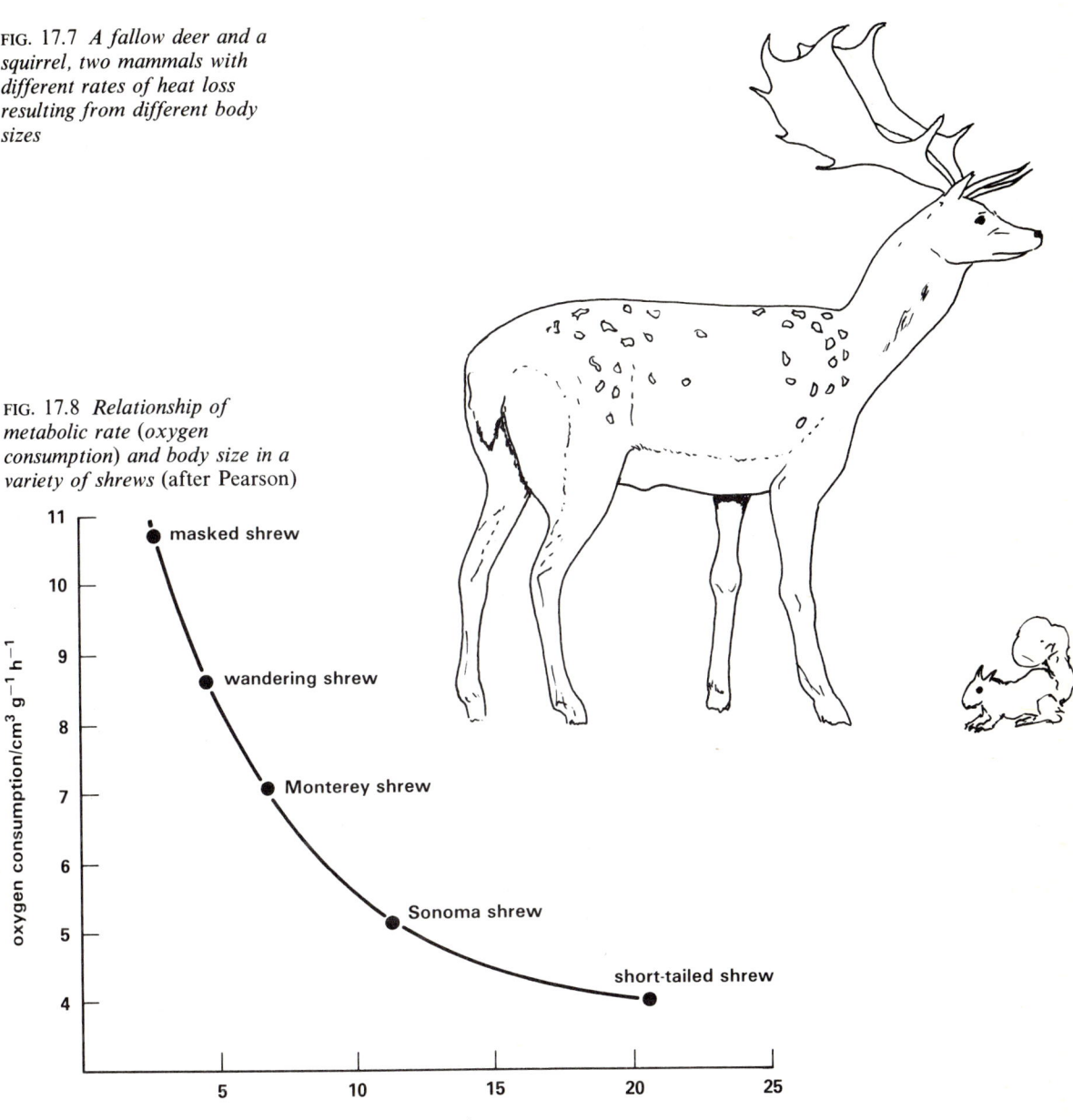

FIG. 17.7 *A fallow deer and a squirrel, two mammals with different rates of heat loss resulting from different body sizes*

FIG. 17.8 *Relationship of metabolic rate (oxygen consumption) and body size in a variety of shrews* (after Pearson)

Since metabolic rate and surface area can be related in this way, measurements of BMR are usually expressed in units which take account of surface area. Expressing BMR in this way allows comparison to be made between animals of different sizes. In man, for example, the BMR is expressed in $kJ\ m^{-2}h^{-1}$. The surface area of a human is difficult to measure directly but it can be calculated from measurements of height and mass (Fig. 17.9).

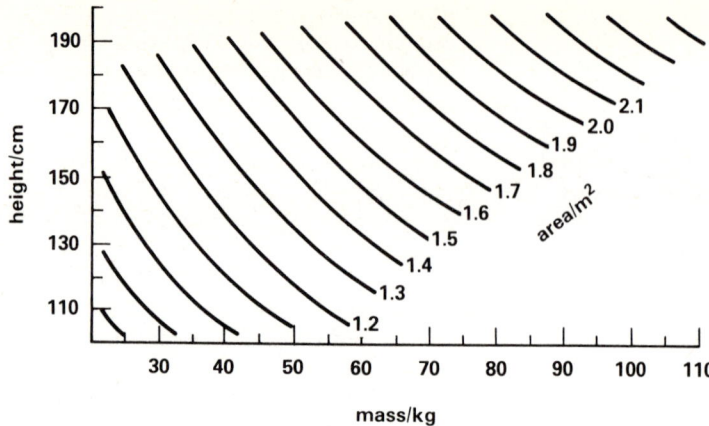

FIG. 17.9 *Relationship between height, mass and surface area in man*

17.3 Thermo-regulation

The regulation of body temperature in mammals is brought about by balancing heat production and heat loss.

17.3.1 The thermo-regulation centres

Thermoregulation takes place in response to changes in body temperature, but the changes must first be detected. This function is performed by specific parts of the hypothalamus in the forebrain (Chapter 15). In the hypothalamus are two **thermoregulation centres**. The **cold centre** responds to blood which has a temperature less than normal by triggering off responses which increase heat production and decrease heat loss. The **heat centre** responds to blood which has a temperature higher than normal by triggering off responses which reduce heat production and increase heat loss (Fig. 17.10).

FIG. 17.10 *Action of thermoregulation centres in the hypothalamus*

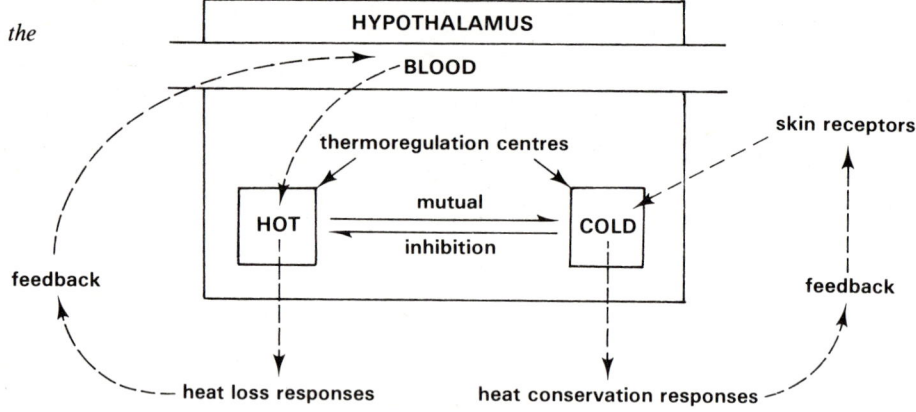

The heat centre acts like a **thermostat**. It switches on heat loss mechanisms when the temperature of the blood is higher than normal. Conversely it switches on heat conservation mechanisms when the temperature of the blood is lower than normal. The cold centre is inhibited by the heat centre, but becomes active when receptors in the skin signal that the environment is getting cooler. Heat production mechanisms are then switched on and heat conservation mechanisms are switched off.

In this way the body temperature is kept constant even though the environmental temperature varies. The cold centre also receives information about potential body temperature changes. The information travels along sensory neurons from the receptors in the skin which are sensitive to temperature changes outside the body. It is an early warning system which enables the thermoregulation centres to trigger off appropriate responses before external changes alter the internal body temperature too much.

17.3.2 Regulating heat production

Heat release in the body is ultimately limited by the availability of food. Assuming food to be adequate, other factors become important. Among these is production of the hormone **thyroxin** which controls the BMR (Chapter 18). Increased thyroxin output can double the BMR in man, but the response takes several days to come into effect. A similar but more immediate response is brought about by **adrenalin** made in the adrenal glands and **noradrenalin** secreted by sympathetic nerve endings (Chapters 15 and 18).

When the environmental temperature is very low considerable heat can be generated by **shivering**. Shivering is very rapid alternate contraction and relaxation of the skeletal muscles.

17.3.3 Regulating heat loss

There are several means by which mammals can alter the rate at which heat is lost from the body. Nearly all involve the **skin**. The mammalian skin acts as a physical barrier which prevents excessive loss of heat and water from the body and stops foreign matter and pathogenic microbes gaining access to the underlying tissues and organs. The skin has a thin outer layer called the **epidermis**, one of the main functions of which is to replace cells which are constantly lost from the surface. Below the epidermis is a much thicker layer, the **dermis**. The dermis contains blood vessels and a variety of receptors sensitive to heat, cold, pressure, touch and pain. Also in the dermis of many mammals are the **sudorific glands** which produce sweat.

FIG. 17.11 (a) *Photomicrograph of a vertical section through the skin from the human scalp,* × 20

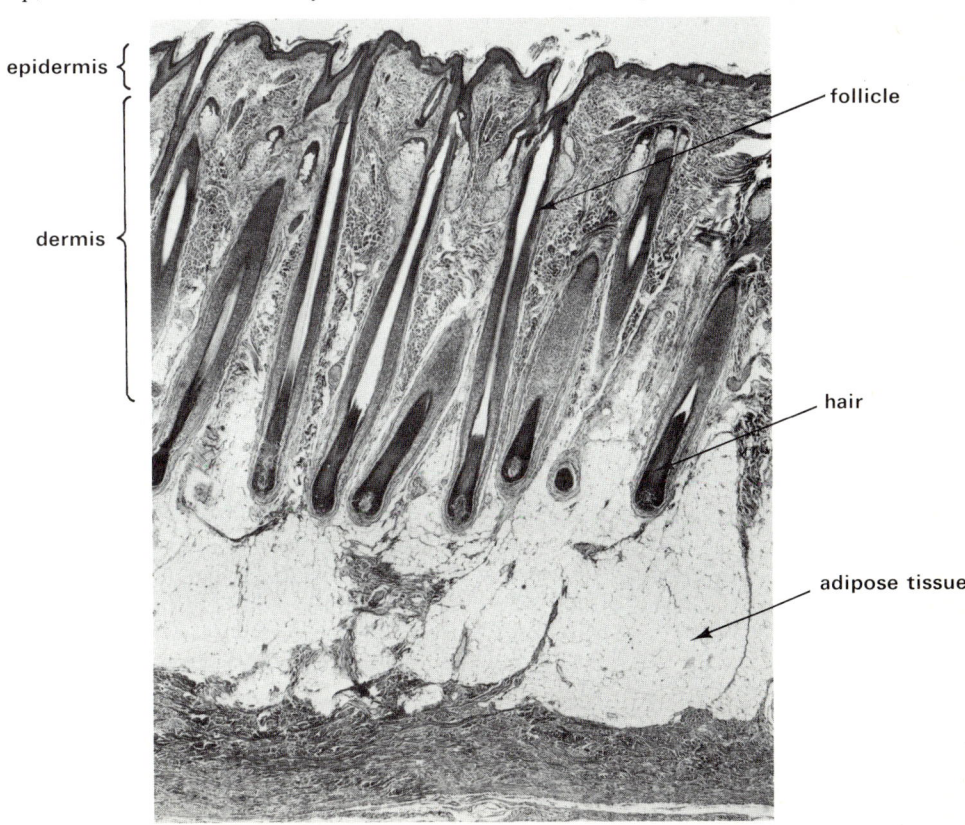

epidermis {

dermis {

follicle

hair

adipose tissue

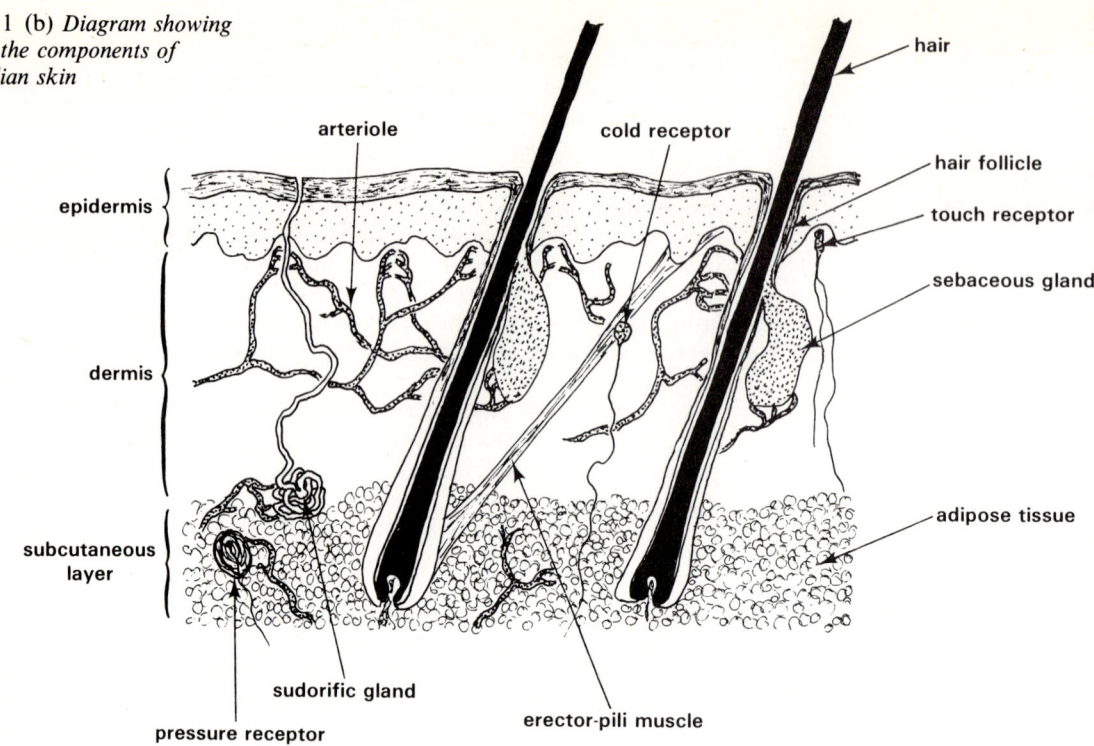

FIG. 17.11 (b) *Diagram showing some of the components of mammalian skin*

Labels on the diagram:
arteriole · cold receptor · hair · hair follicle · touch receptor · sebaceous gland · adipose tissue · epidermis · dermis · subcutaneous layer · sudorific gland · pressure receptor · erector-pili muscle

The sweat glands are coiled structures, each with a duct leading through the epidermis to the surface.

Hairs originate in the dermis and grow in pits called **follicles**. The bases of hairs are attached to small **erector-pili muscles**, the contraction of which can raise the hairs projecting from the surface. Beneath the dermis are fat deposits contained in **adipose tissue** (Fig. 17.11).

1. ADIPOSE TISSUE

Fat is a poor conductor of heat and **adipose tissue** is an effective thermal insulator. Heat is taken into the dermis by blood in numerous arterioles and capillaries. When the flow of blood is rapid, heat is lost quickly through the epidermis. Parasympathetic stimulation of the arterioles leading to the capillaries causes **vasodilation**, increasing blood flow in the skin and promoting heat loss (Fig. 17.12(a)). The nerve impulses which bring about vasodilation originate in the heat centre of the brain. Alternatively, **vasoconstriction** of the arterioles directs blood away from the skin, thereby reducing heat loss. The blood flow in human skin can be reduced by vasoconstriction to about one hundredth of the volume which flows when the arterioles are dilated (Fig. 17.12(b)). Sympathetic stimulation of the arterioles by nerve impulses from the cold centre of the brain causes vasoconstriction. Prolonged constriction of the arterioles resulting from long exposure to intense cold, deprives the dermis of the oxygen and nutrients it requires to maintain its metabolic functions. The tissues in the skin may then die or degenerate. This is the cause of frost bite.

2. HAIR

Most mammals are covered with **hair** which helps thermoregulation by keeping a layer of air next to the skin. Since it is kept in contact with the

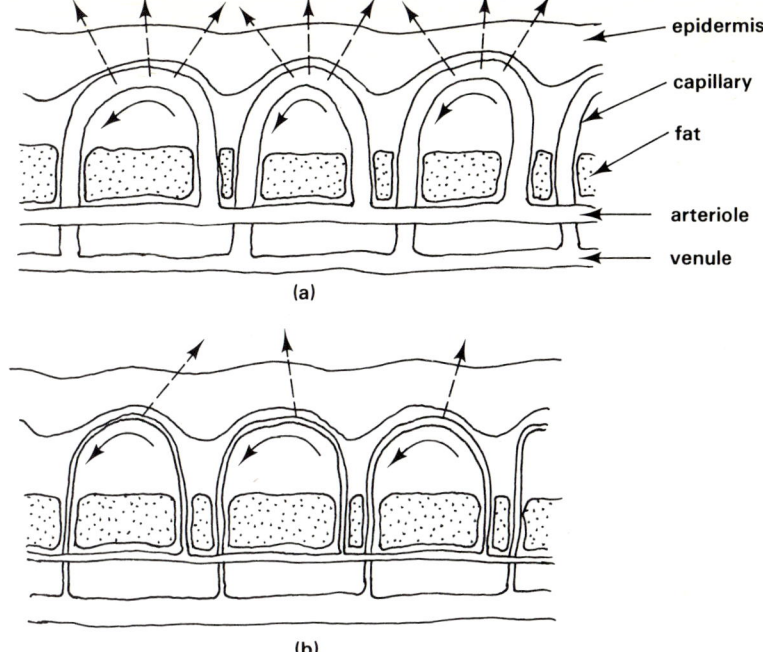

FIG. 17.12 (a) *Vasodilation in the skin leading to heat loss through the epidermis;* (b) *vasoconstriction in the skin reducing epidermal heat to a minimum*

epidermis

capillary

fat

arteriole

venule

(a)

(b)

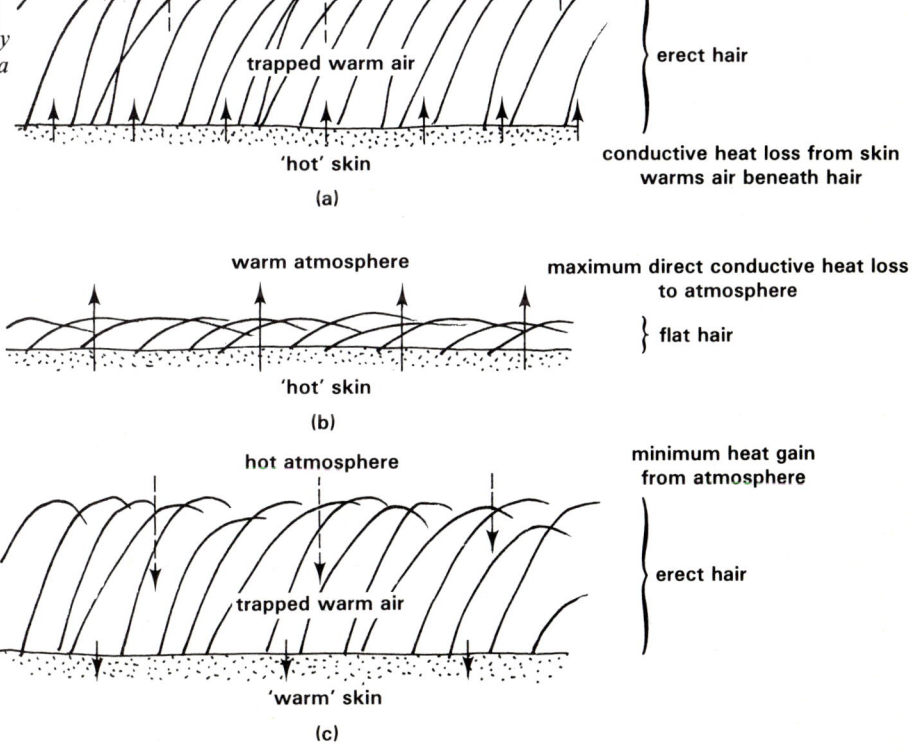

FIG. 17.13 *Variation in heat exchange with the atmosphere caused by elevation and depression of the hair on the skin in* (a) *a relatively cold environment,* (b) *a relatively warm environment and* (c) *a relatively hot environment*

cold atmosphere

minimum heat loss to atmosphere

erect hair

trapped warm air

conductive heat loss from skin warms air beneath hair

'hot' skin

(a)

warm atmosphere

maximum direct conductive heat loss to atmosphere

flat hair

'hot' skin

(b)

hot atmosphere

minimum heat gain from atmosphere

erect hair

trapped warm air

'warm' skin

(c)

343

skin and not readily replaced by cold air from the surroundings, the temperature gradient between the skin and the trapped air is relatively small. Consequently, although some of the heat in the trapped air is lost to the atmosphere, the total heat loss from the skin is reduced (Fig. 17.13(a)). The volume of warm air trapped by the hair depends on whether the hairs are erect or flat. In cold conditions, the hairs stand on end providing maximum heat conservation. In warm conditions the hairs lie flat on the skin allowing maximum heat loss (Fig. 17.13(b)). When the atmosphere is hotter than the body, the air trapped under the hair prevents excessive inward heat transfer (Fig. 17.13(c)). This is particularly important to mammals living in hot climates.

Some mammals do not have abundant hair and rely more on other methods of thermoregulation. In cetaceans such as whales and dolphins, the skin is in constant contact with cool water rather than air. Such animals have a thick fat layer called **blubber** which prevents excessive heat loss. Many aquatic mammals do have hair, however, including otters, beavers and water rats. Their fur is oily, preventing water intruding into the air spaces beneath the hairs. The fur thus acts in the usual way as an insulating layer.

3. SWEATING

Evaporation of **sweat** from the skin's surface is a means of increasing heat loss. When it is necessary to conserve heat, the flow of sweat can be stopped. Activity of the sweat glands is controlled by autonomic nerves. The impulses come from the thermoregulation centres of the brain.

Benzinger in 1961, experimenting on human subjects, found strong correlations between fluctuations in evaporation from the body, skin temperature and body temperature following ingestion of ice (Fig. 17.14). Ice in the gut removes heat from the blood. Within a few minutes, a lower blood temperature is recorded in the cold centre of the hypothalamus and evaporation from the skin decreases almost at once. The skin temperature rises as less heat is lost from it by evaporation. Within minutes the internal body temperature begins to rise, having dropped by only 0·35°C. After about twenty-five minutes the internal body temperature returns to normal.

Many mammals, such as dogs, have few sweat glands. Nevertheless evaporation is an important means of heat loss in dogs. In hot weather dogs **pant**. Panting is rapid breathing usually with the mouth open. Breathing is shallow and air in the mouth, trachea and bronchi is exchanged at a faster rate than normal. The rapid exchange of air is accompanied by increased evaporation. Lung ventilation is not significantly increased, an important point as this could change the balance of blood gases (Chapter 8).

Camels and many other mammals living in the desert sweat very little, if at all. They conserve water at the expense of sweating. Because of this, a camel's body temperature can fluctuate by as much as 6°C. Such a fluctuation is **tolerated** by a camel yet would be fatal to other mammals. Other desert mammals such as the kangaroo rat avoid the heat by sleeping in burrows during the day. They are active at night, when the desert can be very cold and less likely to encourage evaporative losses.

Humans are far less tolerant of fluctuations of body temperature than camels (Fig. 17.15). However, the temperature of the human body often rises following an infection. In these circumstances the heat centre responds to blood which is much hotter than normal, by triggering off thermoregulatory mechanisms at about 40°C compared with 37°C the normal body

FIG. 17.14 *Relationship between evaporative heat loss, skin and hypothalamus temperatures in man following experimental ice meals* (after Benzinger)

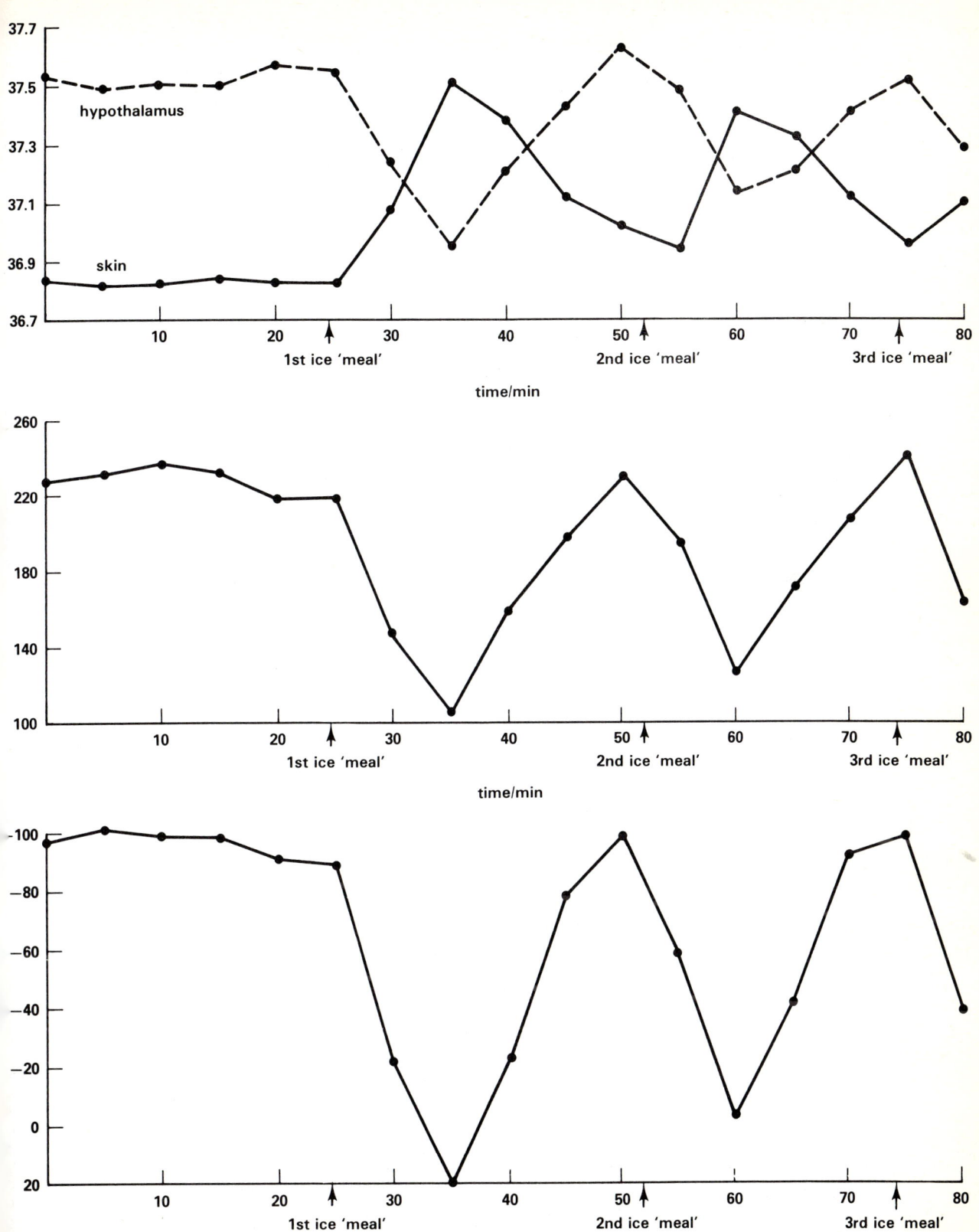

temperature. The maximum upper limit which the human body can survive is about 45°C and the lower limit is about 24°C. Nevertheless, permanent damage can be done to the tissues near to the upper and lower critical temperatures.

Figure 17.16 summarises the mechanisms of thermoregulation. Reference is made in Chapter 15 to the reflexes involved.

FIG. 17.15 *Average daily fluctuations in human body temperature*

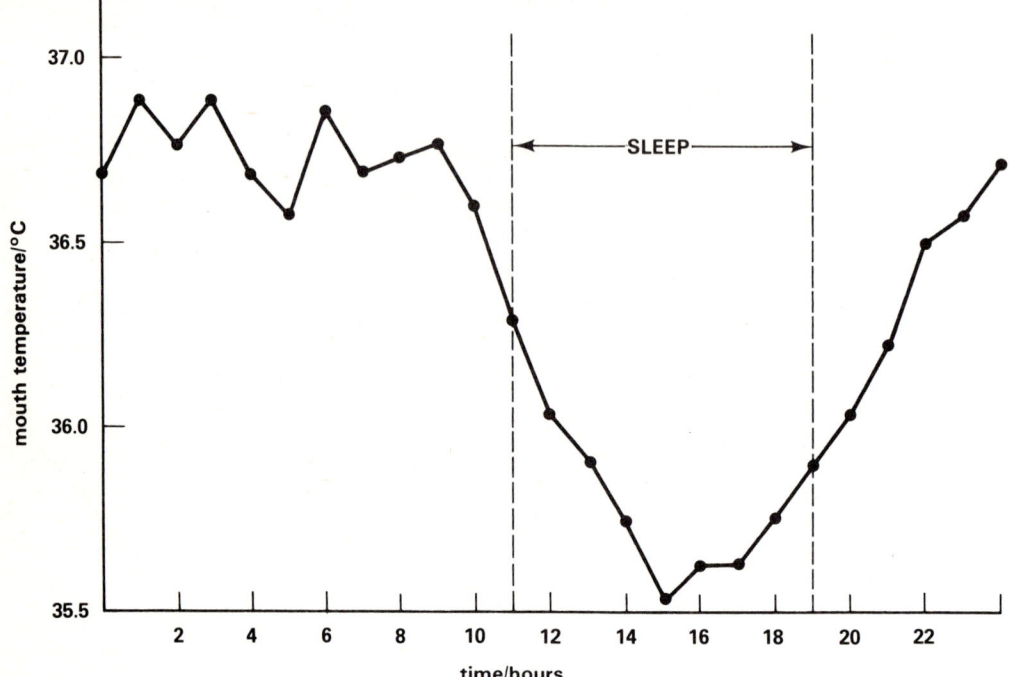

FIG. 17.16 *Summary of the main thermoregulatory mechanisms in mammals (A = hot centre; B = cold centre)*

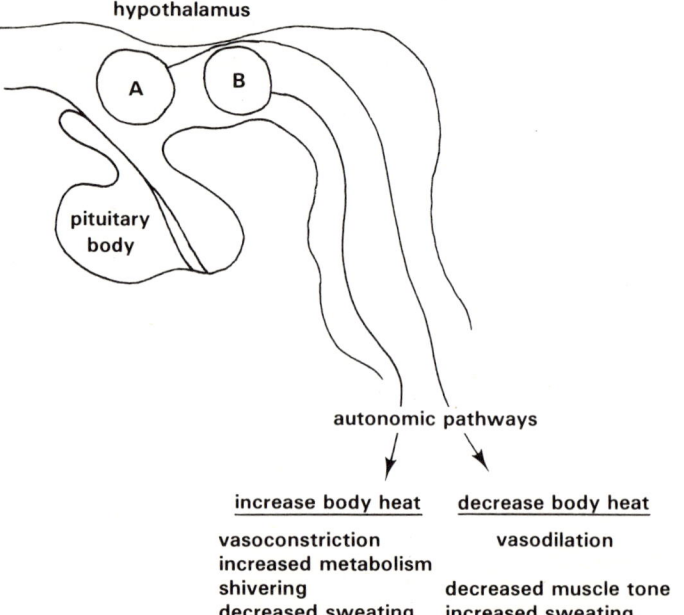

17.3.4 Hibernation

Many endothermic animals become inactive in cold weather and enter a prolonged sleep. In this condition, called **hibernation**, the metabolic rate, oxygen consumption, heart and breathing rates are very low. The body temperature falls to only a few degrees above that of the surroundings, yet it is still kept fairly constant (Table 17.1). Hibernation is a means of surviving long periods of cold weather when food is scarce. Reserves of food in the body provide the energy needed to maintain the slow rate of metabolism during hibernation. When the climate becomes warmer, hibernating animals become aroused very rapidly. In some species, the oxygen consumption necessary to increase their metabolic rate can increase several thousand times in a few hours. Some large mammals, including grizzly bears, have a period of hibernation sometimes called **winter sleep**. The body temperature remains high, even when the surroundings are intensely cold. It is thought that hibernation of some mammals is brought on by a peptide hormone secreted by the hypothalamus. During hibernation, urination and defaecation stop, so helping to conserve body heat.

Several desert-living mammals have a form of hibernation which is of value in surviving long periods of drought rather than cold. Hibernation of this kind is called **aestivation**. During severe drought food is not likely to be plentiful. The reduced rate of breathing during aestivation is of survival value, since even a moderate rate of breathing results in considerable water loss which could lead to desiccation (Chapter 8).

Table 17.1 Comparison of some body functions in active and hibernating marmots (from Hughes 1965)

FUNCTIONS	ACTIVE	HIBERNATING
body temperature/°C	34–39	3–8
basal metabolism/kJ m^{-2}day^{-1}	1714	113
heart rate/beats min^{-1}	80	4–5
breathing rate/ventilations min^{-1}	25–30	0·2

Chapter 18

Endocrine control in mammals

Many functions of the mammalian body are co-ordinated and controlled by **hormones** which are produced by **endocrine glands**. Endocrine glands are **ductless** and secrete hormones directly into the body fluids, mainly the blood. The endocrine glands are distributed throughout the mammalian body (Fig. 18.1). Because they are transported in the blood, hormones can control body functions taking place some distance from the endocrine glands. This distinguishes endocrine glands from exocrine glands which have ducts to channel their products into nearby regions without using blood for transport (Fig. 18.2).

FIG. 18.1 *The main endocrine glands in man*

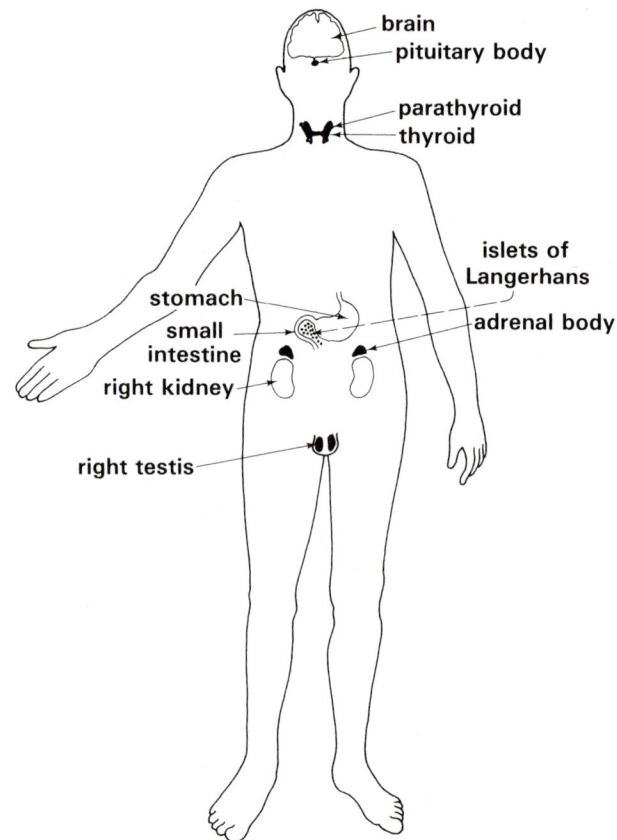

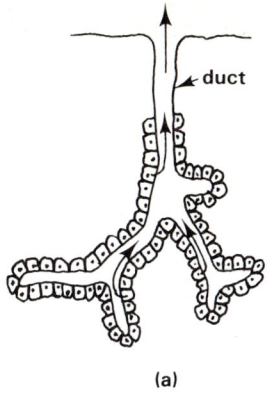

(a)

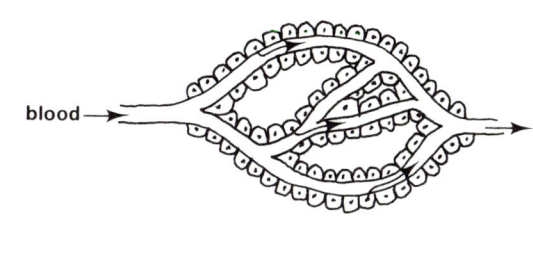

(b)

Although only small amounts of hormones are found in blood their effects on the body tissues are very great. Once in the blood, hormones are bound to plasma proteins which carry them to their sites of activity called **target organs**. Activity of most endocrine glands is controlled by the nervous system (Chapter 15).

18.1 The thyroid and parathyroid glands

18.1.1 The thyroid gland

The mammalian **thyroid gland** is found in the neck close to the larynx (Fig. 18.1). Under the microscope the thyroid gland is seen to be composed of cells arranged in globular groups called **follicles** (Fig. 18.3). The closely packed follicles are made of cubical epithelium and are held together by connective tissue supplied with blood vessels. The epithelial cells secrete thyroid hormones into the cavities of the follicles where they are temporarily stored before they are taken into the blood.

The thyroid gland continuously removes relatively large quantities of iodine from the blood. The iodine ultimately comes from the diet. In the thyroid gland iodine is combined with the amino acid tyrosine, to produce mono-and diiodotyrosine then tetra- and finally triiodothyronine (Fig. 18.4). **Triiodothyronine (T_3)** and **tetraiodothyronine (T_4)** in a ratio of about 1:9 in man are the **thyroid hormones**. T_4 is usually called **thyroxin**.

The thyroid hormones affect all tissues of the body and have two main functions. First, they regulate growth and development by controlling the growth and differentiation of cells. It is thought that the hormones modify the activity of appropriate genes. The second main function of thyroid hormones is to control the basal metabolic rate (Chapter 17). When the body is at rest thyroid hormones adjust the basal metabolic rate to a level suitable for survival.

FIG. 18.3 *Photomicrograph of a thin section of cat thyroid,* × 400

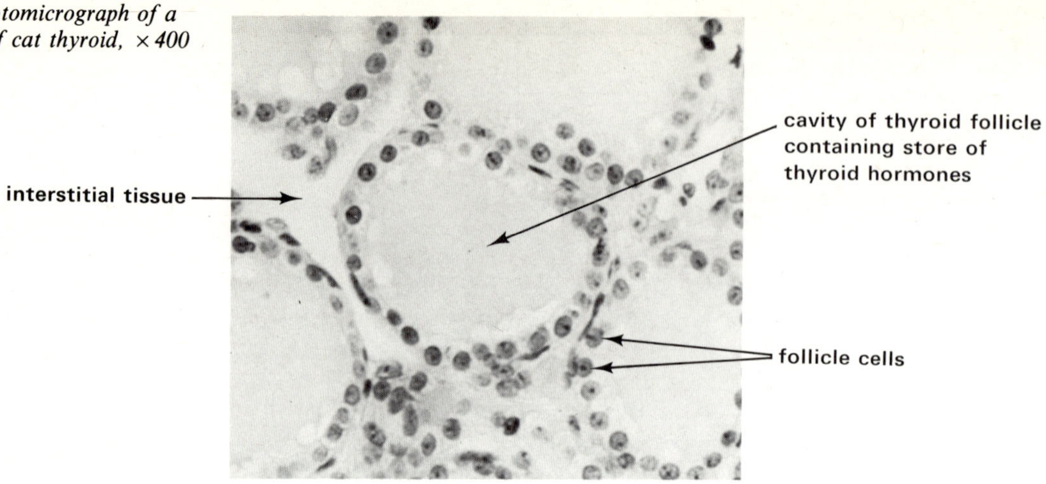

cavity of thyroid follicle containing store of thyroid hormones

interstitial tissue

follicle cells

FIG. 18.4 *Pathway of thyroxin production*

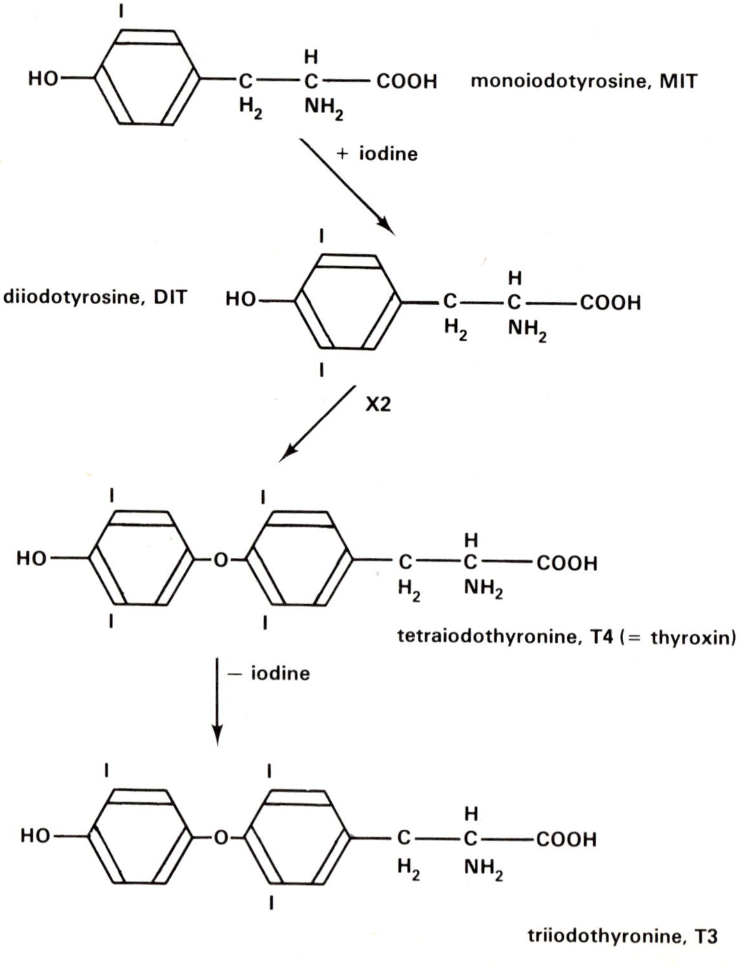

monoiodotyrosine, MIT

+ iodine

diiodotyrosine, DIT

X2

tetraiodothyronine, T4 (= thyroxin)

− iodine

triiodothyronine, T3

18.1.2 Control of thyroid activity

FIG. 18.5 *Mechanism controlling thyroid activity*

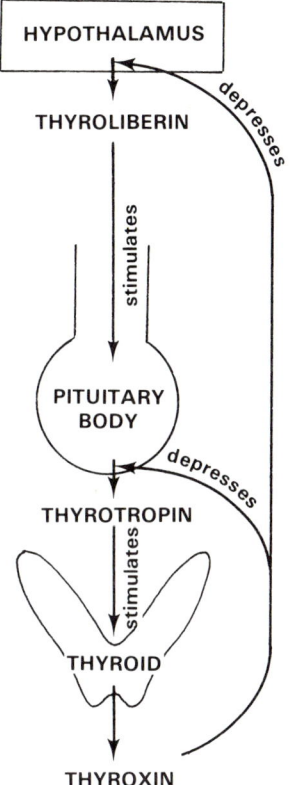

As with several other endocrine glands, growth and secretion of the thyroid gland are controlled by the **pituitary body** (Fig. 18.1). The pituitary body secretes a number of **tropins** which affect other endocrine glands (section 18.5.1). The trophic hormones include **thyrotropin (TSH)**. Growth of the thyroid gland and its output of hormones is stimulated by TSH. Secretion of TSH is in turn depressed by the thyroid hormones. The interaction is a **negative feedback control** mechanism. Another hormone controls the release of TSH. This hormone is called **thyroliberin**. It is made in the **hypothalamus**, the part of the brain immediately above the pituitary body (Fig. 18.5). Such control ensures that growth and activity of the thyroid gland meets the body's needs at any time. It is a homeostatic device which helps maintain the stability of the body's internal environment. Other comparable examples are described in the remainder of the chapter.

FIG. 18.6 *The fate of ingested iodine*

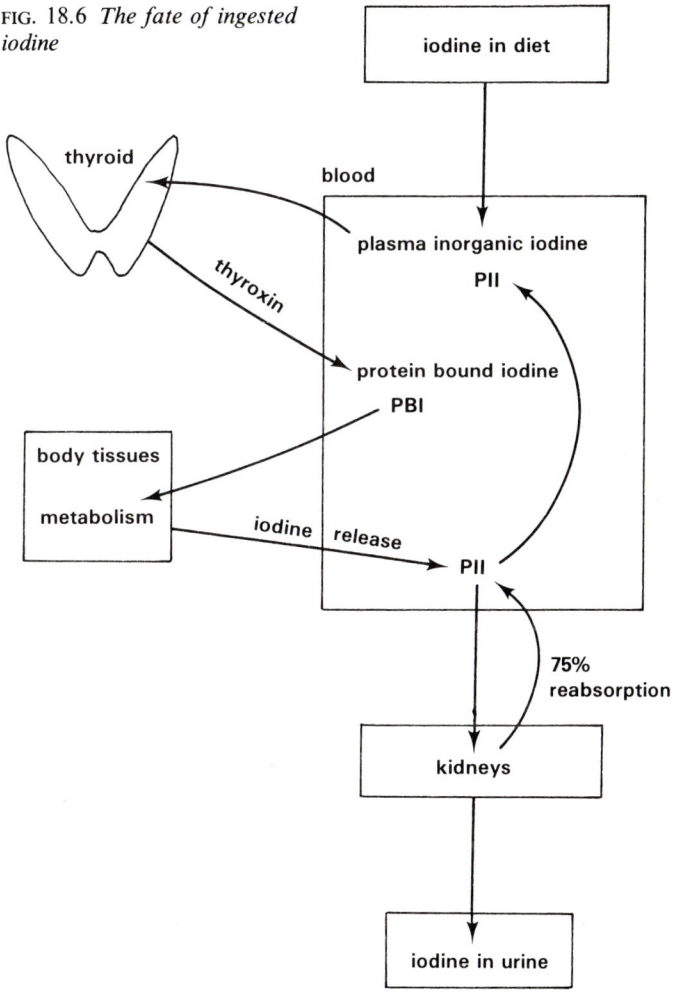

18.1.3 Measurement of thyroid activity

Human thyroid activity can be measured with the use of **radioactive isotopes** of iodine. Sodium iodide containing ^{125}I, ^{131}I or ^{132}I is given orally in doses which produce harmless levels of radiation to the patient. The radioiodine circulates in the plasma as **plasma inorganic iodine (PII)**. Much of the iodine in PII is normally absorbed by the thyroid gland and used to make thyroid hormones. The hormones are secreted into the blood where the radio-iodine is carried as **protein bound iodine (PBI)**. When the hormones have performed their functions in the tissues iodine is released back into the blood to be absorbed once more by the thyroid gland. Some of the body's total **iodine pool** including the radioisotopes is excreted in the urine (Fig. 18.6). There is a well-defined relationship between the amounts of radio-iodine in the plasma, thyroid gland and urine following a dose of the isotope (Fig. 18.7(a)). Consequently, measurements of the radiation in the thyroid gland, samples of blood and urine are useful in detecting abnormal activity of the thyroid gland (Fig. 18.7(b)).

FIG. 18.7 *Relationship between the levels of iodine in the thyroid gland, blood and urine after oral administration of radioactive iodine in man* (after Greig, Boyle and Boyle): (a) *normal* and (b) *thyrotoxic*

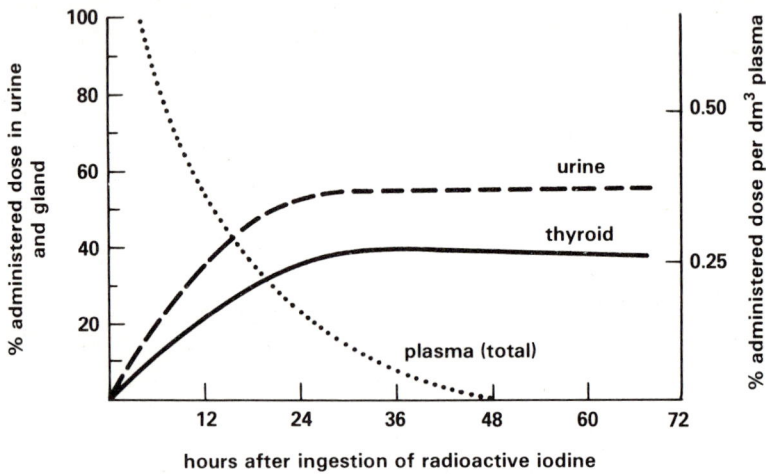

(a)

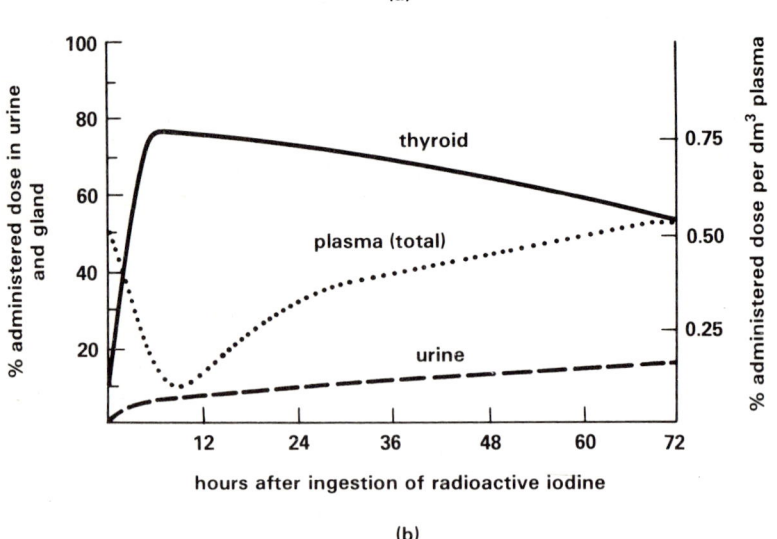

(b)

18.1.4 Malfunction of the thyroid gland

Overactivity of the thyroid gland is called **hyperthyroidism**. The condition can result from a failure of the TSH—thyroid control mechanism resulting in increased thyroxin output. Another cause of hyperthyroidism is secretion of thyroid hormones by a thyroid tumour. Either way an increase in basal metabolism occurs with increased heart rate, extreme irritability and loss of body mass. Removal of part of the thyroid gland may be necessary to check the condition. A common side-effect of an overactive thyroid gland is the secretion of exophthalmos-producing substance by the pituitary body. This substance causes excessive growth of the tissues immediately behind the eyes. Consequently protrusion of the eyes called **exophthalmos** often accompanies hyperthyroidism (Fig. 18.8).

Underactivity of the thyroid gland is called **hypothyroidism**. In adult humans hypothyroidism is the cause of a condition called **myxoedema**. The effects on basal metabolism are the opposite to those of hyperthyroidism. The BMR slows down, a smaller proportion of the energy-rich ingredients

FIG. 18.8 *Facial characteristics of a person with thyrotoxicosis. Note the goitre and protruding eyes*

in the diet are respired, hence body mass increases. Mental activity also slows down so the patient is less alert than normal. At one time hypothyroidism was common in parts of Derbyshire and the Swiss Alps. Here the local soil and water supply are deficient in iodine and the population was thus deprived of an essential requirement for the synthesis of thyroid hormones. Addition of traces of iodine to commercial table salt has largely overcome the problem.

The abnormal growth and development which accompanies an underactive thyroid gland is particularly distressing in infants. **Cretinism** is the name given to the effects of thyroid deficiency in children. The main symptoms are retardation in mental, physical and sexual development.

Both hyper- and hypothyroidism usually result in excessive growth of the thyroid gland. The enlarged thyroid gland is called a **goitre** which causes swelling of the neck.

18.1.5 The parathyroid glands

In man the **parathyroid glands** are four small oval bodies embedded in the thyroid gland (Fig. 18.9). The parathyroid glands make a polypeptide hormone called **parathyrin (PTH)** which has a profound effect on the calcium ion content of the blood and other body fluids. PTH elevates the amount of calcium in the blood which, by negative feedback control, regulates PTH output (Fig. 18.10).

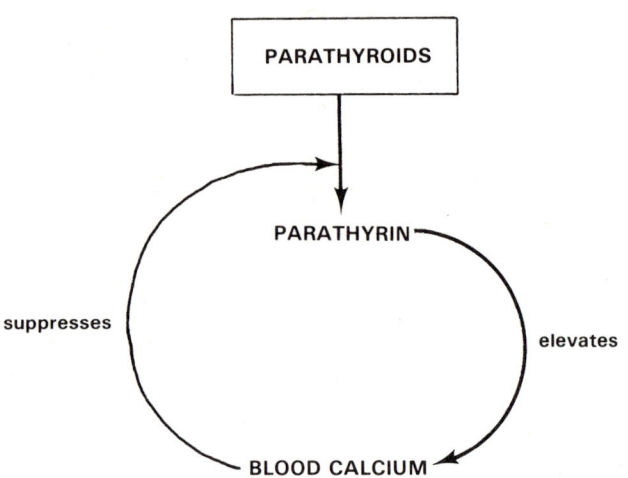

FIG. 18.9 *Position of the parathyroid glands in man*

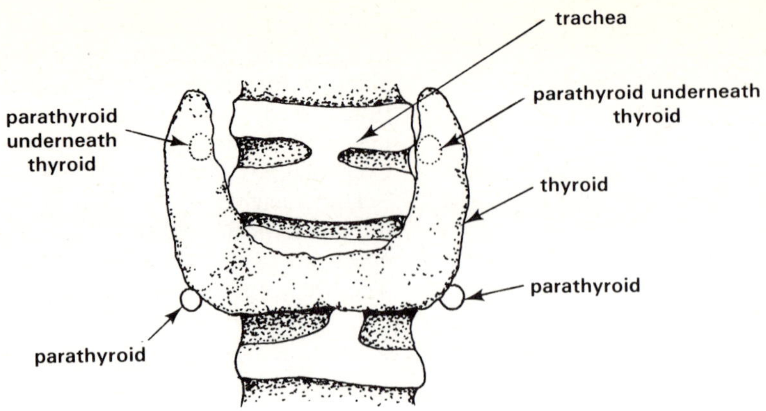

trachea

parathyroid underneath thyroid

parathyroid underneath thyroid

thyroid

parathyroid

parathyroid

FIG. 18.10 *Mechanism regulating the output of parathyrin, PTH*

PARATHYROIDS

PARATHYRIN

suppresses

elevates

BLOOD CALCIUM

Calcium is an important constituent of body fluids. It participates in many body functions including coagulation of the blood (Chapter 9), nerve and muscle activity (Chapter 15) and teeth and bone formation. These functions depend not just on the presence of calcium ions but on a precise concentration of calcium ions. It is therefore important that the amount of calcium is regulated within narrow limits. Calcium is one of the most precisely regulated constituents of the mammalian body. The concentration of calcium ions in normal human blood serum ranges between only 9 and 11 mg per 100 cm^3.

PTH affects blood calcium in several ways:

1. RELEASE OF CALCIUM FROM BONE

Bone contains calcium salts, especially calcium phosphate. Calcium and phosphate(V) ions are continuously released from bones into the tissue fluids and redeposited in bones. Clearly the composition and structure of bone depends on regulating this reversible process (Fig. 18.11). By stimulating the **release of calcium** and phosphate ions from bone, PTH helps maintain normal concentrations of calcium and phosphate ions in blood.

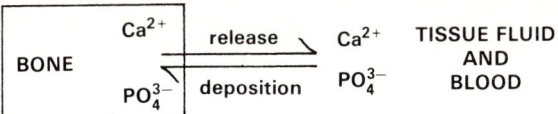

FIG. 18.11 *Dynamic equilibrium between the calcium and phosphate in bone and that in the tissue fluid and blood*

2. REABSORPTION OF CALCIUM FROM THE URINE

Calcium and phosphate ions are filtered from the blood in the kidney nephrons. PTH promotes **reabsorption of calcium** ions from the filtrate in the proximal convolutions at the expense of phosphate ions which are excreted (Chapter 10). PTH thus elevates the concentration of calcium ions in the blood while lowering the concentration of blood phosphate ions (Fig. 18.12). The inverse relationship between calcium and phosphate ions in blood

$$[Ca^{2+}] \propto \frac{1}{[PO_4^{3-}]}$$

prevents the accumulation of unwanted calcium phosphate in the blood and tissues.

FIG. 18.12 *Effect of parathyrin on calcium and phosphate in the kidney nephrons*

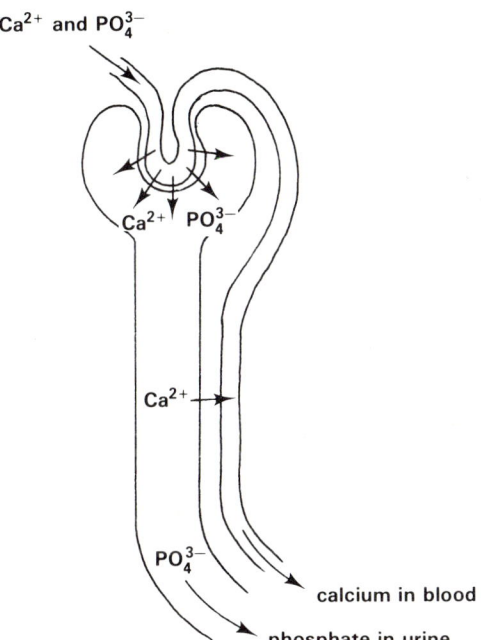

3. ABSORPTION OF CALCIUM FROM THE GUT

PTH is also thought to promote **absorption of calcium** ions from the gut, thus helping to maintain the normal concentration of calcium ions in the blood.

18.1.6 Role of the thyroid gland in calcium control

As well as PTH from the parathyroid glands, another hormone called **calcitonin** affects the calcium ion concentration of blood. Calcitonin is secreted from **C cells** in the thyroid gland. Its effect is to lower the concentration of calcium ions in blood by causing the deposition of calcium phosphate in bone. Calcitonin has been found in mammals and fish.

Salmon calcitonin is one of the most potent of biological compounds with a strength over 40 times that of human calcitonin. In experiments on rats, a dose of salmon calcitonin of 0·000000002 g kg^{-1} of rat lowered the concentration of calcium ions in blood by 10 per cent.

Thus it is the combined effects of PTH and calcitonin which regulate calcium and phosphate ions to the concentrations necessary for many physiological functions to take place normally (Fig. 18.13).

FIG. 18.13 *Summary of the effects of PTH and calcitonin on blood calcium and phosphate*

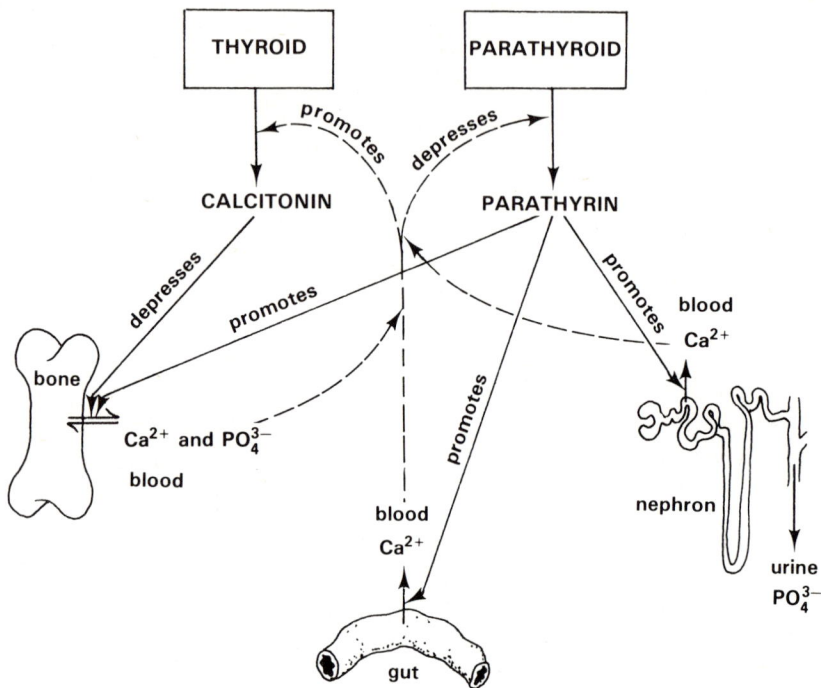

18.1.7 Malfunction of the parathyroid glands

Underactivity of the parathyroid glands, **hypoparathyroidism**, results in a lowering of the concentration of calcium ions and an elevation of the concentration of phosphate ions in blood. The condition is called **hypocalcaemia**. If the concentration of calcium ions in human blood serum drops below 7 mg per 100 cm^3 a condition called **tetany** results. Tetany is characterised by increased excitability of the nervous system. Muscular activity becomes spasmodic and uncontrolled.

Surgical removal of the parathyroid glands from dogs results in tetany within three days. Subsequent injection of parathyroid extract quickly corrects the condition but the effect is short-lived. Maintenance of normal concentrations of calcium ions in the blood of experimental dogs requires continued treatment with PTH (Fig. 18.14).

Overactivity of the parathyroid glands, **hyperparathyroidism**, can result in breakdown of bone structure and consequent elevation of the concentration of calcium ions in blood. The condition is called **hypercalcaemia**. Recent work indicates that the decalcifying activity of PTH increases in post-menopausal women. Before the menopause PTH is inhibited by oestrogen hormones secreted by the ovaries (Chapter 22). After the menopause oestrogen production stops. This explains why the bones of older women are often fragile.

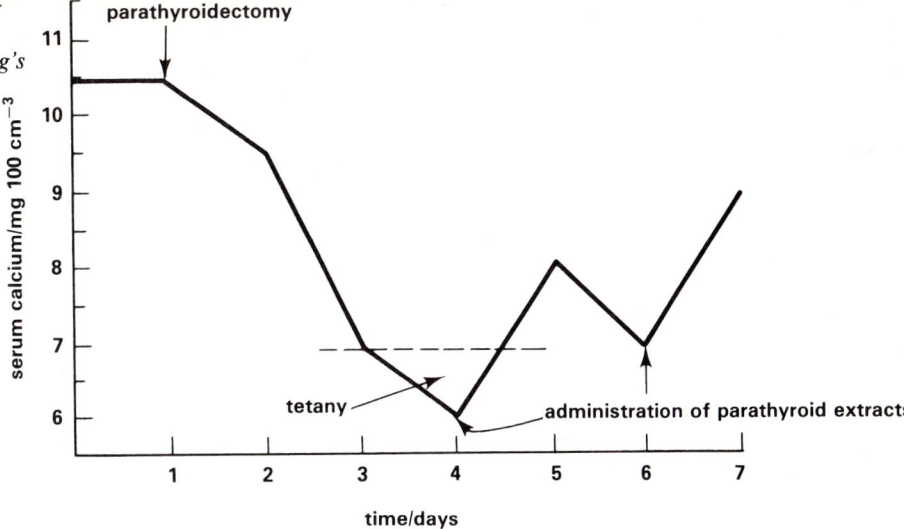

FIG. 18.14 *Effects of removing the parathyroids on the concentration of calcium in dog's blood*

18.2 Control of blood glucose

Carbohydrate is the main source of free energy released in respiration (Chapter 5). Nearly all the carbohydrate absorbed from the gut circulates in the blood as **glucose** and it is this sugar which is the main respiratory substrate. It is therefore vital that the blood maintains a constant and adequate supply of glucose to all the tissues. It is equally important, however, that the concentration of glucose in blood and tissue fluids is not excessive. Abnormally high concentrations of glucose in the tissue fluid would draw water from cells by osmosis (Chapter 2). Furthermore the water content of the body as a whole can be affected by the concentration of glucose in the blood. Normally the concentration of glucose in blood is sufficiently low for all the glucose in the renal filtrate to be reabsorbed into the blood in the kidneys (Chapter 10). Thus glucose is not usually excreted in urine. Since the reabsorption of water from the nephrons depends on the water potential gradient between the renal filtrate and the blood, any glucose remaining in the filtrate would minimise the gradient and reduce water reabsorption. Thus as well as tissue dehydration, a high concentration of glucose in blood could cause total body dehydration.

In the mammalian body mechanisms exist which regulate the concentration of glucose in the blood so that the tissues' metabolic needs are met without adversely affecting the body's water content. A number of hormones help to control the concentration of glucose in blood.

18.2.1 Insulin and glucagon

The mammalian pancreas is an exocrine gland which secretes pancreatic juice into the gut (Chapter 14). It is situated in a loop of the small intestine just below the stomach (Fig. 18.1). Embedded in the pancreas and sometimes also in the wall of the small intestine is a large number of microscopic patches of endocrine tissue called the **islets of Langerhans** (Fig. 18.15). Histochemical studies show that the islets consist of two types of cells called α- and β-cells. They are responsible for the production and secretion of two hormones called **glucagon** and **insulin** respectively. The hormones are discharged directly into the islets' blood capillaries. The rest of the pancreas and its exocrine ducts are not involved in hormone production or secretion. Experimentally tying off the pancreatic ducts prevents secretion of pancreatic digestive enzymes but has no effect on hormone secretion.

357

Insulin lowers the concentration of blood glucose while glucagon raises it. Output of the pancreatic hormones is regulated by the concentration of glucose in blood (Fig. 18.16).

FIG. 18.15 *Photomicrograph of two islets of Langerhans seen in a thin section of monkey pancreas,* × *100*

FIG. 18.16 *Effects of insulin and glucagon on the level of blood glucose*

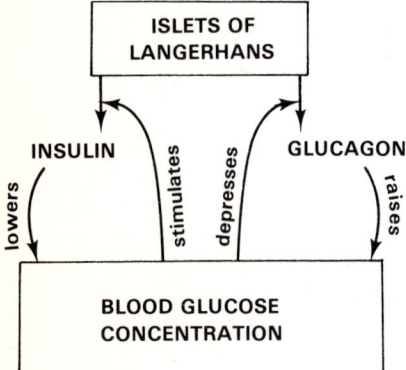

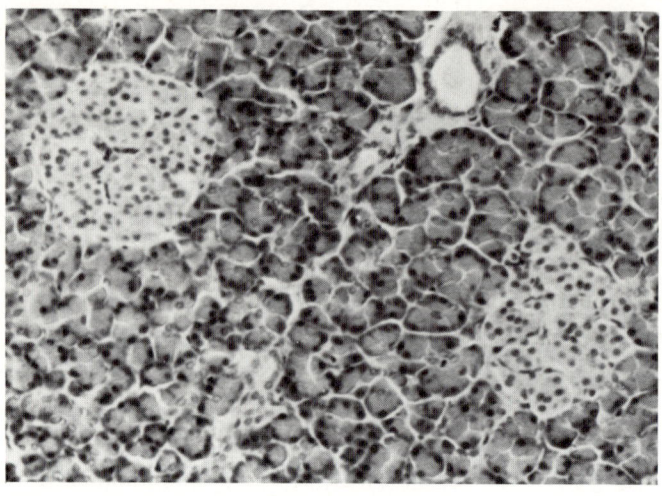

1. THE FUNCTIONS OF INSULIN

Insulin performs two main functions in mammals. The first is to regulate the concentration of glucose in the blood. On the one hand glucose is used by the tissues at varying rates dependent on the rate of metabolic activity. More glucose is used during exercise than when the body is at rest. On the other hand the body's input of glucose varies depending on what has been eaten. The concentration of glucose in the blood draining the gut can double after a meal. The problem then is to reconcile the variable input and respiration of glucose with a relatively constant concentration of glucose in the blood. In dealing with this problem the liver, insulin and glucagon play vital roles.

Blood leaving the gut contains the absorbed products of digestion and passes through the hepatic portal system into the liver (Chapter 14). The liver cells contain enzymes which under the control of insulin promote the synthesis of **glycogen**, a polymer of glucose. It is as glycogen that much of the glucose absorbed from the gut is stored in the liver. Some glycogen is also stored in the skeletal muscles. Insulin therefore prevents any undue rise in the concentration of glucose in the blood. Glucose removed from the blood for respiration is replaced from the glycogen stores. Glucagon has the opposite effect to insulin. It stimulates the conversion of glycogen to

FIG. 18.17 *Summary of the fate of glucose in the mammalian body*

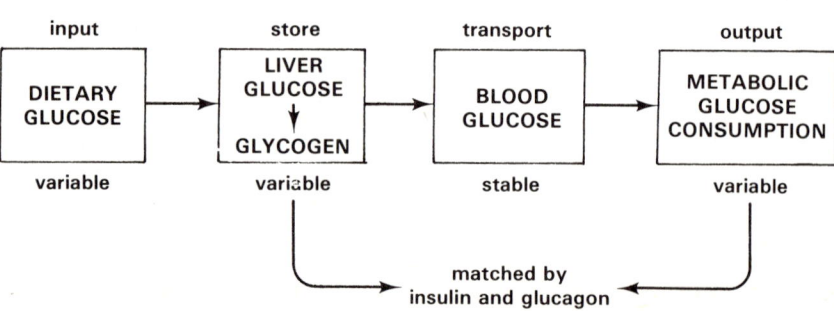

glucose. The balance between the effects of the two hormones results in regulation of the concentration of glucose in blood (Fig. 18.17). The **negative feedback relationship** between the hormones and glucose in blood ensures that glucose is released from the glycogen stores at a rate sufficient to match its uptake from the blood by respiring tissues.

The second main function of insulin is to affect the **rate of entry** of glucose into respiring cells. Glucose is taken into living cells by **active absorption**. Insulin greatly increases the rate of glucose absorption, possibly by triggering off a membrane carrier mechanism or by acting as a carrier itself. In a normal healthy man a concentration of about 50 to 80 mg glucose per 100 cm³ blood is sufficient to meet the requirements of respiring cells. In the absence of insulin, this concentration would have to increase by between ten and twenty times for glucose to enter the cells at the same rate by diffusion (Fig. 18.18).

FIG. 18.18 *Insulin promotes the entry of glucose into body cells. In the absence of insulin the amount of intercellular glucose must rise to about 10 times its normal concentration for glucose to enter the cells at the same rate*

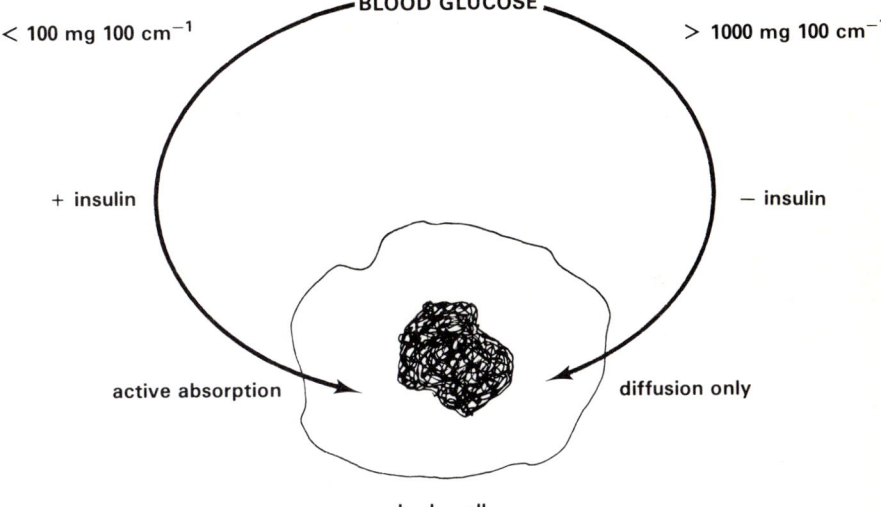

BLOOD GLUCOSE

< 100 mg 100 cm⁻¹   > 1000 mg 100 cm⁻¹

+ insulin   − insulin

active absorption   diffusion only

body cell

2. MALFUNCTION OF THE ISLETS OF LANGERHANS

Underactivity of the islets of Langerhans results in reduced secretion of insulin and is the cause of **diabetes mellitus**. This should not be confused with diabetes insipidus (Chapter 10). In diabetes mellitus the glucose concentration in blood rises. The condition is called **hyperglycaemia**, and glucose exceeds the maximum concentration which can be totally reabsorbed from the renal filtrate in the kidneys. Consequently glucose is excreted in the urine, a condition called **glycosuria**. The presence of glucose in the urine disturbs the water potential gradient which normally results in water reabsorption from the nephrons. Large volumes of dilute urine are thus produced, a condition called **diuresis**. Diuresis is dangerous because it brings about dehydration of the body. Since the breakdown of glycogen is uninhibited in diabetes mellitus the stores of glycogen in the liver and muscles are quickly used up. Body fats and proteins are then used as respiratory substrates causing a rapid loss of body mass.

The condition can be rectified by regular doses of insulin and by eating a carefully controlled diet. A clinical test often used in hospitals to assess whether insulin production is normal involves measurement of **glucose tolerance**. Patients are made to fast for several hours before ingesting 50 g of glucose in 150 cm³ water. The concentration of glucose in the patient's

359

blood is then measured immediately and at 30-minute intervals over a period of two to three hours. If necessary the urine is also analysed for glucose at hourly intervals. The concentration of glucose in the blood is plotted against time and a glucose tolerance curve is obtained (Fig. 18.19). In a healthy individual there is a slight rise in blood glucose concentration after the glucose drink. Insulin then brings the concentration of glucose in the blood down to its original value within about two hours and no glucose is excreted in the urine.

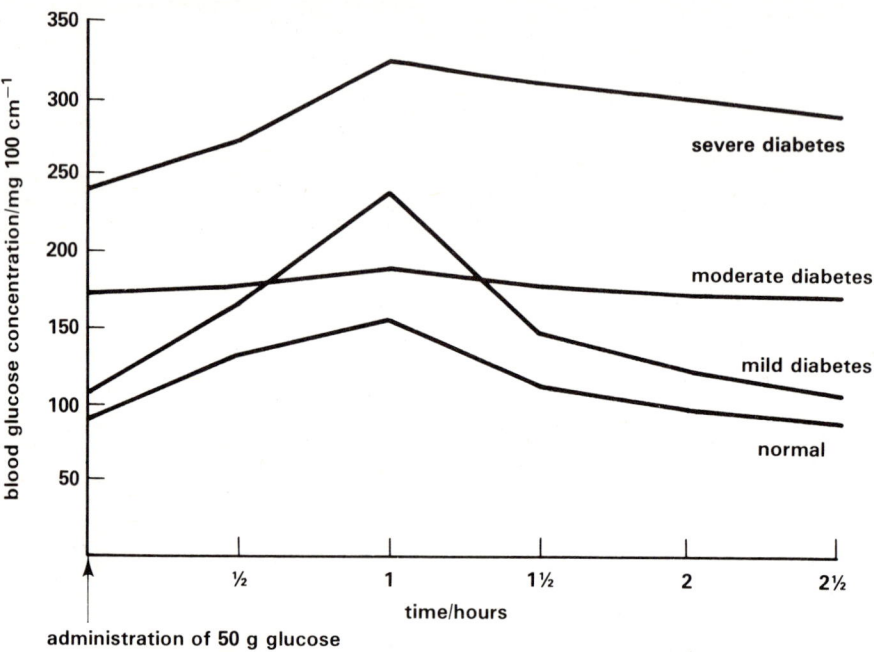

FIG. 18.19 *Glucose tolerance curves*

During late pregnancy the renal threshold for glucose may be lower than normal. Thus even though the concentration of glucose in the blood is normal and only slightly rises after a meal, the rise may be sufficient for some glucose to be excreted in the urine. Glycosuria during pregnancy therefore does not necessarily mean that the mother is suffering from diabetes mellitus.

18.2.2 Cortisol

The reserves of glycogen in the mammalian body are limited and in certain conditions fats and proteins can be converted to glucose to supply metabolic demands. The conversion is influenced by a number of steroid hormones called **glucocorticoids** secreted by the **adrenal bodies**, a pair of glands lying close to the kidneys (Fig. 18.1). Each adrenal body is divided into an inner **medulla** and an outer **cortex** (Fig. 18.20(a)). It is in the cortex that glucocorticoids are made. The most abundant of these hormones is **cortisol** (Fig. 18.20(b)).

Cortisol stimulates the conversion of fats and proteins to glucose and is thus involved in regulating the glucose concentration in blood. Output of cortisol is controlled by a trophic hormone from the pituitary body. The hormone is called **corticotropin (ACTH)**. Secretion of ACTH is affected by the concentration of cortisol in the blood, another example of a negative feedback control mechanism (Fig. 18.21). The combined effects of insulin, glucagon and cortisol in regulating the glucose concentration in blood are summarised in Figure 18.22.

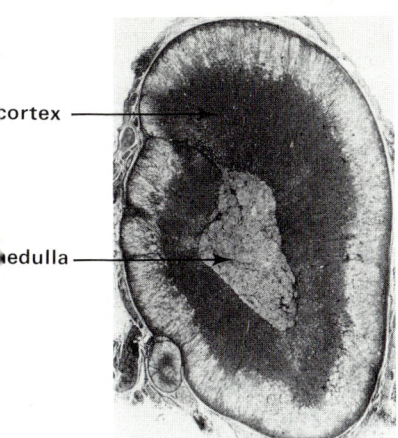

FIG. 18.20 (a) *Photomicrograph of a transverse section of an adrenal body from a cat,* ×11

cortex —

medulla —

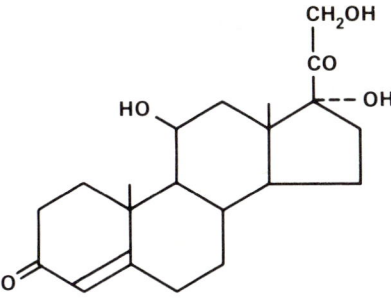

FIG. 18.20 (b) *Molecular structure of cortisol*

FIG. 18.20 (c) *The basic structure of a steroid*

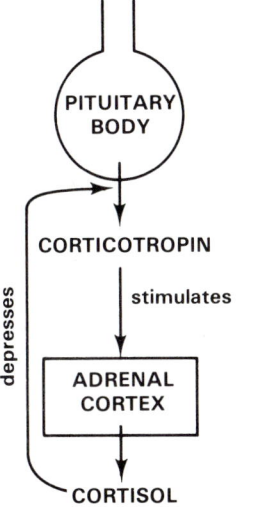

FIG. 18.21 *Mechanism regulating cortisol production*

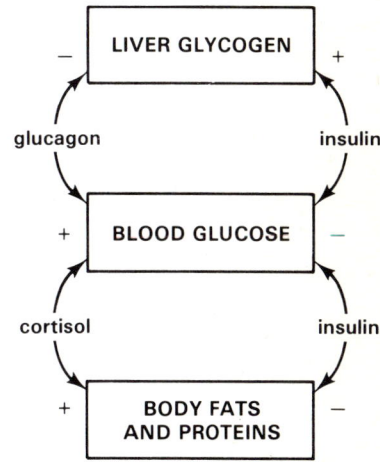

FIG. 18.22 *Combined effects of insulin, glycogen and cortisol on blood glucose*

Another important function of cortisol is its ability to depress autolysis of damaged body cells by lysosomes (Chapter 1). Cortisol-like drugs are valuable in promoting the repair of body organs damaged in degenerative diseases such as arthritis. One of the effects of adrenalin, secreted by the medulla of the adrenal glands, is to increase ACTH output (section 18.5.1). This partly explains the increased secretion of cortisol which normally accompanies mental and physical stress.

18.3 Control of blood sodium and potassium

Sodium and potassium salts, especially chlorides, participate in a variety of body functions, particularly nerve and muscle activity. Consequently the maintenance of stable concentrations of sodium and potassium ions in the blood is vital for normal physiological activity (Chapter 15). Furthermore, because sodium and potassium salts are soluble in water they affect the osmotic potential of the body fluids. The concentrations of sodium and potassium ions in the body are regulated by **mineralocorticoid hormones** made in the adrenal cortex.

18.3.1 Aldosterone

FIG. 18.23 *Molecular structure of aldosterone*

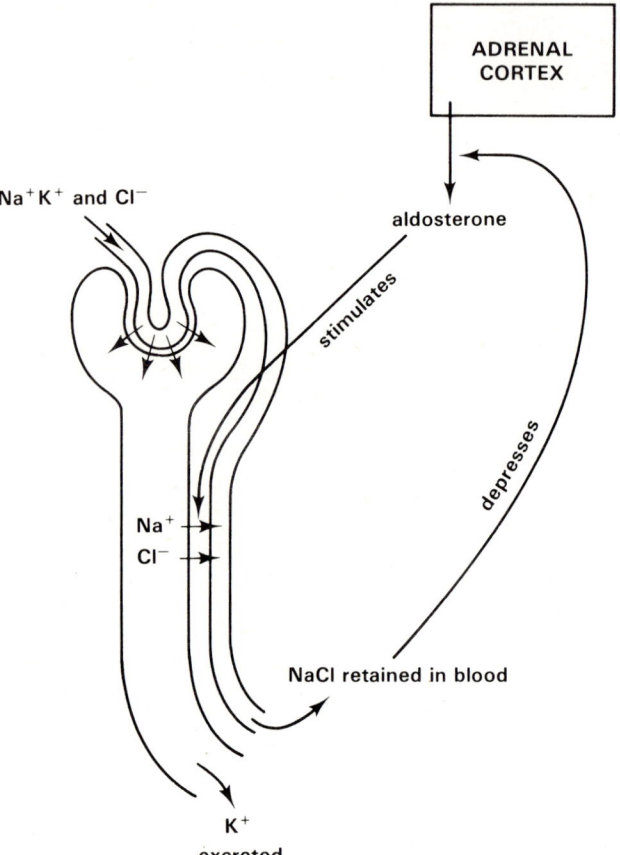

The most important mineralocortoid hormone is **aldosterone** (Fig. 18.23). This steroid is strikingly similar in molecular structure to cortisol (Fig. 18.20(b)). Nevertheless their functions are very different. Aldosterone promotes reabsorption of sodium ions (Na^+) into the blood from the filtrate in the renal nephrons (Chapter 10). Chloride ions (Cl^-) usually accompany the sodium ions, probably because of electrostatic attraction. Uptake of sodium ions depresses reabsorption of potassium ions (K^+). This is why potassium ions are found in urine. Output of aldosterone is inhibited by a high concentration of sodium ions in the blood, yet another example of a feedback mechanism which assists homeostasis (Fig. 18.24).

FIG. 18.24 *Aldosterone stimulates the reabsorption of sodium chloride by the nephrons*

Depressed output of aldosterone can cause excessive excretion of water by the kidneys resulting in a fall in the blood volume. One effect of the fall is the secretion into the blood of the enzyme **renin** from special kidney cells. Renin activates the conversion of a plasma protein called **proangiotensin** into an active derivative **angiotensin**. Angiotensin stimulates secretion of aldosterone thereby aiding water reabsorption by the kidneys. Angiotensin also stimulates the brain to trigger off impulses which create the sensation of **thirst** (Fig. 18.25). These are appropriate responses to too little water in the body fluids, and providing water is consumed the body's state of hydration is rectified.

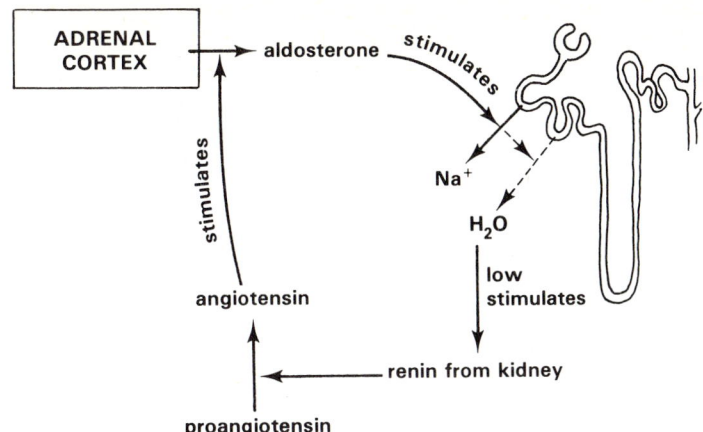

18.4 The adrenal medulla

The centre of each adrenal gland, the **medulla**, is derived from the same embryonic tissue which gives rise to sympathetic nerve ganglia. For this reason the medulla of the adrenal glands can be regarded as a modified part of the sympathetic nervous system. Sympathetic nerves, as part of the autonomic system play an important role in regulating many body functions (Chapter 15). Hormones secreted into the blood by the adrenal medulla have similar effects to stimulation of sympathetic nerves.

18.4.1 The medullary hormones

FIG. 18.26 *Structure of (a) noradrenalin and (b) adrenalin. These hormones belong to a group of chemicals called catecholamines*

(a)

(b)

Two hormones are made in the adrenal medulla of mammals, **adrenalin** and **noradrenalin** (Fig. 18.26). Noradrenalin is the neuroeffector transmitter secreted by sympathetic nerve endings. However, adrenalin is the main secretion of the medulla. The effects of both hormones are similar and like sympathetic activity generally, they prepare the body to expend a lot of energy quickly. In fact, stimulation of the sympathetic nerves activates the medulla and promotes secretion of adrenalin.

Normally the adrenal medulla secretes small amounts of adrenalin and noradrenalin into the blood. However, following increased activity of the sympathetic nerves which usually accompanies physical and mental stress, larger quantities are released. Because the medullary hormones prepare mammals to run away from or to face an enemy, they are sometimes called the **flight or fight hormones**. In many respects this is an appropriate description. The effects of adrenalin and noradrenalin can be summarised as follows:

1. EFFECTS ON THE GUT AND RESPIRATORY SYSTEM
The smooth muscle of the gastro-intestinal tract relaxes and the bronchi become dilated. The thorax is enlarged because the diaphragm can now be pushed down further into the abdomen. The net result is that volumes of air larger than normal can be drawn in and out of the lungs, thereby increasing the rate of oxygen uptake by the blood.

2. EFFECTS ON THE CARDIOVASCULAR SYSTEM
The heart rate and the power of cardiac contractions increase with consequent rise of blood pressure. Arterioles in the skin and gut become constricted while the vessels supplying the skeletal muscles dilate. These effects, together with increased ventilation of the lungs, ensure an adequate supply of oxygenated blood to organs such as muscles which produce energy for movement.

3. EFFECTS ON BLOOD GLUCOSE

Glycogen stored in liver and skeletal muscles is converted to glucose, thus providing the source of energy required for increased muscular activity. Adrenalin stimulates secretion of ACTH and hence cortisol. Some of the extra glucose may therefore come from fats and proteins (section 18.2.2).

4. EFFECTS ON THE NERVOUS SYSTEM

Adrenalin increases sensitivity of the nervous system thereby increasing the speed with which the body reacts to environmental stimuli. This is of obvious advantage to a mammal in flight or fight. The various effects of the medullary hormones are illustrated in Figure 18.27.

FIG. 18.27 *Summary of the effects of adrenalin*

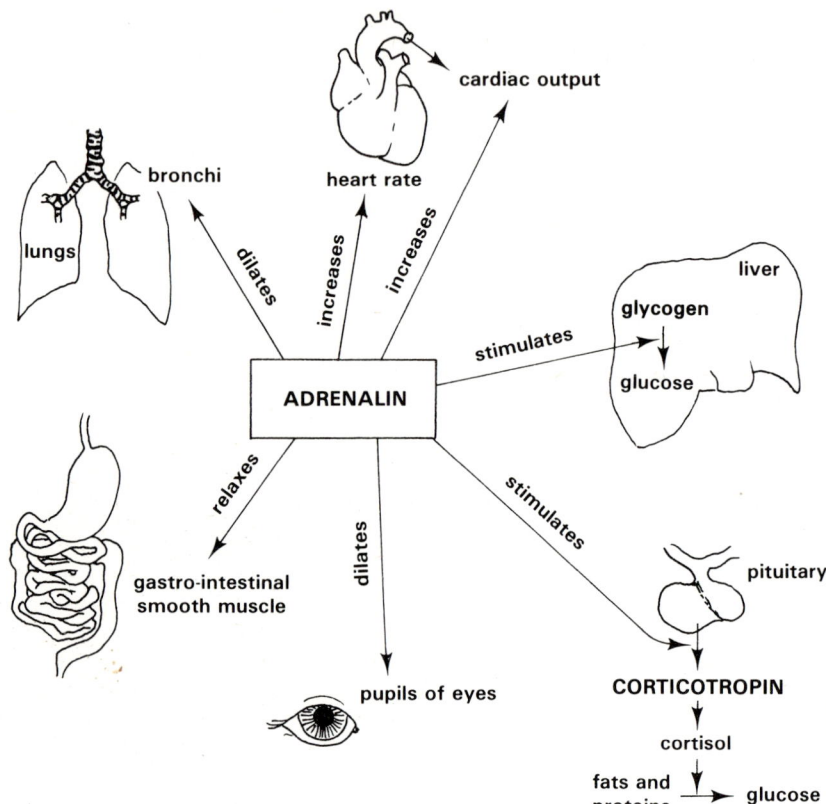

18.5 The pituitary body

Projecting downward from the base of the forebrain and almost totally enclosed by bone is the **pituitary body**, also called the **hypophysis**. It consists of two functional glands called the **anterior pituitary**, derived from the embryonic pharynx and the **posterior pituitary**, derived from the hypothalamus in the forebrain (Fig. 18.28).

18.5.1 The anterior pituitary

The anterior pituitary secretes several hormones including the various **tropins** referred to earlier in the chapter. **TSH** controls the secretion of thyroid hormones while **ACTH** controls the output of cortisol from the adrenal cortex. Other tropins affecting mainly the testes and ovaries are called **gonadotropins**. They are also produced by the anterior pituitary. The activities of **follitropin (FSH)** and **lutropin (LH)** are described in Chapter 22.

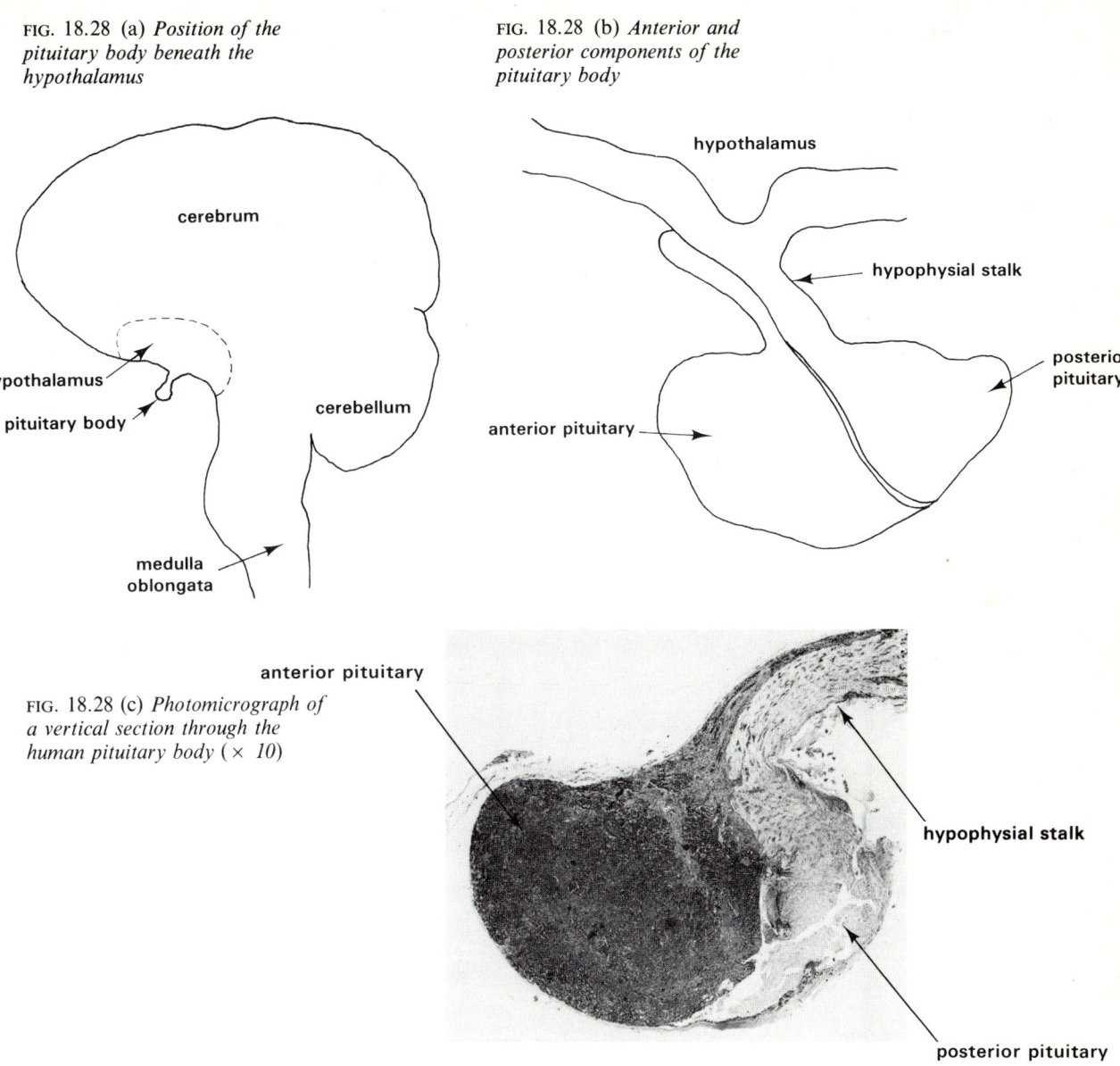

FIG. 18.28 (a) *Position of the pituitary body beneath the hypothalamus*

cerebrum

hypothalamus

pituitary body

cerebellum

medulla oblongata

FIG. 18.28 (b) *Anterior and posterior components of the pituitary body*

hypothalamus

hypophysial stalk

posterior pituitary

anterior pituitary

FIG. 18.28 (c) *Photomicrograph of a vertical section through the human pituitary body (× 10)*

anterior pituitary

hypophysial stalk

posterior pituitary

Another hormone made in the anterior pituitary is called the growth hormone or **somatotropin (STH)**. Growth is also controlled by thyroxin (section 18.1.1). STH is not strictly a tropin since it affects the tissues directly rather than through secretions of other endocrine glands. However, STH enhances the activities of the other hormones from the anterior pituitary. The main function of STH is to promote the growth of the body's tissues and organs by stimulating the synthesis of macro-molecules, especially proteins. The precise mechanism of the stimulation is unknown but it may involve promoting the absorption of vital nutrients, especially amino acids by body cells from tissue fluid, and activation of the genes which control growth.

Secretion of STH occurs throughout life but diminishes after the growing period. The way in which its secretion is controlled is unknown.

Oversecretion of STH leads to **gigantism**, while undersecretion causes **dwarfism** (Fig. 18.29). In man, oversecretion of STH during adulthood when normal growth is complete leads to **acromegaly**. Many of the internal organs become enlarged as do the hands and feet. The most striking characteristic is the lower jaw which often grows to protrude forward rather noticeably (Fig. 18.30). The overall body height does not usually increase since after adolescence the limb bones are normally incapable of further growth in length.

FIG. 18.29 *Diagram showing gigantism* (left) *and dwarfism* (right) *contrasted with normal growth* (centre)

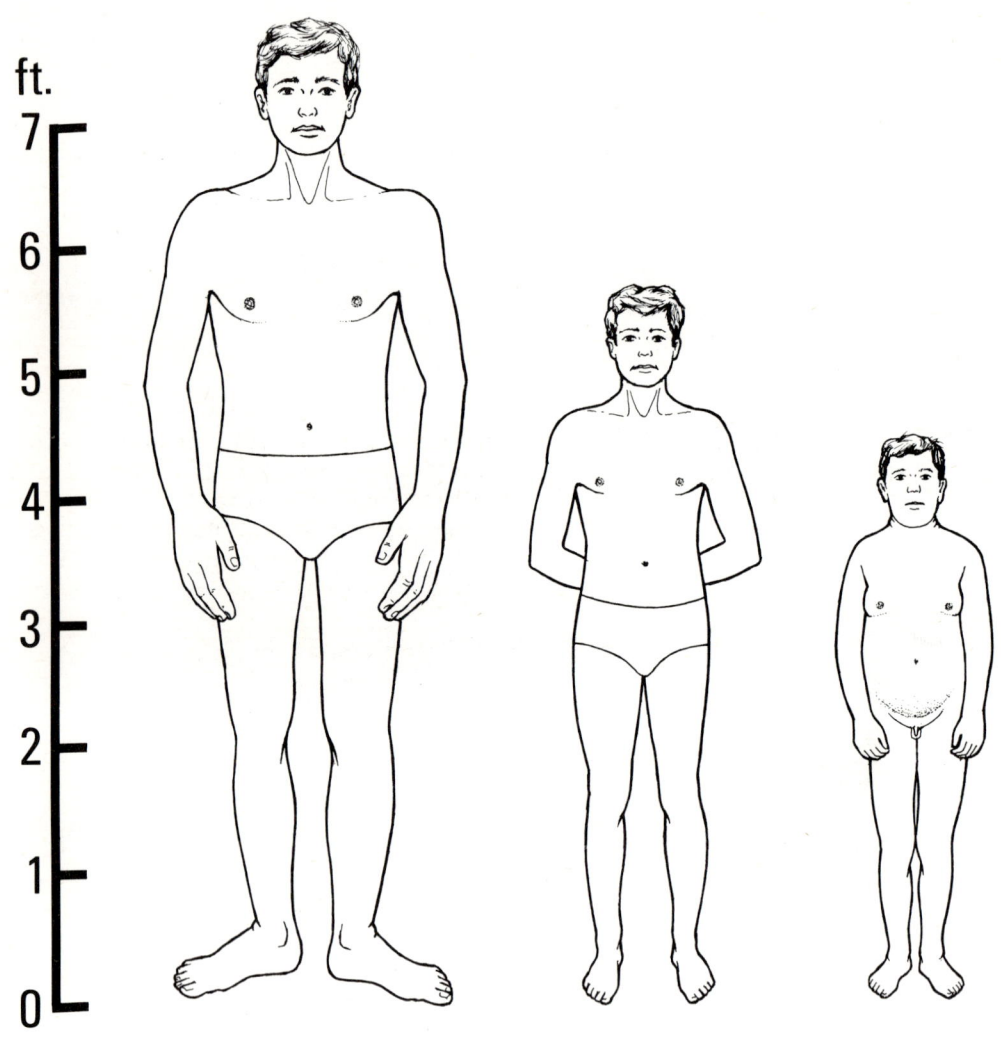

FIG. 18.30 *Facial characteristics of acromegaly. Note the protruding jaw and forehead*

Secretion by the anterior pituitary is partly controlled by a variety of substances from the hypothalamus. The secretions are carried from the hypothalamus into the anterior pituitary in blood vessels (Fig. 18.31). One secretion is **thyroliberin**. Thyroliberin stimulates the release of TSH which in turn stimulates secretion of the thyroid hormones (Fig. 18.5). Other hormones made in the hypothalamus control the release of the rest of the hormones made in the anterior pituitary.

FIG. 18.31 *The hypothalamic–hypophysial portal system responsible for transporting tropin hormones made in the hypothalamus down into the anterior pituitary*

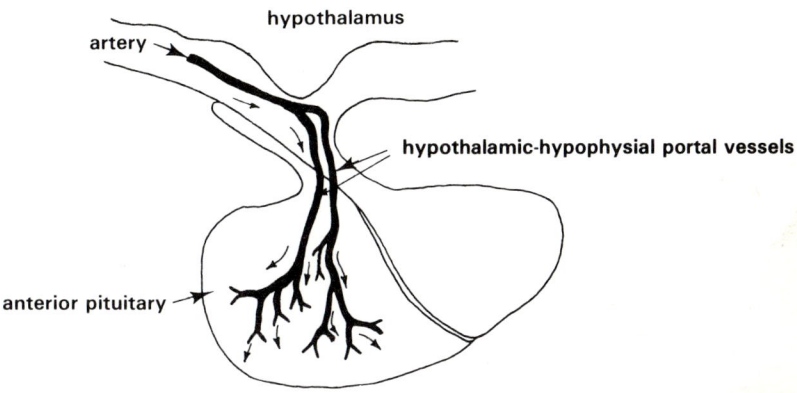

18.5.2 The posterior pituitary

The **posterior pituitary** produces two hormones. These are **vasopressin** and **ocytocin**. Both are polypeptides consisting of nine amino acids. Vasopressin (VP) and ocytocin are made by neurons in the hypothalamus and migrate in the axons of nerve fibres to the posterior pituitary where they are stored. When the neurons connecting the hypothalamus and posterior pituitary are stimulated, VP and ocytocin are released into the blood (Fig. 18.32).

Ocytocin affects smooth muscle and is particularly important in female mammals at the end of pregnancy when it brings about rhythmical contraction of the uterine wall during birth. This function and its role in controlling the ejection of milk from the mammary glands are described in Chapter 22. Vasopressin plays an important role in the regulation of water reabsorption by the renal nephrons (Chapter 10).

FIG. 18.32 *Hypothalamic–hypophysial neurons responsible for transporting vasopressin and ocytocin from the hypothalamus down into the posterior pituitary*

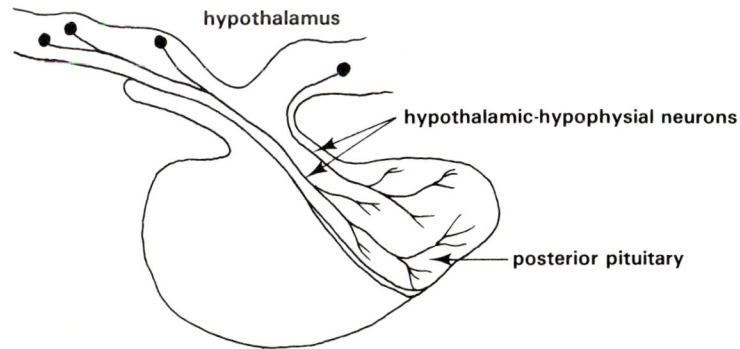

Chapter 19

Reproduction in flowering plants

Reproduction is the formation of new individuals. It is the means whereby successive generations are produced resembling their parents to a greater or lesser extent. The transmission of hereditary material from parent to offspring is an essential feature of reproduction. In flowering plants the transmission can occur in one of two main ways. In **sexual reproduction** male and female sex cells called **gametes** are produced. The gametes come together and nuclear fusion takes place. The genome of a sexually produced organism is a mixture of genes from two sources. In **asexual reproduction** offspring are formed in a number of ways but fusion of nuclear material does not take place. Offspring produced asexually nearly always have an exact replica of the genome of their parent. Many species of flowering plants reproduce both sexually and asexually. Most flowering plants are **hermaphrodite** yet most species cross-breed.

Man is highly dependent on the end products of flowering plant reproduction. Seeds and fruits, tubers and bulbs figure prominently in our diet and in the diets of our domesticated stock.

19.1 Sexual reproduction

19.1.1 Flower structure

In flowering plants male gametes develop in **pollen grains**, female gametes in **ovules**. Pollen grains and ovules are produced in **flowers**.

Following a period of vegetative growth, most flowering plants produce flowers. Annual plants flower in their first season of growth, biennials in their second season. After flowering both types die leaving seeds to germinate the following year. Perennial species generally flower every year when established. Some angiosperms, for example the tulip, produce large **solitary flowers** but others such as the dandelion have small flowers massed together in an **inflorescence**.

Flowers develop from apical meristems which previously gave rise to leafy shoots. Some of the factors which regulate the switch from a vegetative to a reproductive function are described in section 19.5. As a flower bud opens it is possible to see four kinds of floral organs in most flowers. On the outside are the **sepals**, usually green in colour. Collectively the sepals of a flower are called the **calyx**. Immediately inside the calyx are the **petals**, collectively called the **corolla**. The petals of many species of flowering plants are coloured. Yellow, blue and red are the most common colours in wild flowers. Next inwards are the **stamens**, each consisting of a **filament** and an **anther** in which pollen is produced. The term **androecium** is used to describe the collection of stamens in a flower. At the centre of the flower are found one or more **carpels**, collectively known as the **gynoecium** or **ovary**. Each carpel or group of carpels contains one or more ovules and has a **stigma** and a **style**.

The apex of the flower stalk from which the floral organs arise is called the **receptacle**. The numbers, size, colour and arrangement of the various organs vary from one species to another, although the flowers of closely related species are very similar in appearance. Figure 19.1 shows a range of flower structures of a few common British species.

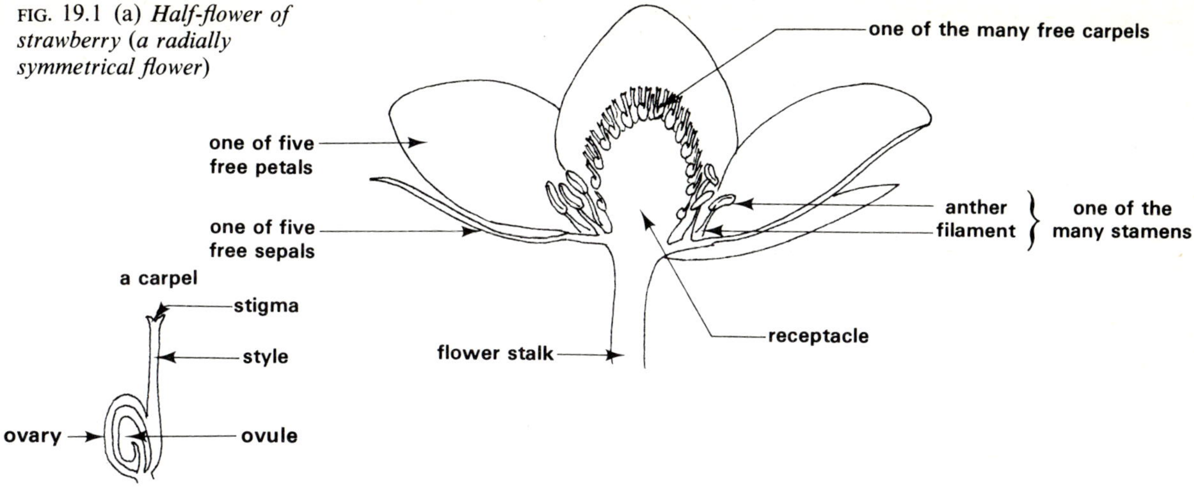

FIG. 19.1 (a) *Half-flower of strawberry (a radially symmetrical flower)*

one of the many free carpels

one of five free petals

one of five free sepals

anther
filament
} one of the many stamens

a carpel

stigma

style

ovary →

ovule

receptacle

flower stalk

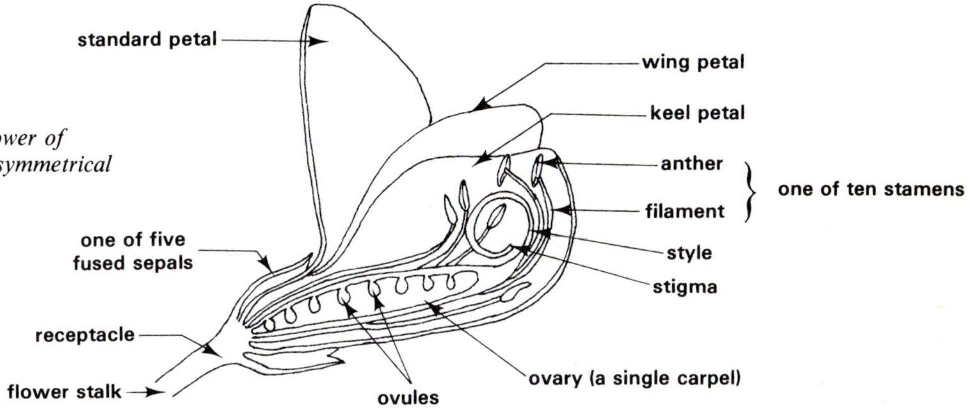

FIG. 19.1 (b) *Half-flower of broom (a bilaterally symmetrical flower)*

standard petal

wing petal

keel petal

anther

filament
} one of ten stamens

one of five fused sepals

style

stigma

receptacle

ovary (a single carpel)

flower stalk →

ovules

FIG. 19.1 (c) *Flowers of dandelion*

a floret

inflorescence of many small flowers (florets)

stigma

style →

corolla of five fused petals

tube of five fused anthers through which the style grows

calyx (hairs)

ovary of two fused carpels containing one ovule

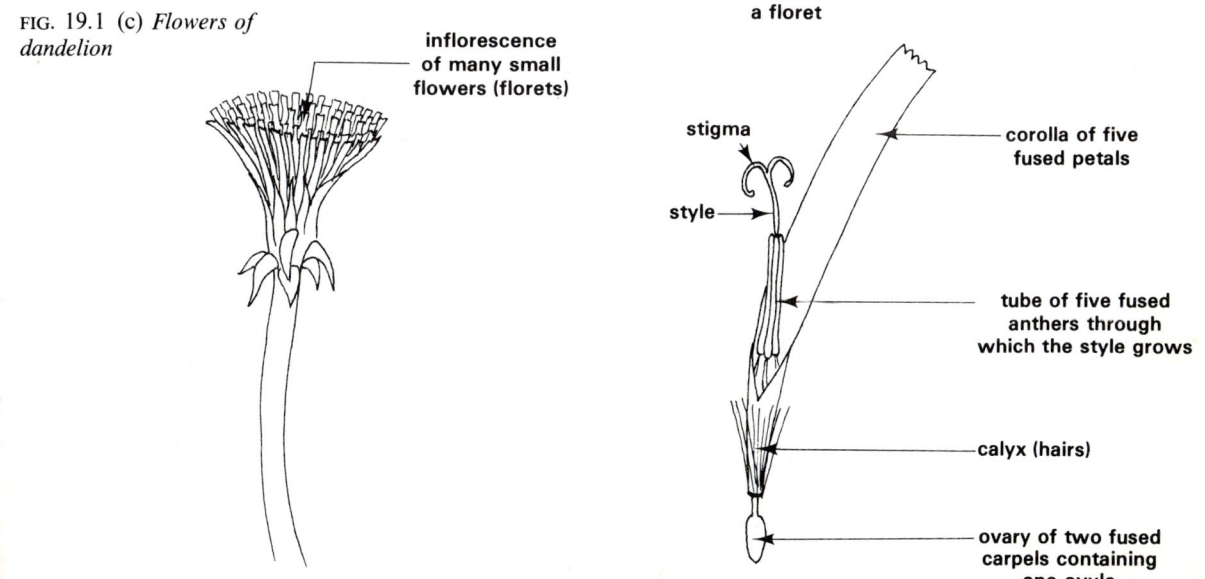

19.1.2 Development of pollen and ovules

Sexual reproduction cannot occur without the formation of sexual cells called **gametes**. In flowering plants male gametes are formed in the **pollen grains**. Female gametes are formed in the **ovules**.

1. POLLEN FORMATION

Pollen production takes place in the anthers (Fig. 19.2). Early in their development anthers are elongate masses of parenchyma tissue enclosed in an epidermis. In transverse section a young anther is seen to have two or four lobes according to the species. A row of cells extending the whole length of each lobe becomes organised into a central mass of **pollen mother cells** surrounded by several layers of flattened cells. The layer immediately surrounding the pollen mother cells is called the **tapetum**. Soon the mother cells undergo meiosis to produce tetrads of haploid **pollen grains**, also called **microspores**. Each pollen grain secretes a thick wall and its haploid nucleus divides by mitosis to form a **generative cell** and a **tube nucleus**.

FIG. 19.2 *Development of anther and pollen*
(a) *T.s. young anther*

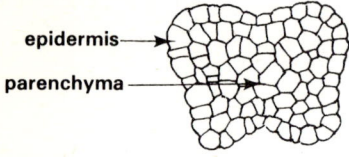

epidermis
parenchyma

FIG. 19.2 (b)

(i) *transverse section through a young anther of lily,* × 40

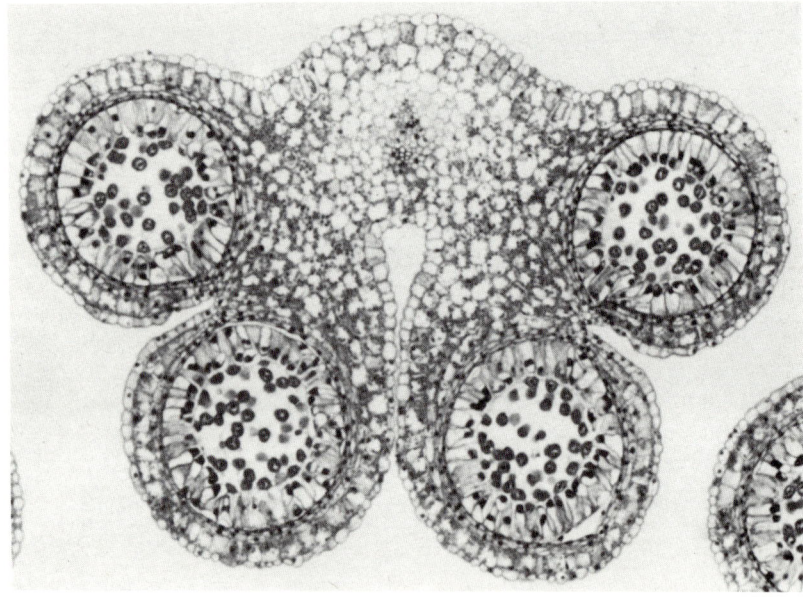

While these changes are going on the anther grows considerably in size, and depending on the species, two or four pollen sacs become prominent. The cells around each sac become thickened with bands of lignin forming a **fibrous layer**. On drying out the fibrous layer shrinks causing the pollen sacs to rupture. Usually they open by splitting lengthwise down each side of the anther, though in some species pores or valves are formed instead.

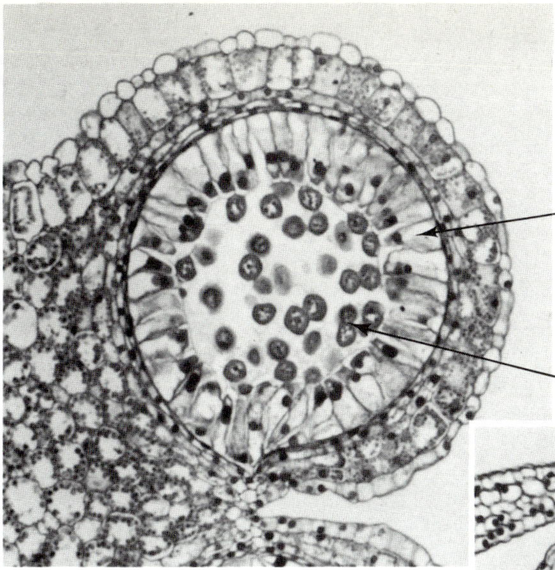

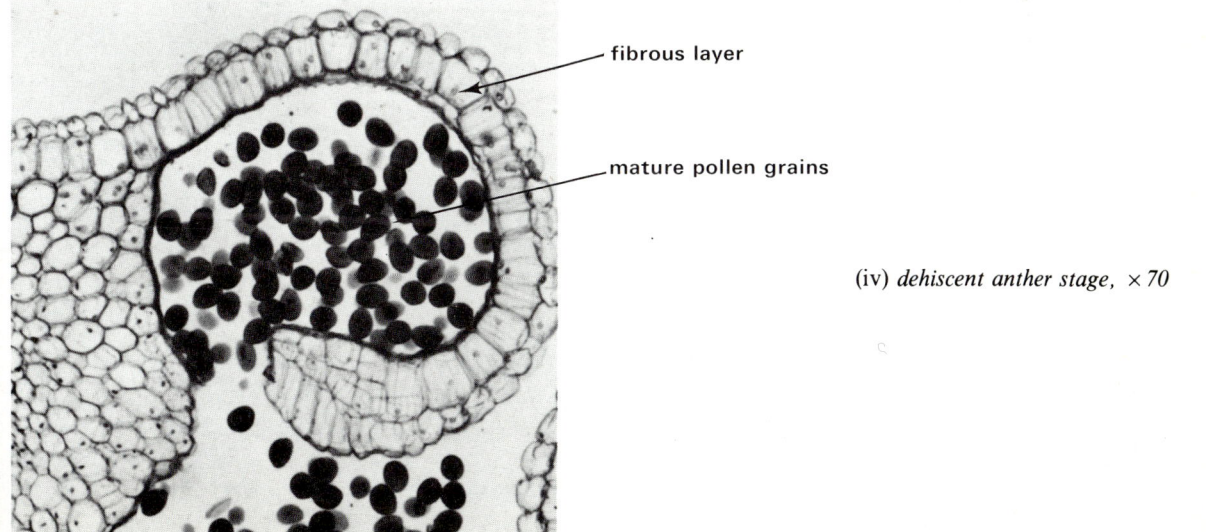

tapetum

(ii) *pollen mother cell stage,* × 70

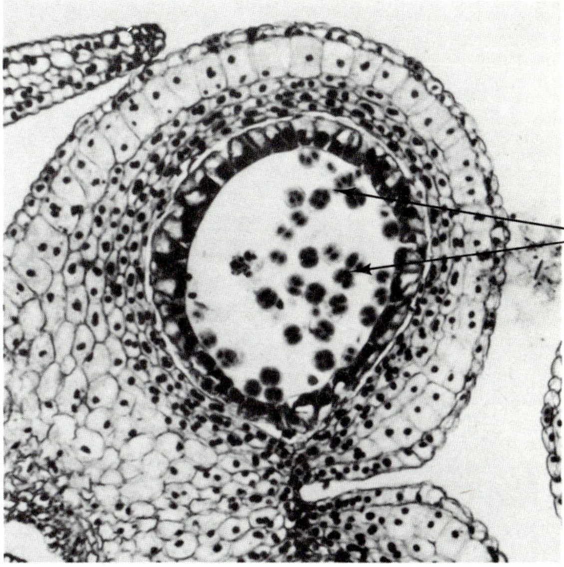

pollen mother cells

tetrads of
pollen grains

(iii) *tetrad stage,* × 70

fibrous layer

mature pollen grains

(iv) *dehiscent anther stage,* × 70

FIG. 19.2 (c) *Development of pollen grains*

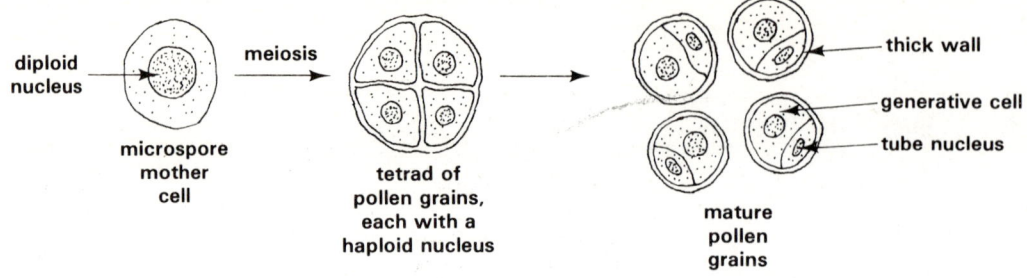

diploid nucleus → microspore mother cell

meiosis →

tetrad of pollen grains, each with a haploid nucleus

→

mature pollen grains

thick wall
generative cell
tube nucleus

(vi) *section through mature pollen grains,* $\times 400$

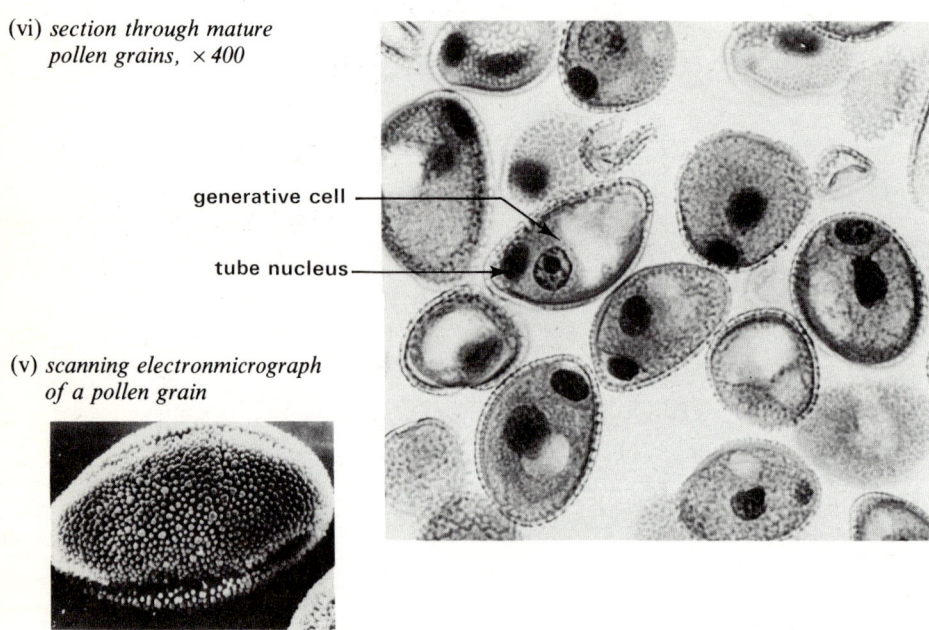

generative cell

tube nucleus

(v) *scanning electronmicrograph of a pollen grain*

FIG. 19.2 (d) *T.s. dehisced anther*

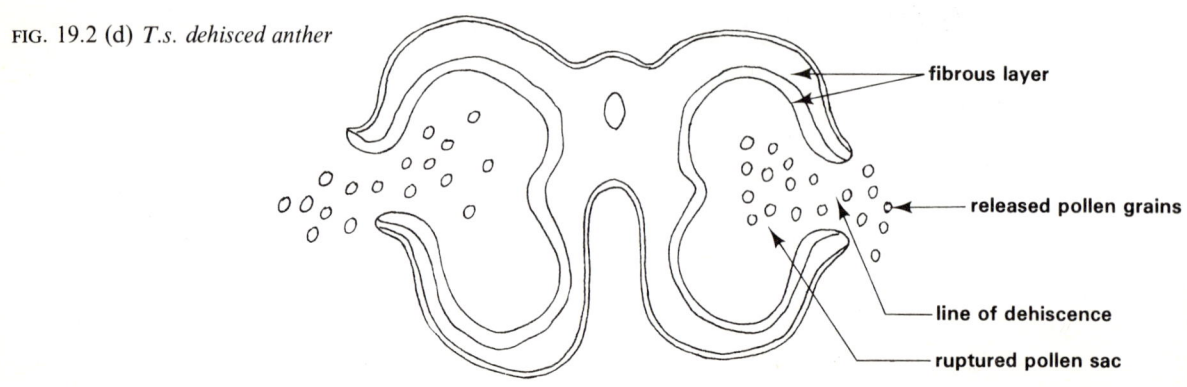

fibrous layer

released pollen grains

line of dehiscence

ruptured pollen sac

2. OVULE FORMATION

Ovules are formed inside the carpels. Depending on the species, one or more ovules develop in each carpel. In all species, however, an ovule grows from a pad of carpellary tissue called the **placenta** to which it is attached by a short stalk called a **funicle**. At first the young ovule is just a bulge of parenchyma tissue called the **nucellus**. As it enlarges the nucellus is enclosed by two layers called **integuments** which grow up from the funicle. When fully developed the integuments completely surround the nucellus except for a tiny pore called the **micropyle**. Meanwhile, near the tip of the nucellus a cell divides by mitosis to form an outer **parietal cell** and an inner **megaspore mother cell**. The latter divides by meiosis to form a row of four haploid **megaspores**. Only the innermost megaspore usually develops further. It grows considerably in size and its nucleus undergoes three mitotic divisions. The eight haploid nuclei which are formed become arranged in a characteristic fashion (Fig. 19.3). The resulting structure is called an **embryo sac**.

FIG. 19.3 (a) *Development of an ovule*

(i) l.s. young ovule in carpel

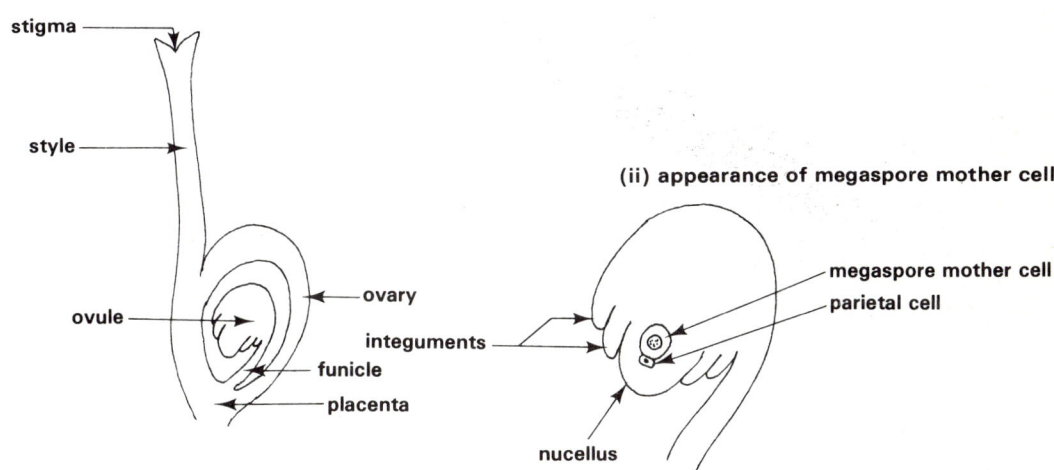

(ii) appearance of megaspore mother cell

(iii) formation of embryo sac

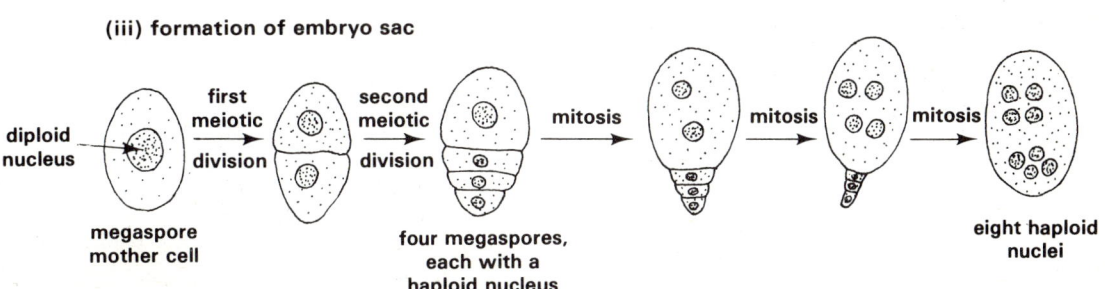

(iv) l.s. mature ovule

FIG. 19.3 (b) *Longitudinal sections through ovules at various stages of development*

embryo sac

antipodal cells

egg cell

polar nuclei

synergids

micropyle

vascular bundle

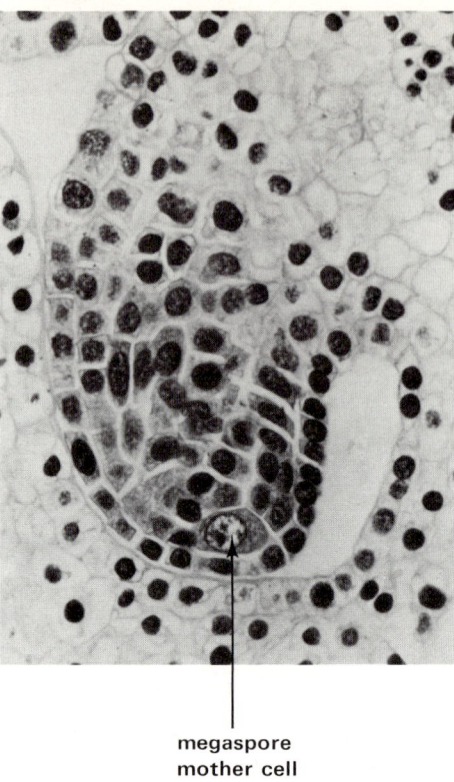

(i) *megaspore mother cell stage, ×250*

megaspore
mother cell

(ii) *embryo sac stage, ×125*

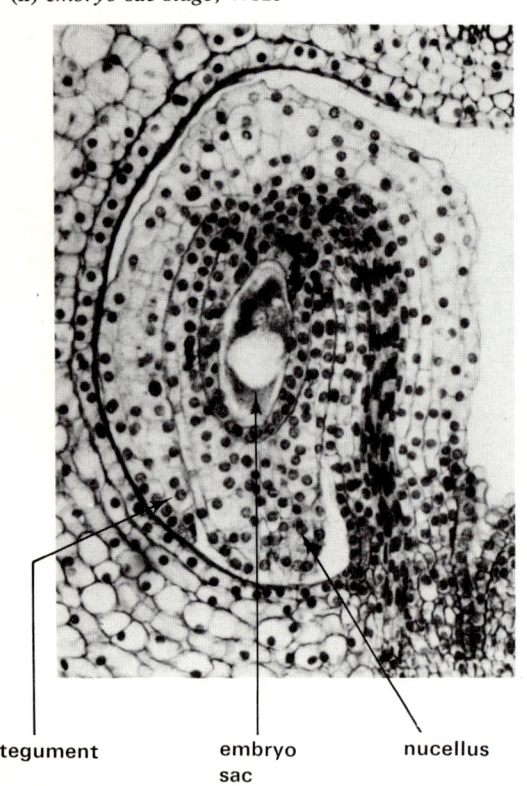

integument

embryo
sac

nucellus

(iii) *young embryo sac, ×400*

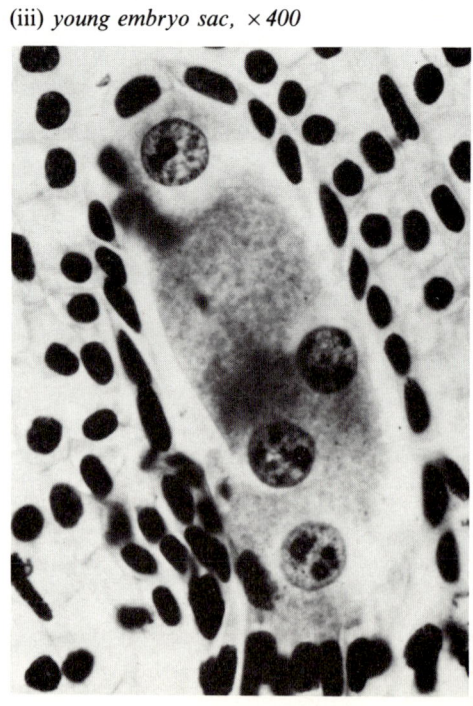

(iv) mature embryo sac, × 300

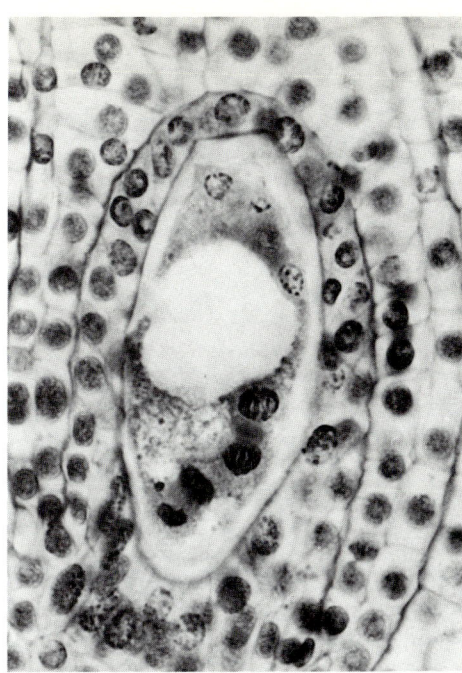

19.1.3 Fertilisation and its consequences

If fertilisation is to occur the male and female sexual cells must be brought together. The first step in bringing the gametes near to each other is **pollination** (section 19.3). Wind and insects are the main pollinating agents. They carry pollen from the anthers to the stigmas.

FIG. 19.4 (a) *Germinating pollen grains*

FIG. 19.4 (b) *Fertilisation*

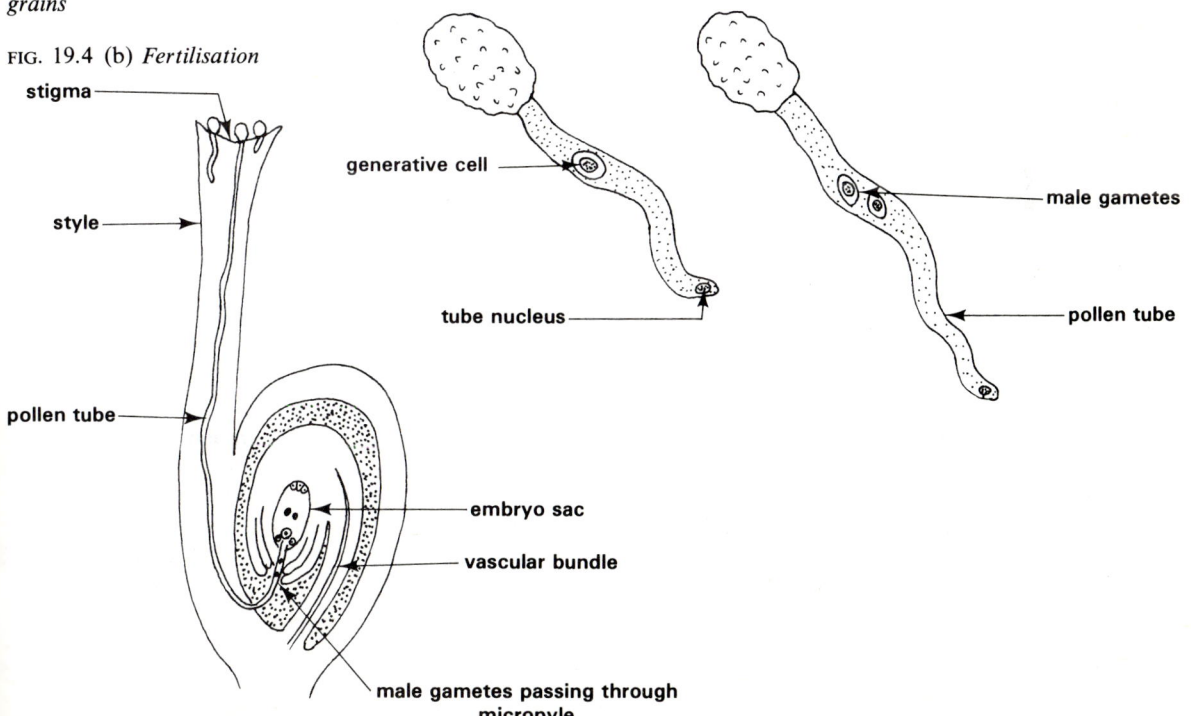

FIG. 19.4 (c) *Pollen grains germinating on the stigma of evening primrose,* ×*400*

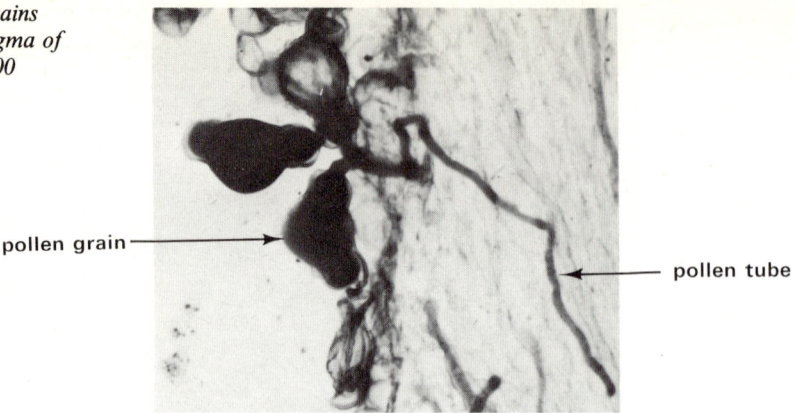

pollen grain

pollen tube

The **egg cell** of each embryo sac is a **female gamete**. Following the transfer of pollen to the stigma a **pollen tube** is formed from each pollen grain. As the tube grows down into the style the tube nucleus is situated at its tip. The generative cell then enters the tube and divides by mitosis to form two cells which are **male gametes** (Fig. 19.4). In some species the male gametes are formed before the generative cell enters the pollen tube. On reaching the ovule the pollen tube grows through the micropyle and nucellus to the embryo sac where its tip opens out. The tube nucleus disintegrates and the two male gametes enter the embryo sac. One of the gametes fuses with the nucleus of the egg cell to form a **diploid zygote**. The other fuses with the diploid nucleus formed by the fusion of the two polar nuclei. The result is a **triploid fusion nucleus**. This **double fertilisation** is unique to flowering plants. Following fertilisation the ovules develop into **seeds** and the ovary of the flower into a **fruit**.

1. SEED FORMATION

The subsequent events leading to seed production vary from one species to another. Essentially what happens is that the zygote develops into an **embryo plant** which is nourished as it grows by **endosperm** formed from the triploid fusion nucleus.

In shepherd's purse, mitosis of the zygote produces a short chain of cells. The first division is at right angles to the long axis of the ovule. The cell furthest away from the micropyle eventually becomes the embryo. The rest of the chain called the **suspensor** elongates, pushing the **embryonic cell** deeper into the embryo sac. Successive divisions of the triploid fusion nucleus produce a large number of triploid nuclei lying in the embryo sac. The nuclei may or may not become separated by cell walls forming a tissue called **endosperm** which surrounds the developing embryo. The embryonic cell divides to form a multicellular embryo with its **radicle**, the embryonic root, pointing towards the micropyle. In shepherd's purse the long axis of the mature embryo is bent. The two cotyledons with the **plumule**, the embryonic shoot, tucked between them point in the same direction as the radicle. During its development the embryo goes through globular, heart-shaped and torpedo stages. As the embryo grows the endosperm gradually disappears. Ultimately no endosperm remains so the seeds of shepherd's purse are **non-endospermic** (Fig. 19.5). However, in some flowering plants such as grasses, some of the endosperm is still present when the embryo is fully developed. These seeds are described as **endospermic** (Chapter 20).

During the final stage in the formation of a seed some of the integument cells usually become lignified, forming a tough protective seed coat called the **testa** perforated only by the micropyle. Gradual dehydration of the entire seed takes place so that eventually as little as 5–10 per cent of its mass is water. The embryo is now in a state of **dormancy** (Chapter 20).

FIG. 19.5 (a) *Stages in the development of an embryo of shepherd's purse*

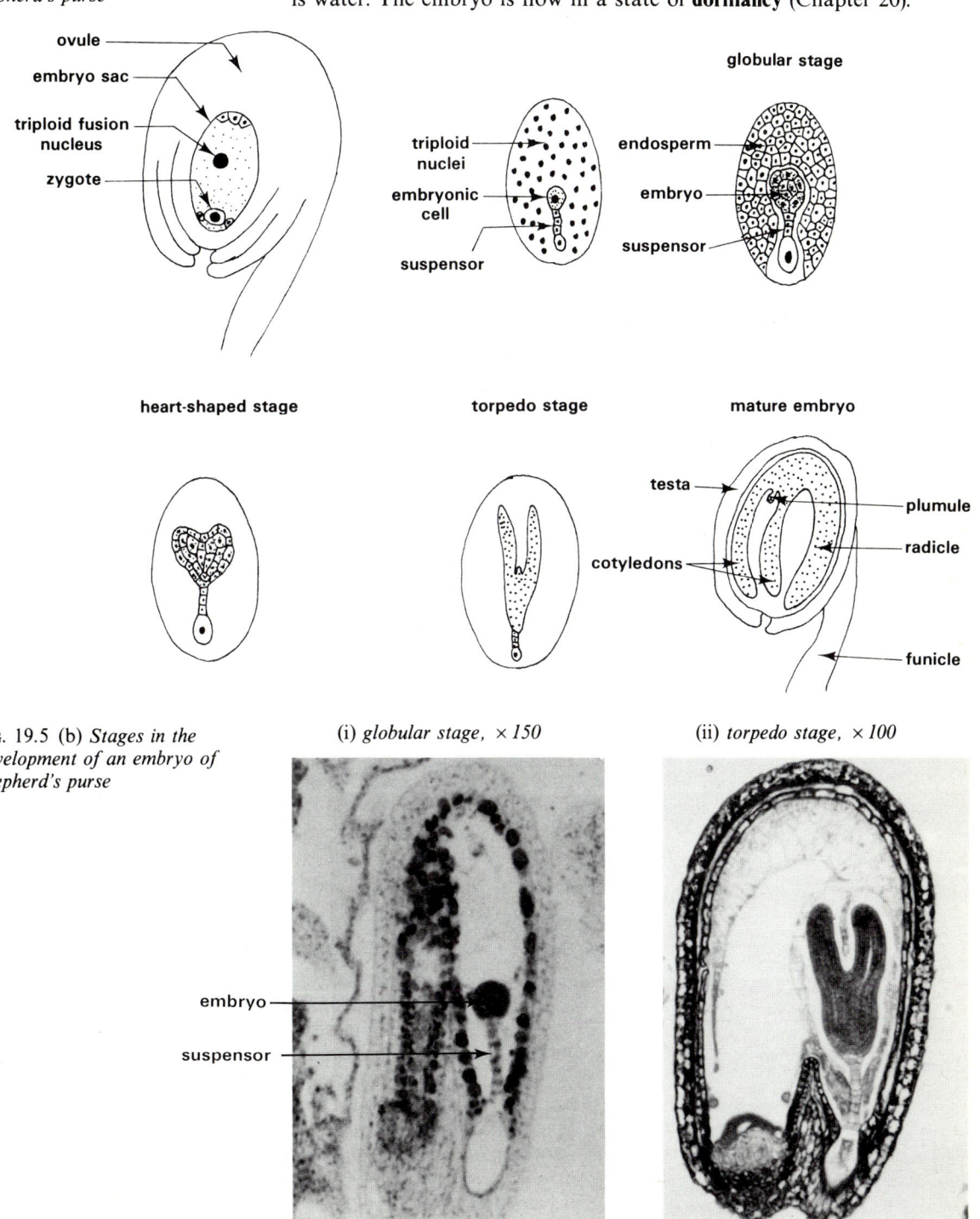

FIG. 19.5 (b) *Stages in the development of an embryo of shepherd's purse*

(i) *globular stage, × 150*

(ii) *torpedo stage, × 100*

(iii) *mature embryo,* × *100*

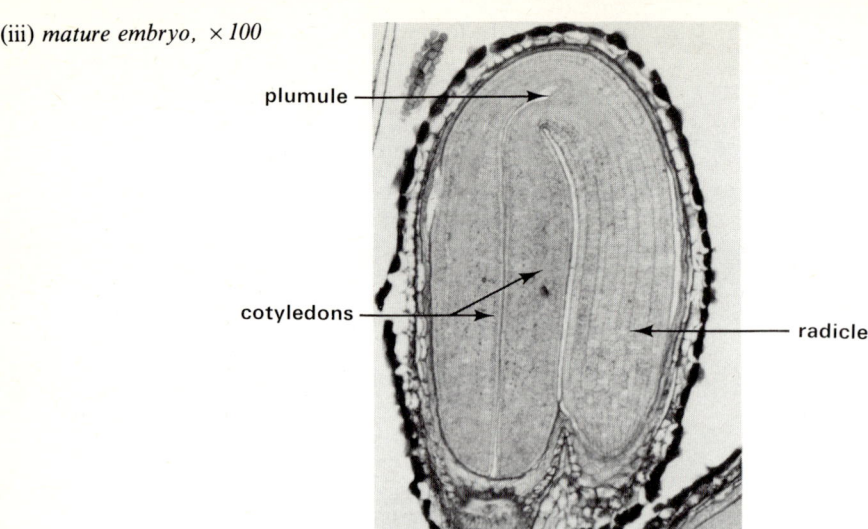

plumule

cotyledons

radicle

2. FRUIT FORMATION AND SEED DISPERSAL

While the seeds are developing, changes take place in the surrounding carpels. Growth of the carpels keeps pace with enlargement of the developing seeds. At maturity the seeds are completely enclosed and protected by carpellary tissue called the **pericarp**. The gynoecium has now developed into a **fruit**, the structure of which depends on the number and arrangement of carpels in the flower and on the changes taking place after fertilisation.

In many species the pericarp dries out at the same time as the seeds. The **dry fruits** which result often open by valves or pores through which the seeds are released when the fruit is shaken by wind. The pods of leguminous plants such as gorse and broom open by splitting along the dorsal and ventral edges and the seeds are shot out (Fig. 19.6(a)). In small-seeded dry fruits the pericarp often has an extension enabling the fruit to remain airborne for some time after its release from the parent. Dandelion fruits, for example, have a parachute-like pappus (Fig. 19.6(b)). In the dandelion and in many other flowering plants the seeds are not released. The pericarp splits open when the seeds germinate in the soil.

Some flowering plant species produce fleshy, **succulent fruits**. The fleshiness is sometimes due to growth of the pericarp as in plums and cherries. In other species another part of the flower enlarges instead. In the strawberry for example the receptacle swells considerably after fertilisation, pushing the tiny one-seeded carpels apart so that they are eventually scattered over the surface of the fruit (Fig. 19.6(c)). Fleshy fruits are often part of the diets of many animals. Although the seeds may be eaten they come to no harm unless crushed by the animals' teeth. This is because they are protected by the tough pericarp which is undigested in its passage through the animal's gut. By the time the animal has egested the seeds in its faeces they have probably been carried some distance from the parent plant.

There are many other patterns of fruit development but the end-product is generally structured so that the seeds are self-dispersed or are dispersed by wind, animals, or, occasionally, water. Dispersal helps to avoid overcrowding and competition which occurs if a large number of seeds germinate in a confined area. It also provides opportunities for species to colonise territories where previously they have not grown and where they may thrive.

FIG. 19.6 *Fruits*

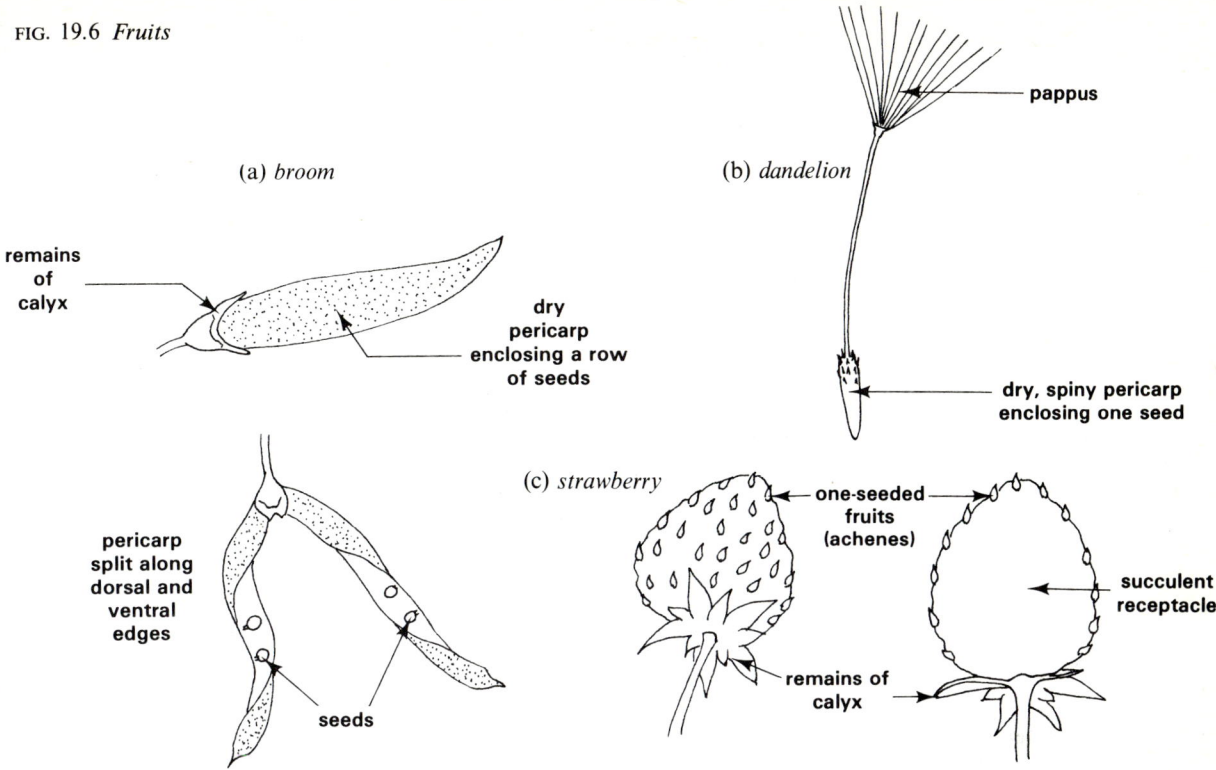

(a) *broom*

remains
of
calyx

dry
pericarp
enclosing a row
of seeds

(b) *dandelion*

pappus

dry, spiny pericarp
enclosing one seed

pericarp
split along
dorsal and
ventral
edges

seeds

(c) *strawberry*

one-seeded
fruits
(achenes)

succulent
receptacle

remains of
calyx

whole

l.s.

19.2 The life cycle of a flowering plant

The reproductive function of flowers involves the formation of two kinds of spores. Pollen grains are sometimes called **microspores** and the anthers in which they are produced can be called **microsporangia**. The larger **megaspores** are formed in the ovules or **megasporangia**. Thus a diploid spore-forming flowering plant is a **sporophyte** and produces two types of haploid spores. On developing further the spores give rise to gamete-bearing structures called **gametophytes**. The germinated pollen grain is a **male gametophyte** from which two male gametes arise whilst an embryo-sac is a **female gametophyte** in which a female gamete is formed. The life cycle is completed when an embryo sporophyte plant is produced in the seed following fertilisation (Fig. 19.7).

In the life cycle of a flowering plant there is thus an **alternation of generations.** Such an alternation of sporophyte and gametophyte is also found in other groups of terrestrial plants. In flowering plants, however, the gametophytes are vestigial and rely entirely on the sporophyte for their development. The situation is rather different in ferns where the sporophyte and gametophyte are independent plants and where the gametophyte has distinctive sexual organs. It is also different from the life cycle of a moss where the gametophyte is the free-living plant on which the sporophyte develops.

A feature of flowering plant reproduction is the development of the young sporophyte inside the carpel of the parent sporophyte. The young sporophyte thus has ideal conditions in which to develop. Because of this the chances that the next generation will be produced are very good. The development of seeds may be one of the reasons for the success of flowering plants compared with other groups of land plants.

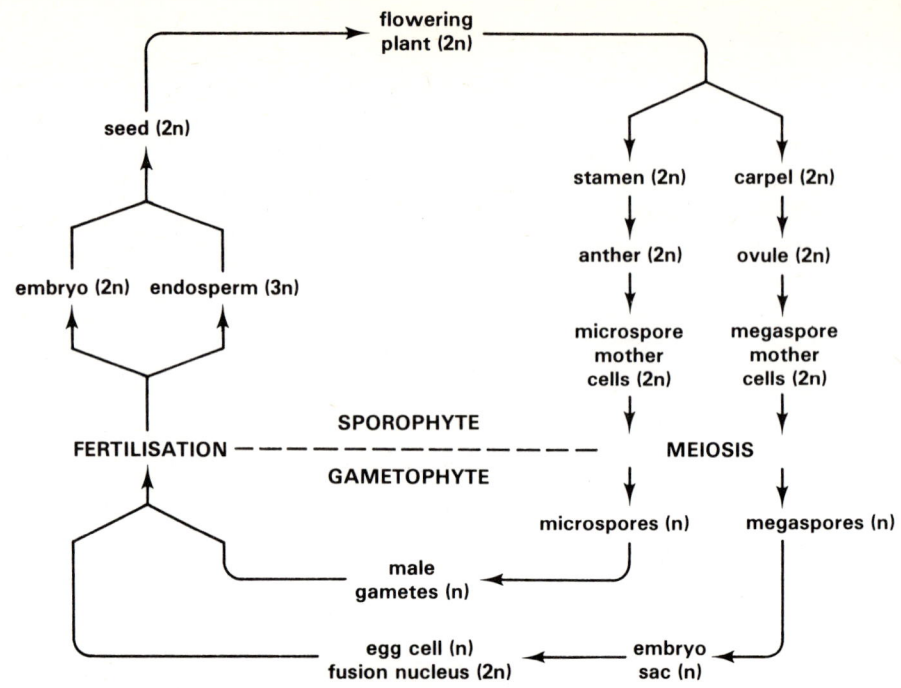

FIG. 19.7 *Life-cycle of a flowering plant*

19.3 Pollination

The transfer of ripe pollen from an anther to the stigma of a flower of the same species is called **pollination**. **Self-pollination** occurs when pollen is carried from an anther to the stigma of the same flower or to the stigma of another flower on the same plant. When pollen is transferred from one plant to another of the same species, **cross-pollination** has taken place.

The two forms of pollination have very different genetic consequences. Self-pollination leads to **self-fertilisation**, cross-pollination to **cross-fertilisation**. Self-fertilised species depend on random assortment and crossing over (in meiosis leading to pollen grain and embryo sac production) and on mutation to bring about variation in the genomes of male and female gametes. Self-fertilised species therefore display less **genetic variation** than cross-fertilised species which are produced from gametes from two different individuals. Plants which are self-pollinated are called **inbreeders** while cross-pollinated plants are called **outbreeders**.

Inbreeding has its virtues because it can preserve particularly good genomes which may be suited to a relatively stable environment. Outbreeding is of greater evolutionary significance because it continually produces a variety of genomes. In the struggle for survival some genomes are more successful than others. Although a few species never outbreed it is interesting, in view of its potential evolutionary advantage, to find that most flowering plants do. In fact many flowering plant species have evolved a variety of mechanisms to ensure that cross-fertilisation usually takes place.

19.3.1 Mechanisms for ensuring outbreeding

When the stamens and ovules of flowers do not ripen at the same time pollen has to come from another plant of the same species if fertilisation is to occur. In the wood-sage the anthers discharge their pollen before the style has completed its growth and the stigma is receptive. The condition is called **protandry** (Fig. 19.8(a)). It contrasts with the behaviour of the rib-wort plantain in which the styles protrude from the flowers and the stigmas are ready to be pollinated before the stamens have grown. The latter condition is called **protogyny** (Fig. 19.8(b)). Some angiosperms such as willows and poplars have separate plants for the production of pollen and ovules. They have no option but to cross-fertilise. The primrose has two types of flower, pin-eyed and thrum-eyed (Fig. 19.8(c)). As you can see the structure of the pin-eyed flowers prevents them being self-pollinated. If pollen from thrum-eyed flowers falls on to their own stigmas the pollen will not germinate.

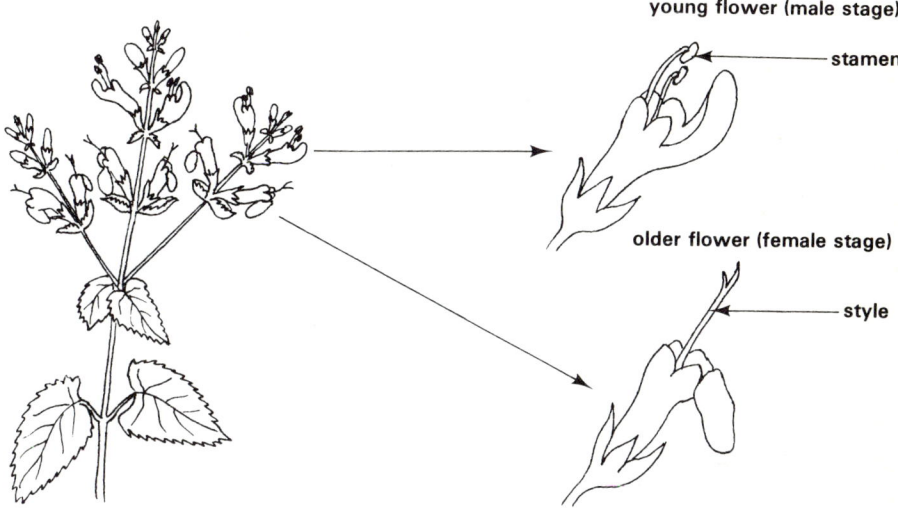

FIG. 19.8 (a) *Protandrous flowers of wood-sage*

young flower (male stage)

stamen

older flower (female stage)

style

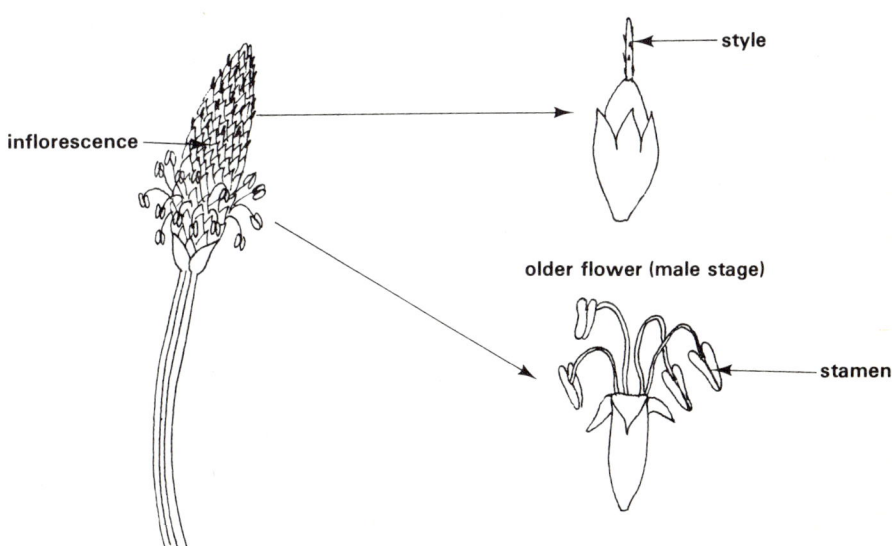

FIG. 19.8 (b) *Protogynous flowers of ribwort plantain*

young flower (female stage)

style

inflorescence

older flower (male stage)

stamen

Other flowers, although producing both pollen and ovules, are **self-sterile**. When transferred to the stigma pollen will not germinate unless it comes from another plant of the same species. Self-sterility is a genetic trait. It is the reason why many varieties of apple, pear and cherry will not set fruit unless another variety is growing nearby to act as a **pollinator**.

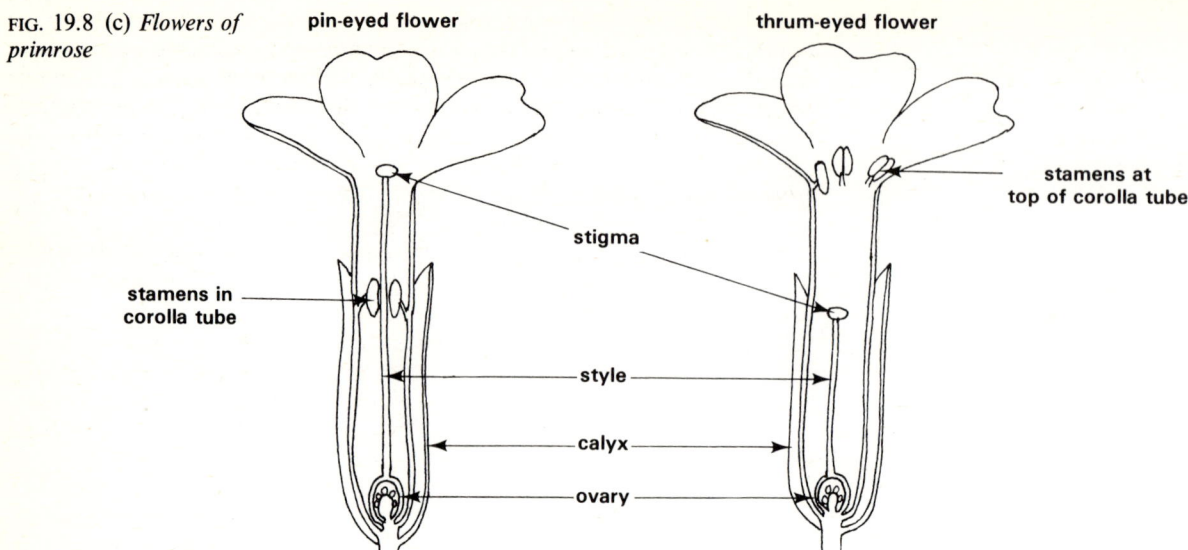

FIG. 19.8 (c) *Flowers of primrose*

pin-eyed flower

thrum-eyed flower

stamens at
top of corolla tube

stigma

stamens in
corolla tube

style

calyx

ovary

19.3.2 Pollination mechanisms

A number of agencies bring about pollination, the two most common being **wind** and **insects**.

1. WIND-POLLINATED FLOWERS

The flowers of **wind-pollinated** species often show a complete lack of petals and sepals. They have no means of attracting insects (Fig. 19.9). The flowers are often borne in inflorescences which either appear before the leaves as in trees such as ash, and shrubs such as hazel, or well above the leaves as in grasses and some plantains. In this way the leaves do not interfere with air movements around the flowers.

FIG. 19.9 (a) *Wind-pollinated flowers of rye-grass*

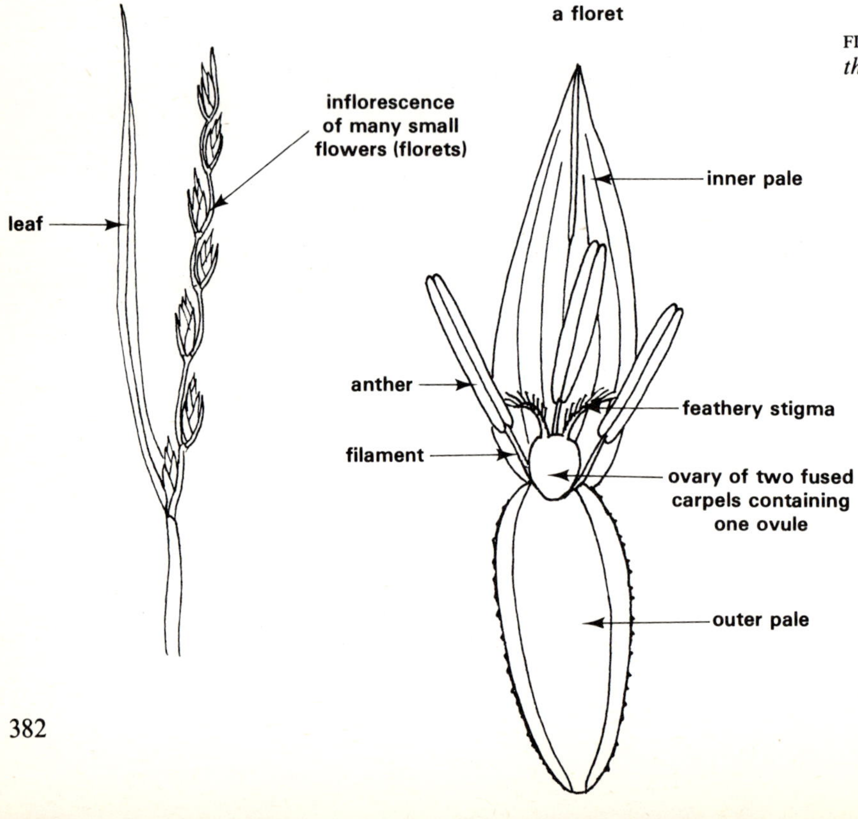

a floret

FIG. 19.9 (b) *Grass flowers. Notice the pendulous stamens*

inflorescence
of many small
flowers (florets)

inner pale

leaf

anther

feathery stigma

filament

ovary of two fused
carpels containing
one ovule

outer pale

The long, feathery and often sticky stigmas which protrude into the air improve the chances of pollination. The anthers are often hinged on to the filaments and hang out of the flower where the slightest air movement shakes out the pollen. The pollen grains are extremely light, often with air bladders, so that they remain suspended for long periods in the atmosphere. Many people suffer from hay-fever and similar allergies during the summer when the density of airborne pollen is high (Chapter 9). Grass pollen makes a large contribution to the pollen count.

2. INSECT-POLLINATED FLOWERS

One of the most remarkable facets of plant biology is the way in which certain insects and flowers of many species have evolved to their mutual advantage. **Insect-pollinated** flowers offer food. While taking it insects transfer pollen. Insects, like all animals, must have a balanced diet and many flowers provide carbohydrate, protein and lipids. The carbohydrate is found in **nectar**, a sugary solution secreted by special glands called nectaries which are often situated at the base of the petals. Pollen contains the other main dietary ingredients. Some insect-pollinated flowers such as poppies offer pollen only. Nectar and pollen have very little smell so are unlikely to be found by insects, except by chance. Flowers pollinated by insects have powerful insect attractants. Two senses which are highly developed in insects are smell and sight. It is not surprising therefore that insect-pollinated flowers are often highly scented. They are often relatively large and so easily seen. If small, they are grouped into conspicuous inflorescences. They are also brightly coloured. Significantly, the most common flower colours, yellow, blue and red, are those to which insects respond most readily.

The mouth-parts of the insects involved, such as bees and wasps, are adapted for sucking up nectar. Their legs are often modified to collect pollen (Fig. 19.10). Often the insects are social creatures and in co-operating in their search for food they increase the chances of pollination. Although pollen is collected as food by insects, insect-pollinated flowers are generally so structured that some pollen is transferred to the stigma by a foraging bee or wasp. This condition is seen at its extreme in zygomorphic flowers. The structure of zygomorphic flowers ensures that insects stand only in one position while feeding. Consequently pollen is dusted on to a specific site on the insect's body. In the white dead nettle it is the insect's back which receives the pollen. In gorse and broom it is its belly. Often the heavier bees operate a lever mechanism which brings the anthers down on to the body (Fig. 19.11). When visiting another flower of the same species the pollen is then at just the right position to make contact with the stigma.

FIG. 19.10 *Head and legs of worker bee*

front view of head

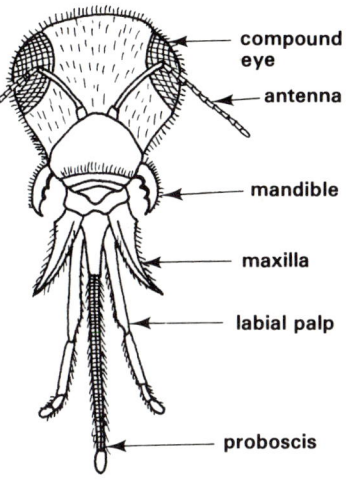

- compound eye
- antenna
- mandible
- maxilla
- labial palp
- proboscis

The proboscis is used as a suction tube to draw nectar into the mouth.

legs

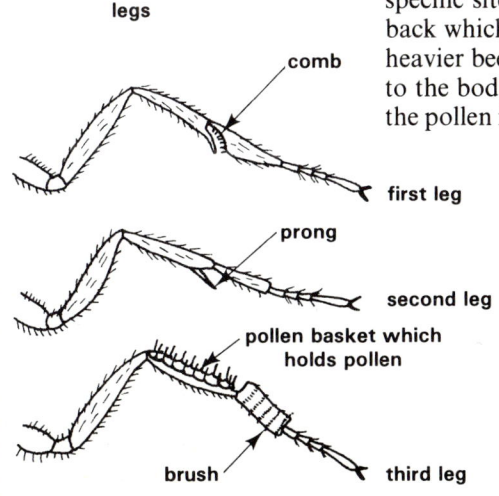

- comb
- first leg
- prong
- second leg
- pollen basket which holds pollen
- brush
- third leg

The comb removes pollen from the hairs on the front end of the body to the pollen basket. The brush transfers pollen from hairs on the rest of the body to the basket. The prong is used to dig out pollen from the basket when the bee returns to the hive.

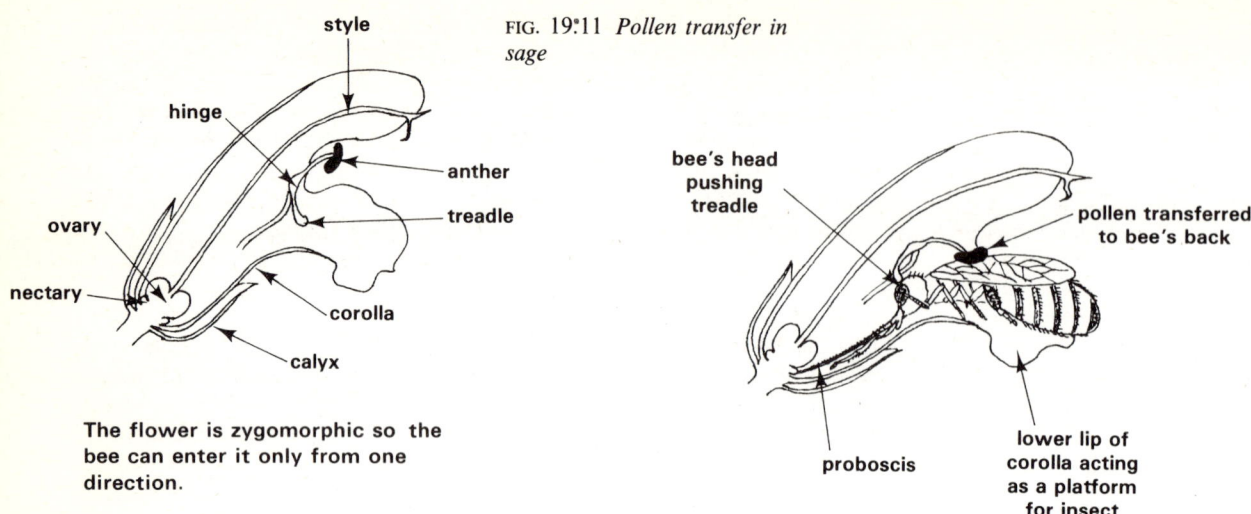

FIG. 19.11 *Pollen transfer in sage*

style

hinge

anther

treadle

ovary

nectary

corolla

calyx

The flower is zygomorphic so the bee can enter it only from one direction.

bee's head pushing treadle

pollen transferred to bee's back

proboscis

lower lip of corolla acting as a platform for insect

19.4 Asexual reproduction

Some plants which produce flowers can produce seeds without sexual fusion taking place as previously described. The production of seeds without fertilisation is called **parthenogenesis**. It occurs in a number of ways.

19.4.1 Diploid parthenogenesis

Diploid parthenogenesis gives rise to seeds with normal diploid embryos. In the dandelion the megaspore mother cell invariably does not undergo meiosis and develops into an embryo without fertilisation. In the blackberry, embryos develop from a diploid cell in the nucellus.

19.4.2 Haploid parthenogenesis

Haploid parthenogenesis results in the formation of seeds with haploid embryos. This occurs when an unfertilised egg cell grows into an embryo, or more rarely when a male gamete on reaching the embryo sac does so. Examples of this form of reproduction have been reported in the thorn apple and the evening primrose. However, because a complete set of chromosomes is lost the plants which grow from the seeds are sterile.

19.4.3 Vegetative propagation

Many angiosperms can produce offspring by means other than seed formation. In these plants progeny are formed by proliferation of part of the vegetative body of the parent. Species which have **vegetative propagation** usually flower too. The garden strawberry, for example, flowers in early summer and produces fruit. Later, axillary buds on the parent plant grow over the soil surface. At the tip of each **runner** is a terminal bud. Along the length of the runner are scale leaves in the axils of which are axillary buds. One or more of the axillary buds produces roots which grow into the soil, and leaves grow up into the air. When the internodes of the runner decay several new plants are independent of the parent (Fig. 19.12).

FIG. 19.12 *Vegetative propagation in strawberry*

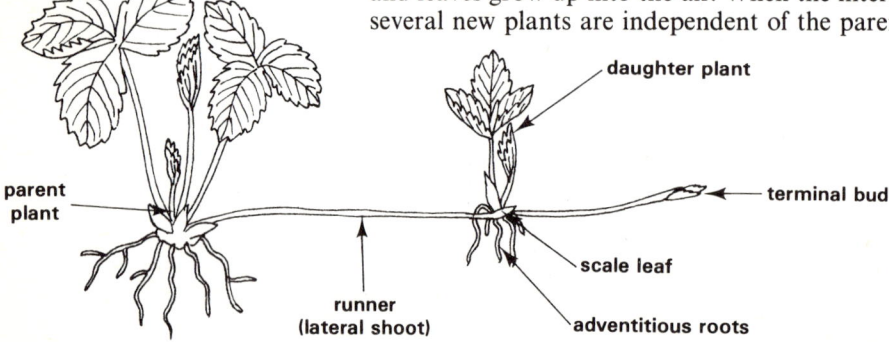

daughter plant

terminal bud

parent plant

scale leaf

runner (lateral shoot)

adventitious roots

FIG. 19.13 *Organs of perennation and vegetative propagation*

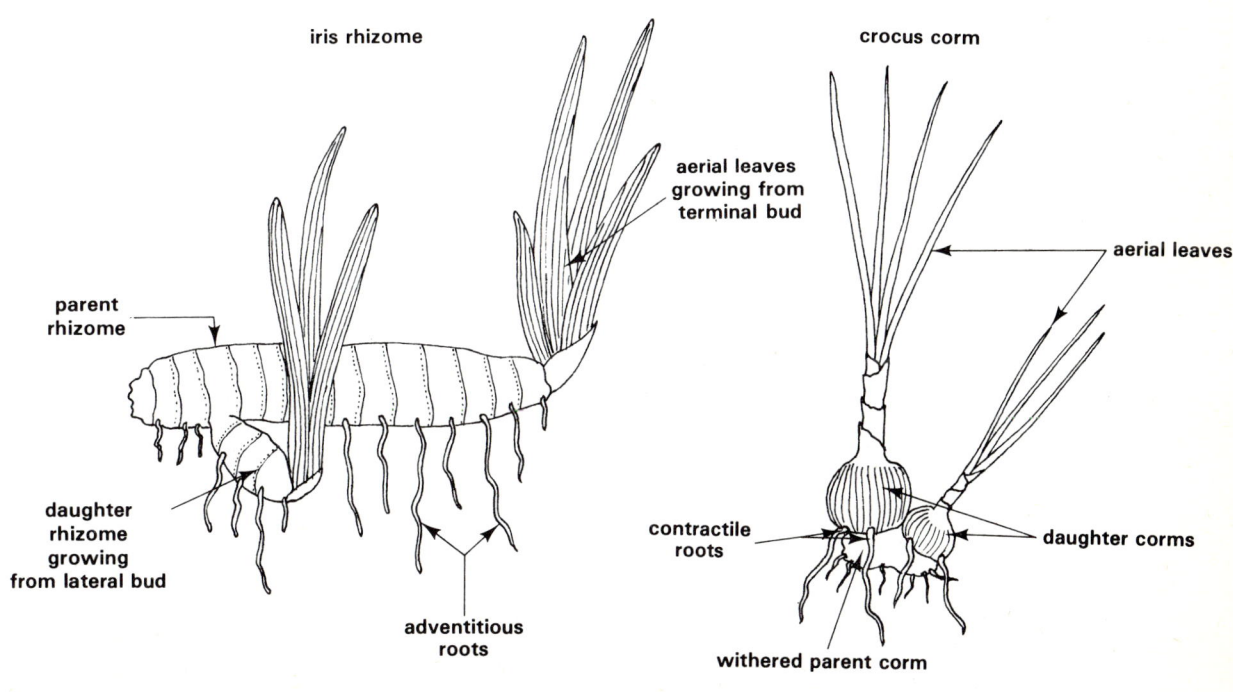

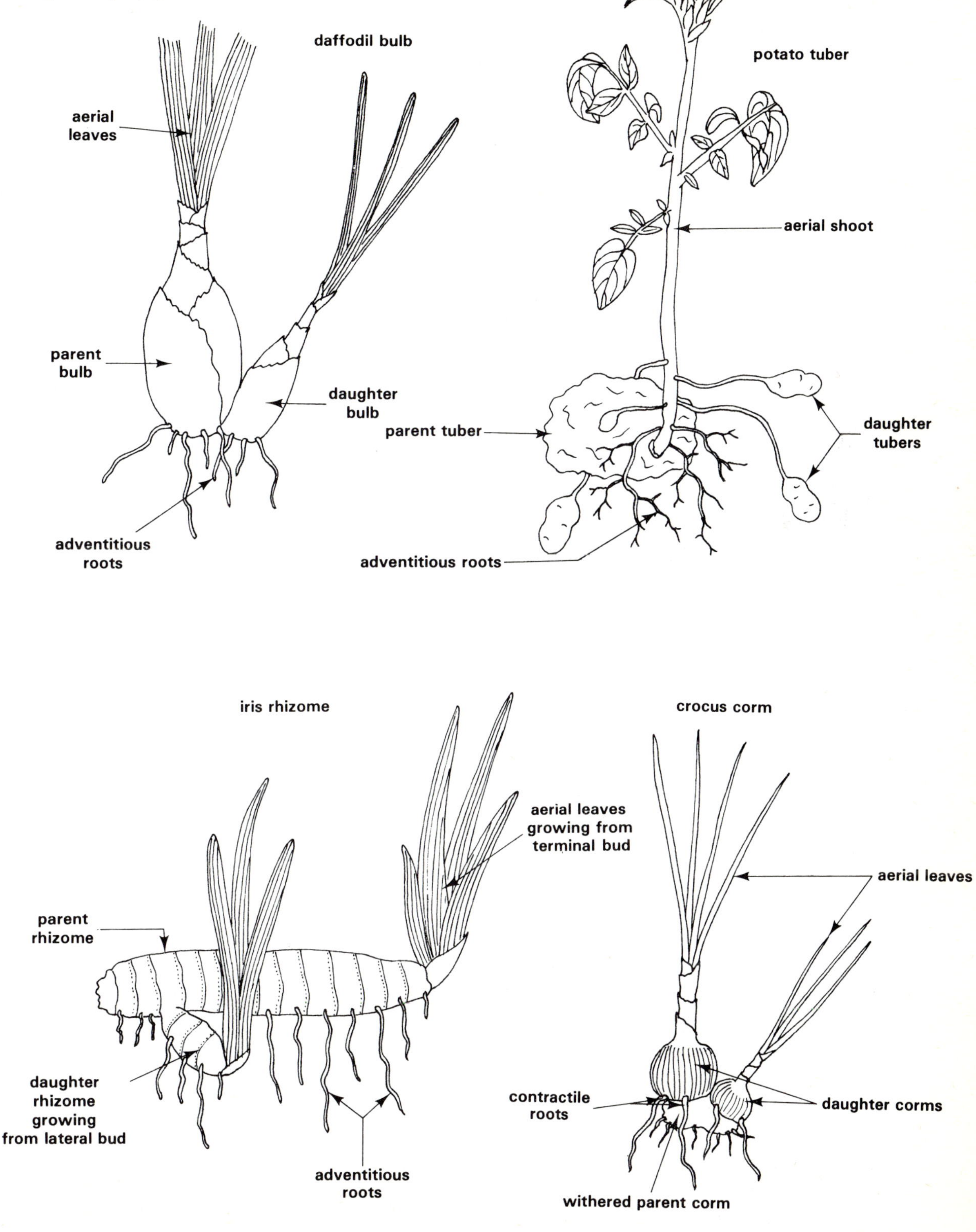

daffodil bulb

aerial leaves

parent bulb

daughter bulb

adventitious roots

potato tuber

aerial shoot

parent tuber

daughter tubers

adventitious roots

iris rhizome

aerial leaves growing from terminal bud

parent rhizome

daughter rhizome growing from lateral bud

adventitious roots

crocus corm

aerial leaves

contractile roots

daughter corms

withered parent corm

Many other methods of vegetative propagation are seen in wild and cultivated plants. Often propagation is coupled with a means of survival in adverse seasons. **Rhizomes**, **corms**, **bulbs** and **tubers**, for example, are frequently swollen with reserves of food which are used to establish new growth when the weather is suitable. As growth proceeds, lateral buds develop into daughter organs which later give rise to new plants (Fig. 19.13).

The various forms of asexual reproduction observed in flowering plants are often collectively called **apomixis**.

19.4.4 The consequences of asexual reproduction

In section 19.3 the differences between inbreeding and outbreeding were described and the genetic consequences were explained. All forms of asexual reproduction are examples of **inbreeding**. They are a means whereby genomes are passed with little or no change from parent to progeny. In fact asexual reproduction gives less scope for genetic variation than self-fertilisation where some change in the genomes of gametes will result from random assortment and crossing-over in meiosis. There is a tendency therefore for asexual reproduction to give rise to progeny of **genetic uniformity**. Such a group of plants is called a **clone**. Even so, the ability of many cultivated plants to reproduce by vegetative propagation is of great advantage to humans. It ensures that the genomes of useful varieties of plants are kept unchanged. Nevertheless occasional mutants, called **sports** by gardeners, appear among plants which propagate vegetatively.

Very few species of wild plants reproduce by asexual methods only. However, plants which can produce asexually are more versatile than those which rely entirely on sexual reproduction. It means that progeny can be produced even when seed production fails.

19.5 The physiology of angiosperm reproduction

Earlier in the chapter an orderly sequence of events beginning with the development of flowers and ending with the formation of seed-containing fruits is described. We will now examine some of the reasons for the changes. What causes a vegetative shoot to produce flowers and what controls the development of fruits and seeds? Why does a pollen tube grow towards an embryo sac when a pollen grain germinates on the stigma?

19.5.1 The physiology of flowering

The flowers of wild and cultivated plants grown outdoors appear at a particular time of the year. Daffodils, snowdrops, crocuses and hyacinths produce flowering shoots in early spring. Grasses and roses flower in summer, chrysanthemums and Michaelmas daisies flower in the autumn. This suggests that climate controls the times of flowering in different species. Among the more obvious seasonal changes of climate in temperate zones are **day-length** and **temperature**.

1. EFFECT OF DAY-LENGTH
The number of hours of light each day varies from one time of the year to another. In mid-winter the sun rises at about 8 am in southern England and dusk occurs at about 4.30 pm. This contrasts with mid-summer when dawn is at 5 am and sunset at 10 pm. British vegetation is thus exposed to little more than 8 hours of sunlight each day in winter and as much as 17 hours in summer. From December to June the number of daylight hours gradually increases while from June to December they gradually decrease.

The effect of day-length in controlling the time of flowering was discovered in America in the 1920s. It was found that a variety of tobacco flowered much more readily if the plants were given as little as five hours light a day compared with longer periods of light. The light had to be above a certain intensity for it to be effective. Since then it has been shown that many other species behave in the same way. They flower after exposure to several days when the number of light hours or **photoperiod** is small. Other species flower only after exposure to a number of days with long photoperiods. A third group are unaffected by daylength. Thus we have three categories of **photoperiodic response**:

(i) SHORT-DAY PLANTS which flower after a number of photoperiods of less than twelve hours, as occurs in Britain during autumn and winter. This group of plants includes chrysanthemums, rice and cotton.

(ii) LONG-DAY PLANTS which flower only after a succession of long photoperiods of more than twelve hours. In this group are many important crop plants such as wheat, clover and lettuce.

(iii) DAY-NEUTRAL PLANTS which produce flowers regardless of the length of photoperiod to which they are exposed. They include tomato, cucumber and dandelion.

The number of photoperiods required to induce the conversion of vegetative apices to floral apices varies from one species to another. The cocklebur requires only one **inductive photoperiod**. Most plants need about ten consecutive photoperiods of suitable length. It may then take several days or even weeks for vegetative apices to change into floral apices. A plant must be at a certain stage in its development when it receives the photoperiod treatment if eventually it is to flower. In some species the plant can be at the seedling stage. In others several mature leaves must be present.

It is now known that the length of the photoperiod is less critical than the length of the **dark period** in a daily cycle. If the photoperiods are interrupted with short periods of darkness, flowering still follows. However, if the dark period is interrupted by as little as one minute's exposure to light, flowering is prevented. Red-light is most effective in this respect. Yet the effect of red-light treatment can be overcome if the plant is immediately exposed to infra-red light. This suggests that a flower-inducing substance is made in darkness and is broken down by red light. Flowering plants produce a pigment, **phytochrome**, which is affected in this way (Chapter 20). Phytochrome may therefore play a part in regulating flowering. In what way is still unknown, but phytochrome may stimulate production of gibberellic acid (GA). GA has been shown to enhance flowering when applied to long-day plants grown in short-days. Grafting experiments strongly indicate that hormones are involved in regulating flowering and that the leaves perceive the stimulus of day length (Fig. 19.14).

2. EFFECT OF TEMPERATURE

As well as a suitable photoperiod treatment some plants require an appropriate temperature treatment if they are to flower. Winter varieties of wheat and barley sown in autumn in Britain will flower the following year, but if sown in spring they do not flower unless the seedlings are exposed to a temperature of between 1 and 5°C for several days. This shows that cold weather has an effect in regulating flowering of some species. Low temperature treatment induces flowering in many biennial species. In the first year of growth biennial plants such as the carrot and parsnip grow vegetatively and build up a reserve of food which is stored in their swollen taproots.

FIG. 19.14 *Some aspects of photoperiodism in a flowering plant*

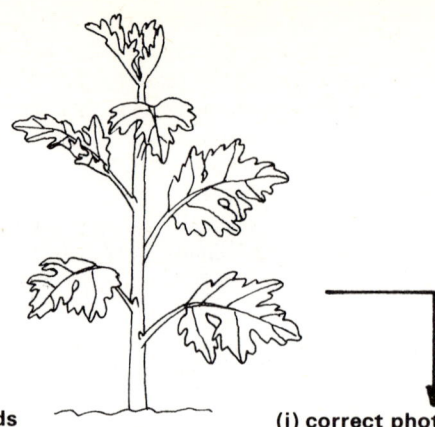

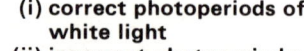

(i) **incorrect photoperiods of white light**
(ii) **correct photoperiods of IR light**
(iii) **flashes of light given in dark periods**
(iv) **all leaves removed then given correct photoperiods**

(i) **correct photoperiods of white light**
(ii) **incorrect photoperiods; leaf of same species given correct photoperiods grafted on**
(iii) **all but one leaf removed then given correct photoperiods**

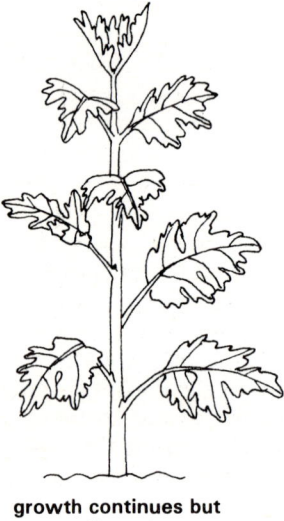

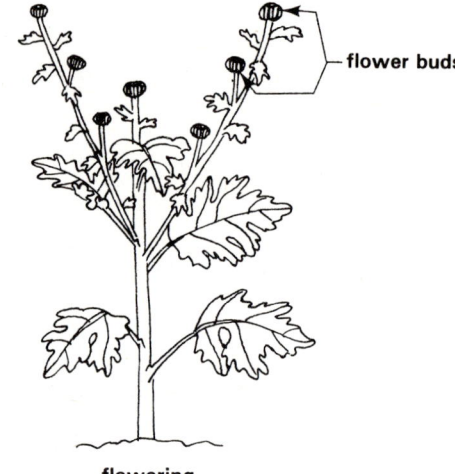

flower buds

growth continues but no flowering

flowering

Next year, if the plants are left out of doors, a flowering shoot is produced using the stored food (Fig. 19.15). However, if they are kept over winter in a warm greenhouse they continue to grow vegetatively.

The cold-requirement for initiation of flowers is called **vernalisation**. How exposure to low temperatures promotes flowering is still poorly understood. It is likely, however, that it changes the hormone balance in the plant, possibly by promoting gibberellic acid production. Biennial plants treated with gibberellic acid will flower without vernalisation.

It is possible to investigate the involvement of hormones in flowering using tissue culture techniques. When floral apices are grown in a culture medium, floral organs differentiate if auxin, gibberellic acid and cytokinin are added (Chapter 21). The cucumber has been much studied in this respect. Application of auxin to male flower buds causes them to produce female flowers. Gibberellic acid has the opposite effect. Studies of this kind suggest that protandry, protogyny and the existence of separate male and female flowers on the same plant may be due to changes in hormone balance at different stages of growth.

FIG. 19.15 *A beetroot plant in the first and second year of its growth*

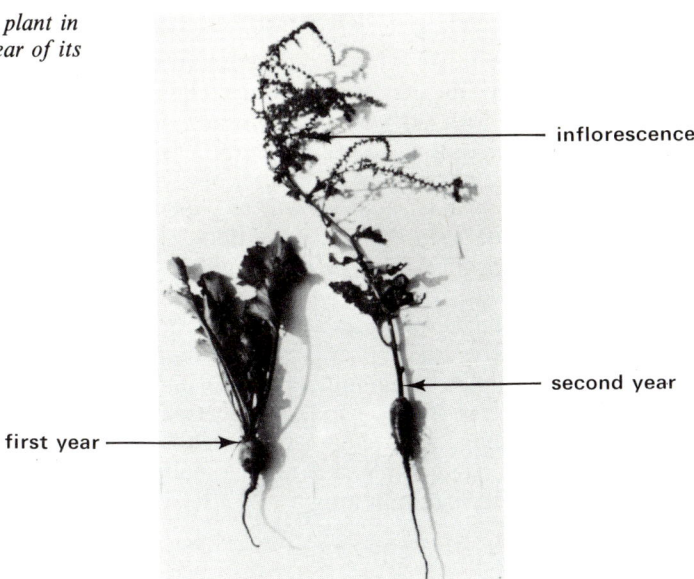

inflorescence

second year

first year

19.5.2 Physiology of pollen growth

A lot has yet to be learned about the reason why pollen tubes transfer male gametes with such accuracy into embryo sacs. The very fact that the tips of pollen tubes so frequently find the micropyle suggests that the nucellus or embryo sac secretes substances to which the pollen tube is sensitive. Growth of the tube is probably an example of **positive chemotropism**. However, there has been little success in identifying the substance or substances involved. More is known of the conditions which encourage germination of pollen grains. Whereas the pollen of some species will germinate in pure water, others need the presence of sugars, amino acids and other ingredients all of which are present in stigmatic and stylar tissues. The required concentrations of these substances is often quite high. Up to 15 per cent sucrose is needed in some cases. The trace element boron also stimulates pollen tube growth (Chapter 13).

Early growth of the pollen tube uses materials stored in the pollen grain. There is evidence that digestion of stylar and carpellary tissues by pectinase and cellulase enzymes takes place when the pollen tube grows further. Even so, pollen may fail to germinate despite successful pollination. Incorrect amounts of stimulants in the stigma and style, secretion of inhibitory materials by the stigma, style and ovules and inability of the pollen tube to produce enzymes which help the pollen tube penetrate the stigmatic surface are among the reasons for this.

A better understanding of the physiology of pollen tube growth could enable plant breeders to break down sterility barriers between varieties of flowering plant species. By bringing together genetic material from two different sources some of the hybrid progeny could be superior to existing varieties. This could mean higher yields of crop and go some way to alleviate the existing world food shortage.

19.5.3 Physiology of seed and fruit formation

FIG. 19.16 *Effect of seed growth on growth of pericarp. In the small pod only one ovule was fertilised. Several ovules were fertilised in the larger pod.*

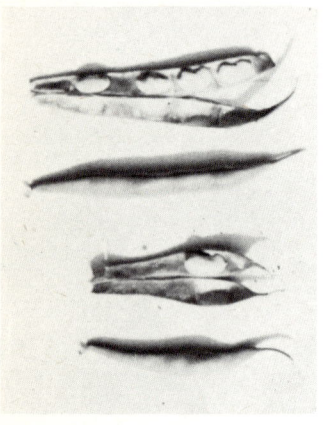

Some information of the conditions required for embryo growth has been obtained by analysing the endosperm which surrounds the embryo during its natural development. The coconut produces a large volume of liquid endosperm, commonly called coconut 'milk'. The milk is rich in sugars, minerals and plant hormones such as cytokinins and auxin. Recently young embryos of shepherd's purse have been grown successfully in a synthetic mixture of sucrose, various minerals, indole-3-acetic acid, kinetin and adenine. As the embryos develop they become less reliant on organic substances in their surroundings. Older embryos continue to develop if supplied with carbon dioxide, water, minerals and light. Other than this and the fact that embryos respire actively as they grow, little is known about the physiology of embryo development.

Simultaneous rapid growth of ovules and pericarp occur immediately after fertilisation. This indicates that growth of the various parts of the fruit are synchronised. Expansion of the fruit is stimulated by pollination in some species (Fig. 19.16). If auxin or gibberellic acid is sprayed on to unpollinated flowers, fruit will form. Yet the pollen of most species contains only a little auxin. It appears that pollination triggers off hormone synthesis in the ovules and carpels so that growth of seed and pericarp go 'hand in hand' (Fig. 19.17). Production of hormones in the growing fruit also inhibits the development of an abscission layer in the fruit stalk. The ovaries of unpollinated flowers usually fall away like the leaves of deciduous trees in autumn.

FIG. 19.17 *Growth of a plum*

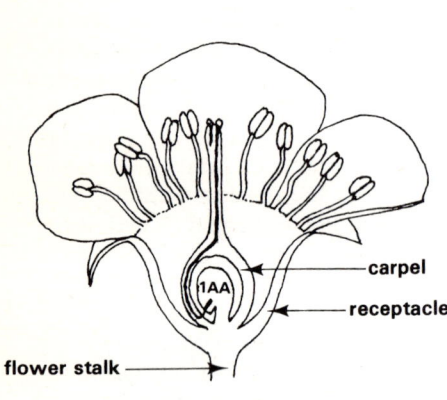

flower

pollination triggers off production of auxin (IAA) in ovule

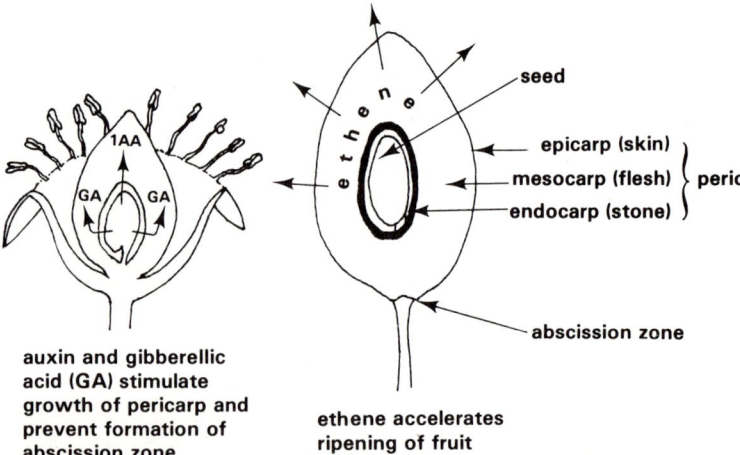

growing fruit

auxin and gibberellic acid (GA) stimulate growth of pericarp and prevent formation of abscission zone

ripe fruit

ethene accelerates ripening of fruit

seed
epicarp (skin)
mesocarp (flesh)
endocarp (stone)
} pericarp
abscission zone

What is known about the physiology of fruit development has been put to commercial use. A large proportion of the grape crop in California, for example, is sprayed with gibberellic acid. Spraying does away with dependence on insects for pollination and ensures that more fruit develop. Seedless fruit develop because fertilisation does not occur. The development of seedless fruit is called **parthenocarpy**. Parthenocarpic fruits are produced naturally by some cultivated plants such as the banana and pineapple. Here fertilisation generally does occur but soon after their formation the embryos abort. Not, however, before they have supplied the necessary stimulus to the pericarp for its expansion.

390

19.5.4 Physiology of fruit ripening

The pericarps of dry fruits quickly lose water when their growth is finished. Drying is usually accompanied by lignification of the pericarp, forming a structure which helps disperse the seeds and protects them until they germinate. The ripening of fleshy fruits is accompanied by many chemical changes in the swollen pericarps or receptacles, making them attractive and palatable to animals. Unripe fruits are generally sour because they contain high concentrations of carboxylic acids. On ripening, the acids together with starch are converted to sweet-tasting sugars. Simultaneously the fruit becomes softer as pectinase and cellulase enzymes digest cell wall materials such as pectin and cellulose. The extent to which this occurs varies between species. In ripe plums for instance there is a lot of unchanged pectin while in cherries there is little. For this reason jam made from plums usually sets readily, the pectin acting as a gelatinising agent, but cherry jam does not set unless extra pectin is added.

Pigment changes are also common. Green unripe fleshy fruit become red, yellow and orange as chlorophyll is broken down and xanthophylls and carotenes predominate. Volatile aromatic oils may be synthesised giving the fruit a pleasant odour. The changes make the fruit attractive to some animals which help in dispersal of the seeds inside the fruit.

The many chemical conversions use energy from ATP made during respiration. Ripening fruit thus show a marked increase in respiratory rate. They also exhibit an increased output of ethene gas which accelerates ripening.

Knowledge of some aspects of fruit ripening is put to practical use. The ability to delay ripening has obvious advantages because it enables us to store fruit for long periods of time after it has been picked, thus avoiding wastage. Ripening can be delayed in a number of ways, all of which are aimed at minimising the respiration rate of the fruit. Storage at low temperatures or in an atmosphere containing a low concentration of oxygen and a high concentration of carbon dioxide is usually effective. It may be necessary to ripen fruit when it is taken out of storage. Ripening can be stimulated by placing the fruit in an atmosphere containing ethene.

Chapter 20

Seed structure and germination

Few species of flowering plants fail to produce seeds. Indeed, the evolutionary success of flowering plants can to some extent be attributed to seed production. New genomes arise when seeds are produced. Seeds also provide a resting stage which can be maintained for long periods. It is not uncommon for seeds to germinate after being stored for up to fifty years. The longest life span for seeds recorded to date is that of the Arctic lupin, *Lupinus arcticus*. Seeds of this species, discovered in frozen silt at Miller Creek, Yukon, Canada, in 1954, were later successfully germinated. Radiocarbon dating methods showed the seeds to be 10 000 to 15 000 years old. The longevity of seeds is due mainly to their extremely low metabolic rate and to the presence of a reservoir of energy in the form of stored food. Humans have long known of the food reserves in seeds and a great deal of time and effort has been spent in improving and cultivating crops such as cereals and legumes which produce edible seeds.

Although biologists are a long way from a complete understanding of many aspects of seed germination we shall see that the physiology of seeds is largely controlled by their environment.

20.1 Seed structure

A **seed** is an **embryo plant** together with its **store of food**. The whole structure is encased in a seed coat called the **testa**. The embryo consists of an immature shoot, the **plumule**, and an undeveloped root, the **radicle**. In the seeds of monocotyledonous plants such as grasses and maize the embryo is attached to a single seed leaf called the **cotyledon**. There are two seed leaves in dicotyledonous plants. The food reserve is stored in swollen cotyledons in sunflower and bean seeds. In maize, food is stored in a separate tissue called **endosperm** (Figs. 20.1, 20.2 and 20.3).

20.2 Morphological aspects of seed germination

Germination is the first step in the development of a mature plant from the embryo of the seed. The appearance of a seedling plant marks the end of germination. However, the way in which this happens differs from one species to another. The various patterns of germination can best be appreciated by studying the germination of seeds of several species.

20.2.1 Germination of broad bean

The seeds of the broad bean develop as a single row inside an elongated pod-shaped **pericarp**. Each seed is attached to the pod by a stalk, called the **funicle**, at the base of which is a tiny hole, the **micropyle**. When a seed is detached from its funicle a scar called the **hilum** is seen at one end of the testa. Each seed has two very large cotyledons packed with starch and protein. The embryo is tucked between the cotyledons to which it is attached by two short stalks (Fig. 20.1(a)).

If placed in moist soil at a suitable temperature the seed quickly swells as it absorbs water. Shortly afterwards the radicle elongates and pushes its way through the micropyle, bursting open the testa as it does so. As the

FIG. 20.1(a) *Seed of broad bean*

testa removed

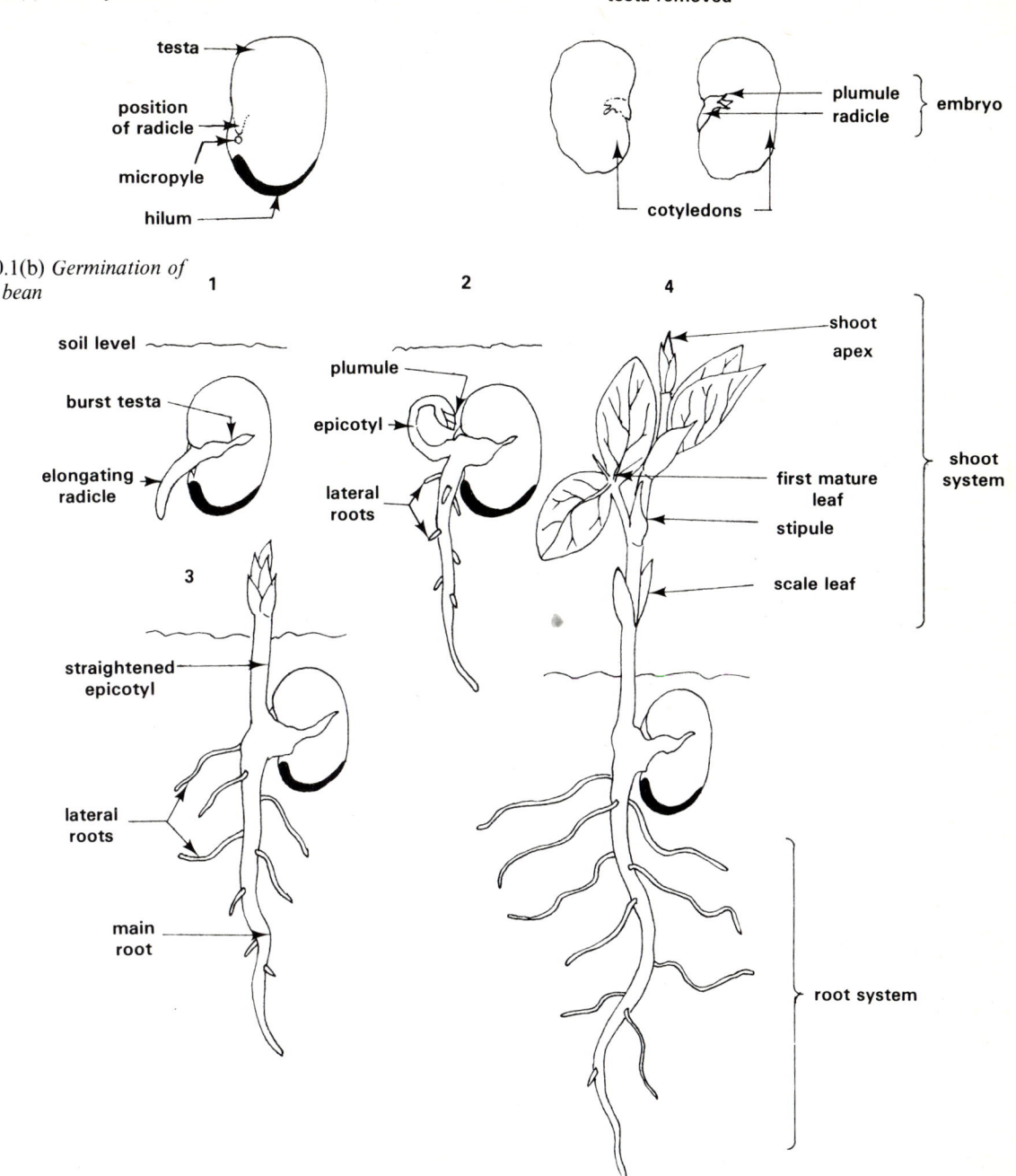

FIG. 20.1(b) *Germination of broad bean*

radicle extends down into the soil a number of lateral roots appear, forming a branched root system. While the radicle emerges the base of the plumule called the **epicotyl** also elongates and becomes hook-shaped as it grows between the cotyledons into the surrounding soil. Eventually the plumule reaches the soil surface where the epicotyl straightens and grows vertically. The embryonic leaves expand and develop into the large green compound leaves typical of the broad bean.

The cotyledons gradually shrink as the food reserves are used up. Such a pattern of germination where the cotyledons remain below the surface of the soil is called **hypogeal** (hypo = below; ge = earth) (Fig. 20.1(b)).

20.2.2 Germination of sunflower

Each sunflower seed is surrounded by a testa and a leathery pericarp and is therefore a one-seeded fruit called an **achene**. At the tapering end of the achene there is a small scar showing where it was attached to the flower stalk during its development. As in the broad bean there is no endosperm. The food reserve mainly of starch and oil is stored in the swollen cotyledons (Fig. 20.2(a)).

FIG. 20.2 (a) *Seed of sunflower*

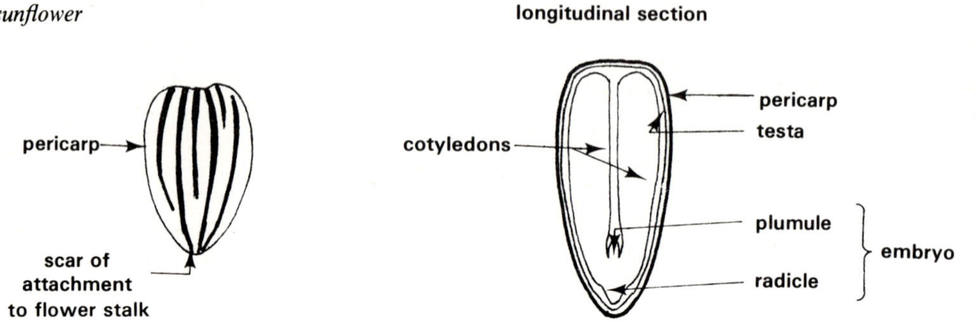

longitudinal section

FIG. 20.2 (b) *Germination of sunflower seed*

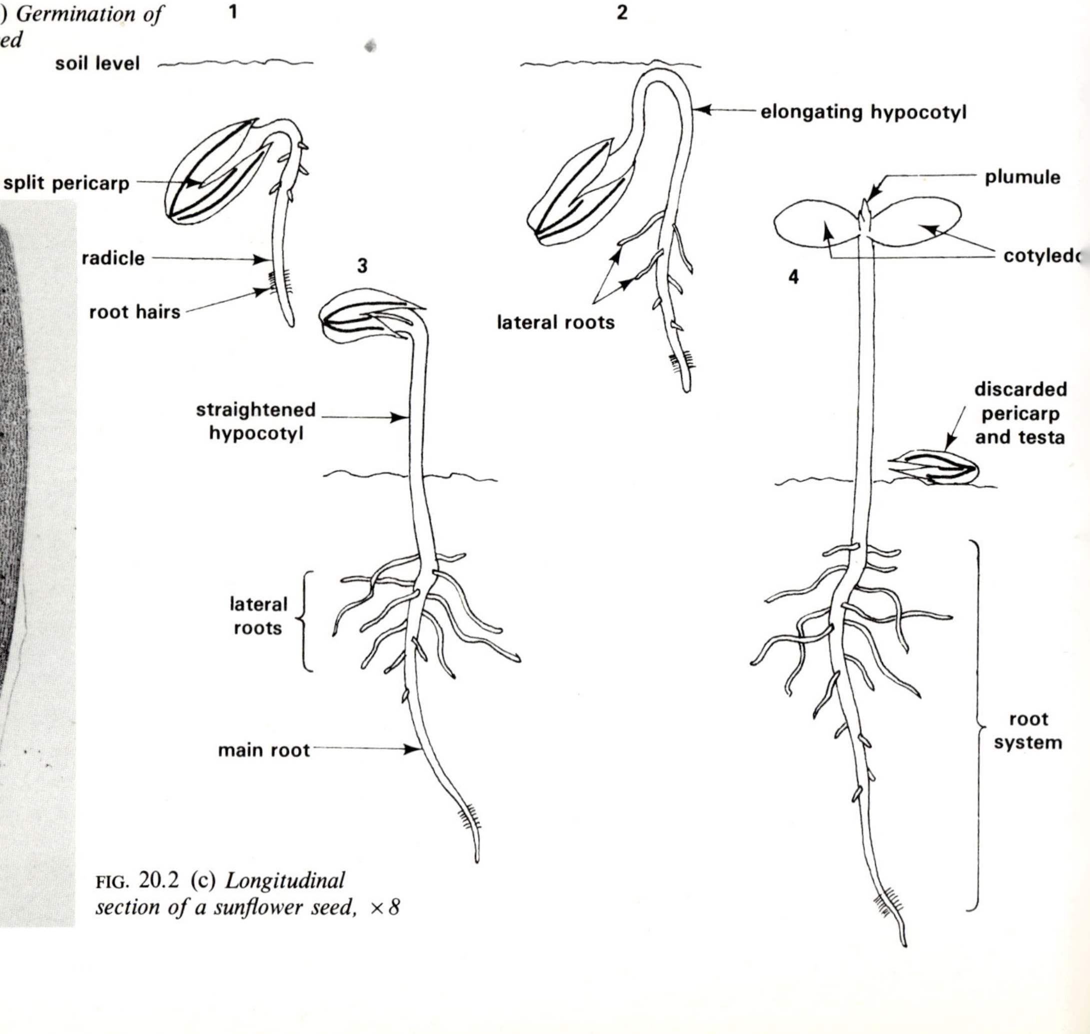

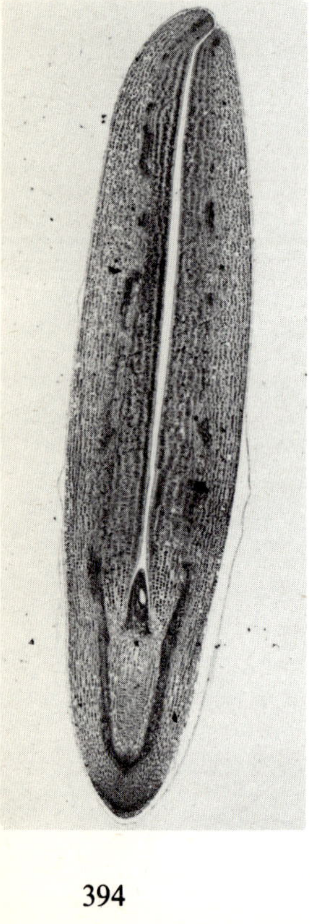

FIG. 20.2 (c) *Longitudinal section of a sunflower seed, × 8*

394

Despite the basic similarity in structure between the sunflower and broad bean seeds there are differences in the patterns of germination. After the absorption of water and subsequent emergence of the radicle, the **hypocotyl**, the part of the embryo between the cotyledon stalks and the radicle, begins to elongate. As it grows it becomes hook-shaped and the cotyledons with the tiny plumule tucked between them are dragged above soil level where the hypocotyl straightens. The cotyledons, by now shrunken because their food reserves have been used up, become green and start to photosynthesise. At this stage the plumule elongates and the first simple mature leaves appear. This type of germination where the cotyledons appear above soil level is called **epigeal** germination (epi = above) (Fig. 20.2(b)).

20.2.3 Germination of maize

A maize seed is enclosed in a tough pericarp which is fused with the testa. This type of one-seeded fruit is called a **caryopsis** and develops with several hundred others on a cob. At the tapered base of the fruit is a scar showing its point of attachment to the cob. On one of the broader sides of the fruit a lighter oval area can be seen under which the embryo is found. The radicle is enclosed in a hollow tube called the **coleorhiza**. The plumule is surrounded by a similar structure called the **coleoptile**. There is a small triangular cotyledon, the **scutellum**, to one side of which the embryo is attached. At its other side the scutellum is fused to the endosperm which stores starch, protein and oil (Fig. 20.3(a)).

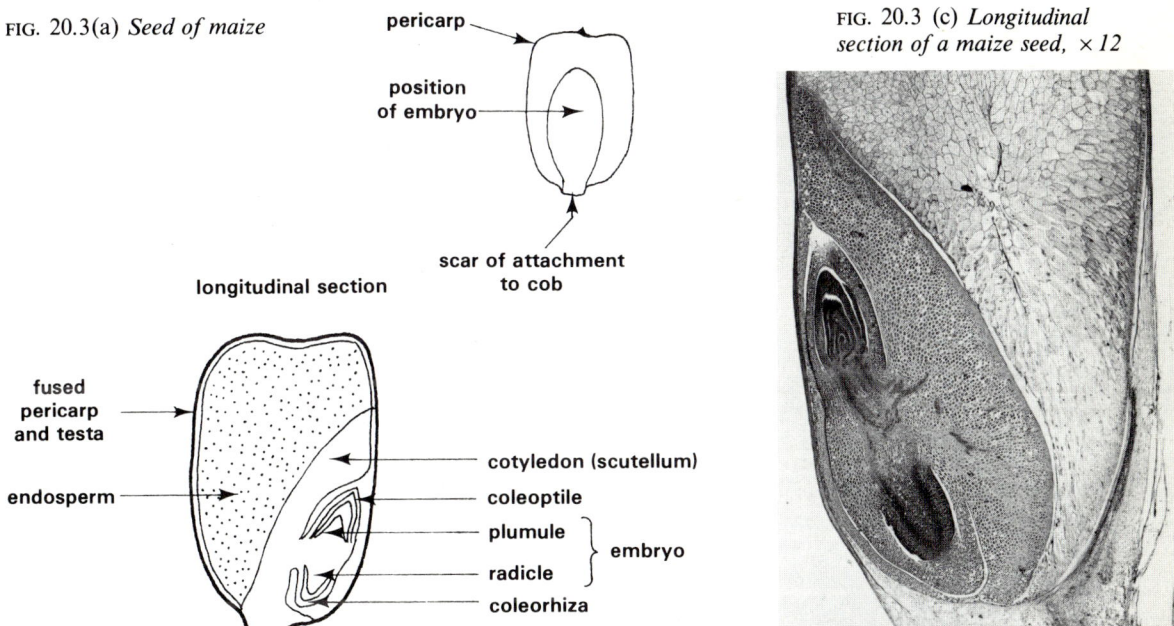

FIG. 20.3(a) *Seed of maize*

pericarp

position of embryo

scar of attachment to cob

longitudinal section

fused pericarp and testa

endosperm

cotyledon (scutellum)

coleoptile

plumule ⎤
          ⎬ embryo
radicle ⎦

coleorhiza

FIG. 20.3 (c) *Longitudinal section of a maize seed, ×12*

Soon after water uptake the radicle emerges, piercing the coleorhiza as it grows. Shortly afterwards the coleoptile appears, enclosing the plumule as it grows upwards through the soil. Growth of the coleoptile is mainly from its base which is called the **mesocotyl**. Once through the soil surface the coleoptile soon stops growing and the first long, strap-shaped leaf bursts through its tip. Adventitious roots grow from the mesocotyl and form a fibrous root system. Because the cotyledon remains below the soil level the germination of maize is **hypogeal** (Fig. 20.3(b)).

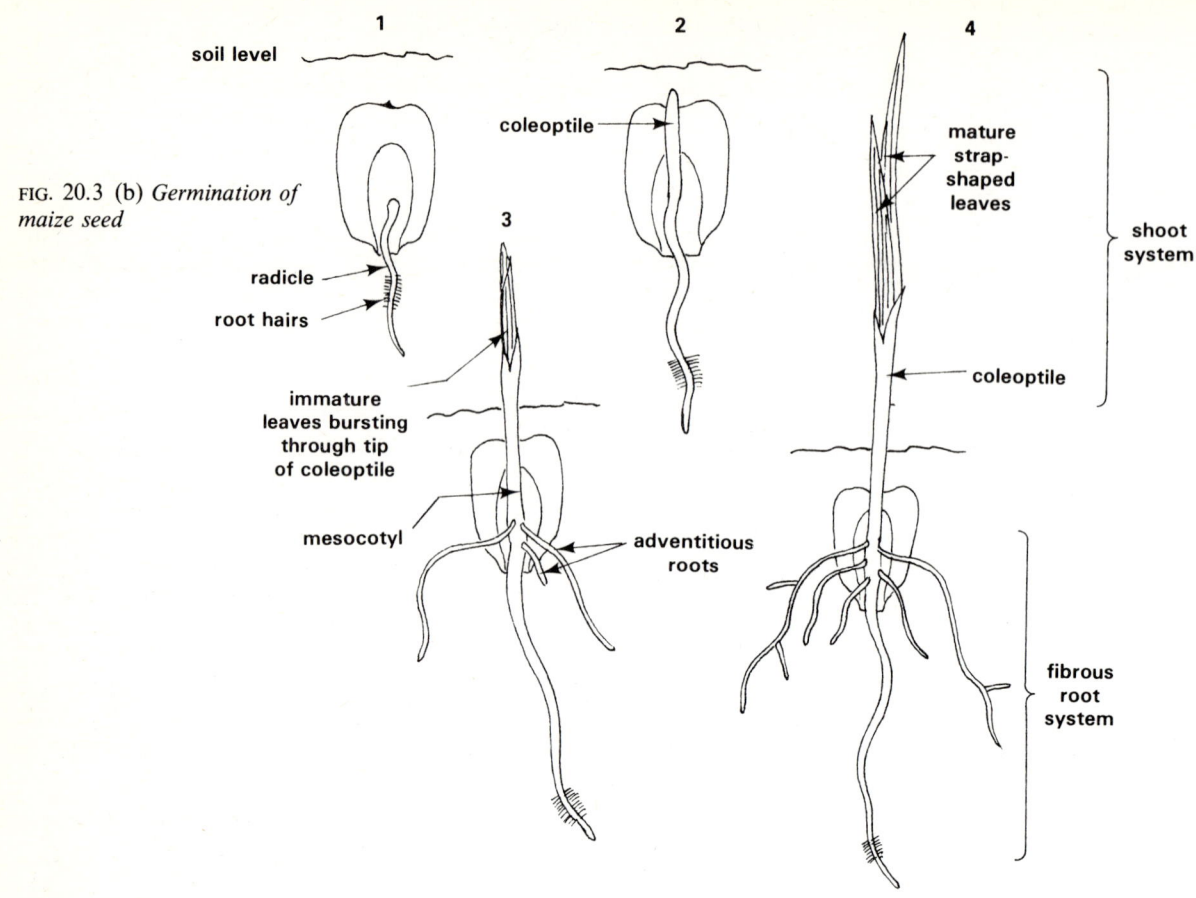

FIG. 20.3 (b) *Germination of maize seed*

soil level

1

2

4

coleoptile

mature
strap-
shaped
leaves

shoot
system

radicle

root hairs

3

immature
leaves bursting
through tip
of coleoptile

coleoptile

mesocotyl

adventitious
roots

fibrous
root
system

20.3 Physiology of germination

The structural changes described so far are the consequences of physiological changes occurring inside the cells and tissues of the germinating seed. The first changes are mainly the result of **cell expansion** following the uptake of water (Section 20.3.2). Later changes involve the **growth** of new cells and tissues at the apices of the radicle and plumule. Growth requires raw materials and energy and initially it occurs at the expense of energy-rich molecules such as starch and lipids stored in the seed. Once the leaves of the young plant have expanded the seedling photosynthesises.

20.3.1 Mobilization of stored food

Table 20.1 gives an indication of the main materials stored in a variety of seeds. Because of their richness in **carbohydrates**, **lipids** and **proteins**, seeds feature prominently in the diets of many animals including man. Food reserves in seeds are insoluble in water and cannot be transported in the seedling. They must be broken down into relatively simple, soluble substances which dissolve in water to be moved to the growing apices of the plumule and radicle. As in the mammalian gut (Chapter 14), hydrolytic enzymes catalyse the breakdown of proteins, lipids and polysaccharides such as starch.

Table 20.1 Chemical composition of seeds

| SPECIES | PERCENTAGE DRY MASS | | |
	CARBOHYDRATES	LIPIDS	PROTEINS
maize	50–75	5	10
wheat	60–75	2	13
rice	65–70	2	10
broad bean	57	2	36
sunflower	2	45–50	25
peanut	12–33	40–50	20–30
pea	34–46	2	20

1. CARBOHYDRATES

The hydrolysis of **starch** into the soluble disaccharide sugar **maltose** is catalysed by a complex of enzymes called **amylase**. The conversion of starch to maltose is used to produce malt for the brewing industry. Barley seeds are germinated for a few days and the sprouted grain then dried and ground into powder. When later steeped in water the maltose dissolves and provides the source of energy for yeast cells in the fermentation stage of brewing. In seeds germinating under natural conditions maltose is further hydrolysed by the enzyme **maltase** to **glucose** which is converted to **sucrose** for transport to the growing apices of the embryo. At the apices sucrose is used for the synthesis of cellulose, hemicelluloses and pectic compounds, the main components of plant **cell walls**. Some sucrose is **respired** to provide free energy for growth.

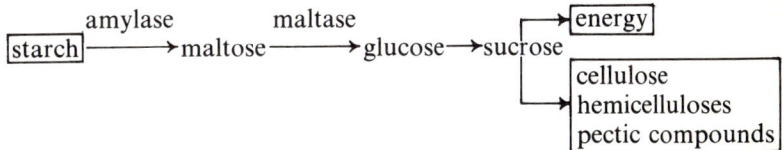

2. PROTEINS

Proteins are hydrolysed to polypeptides and **amino acids** by **peptidase enzymes**. Some amino acids are moved in solution to the embryo. Most are transported as **amides**. At the growing points of the plumule and radicle the amides are de-aminated and the amino acids are used to synthesise structural and enzymic **proteins**.

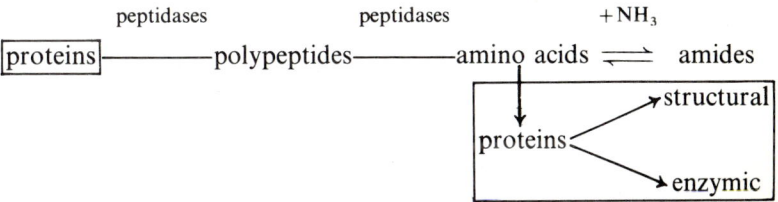

3. LIPIDS

Fats and oils are first hydrolysed to **glycerol** and **fatty acids** by **lipase** enzymes. The fatty acids may be oxidised to release energy or converted to sucrose ready for transport. Glycerol, which is only a small part of a lipid molecule, is also converted to transportable sugars.

Some of the changes described above are shown for wheat seeds in Table 20.2. The general picture is a gradual depletion of reserves in the food storage areas and an increase in dry mass of the embryo (Fig. 20.4).

Table 20.2 Changes in dry mass of maize seedlings during germination

TIME/DAYS	DRY MASS/mg g^{-1}		
	WHOLE SEEDLING	ENDOSPERM	EMBRYO
0	225	200	2
1	210	189	3
2	208	188	5
3	206	155	5
4	175	115	15
5	155	84	23

FIG. 20.4 *Changes in dry mass of embryo and endosperm during germination*

FIG. 20.5 (b) *Summary of the physiology of food mobilization in a barley seed*

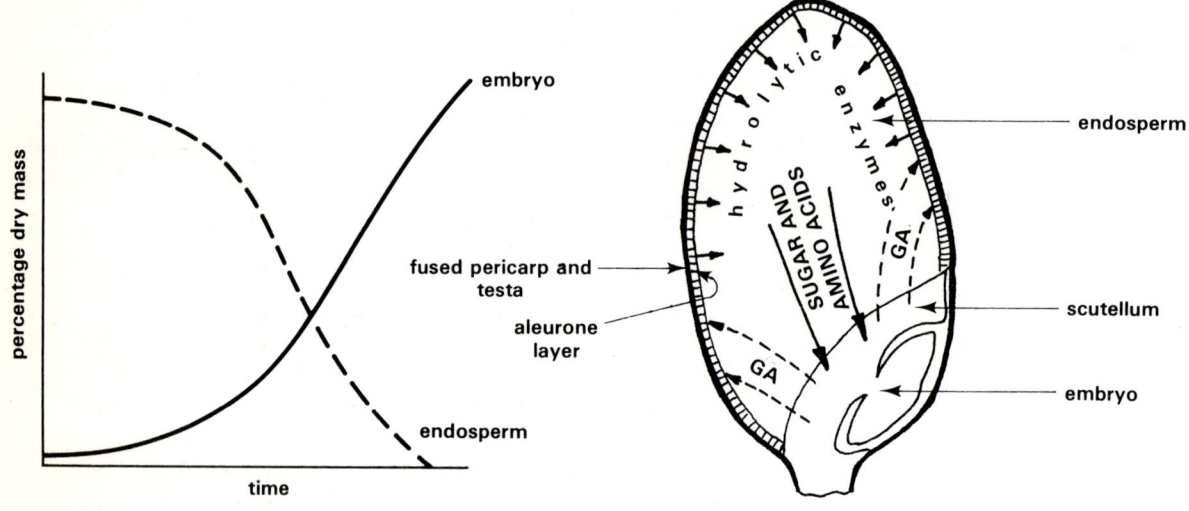

FIG. 20.5 (a) *An experiment to investigate the effect of the embryo on the release of the starch-hydrolysing enzyme amylase by the aleurone layer of barley seeds*

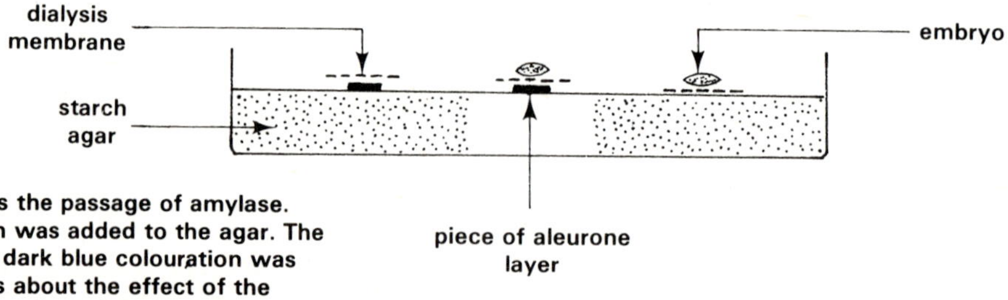

The dialysis membrane prevents the passage of amylase. After a few days iodine solution was added to the agar. The stippled area indicates where a dark blue colouration was observed. What does this tell us about the effect of the embryo on the release of amylase from the aleurone layer?

What causes such changes to occur? Studies with cereal grains show that the secretion of hydrolytic enzymes is triggered off by the hormone **gibberellic acid (GA)** made by the embryo (Fig. 20.5(a)). In cereals the release of enzymes takes place from the **aleurone layer** surrounding the endosperm. After absorption of water by the grain there is a marked rise in embryonic production of GA which diffuses into the endosperm where food reserves are hydrolysed (Fig. 20.5(b)). In some seeds the hydrolytic enzymes are found in lysosomes in the food-storing cells. Other hormones, notably **cytokinin** and **indole-3-acetic acid (IAA)** promote cell division and enlargement at the growing apices of the embryo. IAA also controls differentiation of vascular tissue in the developing shoot and root of the seedling (Chapter 21).

20.3.2 Factors affecting seed germination

If a gardener or farmer sows seeds at the wrong time of the year, few if any will germinate. **Environmental factors** can thus influence seed germination. Instructions concerning the best time of year to plant seeds are usually given on seed packets. For many outdoor crops of vegetables the most suitable time for sowing seeds in Britain is spring and early summer when the temperature is higher than in winter. In preparing a seed bed an experienced gardener ensures that the soil has a good crumb structure and is moist but not waterlogged. Too much water fills pore spaces in soil which are normally filled with air which contains oxygen. Thus a suitable **temperature**, **water** and **oxygen** have to be available for successful seed germination. Some seeds also require **light** before they germinate.

1. TEMPERATURE

Many of the physiological changes which take place in germination are catalysed by **enzymes**. The effect of temperature on enzyme activity is described in Chapter 4. For the seeds of each species of flowering plant there is an **optimum temperature** at which the germination rate is at its highest, providing no other factor is limiting germination (Fig. 20.6). The fact that seeds of most plants germinate early in the year if sown in a warm greenhouse is used by nurserymen in preparing trays of bedding plants for sale in early summer.

The seeds of many species such as red fescue grass, peach and apple germinate only if they have been subjected to a period of low temperature after they have absorbed water. A temperature between 0–5°C is normally

FIG. 20.6 *Effect of temperature on the rate of seed germination*

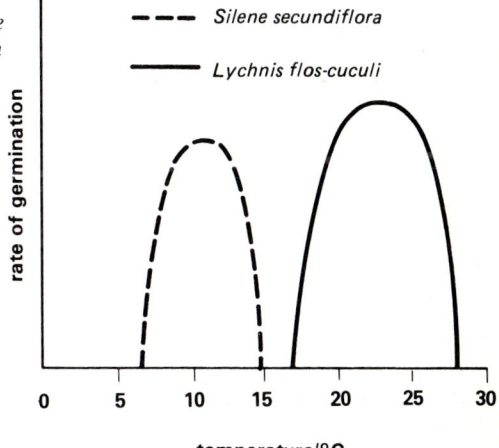

S. secundiflora inhabits the Mediterranean coastal area. The seedlings of this species avoid the hot, dry summer weather because the seeds germinate in the cooler, moister winter months. This is reflected in its relatively low optimum germination temperature.

L. flos-cuculi is widespread in temperate deciduous forest areas including Britain. How does the higher optimum germination temperature for this species reflect the climate of its habitat?

FIG. 20.7 *Effect of chilling on the rate of germination of apple seeds*

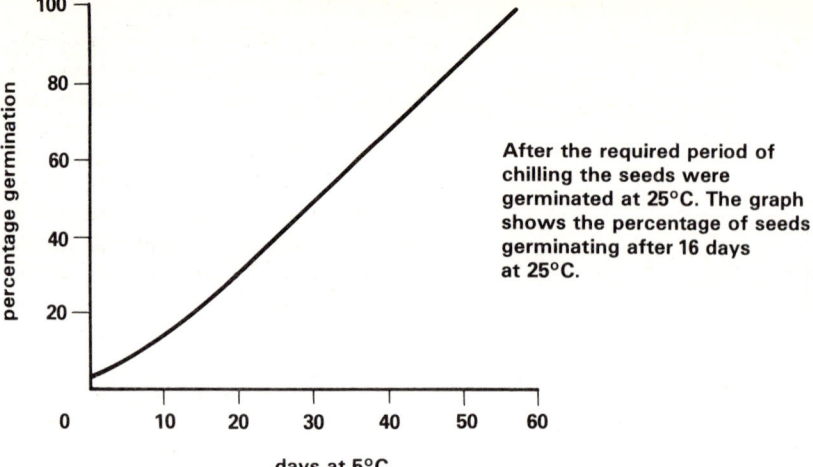

After the required period of chilling the seeds were germinated at 25°C. The graph shows the percentage of seeds germinating after 16 days at 25°C.

sufficient, the period of time varying from a few days to several weeks (Fig. 20.7). **Chilling** reduces the abscissic acid content of the seed coat and enables the embryo to make gibberellic acid (Fig. 20.8). In Britain the cold weather of winter provides the chilling requirement of seeds in the wild.

FIG. 20.8 *Effect of chilling on hormone concentrations in the seeds of sugar maple*

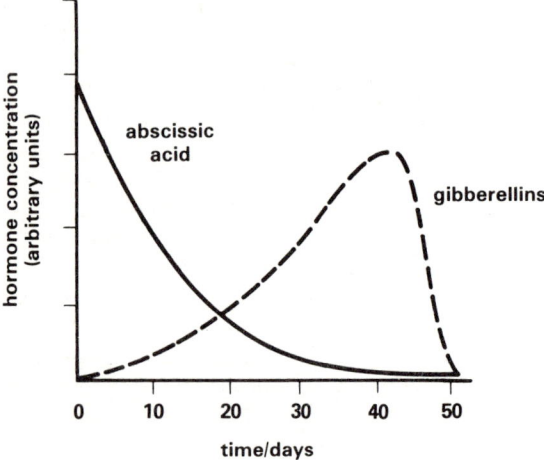

2. WATER

The water content of many types of seeds is as little as 5–10 per cent of their dry mass, far too low for rapid **metabolic activity**. Water is also needed for **vacuolation** of cells and for **transport** during germination.

It is normal practice to store seeds in an atmosphere which keeps them dry and prevents germination. When placed in water, seeds quickly swell enormously, often doubling their mass in a few hours. The initial absorption of water is largely by **imbibition**, the attraction between hydrophilic colloids such as membrane-bound proteins in the seed and water molecules (Chapter 2). Later the roots absorb water by **osmosis**.

In natural circumstances the water content of many seeds does not remain low for long. Most seeds are dispersed as soon as they are formed and on reaching the soil they absorb water. Even so, few seeds germinate at this stage. It is therefore evident that factors other than water content normally control germination.

3. OXYGEN

The synthesis of new cellular components at the apices of plumule and radicle requires free energy. Energy is supplied by ATP produced in **respiration** from stored energy-rich compounds such as carbohydrates and lipids. The manufacture of ATP occurs most efficiently when oxygen is available and aerobic respiration can occur (Chapter 5). As germination proceeds there is rapid intake of oxygen through the seed coat (Fig. 20.9).

Some of the energy in the reserves of food is liberated as heat, so it is possible to detect a rise in temperature of germinating seeds, particularly if kept in bulk. This is a problem farmers frequently encounter when grain with too high a water content has been stored. In these conditions the seeds begin to germinate, and **self-heating** occurs as heat energy is released. The temperature reached is so high that enzyme denaturation occurs and the

FIG. 20.9 *Intake of oxygen by dormant and non-dormant seeds*

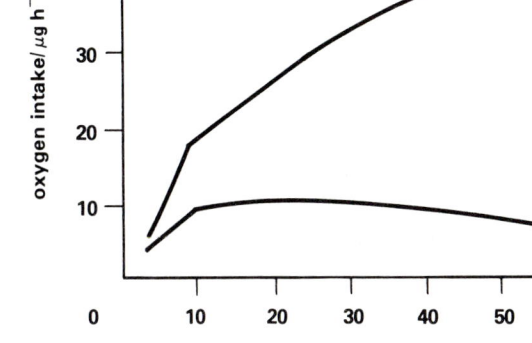

Both sets of seeds were first soaked in water. An initial increase in oxygen intake is observed even in dormant seeds. However the rate of oxygen intake continues to increase only in the non-dormant seeds.

FIG. 20.10 *Respiratory quotients of seeds of pea and castor oil*

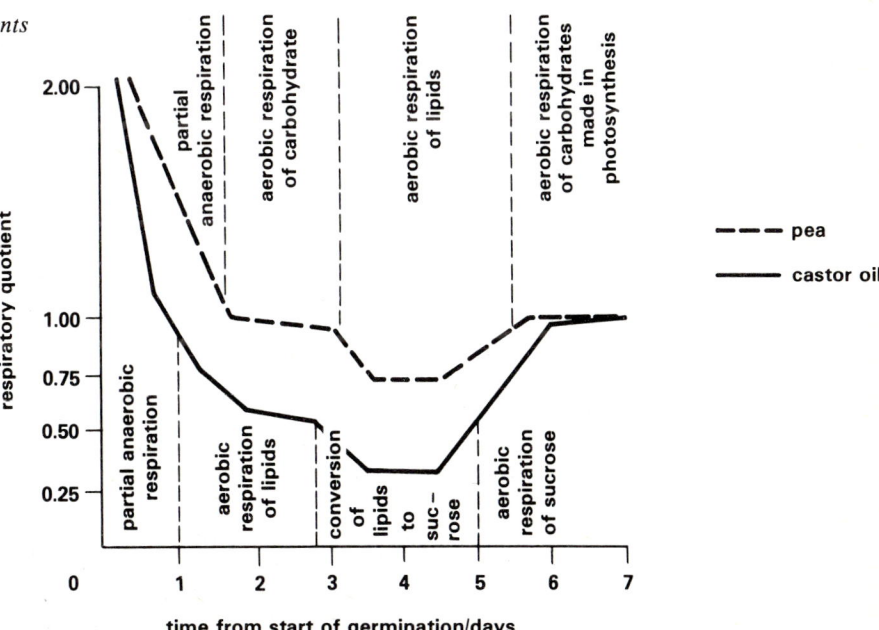

embryos die. Self-heated grain is useless as seed for a potential crop and its value as fodder is considerably reduced because much of the food reserve has been spent.

Carbon dioxide is a product of respiration. The ratio of carbon dioxide given off to oxygen used in a given period of time is called the **respiratory quotient** (Chapter 5). Figure 20.10 shows the respiratory quotients of two types of seeds at various stages in germination. The quotients give an indication of the type of substrate oxidised and whether aerobic respiration, anaerobic respiration, or both is taking place.

The testa of some seeds is relatively impermeable to oxygen so they may respire anaerobically at the start of germination. Large, bulky seeds such as broad beans have a small surface area to volume ratio which may result in an inadequate supply of oxygen to respiring cells and consequently some anaerobic respiration. For these reasons the RQ can be greater than 1·0 during the first 24–48 hours of germination. The RQ of most seeds then falls to 0·7. This is because lipids are now respired aerobically. If large quantities of lipids are stored, as in castor-oil seeds, the RQ at this stage may drop to as little as 0·35 owing to the conversion of lipids to sucrose.

$$C_{16}H_{40}O_6 + 11O_2 \longrightarrow C_{12}H_{22}O_{11} + 4CO_2 + 9H_2O$$
$$\text{lipid} \qquad\qquad\qquad \text{sucrose}$$

$$RQ = \frac{4}{11} = 0·35$$

In carbohydrate-rich seeds the RQ soon climbs to 1·0 as aerobic respiration of sugar gets under way. When the seedling eventually photosynthesises the RQ is 1·0 as in the mature plant. Aerobic respiration of carbohydrates made in photosynthesis now takes place. At this stage RQ measurements have to be made in darkness. In the presence of light, carbon dioxide released in respiration is used for photosynthesis, while some of the oxygen made in photosynthesis is used for respiration.

4. LIGHT

Although the seeds of many plants can germinate in total darkness, others such as tobacco and some varieties of lettuce germinate only after exposure to **light**. The time of exposure varies from several seconds to a few days. Light-treatment is effective only if the seeds have absorbed water. The light need only be of low intensity and it is the red part of the spectrum which is required. Infra-red light inhibits germination, even though the seeds have previously been subjected to red light. The cause of this behaviour is the effect of light on the pigment **phytochrome**. Phytochrome exists in two forms, P_r which absorbs mainly red light and P_{ir} which absorbs mainly infra-red light (Fig. 20.11). Both P_r and P_{ir} are present in seeds. On absorbing red light, P_r is converted to P_{ir} while P_{ir} is reconverted to P_r when it absorbs infra-red light.

$$P_r \underset{\text{infra-red}}{\overset{\text{red}}{\rightleftharpoons}} P_{ir}$$

When a seed contains more than a certain amount of P_{ir} germination can proceed.

Sunlight, to which seeds are naturally exposed, contains both red and infra-red light. The proportion varies from time to time and place to place. Generally there is sufficient red light for the formation of enough P_{ir} to trigger off germination. The effect of P_{ir} accumulation is not certain but it

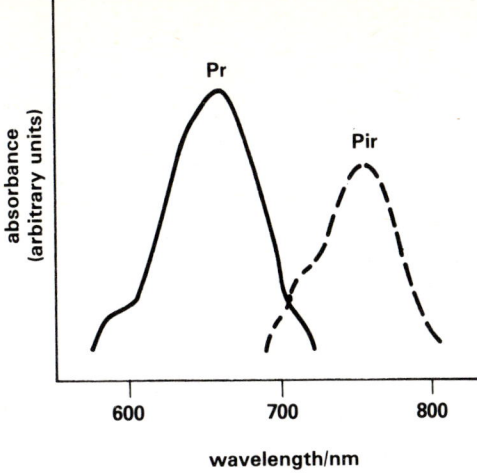

FIG. 20.11 *Absorption spectra of phytochromes*

may stimulate the production of gibberellic acid. Seeds which can germinate in darkness seem to have sufficient P_{ir} and thus do not require light.

Many light-dependent seeds are small. If they germinated too deeply in soil the embryonic shoot system would not reach the soil surface. Their dependence on light ensures that they germinate only when at or near the soil surface where there is a reasonable chance of the seedlings' aerial system becoming established. Ploughing and digging bring the light-dependent seeds of weeds to the soil surface where their dormancy is broken. Another ecological aspect of light-dependence is seen in deciduous forests and woodlands where in summer the canopy of leaves on the trees absorbs much red light. The light reaching the ground is then proportionally richer than normal daylight in infra-red light. In these conditions seeds accumulate P_r which inhibits germination. Germination of the seeds of woodland herbs therefore normally takes place in spring before the leaf canopy appears. The intensity of light in the wood is then also favourable for seedling growth.

20.3.3 Seed dormancy

Even when provided with water, a suitable temperature, oxygen and light, the seeds of many species fail to germinate and remain dormant. There are a number of reasons for **seed dormancy**.

1. IMMATURITY OF THE EMBRYO
The embryo may be **structurally immature** when the seeds are dispersed from the parent plant. This has been shown to be so in the ash and the wood anemone. In some species such as hazel the embryo is unable to synthesise gibberellic acid. This means that the hydrolytic enzymes needed for mobilising food reserves are not produced. This is an example of **physiological immaturity**. Figure 20.12 shows the effect of gibberellic acid in stimulating germination of dormant hazel seeds.

2. IMPERMEABILITY OF THE SEED COAT
The seed coat may be impermeable to oxygen and water or is too hard for the swelling embryo to burst through. This is so in many leguminous species such as lupin, clover and sweet pea. If the testa is chipped or removed, dormancy is broken if the seeds are given conditions suitable for germination. In bulky fruits such as apples the thick wall of the fruit restricts the diffusion of oxygen to the seeds. In natural conditions the seed coat and fruit wall are gradually broken down by microbial activity in the soil.

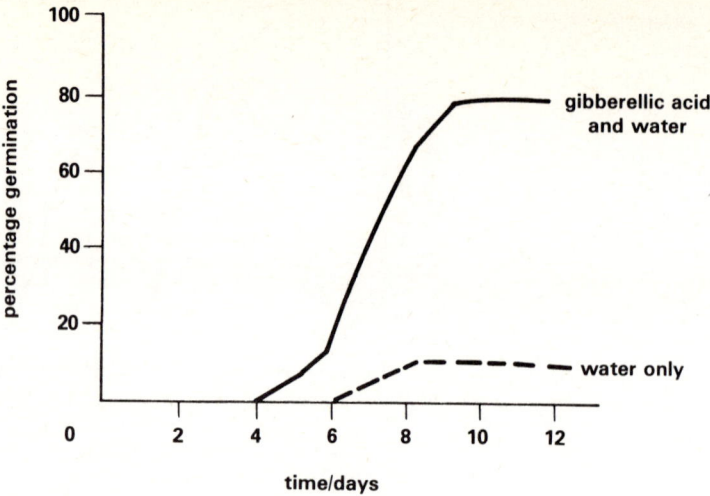

FIG. 20.12 *Effect of gibberellic acid on germination of hazel seeds*

3. GERMINATION INHIBITORS

Seeds of plants with succulent fruits such as tomatoes develop in an environment which seems ideal for germination. They fail to germinate because of **inhibitor** chemicals in the sap of the fruit. In other species such as birch and rose the inhibitor is in the testa and can be removed by immersing the seeds in several changes of water. In natural conditions rainfall probably slowly leaches out the inhibitor. The most important natural inhibitor so far discovered is **abscissic acid** (Fig. 20.13).

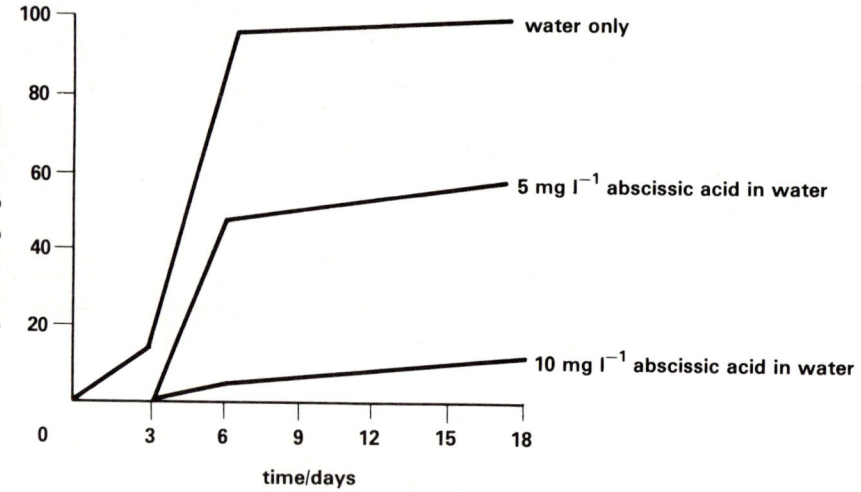

FIG. 20.13 *Effect of abscissic acid on germination of rye grass seeds*

Although the various factors responsible for seed dormancy have been considered separately it is often a combination of factors which generally keeps the seeds of most species dormant. Figure 20.14 shows the combination of dormancy-imposing mechanisms which operate in the European ash.

The few aspects of dormancy considered here demonstrate the subtle interactions between seed and environment and show how germination is controlled so that it usually takes place when there is a reasonable chance that successful growth will follow. In Britain most flowering plants produce their seeds in summer. If germination took place immediately the seeds are

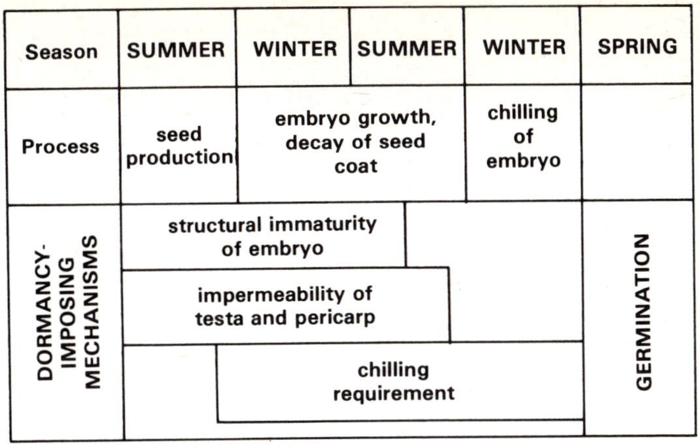

FIG. 20.14 *Combination of dormancy-imposing mechanisms in the European ash*

Season	SUMMER	WINTER	SUMMER	WINTER	SPRING
Process	seed production	embryo growth, decay of seed coat		chilling of embryo	
DORMANCY-IMPOSING MECHANISMS	structural immaturity of embryo				**GERMINATION**
	impermeability of testa and pericarp				
		chilling requirement			

dispersed, large numbers of seedlings would compete for the available space and light during autumn. Few would become established properly before the cold weather of winter. Because germination is delayed until the following spring or summer, subsequent growth is much more likely to be successful with a long spell of favourable weather about to begin. The reasons for dormancy vary between species and the time required for inhibitory factors to disappear also varies. Thus germination is usually spread out over a long period of time whereby the effects of competition between seedlings is diminished and adverse climatic conditions are avoided.

Chapter 21

Growth and development of flowering plants

Growth, which can be thought of as an increase in mass, and development, an increase in complexity, go hand in hand in all living organisms. In flowering plants growth and development begin when a seed germinates and ends when a mature plant has grown. Trees can take many years to reach maturity yet a few weeks is sufficient for some annual plants. It is well known that plants kept in greenhouses grow more quickly than similar plants growing outdoors. It is also known that oak trees and not dandelions appear if we sow acorns. This is because environment and heredity control the pattern and the rate of growth and development.

Over the past twenty years especially, a good deal of effort has been made in an attempt to unravel the physiological basis of plant growth. It is now apparent that growth and development of plants are the consequences of subtle interactions between many external and internal factors. With this knowledge it may soon be possible for humans to control plant growth more effectively than at present. As a result there could be significant improvements in the production of food, timber and other essential plant products.

21.1 Morphological and anatomical aspects of growth and development

In the previous chapter, seed germination and the subsequent establishment of a small immature plant called a seedling is described. How does a seedling grow and develop into a much larger and more complex mature plant? Mature plants have many more cells than seedlings. It is therefore evident that somatic cell division, mitosis (Chapter 7), is an integral part of growth. But the adult plant is more than a large number of identical cells. It has a variety of cell types, each type adapted to perform specific physiological and mechanical roles. The changes which take place to produce several types of cells is called differentiation.

21.1.1 Primary growth and development

In seedling plants mitosis is mainly confined to groups of cells at the tips of radicles and plumules. These areas of somatic cell division are called apical meristems. They produce the cells which differentiate into the tissues of the primary plant body.

1. THE PRIMARY SHOOT SYSTEM
A thin longitudinal section cut through the middle of a shoot tip reveals, at the apex, a mass of closely packed cells with conspicuous nuclei. The thin cell walls consist largely of pectic substances and the cytoplasm lacks any large vacuoles. Some of the cells are undergoing mitosis (Fig. 21.1). This is the apical meristem of the shoot.
(i) THE APICAL MERISTEM OF THE SHOOT. It is usually possible to distinguish an outer zone of cells called the tunica and an inner mass of cells called the corpus in the apical meristem of a shoot. The tunica, which can be one to several cells thick, is characterised by the fact that during mitosis cross walls are laid down at right angles to the surface of the shoot apex. Its main task is to enlarge the surface area of the shoot. The outermost layer of cells derived from the tunica later differentiates into the epidermis of the primary shoot system. Division of the corpus cells occurs

FIG. 21.1 (a) *Longitudinal section through the shoot apex of a dicotyledonous plant,* ×30

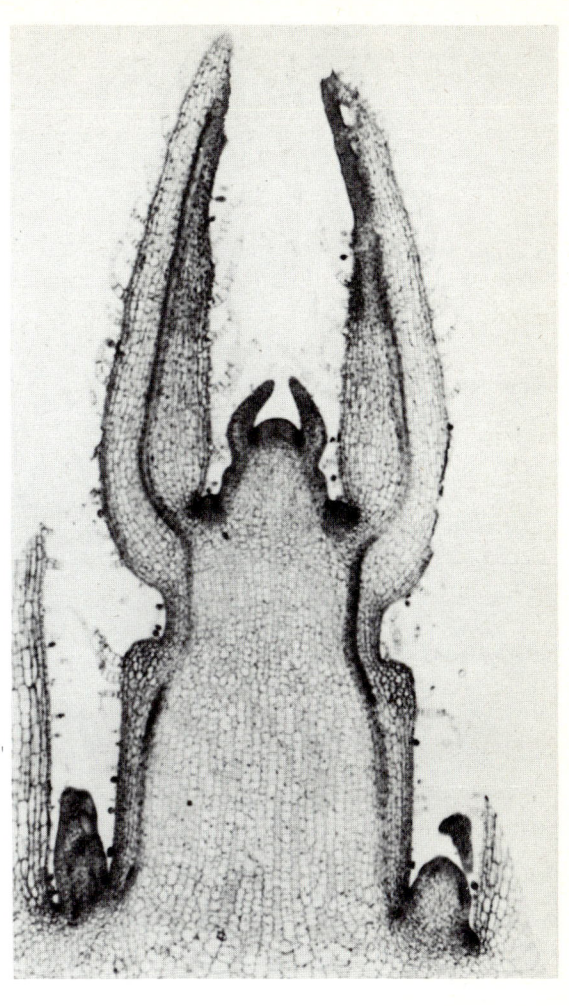

FIG. 21.1 (b) *Close-up view of apex,* ×160

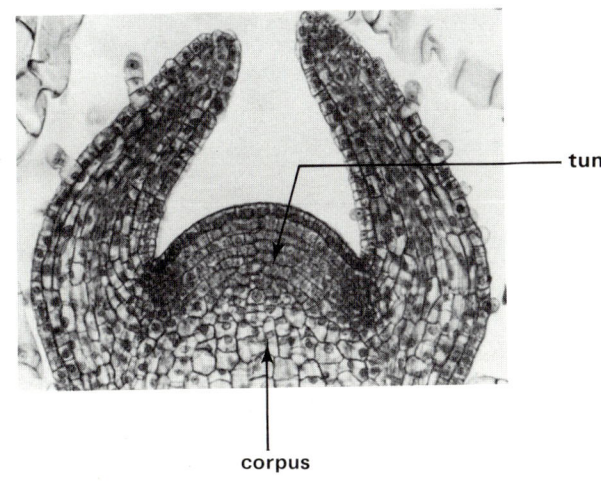

tunica

corpus

in several planes. An internal mass of cells is thus produced, the size of which keeps pace with the ever-increasing surface layer formed by the tunica. Cells derived from the corpus differentiate into the tissues of the **cortex**, **vascular bundles** and **pith** of the primary stem.

Both tunica and corpus take part in the formation of swellings which arise at regular intervals on the surface of the shoot apex. The swellings are called **leaf primordia**. Their growth is so rapid that the earlier formed primordia soon enclose the apical meristem at the tip of the shoot. Mature leaves gradually differentiate from the older primordia, new primordia constantly being formed at the tip of the shoot as apical growth continues. The spatial arrangement of the leaf primordia determines the positions of mature leaves on the shoot. Paired primordia for example give rise to pairs of leaves on the mature stem. **Axillary bud primordia** which have the potential to grow into lateral shoots are generally seen in the axils of developing leaves near the shoot apex (Fig. 21.2). The tunica and corpus again produce the axillary bud primordia. The superficial origin of lateral shoots contrasts with the deep-seated origin of lateral roots Fig. 21.9).

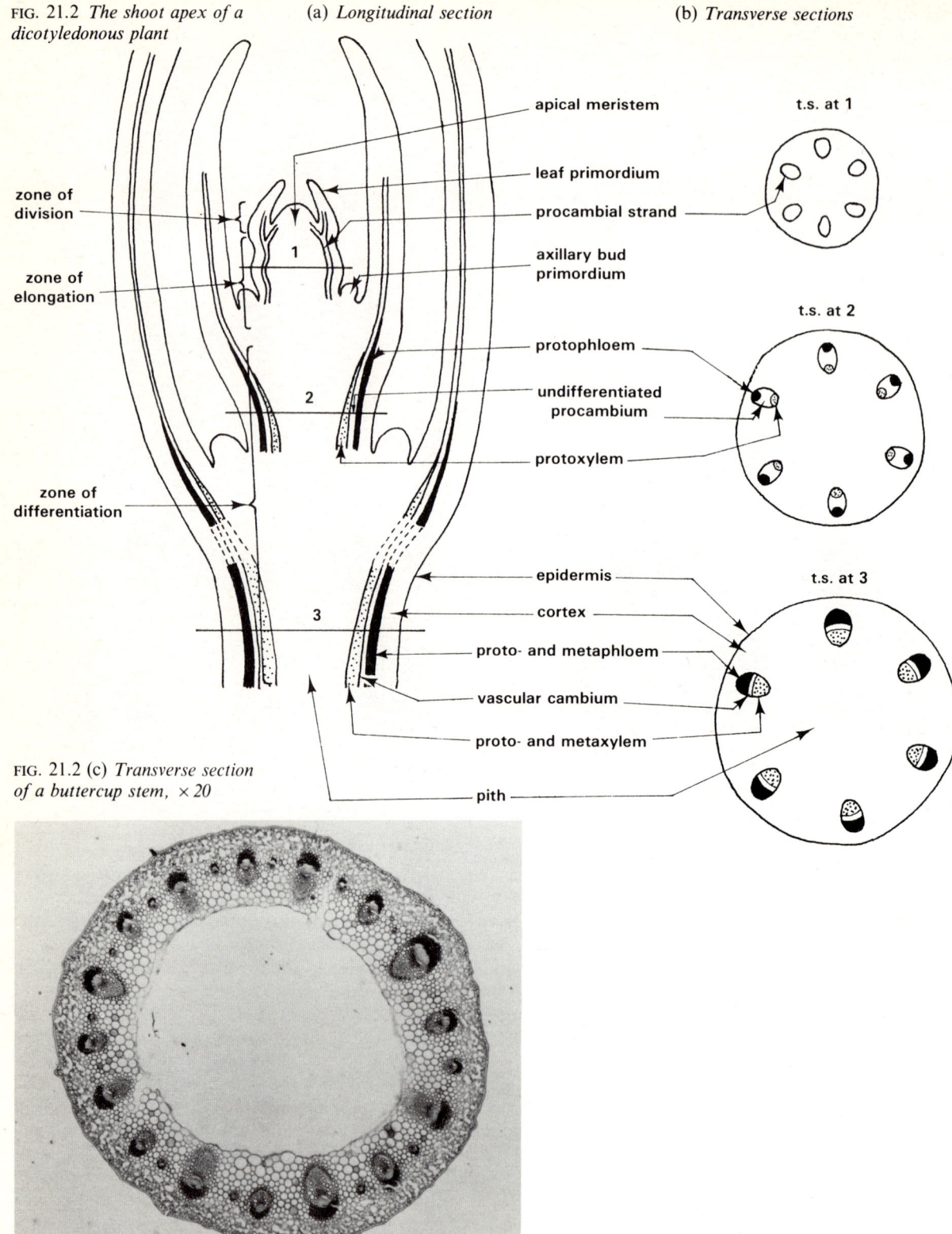

FIG. 21.2 *The shoot apex of a dicotyledonous plant*

(a) *Longitudinal section*

(b) *Transverse sections*

apical meristem

leaf primordium

procambial strand

axillary bud primordium

zone of division

zone of elongation

protophloem

undifferentiated procambium

protoxylem

zone of differentiation

t.s. at 1

t.s. at 2

t.s. at 3

epidermis

cortex

proto- and metaphloem

vascular cambium

proto- and metaxylem

pith

FIG. 21.2 (c) *Transverse section of a buttercup stem, × 20*

FIG. 21.2 (d) *Enlarged view of one of the vascular bundles,* × 120

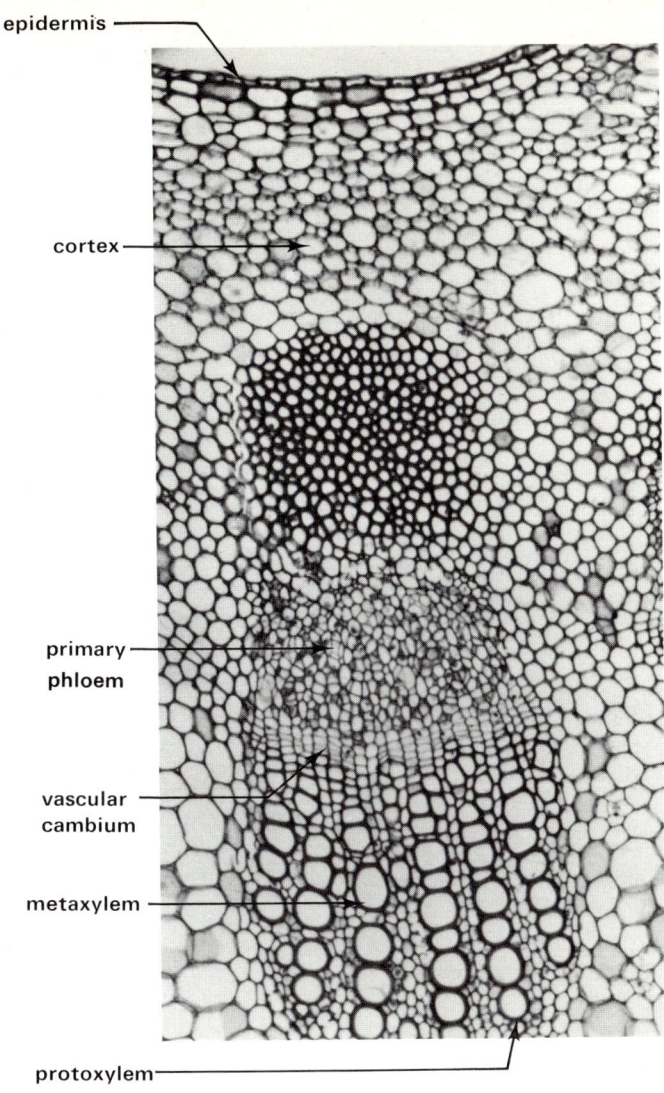

epidermis

cortex

primary phloem

vascular cambium

metaxylem

protoxylem

(ii) TISSUE DIFFERENTIATION IN THE STEM. Just behind the apical meristem the first signs of **tissue differentiation** are seen. Conspicuous vacuoles appear in the cytoplasm of cells cut off from the tunica and corpus. The effect is to cause considerable increase in length and girth of the cells. **Vacuolation** accounts mainly for elongation of the shoot at this stage. However, there is an increase in the protein and nucleic acid content of the cells. Cellulose is also laid down in the expanding cell walls.

Not all of the apically-derived cells undergo the changes outlined above. Several strands of cells extending backwards from the apex remain narrow with pointed ends. Although they elongate they do not develop large vacuoles. They also retain the power to divide even when the surrounding cells can no longer do so. Thus a number of discrete bundles of tissue called **procambial strands** appear among the vacuolate ground tissue. The positions of the strands are determined by the leaf primordia, one strand developing below each primordium.

The outermost cells of the procambial strands differentiate into the first formed cells of primary phloem called **protophloem**. First there is an unequal longitudinal division of each procambial cell. The smaller of the derivatives differentiates into a companion cell. At the same time the larger cell differentiates into a sieve-tube element. Sieve tubes transport organic substances in solution to all parts of the plant including the growing shoot apex (Chapter 12). Shortly afterwards the first formed water-carrying tissue called **protoxylem** differentiates from the innermost cells of each procambial strand. The main changes in protoxylem differentiation are loss of the protoplast, breakdown of the end walls of adjacent cells and secretion of a lignified secondary wall as annular or spiral bands inside the primary cell wall (Fig. 21.3(a)). The primary walls of protoxylem cells can continue growing for some time keeping pace with elongation of surrounding tissues. The secondary wall gives the necessary support to keep the lumen of each xylem element open for water transport.

While the changes outlined above are taking place the shoot apex continues to grow upwards. The part of the shoot where protophloem and protoxylem have formed now stops elongating and further differentiation occurs in the procambial strands. The procambial cells immediately inside the protophloem differentiate into **metaphloem** sieve-tubes and companion cells. The procambial cells immediately inside the protoxylem differentiate into **metaxylem** cells. The main changes in the formation of metaxylem cells are loss of end walls. The primary walls of metaxylem cells become covered internally by scalariform, reticulate or pitted, lignified secondary walls (Fig. 21.3(b)). The lack of plasticity compared with protoxylem cells reflects the cessation of elongation in this part of the shoot.

FIG. 21.3 (a) *Differentiation of protoxylem*

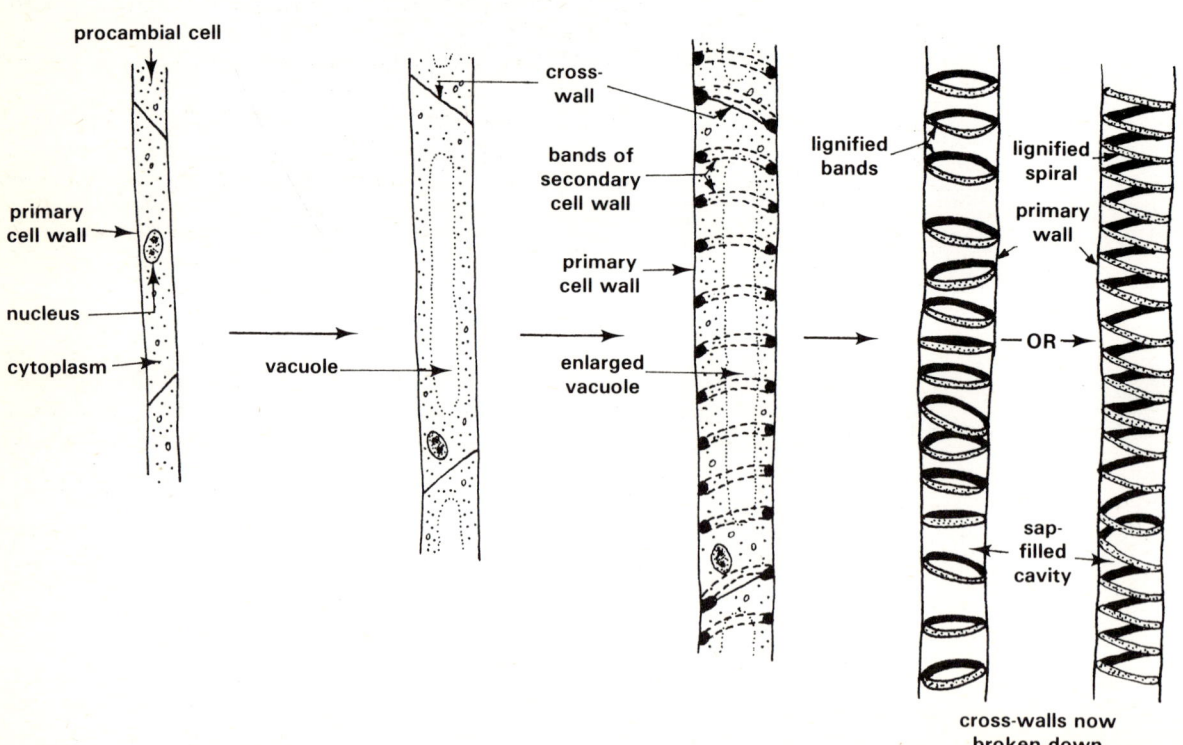

410

FIG. 21.3 (b) *Metaxylem cells*

FIG. 21.3 (c) *Longitudinal section of protoxylem, × 120*

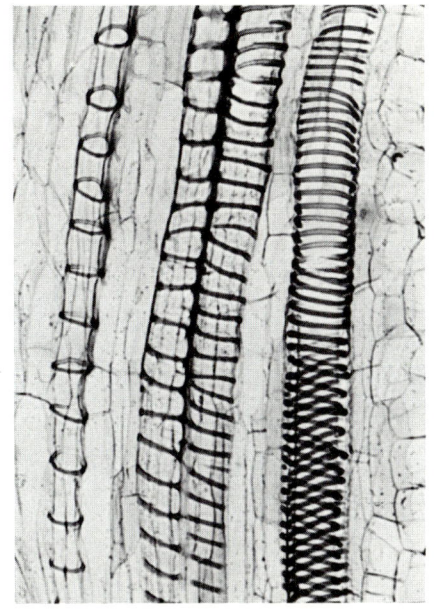

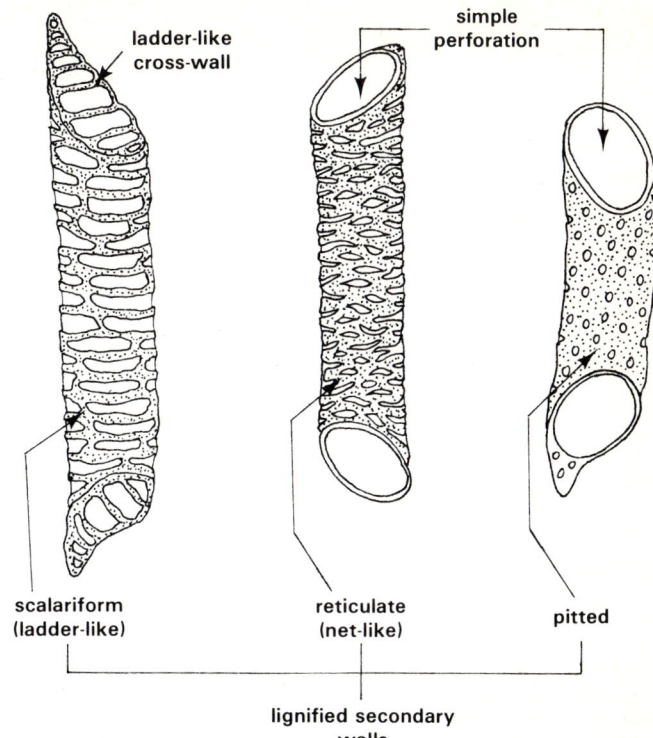

ladder-like cross-wall

simple perforation

scalariform (ladder-like)

reticulate (net-like)

pitted

lignified secondary walls

The cells of much of the ground tissue of the stem, (the cortex and pith) show less obvious changes as they differentiate into a tissue called **parenchyma**. The main changes are enlargement in girth and length, the development of thin cellulose walls and extensive vacuolation of the cytoplasm. The changes prepare the parenchyma cells for the storage of large quantities of sap which helps maintain turgidity of the young shoot. Starch grains too are frequently stored in parenchyma. In the outer cortex the cell walls often become thickened with stiff, yet plastic, deposits of cellulose

FIG. 21.4 (a) *Parenchyma*

l.s.

t.s.

(i) *transverse section of parenchyma, × 300*

thin cellulose wall

cytoplasm

sap-filled vacuole

nucleus

Parenchyma cells near the surface of the stem usually contain chloroplasts

411

FIG. 21.4 (b)

l.s.

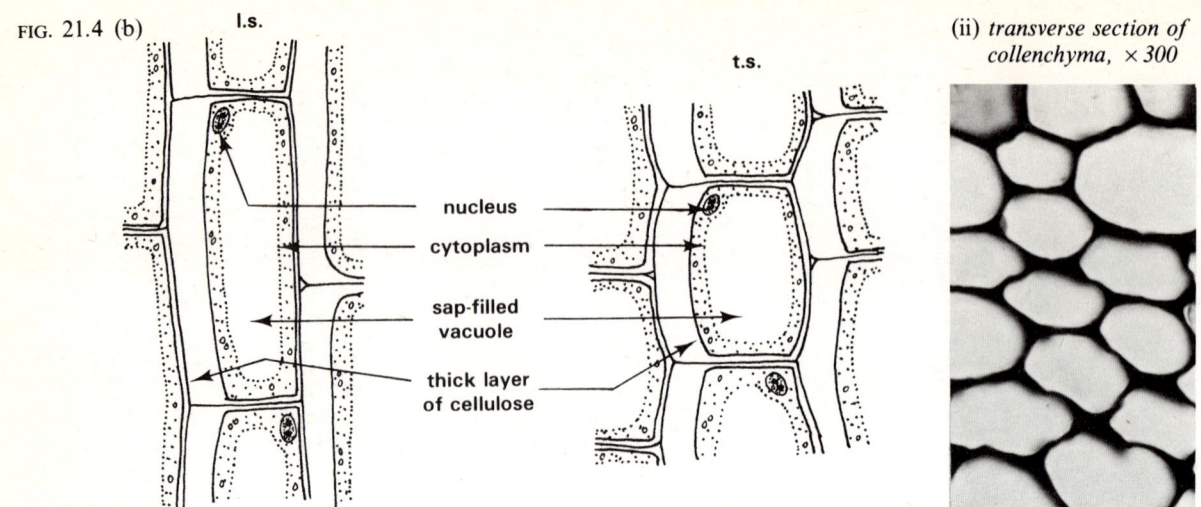

t.s.

(ii) *transverse section of collenchyma, × 300*

nucleus

cytoplasm

sap-filled vacuole

thick layer of cellulose

The position of the thick deposit of cellulose varies from one type of collenchyma to another

FIG. 21.4 (c) *sclerenchyma fibres*

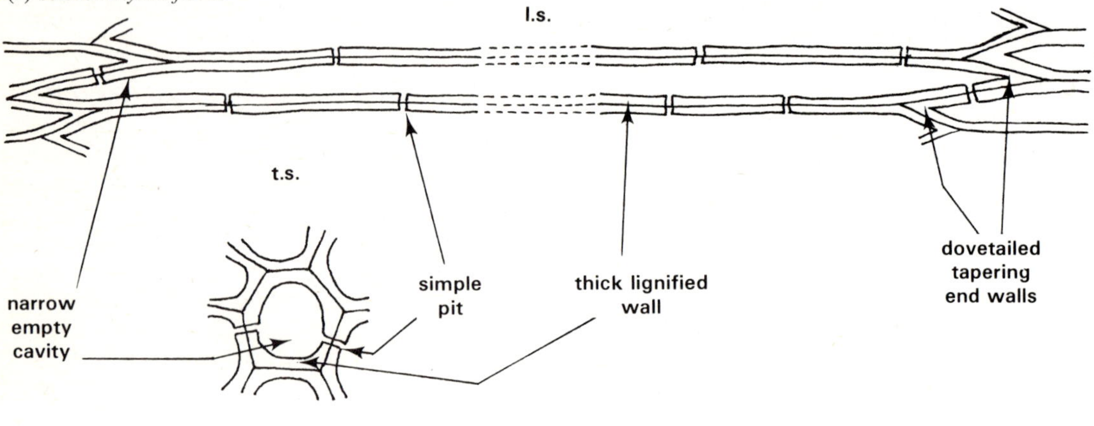

l.s.

t.s.

narrow empty cavity

simple pit

thick lignified wall

dovetailed tapering end walls

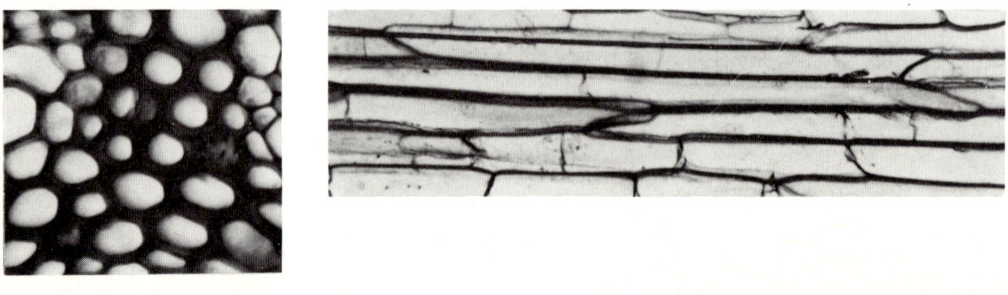

(iii) *sclerenchyma: transverse and longitudinal sections, both × 300*

forming a supporting tissue called **collenchyma**. Parenchyma and collenchyma cells retain the ability to divide and often contain chloroplasts. Some cortical cells, particularly those immediately around each vascular bundle, become extremely elongated and have tapering end walls. They develop extremely thick, lignified secondary walls. These are **sclerenchyma fibres**, the ends of which dovetail together and form a very efficient supporting tissue (Fig. 21.4). The epidermal cells derived from the tunica secrete a waxy material called **cutin** over their outer surface making the **epidermis** waterproof. Some epidermal cells become modified into **guard cells** of **stomata** which are used for gas exchange (Chapters 11 and 12).

(iii) DEVELOPMENT OF LEAVES AND LATERAL SHOOTS. A leaf primordium grows initially into a peg-like outgrowth with an **adaxial meristem** on its lower surface. Division of the cells of the adaxial meristem produces a thicker peg containing procambial strands continuous with those in the stem. The peg later differentiates into the tissues of the leaf midrib. **Marginal meristems** appear on either side of the peg and divide to produce cells which differentiate into the epidermis, mesophyll and vascular tissue of the leaf blade (Fig. 21.5). The spatial arrangement or **phyllotaxis** of leaves on the mature stem is therefore determined by the positions of the leaf primordia at the stem apex. In Figure 21.5 it can be seen that the developing leaves are so close to each other at the shoot apex that they overlap and form a terminal **bud**. Growth and differentiation of the primary stem separates the older leaves as elongation of the **internodes**, the length of stem between successive leaves, occurs. Separation of the mature leaves allows efficient gas exchange and light absorption, important for photosynthesis, the main physiological function of leaves (Chapter 12).

FIG. 21.5 (a) and (b) *Stages in the early development of a leaf*

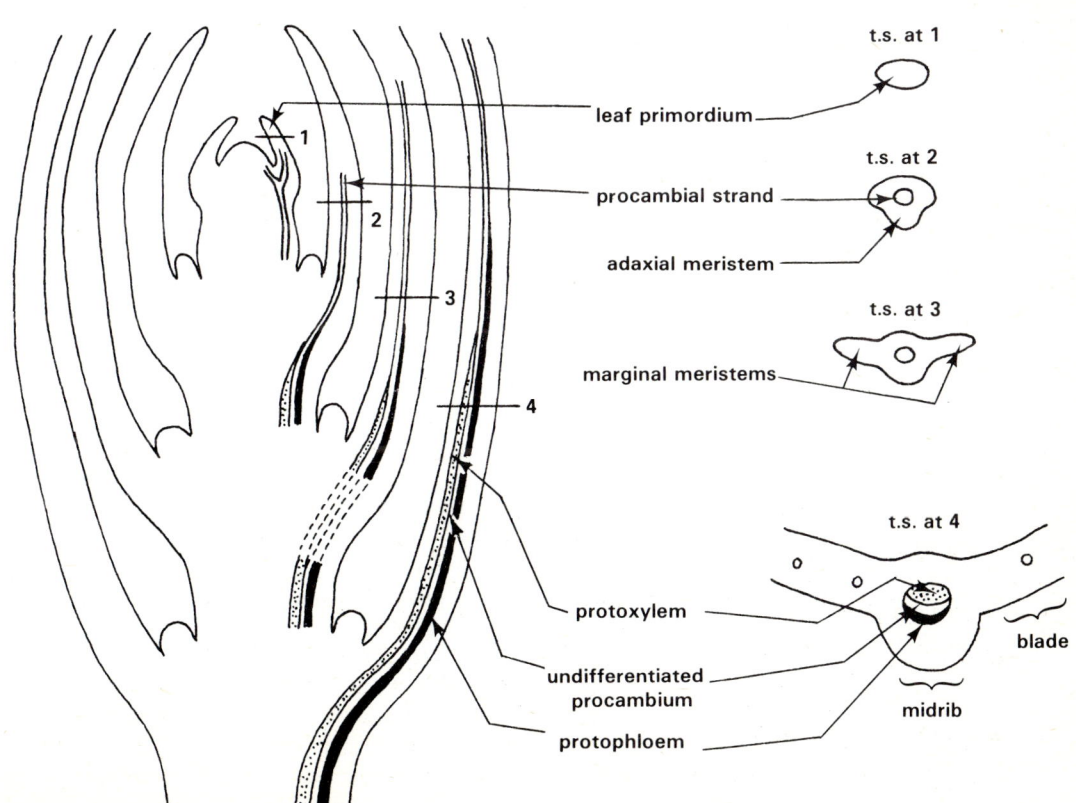

FIG. 21.5 (c) *Transverse section of a mature leaf,* ×30

Lateral buds in the axils of leaves may grow into side shoots. Development of side shoots is the same as described earlier for the main shoot.

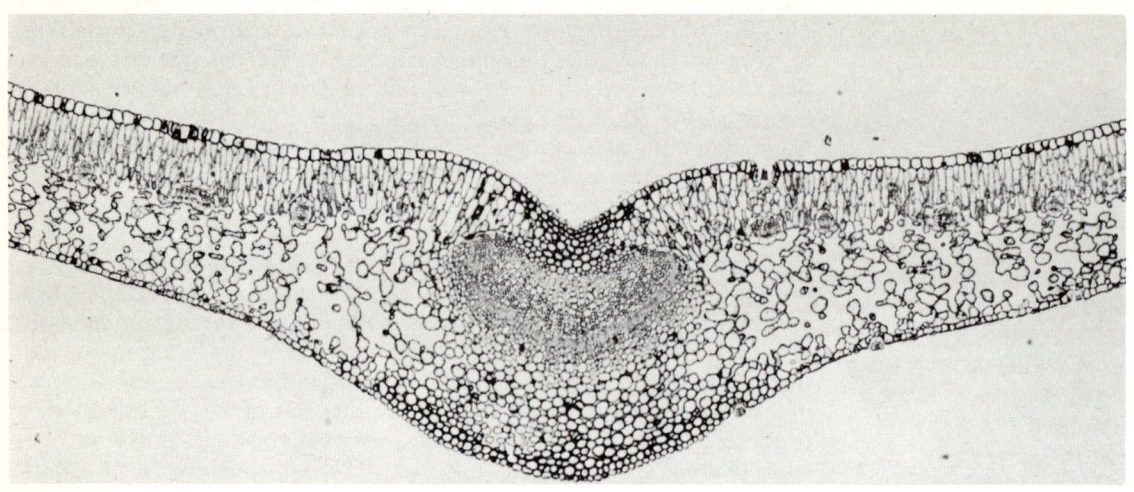

FIG. 21.6 *Longitudinal section through the apex of an onion root* × 100

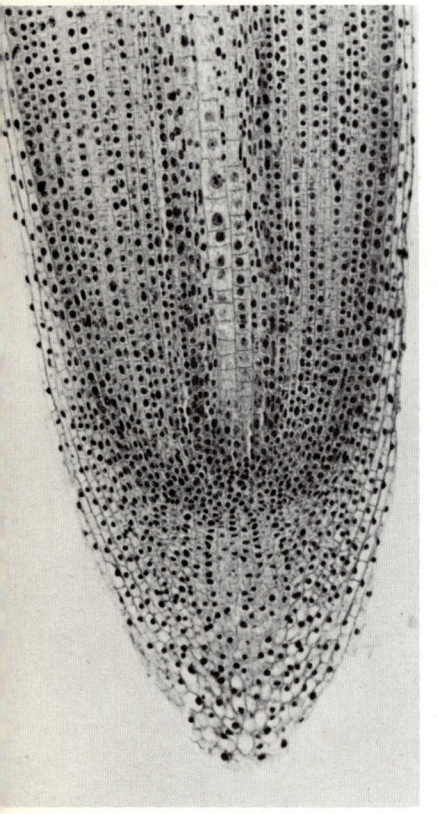

2. THE PRIMARY ROOT SYSTEM

(i) THE APICAL MERISTEM OF THE ROOT. The most obvious difference in appearance between longitudinal sections of stem and root apices is the absence of bulges comparable to leaf and bud primordia on the root apex. The root apex is also covered by a **root cap** (Fig. 21.6). There is, however, a marked similarity in appearance and behaviour of the apical cells which constantly divide by mitosis. In most roots it is possible to distinguish a number of zones of cells at the apex. The outermost zone is called the **protoderm**. It produces cells which differentiate into the root epidermis and root cap. Inside the protoderm is the **ground meristem** which produces cells which differentiate into the root cortex. Just behind the root apex a single **procambial strand** can be seen at the centre of the root. Some roots have an additional meristematic layer, the **calyptrogen**, which gives rise to the cells of the root cap. The meristematic zones radiate from a clump of cells called the **quiescent centre** situated immediately behind the root cap. The significance of the quiescent centre is not as yet fully understood. Its cells divide slowly and it is probably the site from which the meristematic layers arise (Fig. 21.7).

(ii) TISSUE DIFFERENTIATION IN THE ROOT. Differentiation of vascular tissue begins near the root apex. Several strands of sieve-tube elements and companion cells appear near the outside of the procambial strand. Shortly afterwards a similar number of strands of protoxylem cells alternating with the primary phloem strands differentiate. Metaxylem cells differentiate last of all at the centre of the procambial strand. The outermost procambial cells undergo little change and retain their ability to divide. They become the **pericycle** which may later produce lateral roots.

The derivatives of the ground meristem differentiate into parenchyma cells of the **root cortex** which are used mainly for the storage of starch and the transport of water and dissolved minerals from the soil solution to the root vascular tissue. The cortical cells of the root do not usually develop chloroplasts as do the outer cortical cells of the stem. The innermost layer of the root cortex differentiates into a special band of cells called the **endodermis**. Early in development the endodermal cells secrete a suberised

FIG. 21.7 *The root apex of a dicotyledonous plant*

(b) *transverse sections*

(a) *longitudinal section*

epidermis

cortex

endodermis

proto- and metaxylem

proto- and metaphloem

protoxylem

protophloem

ground meristem

procambial strand

protoderm

quiescent centre

root cap

t.s. at 3

t.s. at 2

t.s. at 1

3

2

1

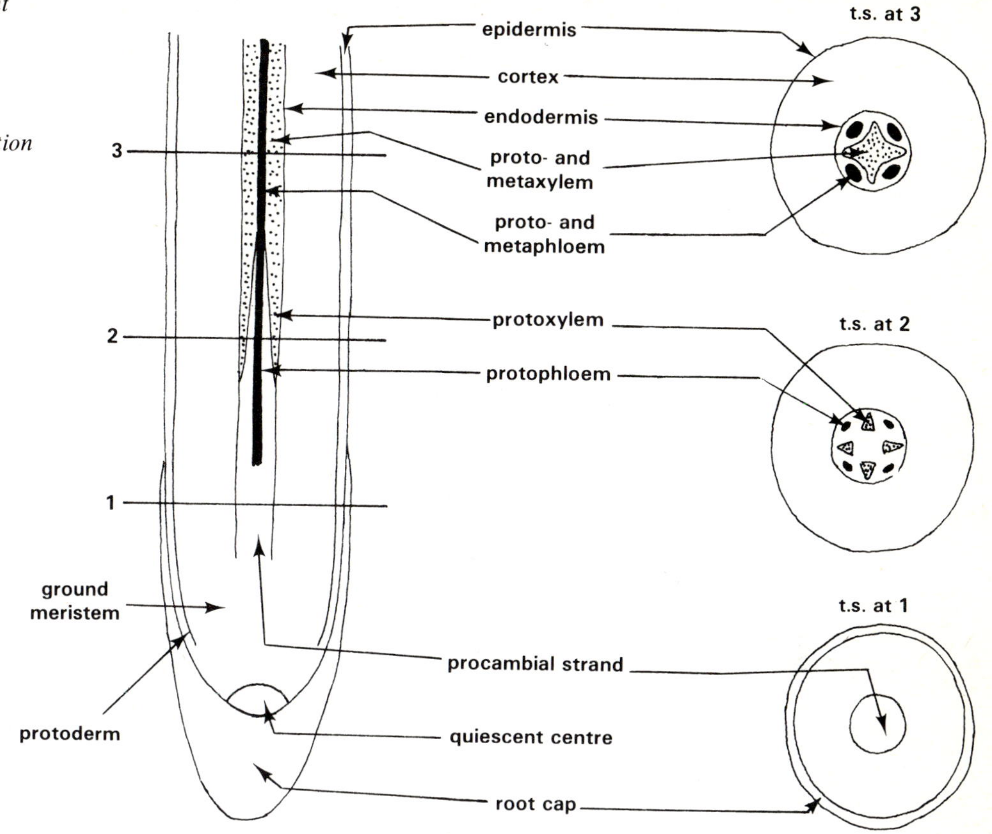

FIG. 21.7 (c) *Transverse section of a mature primary root of a dicotyledonous plant,* ×70

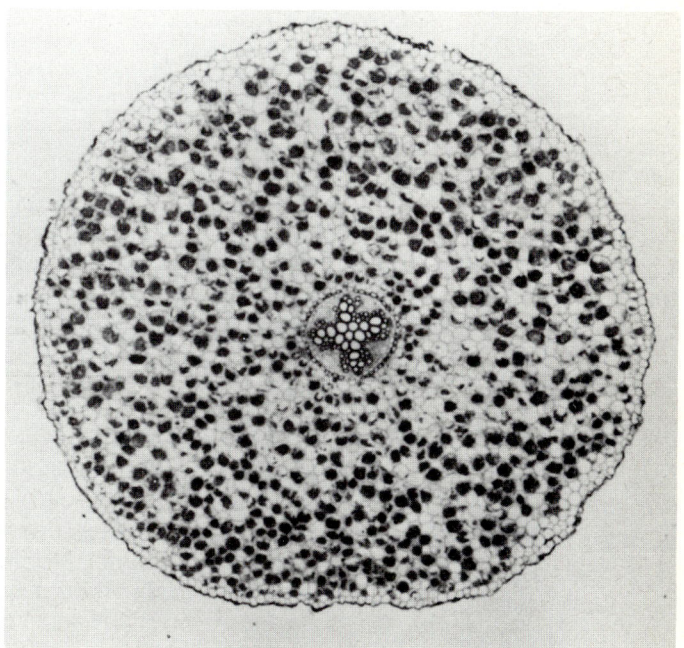

Casparian strip on their radial wall. Later, all but the outer of the endodermal cell walls become thickened and suberised (Fig. 21.8). The endodermis is thought to control the passage of absorbed water and dissolved minerals from the cortex into the root vascular tissue (Chapters 11 and 13).

The root epidermis derived from the protoderm is very different from that of the shoot system. There is no cuticle as this would prevent one of the most important tasks of the root, the absorption of water and dissolved minerals. Just behind the root apex the epidermal cells develop tubular extensions called **root hairs** which vastly enlarge the root surface area through which absorption from the soil can occur. Root cap cells develop thick gelatinous walls of pectic substances which lubricate and protect the root apex as it constantly probes its way through the soil.

FIG. 21.8 *Transverse section of young endodermis*

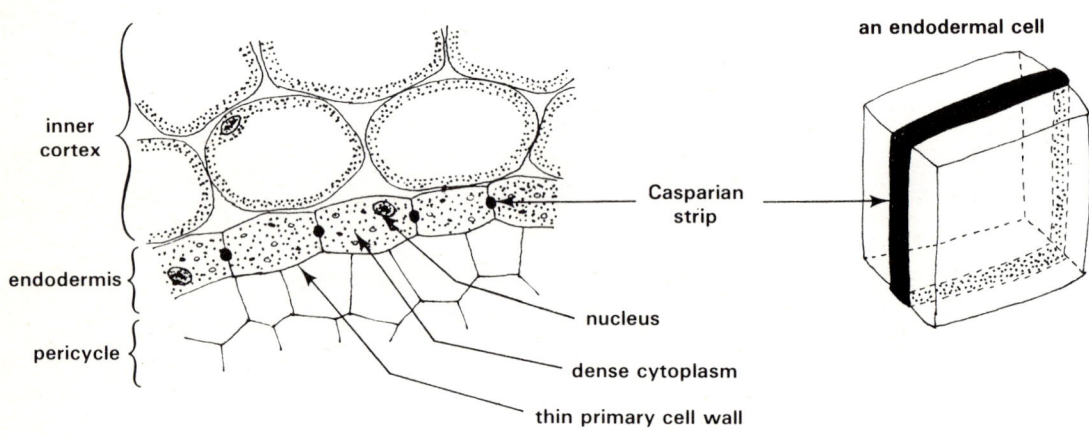

(iii) LATERAL ROOT FORMATION. The formation of **lateral roots** shows two main points of difference compared with the formation of lateral shoots. Firstly, lateral roots normally arise some distance from the root apex. Secondly, they originate deeply in the root from the pericycle. For this reason the origin of lateral roots is described as **endogenous** whereas lateral shoots are exogenous in origin.

Mitosis of the pericycle cells produces bulges of cells (Fig. 21.9). The bulges are the beginnings of lateral roots. They normally arise opposite the groups of protoxylem cells. As they grow, each of the bulges develops an apical meristem and a root cap. The growing lateral roots, helped by enzyme secretion, force their way through the cortex of the main root. Eventually the lateral roots emerge through the ruptured epidermis. By this time a procambial strand has usually differentiated in each lateral root. Further differentiation takes place as in the main root, and the vascular tissues of the lateral roots become joined to the xylem and phloem of the main root.

FIG. 21.9 Stages in the early
development of a lateral root
(a) *pericycle cells begin to divide*

(b) *lateral root primordium formed*

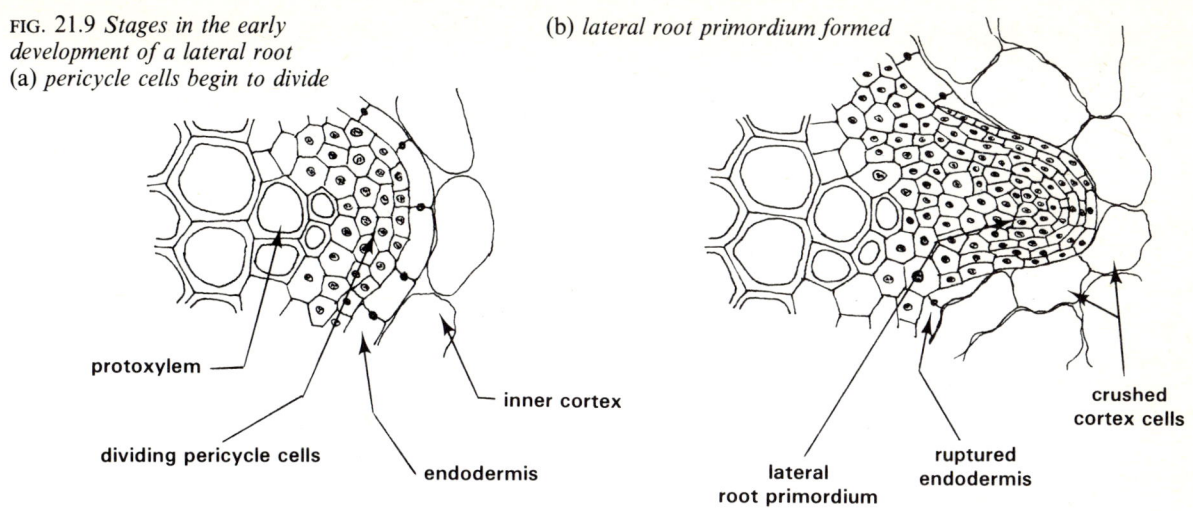

protoxylem —

inner cortex

dividing pericycle cells

endodermis

lateral
root primordium

ruptured
endodermis

crushed
cortex cells

(c) *tissue differentiation*

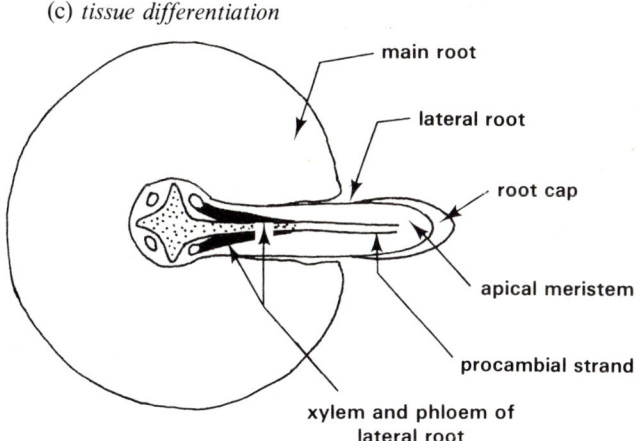

main root

lateral root

root cap

apical meristem

procambial strand

xylem and phloem of
lateral root

FIG. 21.9 (d) *Transverse section
of a dicotyledonous root showing
the early growth of a lateral
root, × 30*

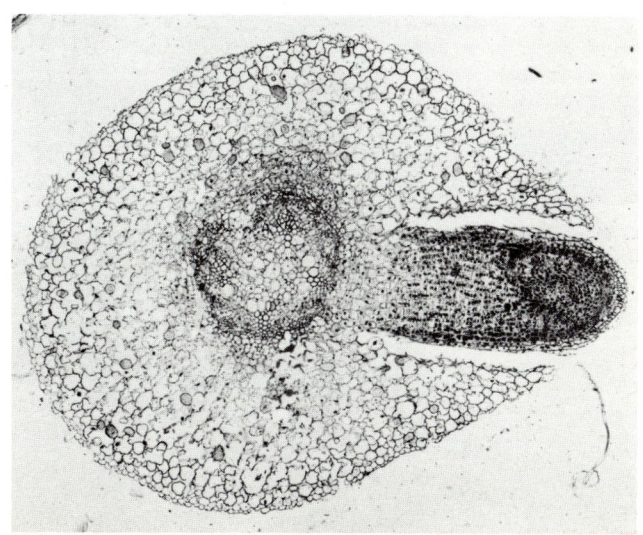

21.1.2 Secondary growth and development

Whereas some plants do not develop further than the primary stage, others can add secondary tissues to the primary body. The **secondary plant body** is derived from **lateral meristems**, the most important of which are the **vascular cambium** and **cork cambium**.

1. THE VASCULAR CAMBIUM

In plants which undergo secondary growth not all the procambial cells in the primary body differentiate into vascular tissue. What remains is called **vascular cambium** and is situated between the primary xylem and phloem of stems and roots. The vascular cambium consists of two types of cells, **ray and fusiform initials** (Fig. 21.10), and usually begins its activity during the first year of the plant's life.

FIG. 21.10 *The vascular cambium*

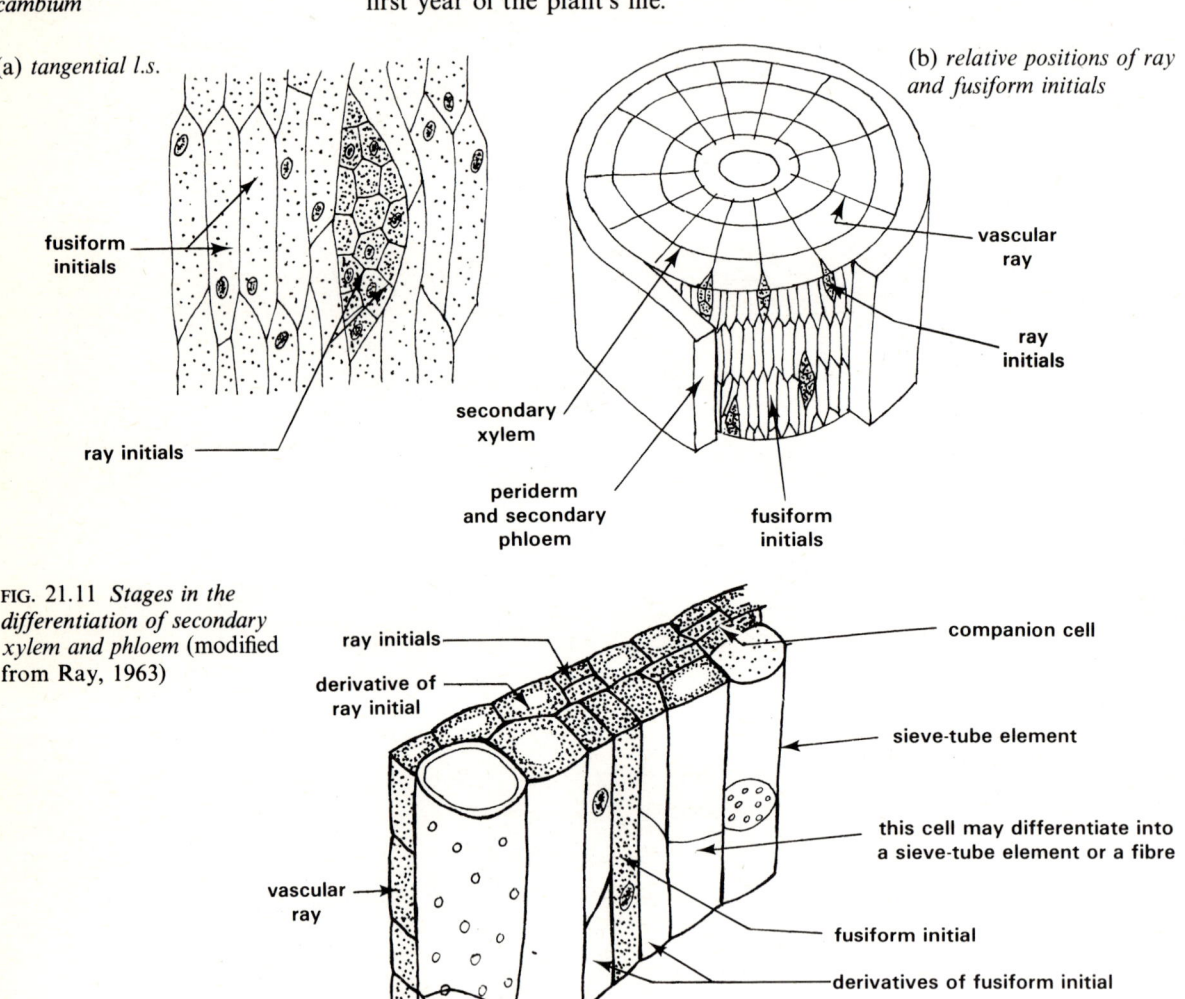

(a) *tangential l.s.*

fusiform initials

ray initials

(b) *relative positions of ray and fusiform initials*

vascular ray

ray initials

secondary xylem

periderm and secondary phloem

fusiform initials

FIG. 21.11 *Stages in the differentiation of secondary xylem and phloem* (modified from Ray, 1963)

ray initials

derivative of ray initial

vascular ray

companion cell

sieve-tube element

this cell may differentiate into a sieve-tube element or a fibre

fusiform initial

derivatives of fusiform initial

this cell may differentiate into a vessel element, a tracheid or a fibre

vessel element

In the stem of a dicotyledonous plant the vascular cambium is at first confined to the vascular bundles. However, a complete cylinder of vascular cambium is formed as parenchyma cells between the bundles become meristematic. The main plane of division of the ray and fusiform initials is tangential. Derivatives cut off from the fusiform initials to the inside of the cylinder of vascular cambium **differentiate** into xylem vessel elements, tracheids and fibres. Those cut off to the outside become sieve-tube elements, companion cells and phloem fibres. Derivatives of the ray initials differentiate into parenchyma cells which cross the secondary xylem and phloem as radial **vascular rays** (Fig. 21.11). Occasionally the ray and fusiform initials divide by radial walls so that the circumference of the cylinder of vascular cambium increases, accommodating the growing amount of secondary xylem it encloses.

FIG. 21.12 *Stages in the secondary growth of a dicotyledonous stem as seen in transverse section*

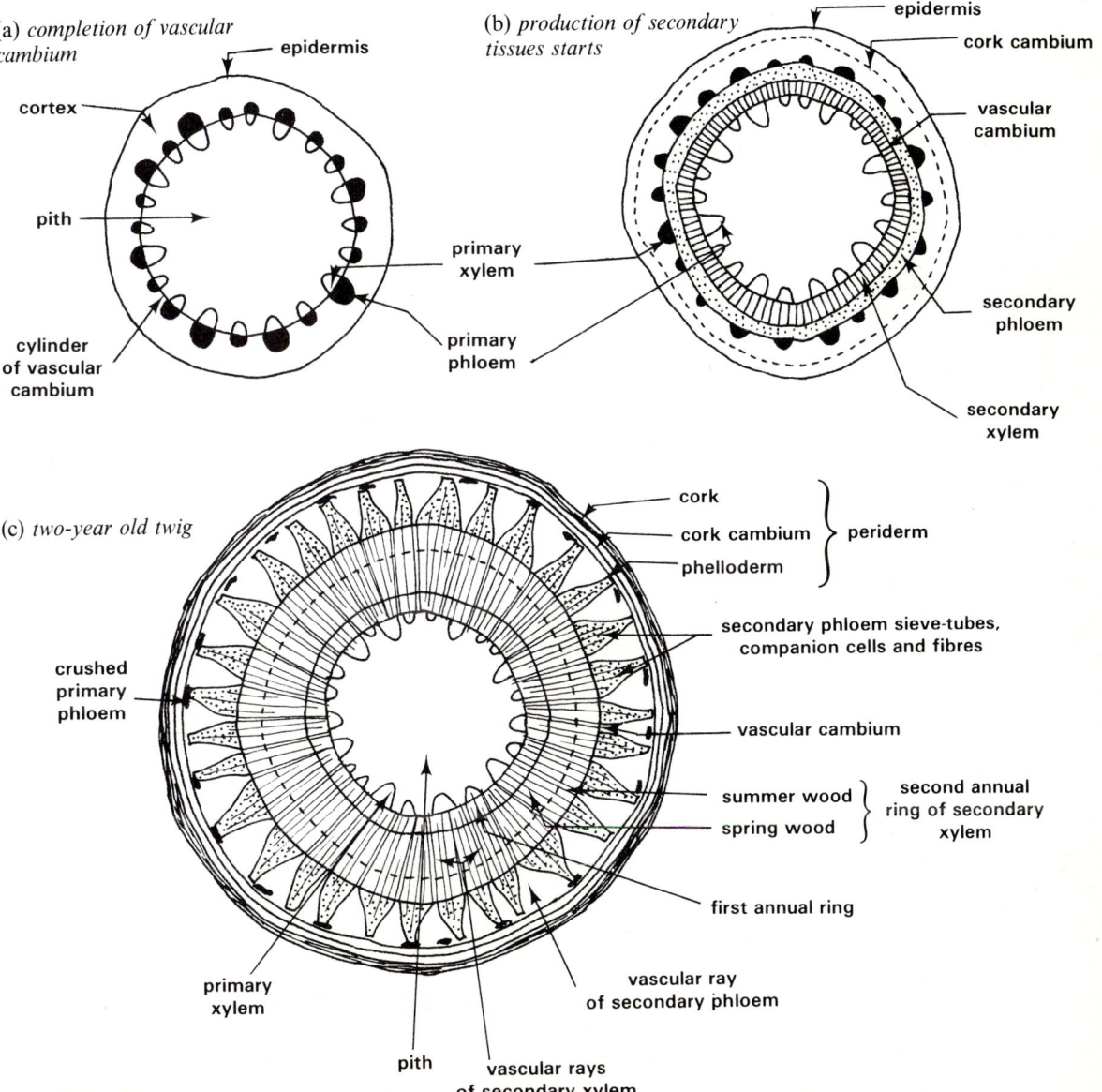

(a) *completion of vascular cambium*

epidermis

cortex

pith

cylinder of vascular cambium

primary xylem

primary phloem

(b) *production of secondary tissues starts*

epidermis

cork cambium

vascular cambium

secondary phloem

secondary xylem

(c) *two-year old twig*

cork
cork cambium } periderm
phelloderm

secondary phloem sieve-tubes, companion cells and fibres

vascular cambium

summer wood } second annual ring of secondary xylem
spring wood }

first annual ring

crushed primary phloem

primary xylem

vascular ray of secondary phloem

pith

vascular rays of secondary xylem

419

FIG. 21.12 (d) *Transverse section through a 5-year-old twig of lime, × 15*

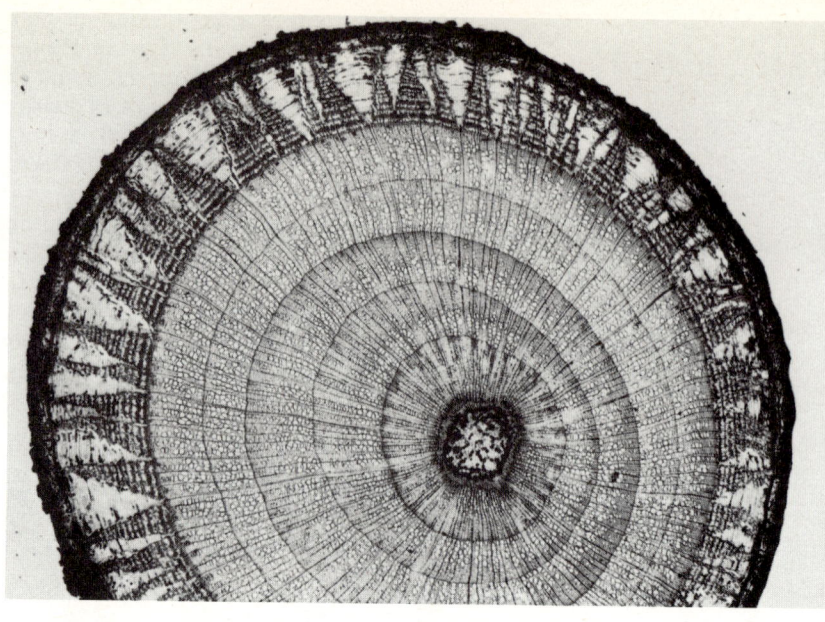

The vascular cambium is seasonal in its activity. It is dormant in winter and active in spring and summer. A cylinder of secondary xylem called an **annual ring** is added each year to the existing plant body. Vessels produced early in the growing season are generally larger in diameter than those arising later. Each annual ring thus consists of two distinct regions, **spring wood** on the inside and **summer wood** on the outside. The age of a shrub or tree can be found by counting the number of annual rings of secondary xylem in a cross-section of the main stem or trunk. There is no obvious layering in the secondary phloem. Here the vascular ray cells often undergo further division by radial walls so that wide, wedge-shaped phloem rays separate groups of sieve-tubes, companion cells and fibres (Fig. 21.12).

2. THE CORK CAMBIUM

The increase in girth resulting from the addition of secondary vascular tissue imposes a strain on the outer tissues of the stem. If the primary cortex and epidermis remained, the stem would split open with the risk of desiccation. This is prevented by the development of tissues called the **periderm** which replace the outer tissues of the primary stem.

A cylinder of **cork cambium** arises, usually from parenchyma cells in the outer cortex. In transverse section the cells of the cork cambium are uniformly rectangular in shape. As in the vascular cambium, mitotic division in a tangential plane cuts off derivatives to the outside and to the inside. The outer derivatives secrete a layer of waxy material called suberin in their secondary walls after which their protoplasts disappear. They are now **cork cells** and form an effective waterproof barrier which prevents dehydration of internal tissues. Here and there the cork cells are loosely arranged, forming **lenticels** which allow gas exchange between the outer tissues and the atmosphere. The small number of derivatives cut off internally undergo little change and remain as a narrow band of uniform cells called **phelloderm** (Fig. 21.13). The cells of the cork cambium now and again divide by radial walls to enlarge the circumference of the meristem, thereby keeping pace with the increase in girth caused by the activity of the vascular cambium.

A comparable pattern of secondary growth also happens in the roots of many dicotyledonous plants. The vascular cambium originates from strands of undifferentiated procambium lying between the primary xylem and phloem. It then develops into a cylinder from which annual rings of secondary vascular tissue are produced as in the stem. The cork cambium often originates from the pericycle so that early in its secondary growth the root becomes temporarily much thinner when the endodermis, cortex and epidermis are lost.

FIG. 21.13 (a) *Transverse section through the periderm of elder, ×360*

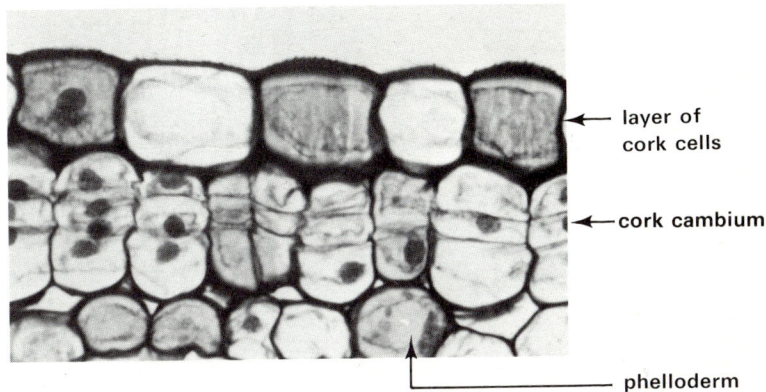

← layer of cork cells

← cork cambium

phelloderm

FIG. 21.13 (b) *Transverse section through a lenticel of elder, ×80*

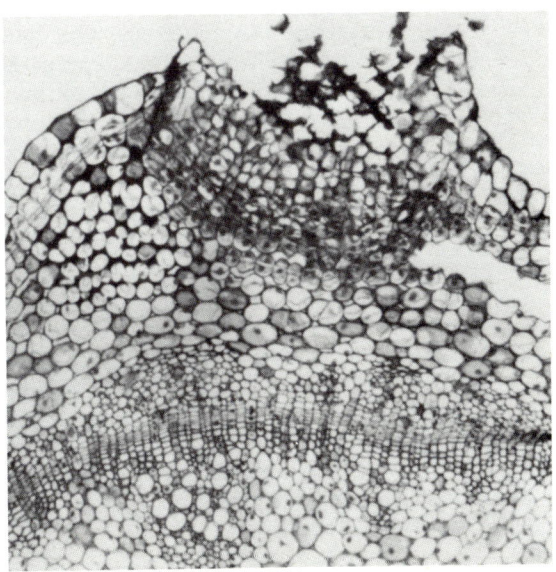

Few monocotyledonous species undergo secondary growth. However, in plants of the grass family shoot growth is not entirely dependent on the apical meristem. In grasses, tissues at the bases of the internodes remain meristematic even though the rest of the internode is fully differentiated. Each internode thus has its own growth zone called an **intercalary meristem** (Fig. 21.14). The activity of the intercalary meristem is one of the reasons why grass shoots regenerate rapidly after being severely damaged in haymaking, lawnmowing and animal grazing.

FIG. 21.14 *Intercalary meristems in a grass shoot*

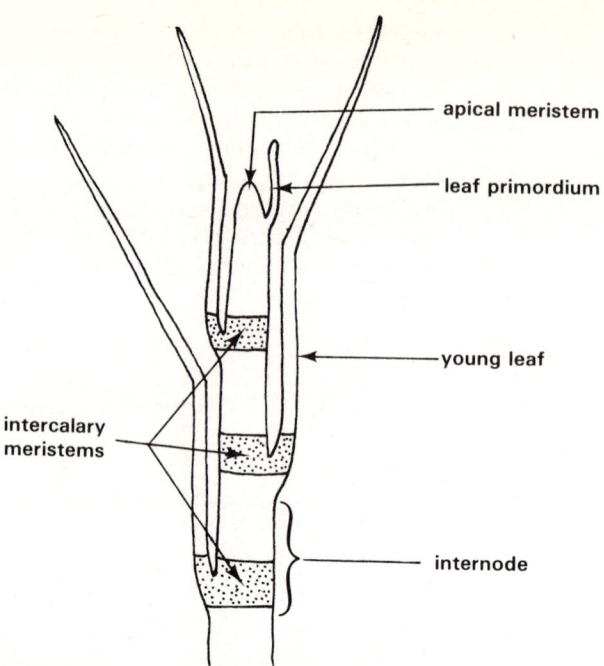

apical meristem

leaf primordium

young leaf

intercalary meristems

internode

21.2 Physiological aspects of growth and development

A seedling becomes established using the reserve of food stored in the seed (Chapter 20). When the food reserve has been used up continued growth and development depend on the young plant obtaining alternative sources of energy and raw materials. The capture of energy in sunlight and its conversion to chemical bond energy is described in Chapters 5 and 12, while the roles of minerals in plant metabolism are dealt with in Chapter 13. There are, however, many other physiological aspects of plant growth. What for example causes roots to grow down into the soil and shoots to grow up into the air? What controls the orderly sequence of cell division and differentiation at the apices of roots and shoots? Why do deciduous trees shed their leaves in autumn, why do buds spurt into growth in spring and summer and why is the vascular cambium seasonal in its activity?

Climate plays some part in controlling plant growth. The metabolism of green plants occurs efficiently only if such factors as temperature, light intensity and water availability are at an optimum. Nevertheless many species of flowering plants stop growing even when environmental factors are ideal. This suggests that there is an inbuilt growth control mechanism which causes plants to enter a period of dormancy at certain times of the year. Thus **external** and **internal factors** must be examined if we are to find a physiological explanation for the morphological and anatomical changes described earlier.

21.2.1 External factors

1. LIGHT

When the **compensation point** (Chapter 12) has been passed a green plant begins to increase in dry mass as it stores the synthesised energy-rich substances, mainly carbohydrates like starch. Some of the energy released from the food reserves as high energy phosphate bonds of ATP is used in growth for building up the many complex constituents of the protoplasm and walls of new cells. The **intensity of light** largely determines the time required to reach the compensation point each day. If a plant cannot build up sufficient food reserves in a day to provide for energy-requiring processes such as growth, it cannot remain active for long and must either die or undergo a period of dormancy.

In a temperate climate the length of time a plant is subjected to light each day varies from season to season. Flowering plants respond to changes in **daylength**. Growth takes place in seasons of favourable climate and stops when winter approaches. This response is called **vegetative photoperiodism**. It accounts for leaf fall and bud dormancy in deciduous trees and the production of storage organs, such as bulbs, tubers, corms and taproots by herbaceous perennials.

The pattern of growth is affected by the **wavelength of light** to which green plants are exposed. Kept in total darkness shoot systems soon begin to show signs of **etiolation**. The internodes become abnormally long and the leaves remain embryonic and do not develop chlorophyll (Fig. 21.15). Red light alone can prevent etiolation. Red light converts the inactive form of phytochrome P_r to the active form P_{ir} (Chapter 20). P_{ir} promotes leaf expansion and inhibits internode elongation and the normal growth pattern of the shoot is established.

Both vegetative photoperiodism and prevention of etiolation are good examples of the ways in which light controls the balance of growth promotors and inhibitors in plants. The **direction** from which light is received by shoots and roots also affects their direction of growth by altering the distribution of growth promotors (section 21.2.2).

FIG. 21.15 *Etiolation in pea seedlings*

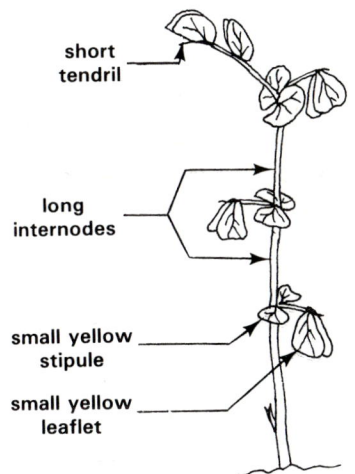

short tendril

long internodes

small yellow stipule

small yellow leaflet

(a) *etiolated seedling grown without light*

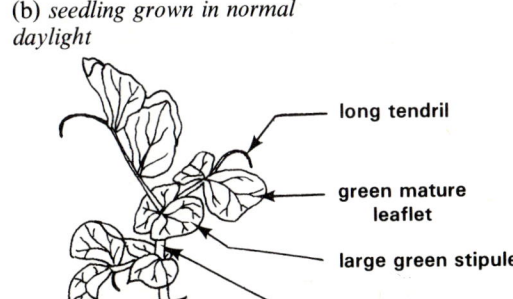

(b) *seedling grown in normal daylight*

long tendril

green mature leaflet

large green stipule

short internode

Both seedlings are of the same variety and of the same age

2. TEMPERATURE

Because growth depends on biochemical reactions catalysed by enzymes, it is affected by **temperature**. In the tropics the average yearly temperature is high enough to allow plants to grow for most of the time. In temperate zones growth is seasonal and often stops altogether in the winter because of low temperatures. In winter annual plants survive as **seeds**, herbaceous perennials as **bulbs**, **corms**, **tubers** and **rhizomes** or by the production of some other kind of fleshy storage organ. Deciduous shrubs and trees develop **dormant winter buds**.

Exposure to a period of **low temperature** is essential to break the dormancy of perennating organs and winter buds (Fig. 21.16). As with seeds, the effect of chilling is thought to reduce the amount of growth inhibitors. This ensures that growth normally occurs when the more suitable weather conditions of spring and summer are about to follow.

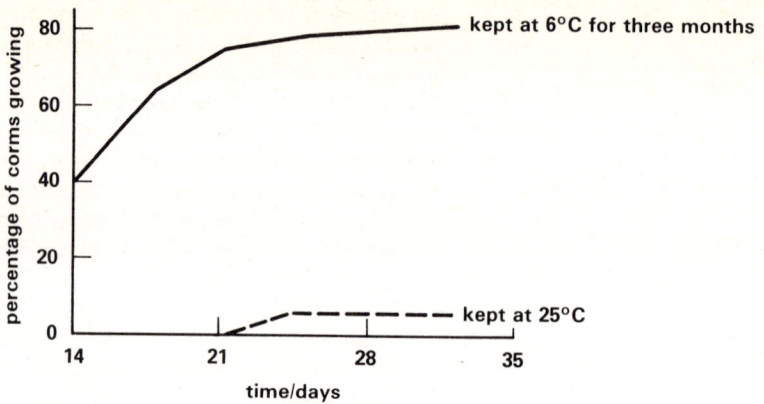

3. GRAVITY
Gravity affects the distribution of growth promotors at root and shoot apices thus influencing the direction in which they grow (section 21.2.2).

4. OXYGEN
Higher plants are aerobic organisms and can produce sufficient free energy for growth only if **oxygen** is available (Chapter 5). Root growth of most plants is inhibited in waterlogged soils where oxygen is scarce. Flowering plants which live totally or partly immersed in water are called hydrophytes. They often have special **aerating tissue** which stores oxygen made in photosynthesis. The storage of oxygen helps hydrophytes to overcome the relative scarcity of dissolved oxygen in water.

5. WATER AVAILABILITY
Elongation of cells during the primary growth of roots and shoots is mainly due to vacuolation. **Water** is essential for **vacuolation** and the subsequent maintenance of **turgidity** of young growing organs. The direction of root growth too is affected by water availability (section 21.2.2).

The synthesis of new cellular components, an essential feature of growth, takes place in water like all enzyme-catalysed reactions (Chapter 4). It is in water that the raw materials for growth are **absorbed** and **circulated** inside plants.

6. CARBON DIOXIDE
Carbon dioxide is essential for photosynthesis (Chapters 5 and 12). If growth is to occur beyond the seedling stage carbon dioxide must be available in sufficient quantities.

7. MINERAL ELEMENTS
Minerals are required for many facets of plant metabolism on which growth depends (Chapter 13). When higher plants are exposed to mineral-rich effluents, growth is not always stimulated even though the minerals are essential for plant growth. Growth is promoted when air containing moderate concentrations of sulphur dioxide comes into contact with flowering plants. However, a high concentration of sulphur dioxide is toxic to higher plants. Copper acts as a trace element in plant metabolism yet land plants do not readily colonise spoil tips rich in copper (Chapter 13).

8. BIOTIC FACTORS

In natural conditions a state of equilibrium exists between the trophic levels of every ecosystem. The balance is necessary for self-perpetuation of ecosystems. There are many well-known examples of the consequences of imbalance in ecosystems. Overgrazing in East Africa has converted productive natural grassland into desert. Removal of forest trees has led to serious soil erosion in China. Significantly, humans have usually been at the heart of these problems. People are the most potent **biotic factor** affecting the growth of plants. Enormous tracts of natural vegetation have been destroyed by man and replaced by farmland and forestry. The growth of plants on a substantial area of the earth's surface is therefore controlled by man.

21.2.2 Internal factors

Plants contain **growth promotors** and **inhibitors**. Because an understanding of the ways in which they work could enable man to control plant growth more efficiently, much effort has been put into investigations concerning these substances which are generally called **plant hormones**. The more important plant hormones so far identified are **auxins**, **gibberellins**, **cytokinins** and **abscissic acid**.

1. AUXINS

Indole-3-acetic acid (IAA) is the main **auxin** made in plants (Fig. 21.17). Produced at the apices of shoots from the amino acid tryptophan, IAA is transported backwards to the zones of elongation in roots and shoots, where it regulates **cell extension** and the **differentiation** of vascular tissues. Auxin transport is uni-directional or **polar** (Fig. 21.18). In coleoptiles IAA is carried in parenchyma tissue but in mature roots and shoots the main transporting tissue is probably the phloem.

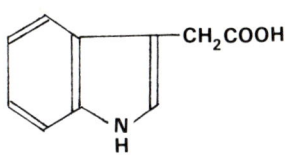

FIG. 21.17 *Indole-3-acetic acid*

FIG. 21.18 *Polar transport of IAA*

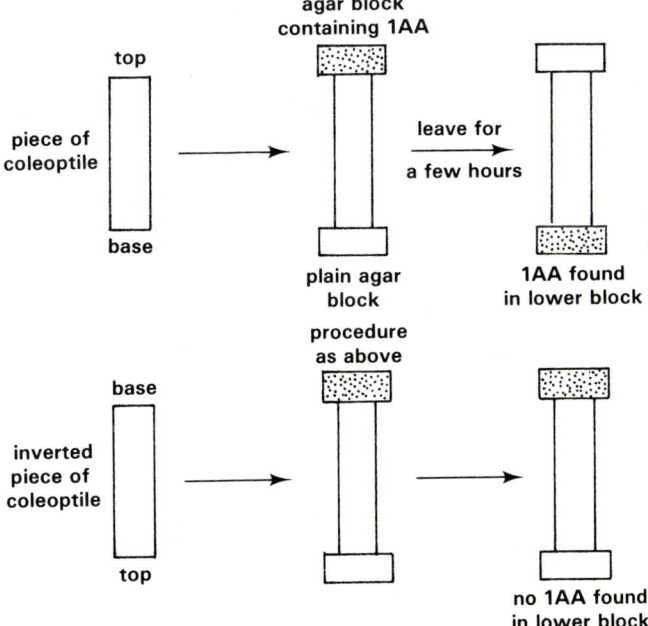

The IAA made by apical buds usually inhibits lateral buds from growing into side shoots, a phenomenon called **apical dominance** (Fig. 21.19). **Leaf expansion** is IAA-controlled (Fig. 21.20) as is the **growth of fruits and seeds** (Chapter 19). The **positions of procambial strands** at shoot apices too are

425

determined by IAA made in leaf primordia (section 21.1.1). IAA released from the expanding buds of woody plants during spring **triggers the vascular cambium** into activity. IAA also **stimulates translocation** of organic substances in the phloem (Chapter 12). It is therefore apparent that the general shape of a plant, its orderly internal growth and development and its rate of growth are very much affected by auxins.

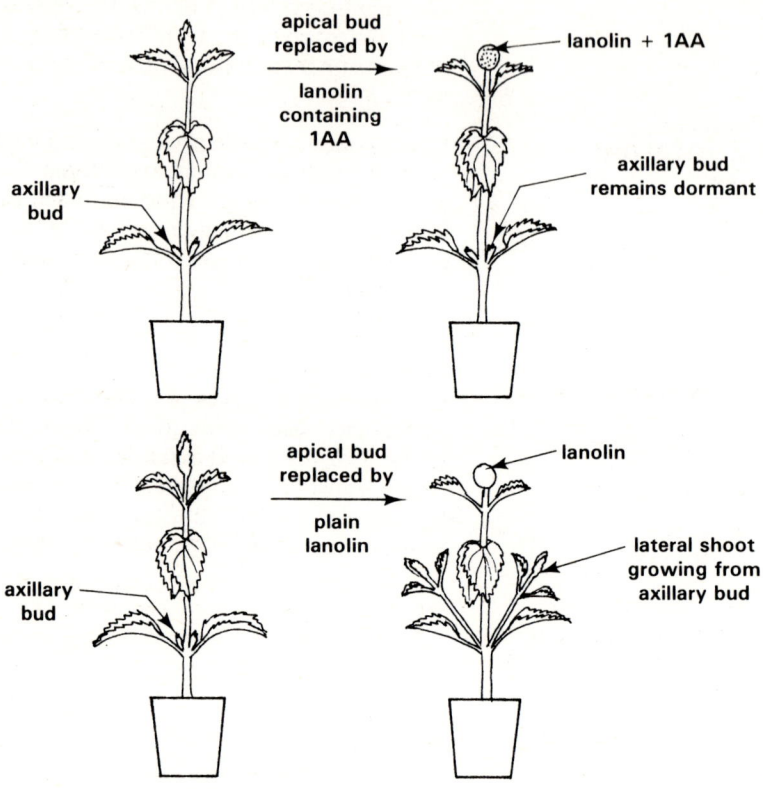

FIG. 21.19 *Demonstration of the role of IAA in apical dominance*

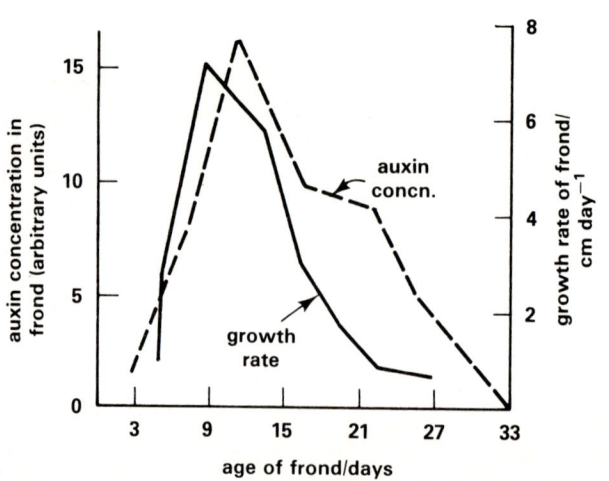

FIG. 21.20 *Correlation between auxin concentration and the growth of* Osmunda *fronds* (after Steeves and Briggs, 1960)

FIG. 21.21 *Some important landmarks in the discovery of auxins* (after Wareing and Philips, 1970)

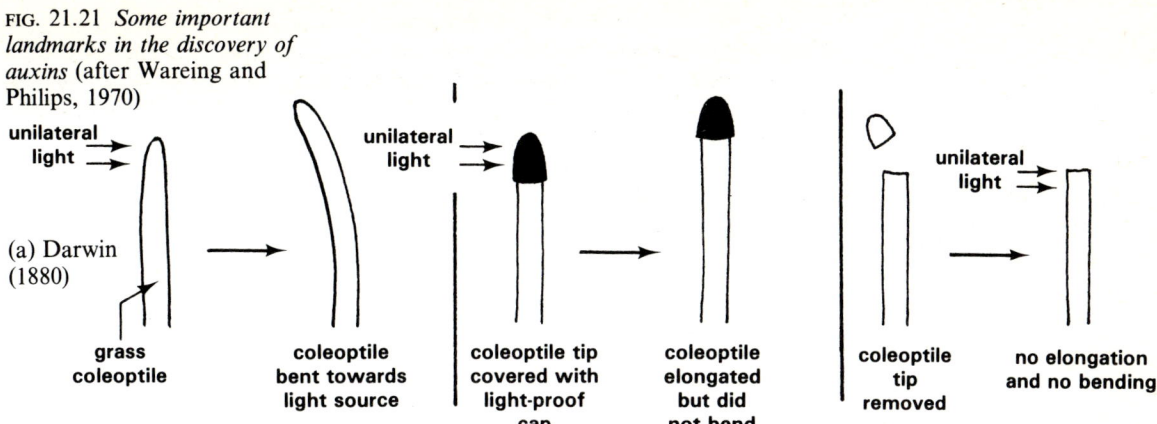

(a) Darwin (1880)

grass coleoptile → coleoptile bent towards light source

coleoptile tip covered with light-proof cap → coleoptile elongated but did not bend

coleoptile tip removed → no elongation and no bending

Darwin's results show that it is the tip of the coleoptile which detects the stimulus of light.

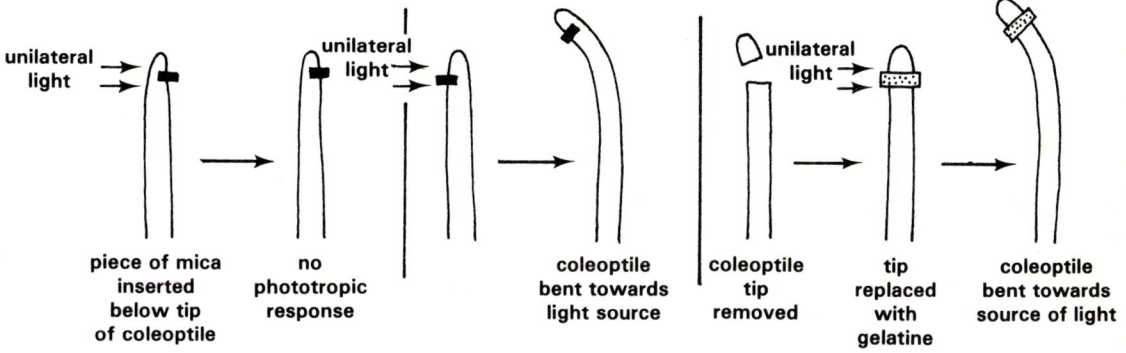

piece of mica inserted below tip of coleoptile → no phototropic response

coleoptile bent towards light source

coleoptile tip removed → tip replaced with gelatine → coleoptile bent towards source of light

(b) Boysen-Jensen (1913)

These results suggest that a stimulus for growth coming from the tip cannot pass through a solid barrier such as mica but can pass through an aqueous medium.

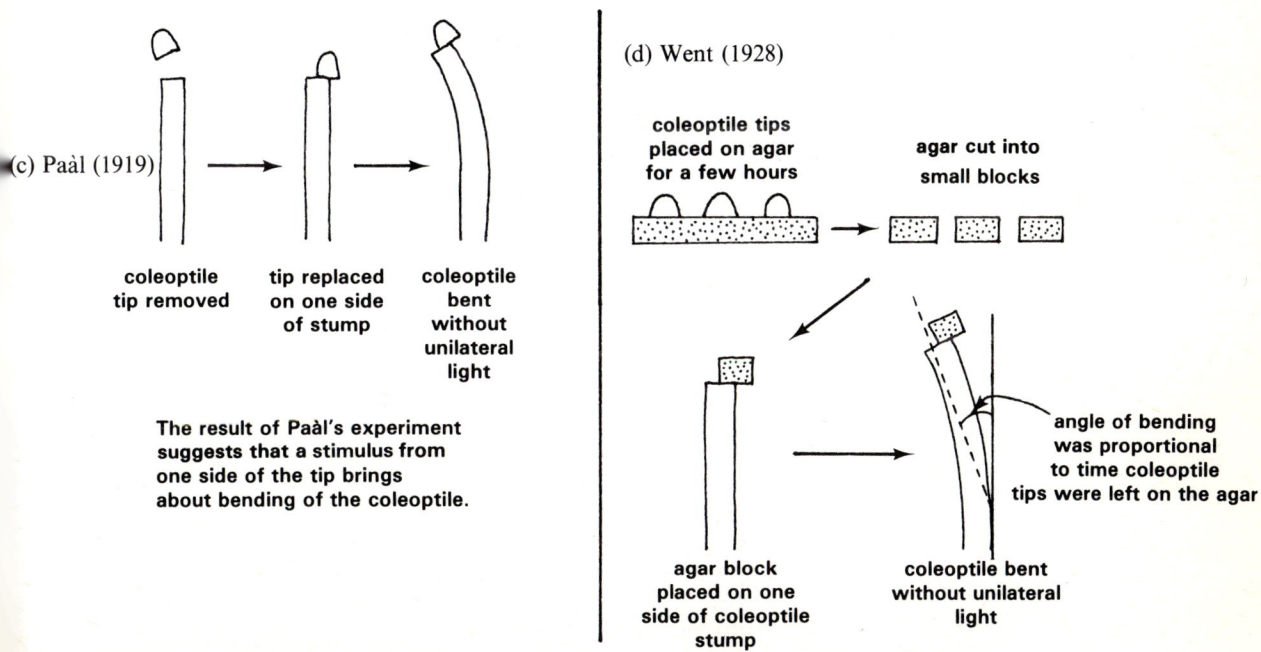

(c) Paàl (1919)

coleoptile tip removed → tip replaced on one side of stump → coleoptile bent without unilateral light

The result of Paàl's experiment suggests that a stimulus from one side of the tip brings about bending of the coleoptile.

(d) Went (1928)

coleoptile tips placed on agar for a few hours → agar cut into small blocks

agar block placed on one side of coleoptile stump → coleoptile bent without unilateral light

angle of bending was proportional to time coleoptile tips were left on the agar

Went's observations indicate that a growth-promoting substance had passed into the agar blocks.

427

The discovery of auxins resulted from investigations into the tendency of shoot systems to grow towards a source of light (Fig. 21.21). The behaviour is called **positive phototropism**. Although there is still much to be understood of the underlying mechanism, auxins almost certainly play a key role. The effect of unilateral light on a shoot apex is to cause more auxin to be present on the shaded side. Thus more IAA is transported to the elongating zone on the shaded side of the shoot. The result is that the shaded side grows more rapidly and the shoot bends towards the source of light (Fig. 21.22).

FIG. 21.22 (a) *Effect of light on distribution of IAA at the apex of an oat coleoptile*

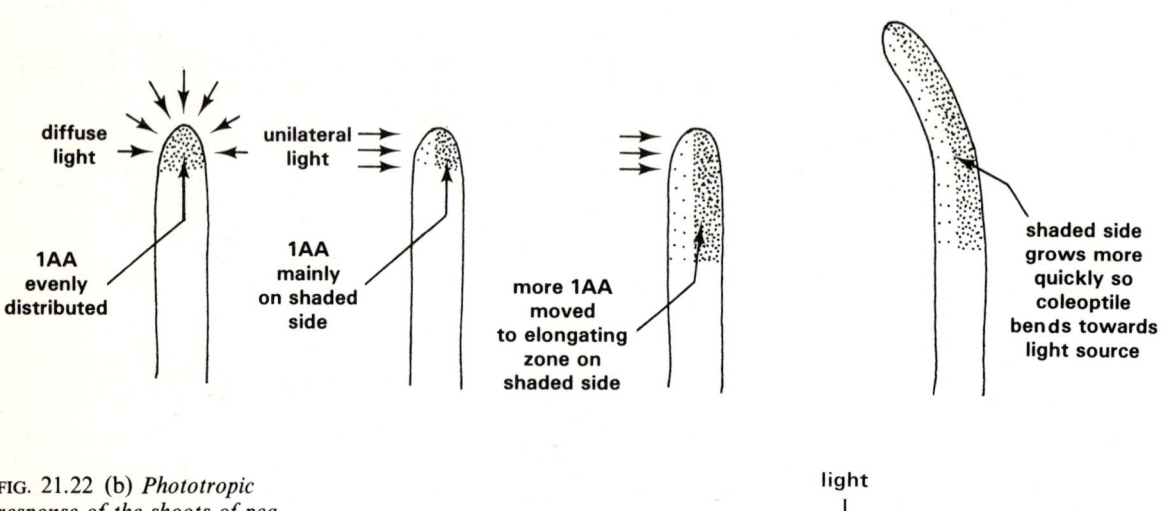

diffuse light

unilateral light

1AA evenly distributed

1AA mainly on shaded side

more 1AA moved to elongating zone on shaded side

shaded side grows more quickly so coleoptile bends towards light source

FIG. 21.22 (b) *Phototropic response of the shoots of pea seedlings*

light

light

The seedlings on this side were subjected to light coming from the left

The seedlings on this side received light from above

Roots are usually **negatively phototropic** and grow away from a light source. The effect of unilateral light on auxin distribution at the root apex is similar to that at a shoot apex. What then causes the shaded side of a root treated in this way to grow less quickly than the side exposed to light? Experiments have shown that whereas relatively high concentrations of auxin stimulate shoot growth, roots grow quickest when given relatively low concentrations of IAA (Fig. 21.23). Thus the small amount of IAA transported to the elongating zone on the illuminated side of a root causes more rapid growth than on the shaded side so the root gradually bends away from the source of light.

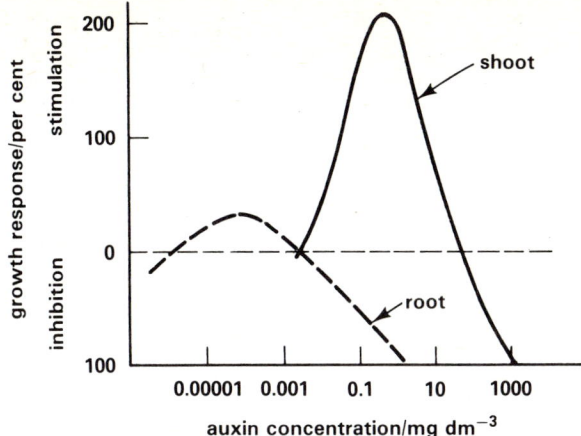

FIG. 21.23 *Effect of IAA on root and shoot growth* (after Audus, 1959)

Having explained the phototropic response in terms of changes in auxin distribution it is much more difficult to determine why uni-directional light has such an effect. Blue light of wavelength 440–460 nm is more effective than any other in stimulating a phototropic response (Fig. 21.24). The **action spectrum** for phototropism also has a minor peak at 360 nm and resembles combined absorption spectra for carotene and riboflavin. If, or how, these substances could affect auxin distribution is not yet known. Little is known too of the way in which IAA affects cell growth. Some workers think that auxins soften the cell wall substances causing rapid cell enlargement by vacuolation as water is absorbed. Auxins may also stimulate cell growth by helping the unloading of metabolites from phloem sieve-tubes.

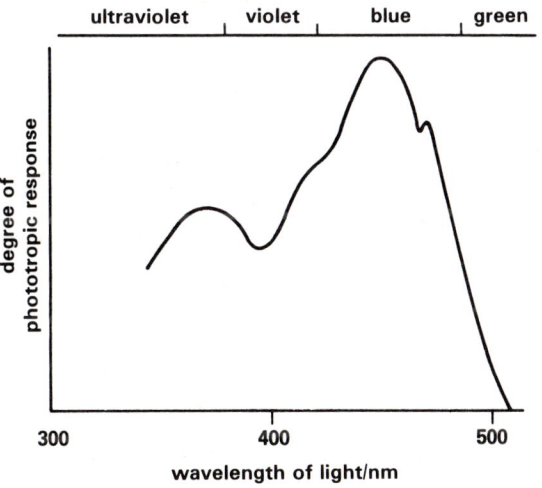

FIG. 21.24 *Action spectrum for phototropism in oat coleoptiles*

Roots and shoots also react to a unilateral stimulus of gravity. Shoots are usually **negatively geotropic**, growing away from a gravitational pull, while roots show **positive geotropism**. Changes in the normal pattern of auxin distribution are again thought to account for the responses (Fig. 21.25). Why gravity should bring about such changes is still not at all clear. Roots are also **positively hydrotropic**, growing towards a source of water, but again the reasons for this are still something of a mystery.

FIG. 21.25 *Geotropism in pea seedling*

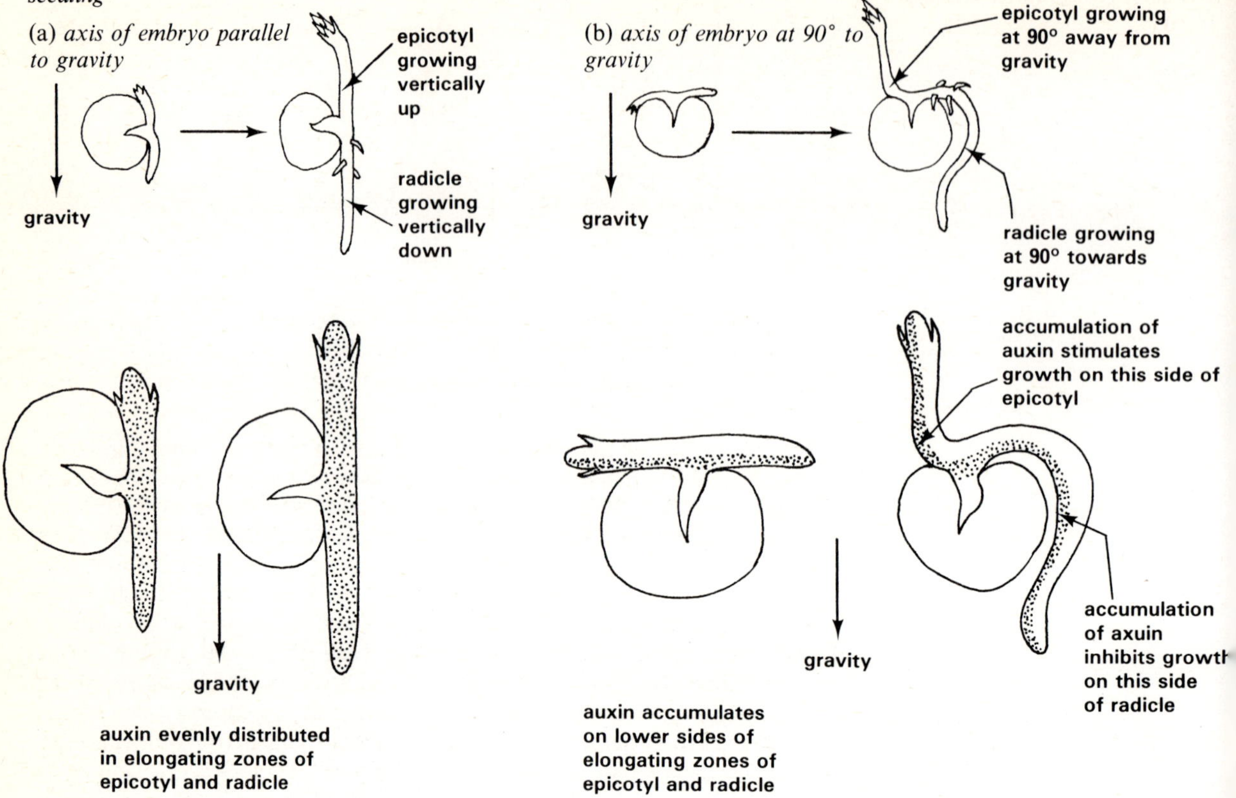

(a) *axis of embryo parallel to gravity*

epicotyl growing vertically up

radicle growing vertically down

gravity

gravity

auxin evenly distributed in elongating zones of epicotyl and radicle

(b) *axis of embryo at 90° to gravity*

epicotyl growing at 90° away from gravity

radicle growing at 90° towards gravity

accumulation of auxin stimulates growth on this side of epicotyl

gravity

auxin accumulates on lower sides of elongating zones of epicotyl and radicle

accumulation of auxin inhibits growth on this side of radicle

gravity

FIG. 21.26 *Gibberellic acid*

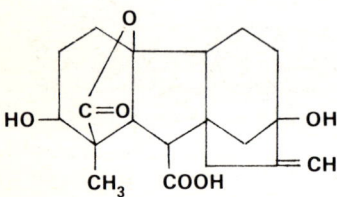

2. GIBBERELLINS

Gibberellins take their name from the fungal parasite *Gibberella fujikuroi* which causes a disease of rice plants known as 'foolish seedling'. Seedlings which are attacked by the fungus develop very long internodes so that the shoot systems are much taller than those of healthy plants. By the late 1920s Japanese workers showed that a growth-promoting substance called gibberellin was present in a liquid medium in which the fungus had been grown in pure culture. However, the chemical nature of gibberellin was unknown until the 1950s when British and American biochemists showed it to be **gibberellic acid** (Fig. 21.26).

Since then a number of closely related substances having similar effects have been isolated from higher plants. They are synthesised at the apices of roots and shoots but little is known of the way in which they are transported. The best known effect of gibberellic acid (GA) is to bring about **internode elongation**. GA also stimulates **enzyme production** during seed germination (Chapter 20) and plays a role in **cell division** and **tissue differentiation**. When applied to woody stems GA promotes division of the vascular cambium, but the derivatives fail to differentiate into mature xylem elements unless IAA is added. However GA on its own brings about differentiation of mature phloem elements. This suggests that the order in which primary xylem and phloem appear in root and shoot may be governed by the ratio of IAA : GA passing back from the apical meristems.

430

FIG. 21.27 *Zeatin*

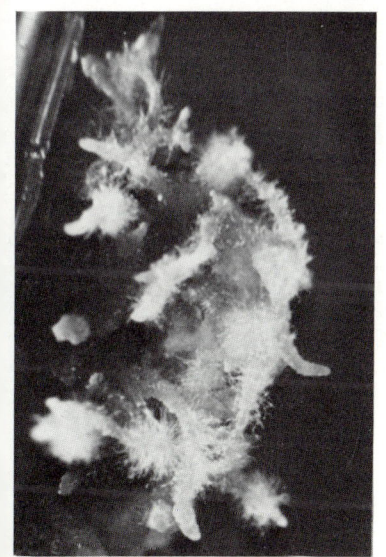

FIG. 21.28 (a) *Root formation from callus tissue of cowpea, ×4*

3. CYTOKININS

Interest in the **cytokinins** began in the 1950s when **kinetin** was found to affect the growth of plant tissue in sterile culture media. Kinetin is DNA which has been autoclaved in an acid solution. Kinetin closely resembles adenine, one of the nitrogenous bases of DNA and RNA, but it has not yet been isolated from higher plants. **Zeatin** is one of several similar compounds having comparable effects which has been extracted from a number of plants in recent years (Fig. 21.27).

Cytokinins are frequently found in xylem sap and are thought to be synthesised in the roots from where they are carried to other parts of the plant in the transpiration stream. Production of cytokinins is at its highest when a plant is growing rapidly and falls off as ageing begins.

Much of what is known about cytokinins comes from tissue culture experiments. Small samples of living cells called **explants** are taken from plant organs and aseptically transferred to sterile growth media containing sugar, minerals, vitamins and hormones. Growth of the transplanted tissue follows, but the pattern of growth depends on the composition of the medium, in particular the plant hormones present. With no cytokinin but IAA present the tissue grows into a mass of similar large multinucleate cells. When a small amount of cytokinin is added, a **callus** of uninucleate cells appears indicating that cytokinins have an effect on cell division. By changing the ratio of IAA to cytokinin the first signs of **organ differentiation** are seen. Roots appear on the explant if the ratio of auxin to cytokinin is high while shoots develop if the ratio is low. With no hormones, growth of the explant does not take place (Fig. 21.28).

FIG. 21.28 (b) *Shoot formation from callus tissue of tobacco, ×2·5*

If these findings reflect what happens in intact plants then cytokinins evidently play a significant role in various aspects of plant growth. Particularly important are the ways in which cytokinins and other plant hormones interact in controlling plant growth.

4. ABSCISSIC ACID

Plant cells contain many substances which inhibit some process or another. One of the best known compounds which inhibits growth is **abscissic acid** (Fig. 21.29). Abscissic acid (ABA) was discovered in 1964 and has also

FIG. 21.29 *Abscissic acid*

been called abscissin and dormin. The quantity of ABA in the leaves of deciduous trees increases in the shorter days of autumn. The accumulation of ABA slows down the growth rate of shoots and stimulates the **formation of dormant winter buds**. Soon leaf fall, **abscission**, occurs (Fig. 21.30). In this way the plant is prepared for winter conditions. The quantity of ABA in the shoot system falls after a period of cold weather. Dormancy of winter buds is then broken and growth follows in spring and summer. Once again it can be seen that hormone balance is controlled by climate and growth generally occurs only when a spell of good weather is about to follow.

FIG. 21.30 (a) *Longitudinal section through an abscission zone of sycamore,* × 10

FIG. 21.30 (b) *Winter twig of sycamore. All the buds are dormant*

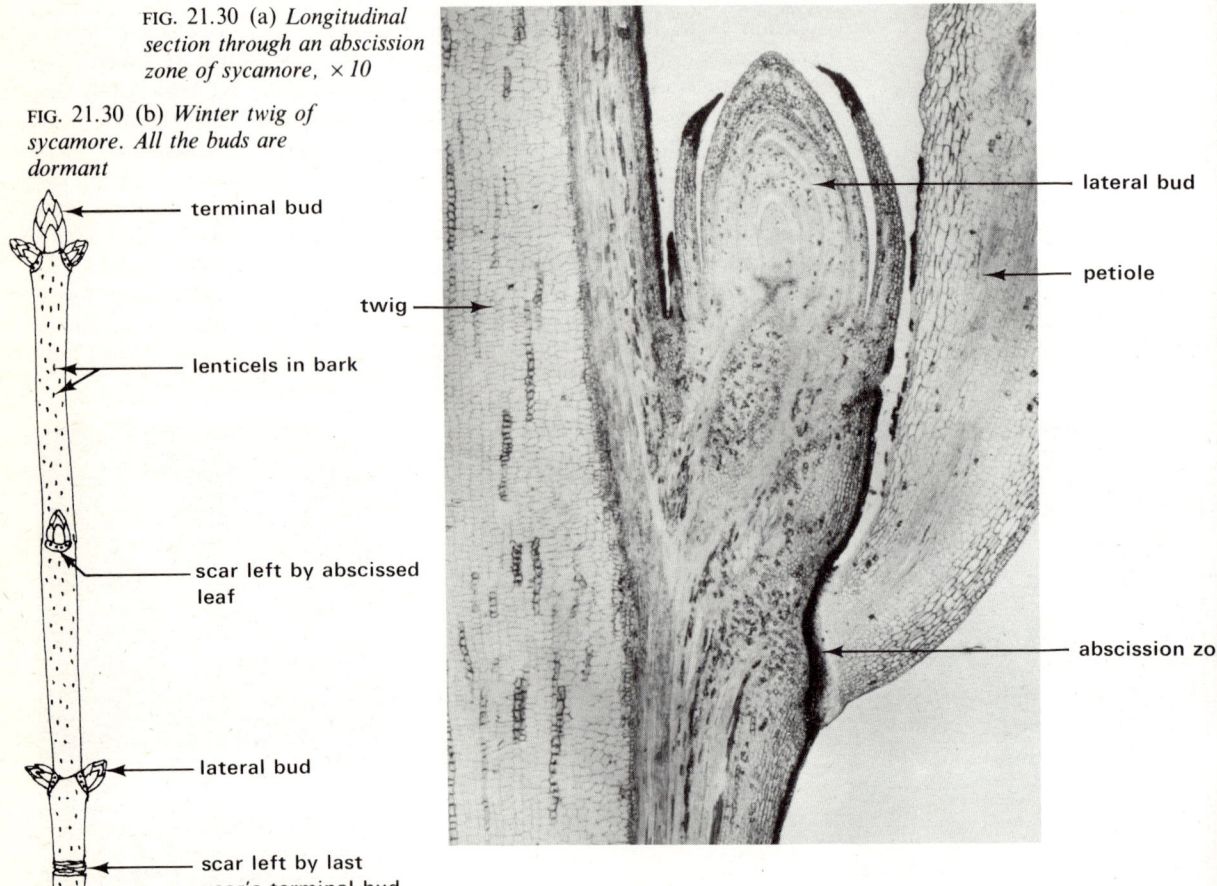

terminal bud

lenticels in bark

scar left by abscissed leaf

lateral bud

scar left by last year's terminal bud

twig

lateral bud

petiole

abscission zo

21.2.3 Practical uses of plant hormones

With the explosion of information obtained about growth-regulating substances over the past twenty years or so it is not surprising that some of the knowledge has been put to practical use. There are many synthetic compounds with auxin-like effects now in common use in agriculture and horticulture. They include various **phenoxyacetic acids** such as 2:4D (Fig. 21.31). 2:4D stimulates the formation of adventitious roots at the bases of cut shoots (Fig. 21.32) and is an invaluable aid to horticulturalists in accelerating the **striking of cuttings**. Many species of ornamental plants such as carnations and chrysanthemums are propagated by means of cuttings. Synthetic auxins also encourage the formation of wound tissue which knits together parts of plants after **grafting**.

FIG. 21.31 *2,4-dichlorophenoxy-acetic acid*

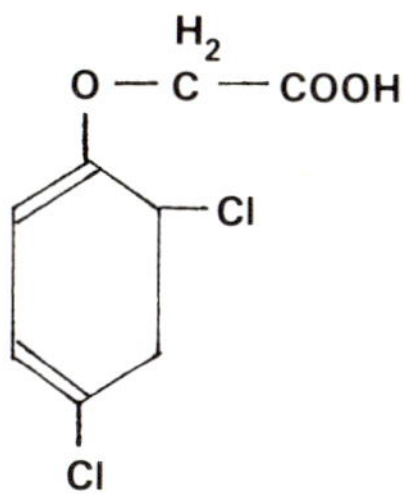

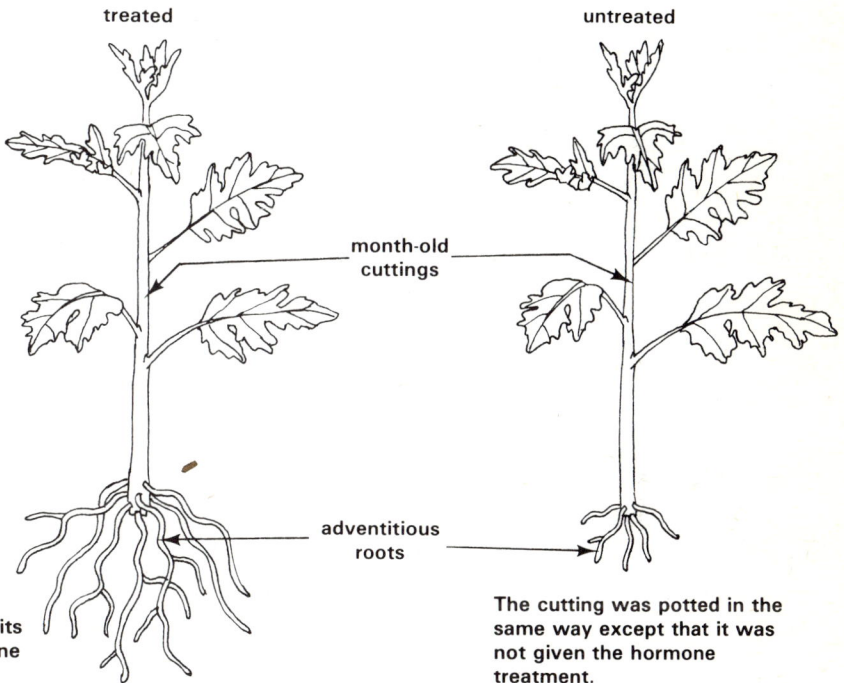

When sprayed on to fruit trees after flowering the phenoxyacetic acids prevent young fruit from dropping off. Large increases of apple and pear crops have been achieved in this way (Chapter 19).

Perhaps the most widespread use of synthetic auxins has been as **weed killers** applied to fields of cereal crops, pasture and lawns. The roots and shoots of the same plant respond differently to a particular concentration of auxins (section 21.2.2). It has also been found that different groups of plants vary in their reaction to applications of similar concentrations of synthetic auxin. Dicotyledonous plants such as dandelions and thistles are more sensitive to 2:4D than are monocotyledons such as grasses, wheat and maize. When sprayed on to an area of mixed species, for example a weedy lawn or a field of wheat infested with thistles, the dicotyledons are stimulated into abnormal growth. The internodes become grossly elongated, sieve-tubes become blocked and cells with unbalanced chromosome numbers arise. Distortion of the root system also occurs with the net result that the growth pattern is thrown completely out of control and death quickly follows. Meanwhile the monocotyledons are unaffected and are thus freed from unwanted competitors.

FIG. 21.32 *Effect of synthetic auxin on the striking of chrysanthemum cuttings*

treated   untreated

month-old
cuttings

adventitious
roots

Before potting the cutting its base was dipped in hormone rooting powder.

The cutting was potted in the same way except that it was not given the hormone treatment.

Despite the many practical benefits which result from their uses, synthetic auxins can cause serious ecological problems if used indiscriminately. Some are toxic to insects and have been known to kill large numbers of useful insects such as bees with considerable economic loss. Many weeds are the food plants of pollinating insects which starve when weeds are destroyed. Finally some weeds have evolved auxin-resistant varieties which are virtually impossible to eradicate by treatment with synthetic auxins.

Chapter 22

Mammalian reproduction

Whereas most flowering plants are hermaphrodite (Chapter 19) all mammals are **unisexual**. This means that mammalian gametes are produced in separate male and female animals and **cross-fertilisation** is inevitable. Consequently genetic variation among the offspring of mammals is ensured.

For sexual reproduction to occur successfully gametes have to be mature at the right time. Males and females have to be in the same locality and attracted to each other. Fertilisation must occur and the resulting embryo must be protected and provided with food for it to survive and develop properly.

Reproduction in mammals is a carefully regulated process although the controlling mechanisms vary in different species. An important feature of mammalian reproduction is that it occurs in such a way that the effects of chance and adverse environmental conditions are minimised. A successful outcome to sexual reproduction is therefore more likely in mammals than it is in many other groups of animals. Mammalian gametes are made in sex organs called **gonads**. The **testes** are the male gonads, the **ovaries** the female gonads. **Gametogenesis** is the process which gives rise to male and female **gametes**.

22.1 Testes

The production and maturation of **spermatozoa** (sperm) which are the male gametes, takes place in a pair of compact **testes**. The testes originate in the abdomen from the embryonic tissue which also gives rise to the urinary system. Indeed the urethra through which urine passes to the outside also provides the route for sperm to leave the male reproductive system (Fig. 22.1). As well as making sperm the testes produce male sex hormones.

FIG. 22.1 *Diagram of the male uro-genital system in man*

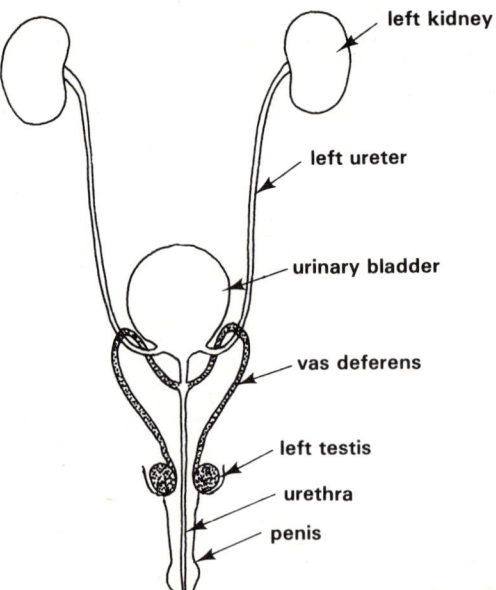

left kidney

left ureter

urinary bladder

vas deferens

left testis

urethra

penis

22.1.1 Structure of the testes

FIG. 22.2 *The human male reproductive organs (side view)*

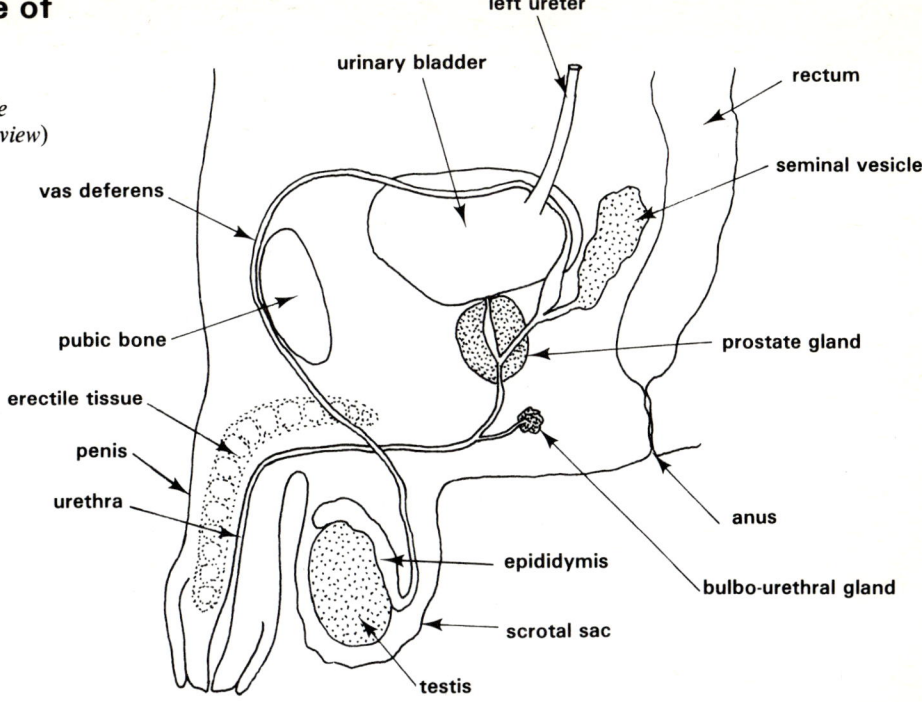

The testes of most mammals descend from the abdomen into two sacs of skin called **scrotal sacs**. Under the microscope a thin section of a testis is seen as thousands of very fine, highly coiled **seminiferous tubules**. It is in the tubules that sperm are produced. The seminiferous tubules are continuous with other tubules called the **vasa efferentia**, **epididymes** and **vasa deferentia**. The sperm travel through the system of tubules on their way to the **seminal vesicles** where they are stored before passing out of the body through the **urethra** (Fig. 22.2). Cutting the vasa deferentia prevents the passage of sperm to the outside. The operation which is called **vasectomy** is now commonly used to sterilise men.

FIG. 22.3 (a) *Vertical section through testis. The seminiferous tubules have only been shown in one lobule*

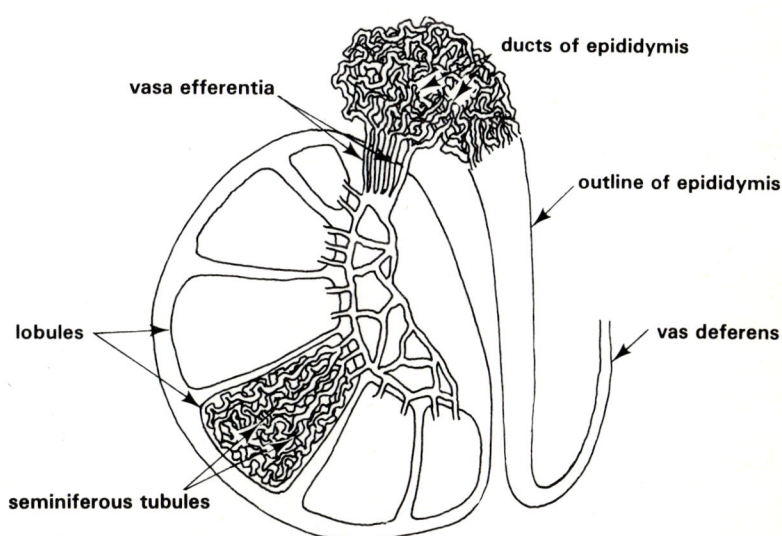

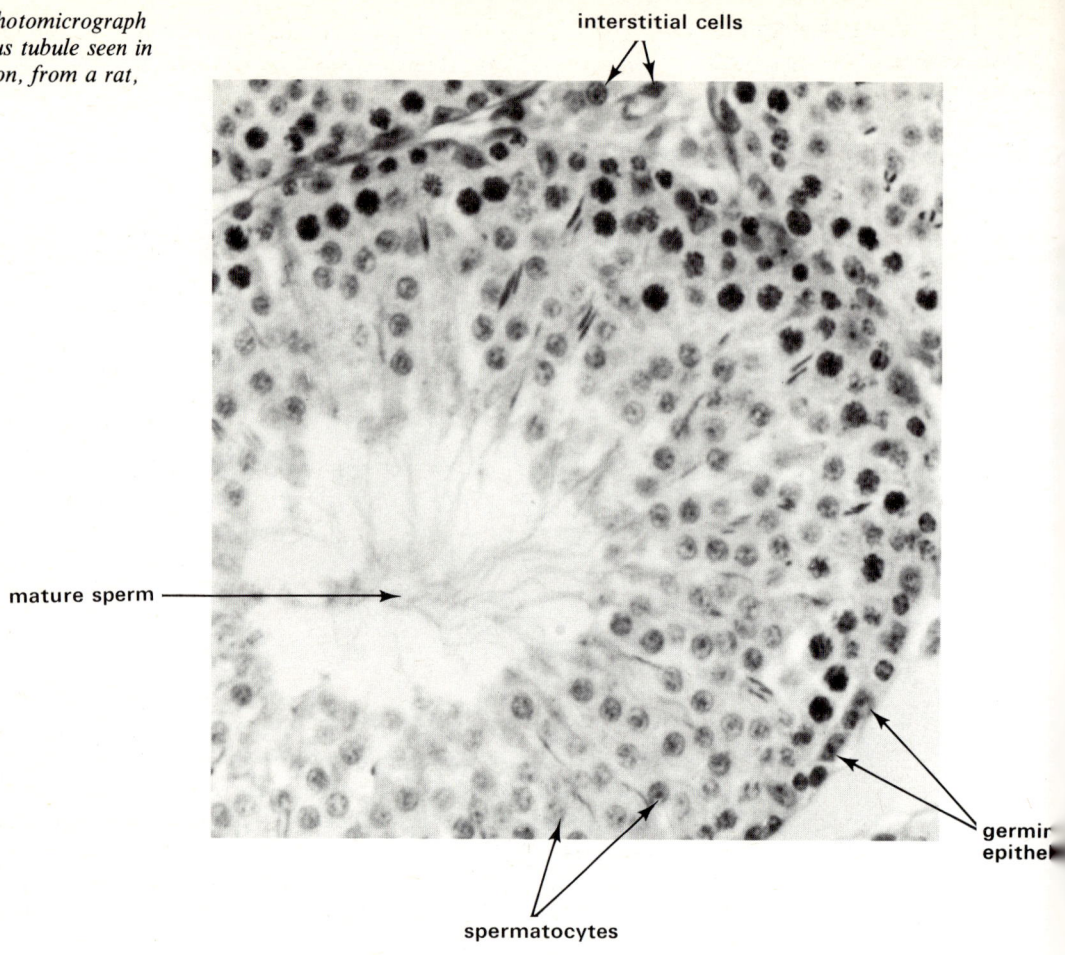

FIG. 22.3 (b) *Photomicrograph of a seminiferous tubule seen in transverse section, from a rat, ×400*

interstitial cells

mature sperm

spermatocytes

germinal epithelium

The walls of the seminiferous tubules consist of a layer of cells called the **germinal epithelium** from which sperm originate (Fig. 22.3). Sperm at different stages of development are found inside the seminiferous tubules. As they mature the sperm become attached to the relatively large **Sertoli cells**. The tubules are held together by connective tissue which contains **interstitial (Leydig) cells**, nerve fibres, blood and lymphatic capillaries.

22.1.2 Spermato-genesis

Spermatogenesis is the process whereby sperm are made. The production of sperm begins with divisions of the **primordial germ cells** which make up the germinal epithelium. The first divisions are mitotic (Chapter 7) and produce many genetically identical cells called **spermatogonia**. Next the spermatogonia increase in size and mass without dividing to become **primary spermatocytes**. The primary spermatocytes then divide by meiosis (Chapter 7). The first meiotic division produces **secondary spermatocytes**, two from each primary spermatocyte. The second meiotic division gives rise to **spermatids**, two from each secondary spermatocyte (Fig. 22.4). The final stage of spermatogenesis is differentiation of **spermatozoa** from the spermatids.

A mature spermatozoon has a **head** which contains a nucleus. Attached to the head is a long **tail** (flagellum) which the sperm uses to swim. Human sperm can swim at a rate of 1–4 mm per minute. At the base of the tail

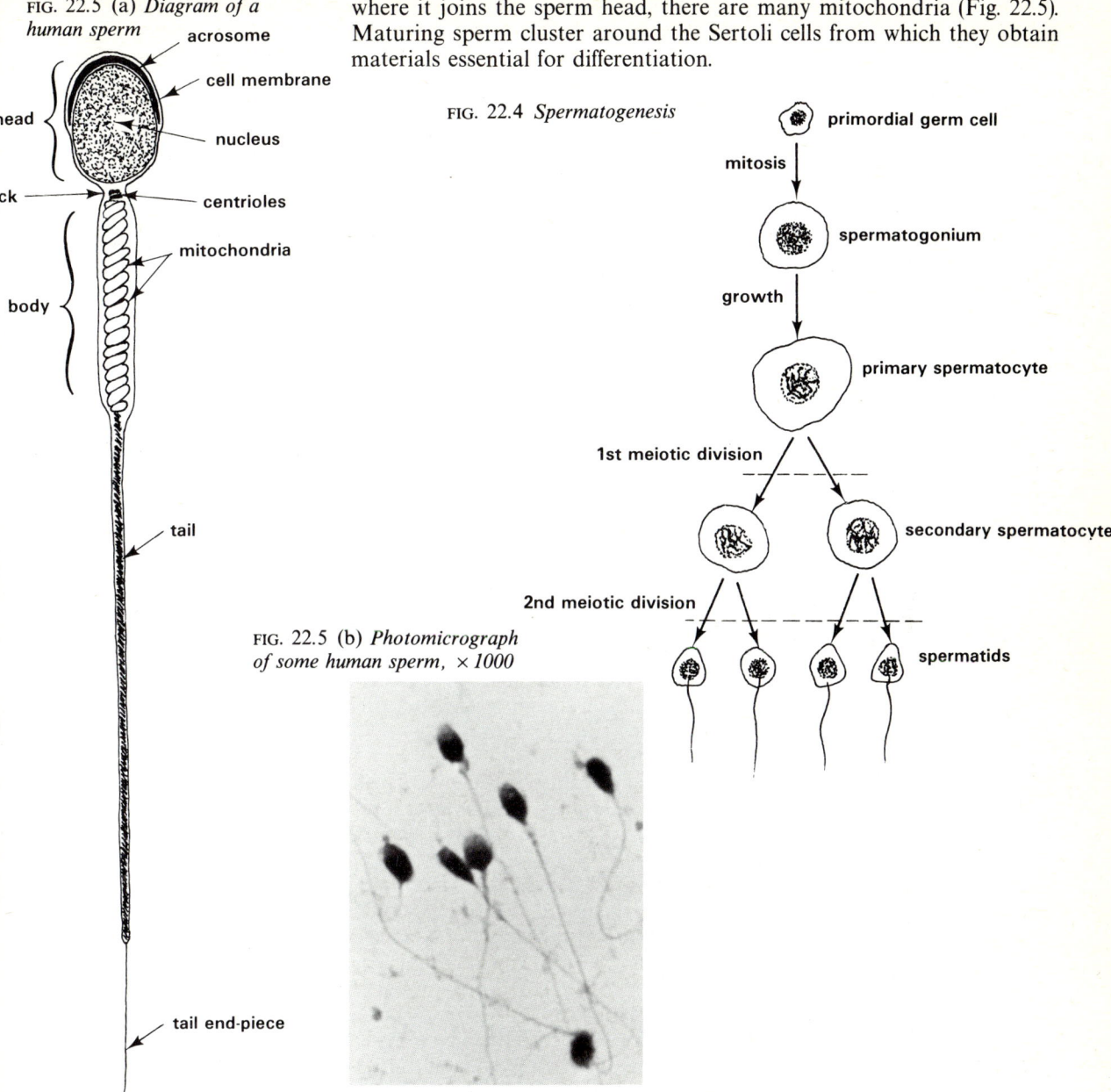

FIG. 22.5 (a) *Diagram of a human sperm*

acrosome

cell membrane

head

nucleus

ck — centrioles

mitochondria

body

tail

tail end-piece

FIG. 22.4 *Spermatogenesis*

primordial germ cell

mitosis

spermatogonium

growth

primary spermatocyte

1st meiotic division

secondary spermatocyte

2nd meiotic division

spermatids

FIG. 22.5 (b) *Photomicrograph of some human sperm,* × 1000

where it joins the sperm head, there are many mitochondria (Fig. 22.5). Maturing sperm cluster around the Sertoli cells from which they obtain materials essential for differentiation.

Spermatogenesis is inhibited by heat. Because the testes are located in the scrotal sacs where the temperature is lower than in the abdomen, production of sperm is normally encouraged. Before the development of efficient methods of contraception, it used to be common for men to take a very hot bath before sexual intercourse. However, it is questionable whether this practice had the desired contraceptive effect.

22.1.3 Male hormones

Spermatogenesis is stimulated by a hormone called **follitropin** (**FSH**) produced by the anterior pituitary (Chapter 18). FSH gains its name from the action it has in the female (section 22.2.3). In both male and female, FSH stimulates gametogenesis.

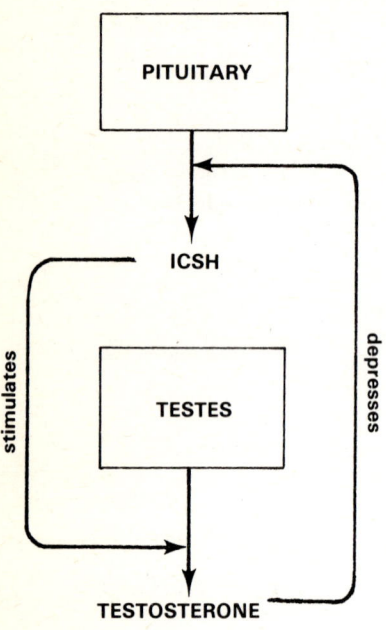

FIG. 22.6 *Mechanism controlling testosterone secretion*

Another gonadotropin hormone made in the anterior pituitary is **lutropin (LH)**. LH is also named because of the effect it has in the female (section 22.2.3). In the male, LH is alternatively called **interstitial cell stimulating hormone (ICSH)**. As the name suggests ICSH stimulates the interstitial cells which respond by secreting a male sex hormone called **testosterone**. ICSH and testosterone interact in a negative feedback control mechanism so that the output of testosterone is kept relatively constant (Fig. 22.6).

Male sex hormones are collectively called **androgens**. The functions of androgens are twofold. First, they regulate the development of the male **accessory sex organs**. They include the vasa efferentia, epididymes, vasa deferentia, penis and scrotal sacs. The second main function of androgens is to control the development and maintenance of the **secondary sexual characteristics** which in man include growth of facial and pubic hair, breaking of the voice and general muscular development. Androgens are also partly responsible for a number of behavioural characteristics such as aggressiveness.

Castration, removal of the testes, does not necessarily lead to loss of secondary sexual characteristics in males. How can this be so when maintenance of the secondary sexual characteristics is an important function of androgens made in the testes? Androgens are steroids (Figs. 22.7 and 18.20(b)). Following castration there is increased output of steroids similar to androgens from the adrenal cortex (Chapter 18). This explains why eunuchs outwardly appear as normal males.

The development of male secondary sexual characteristics and the production of sperm begin in man at a period of life called **adolescence**. The point at which sexual maturity is reached is called **puberty**. The testes also produce female sex hormones, but their function in males is not clear.

22.2 Ovaries

Female gametes are called **ova**. Their production and maturation take place in the **ovaries**. The ovaries also produce **female sex hormones**. In female mammals a pair of ovaries lie in the abdominal cavity (Fig. 22.8).

22.2.1 Structure of the ovaries and oogenesis

Spermatogenesis in the testes (section 22.1.2) is paralleled in the ovaries by **oogenesis**, the production of ova. Unlike the testes, ovaries do not contain tubules but consist mainly of connective tissue. Like the testes, however, ovaries have a **germinal epithelium**. The ovarian germinal epithelium is on the outside of the ovaries. **Primordial germ cells** in the germinal epithelium divide by mitosis to form many **oogonia**. The oogonia become surrounded by **follicle cells**, also derived by mitosis from the germinal epithelium. The follicle cells and enclosed oogonia are called **primary follicles**. They migrate to the centre of the ovary.

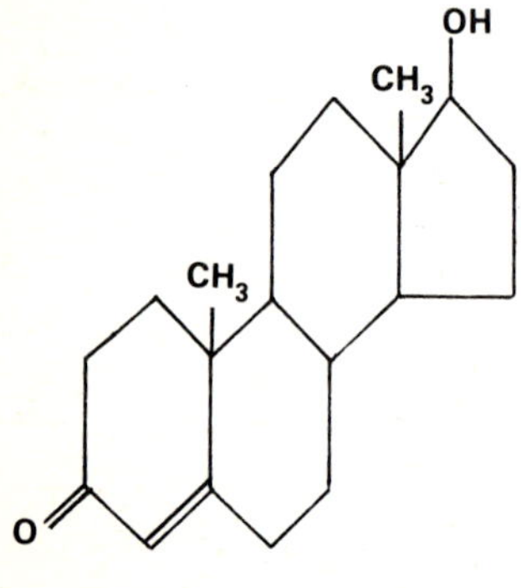

At birth there are up to 400 000 primary follicles in a human ovary where they remain dormant until **puberty**. Hormones from the pituitary gland then start the process of oogenesis. In women only about 400 primary follicles normally mature over a reproductive lifetime of about 30–40 years. The end of this period, when follicle development stops, is called the **menopause**. The remaining primary follicles degenerate into cyst-like **atretic follicles** and remain in the ovaries.

In women, ova usually develop one at a time. Each oogonium grows into a **primary oocyte** which by now is surrounded by several layers of follicle cells. Between the dividing follicle cells cavities appear, filled with **follicular liquid**. The follicle becomes surrounded by two layers derived from the

FIG. 22.7 *Structure of testosterone*

438

ovarian connective tissue. An outer fibrous **theca externa** encloses an inner vascular **theca interna** (Fig. 22.9). The mature follicle is called a **Graafian follicle** and can usually be seen bulging at the surface of the ovary (Fig. 22.10). During development the follicles of women grow from about 0·05 mm to about 12 mm in diameter.

FIG. 22.8 *The human female reproductive organs (ventral view)*

FIG. 22.9 *Oogenesis*

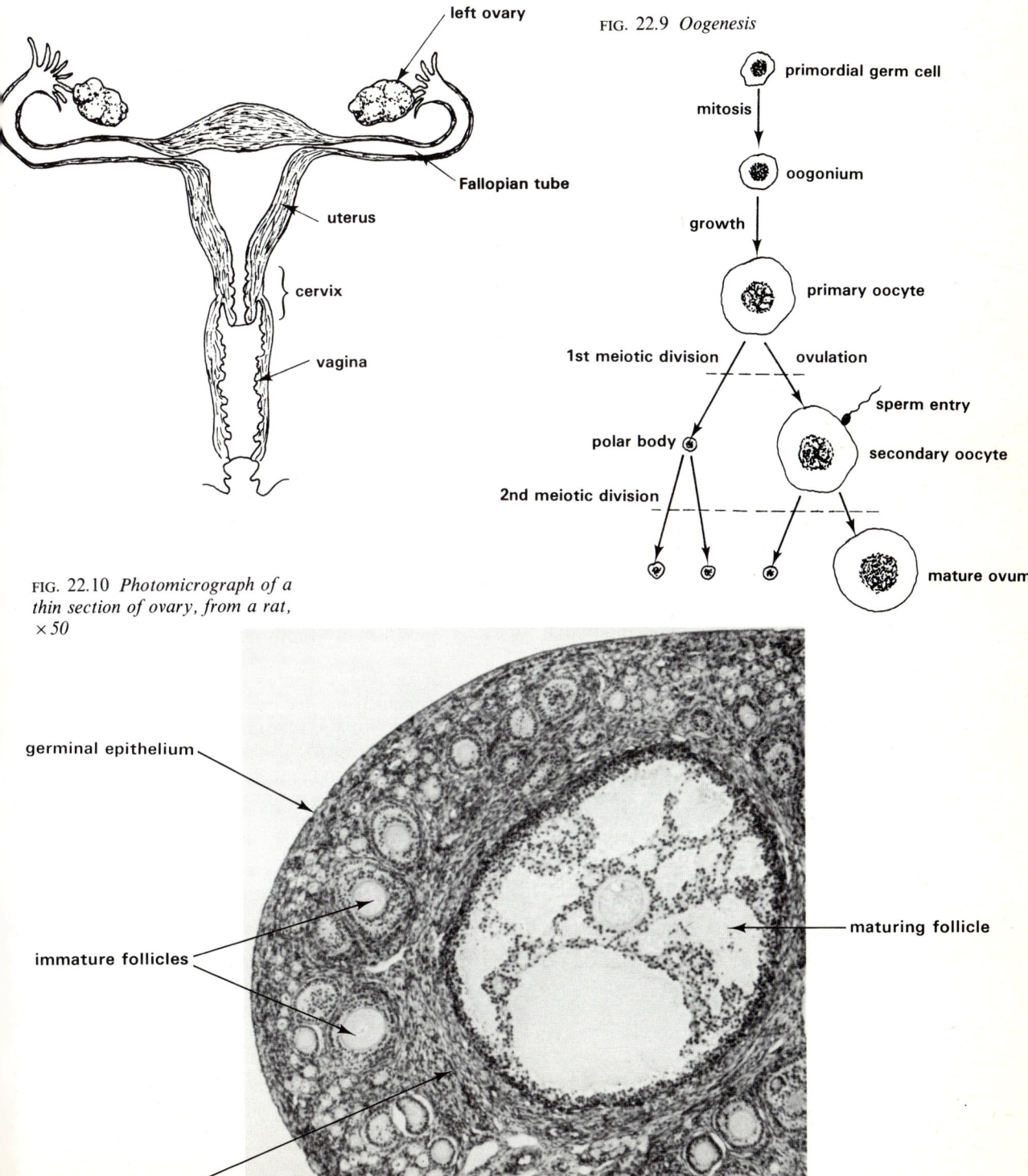

FIG. 22.10 *Photomicrograph of a thin section of ovary, from a rat, ×50*

The primary oocyte inside the follicle now divides by meiosis. The first meiotic division which occurs just before ovulation produces a **secondary oocyte** and attached to it, a small cell called a **polar body**. **Ovulation** is the release of the secondary oocyte from the ovary into the abdominal cavity (sections 22.2.2 and 22.3.2). The second meiotic division usually occurs following fertilisation, just after a sperm nucleus enters the secondary oocyte. The secondary oocyte then produces the **ovum** and another small polar body. The second meiotic division also gives rise to two more polar bodies from the polar body produced in the first meiotic division. The polar bodies have no known function.

During oogenesis only one functional ovum arises from each primary oocyte, whereas spermatogenesis produces four functional sperm from each primary spermatocyte (section 22.1.2).

22.2.2 Hormonal control of ovulation

As in the male, gametogenesis in the female mammal is influenced by hormones from the anterior pituitary, particularly **follitropin (FSH)**. Unlike spermatogenesis, however, oogenesis is not a continuous process. It is a characteristic of mammals that the embryos develop inside the mother's reproductive tract for a period prior to birth. This period is called the **gestation period**. Since there is limited space in the mother for embryo development, there is a limit to the number of embryos which can develop at a time. The number of embryos carried by a mother during the gestation period is usually determined by the number of ova released at a time from the ovaries. In women only one ovum normally leaves the ovaries at each ovulation.

For many mammals living in the wild, it is advantageous to mother and offspring if birth occurs when food is plentiful and the weather is mild. In many species this is ensured by the seasonal secretion of FSH. Secretion of FSH in **seasonal reproducers** is influenced by day-length. In some species such as ferrets, FSH output is stepped up in spring when the number of hours of light in a day increases. In sheep, production of FSH increases in autumn when the day-length shortens. Other mechanisms such as **delayed implantation** (section 22.3.2) can also determine the time of year when birth takes place.

Secretion of FSH in some species is triggered off by the nervous stimulation of copulation. The mechanism is called **induced ovulation** and occurs, for example, in rabbits. Induced ovulation ensures that ripe ova are provided immediately after sperm have been introduced into the female tract (section 22.3.1). Since fertilisation is then likely to occur it is not surprising that rabbits have gained a reputation as prolific breeders.

Many mammals produce ova at regular intervals by a mechanism in which FSH interacts with ovarian hormones to produce a rhythm of activity. The rhythm is called the **oestrus** or **menstrual cycle**. This internal method of controlling ovulation is largely independent of external influences.

22.2.3 The oestrus cycle

Since the oestrus cycle is a sequence of events which repeats itself regularly in a cyclical fashion, there is no fixed beginning or ending. For convenience of description we shall begin with the onset of follicle development. Follicle development is stimulated by FSH from the anterior pituitary. FSH also causes the ovary to secrete a female sex hormone, an **oestrogen**. The most prominent oestrogen is called **estradiol**. Estradiol has several effects in the female body, some of which are described later (section 22.2.4). Among these is a negative feedback effect on FSH (Fig. 22.11).

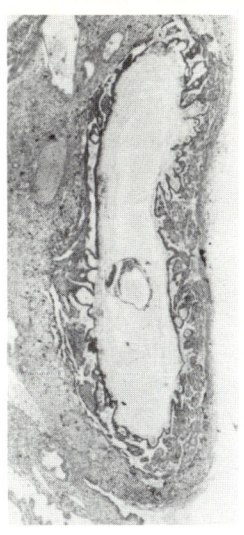

Sometimes the zygote divides into two separate cells from which two individuals develop. These individuals are called **monozygotic twins**. Monozygotic twins, coming from the same sperm and ovum, are almost identical. Small differences appear as the twins develop. **Dizygotic twins** result from two different zygotes produced from two ova fertilised at the same time. Consequently dizygotic twins are usually quite different from one another.

2. IMPLANTATION

The zygote undergoes several divisions, producing a ball of 16 or 32 cells by the time it reaches the uterus. It is in this form that the embryo is **implanted** in the endometrium (Fig. 22.15). Peptidase enzymes from **trophoblast cells** on the outside of the embryo digest part of the endometrium, making space for the embryo. The products of digestion are absorbed by the trophoblasts and are used as nutrients by the embryo. Meanwhile the trophoblasts along with surrounding endometrial cells divide to form the placenta and foetal membranes.

Implantation of the embryo is delayed after fertilisation in some mammals. In deer for example, ova fertilised in the autumn rut are implanted in late December. The delay ensures that the fawns are born in midsummer when the weather favours survival.

22.3.3 The placenta

Considerable growth and development of the embryo, now called the **foetus**, takes place inside the uterus. Growth and development require the provision of oxygen and nutrients and the removal of waste materials. Oxygen and nutrients are supplied to the foetus by the mother's blood which also serves to remove foetal wastes. However, the foetal blood and the mother's blood do not mix. The two circulations come near to each other in the **placenta** (Fig. 22.16). Materials are exchanged between the blood of the mother and foetus across the placenta almost entirely by diffusion. The arrangement of the blood vascular systems of the mother and foetus is described in Chapter 9.

The placenta is formed by growth of some of the **foetal membranes** and part of the uterine wall in which the foetus is implanted. At an early stage several foetal membranes develop. They are the **yolk sac**, **allantois**, **amnion** and **chorion** (Fig. 22.17). The yolk sac is an extension of the embryo's hind

FIG. 22.16 *Ultrasonic B-scan showing a foetus inside the uterus of a pregnant woman*

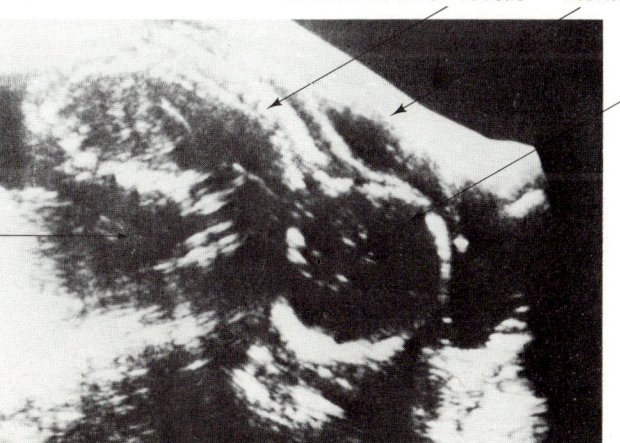

arched back of foetus   uterine wall

head of foetus

placenta

445

gut. In birds and reptiles yolk is the source of nutrients for the embryo. This is also the case in primitive mammals such as the duck-billed platypus which lay reptile-like eggs. In marsupial mammals also, the yolk sac provides nutrients for the embryo. In placental mammals, however, the yolk sac is relatively small. Nevertheless it is important in some mammals because antibodies pass through it into the embryo from the uterus. The antibodies provide the developing foetus with passive immunity (Chapter 9). Like the yolk sac, the allantois projects from the embryonic gut.

FIG. 22.17 *Development of the embryonic membranes in the mammal*

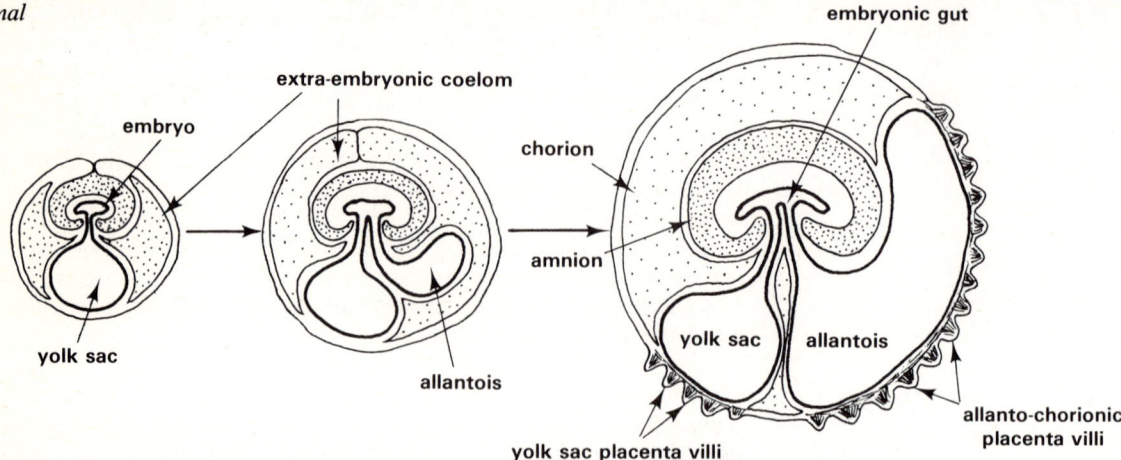

In reptiles, birds and egg-laying mammals, the allantois acts as a bladder, collecting wastes. The allantois is important in these animals, since the embryo, encased in its shell, has no means of disposing of solid wastes. In these animals too the allantois acts like a lung. Respiratory gases are exchanged across the allantoic membrane between the allantoic fluid and the atmosphere (Fig. 22.18). The amnion and chorion are membranes which grow to enclose the whole embryo. They are derived from the outermost layers of the embryo which by now lies in a fluid-filled cavity called the amniotic cavity. The chorion has a great number of **villi** which are attached to the uterine wall. The blood vessels of the allantois project into the chorionic villi and it is this vascular structure which becomes the **placenta** (Fig. 22.19).

FIG. 22.18 *Embryonic membranes of a reptilian egg*

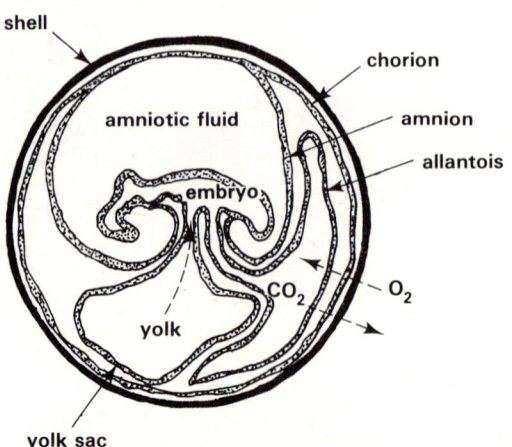

FIG. 22.19 *Blood vessels in the placenta*

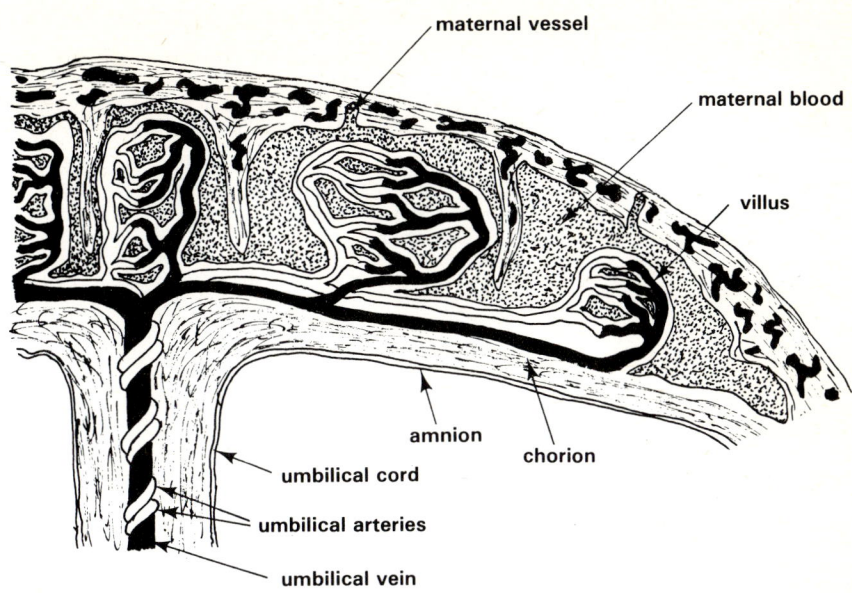

The chorionic villi provide a large surface area across which materials can be exchanged between maternal and foetal blood. Exchange of materials is further aided by the pattern of blood flow in the placental vessels. The maternal and foetal vessels are arranged in the placenta in such a way that maternal blood is close to the blood of the foetus (Fig. 22.20). Only substances of low relative molar mass can cross the placental barrier. Essential metabolites such as oxygen, sugars, amino acids, minerals, water and vitamins are transported from the mother's blood into that of the foetus. Antibodies which provide passive immunity in the foetus also pass from the mother across the placenta. Carbon dioxide, urea and other waste products made by the foetus move in the opposite direction. Most substances pass through the placenta by diffusion, though there is evidence that some sugars are moved by active transport.

FIG. 22.20 *Opposite blood flow in maternal and foetal vessels in the placenta*

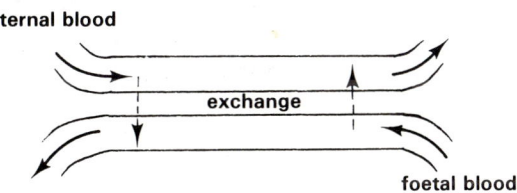

As well as acting as a route by which materials are exchanged between maternal and foetal blood, the placenta performs other vital functions. Among its other functions is the secretion of hormones. By about the second month of pregnancy in women the chorion secretes **choriogonadotropin**. This hormone is present in the urine of pregnant women and its detection is the basis of modern pregnancy tests. Small latex beads which are coated with an antibody specific to **human choriogonadotropin (HCG)** are added to a sample of urine. If the urine contains HCG the beads clump together due to the binding of HCG with the antibody. Some tests are more complicated and work by preventing the clumping of latex beads (Fig. 22.21).

FIG. 22.21 *Principle of the HCG agglutination inhibition test*

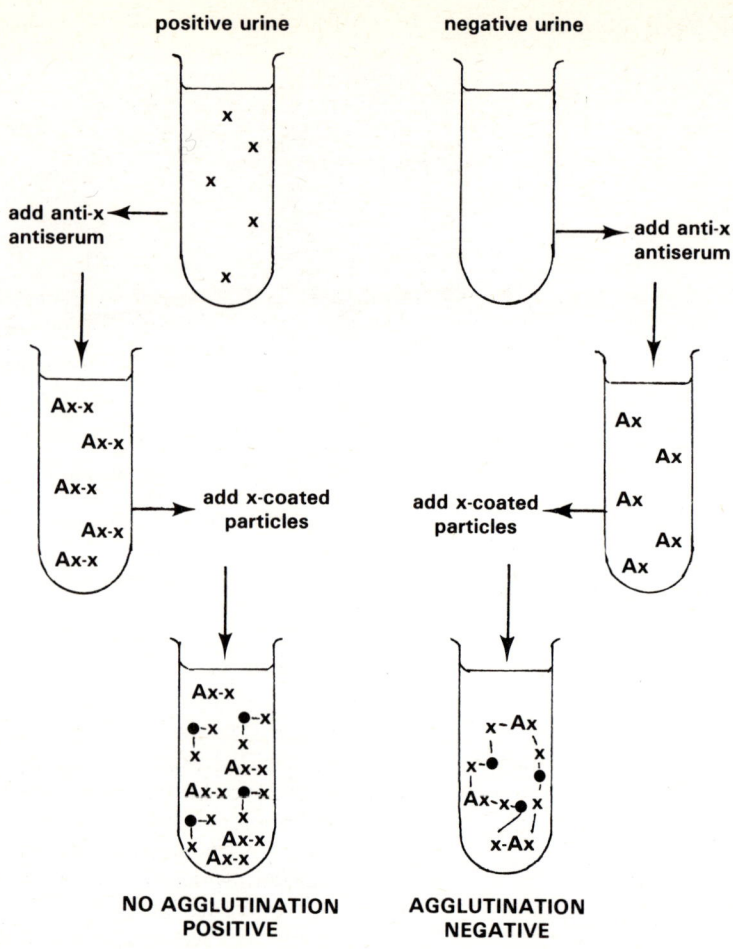

Estradiol and progesterone are also secreted from the placenta (section 22.2.3). These hormones prevent output of FSH from the anterior pituitary, thus stopping the development of new follicles in the ovaries. They also maintain the endometrium during pregnancy. Relatively high concentrations of progesterone in the uterus also keep the smooth muscle of the uterus relaxed, so protecting the developing embryos from premature birth.

22.3.4 Birth and lactation

Birth or **parturition** normally occurs when the foetus is fully developed at the end of the gestation period. Sudden and powerful intermittent contractions of the uterine wall mark the start of parturition. The smooth muscle of the uterus responds to at least two hormonal signals. One is the sudden drop in progesterone output from the placenta. Progesterone inhibits smooth muscle contraction. The other signal is a sudden increase in output of **ocytocin** from the posterior pituitary (Chapter 18). Ocytocin causes contraction of smooth muscle. Birth can be induced artificially by injecting the mother with ocytocin.

Ocytocin also brings about expulsion of milk from the mammary glands. The release of milk is an active muscular process and is a reflex response. Tactile stimulation of the nipples by a suckling infant generates nerve impulses which are transmitted in sensory neurons from the mammary glands to the hypothalamus in the brain. Ocytocin is released into the blood from the posterior pituitary. The hormone causes contraction of the smooth muscle lining the milk ducts in the mammary glands. Constriction of the ducts causes the milk to be ejected actively from the nipples into the infant's mouth.

FIG. 22.22 *The milk-expulsion reflex*

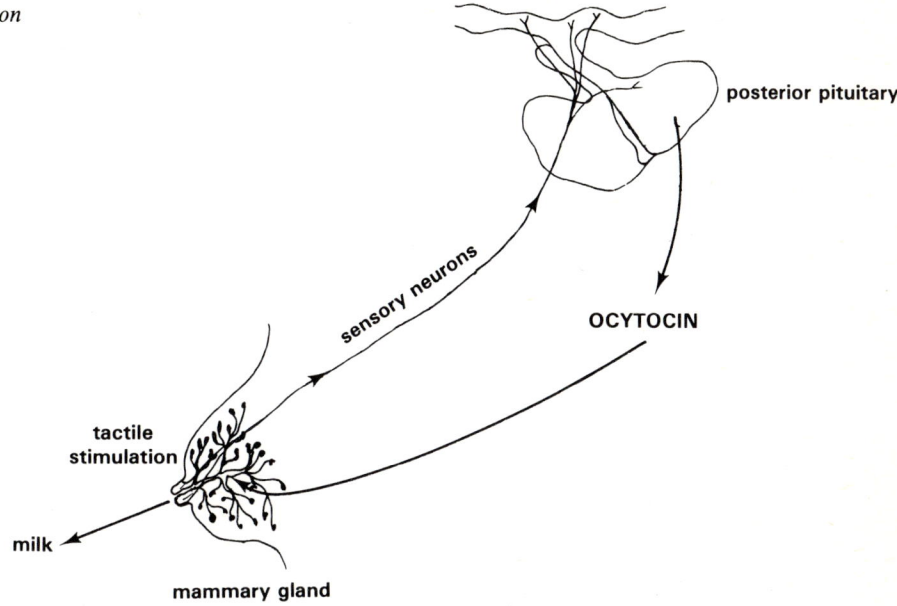

posterior pituitary

sensory neurons

OCYTOCIN

tactile stimulation

milk

mammary gland

Milk production called **lactation**, is caused by the action of other hormones. During pregnancy, estradiol and progesterone from the placenta bring about growth of the milk ducts and secretory tissue in the mammary glands. At birth the placenta is released from the uterus as the **afterbirth**. The amounts of estradiol and progesterone in the mother's blood then quickly fall, causing the anterior pituitary to secrete the hormone **prolactin**. Prolactin stimulates the secretory tissue to produce milk. Milk is the normal source of nutrients for newly-born mammals. It also contains antibodies which give passive immunity in the digestive system of newly-born mammals. When sufficiently mature young mammals are weaned on to solid food.

Units, Symbols and Quantities

SI and SI derived units, symbols and quantities

Système International (SI) and SI-derived units, symbols and quantities have been used throughout the text. Non-SI units have been given in addition where it is anticipated that the SI system may not be adopted in the near future.

Non-SI units and their conversion to SI units are given in brackets under each section.

LENGTH

> 1 metre (m) = 1000 millimetres (mm)
> 1 mm (10^{-3}m) = 1000 micrometres (μm)
> 1 μm (10^{-6}m) = 1000 nanometres (nm)
> (1 in = 25·4 mm)

ENERGY

> 1 kilojoule (kJ) = 1000 joules (J)
> (1 calorie = 4·187 J)

PRESSURE

> 1 kilopascal (kPa) = 1000 pascals (Pa)
> (1 atm = 760 mm Hg = 101·325 kPa)

MASS

> 1 kilogram (kg) = 1000 grams (g)
> 1 g (10^{-3}kg) = 1000 milligrams (mg)
> 1 mg (10^{-6}kg) = 1000 nanograms (ng)
> 1 ng (10^{-9}kg) = 1000 picograms (pg)
> (1 lb = 0·454 kg)

VOLUME

> 1 cubic decimetre (dm^3) = 1000 cubic centimetres (cm^3)
> 1 cubic centimetre $(10^{-3}\ \text{dm}^3)$ = 1000 cubic millimetres (mm^3)
> (1 in^3 = 16·38 cm^3)

ELECTRICAL POTENTIAL DIFFERENCE

1 volt (V) = 1000 millivolts (mV)

TIME

1 second (s) = 1000 milliseconds (ms)

TEMPERATURE

> thermodynamic temperature   kelvin (K)
> degree Celsius = °C
> (t °Fahrenheit = $\dfrac{5}{9}$ (t − 32) °C)

Chemical Nomenclature

In recent years the International Union of Physics and Chemistry (IUPAC) and the International Union of Biochemistry (IUB) have reached agreement on the rules of nomenclature for many compounds of biological interest. The systematic names of most biochemicals are extremely cumbersome and require a knowledge of chemistry beyond what is expected of GCE Advanced Level Biology students. We have used systematic names where students should reasonably be able to cope with them or where they present little or no more difficulty than the trivial names. Trivial names have been given where it is not anticipated that the systematic names will be used at this level of study in the near future. Because some students may wish to know the systematic names of some of the biochemicals given trivial names in the text the following list has been drawn up. The list may also be useful for those who are not familiar with the systematic names used in the text.

Table A2.1

Trivial name	Systematic name
acetaldehyde	ethanal
acetic acid	ethanoic acid
acetone	propanone
alanine	2-aminopropionic acid
aniline	phenylamine
benzoic acid	benzenecarboxylic acid
carbon tetrachloride	tetrachloromethane
chloroform	trichloromethane
citric acid	2-hydroxypropane, 1,2,3-tricarboxylic acid
ethylene	ethene
formaldehyde	methanal
fumaric acid	trans-butenedioic acid
glyceraldehyde	2,3-dihydroxypropanal
glycerol	propane-1,2,3 triol
glycine	aminoethanoic acid
glycollic acid	hydroxyethanoic acid
hippuric acid	N-benzoylglycine
lactic acid	2-hydroxypropanoic acid
malic acid	2-hydroxybutanedioic acid
malonic acid	propanedioic acid
ornithine	2,5-diaminovaleric acid
phosphoenolpyruvic acid	2-hydroxy-2-propanoic acid
pyruvic acid	2-oxopropanoic acid
succinic acid	butanedioic acid
urea	carbamide
xylene	dimethylbenzene

The names of mammalian hormones have recently been revised by the IUB and the revised names have been used in the text. Abbreviations for the old names have also been included as these are in such wide use.

Appendix 3

Questions

The questions have been arranged so that they can often be answered after the appropriate chapter has been studied. In some instances it may be necessary to use information from more than one chapter.

JMB = Joint Matriculation Board Examinations Council

London = University of London Examinations Council

ONC and OND = Joint Committee for Ordinary National Diplomas and Certificates

B = Biology   Bot = Botany   Z = Zoology

Chapter 1

1 Write concise notes on each of the following:

    (a) mitochondria,
    (b) nucleoli,
    (c) endoplasmic reticulum.

How has electron microscopy changed our views on the structure of cells?

(London: June 1975)   B

2 Describe how tissues are prepared for observation under either (a) the light microscope or (b) the electron microscope. Include each step of the procedure, and explain why it is necessary.

(London: January 1974)   Z

3 (a) Make a labelled diagram to illustrate chloroplast structure as revealed by the electron microscope.
  (b) Describe, with the aid of diagrams, the structural and functional characteristics of

    (i) the endoplasmic reticulum, and
    (ii) the Golgi apparatus.

  (c) (i) What are lysosomes, where are they formed, why are they sometimes referred to as 'suicide bags'?
    (ii) Suggest a possible reason for the occurrence of large numbers of lysosomes in phagocytic cells.

(JMB: June 1976)   B

4 (a) (i) Give three differences, in working, between the light microscope and the electron microscope.

    (ii) What are the relative ranges of magnification and resolving power of a light microscope and an electron microscope?
    (iii) Give two disadvantages in the use of an electron microscope as compared with that of a light microscope.
  (b) Using diagrams, compare what is revealed of the structure of any three components of the cytoplasm of a given cell seen under the light microscope with that seen under the electron microscope.

(JMB: June 1974)   B

5 Give a brief account of the distribution of membranes within a generalised cell. Describe the structure and functions of the surface membrane (the plasmalemma) of cells.

(JMB: June 1973)   B

Chapter 2

1 Consider this statement: 'When a plant cell is placed in a sugar solution which is stronger than the cell sap, plasmolysis occurs.'

  (a) What does the word 'stronger' mean in this context?
  (b) Draw a fully labelled diagram of a plant cell which has undergone plasmolysis.
  (c) Describe a method involving plasmolysis and using sucrose solution by which you could determine the osmotic potential (or osmotic pressure) of the cell sap of the cells of a plant. Use the outline below for your answer.

    (i) The preparation of suitable tissue of a named plant for investigation.
    (ii) The procedure to be followed.
    (ii) The results you would expect. State how these may be used to determine the osmotic potential (or osmotic pressure) of the cell sap.

(JMB: June 1978)   B

2 (a) What is meant by
    (i) osmotic pressure,
    (ii) turgor pressure,
    (iii) suction pressure or diffusion pressure deficit or water potential,
    (iv) plasmolysis.

  (b) Discuss whether a plant membrane should be considered to be semi-permeable or differentially permeable.

(c) Explain what you would expect to happen when a plant cell having an osmotic potential equivalent to 5 atmospheres and a suction pressure of 2 atmosperes is

    (i) immersed in pure water,
    (ii) immersed in a hypertonic sucrose solution,
    (iii) immersed in strong acid.
                          (London: January 1977) B

3 Discuss the importance of water to living organisms.
                          (London: June 1974) B

4 Mammalian red blood cells are sensitive to a change in salt concentration of the external solution. If they are transferred from plasma to a less concentrated solution they swell and, if they swell sufficiently, may burst, in which case they are said to be haemolysed. In an experiment to find the percentage of human red cells haemolysed at different concentrations of salt solution, the following results were obtained:

percentage salt concentration/ g 100 cm^{-3}	0·33	0·36	0·38	0·39	0·42	0·44	0·48
percentage red cells haemolysed	100	90	80	68	30	16	0

(a) Plot the results on graph paper, using the horizontal axis for the percentages of salt concentration.
(b) Explain why red cells swell and burst when placed in a dilute salt solution.
(c) At what percentage salt solution is the proportion of haemolysed to nonhaemolysed cells equal?
(d) Suggest a hypothesis to account for the red cells haemolysing over a range of salt concentrations rather than at one particular salt concentration.
(e) What is the significance of these observations in relation to the functioning of the human body?
                                (ONC: 1978)

4 What is the chemical nature of fats and oils that occur in plant tissues? How do (a) protein and (b) starch differ from fats and oils? Describe precisely how you would investigate whether sucrose and protein are present in carrot root tissue.
                      (London: June 1975) Bot.

5 (a) Distinguish

    (i) between sucrose and maltose, and
    (ii) between starch and glycogen.

(b) How could you obtain from a powder containing sucrose, starch and cellulose, cellulose with minimum amounts of the others as contaminants?
(c) (i) What are the two main products formed by the hydrolysis of fats?
    (ii) What general term is given to the reaction leading to the formation of fats?
    (iii) Give two reasons why fats (oils) are important as storage compounds.
    (iv) How would you demonstrate that a castor-oil seed contains an appreciable amount of fat?
(d) (i) Proteins have been called 'biological polymers'. Explain what this means.
    (ii) Name two elements likely to be found in proteins but not in carbohydrates or fats.
    (iii) Give one simple physical or chemical test which would suggest that an unknown substance was a protein.
    (iv) Proteins form what are termed 'hydrophilic colloids'. Explain what this means.
    (v) What is imbibition? Give one biological process in which imbibition occurs.
(e) Name the two main chemical constituents of the plasmalemma and show by means of a diagram, how they are thought to be arranged.
                         (JMB: June 1975) B

Chapter 3

1 Discuss the importance of proteins within cells.
                        (London: January 1977) B

2 Give an account of the ways in which reserve food materials are stored and utilised in plants and animals.
                        (London: January 1973) B

3 Discuss the ways in which the use of each of the following has contributed to advances in biological knowledge and understanding:

(a) the electronmicroscope,
(b) chromatography,
(c) radioactive isotopes.
                        (London: June 1978) B

Chapter 4

1 Discuss the role of enzymes in the regulation of cellular activities.
                  (JMB: Special Paper June 1978) B

2 (a) What is an enzyme?
(b) List the factors which affect the rate of an enzyme-controlled reaction.
(c) Describe an experiment to investigate the effect of one factor on the rate of an enzyme-controlled reaction.
(d) What is meant by a co-enzyme? Discuss briefly the role of co-enzymes in metabolism.
                      (London: January 1977) B

3 (a) What is meant by (i) specificity and (ii) denaturation of an enzyme?

(b) Discuss the ways in which (i) temperature and (ii) pH may affect the activity of an enzyme.

(c) How may starch be converted to glucose? Would you expect the same enzymes to be involved in the synthesis of starch? Give reasons for your answer.
(London: June 1976)   B

4 Enzymes may be described as biological catalysts. Give an account of (a) their catalytic and (b) their other properties. Explain briefly why enzymes are essential to living organisms.
(London: January 1974)   B

5 (a) Give an account of the factors affecting the rate of enzyme reactions, relating the action of these factors to the nature and properties of enzymes.

(b) Outline a method you might use to determine the effect of one of these factors on the rate of a named reaction.
(JMB: June 1975)   B

6 (a) State four properties you would expect an enzyme to possess.

(b) What is meant by:

(i) an enzyme activator?
(ii) an enzyme inhibitor?

Give a named example of each.

(c) (i) For each numbered enzyme 1–6 below write down the letter of the reaction A–F which it catalyses:

1. diastase (α-amylase)   2. cytochrome oxidase
3. a dehydrogenase   4. phosphorylase
5. invertase   6. phosphatase.

Reactions:

A. glucose-phosphate + water⇌starch + inorganic phosphate.
B. starch⇌dextrins.
C. sucrose⇌glucose + fructose.
D. glucose-1-phosphate⇌glucose + inorganic phosphate.
E. acid I $(C_4H_6O_4)$⇌acid II $(C_4H_4O_4)$.
F. transfer of electrons to molecular oxygen during aerobic respiration.

(ii) Briefly indicate the physiological significance of any four of these reactions.
(JMB: June 1973)   B

Chapter 5

1 In aerobic respiration the energy contained in a molecule of glucose is made available to the cells by three main processes: glycolysis, tricarboxylic acid cycle, electron-transfer system.

(a) Explain concisely the essential features of each of these processes.

(b) In what ways does the anaerobic respiration of glucose differ from aerobic respiration of the same substrate?
(JMB: June 1977)   B

2 Give an account of the fixation and utilisation of carbon in plants which is sufficiently detailed to explain the following statements.

(a) (i) The first product of carbon dioxide fixation is 3-phosphoglyceric acid (PGA).

(ii) Light energy is required only to provide the ATP and electron donors (reduced coenzymes) needed to reduce PGA and maintain a supply of ribulose diphosphate.

(iii) There are alternative uses for ATP and the electron donors in addition to those of carbon dioxide fixation.

(b) (i) Starch grains begin to grow close to grana.

(ii) The formation of cellulose and pectins occurs near to the plasmalemma.
(JMB: June 1978)   B

3 (a) Give an outline of the biochemical processes which occur during aerobic respiration in a cell.

(b) Briefly differentiate between aerobic and anaerobic respiration.

(c) Explain why (i) a yeast cell and (ii) a muscle fibre need not take up oxygen during respiration.
(London: January 1978)   B

4 (a) What is the main function of ATP in cell metabolism? State briefly how it fulfils this function.

(b) Explain, in outline only, how ATP is generated by the cells of a green plant in the light and in darkness.
(JMB: June 1976)   B

5 In an experiment with mitochondria a medium which contained inorganic phosphate and oxidisable substrates was used. The medium was saturated with air at

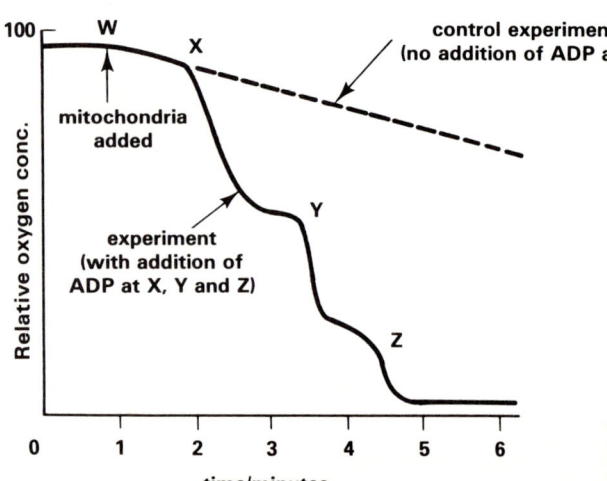

the start. The experiment determined the fall in oxygen concentration in the medium when firstly mitochondria, and then ADP, were added.

The letters used below refer to those on the graph. In the experiment, mitochondria were added at W. A standard amount of ADP was added at X, a similar amount at Y, and at Z.

In a control experiment, mitochondria were added at W. No ADP was added.

The results are shown in the graph.

(a) State one function of the phosphate in the medium.
(b) Account for

   (i) the slight drop in oxygen concentration when the mitochondria were added at W,
   (ii) the shape of the graph when the ADP was added at X and Y.

(c) Why does the graph become horizontal shortly after the addition of ADP at Z?
(d) Briefly explain what effect the following additions might have.

   (i) The addition of excess ADP at X.
   (ii) The addition of cyanide at X as well as the excess ADP.

(JMB: June 1977) B

6 The apparatus drawn below can be used to measure the rate of respiration of small organisms.

(a) Name any two organisms which could be used in this apparatus.
(b) To measure oxygen consumption, you would need to put a chemical into the specimen tube. Write down
   (i) which chemical you would use,
   (ii) the function it performs in the investigation,
   (iii) how you would place it in the tube, keep it in position, and isolate it from the organisms.

(c) Explain how you would proceed to use the apparatus to measure the uptake of oxygen by an organism.
(d) Could this apparatus be used to measure the rate of anaerobic respiration? Explain your answer.
(e) The apparatus provides a very simple device to measure the rate of respiration. Give any two sources of error which might arise when using it.

(JMB: June 1977) B

Chapter 6

1 (a) Explain concisely, those features of the structure of DNA which enable it to

   (i) serve as a store of genetic information, and
   (ii) transmit identical information to the cells produced as a result of mitosis.

  (b) (i) What is a gene and what is a gene mutation?
     (ii) Explain why mutations may appear
       (A) immediately (somatic),
       (B) in the first generation,
       (C) not until several generations of offspring have appeared.

(JMB: June 1978) B

2 Give an illustrated account of the structure of DNA. How is DNA involved in the synthesis of proteins? Indicate the genetic significance of changes that may occur in the structure of a DNA molecule.

(London: January 1976) B

3 (a) Describe the structure of DNA.
  (b) Bacterial cells were grown in a medium containing radioactive nitrogen. On division the DNA strands of the daughter cells were radioactive. The daughter cells were placed in a normal medium and when division occurred half of the DNA contained labelled nitrogen. How can this be explained?

(OND: 1974)

4 (a) Mouse cells grown in culture were exposed for 30 minutes to radioactive uracil, then either (A) fixed or (B) grown on for an additional 4 hours in a medium containing non-radioactive uracil and then fixed. The distribution of the radioactivity was determined by autoradiography and the results of such an experiment are given below.

	PERCENTAGE OF RADIOACTIVITY OVER		
	NUCLEOLUS	REST OF NUCLEUS	CYTOPLASM
(A) 30 minutes	54	41	5
(B) 30 minutes + 4 hours	14	19	67

   (i) Give a brief outline of the principles of the technique of autoradiography.
   (ii) From these data and your knowledge of the functioning of the nucleus, account for the distribution of radioactivity after 30 minutes and after 30 minutes plus 4 hours.

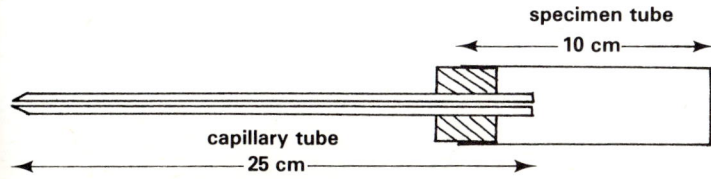

specimen tube
10 cm
capillary tube
25 cm

(iii) Supposing the cells had been growing continuously for $4\frac{1}{2}$ hours in radioactive medium, what then would have been the likely distribution of radioactivity within the cell?

(b) The following experiment was devised to see whether newly-synthesised protein molecules go directly from the endoplasmic reticulum-bound ribosomes into the cavity of the endoplasmic reticulum or whether they are first freed into the cytoplasm and later picked up by the endoplasmic reticulum.

A microsome fraction was prepared by differential centrifugation of homogenised tissue and was suspended in a medium which can be regarded as being equivalent to the cytoplasm of the cell. The microsome fraction consists chiefly of fragments of the rough endoplasmic reticulum. A radioactive amino acid was added to the medium containing the microsomes. At successive intervals samples were taken and treated with a detergent which frees the ribosomes and disperses the membranes. Radioactivity in protein was determined separately for the ribosomes, the detergent-dispersed fractions, and the medium. The results were as follows.

TIME IN MINUTES	RADIOACTIVITY IN C.P.M.		
	DETERGENT-DISPERSED FRACTION	RIBOSOMES	MEDIUM
1	10	65	30
2	40	100	50
3	80	116	60
5	150	115	70
10	200	115	70

(i) Plot a graph of these data.
(ii) Interpret these results as fully as you can, indicating whether they show that newly-synthesised protein follows one or other of the routes described in the aims of the experiment.
(JMB: Special Paper June 1977) B

5 (a) Briefly describe one piece of experimental evidence which directly supports the view that DNA is the molecule of heredity.
(b) Give three chemical features of DNA which render this molecule particularly suitable to be the vehicle of heredity.
(c) By using the following keywords in an appropriate context construct a simple account of protein synthesis in living cells. The account need be no longer than is necessary to include all of the keywords which may be used in any order. Keywords: codon (triplet), messenger RNA, transcription, enzyme, polypeptide, ribosomes, nuclear membrane, amino-acids, proteins.
(JMB: June 1974) B

Chapter 7

1 (a) Give an illustrated account of mitosis.
(b) Where does meiosis occur in a mammal and in a flowering plant? What is its significance?
(c) List four differences between mitosis and meiosis.
(London: January 1978) B

2 'It is sometimes said that meiosis consists of two mitotic divisions.' Discuss the validity of this statement.
(London: January 1974) Z

3 What are the functions of a nucleus in a non-dividing cell? Make fully labelled diagrams showing the stages of mitosis in a plant cell, the nucleus of which contains four chromosomes.
(London: January 1975) Bot.

4 In vegetative cells the cycle of the main stages of cell division can be represented by the following scheme:

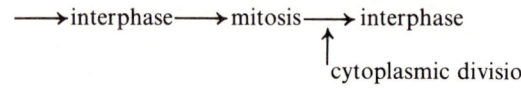

(a) (i) At what stage does DNA synthesis occur?
(ii) Give an example of nuclear division proceeding in the absence of cell division.
(b) Give an account of the structural changes which take place as a cell passes through this cycle and note any difference between plant and animal cells.
(JMB: June 1974) B

5 (a) Suppose you are given an onion bulb. Outline the procedure you would follow in order to be able to see its chromosomes.
(b) The relative amounts of DNA were measured in the cells of the plant Tradescantia during meiosis and mitosis. The results are given below:

PHASE OF NUCLEAR CYCLE	DNA CONTENT PER CELL
Mitosis: Early interphase	2·0
Prophase	4·0
Late telophase	2·0
Meiosis 1: Leptotene	4·0
Diplotene	4·0
Meiosis 2: Telophase	1·0

(i) Account for the differences in DNA content between
(A) mitosis early interphase and mitosis prophase,
(B) mitosis prophase and mitosis late telophase,
(C) meiosis 1 diplotene and meiosis 2 telophase,
(D) meiosis 2 telophase and mitosis early interphase.

(ii) What would you estimate the DNA content of cells at meiosis 1 telophase to be?

(c) A diploid cell contains three pairs of chromosomes in the nucleus. Draw a labelled diagram to show the appearance of the cell at anaphase of a mitotic division.

(JMB: June 1977)   B

Chapter 8

1 Describe the mechanism of (a) ventilation and (b) gaseous exchange in a mammalian lung. Explain how the oxygen concentration of the blood is maintained under varying degrees of physical activity.

(London: June 1976)   B

2 The table below gives the percentage saturation with oxygen of the blood of three animals at different pressures of oxygen:

OXYGEN PRESSURE (mm OF MERCURY)	% SATURATION OF BLOOD WITH OXYGEN		
	HORSE	FOETAL HORSE	LLAMA
10	20	10	20
20	40	50	60
30	50	70	80
40	70	80	90
50	80	90	90
60	80	90	—

Plot these data in such a way that the graphs (oxygen dissociation curves) can be compared. Given that the llama lives at a high altitude, in the Andes, explain the differences between these graphs. What do you think the oxygen dissociation curves might be like for (a) horses normally kept at sea level but used to take travellers across high mountains, and (b) horses used for ploughing, and (c) race horses, and (d) llamas in zoos, and (e) foetal llamas?

(London: June 1971)   Z

3 (a) Describe the mechanism involved in the ventilation of the lungs in a mammal.
(b) Explain how these breathing movements are controlled.

(JMB: June 1974)   B

4 Write a detailed account of the carriage of the blood gases in mammals, including reference to environmental effects.

(JMB: Special Paper June 1978)   B

5 (a) State briefly what you consider to be the main properties and requirements of surfaces used for gaseous exchange.
(b) How are these requirements achieved in

(i) a terrestrial flowering plant, and
(ii) a terrestrial mammal?

(JMB: June 1976)   B

Chapter 9

1 Describe the structure, mode of action and function of the heart in a mammal.

(London: June 1977)   B

2 Explain how the structure and operation of the mammalian heart enable it to function as a pump in the circulatory system. What effects do the aorta, other arteries, capillaries and veins have on the flow of blood?

(London: June 1968)   Z

3 (a) Describe the mode of action of a mammalian heart.
(b) Explain how the heart beat is regulated according to the demands of the body.

(JMB: June 1978)   B

4 (a) What do you understand by the terms antigen and antibody? State their general importance in mammals.
(b) Explain, with *one* example, how antigen/antibody responses in man have been used in the fight against infectious disease.

	DIAMETER OF LUMEN	THICKNESS OF WALL	APPROX. TOTAL CROSS-SECTIONAL AREA FOR ALL VESSELS OF EACH TYPE (cm^2)	PERCENTAGE OF TOTAL BLOOD VOLUME CONTAINED IN ALL VESSELS OF EACH TYPE
Aorta	25 mm	2 mm	4·5	2
Artery	4 mm	1 mm	20	8
Arteriole	30 μm	20 μm	400	1
Capillary	6 μm	1 μm	4500	5
Venule	20 μm	2 μm	4000	50
Vein	5 mm	500 μm	40	50
Vena cava	30 mm	1·5 mm	18	50

(c) Explain how antigen/antibody responses in man (other than that given in answer to (b)) have led to medical problems. Describe how these problems have been or are being overcome.

(JMB: June 1978) B

5 The table on p. 457 gives some of the characteristics of various types of blood vessels in man.

(i) Comment on any relationships you observe in this table, and explain them as far as you can.
(ii) Discuss the distribution of the total blood volume in the light of the figures given in the right-hand column above.

(b) The figure below gives diagrammatic curves, drawn to different scales, for total cross-sectional area, mean velocity of blood flow, and mean pressure in each type of blood vessel. Deduce which curve is which.

(c) (i) Make a diagram of the heart labelled to show the route of the excitation wave which accompanies the heart beat.
(ii) Give two pieces of evidence which show that the sino-atrial node is the pacemaker of the heart.
(iii) Give two pieces of evidence which show that the origin of the heart beat is myogenic.

(JMB: Special Paper June 1975) B

Chapter 10

1 Make large, fully labelled diagrams to show

(a) the general internal structure of a mammalian kidney, and
(b) the detailed structure of a nephron.

Explain how the kidney controls the water content and the urea content of the body.

(London: June 1975) B

2 What are the functions of the mammalian kidney? Explain how the structure of the kidney is specialised for the performance of the functions you have described.

(London: June 1967) Z

3 Give an account of the histology of the mammalian kidney, with special reference to high-power detail, to show how structure is related to function.

(London: June 1975) Z

4 The graph shows the effects of a temporary constriction of the renal artery in a mammal.

(a) (i) On which side of the constriction would these effects be shown?
(ii) What corresponding changes would you expect to find in the blood pressure on the opposite side of the constriction?
(iii) What immediate effect would this constriction have on the rate of glomerular filtration?
(iv) Predict, giving your reasons, the effect on the composition of the urine of lowering the temperature of the kidney from 37°C to 25°C.

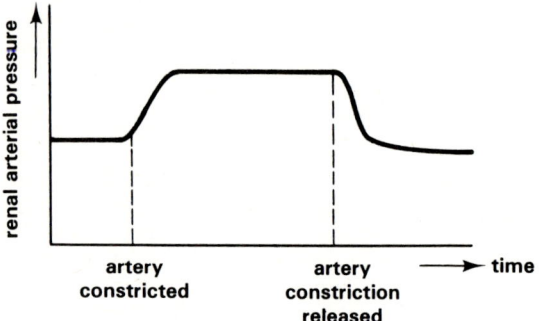

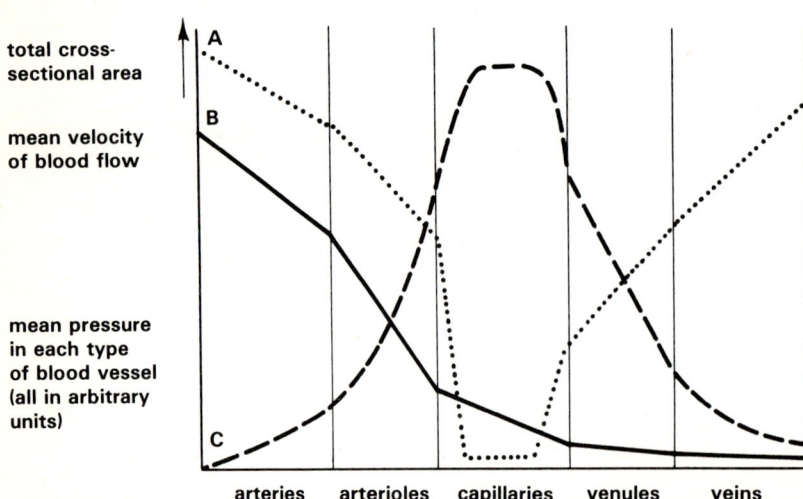

(v) Explain why the total output of urine changes when the environmental temperature of the body is decreased.

(b) In what respect does the kidney function as a homeostatic device?

(JMB: June 1974)   B

5 (a) Make a labelled diagram to show the form of a typical nephron in the kidney of a mammal, including its blood supply.

(b) State the main functions performed by the nephron and relate these, as far as possible, to specific regions. (N.B. Physical and chemical details of how these functions are carried out are not needed).

(c) The figure below indicates the osmotic pressures of dissolved mineral salts and of urea in the blood of three different fishes, compared with the osmotic pressures of fresh water and of sea water. Assuming that the structure of the fish kidney is essentially the same as that of the mammalian kidney, what differences, if any, would you expect to find between:

(i) the amount of urine produced,
(ii) the concentration of urine produced, and
(iii) the form of the nephrons in these three fishes?

Explain your answer.

(JMB: June 1975)   B

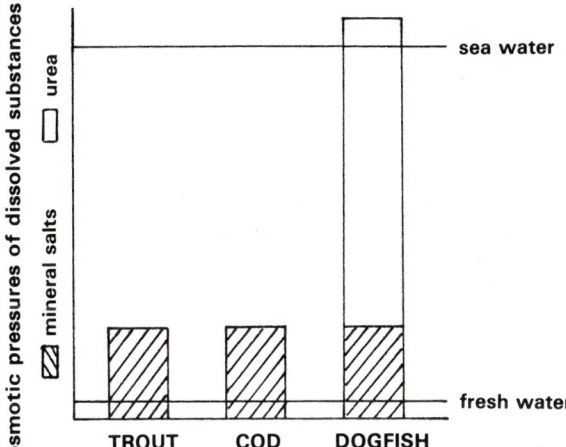

Chapter 11

1 (a) Explain, concisely, the basis of the 'root pressure' and 'cohesion-tension' hypotheses which have been proposed to explain the transport of water from roots to leaves. For each hypothesis, outline the supporting experimental evidence.

(b) (i) Give an illustrated account of the structure of the tissue in which the upward movement of water occurs in the stem.

(ii) Describe one experiment which would show that the upward movement of water does occur in this tissue.

(JMB: June 1978)   B

2 The graph below shows the relationship between stomatal aperture and water loss from a leaf, for a plant in still air.

(a) Describe the relationship shown by the graph.

(b) What is the effect on the rate of water loss when the stomatal aperture reduces from 20 to 10 μm?

(c) How do other factors influence the loss of water vapour from a leaf?

(d) Discuss which of these factors you would expect to have the greatest influence over a 24-hour period in hot summer weather.

(e) Draw a curve to show how the shape of the graph would change if the plant was transferred to and left in moving air.

(London: June 1976)   B

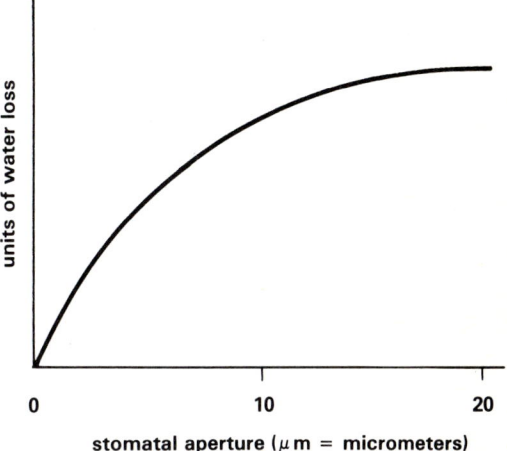

3 By means of a fully labelled diagram describe the movement of water from the soil solution into the vacuole of a parenchyma cell of the root cortex. What effect does the endodermis of the root have on water movement into the stele?

Explain, as fully as you are able, how water enters xylem vessels in the root.

(London: June 1975)   Bot.

4 Briefly compare the functions of phloem and xylem in a flowering plant. What kinds of cells are found in the xylem and how is their structure related to the functions of the tissue?

(London: June 1975)   B

5 (a) Describe one method of measuring the velocity of flow of sap in the xylem tissue of either an intact plant or a cut shoot.

(b) The efficiency of the wood of different species in water conduction can be compared by measuring their specific conductance. This is done by measuring the quantity of water which passes through a standard length (15 cm) of stem under a pressure of 30 cm of mercury in 15 minutes. This quantity divided by the area of cross-section of the conducting tract gives a value which is called the specific conductance.

For a stem of a larch tree, in which the xylem is largely composed of tracheids, the specific conductance was 14·5 and for a beech tree it was 65.

Comment upon this difference as fully as you can.

(c) The graph below records the variation in the velocity of flow of sap during the course of a single day in the xylem of intact trees of the same two species.

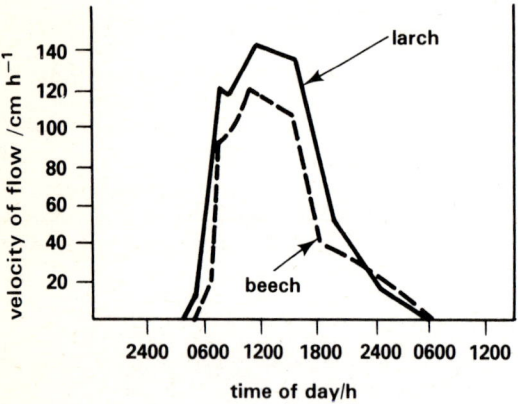

(i) What is the force which moves the sap through the plant?

(ii) Why does the rate of flow rise to a peak at noon and then decline?

(iii) How do you explain the continued flow of sap in late evening?

(d) The values recorded in the graph appear not to agree with the values for specific conductance given in (b). Explain why both sets of values can be correct and compatible with each other.

(JMB: Special Paper June 1977)   B

Chapter 12

1 A shoot of pond weed was submerged in pond water at 15°C and placed on a bench in daylight. The bubbles of gas which escaped from the cut end of the shoot were collected and their volume was measured at intervals of time. The experiment was repeated with the shoot submerged in pond water held at temperatures of 25°C, 35°C and 45°C. The results are given below.

TEMPERATURE °C	TOTAL VOLUME OF GAS IN mm³ COLLECTED AFTER			
	15 MINUTES	30 MINUTES	45 MINUTES	60 MINUTES
15	0·7	1·5	2·2	3·0
25	1·4	3·0	4·3	5·7
35	2·5	4·8	6·7	8·3
45	4·0	6·3	7·1	7·3

(a) Express these results graphically in what you consider to be the most satisfactory way.

(b) (i) Calculate a typical Q_{10} for photosynthesis as shown in this experiment.

(ii) Explain your choice of data for the calculation.

(c) (i) Give one possible reason for the levelling-off of gas production with time at 45°C.

(ii) Suggest one experiment, or modification of the experiment described here, to test if this reason is valid.

(JMB: June 1978)   B

2 (a) The following table shows the rates of carbon dioxide diffusion through membranes perforated with holes of different diameters. In each case the area of membrane perforated by the holes was the same and the number of perforations per unit area of membrane was also similar.

Diameter of holes (mm)	22·7	12·06	6·03	3·23	2·00
cm³ CO_2 diffusing h⁻¹	0·24	0·10	0·06	0·04	0·02

Use the data to construct a graph showing the rate of diffusion of the gas per unit area of hole.

(b) Relate your findings to the exchange of gases between the leaves of a terrestrial plant and its surrounding atmosphere.

(c) Write a brief account of the mechanism of stomatal movement.

(OND: 1977)

3 Explain the idea that stomatal movement is a turgor phenomenon and describe how this can be tested by experiment. Describe and explain the effect on stomatal aperture of changes in the rate of photosynthesis in (a) the leaf mesophyll and (b) the guard cells.

(JMB: Special Paper June 1978)   B

4 Write an account of the ways in which external factors may influence the rate of photosynthesis in a green plant. Describe carefully how you would measure

(a) the rate of photosynthesis, and

(b) the effect of one external factor on this rate.

(London: January 1976)   B

5 Compare the structure of a xylem vessel with that of a sieve-tube. Show how differences between them may be related to their different functions in the plant.

(London: January 1977)   B

6 Leaf discs (1 cm diameter) were punched from the mature leaves of a shrub at two-hour intervals during a warm, sunny day. The dry mass of each sample was determined. The results are given below. From the measurements, the changes in dry mass per two hours were calculated.

1	2	3
TIME OF DAY	DRY MASS OF LEAF DISCS (g)	CHANGE IN DRY MASS SINCE PREVIOUS SAMPLE (g)
5·00	3·006	—
7·00	3·028	0·022
9·00	3·148	0·120
11·00	3·390	0·242
13·00	3·650	0·260
15·00	3·782	0·132
17·00	3·936	0·154
19·00	4·008	0·072
21·00	3·992	−0·016

(a) Make a graph of the changes in dry mass (column 3) during the course of the investigation.

(b) Briefly account for the changes in dry mass recorded in each of the periods below.

   (i) From 07.00 hours to 13.00 hours.
   (ii) From 13.00 hours to 15.00 hours.
   (iii) From 15.00 hours to 17.00 hours.
   (iv) From 17.00 hours to 21.00 hours.

(JMB: June 1976) B

Chapter 13

1 A green plant synthesises hexose sugar in its leaves and absorbs ammonium ions by its roots. What further processes of transport and synthesis must occur, and in what tissues, before some of the organic nitrogen compounds synthesised from these raw materials will be exported from the roots?

(JMB: June 1977) B

2 Radioactive ^{32}P was simultaneously supplied to the roots of two similar plants, one of which had a ring of phloem removed as shown on the diagram. At the same

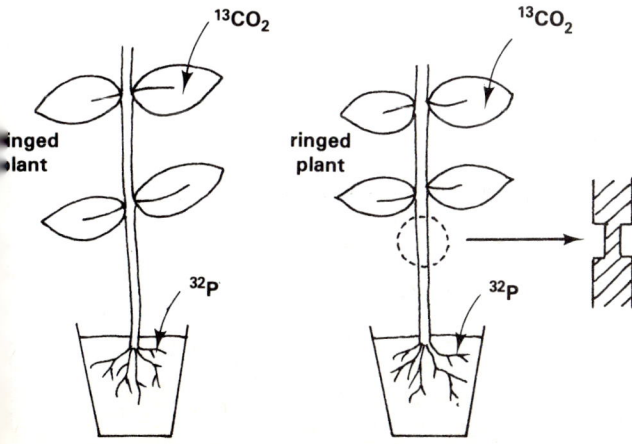

time the leaves of both plants were in an atmosphere containing $^{13}CO_2$.

The concentrations of ^{32}P and ^{13}C after 24 hours are given in the table.

	UNRINGED PLANT	RINGED PLANT
^{32}P in roots	100·0	100·0
^{32}P in leaves	2·2	1·7
^{13}C in roots	0·09	0·00
^{13}C in leaves	0·10	0·99

What do these results show? Comment on the physiological processes involved.

(London: January 1976) Bot.

3 Give an account of the mechanism of mineral ion absorption by the roots of higher plants. For three different mineral ions, discuss their importance in nutrition and indicate the deficiency symptoms which develop in their absence.

(London: June 1978) Bot.

4 An experiment was carried out to measure the uptake of water, potassium and oxygen by different zones of the root in a barley plant. The root of an intact plant was immersed in dilute potassium nitrate solution and measurements were made when the plant was transpiring at a high rate. The results (in arbitrary units) are shown below:

DISTANCE FROM ROOT TIP (mm)	UPTAKE OF		
	WATER	POTASSIUM	OXYGEN
0–25	16	5	7
25–50	29	13	3
50–75	24	10	2

(a) Plot, by drawing three lines on one graph, the uptake of (i) water (ii) potassium (iii) oxygen.

(b) Suggest explanations to account for the differences in uptake of

   (i) water in the different zones of the root,
   (ii) oxygen in the different zones of the root.

(c) What is meant when an uptake process is said to be

   (i) active,
   (ii) passive?

(d) (i) Do the data suggest that the uptake of potassium is

     (A) mainly active
     (B) mainly passive
     (C) neither A nor B?

   (ii) Give two reasons for your answer to (i).

(JMB: June 1977) B

5 The diagrams (Figs. A and B) are of the aerial parts of kidney bean plants grown in culture solution. Ten days after germination a radioactive isotope, either ^{45}Ca or ^{32}P, was placed in the culture solution. After one hour the plants were removed and the roots washed. They were then returned to a normal culture solution and allowed to grow for a number of hours; this is shown on each diagram. After this time they were taken from the culture solution, dried quickly and autoradiographed. In the diagrams the intensity of shading is proportional to the concentrations of radioactivity found in each leaf of the plant. Note that in this plant the first leaves are a pair of heart-shaped primary leaves, the trifoliate (3-leaflets) leaves subsequently formed being borne singly. Losses by radioactive decay in the course of the experiment can be ignored.

(a) Give two assumptions which must be made concerning the absorption and distribution of radioactive isotopes in plants when interpreting data such as these.

(b) Which leaves of the plant show the greatest concentration of

   (i) calcium after 6 hours, and after 96 hours?
   (ii) phosphorus after 6 hours, and after 96 hours?

(c) (i) It was suggested that when calcium had been translocated into a given leaf it remained there. Does Fig. A support this hypothesis? Give your reasons.

(ii) How do you account for the apparent decrease in radioactivity in leaf X (Fig. A) during the time of the experiment?

(d) Give an explanation for the changes in the distribution of radioactive phosphorus in the plant in Fig. B.

(e) (i) Using these experimental results, explain where you would expect deficiency symptoms to show when (A) calcium and (B) phosphorus is lacking in the culture medium.

(ii) Give (A) one role of calcium and (B) one role of phosphorus in the plant.

(JMB: June 1974) **B**

Chapter 14

1 Describe the following processes in a mammal:

(a) food ingestion and mastication,
(b) digestion in the duodenum,
(c) absorption in the ileum,
(d) production of the faeces.

(London: January 1976) **B**

2 Why do animals require a source of nitrogen in their diet? In what forms do animals obtain nitrogen? Describe how animals process nitrogen-containing compounds and make the products available to the cells of the body.

(London: June 1978) **Z**

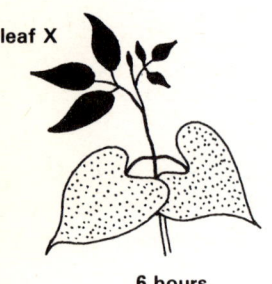
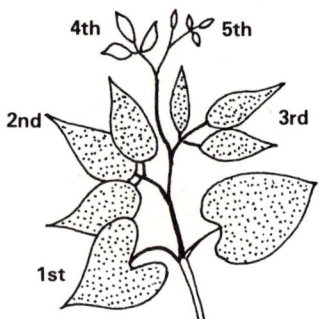

6 hours   24 hours   96 hours

Fig. A. Distribution of ^{45}Ca.

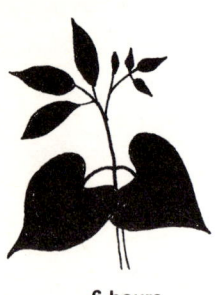

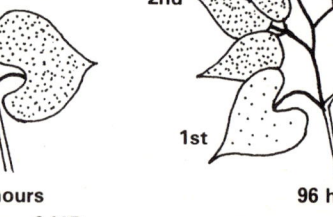

6 hours   24 hours   96 hours

Fig. B. Distribution of ^{32}P.

3 What is an enzyme? Describe the roles of enzymes in the digestion of proteins in a mammal making reference to those factors which are likely to influence enzyme activities.

(London: January 1978) Z

4 'The liver has a strategic role in the metabolism of carbohydrates, fats and proteins and, in addition, performs several other key functions.' Discuss this statement.

(JMB: Special Paper June 1977) B

5 Give a full account of any four proteolytic enzymes of the human digestive system. For each state:

   (i) its site of production (be as precise as you can in locating this),
   (ii) the form in which it is secreted (actual chemical formulae are not required),

(iii) its effect on the substrate molecule on which it acts,
(iv) the particular event or condition which stimulates its production and release.

(JMB: June 1977) B

6 (a) The graph above shows the results of six experiments on the nutritional requirements of young rats in which the following modifications of a complete diet were used:

Curve A. The only protein given was zein, a protein found in maize.
Curve B. The only protein given was zein and the amino acids lysine and tryptophane were added.
Curve C. The only protein given was zein and the amino acid lysine was added.
Curve D. The amino acid cystine was omitted.
Curve E. The amino acid methionine was omitted.
Curve F. The level of methionine was reduced and cystine was omitted. At day 4 cystine was added.

  (i) What conclusions can you make regarding

    (A) zein as a source of amino acids for rats?
    (B) the nutritional importance to rats of the amino acids lysine, tryptophane, cystine and methionine?
    (C) the ability of rats to synthesise these amino acids?

  (ii) Name and describe the process by which mammals can synthesise amino acids.

(b) Measurements were taken of (A) rate of production of urine, (B) concentration of urine, of a dog which had drunk its fill of water. Ten minutes after these observations had commenced a saline solution, hypertonic to the blood plasma, was injected into the carotid artery of the dog. The results of the investigations are shown in the graphs below.

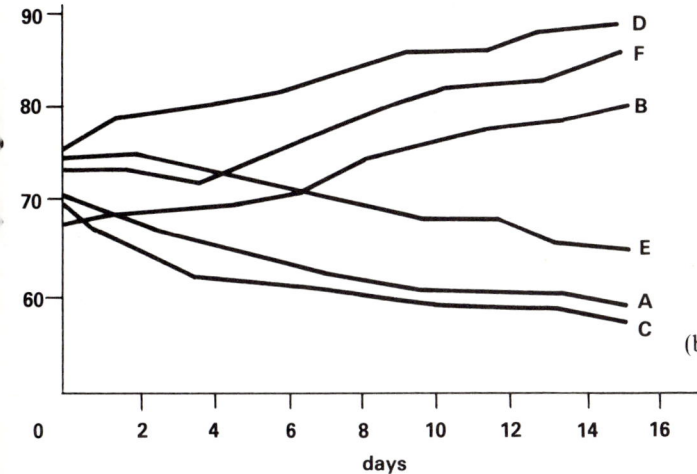

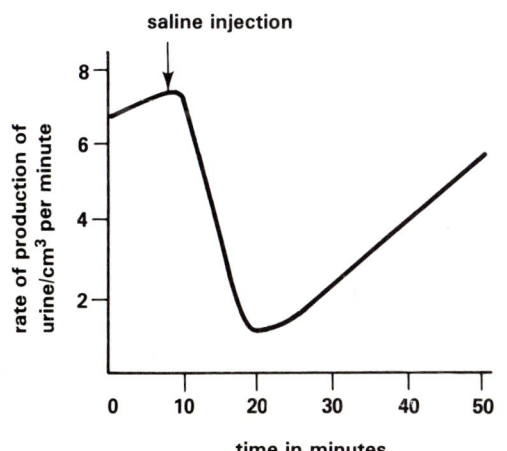

A. Rate of urine production

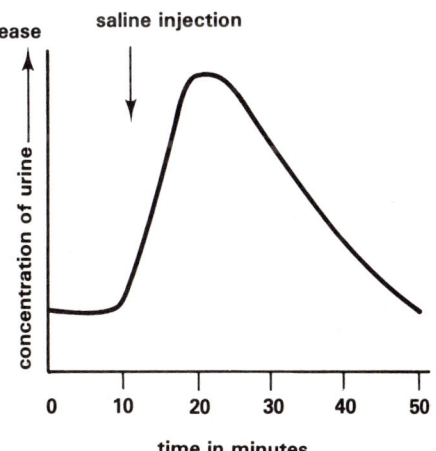

B. Concentration of urine

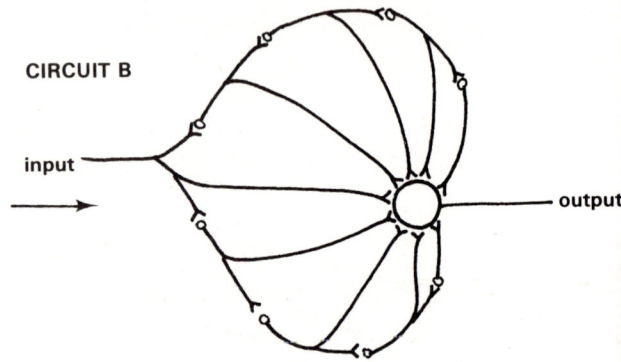

CIRCUIT A

input

output

CIRCUIT B

input

output

(i) Explain (A) the change in rate of production of urine,
  (B) the change in concentration of the urine.
(ii) When the experiment was repeated but with the saline solution injected into a vein in the neck of the dog, it had no effect on the amount of urine produced. Explain this observation.

(JMB: Special Paper June 1975)   B

Chapter 15

1 (a) Make a fully labelled diagram to show the main parts of a mammalian brain.
  (b) Briefly explain the functions of the main parts of the brain in a mammal.
  (c) Give an illustrated account of the way in which nerve impulses are conducted in a mammalian nerve fibre.

(London: June 1977)   B

2 Outline the major pathways of the reflex arc. Describe the way in which a nerve impulse is transmitted along a fibre and across a synapse.

(London: June 1975)   Z

3 What is the nature of the nerve impulse? Describe

  (a) how an impulse is transmitted between neurons, and
  (b) what happens when an impulse reaches a striated muscle.

(London: June 1973)   Z

4 Give a comparative account of the histology of striated (skeletal) and smooth muscle.

  By reference to named organs explain how each type of muscle is related to the particular function of the organ concerned.

(London: June 1978)   Z

5 (a) Write an account of the nature, initiation and transmission of the nerve impulse.
  (b) What are the main points of difference in function between the parasympathetic and sympathetic parts of the autonomic nervous system?

(JMB: June 1975)   B

6 The processing of information in the mammalian brain is assisted by special types of neuronal circuitry. Read the following information carefully, then answer the questions relating to the two diagrams. You should assume that a single impulse arriving at a synapse is sufficient to generate an impulse in the neuron or neurons on the other side of the synapse. If further inputs maintain excitation above threshold, repetitive discharges will occur until excitation falls. Remember that transmission across a synapse involves a short delay.

  (a) (i) If the input to Circuit A were to be a single impulse, what would be the output?
    (ii) Explain how this effect is achieved.

  (b) (i) If the input to Circuit B were to be a single impulse, what would be the output?
    (ii) Explain how this effect is achieved.

(JMB: June 1976)   B

Chapter 16

1 Make a large, fully labelled diagram of the eye of a mammal in section. Write explanatory notes on each of the following:

  (a) retina
  (b) accommodation.

(London: January 1974)   B

2 Make a large, clearly labelled diagram of the ear of a mammal. Describe how those parts of the ear concerned with (a) balance and (b) hearing carry out their functions.

(London: June 1974)   B

3 (a) What are the biological advantages of (i) sight, and (ii) colour vision, to animals?
  (b) Make a large, clearly labelled diagram of the mammalian eye as seen in section.
  (c) Briefly describe how accommodation for distance and light intensity is brought about.

(London: January 1977)   B

4 (a) Make a labelled drawing of a representative portion of the cochlea of a mammal to show the internal structure of the channels in cross section.
  (b) Give a brief account of how

(i) the vibrations on the oval window of the inner ear are transmitted to the sensory cells of the cochlea,

(ii) pitch discrimination may be achieved.

(c) Explain briefly why

(i) in nocturnal mammals the eyes often have

(A) a wide pupil
(B) a tapetum (a reflecting structure situated behind the visual cells)
(C) a retina containing numerous rod cells which function in groups,

(ii) in tree-dwelling mammals the eyes are usually located at the front of the head.
(JMB: June 1974) B

5 This question refers to the mammal.

(a) Make a labelled diagram of a section through the cochlea to show the arrangement of the sensory cells and their surrounding structures and canals.

(b) Explain how mechanical vibrations at the tympanic membrane are transmitted to the sensory cells in the cochlea.

(c) Briefly describe how the direction of the source of a sound of short duration may be located.
(JMB: June 1976) B

6 (a) Describe the structure and functions of the human retina.

(b) What is colour blindness and why does it occur?
(ONC: 1978) B

Chapter 17

1 What is meant by the statement that mammals are warm-blooded while all invertebrates are cold-blooded? How is warm-bloodedness maintained, and what advantages does it confer on a mammal?
(London: January 1969) Z

2 The following results were obtained when a group of students measured their mouth temperatures.

TEMPERATURE IN °C	NUMBER OF STUDENTS
35·0	1
35·5	4
36·0	4
36·5	7
37·0	3
37·5	1

(a) (i) Calculate the average temperature of this group.
(ii) Give two possible reasons why the average differs from the quoted national average of 36·9°C.

(iii) Give two possible reasons why the mouth temperatures of the individuals differ.

(b) Describe briefly how the body detects and reacts to an increase in the external temperature.

(c) What are the advantages of having a constant body temperature?

(d) Explain the causes of the following observations:

(i) The application of heat to a man's unclothed arm may cause a rise in the temperature of the skin of the feet.

(ii) There can be a skin temperature gradient from 35°C to 15°C from the unclothed shoulder to the finger tip when the environmental temperature is 15°C.
(JMB: June 1973) B

3 (a) The graphs below show how the human body responds to an increase in the air temperature. The measurements were made on unclothed, resting male subjects in still air. The relative humidity of the air was low.

(i) Account for the changes in slope in the curves in Graph 1 which occur:

(A) at about 22°C for Metabolic rate and
(B) at about 28°C for Heat loss by evaporation.

(ii) Account for the differences, in Graph II, between the curves for Internal body temperature and Average skin temperature.

(iii) State two ways, other than by evaporation of sweat, by which heat is lost via the skin.

(iv) Name two DIFFERENT external factors: one of which (A) could decrease loss of heat by evaporation of sweat and the other (B) which could increase loss of heat by evaporation of sweat.

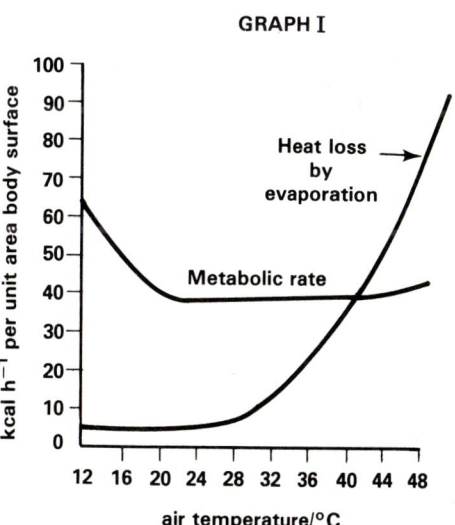

GRAPH I

465

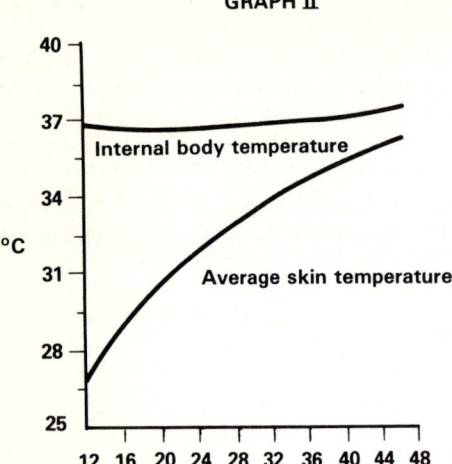

GRAPH II

Internal body temperature

Average skin temperature

°C

air temperature/°C

(b) (i) One arm of a human subject was immersed in hot water while the other arm was left exposed to the air. The exposed arm soon began to sweat and to turn pink.

State two ways in which the responses of the exposed arm could have been brought about.

(ii) When the blood flow from the heated arm was stopped, the exposed arm ceased sweating and lost its pink colour. What does this allow you to deduce about the method of thermo-regulation of the exposed arm?

(JMB: June 1975) B

4 (a) (i) Give a fully-labelled diagram of a generalised vertical section of the skin of a mammal to show the structures visible under a light microscope.

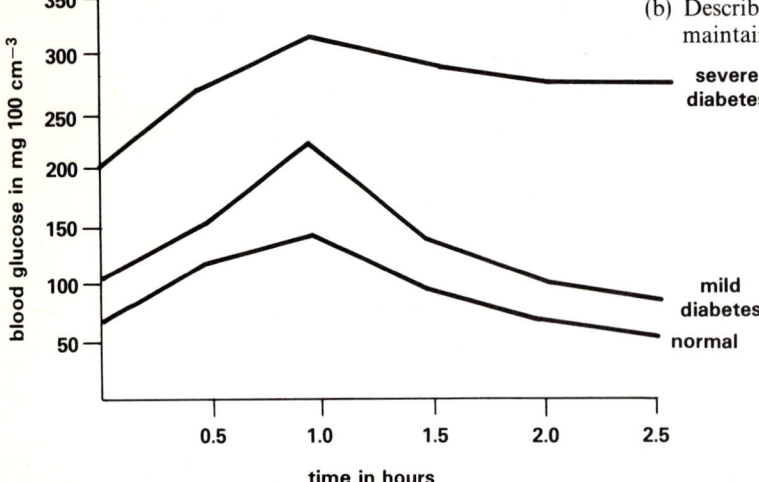

(ii) The skin is not a uniform structure. Describe three different ways in which the skin is specialised at named locations in the human body.

(b) (i) Name the physical processes by which heat can be lost from exposed skin and comment briefly on their relative importance.

(ii) Discuss the role of the skin in the regulation of overall body temperature in mammals.

(JMB: June 1978) B

Chapter 18

1 What is homeostasis? Indicate the part played by each of the following in the homeostatic mechanisms of a mammal:

(a) skin, (b) pancreas and (c) heart and blood vessels.

(London: June 1974) B

2 (a) Show the position of, and label, eight different endocrine glands on a large outline drawing of a human male.

(b) Give an account of the role of two hormones which play a part in carbohydrate metabolism in the mammal.

(JMB: June 1974) B

3 The pituitary is said to be a 'master gland' within the endocrine system of a vertebrate. Using specific examples of the hormones which it produces, describe instances of its activities which you feel reflect this description of the gland.

To what extent do you consider that the statement might misrepresent the pituitary gland in relation to other endocrine organs?

(London: June 1978) Z

4 This question refers to the mammal.

(a) What are the sources of the glucose found in the blood? How and where does glucose enter the blood? What processes lead to a loss of glucose from the blood?

(b) Describe how a constant level of blood glucose is maintained.

(JMB: June 1977) B

5 A person suspected of having diabetes mellitus may take a glucose tolerance test as follows. After several hours of fasting he or she ingests 50 g glucose dissolved in 150 cm³ water. The blood glucose is estimated at the same time, and subsequently every half hour for two and a half hours.

The graph gives glucose tolerance data for three people who had undergone the test.

(a) Using the graph suggest

    (i) why the blood glucose is estimated as frequently as every half hour,

    (ii) why the testing was continued for only two and a half hours.

(b) The kidney starts to excrete glucose when the blood glucose level exceeds about 180 mg per 100 cm³. Glucose in the urine (glycosuria) is therefore a common symptom of diabetes mellitus.

    Using the graph, explain whether a test would or would not reveal glucose in the urine of a mild diabetic

    (i) at half an hour after the start of the test.

    (ii) at two hours after the start of the test.

(c) Treatment of experimental animals with the drug alloxan produces similar symptoms to those of diabetes mellitus. Alloxan selectively destroys particular secretory cells of the body. Give the location and function of the cells which it destroys.

(JMB: June 1978) B

Chapter 19

1 (a) Make a large labelled drawing of a longitudinal section of the mature ovule of a flowering plant.

(b) Describe the structural and physiological events which occur after pollination up to and including the formation of the fruit in a flowering plant.

(JMB: June 1977) B

2 Distinguish between pollination and fertilisation in a flowering plant. Describe the pollination mechanism of a named insect-pollinated flower. To what extent do insects and flowering plants depend on each other?

(London: January 1977) B

3 It is increasingly the practice to grow genetically-uniform plants; breeds of farm animals are also tending to become more uniform genetically.

(a) Explain, with examples, the various means by which such genetic uniformity is obtained.

(b) Discuss, with examples, the advantages and disadvantages of this uniformity.

(JMB: Special Paper June 1977) B

untreated flower   **treated flower**

 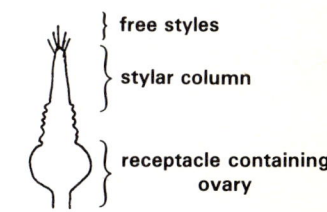

} free styles

} stylar column

} receptacle containing ovary

4 The diagram above shows the design of an experiment to assess the relative effectiveness of insect and wind pollination in apple trees. The weather during the experiment was suitable for both wind and insect pollination. The variety of apple tree used was self-incompatible (when self-pollinated the pollen tubes made little growth in the stigma and style). A large number of untreated and treated flowers were exposed for two days for pollination to occur. The flowers were then protected from further pollination. The extent of pollination was evaluated seven days later by considering the number and penetration of pollen tubes. These data are shown in Table I.

(a) (i) Give two reasons which could explain why there are so many more pollen tubes in the free styles than in the ovary.

  (ii) What is the significance of the presence of pollen tubes in the ovary?

 (iii) Give two ways in which treatment (shown in the diagram) may affect the pollination of the flower.

 (iv) From the data, state, giving two reasons for your answer, whether wind or insect pollination is more effective.

Table I

REGION	AVERAGE NUMBER OF POLLEN TUBES		FLOWERS (AS % OF TOTAL INVESTIGATED) IN WHICH REGION LISTED CONTAINED THE POLLEN TUBES	
	UNTREATED	TREATED	UNTREATED	TREATED
Free styles	158	25	40	21
Stylar column	86	8	15	4
Ovary	38	4	12	3

(b) The data in Table II refer to two varieties of apple.

Table II

	Variety A	Variety B
Longevity of unfertilised ovule measured from the time of flower opening	8 days	12 days
Time required for the growing pollen tube to reach the ovule	5 days	5 days

(i) Calculate the number of days from flower opening during which pollination can be effective for each of the two varieties.
(ii) Which variety is likely to be the most fruitful (taking no account of other factors)?

(JMB: June 1978) B

5 (a) What structure gives rise to the flower in higher plants?
(b) A short-day plant, species P, and a long-day plant, species Q, both flower when grown in cycles of 13 hours light/11 hours dark. Copy the table below (the three columns shown in bold type on the right only) and indicate the likely flowering responses of these plants in the daylengths shown. Use a tick for flowering and a cross for non-flowering.

Daylength (hr)			Flowering response	
Light	Dark		Species P	Species Q
9	15	A		
11	13	B		
13	11	C	✓	✓
15	9	D		
17	7	E		

(c) The critical daylength is that daylength which lies at the boundary between inductive and non-inductive daylengths. The critical daylength for flowering in species X is 14 hours light/10 hours dark and in species Y is 13 hours light/11 hours dark. Do these critical daylengths show that

A species X is a long-day plant and species Y a short-day plant?
B species X is a short-day plant and species Y is a long-day plant?
C both species are long-day plants?
D both species are short-day plants?
E each species could be either a long-day plant or a short-day plant?
(Give one letter only).

(d) Cocklebur is a short-day plant which requires only one inductive cycle for flower formation. A plant grown under long days (18 hours light, 6 hours dark) does not flower. When subjected to a single inductive short day (10 hours light, 14 hours dark) it subsequently flowers.

(i) When an additional 0·5 hours exposure to white light is given in the middle of the dark period to a plant grown under 9·5 hours light, 14 hours dark, the plant does not flower. What does this suggest about the relative importance of the light and dark periods?
(ii) Wavelengths of light in the red region of the spectrum are more effective in inhibiting flowering when given as a light treatment in the middle of the dark period. What pigment is likely to be the receptor for this response?
(iii) Plants maintained under long daylengths flower when the lamina of a single leaf attached to the plant is enclosed and given short-day conditions. How does this observation support the concept of a flowering hormone?

(JMB: June 1974) B

Chapter 20

1 (a) Describe the chemical interactions which exist
(i) between the seed coat and the embryo, and
(ii) between the endosperm and the embryo.
For each case, describe one such chemical interaction and show how the interaction could be demonstrated practically.
(b) (i) How, in nature, are the barriers to germination removed?
(ii) How does the possession of mechanisms of seed dormancy benefit plant species?

(JMB: June 1978) B

2 Write a concise account of the following aspects of the seeds of flowering plants:

(a) the significance of seed size,
(b) mechanisms and significance of dormancy,
(c) mobilisation of storage materials at germination.

(JMB: June 1976) B

3 (a) (i) Give one external factor which affects the rate of respiration of a plant and say how it affects the rate.
(ii) Give two places in, or parts of, a higher plant where you would expect the rate of respiration to be high, and two where you would expect it to be low.
(iii) Suggest one method, giving reasons, by which ripe apples could be stored for as long as possible.
(b) (i) Give a labelled diagram to show the apparatus you would use to measure the respiratory quotient (RQ) of germinating seeds.
(ii) Describe how you would use this apparatus to measure the oxygen intake and carbon dioxide production of the seeds. State what control you would use.
(iii) The following results were given by germinating seeds under standard time and conditions in such an experiment.

Volume of O_2 used $= 1·60$ cm^3
Volume of CO_2 given out $= 1·15$ cm^3.

Calculate the RQ. What type of substrate was being respired?

(iv) What other measurements would you need in order to calculate (A) the respiratory rate and (B) the Q_{10} for respiration of germinating seeds?

(JMB: June 1974)   B

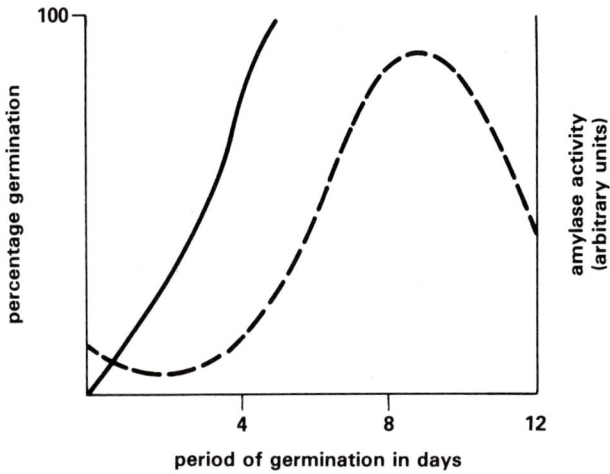

period of germination in days

4 Examine the graph of amylase activity in germinating barley grains and answer the questions that follow:
(a) What is the time interval in days between all the barley grains having commenced germination and the peak in amylase activity?
(b) What reaction does amylase catalyse and what is the product of the reaction?
(c) Suggest reasons for the change in amylase activity after the eighth day of germination.

(d) Make a large, labelled drawing of a longitudinal section of a cereal grain.
Mark on your drawing one site of amylase activity, and one destination for the product of the reaction.
(e) Explain the effect of an increase in temperature from 15° to 45°C on amylase activity.

(London: January 1976)   Bot.

5 The graphs below show changes which occur in (I) dry matter, (II) area and (III) photosynthetic activity, with time from the start of germination, in the cotyledons of the seeds of three species of plant, Species P, Species Q and Species R.
(a) Distinguish briefly between the meanings of the terms hypogeal and epigeal, as used to describe germination. Give one named example of each.
(b) Of the three species, P, Q and R, which one shows hypogeal germination? Give two reasons for your choice.
(c) Using the information in the graphs, together with your knowledge of germination, explain what processes may account for the following:

(i) the fall in cotyledon weight with time in Species P,
(ii) the fact that the fall in cotyledon weight with time in Species Q is proportionately less than that in Species P,
(iii) the fact that cotyledon weight in Species R
(A) increases
(B) that the increase is very small.

(iv) the fact that the photosynthetic activity of the cotyledons of Species R

(A) rises sharply between day 5 and day 10 and
(B) falls after day 15.

(d) Is it reasonable to suggest that, of the two species which have epigeal germination, one is better adapted to this condition than the other? Briefly explain your answer.

(JMB: June 1975)   B

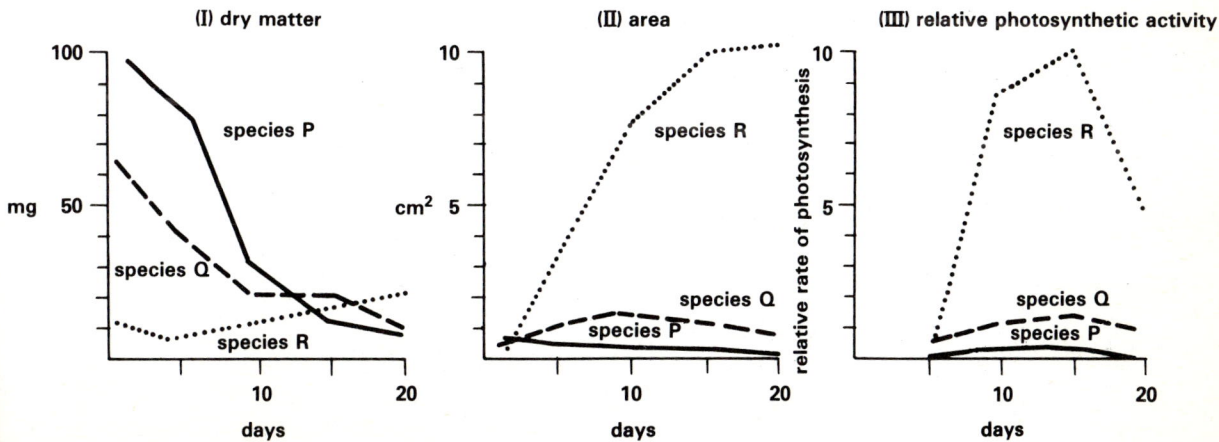

469

Chapter 21

1 Give an account of shoot elongation in flowering plants, showing how it is regulated by external and internal factors.

(JMB: Special Paper June 1977)   B

2 (a)   (i) What is bioassay?
    (ii) Outline a technique which can be used to bioassay auxin.
   (iii) State four effects which auxins may produce in living cells, tissues or organs.
  (b)   (i) Describe how you would attempt to propagate a plant from a stem cutting, outlining the optimum conditions and the ideal procedure for success.
    (ii) What anatomical changes occur in the vicinity of the wounded surface of a stem cutting during successful propagation?
   (iii) What is meant by polar transport of auxin?
   (iv) What, for stem cuttings, are the consequences of the polar transport of auxin?
    (v) With what other category of plant hormones do the auxins which are important for the establishment of the cutting interact and what is the outcome of such interaction?

(JMB: June 1976)   B

3 Describe the various ways in which light affects the growth and development of plants (omitting photosynthesis).

(London: June 1974)   B

4 Give an explanation of each of the following:
  (a) Trees near a street lamp sometimes flower earlier than similar trees elsewhere.
  (b) The shoots of plants placed on a window sill grow towards the light.
  (c) The removal of the terminal bud of a plant may lead to the development of the lateral buds.

(London: January 1976)   B

5 Write about the root and shoot system of flowering plants under the following headings:

  (a) A comparison of the structure of the root and shoot apical regions.
  (b) The sites of initiation and early development of lateral roots and lateral buds.
  (c) The factors which affect the direction of growth of roots and shoots.

(JMB: June 1973)   B

Chapter 22

1 Describe how the following intermittent actions are initiated and controlled in a mammal:

  (a) secretion of gastric juice,
  (b) ovulation.

Discuss the main control principle shown by both of these examples.

(JMB: June 1973)   B

2 (a) For mammals:

    (i) state three features of the placenta which facilitate the exchange of materials,
    (ii) give a brief account of the endocrine functions of the placenta.
  (b) In the foetus, oxygenated blood from the placenta passes via the umbilical vein and posterior vena cava to the right atrium (auricle). There is an opening in the septum between the left and right atria, the foramen ovale. This usually closes over at birth.

    (i) What function may be served in the foetus by the foramen ovale?
    (ii) What malfunction might be expected in the circulatory system of the infant if this opening did not seal?
   (iii) Mammalian foetal haemoglobin usually takes up oxygen at partial pressures at which the maternal haemoglobin gives up oxygen. What is the advantage of this?
  (c)   (i) The table below shows the calcium (Ca) requirements of three women:

	Ca requirements (mg day^{-1})
Adult female	680
Pregnant female	1600
Nursing mother	2000

Explain these differences in the daily requirement of calcium.
    (ii) Briefly explain why the enzyme rennin is found predominantly in young mammals.
   (iii) The young of grazing mammals open their eyes shortly after birth and can soon move around independently. In many carnivorous mammals the young do not open their eyes for some time after birth, they are often quite helpless and have a much longer period of dependence and association with their parents. What relationship does there appear to be between the situation given above and the ways of life of the animals?

(JMB: June 1975)   B

3 (a) Give an account (using graphs and or diagrams if you wish) of the events which occur in

    (i) the ovary, and
    (ii) the uterus

of a non-pregnant human female during the monthly reproductive cycle.
  (b) Describe how these events are controlled by hormones, stating, for each hormone, its precise effect and its site of production.

(JMB: June 1976)   B

4 (a) Make a fully-labelled plan diagram of a transverse section of a mammalian testis and indicate where and how spermatogenesis occurs.

(b) Explain the importance of mitosis and meiosis in spermatogenesis.

(ONC: 1978)

5 What are the essential features of the process of fertilisation in plants and animals? Consider the variety of means by which fertilisation is accomplished.

(JMB: Special Paper June 1977)   B

6 Compare and contrast the processes of gametogenesis in a male and female vertebrate. Show how the differences described are related to the nature of the gametes produced.

(London: June 1976)   Z

Index

480